Managing Global Resources and Universal Processes

Environmental Management Handbook, Second Edition

Edited by
Brian D. Fath and Sven E. Jørgensen

Volume 1
Managing Global Resources and Universal Processes

Volume 2
Managing Biological and Ecological Systems

Volume 3
Managing Soils and Terrestrial Systems

Volume 4
Managing Water Resources and Hydrological Systems

Volume 5
Managing Air Quality and Energy Systems

Volume 6
Managing Human and Social Systems

Managing Global Resources and Universal Processes

Second Edition

Edited by
Brian D. Fath and Sven E. Jørgensen

Assistant to Editor
Megan Cole

CRC Press
Taylor & Francis Group
Boca Raton London New York

CRC Press is an imprint of the
Taylor & Francis Group, an **informa** business

Cover photo: Paphos Archeological Park, Cyprus, N. Fath

Contents

SECTION I Anthropogenic Chemicals: Human Manufactured and Activities

SECTION II   Natural Elements and Chemicals

SECTION III Basic Environmental Processes

Preface

Given the current state of the world as compiled in the massive Millennium Ecosystem Assessment Report, humans have changed ecosystems more rapidly and extensively during the past 50 years than in any other time in human history. These are unprecedented changes that need certain action. As a result, it is imperative that we have a good scientific understanding of how these systems function and good strategies on how to manage them.

In a very practical way, this multivolume *Environmental Management Handbook* provides a comprehensive reference to demonstrate the key processes and provisions for enhancing environmental management. The experience, evidence, methods, and models relevant for studying environmental management are presented here in six stand-alone thematic volumes, as follows:

VOLUME 1 – Managing Global Resources and Universal Processes
VOLUME 2 – Managing Biological and Ecological Systems
VOLUME 3 – Managing Soils and Terrestrial Systems
VOLUME 4 – Managing Water Resources and Hydrological Systems
VOLUME 5 – Managing Air Quality and Energy Systems
VOLUME 6 – Managing Human and Social Systems

In this manner, the handbook introduces in the first volume the general concepts and processes used in environmental management. The next four volumes deal with each of the four spheres of nature (biosphere, geosphere, hydrosphere, and atmosphere). The last volume ties the material together in its application to human and social systems. These are very important chapters for a wide spectrum of students and professionals to understand and implement environmental management. In particular, features include the following:

- The first handbook that demonstrates the key processes and provisions for enhancing environmental management.
- Addresses new and cutting-edge topics on ecosystem services, resilience, sustainability, food–energy–water nexus, socio-ecological systems, etc.
- Provides an excellent basic knowledge on environmental systems, explains how these systems function, and gives strategies on how to manage them.
- Written by an outstanding group of environmental experts.

Since the handbook covers such a wide range of materials from basic processes, to tools, technologies, case studies, and legislative actions, each handbook entry is further classified into the following categories:

APC: Anthropogenic chemicals: The chapters cover human-manufactured chemicals and their activities

COV: Indicates that the chapters give comparative overviews of important topics for environmental management

CSS: The chapters give a case study of a particular environmental management example

DIA: Means that the chapters are about diagnostic tools: monitoring, ecological modeling, ecological indicators, and ecological services

ELE: Focuses on the use of legislation or policy to address environmental problems

ENT: Addresses environmental management using environmental technologies

NEC: Natural elements and chemicals: The chapters cover basic elements and chemicals found in nature

PRO: The chapters cover basic environmental processes.

Overall, these volumes will be a valuable resource for all libraries supporting programs in environmental science and studies, earth science, geography, and policy.

In this volume, #1, the collection of over 50 entries provides an overview of global resources and universal processes. This serves as a good introduction to the key aspects of environmental management and includes descriptions of elements of the periodic table as well as organic and inorganic processes leading to pollution and alteration of natural conditions. A new chapter on telecoupling shows the long distance relations and interactions that mark most environmental systems.

Brian D. Fath
Brno, Czech Republic
December 2019

Editors

Brian D. Fath is Professor in the Department of Biological Sciences at Towson University (Maryland, USA) and Senior Research Scholar at the International Institute for Applied Systems Analysis (Laxenburg, Austria). He has published over 180 research papers, reports, and book chapters on environmental systems modeling, specifically in the areas of network analysis, urban metabolism, and sustainability. He has co-authored the books *A New Ecology: Systems Perspective* (2020), *Foundations for Sustainability: A Coherent Framework of Life–Environment Relations* (2019), and *Flourishing within Limits to Growth: Following Nature's Way* (2015). He is also Editor-in-Chief for the journal *Ecological Modelling* and Co-Editor-in-Chief for *Current Research in Environmental Sustainability*. Dr. Fath was the 2016 recipient of the Prigogine Medal for outstanding work in systems ecology and twice a Fulbright Distinguished Chair (Parthenope University, Naples, Italy in 2012 and Masaryk University, Czech Republic in 2019). In addition, he has served as Secretary General of the International Society for Ecological Modelling, Co-Chair of the Ecosystem Dynamics Focus Research Group in the Community Surface Modeling Dynamics System, and member and past Chair of Baltimore County Commission on Environmental Quality.

Sven E. Jørgensen (1934–2016) was Professor of environmental chemistry at Copenhagen University. He received a doctorate of engineering in environmental technology and a doctorate of science in ecological modeling. He was an honorable doctor of science at Coimbra University (Portugal) and at Dar es Salaam (Tanzania). He was Editor-in-Chief of *Ecological Modelling* from the journal inception in 1975 until 2009. He was Editor-in-Chief for the *Encyclopedia of Environmental Management* (2013) and *Encyclopedia of Ecology* (2008). In 2004, Dr. Jørgensen was awarded the Stockholm Water Prize and the Prigogine Medal. He was awarded the Einstein Professorship by the Chinese Academy of Sciences in 2005. In 2007, he received the Pascal Medal and was elected a member of the European Academy of Sciences. He had published over 350 papers, and has edited or written over 70 books. Dr. Jørgensen gave popular and well-received lectures and courses in ecological modeling, ecosystem theory, and ecological engineering worldwide.

Contributors

Imad A.M. Ahmed
Lancaster Environment Center
Lancaster University
Lancaster, United Kingdom

Ronald G. Amundson
College of Natural Resources
University of California—Berkeley
Berkeley, California

Jirapat Ananpattarachai
Faculty of Engineering
Center of Excellence for Environmental Research
    and Innovation
Naresuan University
Phitsanulok, Thailand

Lars Bergström
Department of Soil Science
Swedish University of Agricultural
    Sciences (SLU)
Uppsala, Sweden

Angelika Beyer
Department of Analytical Chemistry
Chemical Faculty
Gdansk University of Technology
Gdansk, Poland

Marek Biziuk
Department of Analytical Chemistry,
    Chemical Faculty
Gdansk University of Technology
Gdansk, Poland

Frederick Paxton Cardell Blamey
School of Agriculture and Food Sciences
The University of Queensland
St. Lucia, Queensland, Australia

Elke Bloem
Institute for Crop and Soil Science
Julius Kuhn Institute (JKI)
Braunschweig, Germany

Pascal Boeckx
Faculty of Agricultural and Applied
    Biological Sciences
University of Ghent
Ghent, Belgium

John Borden
Department of Biological Sciences
Simon Fraser University
Burnaby, British Columbia, Canada

Céline Boutin
Science and Technology Branch
Environment Canada
Carleton University
Ottawa, Ontario, Canada

Dominic A. Brose
University of Maryland
College Park, Maryland

Ray Correll
Commonwealth Scientific and Industrial
    Research Organization (CSIRO)
Adelaide, South Australia, Australia

Marianna Czaplicka
Institute of Non-Ferrous Metals
and
Department of Analytical Chemistry
Silesian University of Technology
Gliwice, Poland

Vera Lucia S.S. de Castro
Ecotoxicology and Biosafety Laboratory
Brazilian Agricultural Research Corporation
    (Embrapa Environment)
São Paulo, Brazil

Malcolm Devine
Aventis CropScience Canada Co.
Saskatoon, Saskatchewan, Canada

Peter Dillon
Commonwealth Scientific and Industrial
    Research Organization (CSIRO)
Adelaide, South Australia, Australia

J.K. Dubey
Department of Entomology
Dr. Y.S. Parmar University of Horticulture
    and Forestry
Solan, India

Renata Gaj
Institute of Soil Science
Agricultural University
Poznan, Poland

Ardell D. Halvorson
U.S. Department of Agriculture (USDA)
Fort Collins, Colorado

Silvia Haneklaus
Institute for Crop and Soil Science
Julius Kuhn Institute (JKI)
Braunschweig, Germany

Philippe Hinsinger
Sun and Environment Unit
National Institute for Agricultural Research
    (INRA)
Montpellier, France

Zhengyi Hu
Institute of Soil Science
Chinese Academy of Sciences
Nanjing, China

Hayriye Ibrikci
Soil Science and Plant Nutrition Department
Cukurova University
Adana, Turkey

Alexandra Izosimova
St. Petersburg Agricultural Physical
    Research Institute
St. Petersburg, Russia

Bruce R. James
University of Maryland
College Park, Maryland

Philip M. Jardine
Oak Ridge National Laboratory
Oak Ridge, Tennessee

Sven Erik Jørgensen
Institute A, Section of Environmental
    Chemistry
Copenhagen University
Copenhagen, Denmark

Puangrat Kajitvichyanukul
Faculty of Engineering
Center of Excellence for Environmental Research
    and Innovation
Naresuan University
Phitsanulok, Thailand

Gabriella Kakonyi
Kroto Research Institute
Sheffield University
Sheffield, United Kingdom

Keith A. Kelling
Professor Emeritus, Department of
    Soil Science
University of Wisconsin—Extension
Madison, Wisconsin

Rami Keren
Agricultural Research Organization of Israel
Bet-Dagan, Israel

Peter Kleinman
Pasture Systems and Watershed Management
   Research Unit
U.S. Department of Agriculture (USDA)
University Park, Pennsylvania

Rai Kookana
Commonwealth Scientific and Industrial
   Research Organization (CSIRO)
Adelaide, South Australia, Australia

Peter Martin Kopittke
School of Agriculture and Food Sciences
The University of Queensland
St. Lucia, Queensland, Australia

David P. Kreutzweiser
Canadian Forest Service
Natural Resources Canada
Sault Sainte Marie, Ontario, Canada

Tore Krogstad
Department of Plant and Environmental
   Sciences
Norwegian University of Life Science
Aas, Norway

Witold Kurylak
Institute of Non-Ferrous Metals
Gliwice, Poland

Gamini Manuweera
Chemicals and Health Branch
United Nations Environmental Program
Geneva, Switzerland

Richard McDowell
AgResearch Ltd., Invermay Agricultural Center
Mosgiel, New Zealand

Leslie D. McFadden
Department of Earth and Planetary Sciences
University of New Mexico
Albuquerque, New Mexico

Ronald G. McLaren
Soil, Plant, and Ecological Sciences Division
Lincoln University
Canterbury, New Zealand

Mike J. McLaughlin
Land and Water, Commonwealth Scientific and
   Industrial Research Organization (CSIRO)
Glen Osmond, South Australia, Australia

Mallavarapu Megharaj
Commonwealth Scientific and Industrial
   Research Organization (CSIRO)
Adelaide, South Australia, Australia

Neal William Menzies
School of Agriculture and Food Sciences
The University of Queensland
St. Lucia, Queensland, Australia

J. David Miller
Department of Chemistry
Carleton University
Ottawa, Ontario, Canada

H. Curtis Monger
Department of Agronomy and Horticulture
New Mexico State University
Las Cruces, New Mexico

Ravendra Naidu
Commonwealth Scientific and Industrial
   Research Organization (CSIRO)
Adelaide, South Australia, Australia

Niels Erik Nielsen
Plant Nutrition and Soil Fertility Laboratory
Department of Agricultural Sciences
Royal Veterinary and Agricultural University
Frederiksberg, Denmark

David R. Parker
Department of Soil and Environmental Sciences
University of California—Riverside
Riverside, California

Judith F. Pedler
Department of Environmental Sciences
University of California—Riverside
Riverside, California

Mark B. Peoples
Agriculture and Food, Commonwealth Scientific
    and Industrial Research Organization (CSIRO)
Canberra, Australian Capital Territory, Australia

Robert Pietrzak
Department of Chemistry
Adam Mickiewicz University
Poznan, Poland

Paola Poli
Department of Genetics, Biology of
    Microorganisms, Anthropology,
    and Evolution
University of Parma
Parma, Italy

Abdul Rashid
Pakistan Atomic Energy Commission
Islamabad, Pakistan

Lisa A. Robinson
Independent Consultant
Newton, Massachusetts

Alexandru V. Roman
School of Public Administration
Florida Atlantic University
Boca Raton, Florida

John Ryan
International Center for Agricultural Research
    in the Dry Areas (ICARDA)
Aleppo, Syria

Vilma Sandström
Sustainability Science
LUT-University
Lahti, Finland

William H. Schlesinger
Department of Geology and Botany
Duke University
Durham, North Carolina

Ewald Schnug
Institute for Crop and Soil Science
Julius Kuhn Institute (JKI)
Braunschweig, Germany

A. Paul Schwab
Department of Agronomy
Purdue University
West Lafayette, Indiana

Andrew N. Sharpley
University of Arkansas
Fayetteville, Arkansas

Paul K. Sibley
School of Environmental Sciences
University of Guelph
Guelph, Ontario, Canada

Bogdan Skwarzec
Faculty of Chemistry
University of Gdansk
Gdansk, Poland

Rolf Sommer
International Center for Agricultural Research
    in the Dry Areas (ICARDA)
Aleppo, Syria

Gerd Sparovek
College of Agriculture
Graduate School of Agriculture Luiz de Queiroz
    (ESALQ)
University of São Paulo
São Paulo, Brazil

Scott J. Sturgul
Nutrient and Pest Management Program
University of Wisconsin
Madison, Wisconsin

Donald L. Suarez
U.S. Salinity Laboratory
Agricultural Research Service (USDA-ARS)
U.S. Department of Agriculture
Riverside, California

Meena Thakur
Department of Entomology
Dr. Y.S. Parmar University of Horticulture
    and Forestry
Solan, India

Oswald Van Cleemput
Faculty of Agricultural and Applied
   Biological Sciences
University of Ghent
Ghent, Belgium

Doug Walsh
Washington State University
Prosser, Washington

Leszek Wachowski
Department of Chemistry
Adam Mickiewicz University
Poznan, Poland

Johannes Bernhard Wehr
School of Agriculture and Food Sciences
The University of Queensland
St. Lucia, Queensland, Australia

W.W. Wenzel
Institute of Soil Research
University of Natural Resources and
   Life Sciences
Vienna, Austria

Larry P. Wilding
Department of Soil and Crop Sciences
Texas A&M University
College Station, Texas

He Zhong
Pesticide Environment Impact Section
Public Health Entomology Research and
   Education Center
Florida A&M University
Panama City, Florida

I

Anthropogenic Chemicals: Human Manufactured and Activities

Acaricides

Doug Walsh

Introduction

Approximately 45,000 species of mites and hundreds of species of ticks are described worldwide. Many thousands of species still remain unidentified. About half are plant-feeding species, and among these, about half are in the superfamily Eriophyoidea (gall, bud, and rust mites). Most of the other plant-feeding mites are classified into the superfamilies Tetranychoidea and Tarsonemidae. The superfamily Tetranychoidea includes the economically important spider, flat, and fowl mites, and the superfamily Tarsonemidae includes the economically important broad, cyclamen mites and Varroa mites. Over another 3000 mite species are loosely classified in the order Astigmata. Economically important species include feather and scabies mites. Ticks are placed in the superfamily Ixodoidea and all are ectoparasites (blood feeders) of vertebrate animals.[1]

Most mites are small to minute and mites are universally cryptic, making them difficult to detect. Often infestations are overlooked. Mites are often colonizers of new or disturbed habitats, and once established on a new host, mites possess biological characteristics that permit rapid increases in population abundance. Factors in most mites' lifestyle that lead to rapid population buildup include high egg production, various modes of reproduction (parthenogenesis, pedogenesis, and sexual), short life cycles, a myriad of dispersal techniques, and adaptability to diverse ecological conditions.[1] These traits combined with an exponential increase in worldwide transport of humans and plant and animal products will likely contribute to increased concerns over mite pests in the future.

In plant-based agriculture, Van de Vrie et al.[2] observed that outbreaks of mite populations were uncommon historically in systems where productivity languished far below the levels achieved in modern production agriculture. Spider mite populations stayed below observable levels due to natural regulation by predators, disease, and poor nutrition from low quality host plants. Van de Vrie et al.[2] went on to observe that mite populations often experienced outbreaks in agroecosytems where production levels were bolstered by the use of synthetic inputs including fertilizers and pesticides. When crop production is optimized (i.e., not limited by water, nutrients, competition from weeds, or predatory mites and insects), the plants in production become an excellent food source for mite pests. Under these conditions, the developmental rate, fecundity, and life span of mites are increased and contribute to population outbreaks.

Spider Mite Pests

A number of mite species can achieve pest status at high population abundance. Spider mites develop through several stages: egg, six-legged larva, eight-legged protonymph, deutonymph, and adult. Males typically reach maturity before females and will position themselves near developing quiescent females. When an adult female emerges, copulation will often occur immediately. Under optimal conditions, most mite species can develop from egg to adult in 6 to 10 days. Egg laying can begin as soon as one or two days after maturing to adults. Most spider mite species overwinter as mated adult females. A notable exception is the European red mite that overwinters as eggs.[3]

A Big Drain from the Feeding of Such Small Pests

At the microscopic level, significant quantities (relative to mite size) of plant juices pass through the digestive tract of spider mites as they feed on leaf tissues. McEnroe[4] estimated this volume at 1.2×10^{-2} microliters per mite per hour. This quantity represents roughly 50% of the mass of an adult female spider mite. Leisering and Beitrag[5] calculated that the number of photosynthetically active leaf cells that are punctured and emptied per mite is 100 per minute. In gut content studies of two-spotted mites, Mothes and Seitz[6] observed only thylakoid granules inside their digestive tract following feeding. The thylakoid grana on which spider mites focus their feeding are key photosynthetic engines in plant cells. The grana consist of 45%–50% protein, 50%–55% lipid, and minute amounts of RNA and DNA.[7] Water and other low-density plant cell contents are directly excreted.[4]

At the macroscopic level, damage from mite feeding can cause leaf bronzing, stippling, or scorching. For most horticultural crops, economic loss is caused by a drop in yield or quality due to reduction in photosynthesis.

Spider Mite Outbreaks Are Promoted by Hot, Dry Weather

Water stress, wind, and dust all contribute to the potential for mite outbreaks. When mite outbreaks occur, acaricide treatments are often used for suppression.

Varroa mites *Varroa jacobsoni* provide an ideal example of how rapidly a mite species can spread and exploit a new habitat. First recorded in honeybee colonies in Southeast Asia in 1904, Varroa mites are now pandemic. Varroa mites feed parasitically on an individual bee's hemolymph fluid, weakening the bee and often causing premature death. Mites attach themselves to foraging workers in order to spread themselves from one hive to another. This mite can severely weaken bees, and an unchecked mite population will almost certainly lead to the premature death of a honeybee colony. Apiculturists speculate that Varroa mite has contributed substantially to the collapse of feral honeybee populations worldwide.[8]

The northern fowl mite *Ornithonyssus sylviarum* is a common pest of domestic fowl and other wild birds commonly associated with human settlements. The nymphs and adults have piercing mouthparts and seek blood meals. Mite populations build up rapidly and a generation can be completed in 5 to 12 days. Several generations occur each year. Northern fowl mite spends virtually its entire life on the host bird.[9]

Deer and dog ticks *Ixodes scapularis* and *Dermacentor variabilis* are two common ticks to which acaricides are applied for on a consistent basis, especially since both are parasitic feeders on mammals. Deer ticks are a significant concern since they are the primary vector for Lyme disease.[10]

Mange or scabies in livestock is a skin condition caused by microscopic mites in or on the skin. The mites cause intense itching and discomfort that is associated with decreased feed intake and production. Scratching and rubbing result in extensive damage to hides and fleece. Mange mites are able to cause mange on different species of livestock but are somewhat host specific, thus infecting some species more severely than others. The three most important types of mange are as follows: sarcoptic mange, caused by *Sarcoptes scabiei* feeding; psoroptic mange, caused by *Psoroptes ovis* feeding; and chorioptic mange, caused by *Chorioptes bovis* feeding.[11] Infestations of these mites on their respective livestock, domestic pet, or human host will cause skin irritation and itching and leave entry points for secondary infections. Weight gain can be reduced in livestock, pets can lose hair and itch persistently, and disfigurement can occur in humans. Acaracides are often applied to suppress mite populations parasitizing humans, pets, and livestock.

Smothering Agents

Solutions containing petroleum-based horticultural oils, vegetable oils, or agricultural soaps are applied to many crops and, occasionally, livestock. Application of these types of products kills spider mites through suffocation. Unfortunately, oils and soaps can prove phytotoxic to crop plants and are typically not effective on mites or ticks infesting livestock, pets, or humans. Mites on animal hosts are typically cryptic or subcutaneous, so acaricide coverage is an impediment to effective control.

Organochlorines

Endosulfan and dicofol are organochlorine miticides registered for use on many crops. Unlike many other organochlorine pesticides, endosulfan and dicofol are relatively non-persistent in the environment. Organochlorine acaricides interfere with the transmission of nerve impulses and disrupt the nervous system of pest mites. Organochlorine acaricides are more effective at killing mites at warmer temperatures. Overuse of organochlorine acaricides in commercial situations has resulted in the development of tolerance in many pest mite populations. Organochlorines were used substantially in the mid-20th century, but regulatory actions and public health and environmental concern have eliminated their use in most developed countries (though some continue to use them). Lindane was commonly used for mange mite in pets, livestock, and humans. Only in limited circumstances is lindane still permitted as a pharmaceutical second-line treatment. However, use of lindane continues in developing countries due to its low cost, effectiveness, and persistence.

Organophosphates and Carbamates

Many organophosphate and carbamate pesticides have acaricidal activity. Studies have demonstrated significant mite control with applications of parathion, TEPP (tetraethyl pyrophosphate), and aldicarb. Spider mites are listed as target pests on many organophosphate and carbamate products. However, many mite populations following long-term exposure have developed resistance to the toxic effects of organophosphates.[12] Carbaryl, a common carbamate, continues to be a mainstay for mite control on livestock and poultry, but its use on domestic pets and households is no longer permitted in most developed countries. The use of carbaryl continues extensively in many developing countries in domestic settings.

Organotins

Miticides in this category were synthesized in the 1960s and 1970s and registered for commercial use in the United States in the 1970s. They have been used extensively for their ability to quickly knock down spider mite populations through contact activity. Fenbutatin-oxide has been used extensively since the 1970s. Cyhexatin was used extensively in the 1970s and 1980s, but regulatory actions have now limited its use. Efficacy of the organotin acaricides is improved with warmer weather. Overuse of cyhexatin during the mid-1980s led to the development of resistance in several cropping and livestock production systems. However, populations of pest mites can regain susceptibility to organotins following a period of non-exposure.[13]

Propargite

This acaricide has been used since the 1960s. It provides effective suppression of pest mites on many crops. Regulatory constraints have resulted in the cancellation of a number of uses. Identification of propargite as a dermal irritant has led to substantial increases in time required following application before re-entry is permitted into the treated site.

Amidines

Amitraz is a miticide that once had significant use in plant and animal agriculture. At present, its use is restricted to only a small subset of the domestic pet care market.

Ovicides

Clofentazine and hexythiazox are selective carboxamide ovicidal acaricides. Spider mite eggs exposed to either compound fail to hatch. These acaricides are selective and aid in the conservation of populations of beneficial arthropods. These acaricides are typically used relatively early in the production season before mite populations reach outbreak conditions.

Antimetabolites

A number of miticidal compounds have been developed within the past 30 years. These include avermectins, pyridazinones, carbazates, and pyrroles. Pest mortality results from disruption of metabolic pathways typically within the mitochondria of nerve cells of spider mites.[14] Avermectins, ivermectins, and related compounds are fermentation products derived from mycelial extracts of *Streptomyces* species (reviewed by Burg and Stapley).[17] Avermectins are locally systemic (translaminar) in plant tissues, [15] and ivermectins can be applied dermally, by injection, suppository, or in a bolus to livestock and domestic pets. The ivermectins are the predominant parasiticide used in livestock production today. A number of products have been commercialized in recent years. Pyridaben is a pyridazinone recently registered for use on ornamentals and some tree crops. Bifenazate is a carbazate acaricide. It has a new mode of action that is not clearly understood, but it has proven toxicologically safe in mammalian studies. Bifenazate is registered on ornamentals and food products. Chlorfenapyr is a synthetic pyrrole that has been commercially available on cotton. Other uses are pending.

Synthetic Pyrethroids

Fenpropathrin and bifenthrin are two synthetic pyrethroid insecticides registered for control of spider mites in plant agriculture. Permethrin is registered for mite control on livestock and poultry. Mites have a well-documented history of rapidly developing resistance to pyrethroid insecticides in both plant and animal production systems, and resurgence of spider mite populations following pyrethroid application is typical.[18]

Tetronic Acids

Spiromesifen and spirotetramat are acaricides in a recently introduced class of selective chemistry tetronic acids that exhibit a broad-spectrum insecticidal acaricidal activity against mites. Their mode of action is by inhibition of lipid biosynthesis that affects the egg and immature stages of mites. Foliar sprays of spiromesifen are translaminar in plants and effective against mites in many cropping systems. Spirotetramat has a relatively unique property among currently registered acaricides in that it is phloem systemic within the plant it is applied to. These two acaricides have recently entered the acaricide market and are quickly gaining in use in production agriculture for mite control.

Application Technology

Mite pests can prove difficult to control with acaricides due to their potential for high population abundance, small size, and propensity to live on the bottom surfaces of leaves or within the folds of plant tissues. Good acaricide spray coverage is essential for mite control, particularly for acaricides that kill on contact with the pest mite.

Combating Miticide Resistance

Following repeated exposure, spider mite populations have a history of rapidly developing resistance to acaricides.[16] Alternating acaricides that have different modes of action reduces the potential for development of resistance to acaricides within specific modes of activity. Other techniques to discourage resistance development include spraying only when necessary and treating only infested portions of the crop. Organophosphate, carbamate, and pyrethroid insecticide applications can induce spider mite outbreaks. If possible, avoid early-season insecticide application or apply insecticides that are less disruptive to beneficial arthropods. Careful selection and use of insecticides can potentially reduce the number of miticide applications required later in the season.

References

1. Krantz, G.W.; Walter, D.E., Eds. *A Manual of Acarology,* 3rd Ed.; Texas Tech University Press: Lubbock, Texas, 2009; 807 pp.
2. Van de Vrie, M.; McMurtry, J.A.; Huffaker, C.B. Ecology of mites and their natural enemies. A review. III Biology, ecology, pest status, and host plant relations of tetranychids. Hilgardia **1972,** *41,* 345–432.
3. Bostanian, N.J. The relationship between winter egg counts of the European red mite *Panonychus ulmi* (Acari: Tetranychidae) and its summer abundance in a reduced spray orchard. Exp. Appl. Acarol. **2007,** *42,* 185–195.
4. McEnroe, W.D. The role of the digestive system in the water balance of the two-spotted spider mite. Adv. Acarol. **1963,** *1,* 225–231.
5. Leisering, R.; Beitrag, O. Beitrag zum phytopatologischen Wirkungsmeechanismus von Tetranychus urticae. Pflanzenschutz **1960,** *67,* 525–542.
6. Mothes, U.; Seitz, K.A. Functional microscopic anatomy of the digestive system of *Tetranychus urticae* (Acari: Tetranychidae). Acarologia **1981,** *22,* 257–270.
7. Noggle, G.R. The organization of plants. In *Introductory Plant Physiology;* Noggle, G.R., Fritz, G.J., Eds.; Prentice Hall: Englewood Cliffs, New Jersey, 1983; 9–38.
8. Mangum, W.A. Honey bee biology: The third annual report on the coexistence of my North Carolina bees with varroa mites. Am. Bee J. **2009,** *149,* 63–65.
9. Mullens, B.A. Temporal changes in distribution, prevalence and intensity of northern fowl mite (*Ornithonyssus sylviarum*) parasitism in commercial caged laying hens, with a comprehensive economic analysis of parasite impact. Vet. Parasitol. **2009,** *160,* 116–133.

10. Diuk-Wasser, M.A. Field and climate-based model for predicting the density of host-seeking nymphal *Ixodes scapularis,* an important vector of tick-borne disease agents in the eastern. U. S. Global Ecol. Biogeogr. **2010,** *19,* 504–514.

11. Vercruysse, J. World Association for the Advancement of Veterinary Parasitology (W.A.A.V.P.) guidelines for evaluating the efficacy of acaricides against (mange and itch) mites on ruminants [electronic resource]. Vet. Parasitol. **2006,** *136,* 55–66.

12. Smissaeret, H.R.; Voerman, S.; Oostenbrugge, L.; Reenooy, N. Acetylcholinesterases of organophosphate-susceptible and resistant spider mites *Tetranychus urticae.* J. Agric. Food Chem. **1970,** *18,* 66–75.

13. Hoy, M.A.; Conley, J.; Robinson, W. Cyhexatin and fenbutatin-oxide resistance in Pacific spider mite (Acari: Tetranychidae) stability and mode of inheritance. J. Econ. Entomol. **1988,** *81,* 57–64.

14. Hollingsworth, R.M.; Ahammadsahib, K.I.; Gadelhak, G.; McLaughlin, J.L. New inhibitors of Complex I of the mitochodrial electron transport chain with activity as pesticides. Biochem. Soc. Trans. **1994,** 22, 230–233.

15. Walsh, D.B.; Zalom, F.G.; Shaw, D.V.; Welch, N. C. Effect of strawberry plant physiological status on the translaminar activity of avermectin B1 and its efficacy on the two-spotted spider mite *Tetranychus urticae* Koch (Acari: Tetranychidae). J. Econ. Entomol. **1996,** *89* (5), 1250–1253.

16. Leeuwen, T.V.; Dermauw, W.; Tirry, L.; Vontas, J.; Tsagkarakou, A. Acaricide resistance mechanisms in the two- spotted spider mite *Tetranychus urticae* and other important Acari. Insect Biochem. Mol. Biol. **2010,** *40,* 563–572.

17. Burg, R.W., and E.O. Stapley. Isolation and characterization of the producing organism. In W.C. Campbell (ed) Ivermectin and Abamectin. Springer-Verlag, New York, N.Y. 1989. pp. 24–32.

18. Leigh, T.F. Cotton. In W. Helle and M.W. Sabelis (eds.) World Crop Pests: Spider Mites. Elsevier Press, Amsterdam, the Netherlands. 1990. pp. 349–358.

2

Endocrine Disruptors

Vera Lucia S.S.
de Castro

Introduction

Endocrine-disrupting chemicals (EDCs) refer to anthropogenic compounds that are able to mimic, antagonize, alter, or modify normal hormonal activity. Dichlorodiphenyl-trichloroethane (DDT), an insecticide first produced on a wide scale in 1945, was used extensively during the 1960s and 1970s and was the first chemical found to be estrogenic. Subsequently, other organochlorine insecticides such as dieldrin, endosulfan, and methoxychlor were found to be estrogenic. Endocrine-disrupting chemicals include environmental estrogens such as o,p-DDT, endosulfan, non-planar polychlorinated biphenyl (PCB), octyl-and nonylphenols, the antiandrogens such as vinclozolin and DDE, and the thyroid hormone disrupters such as fenvalerate and benzene hexachloride.[1]

Endocrine-disrupting chemicals are a significant public health concern since these compounds interfere with normal function of pathways responsible for both reproduction and development and can affect the endocrine system, interfering in the production or action of hormones or compromising sexual identity, fertility, or behavior.[2,3] Besides, many of them are persistent in the environment, can be found in waters and sediments, and are easily transported long distances in the atmosphere.[4]

In recent years, numerous studies have suggested that many environmentally persistent chemicals have a potential to disrupt normal functions of the endocrine system. The field of endocrine disrupters, such as the special susceptibility of the developing organism and early induction of latent effects, has come a long way since its initial impetus in 1991.[5] Specially, the possible effects of EDCs on early events of proper gonadal development—which is dependent on intercellular signaling mechanisms— deserve attention since the early steps in mammalian sexual development are vulnerable to genetic and environmental perturbation.[6]

Exposure to EDCs is associated with dysfunctions of metabolism, energy balance, thyroid function, and reproduction, and an increased risk of endocrine cancers. These multifactorial disorders can occur through molecular epigenetic changes induced by exposure to EDCs early in life, the expression of which may not manifest until adulthood.[7] Effects attributed to the EDCs include developmental demasculinization and feminization in reptiles, mammals, amphibians, fish, and birds; reduced fecundity in reptiles, birds, and fish; and possibly increased breast cancer rates and reduced sperm counts in humans.[1]

Since hormones, in synergy with genes, are responsible for sex-related differences in anatomical, physiological, and behavioral traits, even if EDCs are present in minute amounts in environment, their effects in male and female physiology could be greater than before expected. They might also prejudice the sex steroid hormone–induced integrated physiological responses in women and men. In addition, differences in male and female susceptibility to EDCs could be present even if there is still scarce information available on this aspect.[8]

Mechanisms of Action

Several EDCs may work by multiple mechanisms, including uncharacterized mechanisms of action. Because of cross talk between different components of the endocrine systems, effects may occur unpredictably in endocrine target tissues other than the system predicted to be affected. A few modes of action could contribute to the same outcome, including aromatase inhibition, antiestrogenicity, testosterone biosynthesis disruption, and antiandrogens that alter upregulation of aromatase in the target regions within the brain. More complex biological responses to EDCs will generally represent combinations of several physiological processes integrated through multiple biological pathways.[9,10]

Endocrine disrupters may interfere with the functioning of hormonal systems in at least three possible ways: 1) by mimicking the action of a naturally produced hormone, producing similar but exaggerated chemical reactions in the body; 2) by blocking hormone receptors, preventing or diminishing the action of normal hormones; and 3) by affecting the synthesis, transport, metabolism, and/or excretion of hormones, thus altering the concentrations of natural hormones.[11]

The first characterized mechanism of action of EDCs is to act directly as ligands to steroid hormone nuclear receptors (NRs), in particular, estrogen, androgen, and thyroid NRs. Nuclear receptors are a class of proteins found within cells. In response to the presence of hormones, these receptors work in concert with other proteins to regulate the expression of specific genes by a conformation change. Schematically, NRs may be classified into four classes according to their dimerization and DNA-binding proprieties.[10]

Cross talk between NR-mediated and other signal-transduction pathways is an important aspect of NR-based regulation. This so-called genomic or genotropic signaling is normally slow and sustained, taking hours before biological outcomes become manifest. For example, in the classic view of estrogen action, the effects of 17β-estradiol (E2) were thought of as mediated by the NRs estrogen receptor α and β, acting as ligand-dependent transcription factors, thereby regulating gene expression by binding estrogen response elements in the DNA.[12]

Another type of NR cross talk that has recently been recognized is the non-genomic action of several NRs. Some non-genomic actions of NR ligands are apparently mediated through membrane receptors that are not part of the NR superfamily.[10] For example, it has become clear that E2 can also rapidly and transiently trigger a variety of second messenger signaling events, including the induction of cyclic adenosine monophosphate (cAMP) and adenylate cyclase; the mobilization of intracellular calcium; and the stimulation of PI3K, PKB, and Src with consequent activation of the extracellular-regulated kinases Erk1 and Erk2 in the Src/Ras/Erk cascade. All these effects are believed to be mediated through a membrane-associated or cytosolic estrogen receptor (ER) and have, therefore, been termed non-genomic or extranuclear actions of E2.[12]

The cellular activities of estrogens and xenoestrogens are the result of a combination of extranuclear (non-genomic) and nuclear (genomic) events and highlight the need to take non-genomic effects and signaling cross talk into consideration when screening for environmental estrogen.[12]

Disruption of the endocrine system by xenobiotic compounds is consistently reported in humans and wildlife and is a matter of concern worldwide. A great variety of natural or synthetic chemicals such as

EDCs are thought to exert an acute effect at different levels of the thyroid cascade. It is consensual that EDCs probably act by interfering with thyroid hormone (TH) synthesis, cellular uptake, and metabolism, at the level of TH receptors and also TH transport, by binding to thyroid hormone distributor proteins (THDPs). The TH transport system in particular may be quite susceptible to EDCs as many chemicals are structurally related to THs and may bind THDPs and disturb homeostasis of extracellular TH levels or even cellular uptake.[13]

Steroid hormone synthesis is controlled by the activity of several highly substrate-selective cytochrome P450 enzymes and a number of steroid dehydrogenases and reductases. Cytochrome P450 monooxygenases (CYPs) form a large group of enzymes found in most organisms from bacteria to mammals and can be grouped into 281 families. According to their function CYPs can be classified into enzymes metabolizing xenobiotics and enzymes that are part of key biosynthetic pathways, with narrow substrate specificity. Particularly, aromatase (CYP19), the enzyme that converts androgens to estrogens, has been the subject of studies into the mechanisms by which chemicals interfere with sex steroid hormone homeostasis and function, often related to (de)feminization and (de)masculinization processes.[14,15]

After all, several findings suggest that responses to EDCs cannot be assumed to be monotonic across a wide dose range and that change can occur in response to extremely low concentrations. In particular, low-dose effects may be mediated by endocrine-signaling pathways, evolved to act as powerful amplifiers, with the result that large changes can occur in response to extremely low concentrations. Dose–response relationship, however, is perhaps one of the most controversial issues in EDC studies. Reports on non-linearities in dose–response functions are highly controversial and the subject of intense research: non-monotonic, linear, and even threshold responses are all possible outcomes of low-dose exposure.[16] A non-monotonic response decreases testing efficiency and multiplies the time and other resources necessary to understand the potential hazard posed by a chemical. Because the issue of low-dose effects of EDCs was based on unknown and unexpected mechanisms, the actual features of these effects were not readily resolved.[17]

Low-dose effects of EDCs are based on unknown and unexpected mechanisms. Recent developments in the biological sciences, including homeostatic regulatory disturbance and epigenetic response, have aided in clarifying the mechanisms underlying the low-dose issue. Elucidating the xenobiotic effects of EDCs requires development of systems toxicology, i.e., deciphering the toxicity mechanisms underlying homeostatic regulatory disturbances.[17,18]

Hormones, Reproductive Aspects, and EDCs

In vertebrates, the ability to attain reproductive competence in adulthood involves the organization of a complex, steroid sensitive network in hypothalamic–preoptic–limbic brain regions during critical developmental windows. This process includes the establishment of the hypothalamic neural network of gonadotropin-releasing hormone (GnRH) cells, together with their regulatory inputs from other neuronal and glial cells in the brain, which enable feedback effects of steroid hormones on pulsatile GnRH release and the preovulatory GnRH/LH–luteinizing hormone surge in females. The anatomical development of this steroid-sensitive hypothalamic network occurs early in life, typically the late embryonic and early postnatal period in mammals, and its organization is important to the attainment and activation of appropriate reproductive functions in adulthood. Importantly, this same early developmental period is also a critical period for sexual differentiation of hypothalamic–limbic neural networks that must be organized perinatally to enable proper behavioral activation in adulthood. During mammalian development, the fetal organism is exposed to its own gonadal hormones, placental steroids, and maternal hormones that may cross the placental barrier. There are sex differences in exposures to androgens and estrogens that appear to underlie normal reproductive neuroendocrine development. Aberrations in these developmental patterns in females can cause masculinization (acquisition of a male-typical trait) or defeminization (loss of a female-typical trait) and in males may cause feminization or demasculinization (comparably defined). Perinatal hormones have permanent imprinting effects on the hypothalamus, manifested early on as morphological sex differences in the brain and manifested much later on as physiological and behavioral differences between the sexes.[19–21] Besides, androgens and estrogens can play a special role in the development of sexually dimorphic behaviors.[22]

Some populations exposed to chemicals from industrial accidents or chemical misadventures are of particular interest. Data from these select populations with higher levels of exposure than the common population seem to suggest that some of these chemicals have a role in genitourinary development as endocrine disrupters. Furthermore, animal studies of EDC effects on genitourinary development have confirmed that changes occur with exogenous manipulation of steroid levels or hormone receptors. These findings in animals have led to observational and epidemiological studies in humans to document a link between environmental exposure and human disease.[23]

Global declines in semen quality were suggested to be associated with enhanced exposure to environmental chemicals that act as endocrine disrupters as a result of increased use of pesticides, plastics, and other anthropogenic materials. A significant body of toxicology data based upon laboratory and wildlife animals studies suggests that exposure to certain endocrine disrupters is associated with reproductive toxicity, including the following: 1) abnormalities of the male reproductive tract (cryptorchidism, hypospadias); 2) reduced semen quality; and 3) impaired fertility in the adult.[24]

Recently, there has been increasing concern about the potential for environmental EDCs as fungicides to alter sexual differentiation in mammals. In this direction, observations demonstrate that vinclozolin (a systemic dicarboximide fungicide) can affect embryonic testicular cord formation in vitro. This transient in utero exposure to the fungicide increases apoptotic germ cell numbers in the testis of pubertal and adult animals. This effect is correlated with reduced sperm motility in the adult and putative effects on spermatogenic capacity later in adult life. In conclusion, transient exposure to this fungicide during the time of testis differentiation alters testis development and function.[25]

A higher prevalence of cryptorchidism and hypospadias was found in areas with extensive farming and pesticide use and in sons of women working as gardeners. Recently, a relation has been reported between cryptorchidism and persistent pesticide concentration in maternal breast milk.[26]

Other commonly used fungicides, such as the azoles, may also act as endocrine disrupters in vivo. They showed endocrine-disrupting potential when tested for endocrine-disruptive effects using a panel of in vitro assays. Overall, the imidazoles (econazole, ketoconazole, miconazole, prochloraz) were more potent than the triazoles (epoxiconazole, propiconazole, tebuconazole). The critical mechanism in vitro seems to be disturbance of steroid biosynthesis.[27]

Regarding in vivo effects, many of the commonly used azole fungicides act as endocrine disrupters, although the profile of action varies. Common features for azole fungicides are that they increase gestational length, virilize female pups, and affect steroid hormone levels in rat fetuses and/or dams.[28] For example, prochloraz causes reproductive malformations in androgen-dependent tissues of male offspring of exposed rats.[29] Also, tebuconazole has been found to demasculinize the male offspring and to possess some of the same endocrine effects as prochloraz. These effects strongly indicate that one major underlying mechanism for the endocrine-disrupting effects of azole fungicides is disturbance of key enzymes like CYP17 involved in the synthesis of steroid hormones.[28]

Also, triazole-induced male reproductive toxicity includes disruption of testosterone homeostasis. Elevated serum testosterone, increased testis, weights and anogenital distance, and hepatomegaly indicative of altered liver metabolism of steroids are the key events consistent with this mode of action.[30] Developmental exposure to triazole fungicides such as propiconazole, myclobutanil, and triadimefon can adversely impact reproduction in the female rat.[31] In this way, epoxiconazole and ketoconazole may be fetotoxic, increasing postimplantation loss and late resorptions.[32]

Aside from triazoles, the organic fungicide fenarimol possesses estrogenic properties[33] and acts both as an estrogen agonist and as an androgen antagonist.[34] In addition, fenarimol affects rat aromatase activity in vivo, inhibiting estrogen biosynthesis in rat microsomes[35] and in human tissues.[3]

This compound also affects other enzymes of the cytochrome P450 gene family that are involved in the metabolism of steroids.[36]

Induction of reactive oxygen species (ROS) by environmental contaminants and associated oxidative stress also have a role in defective sperm function and male infertility, although there are some controversial data. This is evidence for the existence of a link between endocrine-mediated and ROS-mediated adverse effects of environmental contaminants on male reproduction. Another link is the antioxidant enzyme superoxide dismutase, which has been shown to have a superoxide scavenging effect as well as act as an alternate regulatory switch in testicular steroidogenesis.[37]

Endocrine-disrupting chemicals can also impact female fertility by altering ovarian development and function, purportedly through estrogenic, antiestrogenic, and/or antiandrogenic effects. These compounds may also cause transgenerational effects by targeting oocyte maturation and maternal sex chromosomes.[38]

In girls, earlier age at menarche was reported after exposure to PCBs, polybrominated biphenyls, persistent pesticides (DDT), and phthalate esters. However, several other studies found no effect of these compounds on age at menarche. In boys, exposure to PCBs, PCDFs, or the pesticide endosulfan was associated with delayed puberty or decreased penile length. Much of the results found in population studies are in accordance with experimental studies in animals. However, the mixture of different components with antagonistic effects (estrogenic, antiestrogenic, antiandrogenic) and the limited knowledge about the most critical window for exposure (prenatal, perinatal, and pubertal) may hamper the interpretation of results.[39]

In human and rodent models, EDCs also interfere with the development of cognition and behaviors. In this way, fenvalerate is a potential EDC and is a candidate environmental risk factor for cognitive and behavioral development, especially in the critical period of development.[40] Also, prenatal phthalate exposure was associated with childhood social impairment in a multiethnic urban population.[41]

Recently, the interference of EDCs with receptors regulating metabolism has been proposed especially in relation with the etiology of metabolic diseases such as obesity and diabetes. In particular, the harmful action of EDCs on normal adipocyte development, homeostatic control of adipogenesis, early energy balance, and, in turn, body weight has been demonstrated. Much remains to be studied about the endocrine pathways responding to EDC exposure, especially those controlling feeding behavior, as their impairment represents a real risk factor for metabolic and feeding disorders.[42]

Screening Assays and Biomarkers for EDCs

Environmental stresses as presence of EDCs due to human activities are increasingly likely to pose habitat disturbances that could have potential deleterious effects on physiological function in vertebrates. These effects could result in major impacts on the life cycles of organisms, affecting morphology, physiology, and behavior. However, because animals live in diverse habitats, there is variation in susceptibility to disruption of response systems to environmental cues. While some populations of vertebrates, from fish to mammals, temporarily resist environmental stresses and breed successfully, many others show varying degrees of failure, sometimes resulting in population decline.[43] The development of targeted bioassays in combination with adequate chemical analyses is important for EDC risk assessment.

It is acknowledged that EDCs can affect humans and animals at low exposure levels and that responses to EDCs are in many cases complex, activating a range of different molecular events, e.g., receptor agonism/antagonism and enzyme induction, in multiple hormone systems. As a result, regulatory testing for these effects and evaluating the results is complicated.[44,45] In the typical case of assessing human risk, a scientifically justified validation could only mean an experimentally validated mechanistic link between EDC assay results and human susceptibilities at environmental exposures, sustained by reliable sensitivity and specificity benchmarks. Analytical methods have long been used to determine concentrations of chemical residues that persist in environment and accumulated in biota. Although they are useful, the development of EDC screening and monitoring procedures may help in the establishment of EDC exposure and biological responses. In this way, several in vitro and in vivo procedures have been proposed to screen and monitor individual EDCs or their mixtures.[45]

As an in vitro model, the use of the bovine ovarian follicle has already been recommended as a valuable instrument to unravel reproductive events in women due to the similarities in ovarian follicular dynamics and endocrine control.[46]

As an in vivo test, some external biomarkers of prenatal androgen disruption may be used, including the anogenital distance and the juvenile nipple/areola number. The anogenital distance is defined as the distance between the genital papilla and the anus; male rodents have an anogenital distance that is approximately twice the length as that of females. Areolae are dark areas surrounding the nipple bud and. their presence as measured at postnatal day 2–3 is indicative of adult nipples. Adult female rats typically have 12 nipples, whereas males have none. Both of these biomarkers vary with prenatal exposure to androgens or antiandrogens in females and males, respectively. Reduction of anogenital distance and/or retention of nipples in male rats is indicative of prenatal exposure to antiandrogens.[47]

In this context, a wide spectrum of potential biomarkers also could be applied to the study of endocrine disruption in the aquatic environment. In fish, they include changes in hormone titers (steroid hormones, thyroid hormones), abnormal gonad development, low gamete viability, and alterations in some enzyme activities (i.e., aromatases) and protein levels (i.e., vitellogenin, zona radiata proteins, spiggin). Likewise, evidence is slowly growing that indicates that gamete development and vitellogenesis of marine bivalve mollusks are targets of EDCs.[48]

On the other hand, although it is known that aquatic invertebrates contain different classes of steroids,[49] a clear cause–effect relationship between exposure and specific responses for most EDCs is far from being established.[48]

In crustacean populations, the attribution of endocrine toxicity to observed disturbances requires the identification of definitive biomarkers of such toxicity. Mortality, reduced fecundity, lowered recruitment, and impaired growth all might serve as indicators of endocrine disruption in crustaceans; however, such end points are indicative of adversity involving a variety of mechanisms. An exception to this premise is excess males in parthenogenic branchiopod populations that normally exist predominantly as females.[50]

Other organisms, such as amphibians, may be used to study the endocrine system and can serve as sentinels for detection of the modes of action of EDCs. Recently, amphibians are being reviewed as suitable models to assess (anti)estrogenic and (anti)androgenic modes of action influencing reproductive biology as well as (anti)thyroidal modes of action interfering with the thyroid system.[51]

Biochemical end points can also be useful biomarkers since environmental toxicants can trigger biological effects at the organism level only after initiating biochemical and cellular events. The cellular response to stress is characterized by the activation of genes involved in cell survival to counteract the physiological disturbance induced by physical or chemical agents. As an example, Hsps are suitable as an early-warning bioindicator of environmental hazard by various pollutants such as EDCs, because of their sensitivity to even minor changes in cellular homeostasis and their conservation along the evolutionary scale.[52] In addition, a combined testing strategy, considering both markers of endocrine/hormonal maturation and behavioral end points under hormonal control, may evidence even subtle perturbations of the neuroendocrine homeostasis, which often go undetected.[53]

Guidelines for Regulatory Purposes

In recent years, under the current European Union chemical regulation REACH (Registration, Evaluation, Authorization and Restriction of Chemicals), which revised plant protection product and biocide directives,[54] evaluation of endocrine-disrupting properties of chemicals has become a regulatory effort.

The initial framework for regulatory purposes has been revised by the Endocrine Disrupters Testing and Assessment (EDTA) Task Force at its meetings to reflect the Organization for Economic Cooperation and Development (OECD) member countries' views. The conceptual framework agreed upon by the EDTA6 in 2002 is not a testing scheme but rather a toolbox in which the various tests that can contribute information for the detection of the hazards of endocrine disruption are placed. The toolbox is organized into five compartments or levels, each corresponding to a different level of biological complexity (for both toxicological and ecotoxicological areas). Even though the conceptual framework may be full of testing tools, this does not imply that they all will be needed for assessment purposes. Tools will be added as they are validated in future. The conceptual framework is subject to further elaboration and discussion as the work on endocrine disrupters proceeds.[55] The OECD adopted in 2007 the uterotrophic bioassay as a standardized screening test with international regulatory acceptance. This assay may be used to screen for estrogenic properties of chemicals. However, generally, EDCs are handled as such only if their endocrine-disruption potential has been previously identified via, for example, academic research or is indicated by effects observed in required toxicity tests.[44]

The Endocrine Disruptor Screening Program (EDSP) of the United States Environmental Protection Agency (EPA) has been working to reach a consensus validation on a battery of screens and long-term tests for endocrine disrupters.[45] The Endocrine Disrupter Screening and Testing Advisory Committee (EDSTAC) was established by the EPA in 1996 as a federal advisory committee to provide advice in developing and implementing new screening and testing procedures for endocrine effects as mandated by the U.S. Congress (through the Food Quality Protection Act of 1996) in response to public concern.[56] The ED-STAC assesses the current state of the science and assists the agency in developing an endocrine screening program. The EDSTAC consists of scientists and others representing various interests, including advocates of the endocrine-disruption theory and the regulated community.

The EDSTAC concluded that the assays necessary to determine the potential endocrine activity of chemical substances varied significantly in their degree of development and validation. Several screens had an extensive history, e.g., the uterotrophic and the Hershberger screens, but others were only partially developed or were only hypothetically useful as screens, e.g., the amphibian developmental screen and the fish gonadal recrudescence screen. The fundamental validation principles are to clearly state the purpose and biological basis for the assay and to verify the performance of the assay against validation criteria using a common set of test chemicals across multiple laboratories.[57]

At the same time, EDSTAC recommended that EPA develop an extensive program that would subject all chemicals to screening and testing for estrogenic, antiestrogenic, androgenic, antiandrogenic, and thyroid effects in both humans and wildlife. Specifically, EDSTAC recommended, among other things, that the EPA do the following: 1) adopt a two-tiered, hierarchical testing and evaluation framework; and 2) initiate a research program, composed of both basic and applied research, to develop, standardize, and validate the necessary endocrine test methods. The EPA's EDSP was implemented in 2009–2010 with the issuance of test orders requiring manufacturers and registrants of 58 pesticide active ingredients and 9 pesticide inert/high-production-volume chemicals to evaluate the potential of these chemicals to interact with the estrogen, androgen, and thyroid hormone systems. Despite this great effort, numerous questions and uncertainties remain as to the usefulness and limitations of the specific assays. Understanding the tests' strengths and limitations is critical for interpretation of the screening results and for decision making based on those results.[57,58]

During the time EDSTAC was meeting, OECD began collaborating with its member countries, including the U.S. EPA, to develop internationally harmonized test guidelines.[48] Although the EPA and OECD endocrine screening and testing methods have been substantially harmonized, the framework of OECD's endocrine screening and testing program differs significantly from EPA's two-tiered EDSP. The EPA screening will entail evaluation of responses in the Tier 1 Endocrine Screening Battery, consisting of 11 distinct in vitro and in vivo assays. The OECD framework provides the flexibility to enter and exit at any level depending on information needs and encourages the maximal use of all existing relevant information that may be equally predictive and reduce vertebrate testing. The screening results are collectively intended to identify chemicals for which subsequent Tier 2 testing is necessary. Tier 2 testing uses test methods that encompass reproduction and developmental life stages in several species to provide data on adverse effects and dose response for risk assessment.[57]

In the years that the EPA worked on developing, standardizing, and validating the EDSTAC-recommended assays and implementing the EDSP, significant advances have been made in both computational and molecular technologies for discerning potential endocrine activity.[57] Accordingly, there are efforts to model EDC effects using computational approaches by the development and validation of mechanistically based computational models of hypothalamic–pituitary–thyroid (HPT); hypothalamic–pituitary–gonadal (HPG); hypothalamic–pituitary adrenal (HPA) axes in ecologically relevant species to better predict accommodation and recovery of endocrine systems.[59]

Overview of EDC Exposure

Human Exposure Effects

Models for estimating human exposure to endocrine disruptor (ED) pesticides are an important risk management tool. Many of them are harmful at very low doses, especially if exposure occurs during sensitive stages of development, producing effects that may not manifest for many years or that affect descendants via epigenetic changes. The main requirement for the use of such models is more quantitative data on the sources and pathways of human ED pesticide exposure. Quantifying the risks posed by the different routes of exposure will play an important part in designing and implementing effective risk mitigation for ED pesticides. In fact, it is difficult to assess the relative importance of some routes of exposure because no data sets that would allow these to be calculated are available. Pesticide exposure from the use of pesticides for medicinal purposes and exposure from cut flowers and ornamental plants both need to be quantified, and better data sets are required for pesticide exposure from spray drift, home use, municipal use, and travel.[60]

Food and water are both chronic exposure routes affecting the entire population. Food residues are currently thought to be the most important exposure pathway, for although residue levels present in food tend to be below the maximum residue levels permitted by law, they do result in constant measurable low-level exposure.[51] Food as a major xenobiotic and heavy metal exposure route to humans is studied intensively. More than 100 chemicals have been identified as antiandrogens, including certain phthalates, widely used as plasticizers, pesticides, and various other chemicals found in food and consumer products.[61]

Indeed, typical food contaminants, like pesticides, dioxins, PCBs, methylmercury, lead, etc., are well characterized in food. In contrast, the role of food and beverage packaging as an additional source of contaminants has received much less attention, even though food packaging contributes significantly to human xenobiotic exposure. Especially, EDCs in food packaging are of concern since even at low concentrations, chronic exposure is toxicologically relevant. Thus, non-intentionally added substances migrating from food contact materials need toxicological characterization.[62]

Some chemicals used in food processing have an environmental endocrine-disrupting effect that affects reproduction in wildlife. For example, bisphenol-A is a monomer of polycarbonate plastics and a constituent of epoxy and polystyrene resins, which are used in the food cans and found as a contaminant not only in the liquid of the preserved foods but also in the water autoclaved in the cans. This chemical is also released from polycarbonate flasks during autoclaving. Moreover, it has been reported that significant amounts of bisphenol-A are detected in the saliva of dental patients treated with fissure sealants. The exposure to low doses of this chemical was reported to affect the rate of growth and sexual maturation, hormone levels in blood, reproductive organ function, fertility, immune function, enzyme activity, brain structure, brain chemistry, and behavior.[63]

Another important route of EDC exposure is occupational. Relatively high levels of exposure to environmental endocrine disrupters in the form of pesticides occur among people working in agriculture. Some pesticides are able to influence the synthesis, storage, release, recognition, or binding of hormones, which may lead to alterations in reproductive hormone levels. The issue of male infertility caused by occupational exposure is pertinent worldwide. A significant increase in the incidence of male infertility has been described in the international literature. Part of this effect may result from synthetic toxic substances acting on the endocrine system, many of which are routinely used in work processes. However, progress is needed in the knowledge of possible effects of exposure on male fertility since monitoring these effects requires sufficient time for the manifestations to occur. Such progress will allow the development of preventive measure within the field of workers' health.[64]

Apart from EDC effects on males, several studies on occupational exposure to pesticides and adverse effects on human reproduction have been performed, including end points such as prolonged time to pregnancy, spontaneous abortion or stillbirth, low birth weight, and developmental disorders.[65]

Complex EDC Mixtures

Concerns increase when humans are exposed to mixtures of similar-acting EDCs and/or during sensitive windows of development. It is difficult to predict biological effects directly from the composition of pollutant mixtures. In addition to simple additive effects, interactions between different chemicals in a mixture may result in either a weaker (antagonistic) or a stronger (synergistic) combined effect than would be expected from knowledge about the toxicity and mode of action of each individual compound. Such interactions may take place in the toxicokinetic phase (i.e., processes of uptake, distribution, metabolism, and excretion) or in the toxicodynamic phase (i.e., effects of chemicals on the receptor, cellular target, or organ). A chemical mixture may contain a number of xenoestrogens enhancing the response of endogenous estrogens, or it may contain xenoantiestrogens that inhibit the normal action of endogenous estrogens.[66] Substances of concern include certain phthalates, pesticides and chemicals used in cosmetics and personal care products. A lack of knowledge about relevant exposure scenarios presents serious obstacles for better human risk assessment. A disregard for combination-effect studies may lead to underestimations of risks. In this way, the study of EDC mixture effects by developing biomarkers that capture cumulative exposure to endocrine disrupters is needed.[67]

Doses of endocrine-disrupting pesticides that appear to induce no effects on gestation length, parturition, and pup mortality when alone induced marked adverse effects on these end points together with other pesticides.[68] They can also affect the sexual differentiation of offspring.[68] Chemicals that act on different fetal tissues via diverse cellular mechanism of action may produce additive effects. This fact indicates that the current framework for conducting cumulative risk assessments should not only consider including chemicals from different classes with the same mechanism of toxicity but also include chemicals that disrupt differentiation of the same fetal tissue at different sites in the androgen signaling pathway.[5] Compounds that act by disparate mechanisms of toxicity to disrupt the dynamic interactions among the interconnected signaling pathways in differentiating tissues produce cumulative dose-additive effects, regardless of the mechanism or mode of action of the individual mixture component.[69] Predictive approaches are generally based on the mathematical concepts of concentration addition and independent action, both predicting the toxicity of a mixture based on the individual toxicities of the mixture components.[26]

In this sense, a combination of five pesticides with dissimilar mechanisms of action produced greater androgen sensitive end-point responses than would be expected using response-addition modeling.[70] Deltamethrin, methiocarb, prochloraz, simazine, and tribenuron-methyl are all commonly used for agricultural and horticultural purposes. In vivo, the levator ani/bulbocavernosus muscle and adrenal gland weight changes indicated that the pesticides had an accumulating effect that was not observed for the individual pesticides. Several pesticide-induced gene expression changes were observed, indicating that these may be very sensitive antiandrogenic end points. In another study,[71] dexamethasone appeared to exacerbate the reproductive anomalies induced by in utero exposure of male rats to dibutyl phthalate.

In a recent study, male Sprague Dawley rats were sub-chronically exposed to single doses of dibutyl phthalate, single doses of benzo(a)pyrene, and combined doses of both EDCs. Significant adverse effects were observed on the reproductive system, including decreased sperm count, increased production of abnormal sperm, changes in serum testosterone levels, and irregular arrangements of the seminiferous epithelium. It is also observed that biochemical analyses showed that the activities of superoxide dismutase and glutathione peroxidase decreased after exposure to these EDCs. Therefore, the data suggest that exposure to them, in either separate or combined doses, can affect the reproductive system of male rats adversely via oxidative stress-related mechanisms.[72]

Thus, assessment of risks posed by chemicals causing reproductive effects and protection of future generations are important public health tasks. To determine the levels of significant human exposure to a given chemical and associated health effects, the Agency for Toxic Substances and Disease Registry's (ATSDR's) toxicological profiles examine and interpret available toxicological and epidemiological data. The ATSDR categorizes the health effects according to their seriousness as serious (effects that prevent the organism from functioning normally or can cause death), less serious (changes that will prevent an organ or organ system from functioning in a normal manner but will not necessarily lead to the inability of the whole organism to function normally), or minimal (effects that reduce the capacity of an organ or organ system to absorb additional toxic stress but will not necessarily lead to the inability of the organ or organ system to function normally). The ATSDR uses the highest no-observed-adverse-effect level or the lowest-observed-adverse-effect level (LOAEL) in the available literature to derive a health-based guidance value called a minimal risk level (MRL). An MRL is defined as "an estimate of the daily human exposure to a substance that is likely to be without an appreciable risk of adverse, non-cancer effects over a specified duration of exposure." Minimal risk levels based on reproductive and endocrine effects were described in a review by Pohl et al.[73]

Some Examples of Animal Exposure Effects

There is widespread exposure to EDCs, which can disrupt the reproduction and development of various non-target organisms. Effects of EDCs have been shown by observed adverse reproductive and developmental effects. Indeed, most studies of potential EDC effects are based on indirect evidence of endocrine disruption rather than defined endocrine pathways.

Some domestic mammals may come into contact with EDCs by sewage exposure. As an example, sewage sludge is sometimes recycled to arable land or pasture and contains large amounts of a variety of pollutants, including EDCs and heavy metals, derived from industrial, agricultural, and domestic sources. A demasculinizing effect of exposure to higher pollutant concentrations with respect to exploratory sheep behavior was observed.[74] These observations demonstrate the need to take into account the effects of pollutant combinations, even at very low, environmental concentrations, and further highlight the usefulness of ethotoxicology for the study of biological effects of environmental pollutants.

Endocrine-disrupting chemicals have been found in sewage effluent in low concentrations (ng/L). Some of these estrogens bind with estrogen receptors in exposed organisms and have the potential to exert effects at extremely low concentrations. Data from laboratory experiments support the hypothesis that EDCs in the aquatic environment can impact the reproductive health of various fish species, but evidence in the aquatic environment is still weak and needs a dependable method or indicator to assess reproduction of fish in situ. The link between endocrine disruption and reproductive impairment that cause an ecologically relevant impact on the sustainability of fish populations remains to be better understood.[75]

Surface waters are the main sink of EDCs, which are mainly of anthropogenic origin. Thus, aquatic organisms, especially lower vertebrates such as fish and amphibians, are the main potential targets for EDCs, being at direct or indirect risk via ingestion and accumulation of EDCs via exposure or the food chain. These compounds may play an important role in the decline of the amphibian population.[51] Several incidents in the wildlife population strongly correlated decreased reproductive capacity with exposure to specific industrial chemicals, and the organisms may be viewed as sentinels of human health effects. Reported reproductive disorders in wildlife have included morphologic abnormalities, eggshell thinning, population declines, sex reversal, impaired viability of offspring, altered hormone concentration, and changes in sociosexual behavior.[76]

The ED are prevalent over a wide range of chemicals in the aquatic ecosystems, most of them being resistant to environmental degradation and considered ubiquitous contaminants.[48,77] Some imidazole (prochloraz, imazalil) and triazole (epoxiconazole) agricultural fungicides induced oocyte maturation in rainbow trout. Prochloraz, epoxiconazole, and imazalil strongly potentiated the induction of oocyte maturation by gonadotropin in a dose-dependent manner.[78] Above all, prochloraz caused responses consistent with aromatase inhibition, although there were indications that the fungicide may also be disturbing the balance between estrogens and androgens via effects elsewhere in the steroidogenic pathway.[77]

In U.K. rivers, a widespread feminization of wild fish was observed involving contributions from both steroidal estrogens and xenoestrogens and from other yet-unknown contaminants with antiandrogenic properties. The widespread occurrence of feminized male fish downstream of some wastewater treatment works has led to substantial interest from ecologists and public health professionals. This concern stems from the view that the effects observed have a parallel in humans and that both phenomena are caused by exposure to mixtures of contaminants that interfere with reproductive development.[79] Some authors reported the occurrence of fish feminization as well as reproduction and development interference with other aquatic organisms,[80] although there is no universally accepted bioassay or chemical technique to quantify EDCs in the aquatic environment.[81] Endocrine-disrupting chemicals can also promote disrupting effects in vitro on ovarian follicular cells exposed to environmentally relevant doses of mixtures of persistent organic pollutants extracted from marine and freshwater ecosystems.[66]

Population studies have revealed alterations in crustacean growth, molting, sexual development, and recruitment that are indicative of environmental endocrine disruption. However, environmental factors other than pollution (i.e., temperature, parasitism) also can elicit these effects and definitive causal relationships between endocrine disruption in crustacean field populations, and chemical pollution is generally lacking.[50] Also, temperature and photoperiod are the two most important environmental cues in the regulation of the annual cycles of circulating sex steroid hormones and reproduction in fish. Thus, these variables may alter the endocrine-disruption effects induced by EDCs.[82]

In contrast to mammals and birds, the mechanisms underlying sex determination and differentiation in fishes vary widely and are changeable or labile in response to environmental parameters. These environmental parameters include temperature, behavioral cues and demographic structure of the local population, and EDCs. Understanding the gender similarities and differences in how organisms respond following exposure to environmental chemicals is important to determine the relative risk of these agents to wildlife and human populations. Given the central role of sex steroid hormones in the sex determination and sexual differentiation of fishes, amphibians, and reptiles, future research that includes sex as a factor is recommended. Thus, the risk assessment can address the probable gender differences in effects from exposure to chemicals in the environment.[83]

Dealing with Environment EDC Emission

Municipal wastewater contains a complex mixture of EDCs originating from different sources. A number of organic pollutants, such as polycyclic aromatic hydrocarbons, PCBs, and pesticides, are resistant to degradation and represent an ongoing toxicological threat to both wildlife and human beings. Furthermore, recently, wastewater sludge has been subjected to reuse for production of value-added products. These facts have heightened the need for novel and advanced bioremediation techniques to effectively remove EDCs from a variety of contaminated environmental media including water, wastewater sludge, sediments, and soils. One possibility to solve this problem is the use of microbial potential to degrade or detoxify EDCs and other toxic intermediates.[80,84]

Also, there are physical methods such as absorption by activated carbon and rejection by membranes to remove EDCs. However, pollutant removal from wastewater is a process with high energy consumption, where cost and efficiency are the key considerations for their application. Biodegradation processes have proven to be the most cost-effective.[85]

Water companies became aware of the endocrine-disruption problem when a survey confirmed the observation by anglers of hermaphrodite fish in wastewater treatment plant lagoons after being exposed to significant levels of persistent man-made chemicals. The evolving regulatory context related to micropollutants in the environment may have a decisive impact on wastewater management and requires an increased knowledge of the fate of micropollutants during wastewater treatment. Advanced treatments such as oxidation (ozone) are known to be able to enhance the removal of micropollutants, but technical, economic, and environmental risk/benefit evaluations must be performed before implementing such additional processes. In any case, the reduction of the pollution at the source, i.e., upstream of the wastewater treatment plant, represents the most sensible option, which should be promoted.[86]

Although numerous studies have investigated degradation of individual EDCs in laboratory or natural waters, chemical-based analytical methods cannot represent the combined or synergistic activities between water quality parameters and/or the EDC mixtures at environmentally relevant concentrations since natural variations in water matrices and mixtures of EDC in the environment may confound analysis of the treatment efficiency. In conjunction with standard analytical approaches, bioanalytical assessments of residual estrogenic activity in treated water will enable estimates of the interactions and/or combined estrogenic activity among mixtures of EDCs and the water matrix in natural water.[87]

By contrast, the agricultural sector, a significant user of veterinary pharmaceuticals, has no such treatment—compounds are deposited straight to the ground in dung and urine or washed from hides in the case of topical applications. There has been little research as to whether any of these compounds leach into and persist in local soil and aquatic ecosystems. The extent to which the active ingredients of any of these chemicals (and their metabolites) leach into pastures, soil, runoff, and groundwater is a matter for field research. Also, much spraying of pesticides as herbicides and insecticides is done by ground crews. In such circumstances, it is not known whether they react with each other as well as pesticides and herbicides, forming further compounds which, either acting individually or in combination, could adversely affect bacteria, fungi, and higher organisms.[88]

Above all, a ranking system that could be customized for specific geographical locations will aid public policies in prioritizing EDCs that need monitoring and removal of aquatic sources as drinking water.[89] The establishment of simple but integrative screening assays for regulatory purposes is allowed by a strong correlation between xenoestrogen exposure and reproductive impairment. In fact, molecular screening assays could contain a battery of molecular targets allowing a more comprehensive approach in the identification of endocrine-disrupting compounds in fish and vertebrates in general.[90]

Different assays can be successfully employed as a battery of assays to screen environmental water samples for estrogenicity. The results obtained from this battery of assays should be interpreted as a first-tier screen for estrogenicity. Samples that test positive should be further investigated using second-and third-tier screens with routine sampling in order to monitor rivers for estrogenicity.[91] Complementarily, a fugacity-based model may be applied to simulate the distribution of EDCs in reservoirs of recycled waste-water,[92] or a fugacity-hydrodynamic model may be used for predicting the concentrations of the organic pollutants in surface water.[93]

Conclusion

Endocrine-disrupting chemicals can cause a wide range of reproductive damage and developmental, growth, immune, and behavior effects even in low doses and by different mechanisms of action. They encompass a variety of chemical classes, including hormones, plant constituents, pesticides, compounds used in the plastics industry and in consumer products, and other industrial by-products and pollutants. Some of them are widely dispersed in the environment. Exposure to EDCs can occur through direct contact with these chemicals or through ingestion of contaminated water, sediment, air, soil, and food and consumer products.

In humans, it is difficult to definitively link a particular EDC with a specific effect because the studies have inconclusive results. However, fetuses and embryos, whose growth and development are highly controlled by the endocrine system, are more vulnerable to exposure and may suffer reproductive abnormalities. The timing of exposure is also presumed to be critical, since different hormone pathways are active during different stages of development. Perinatal exposure, in some cases, can lead to permanent alterations that may be overt in adulthood.

Compared with humans, the evidence that wildlife has been affected adversely by exposures to EDCs is extensive. Available evidence seems to indicate that endocrine disruption caused by xenobiotics is primarily an ecotoxicologic problem. These chemicals may be extremely challenging for aquatic organisms and mammals that have a large habitat and that consume fish from many different areas throughout their lives. Low concentrations of endocrine disrupters can have synergistic effects in various organisms as amphibians. For removal of these compounds from aquatic sources, the most cost-effective process is biodegradation.

In spite of the need to manage the environmental, human health, and economic impacts of EDCs, most attention is focused on pharmaceutically active chemicals instead of those for agricultural use. The impact of these latter compounds is understudied.

The legal approach has been improved by new test protocols. Progress has been made in the identification and quantification of a wide array of chemicals with endocrine-active properties, especially those that persist and bioaccumulate in organisms and their environment. Studies with mammals have shown that exposure to endocrine-active compounds during early development may result in adverse health impacts that are not realized until adulthood. However, from a regulatory perspective, the ability of animals to recover from chemical insults is problematic because it complicates efforts to establish acceptable levels of exposure. Consequently, research to define the limits and biological cost recovery using standardized test designs is needed.[94]

However, exposure complexities, including transient and low-concentration exposure to EDCs, maternal metabolism of bioaccumulated EDCs, varying vulnerability and response by developmental stage, poorly understood exposure sources, mixtures and synergies, and cultural, social, and economic patterns, make it difficult for science to make solid exposure determinations. While there has been a great deal of research and effort in context with the hazard assessment and regulation of EDCs, there are also remaining uncertainties and issues. These include animal rights concerns due to significant increases in the use of animals to fulfill testing requirements; associated needs for alternative testing concepts such as *in vitro, in silico,* and modeling approaches; and lack of understanding of the relevance of exposure of humans and wildlife to EDCs.[95] Given the dynamic nature of the endocrine system, future efforts in the study of EDCs need more focus on the timing, frequency, and duration of exposure to these chemicals.

References

1. Kretschmer, X.C.; Baldwin, W.S. CAR and PXR: Xenosensors of endocrine disrupters? Chem.-Biol. Interact. **2005**, *155*, 111–128.
2. Castro, V.; Mello, M.A.; Diniz, C.; Morita, L.; Zucchi, T.; Poli, P. Neurodevelopmental effects of perinatal fenarimol exposure on rats. Reprod. Toxicol. **2007**, *23*, 98–105.
3. Vinggaard, A.; Jacobsen, H.; Metzdorff, S.; Andersen, H.; Nellemann, C. Antiandrogenic effects in short-term in vivo studies of the fungicide fenarimol. Toxicology **2005**, *207*, 21–34.
4. Waissmann, W. Health surveillance and endocrine disruptors. Cadernos de Saude Publica **2002**, *18* (2), 511–517.

5. Hotchkiss, A.K.; Rider, C.V.; Blystone, C.R.; Wilson, V.S.; Hartig, P.C.; Ankley, G.T.; Foster, P.M.; Gray, C.L.; Gray, L.E. Fifteen years after "wingspread"—Environmental endocrine disrupters and human and wildlife health: Where we are today and where we need to go. Toxicol. Sci. **2008**, *105* (2), 235–259.

6. Koopman, P. The delicate balance between male and female sex determining pathways: Potential for disruption of early steps in sexual development. Int. J. Androl. **2010**, *33*, 252–258.

7. Walker, D.M.; Gore, A.C. Transgenerational neuroendocrine disruption of reproduction. Nat. Rev. Endocrinol. **2011**, *7* (4), 197–207.

8. Bulzomi, P.; Marino, M. Environmental endocrine disruptors: Does a sex-related susceptibility exist? Front. Biosci. **2011**, *16* (7), 2478–2498.

9. Andersen, M.; Dennison, J. Mechanistic approaches for mixture risk assessments—Present capabilities with simple mixtures and future directions. Environ. Toxicol. Pharmacol. **2004**, *16*, 1–11.

10. Porcher, J.; Devillers, J.; Marchand-Geneste, N. Mechanism of endocrine disruptions—A tentative overview. In *Endocrine Disruption Modeling*; Devillers, J., Ed.; CRC Press: Boca Raton, 2009; 11–46.

11. Flynn, K. Dietary exposure to endocrine-active pesticides: Conflicting opinions in a European workshop. Environ. Int. **2011**, *37* (5), 980–990.

12. Silva, E.; Kabil, A.; Kortenkamp, A. Cross-talk between non-genomic and genomic signaling pathways—Distinct effect profiles of environmental estrogens. Toxicol. Appl. Pharmacol. **2010**, *245*, 160–170.

13. Morgado, I.; Campinho, M.A.; Costa, R.; Jacinto, R.; Power, D.M. Disruption of the thyroid system by diethylstilbestrol and ioxynil in the sea bream (*Sparus aurata*). Aquat. Toxicol. **2009**, *92*, 271–280.

14. Trosken, E.R.; Adamskab, M.; Arand, M.; Zarn, J.A.; Patten, C.; Volkel, W.; Lutz, W.K. Comparison of lanosterol-14α-demethylase (CYP51) of human and *Candida albicans* for inhibition by different antifungal azoles. Toxicology **2006**, *228*, 24–32.

15. Sanderson, J.T. The steroid hormone biosynthesis pathway as a target for endocrine-disrupting chemicals. Toxicol. Sci. **2006**, *94* (1), 3–21.

16. Belloni, V.; Dessì-Fulgheri, F.; Zaccaroni, M;. Di Consiglio, E.; De Angelis, G.; Testai, E.; Santochirico, M.; All-eva, E.; Santucci, D. Early exposure to low doses of atrazine affects behavior in juvenile and adult CD1 mice. Toxicology **2011**, *279*, 19–26.

17. Hirabayashi, Y.; Inoue, T. The low-dose issue and stochastic responses to endocrine disruptors. J. Appl. Toxicol. **2011**, *31*, 84–88.

18. Calabrese, E.J. Hormesis: A revolution in toxicology, risk assessment and medicine—Re-framing the dose–response relationship. EMBO Rep. **2004**, *5*, S37–S40.

19. Gore, A.C. Developmental programming and endocrine disruptor effects on reproductive neuroendocrine systems. Front. Neuroendocrinol. **2008**, *29*, 358–374.

20. Anway, M.D.; Skinner, M.K. Epigenetic transgenerational actions of endocrine disruptors. Endocrinology **2006**, *147* (6), S43–S49.

21. Crews, D.; McLachlan, J.A. Epigenetics, evolution, endocrine disruption, health, and disease. Endocrinology **2006**, *147* (6), S4–S10.

22. Li, A.A.; Baum, M.J.; McIntosh, L.J.; Day, M.; Liu, F.; Gray L.E., Jr. Building a scientific framework for studying hormonal effects on behavior and on the development of the sexually dimorphic nervous system. Neurotoxicology **2008**, *29*, 504–519.

23. Yiee, J.H.; Baskin, L.S. Environmental factors in genitourinary development. J. Urol. **2010**, *184*, 34–41.

24. Phillips, K.P.; Tanphaichitr, N. Human exposure to endocrine disrupters and semen quality. J. Toxicol. Environ. Health, Part B **2008**, *11* (3–4), 188–220.

25. Uzumcu, M.; Suzuki, H.; Skinner, M.K. Effect of the anti-androgenic endocrine disruptor vinclozolin on embryonic testis cord formation and postnatal testis development and function. Reprod. Toxicol. **2004**, *18* (6), 765–774.

26. Mnif, W.; Hassine, A.I.H.; Bouaziz, A.; Bartegi, A.; Thomas, O.; Roig, B. Effect of endocrine disruptor pesticides: A review. Int. J. Environ. Res. Pub. Health **2011**, *8* (6), 2265–2303.

27. Kjærstad, M.B.; Taxvig, C.; Nellemann, C.; Vinggaard, A.M.; Andersen, H.R. Endocrine disrupting effects in vitro of conazole antifungals used as pesticides and pharmaceuticals. Reprod. Toxicol. **2010**, *30,* 573–582.

28. Taxvig, C.; Hass, U.; Axelstad, M.; Dalgaard, M.; Boberg, J.; Andersen, H.R.; Vinggaard, A.M. Endocrine-disrupting activities in vivo of the fungicides tebuconazole and epoxiconazole. Toxicol. Sci. **2007**, *100* (2), 464–473.

29. Noriega, N.C.; Ostby, J.; Lambright, C.; Wilson, V.S.; Gray, L.E., Jr. Late gestational exposure to the fungicide prochloraz delays the onset of parturition and causes reproductive malformations in male but not female rat offspring. Biol. Reprod. **2005**, *72,* 1324–1335.

30. Goetz, A.K.; Ren, H.; Schmid, J.E.; Blystone, C.R.; Thillai-nadarajah, I.; Best, D.S.; Nichols, H.P.; Strader, L.F.; Wolf, D.C.; Narotsky, M.G.; Rockett, J.C.; Dix, D.J. Disruption of testosterone homeostasis as a mode of action for the reproductive toxicity of triazole fungicides in the male rat. Toxicol. Sci. **2007**, *95* (1), 227–239.

31. Rockett, J.C.; Narotsky, M.G.; Thompsona, K.E.; Thillai-nadarajah, I.; Blystone, C.R.; Goetz, A.K.; Rena, H.; Best, D.S.; Murrell, R.N.; Nichols, H.P.; Schmid, J.E.; Wolf, D.C.; Dix, D.J. Effect of conazole fungicides on reproductive development in the female rat. Reprod. Toxicol. **2006**, *22,* 647–658.

32. Taxvig, C.; Vinggaard, A.M.; Hass, U.; Axelstad, M.; Metzdorff S.; Nellemann, C. Endocrine-disrupting properties in vivo of widely used azole fungicides. Int. J. Androl. **2008**, *31,* 170–177.

33. Grünfeld, H.; Bonefeld-Jørgensen, E. Effect of in vitro estrogenic pesticides on human oestrogen receptor α and β mRNA levels. Toxicol. Lett. **2004**, *151* (3), 467–480.

34. Andersen, H.; Vinggaard, A.; Rasmussen, T.; Gjermandsen, I.; Bonefeld-Jørgensen, E. Effects of currently used pesticides in assays for estrogenicity, androgenicity, and aromatase activity in vitro. Toxicol. Appl. Pharmacol. **2002**, *179* (1), 1–12.

35. Hirsch, K.; Weaver, D.; Black, L.; Falcone, J.; Maclusky, N. Inhibition of central nervous system aromatase activity, a mechanism for fenarimol-induced infertility in the male rats. Toxicol. Appl. Pharmacol. **1987**, *91,* 235–245.

36. Paolini, M.; Mesirca, R.; Pozzetti, L.; Sapone, A.; Can-telli-Forti, G. Molecular non-genetic biomarkers related to fenarimol cocarcinogenesis, organ-and sex-specific CYP induction in rat. Cancer Lett. **1986**, *101,* 171–178.

37. Saradha, B.; Mathur P.P. Effect of environmental contaminants on male reproduction. Environ. Toxicol. Pharmacol. **2006**, *21,* 34–41.

38. Uzumcu, M.; Zachow, R. Developmental exposure to environmental endocrine disruptors: Consequences within the ovary and on female reproductive function. Reprod. Toxicol. **2007**, *23,* 337–352.

39. Hond, E.; Schoeters, G. Endocrine disrupters and human puberty. Int. J. Androl. **2006**, *29,* 264–271.

40. Meng, X.; Liu, P.; Wang, H.; Zhao, X.; Xu, Z.; Chen G.; Xu, D. Gender-specific impairments on cognitive and behavioral development in mice exposed to fenvalerate during puberty. Toxicol. Lett. **2011**, *203* (3), 245–251.

41. Miodovnik, A.; Engel, S.M.; Zhu, C.; Ye, X.; Soorya, L.V.; Silva, M.J.; Calafat, A.M.; Wolff, M.S. Endocrine disruptors and childhood social impairment. Neurotoxicology **2011**, *32* (2), 261–267.

42. Migliarini, B.; Piccinetti, C.C.; Martella, A.; Maradonna, F,; Gioacchini, G.; Carnevali, O. Perspectives on endocrine disruptor effects on metabolic sensors. Gen. Comp. Endocrinol. **2011**, *170,* 416–423.

43. Wingfield, J.C.; Mukai, M. Endocrine disruption in the context of life cycles: Perception and transduction of environmental cues. Gen. Comp. Endocrinol. **2009**, *163*, 92–96.

44. Beronius, A.; Rudén, C.; Hanberg, A.; Håkansson, H. Health risk assessment procedures for endocrine disrupting compounds within different regulatory frameworks in the European Union. Regul. Toxicol. Pharmacol. **2009**, *55*, 111–122.

45. Gori, G. Regulating endocrine disruptors. Regul. Toxicol. Pharmacol. **2007**, *48*, 1–3.

46. Campbell, B.K.; Souza, C., Gong, J.; Webb, R., Kendall, N.; Marsters, P., Robinson, G.; Mitchell, A.; Telfer, E.E.; Baird, D.T. Domestic ruminants as models for the elucidation of the mechanisms controlling ovarian follicle development in humans. Reprod. Suppl. **2003**, *61*, 429–43.

47. Wilson, V.S.; Blystone, C.R.; Hotchkiss, A.K.; Rider, C.V.; Gray, L.E., Jr. Diverse mechanisms of anti-androgen action: impact on male rat reproductive tract development. Int. J. Androl. **2008**, *31*, 178–187.

48. Porte, C.; Janer, G.; Lorusso, L.C.; Ortiz-Zarragoitia, M.; Cajaraville, M.P.; Fossi, M.C.; Canesi, L. Endocrine disruptors in marine organisms: Approaches and perspectives. Comp. Biochem. Physiol., Part C **2006**, *143*, 303–315.

49. Lafont, R.; Mathieu, M. Steroids in aquatic invertebrates. Ecotoxicology **2007**, *16*, 109–130.

50. LeBlanc, G.A. Crustacean endocrine toxicology: A review. Ecotoxicology **2007**, *16*, 61–81.

51. Kloas, W.; Lutz, I. Amphibians as model to study endocrine disrupters. J. Chromatogr. A **2006**, *1130*, 16–27.

52. Morales, M.; Planelló, R.; Martfnez-Paz, P.; Herrero, O.; Cortés, E.; Martfnez-Guitarte, J.L.; Morcillo, G. Characterization of Hsp70 gene in *Chironomus riparius*: Expression in response to endocrine disrupting pollutants as a marker of ecotoxicological stress. Comp. Biochem. Physiol. **2011**, *153*, 150–158.

53. Calamandrei, G.; Maranghi, F.; Venerosi, A.; Alleva, E.; Mantovani, A Efficient testing strategies for evaluation of xenobiotics with neuroendocrine activity. Reprod. Toxicol. **2006**, *22*, 164–174.

54. Danga, Z.C.; Ru, S.; Wang, W.; Rorije, E.; Hakkert, B.; Vermeire, T. Comparison of chemical-induced transcriptional activation of fish and human estrogen receptors: Regulatory implications. Toxicol. Lett. **2011**, doi:10.1016/j.toxlet.2010.12.020.

55. OECO Conceptual Framework for the testing and Assessment of Endocrine Disrupting Chemicals, available at http://www.oecd.org/document/580.3343,en_2649_34377_2348794_1_1_1_1.00.html (accessed May 2011).

56. Juberg, D.R. An evaluation of endocrine modulators: Implications for human health. Ecotoxicol. Environ. Safety **2000**, *45*, 93–105.

57. Borgert, C.J.; Mihaich, E.M.; Quill, T.F.; Marty, M.S.; Levine, S.L.; Becker, R.A. Evaluation of EPA's Tier 1 Endocrine Screening Battery and recommendations for improving the interpretation of screening results. Regul. Toxicol. Pharmacol. **2011**, *59*, 397–411.

58. Vogel, J.M. Perils of paradigm: Complexity, policy design, and the Endocrine Disruptor Screening Program. *Environ. Health: Global Access Sci. Source* **2005**, *4*, 2, doi:10.1186/1476-069X-4-2.

59. Nichols, J.W.; Breen, M.; Denver, R.J.; Distefano, J.J.; Edwards, J.S.; Hoke, R.A.; Volz, D.C.; Zhang, X. Predicting chemical impacts on vertebrate endocrine systems. Environ. Toxicol. Chem. **2011**, *30* (1), 39–51.

60. McKinlay, R.; Plant, J.A.; Bell, J.N.B.; Voulvoulis, N. Calculating human exposure to endocrine disrupting pesticides via agricultural and non-agricultural exposure routes. Sci. Total Environ. **2008**, *398*, 1–12.

61. Kortenkamp A.; Faust M. Combined exposures to anti-androgenic chemicals: Steps towards cumulative risk assessment. Int. J. Androl. **2010**, *33*, 463–474.

62. Muncke, J. Exposure to endocrine disrupting compounds via the food chain: Is packaging a relevant source? Sci. Total Environ. **2009**, *407*, 4549–4559.

63. Narita, M.; Miyagawa, K.; Mizuo, K.; Yoshida, T.; Suzuki, T. Prenatal and neonatal exposure to low-dose of bisphenol-A enhance the morphine-induced hyperlocomotion and rewarding effect. Neurosci. Lett. **2006**, *402*, 249–252.

64. Queiroz, E.K.R.; Waissmann, W. Occupational exposure and effects on the male reproductive system. Cadernos de Saúde Pública **2006**, *22* (3), 485–493.

65. Bretveld, R.W.; Hooiveld, M.; Zielhuis, G.A.; Pellegrino, A.; van Rooij, I.A.L.M.; Roeleveld, N. Reproductive disorders among male and female greenhouse workers. Reprod. Toxicol. **2008**, *25*, 107–114.

66. Gregoraszczuk, E.L.; Milczarek, K.; Wojtowicz, A.K.; Berg, V.; Skaare, J.U.; Ropstad, E. Steroid secretion following exposure of ovarian follicular cells to three different natural mixtures of persistent organic pollutants (POPs). Reprod. Toxicol. **2008**, *25*, 58–66.

67. Kortenkamp A. Low dose mixture effects of endocrine disrupters: implications for risk assessment and epidemiology. Int. J. Androl. **2008**, *31*, 233–240.

68. Jacobsen, P.R.; Christiansen, S.; Boberg, J.; Nellemann C.; Hass, U. Combined exposure to endocrine disrupting pesticides impairs parturition, causes pup mortality and affects sexual differentiation in rats. Int. J. Androl. **2010**, *33*, 434–442.

69. Rider, C.V.; Furr, J.R.; Wilson, V.S.; Gray, L.E., Jr. Cumulative effects of in utero administration of mixtures of reproductive toxicants that disrupt common target tissues via diverse mechanisms of toxicity. Int. J. Androl. **2010**, *33*, 443–462.

70. Birkhoj, M.; Nellemann, C.; Jarfelt, K.; Jacobsen, H.; Andersen, H.R.; Dalgaard, M; Vinggaard, A.M. The combined antiandrogenic effects of five commonly used pesticides. Toxicol. Appl. Pharmacol. **2004**, *201,* 10–20.

71. Drake, A.J.; van den Driesche, S.; Scott, H.M., Hutchison, G.R.; Seckl, J.R.; Sharpe, R.M. Glucocorticoids amplify dibutyl phthalate-induced disruption of testosterone production and male reproductive development. Endocrinology **2009**, *150,* 5055–5064.

72. Chen, X.; An, H.; Ao, L.; Sun, L.; Liu, W.; Zhou, Z.; Wanga, Y.; Cao, J. The combined toxicity of dibutyl phthalate and benzo(a)pyrene on the reproductive system of male Sprague Dawley rats in vivo. J. *Hazard*. Mater. **2011**, *186* (1), 835–841.

73. Pohl, H.R.; Luukinen, B.; Holler, J.S. Health effects classification and its role in the derivation of minimal risk levels: Reproductive and endocrine effects. Regul. Toxicol. Pharmacol. **2005**, *42,* 209–217.

74. Erhard, H.W.; Rhind, S.M. Prenatal and postnatal exposure to environmental pollutants in sewage sludge alters emotional reactivity and exploratory behaviour in sheep. Sci. Total Environ. **2004**, *332,* 101– 108.

75. Mills, L.J.; Chichester, C. Review of evidence: Are endocrine-disrupting chemicals in the aquatic environment impacting fish populations? Sci. Total Environ. **2005**, *343,* 1– 34.

76. Fox, G.A. Wildlife as sentinels of human health effects in the Great Lakes-St. Lawrence Basin. Environ. Health Perspect. **2001**, *109* (Suppl. 6), 853–861.

77. Kinnberg, K.; Holbech, H.; Petersen, G.I.; Bjerregaard, P. Effects of the fungicide prochloraz on the sexual development of zebrafish *(Danio rerio)*. Comp. Biochem. Physiol., Part C **2007**, *145,* 165–170.

78. Monod, G.; Rime, H.; Bobe, J.; Jalabert, B. Agonistic effect of imidazole and triazole fungicides on in vitro oocyte maturation in rainbow trout *(Oncorhynchus mykiss)*. Mar. Environ. Res. **2004**, *58,* 143–146.

79. Jobling, S.; Burn, R.W.; Thorpe, K.; Williams, R.; Tyler, C. Statistical modeling suggests that anti-androgens in effluents from wastewater treatment works contribute to widespread sexual disruption in fish living in English rivers. Environ. Health Perspect. **2009**, *117* (5), 797–802.

80. Barnabé, S.; Brar, S.K.; Tyagi, R.D.; Beauchesne, I.; Surampalli, R.Y. Pre-treatment and bioconversion of wastewater sludge to value-added products—Fate of endocrine disrupting compounds. Sci. Total Environ. **2009**, *407,* 1471–1488.

81. Nelson, J.; Bishay, F.; van Roodselaar, A.; Ikonomou, M.; Law, F.C.P. The use of in vitro bioassays to quantify endocrine disrupting chemicals in municipal wastewater treatment plant effluents. Sci. Total Environ. **2007**, *374,* 80–90.

82. Jin, Y.; Shu, L.; Huang, F.; Cao, L.; Sun, L.; Fu, Z. Environmental cues influence EDC-mediated endocrine disruption effects in different developmental stages of Japanese medaka (*Oryzias latipes*). Aquat. Toxicol. **2011**, *101*, 254–260.

83. Orlando, E.F.; Guillette, L.J., Jr. Sexual dimorphic responses in wildlife exposed to endocrine disrupting chemicals. Environ. Res. **2007**, *104*, 163–173.

84. Robinson, B.J.; Hellou, J. Biodegradation of endocrine disrupting compounds in harbour seawater and sediments. Sci. Total Environ. **2009**, *407*, 5713–5718.

85. Liu, Z.; Kanjo, Y.; Mizutani, S. Removal mechanisms for endocrine disrupting compounds (EDCs) in wastewater treatment—Physical means, biodegradation, and chemical advanced oxidation: A review. Sci. Total Environ. **2009**, *407*, 731–748.

86. Janex-Habibi, M.; Huyard, A.; Esperanza, M.; Bruchet, A. Reduction of endocrine disruptor emissions in the environment: The benefit of wastewater treatment. Water Res. **2009**, *43*, 1565–1576.

87. Chen, P.; Rosenfeldt, E.J.; Kullman, S.W.; Hinton, D.E.; Linden, K.G. Biological assessments of a mixture of endocrine disruptors at environmentally relevant concentrations in water following UV/H_2O_2 oxidation. Sci. Total Environ. **2007**, *376*, 18–26.

88. Fisher, P.M.J.; Scott, R. Evaluating and controlling pharmaceutical emissions from dairy farms: A critical first step in developing a preventative management approach. J. Cleaner Prod. **2008**, *16*, 1437–1446.

89. Kumar, A.; Xagoraraki, I. Pharmaceuticals, personal care products and endocrine-disrupting chemicals in U.S. surface and finished drinking waters: A proposed ranking system. Sci. Total Environ. 2010, *in press*

90. Scholz, S.; Mayer, I. Molecular biomarkers of endocrine disruption in small model fish. Mol. Cell. Endocrinol. **2008**, *293*, 57–70.

91. Swart, J.C.; Pool, E.J.; van Wykb, J.H. The implementation of a battery of in vivo and in vitro bioassays to assess river water for estrogenic endocrine disrupting chemicals. Eco-toxicol. Environ. Saf. **2011**, *74*, 138–143.

92. Cao, Q.; Yu, Q.; Connell, D.W. Fate simulation and risk assessment of endocrine disrupting chemicals in a reservoir receiving recycled wastewater. Sci. Total Environ. **2010**, *408*, 6243–6250.

93. Zhang, Y.; Song, X.; Kondoh, A.; Xia, J.; Tang, C. Behavior, mass inventories and modeling evaluation of xenobiotic endocrine-disrupting chemicals along an urban receiving wastewater river in Henan Province, China. Water Res. **2011**, *45*, 292–302.

94. Nichols, J.W.; Breen, M.; Denver, R.J.; Distefano, J.J., III; Edwards, J.S.; Hoke, R.A.; Volz, D.C.; Zhang X. Predicting impacts on vertebrate endocrine systems. Environ. Toxicol. Chem. **2011**, *30* (1), 39–51.

95. Hecker, M.; Hollert, H. Endocrine disruptor screening: Regulatory perspectives and needs. Environ. Sci. Eur. **2011**, *23*, 15, doi:10.1186/2190-4715-23-15.

3

Herbicides

Malcolm Devine

Herbicide Discovery

Traditionally, herbicides have been discovered through large chemical synthesis and screening programs. Given the high cost of discovery and development of new compounds, efforts have focused on herbicides that target major weeds in major world crops (e.g., corn, rice, wheat, soybean, cotton). However, these compounds often find uses in minor crops, also. The process normally involves a step wise progression from greenhouse screening on a few crop and weed species, to more extensive indoor testing, and eventually to field trials in many locations examining the interactions between different weed species, soil types, weather conditions, etc. Toxicological testing and formulation improvement proceed concurrently, to ensure that regulatory and efficacy requirements are met.

A recent innovation is the use of combinatorial chemistry to identify lead compounds. Rather than being synthesized and evaluated in isolation, compounds are produced and screened as mixtures. When combined with in vitro screens (activity testing at the biochemical or cellular level) rather than whole-plant assays, this can allow the testing of 20- to 50-fold more compounds per year than a traditional herbicide discovery program.

"Rational" discovery of herbicides involves identification of a candidate herbicide target site in the plant, followed by design of inhibitors that specifically block that target. While this has led to the discovery of some novel enzyme inhibitors, to date no commercial herbicides have been discovered through this approach. One difficulty is that compounds predicted to have high activity might not penetrate the tissue satisfactorily or may be rapidly degraded inside the tissue.

Finally, herbicides can be developed from bacterial, fungal, or plant toxins. One commercial herbicide, glufosinate (= phosphinothricin), was developed from the bacterial toxin bialaphos from *Streptomyces hygroscopicus.*

Herbicide Classification

Method or Timing of Application

Preplant incorporated herbicides are applied to the soil surface and mechanically incorporated into the upper 5–10 cm of the soil, in order to minimize photodecomposition and volatilization losses. Preemergence herbicides are applied to the soil surface and often rely on precipitation or soil moisture to transport them to the plant root or shoot for uptake. Postemergence herbicides are applied to exposed foliage after the plants have emerged.

Chemical Structure or Mode of Action

Herbicides within the same structural family usually have the same mode of action, with varying degrees of activity or selectivity depending on the structural variations between compounds. However, herbicides from different chemical families can have the same mode of action (see later).

Components of Herbicide Action

To be effective, herbicides must penetrate the tissue and reach the target site in sufficiently high concentrations to block its activity. The overall process of herbicide action can be separated into the following components:[1]

Absorption: Herbicides can be absorbed by the roots or directly into the leaves. Root uptake occurs through mass flow of herbicide in soil moisture and is a function of root distribution in the soil, soil moisture status, the physical properties of the soil, and the behavior of the herbicide in the soil. Foliar absorption occurs following application to the leaves. To facilitate entry through the cuticle (the waxy layer on most leaf surfaces), herbicides are usually formulated with an array of inert ingredients including surfactants, emulsifiers, etc. Once inside the tissue, further penetration through cell walls and membranes usually occurs by simple diffusion.

Translocation: Long-distance transport of herbicides in the plant can occur in the xylem and/or phloem. In some instances, distribution through the plant is a critical component of overall activity. The amount of translocation depends on the plant stage of development, the physicochemical properties of the herbicide, and the rate at which herbicide injury slows down the movement of endogenous compounds.

Metabolic degradation: Plants have evolved various enzyme systems to detoxify potentially harmful compounds. Fortuitously, some of these enzymes (e.g., cytochrome P450 monooxygenases, glutathione S-transferases) can degrade herbicides to inactive compounds.

Interaction at the target site: Finally, all herbicides must interfere with some critical process in the plant. In most cases this involves binding to a protein (usually a functional enzyme, or a transport or structural protein) so that it cannot carry out its normal function. Over time, through the combined effect of this direct action and other indirect actions, the plant dies.

Herbicide Selectivity

Most herbicides are selective, that is, they kill some plant species but not others. A few herbicides, on the other hand, are nonselective, and kill essentially all species. Selectivity is normally based on one of the following:[1]

- Failure to absorb the herbicide, due to either selective placement (e.g., directed spray on weeds growing between the crop rows) or failure of the herbicide to reach the roots of deep rooted crops, while killing shallow rooted weeds.

- Enhanced rate of metabolic degradation of the herbicide in the crop. This is the most common basis of selectivity of agricultural herbicides. After entry into the plants, the crop metabolizes the herbicide to inactive compounds, whereas degradation does not occur or is slower in the susceptible weeds.
- Differential sensitivity of the target site. This is particularly common in the case of herbicide resistant weeds (see later).

Herbicide Mode of Action

Herbicides kill plants by interfering with an essential process in the plant. The major modes of action of herbicides, the biochemical target sites, and some examples of chemical groups that interfere with those targets, are shown in Table 1.

In addition, many herbicides have been identified that interact with other, unique target sites in plants. However, most of the herbicides that have been developed over the past 50 years target about 15 distinct molecular targets.

Herbicide Resistance in Weeds

The repeated use of herbicides can lead to the development of herbicide-resistant weed populations. The use of herbicides per se does not create herbicide resistance. Rather, the continuous selection pressure through herbicide use creates a niche, allowing resistant individuals to survive and increase in

TABLE 1 Modes of Action of Major Herbicide Groups[a]

Mode of Action	Target Site	Representative Chemical Groups
	Inhibition of amino acid biosynthesis	
Branched chain amino acids	Acetolactate synthase	Sulfonylureas, imidazolinones, triazolopyrimidines
Glutamine synthesis	Glutamine synthetase	Glufosinate
Aromatic amino acid biosynthesis	Enolpyruvylshikimate phosphate synthase	Glyphosate
	Photosynthesis	
Photosynthetic electron transport (PS II)	Q_b or D1 protein	S-Triazines, phenylureas, benzonitriles
Photosynthetic electron transport (PS I)	PS I electron acceptor	Bipyridiliums
	Pigment biosynthesis	
Chlorophyll synthesis	Protoporphyrinogen oxidase	(Nitro) diphenylethers
Carotenoid synthesis	Phytoene desaturase and others	Aminotriazole, clomazone
	Lipid biosynthesis	
Fatty acid synthesis	Acetyl-coA carboxylase	Aryloxyphenoxypropionates, cyclohexanediones
Fatty acid elongation	"Elongase" complex	Thiocarbamates
	Cell division	
Spindle formation	β-Tubulin	Dinitroanilines, carbamates
	Other	
Auxin disruption	Auxin binding proteins (?)	Phenoxyacetic acids, benzoic acids
Homogentisate biosynthesis	4 hydroxyphenylpyruvate dioxygenase	Isoxazoles

[a] From Devine.[1]

the population as susceptible weeds are killed. Thus, starting with an initial population of perhaps one resistant weed in a population of 10^6–10^9, resistance builds up over time until the resistant weeds become predominant in the field.

Herbicide resistance in weeds was first observed in the late 1950s, but was a minor problem until the mid-1970s, when resistance to triazine herbicides became a widespread concern. Since then the occurrence of resistance has increased dramatically, with >200 cases now reported.[2]

In most cases, resistance is due to a point mutation in the gene coding for the herbicide target site.[3] This alters the structure of the target protein in such a way that it can still perform its natural function, but herbicide binding is reduced. This type of resistance often confers cross-resistance to herbicides in the same chemical family or mode-of-action group, although exceptions exist and each case must be analyzed separately. However, target-site mutations do not alter the sensitivity of the weed to herbicides with other mechanisms of action.

Resistance can also be conferred by elevated activity of the enzyme(s) responsible for herbicide degradation. These weeds can be cross-resistant to other herbicides with different mechanisms of action. Again, the possibilities for cross-resistance have to be analyzed on a case-by case basis.

Although herbicide resistance has become widespread, in almost all cases alternative control methods are available, through the use of other herbicides, changes in cropping or tillage practices, or some combination of these. Avoidance and management of herbicide resistance has become an integral part of good farming practices in modern agriculture.

Herbicide-Resistant Crops

A recent development in selective weed control has been to create resistance to certain herbicides in crops where it did not exist previously. While this has been done primarily to extend the market share of certain products, it offers farmers the advantage of broad-spectrum weed control in crops with a single herbicide application.[4] In some cases this has substantially reduced the total amount of herbicide required in a single season.

Herbicide-resistant crops can be produced by three methods:[4]

1. Making crosses between the crop (sensitive to the herbicide) and a related, resistant species. This method was used to develop triazine-tolerant canola (*Brassica napus*).
2. Selecting resistant cells in tissue culture, through random mutation or by selecting somaclonal variants, and regenerating resistant plants from these cells. Corn lines resistant to the herbicide sethoxydim were generated in this way.
3. Transfer of an herbicide-resistance gene through genetic engineering. The gene is identified in an unrelated species (often a bacterium), cloned, and transferred into the crop species of interest. These procedures were used to develop canola varieties resistant to the nonselective herbicides glufosinate and glyphosate.

Herbicide-resistant crops have greatly facilitated weed control, but present some new research questions that have had to be addressed. These include the likelihood and long term ecological consequences of gene flow to related species, and the need to control volunteer plants in the following season(s). These issues do not present insurmountable obstacles but, again, need to be dealt with on a case-by-case basis.

Safety and Environmental Fate of Herbicides

Environmental safety is of prime concern in the development of new herbicides. This includes an understanding of herbicide toxicology, safe handling and application procedures, and environmental behavior and fate. In most countries, approval of herbicides by the relevant decision making bodies is dependent on the registrant satisfying the regulatory requirements imposed by those countries.

Toxicological requirements vary from country to country, but usually include data on oral and dermal toxicity in a range of species, in tests of varying duration. Based on the data collected, maximum residue levels are established in food or food products. Data from these tests and field experiments are used to establish maximum application doses and safe intervals between product application and harvest or grazing. Data may also be required on effects on nontarget organisms and ecosystems that may be exposed to low herbicide doses.

Appropriate handling procedures are an important aspect of herbicide safety. The use of appropriate safety clothing (gloves, coveralls, masks, etc.), more benign formulations (e.g., dispersible granules rather than wettable powders), etc., contribute to reduced applicator exposure. Recently, novel formulations have been developed that further reduce applicator exposure when adding products to the spray tank.

Herbicide drift immediately after spraying can be a source of off-site contamination, resulting in injury to adjacent sensitive crops and other species. Various measures, including spraying only under calm conditions, use of wind deflectors, and avoiding very small droplets, can substantially reduce the risk of spray drift.

Herbicides in soil are lost by a combination of microbial and chemical degradation, plant uptake, and, in some instances, leaching or surface run-off. Stringent environmental regulations have been introduced in many countries to minimize the possibility of groundwater contamination. Herbicide residues in soil can provide extended weed control, but also may limit crop rotation options in future seasons. Field research is conducted to establish the risk of such carryover and the effects on future crops.

References

1. Devine, M.D.; Duke, S.O.; Fedtke, C. *Physiology of Herbicide Action*; Prentice-Hall: Englewood Cliffs, NJ, 1993; 441.
2. Heap, I.M. International survey of herbicide-resistant weeds. http://www.weedscience.com (accessed April 2000).
3. Devine, M.D.; Eberlein, C.V. Physiological, Biochemical and Molecular Aspects of Herbicide Resistance Based on Altered Target Sites. In *Herbicide Activity: Toxicology, Biochemistry and Molecular Biology*; Roe, R.M., Burton, J.D., Kuhr, R.J., Eds.; IOS Press: Amsterdam, 1997; 159–185.
4. *Herbicide-Resistant Crops. Agricultural, Environmental, Economic, Regulatory and Technical Aspects*; Duke, S.O., Ed.; CRC Press: Boca Raton, FL, 1996; 420.

4

Herbicides: Non-Target Species Effects

Céline Boutin

Introduction

Pesticides such as inorganic chemicals (e.g., sulfur, arsenic, or other metal compounds) have been used in agriculture for centuries to protect crops. In more recent times, especially after World War II, organic pesticides have been discovered and increasingly used to suppress unwanted plants, insects, and other organisms that interfere with crop production. The main pesticides used are categorized by their target organisms: insecticides used to kill or suppress insects, fungicides used for pathogens, and herbicides for weeds. In this entry, we are mostly concerned with herbicides because of their considerable use, especially in North America, and their phytotoxic effects. The definition accepted by the Weed Science Society of America is that an herbicide is "a chemical substance or cultured organism used to kill or suppress the growth of plants."[1]

Phytotoxicity refers to the capability of herbicides or other pesticides to exert toxic effect on plant growth, reproduction, and survival. By extension and in addition to terrestrial and aquatic vascular plants, it usually includes other primary producers such as algae and cyanobacteria, which will not be considered in this entry.

This entry will first consider the history, types, and main uses of herbicides. There are advantages and disadvantages of using herbicides in agriculture and forest management. Benefits have long been established; however, the limitations and undesirable effects attributed to herbicide use are still debated. The environmental impact will be examined, including environmental exposure, phytotoxicity, and toxicity to different trophic levels, as well as the multitudinous factors to take into account in risk assessment. Lastly, the techniques and limitations of the phytotoxicity assessments currently used to determine environmental risk evaluations will be discussed. Mitigation measures and alternatives to herbicide use will be discussed in conclusion.

History, Types of Herbicides, and Their Use

DNOC or 4,6-dinitro-*o*-cresol, developed in 1935, was the first organic herbicide used to control weeds. This herbicide was also used as an insecticide, fungicide, and a defoliant and was shown to be toxic to animals.[2] It is no longer used in many countries. In the 1940s, several phenoxy herbicides were discovered and many are still used today, including 2,4-D and MCPA [(4-chloro-2-methylphenoxy)acetic acid]. In the 1950s, other herbicides appeared on the market, including diuron, diquat, paraquat, and triallate. In the 1960s, the tri-azine herbicides were developed and are still considerably used, especially atrazine. Glyphosate, which is still the most widely used herbicide worldwide, was developed in the 1970s. In the 1980s, low-dose high-efficacy herbicides such as sulfonylureas and imidazolinones were found. Since the late 1990s, a new method for controlling weeds has emerged with the development of herbicide tolerant crops, i.e., genetically engineered crops resistant to glyphosate and, to a lesser extent, glufosinate ammonium. More genetically modified herbicide tolerant crops will no doubt be engineered or bred in the future.

Herbicides exhibit different mechanisms of action.[1,3] They include disruption of photosynthesis (ura-cils, substituted ureas, and triazines), inhibition of lipid biosynthesis (carbamothioates such as EPTC [s-ethyl dipropylthiocarbamate], triallate, clethodim, fluazifop, and metolachlor), inhibition of cell divi-sion (dinitroanilines such as trifluralin and pendimethalin), plant hormone mimics (the phenoxy her-bicide 2,4-D, MCPA, the benzoic acid dicamba, and the picolinic acid picloram), inhibition of amino acid biosynthesis (glyphosate, sulfonyl ureas including metsulfuron methyl and chlorsulfuron, imid-azolinones including imazethapyr), blockage of carotenoid biosynthesis (clomazone), and disruption of cell membranes (acifluorfen and bipyridylium compounds such as diquat and paraquat). Other herbi-cides such as glufosinate ammonium act by inhibiting glutamine synthetase, thus leading to a complete breakdown of ammonia metabolism in affected plants. Uncouplers of oxidation phosphorylation, such as the widely used bromoxynil, interfere with plant respiration.

Many herbicides act primarily on systems unique to plants, e.g., photosynthesis, but some herbicides act at more than one site of action. Undoubtedly, the secondary mode of action of some herbicides could explain their relatively high toxicity to animals (see below). In the case of some herbicides, the precise mode of action is unknown and exact molecular sites of action remain to be determined.[4]

The number of herbicides listed in the Weed Science Society of America reached 374 in 2010.[5] In Canada, there are 500 pesticide active ingredients (the ingredient to which the pesticide is attributed) and 7000 pesticide formulated products (mixture containing one or several active ingredients and for-mulants) available since many formulated products contain a mixture of active ingredients.[6] The num-ber of herbicide active ingredients registered in Canada amounts to approximately 125.[6] The majority of herbicide use occurs in agriculture where in modern practices they dominate weed control practice. Approximately 90% of areas planted with corn, cotton, potato, wheat, and soybean were sprayed with herbicides in the United States in 2004.[1]

While the benefits of using herbicides from an agronomic perspective are well known, the undesir-able effects of herbicides on the environment have not always been considered carefully. Regardless of the method of application, it is generally accepted that misplacement will take place through drift at the time of application or through runoff, leaching, and volatilization from soil or plants or from particles moving with contaminated soil after application has occurred.

Benefits to Agriculture and Forest Management

Benefits of herbicide to agriculture, forest management, and other agronomic applications leave no doubt as to their utility, although some applications for cosmetic reasons (e.g., domestic use, horticul-ture, golf courses) are more questionable. The negative relationship between crop yield and weed density is well established.[1] Weeds are well-adapted species that compete with crops in disturbed environments and will reduce crop yield depending on their germination timing and densities, growth patterns, and growth rates. However, there may be cases where weeds can be beneficial to crops. As an example, field experiments were conducted to evaluate the effects of nicosulfuron and imazethapyr, a sulfonyl urea and an imidazolinone herbicide, respectively, for the control of johnsongrass (*Sorghum halepense* L.), a weed difficult to control in corn fields.[7] It was noted that corn vigor was greatly reduced in plots where these two herbicides were applied and where johnsongrass was reduced. In order to verify if there was an unwanted effect of the two herbicides on corn, small plots were sprayed with the two herbicides separately. Two other treatments were also included in the experimental design: control of johnsongrass via mechanical means and no treatment. It was noticed that in plots where johnsongrass was removed, corn was more prone to being attacked by the maize dwarf mosaic and maize chlorotic dwarf viruses, which are transmitted by aphids or leafhoppers. Results revealed that in treated plots, virus disease was increased because the preferred host johnsongrass was suppressed. Furthermore, there was no change in yield between treatments. Crop yield was reduced in treated plots (whether through mechanical or chemical treatments) due to increased virus severity, while in non- treated plots, reduced corn yield was due to more competition from johnsongrass.

Herbicide Use and Exposure to Primary Producers

The amount of herbicide active ingredients used worldwide exceeded 950 million kg in 2007, of which 39% was herbicides.[8] Herbicides are used most extensively in North America. For instance, in 2003, herbicides accounted for 77% of total pesticide sales in Canada, while fungicides represented 9%, insecticides 8%, and other products 6%.[6] In the United States, 47% of pesticides used were herbicides, while insecticides and fungicides amounted to 8% and 6%, respectively.[8] In Europe, fungicides are used extensively mainly due to the use of sulfur fungicides on vineyards by France, Italy, Spain, and Greece.[9] Herbicides constitute less than 35% of pesticides used in Europe, but nevertheless, this is approximately 75,000 tons of applied active ingredients.[10] Of note, it appears that these European data underestimated the actual amount of pesticides used.[9] France is the biggest herbicide user followed by Germany and the United Kingdom. In other parts of the world, insecticides constitute the bulk of the pesticides used.[11]

Large amounts of herbicides are annually used in terms of hectares sprayed and in terms of quantity per hectares. In Canada, in excess of 25 million ha are sprayed with atleast one herbicide on an annual basis.[12] In the United States, 110 million ha were treated with herbicides in the late 1980s,[13] and in 1997, 209 million kg of herbicide active ingredients were used (Olszyk et al.[14] and references therein). Total land area devoted to agriculture worldwide is considerable,[15] especially in Europe where around 70% of the land is cropland or pastureland.[16] As a result, total areas treated with herbicides reach large figures. Herbicide application is highly dependent on the types of crops, the prevailing climate conditions, the land use, and other factors such as topography. In the United Kingdom where 77% of the land is used for agriculture, pesticide use was also intensive at 5.8 kg/ha in 1988[17] but seemed to have declined in 1999 at 2.7 kg/ha (1.4 kg-ai of herbicides), not taking into account the underestimation mentioned previously.[9] By contrast, in France, the intensity of pesticide used increased from 4.4 kg-ai/ha in 1988 to 6.0 kg-ai/ha in 1999, including 2.0 kg-ai/ha of herbicides. This is well above the average for European member states, which used approximately 4.5 kg-ai/ha of pesticides and 1.3 kg-ai/ha of herbicides. In the Netherlands, 11.8 kg-ai/ha of pesticides are used on an annual basis (mostly non-herbicides).[18] On the other hand, in Canada, pesticide use is low at 0.9 kg/ha (largely herbicides).[6] The United States uses 23% of all the pesticides used in the world, and 28% of all herbicides.

These figures, however, dissimulate important facts. In Canada, where less than 10% of the land is devoted to agriculture, two ecoregions are particularly important in this regard, the Prairies of western Canada and the Mixed- wood Plains of the Great Lakes–Saint Lawrence corridor. The Prairies, which constitute the largest agricultural areas in Canada (5.1% of the land),[19] has been almost totally modified to satisfy increasing needs for more cropping areas. Consequently, there are almost no pristine grassland prairies left where native plant and animal communities can survive. The Mixedwood Plains are a smaller area in southern central Canada (1.5% of the land cover) where 50% of the Canadian population lives and where agriculture is also very intense. In this region, a large portion of the Carolinian forest and the mixed-wood forest has disappeared to satisfy human needs. In the United States, agriculture covers 48% of the land and is concentrated in the mid-west regions[20] where original ecosystems have largely vanished. The same pattern is repeated in Europe where mostly seminatural habitats remain in most countries. What remains of these ecosystems interspersed in a sea of intensively cultivated land can be greatly impacted by herbicide use.

Herbicide exposure to non-target environments can occur when application is performed with aircraft, mist-blower, and ground applications through overspray or via spray drift, vapor drift, revolatilization from soil and plants, runoff, or dust particles moved by wind or water.[14] Depending on the equipment and prevailing weather conditions during application, the amount of sprayed herbicide that will deposit in hedgerows and other field edges from multiple consecutive spray tracks can reach 1% to 10% of the application rate within 10 m of a single swath with ground equipment (Boutin and Jobin[21] and references therein), and much more with mist-blower sprayers and aerial applications.[22] Herbicides can travel a considerably longer distance with aerial equipment applications, for example, 500 m downwind from the source.[23,24]

With ground application, it was found that herbicides could cover long distances. A study was undertaken to assess the protection afforded by buffer zones from herbicide drift (Boutin and Baril, unpublished data). Surveys of nontarget plants situated in two small woodlots adjacent to crop fields were conducted prior to (May) and after (May, June, and July) herbicide application. Vegetation was surveyed for community composition and symptoms of herbicidal impact. The experimental work was conducted in southwestern Ontario, Canada, under normal field operation conditions, for soybean sprayed with imazethapyr in 1993, corn with dicamba in 1994, and wheat with MCPA in 1996, following the usual rotation in southwestern Ontario. The buffer zone was defined as a 12 m wide seeded strip of crop, upwind of a woodlot, where herbicides were not applied. Each treatment consisted of four transects, divided into sampling points, at 1 m, 2 m, 4 m, 8 m, 16 m, and 32 m distances into the woodlots. Herbicide application occurred in the early morning or evening, when no precipitation was forecasted, when wind speed was at or less than 8 km/hr, and when the direction of the wind was across the soybean field into the woodlot.

Results revealed that herbicides could move up to 32 m into woodlots (Figure 1). Up to 43% of the vegetation of one quadrat (of the 23 species) in the woodlots showed visual effects characteristics of herbicidal injury: discoloration, bleaching, epinasty, yellow or brown spots, etc. Effects were less pronounced in transects abutted to buffer zones. The plants most affected [e.g., raspberry (*Rubus idaeus* L.), goldenrod (*Solidago canadensis* L.), and ash tree (*Fraxinus* spp.)] were species of open areas growing in the first few meters of the woodlots. In some cases, effects lasted for more than 2 mo.

Vapor drift can also migrate a long way (Franzaring et al.[25] and references therein) causing recurrent sublethal effects on native plants not only in bordering seminatural habitats but also in more remote habitats. Presence of airborne herbicides in the atmosphere has been reported in Europe[26] and North America.[27] In the Netherlands, it was found that non-target vegetation was repeatedly exposed to small amounts of herbicides (Franzaring et al.[25] and references therein). Pesticides used in agriculture in southern Canada, including some herbicides, have been reported in the arctic environments.[28] Of the 10 chemicals surveyed by Hoferkamp et al.,[29] 9 were detected in the arctic. Traveling distances for these chemicals ranged from 55 km to 12,100 km. Persistence in the environment of these herbicides can explain their presence in the arctic where they are transported via the air or by dust.

Undoubtedly, the large number of herbicides available for use, the geographical extent of their use on different crops, and the quantity applied together with the method of application suggest a high probability of exposure to primary producers and other wildlife.

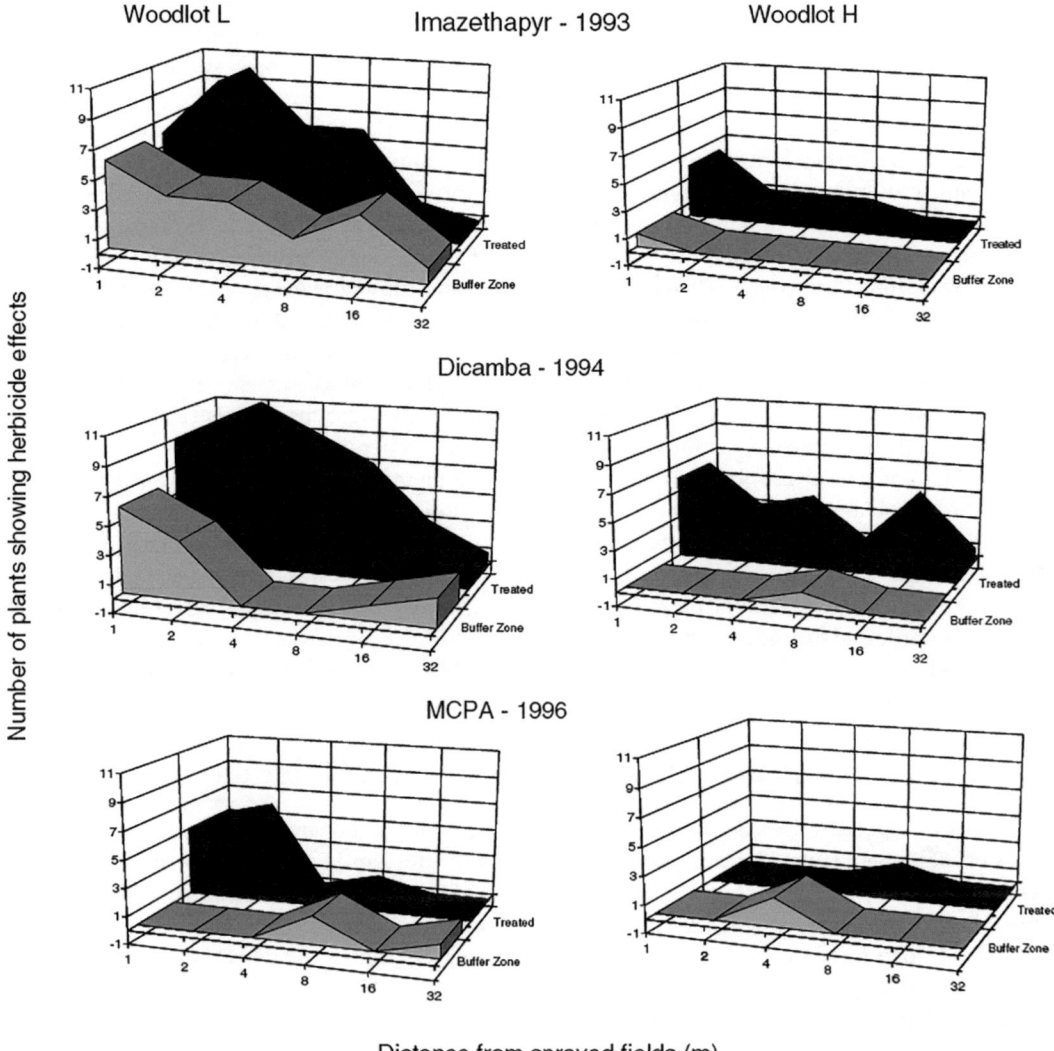

FIGURE 1 Number of plants showing herbicide effects in two woodlots (L and H) surveyed during 3 years prior to and after herbicide spray under normal field operations. Fields were sprayed with the herbicides imazethapyr in 1993, dicamba in 1994, and MCPA in 1996. Plants were surveyed in quadrats situated between 1 and 32 m from crop fields. A 12 m buffer zone within fields was used in half of the fields while the other half was sprayed right to the field edge. Four transects were placed for each treatment. In total, 116 species were inventoried and 35 species were found to be sensitive to herbicides during the course of the 3 years the study lasted.

Herbicide Toxicity to Humans and Animals

Herbicides are generally not very toxic to humans or to animals, including mammals, birds, amphibians, and invertebrates, as well as to microbial organisms. There are some exceptions, however, notably paraquat, which is very toxic to humans. It is still used extensively in the United States[30] and many developing countries. A large epidemiological study suggested that the herbicides pendimethalin and EPTC may be linked with human pancreatic cancer.[31] Sulfonyl ureas, which comprise a large number of low-dose herbicides widely used in agriculture, are known to affect humans as they are used in medicine to treat Type 2 diabetes and other illnesses.[32] They act by increasing insulin release from the beta cells in the pancreas, and side effects such as incidence of hypoglycemia have been reported.

As early as 1976, the herbicide diquat was shown to have an adverse effect on body size and pigmentation of *Xenopus laevis,* the South African clawed frog (family: Pipidae; subfamily: Xenopinae). Moreover, when treated with diquat and the fungicide nabam, the deleterious effect was more pronounced.[33] More recently, the herbicide atrazine has been implicated in a number of reports demonstrating effects on animals as an endocrine disruptor.[34] Similarly, the formulated product containing glyphosate is widely known to affect amphibians.[34] Glyphosate formulation has been implicated in a synergistic effect with a trematode species on fish.[35] Survival of juvenile fish was unaffected by exposure to glyphosate alone or by an infectious trematode parasite alone. However, simultaneous exposure to infection and glyphosate significantly reduced fish survival. Spinal malformations of juvenile fish were also enhanced when both stressors were present. A species of snail acts as the vector between the fish and the trematode. Glyphosate at high concentration killed all the snails, but at moderate concentrations, the snail produced more trematodes than in control and low-concentration groups.[35] This elegant experiment demonstrated the intricate interactions between different components of ecosystems.

Herbicides are categorized for their effect on unwanted plants, yet a number of them demonstrate toxicity to microbial organisms. A recent study revealed the control activities of glyphosate against rust diseases (*Puccinia striiformis* and *Puccinia triticina*) on glyphosate-resistant wheat and soybeans.[36] Control was equivalent to that of registered fungicides. Similarly, diquat was investigated for its potential to control the bacterial infection Columnaris disease (*Flavobacterium columnare*) affecting several fish species and was found to reduce Columnaris infection.[37]

These findings demonstrate that herbicides can be directly toxic not only to plants but also to different types of life form at various trophic levels, including humans.

Toxicity to Plants and Effects to Terrestrial Habitats

Herbicides are especially designed to control weeds and therefore are of particular concern for unwanted effects on desirable plants. However, several types of pesticides can injure plants since pesticides are mostly classified by target organisms. For example, the widely used insecticides chlorpyrifos, diazinon, carbaryl, malathion and others can cause injury to several crops and ornamental species.[38] Several fungicides belonging to the benzimidazole chemical class were shown to be toxic to several plants.[39] Nevertheless, this entry will concentrate on herbicides.

Effects of herbicides on crops and weeds are well documented for obvious agronomic and economic reasons. Unwanted effects on native species is usually assessed through regulatory processes using crops as surrogate species (see below) and via studies performed sporadically by research scientists. By and large, herbicides affect native species at different levels. Herbicides can alter biochemical and developmental processes in plants as well as plant morphology. Habitats within agroecosystems can experience modifications in their species abundance, composition, and diversity when subjected to herbicides. Plants as terrestrial primary producers constitute the basis of terrestrial ecosystems. Herbicide effects on them can have cascading effects at other trophic levels, on overall biodiversity, and on ecosystem functions. These will be addressed in turn.

Effects at the Species Level

Herbicides can modify biochemical processes in plants and as a result can increase plant susceptibility to pests and diseases. Most of the work demonstrating these effects has been performed with crop plants, but there is no reason to believe that native plants would not be affected in the same manner. The incidence of mildew on spring wheat was enhanced by three different herbicides as a result of stress induced by metabolic interference.[40] Wheat treated with 2,4-D was higher in protein content resulting in a proliferation of aphids.[41] The concentration of nitrogenous compounds in crops was enhanced by sublethal doses of several phenoxy, triazine, and uracil herbicides,[42] sometimes leading to an increase in pathogens and pests on corn[43] and rice.[44]

Symbiotic processes with plants such as nodulation and mycorrhizae, which are vital for the biochemical activities of most terrestrial plants, can be greatly modified by herbicides. Nodulation and nitrogen fixation was disrupted by herbicides resulting in deleterious effects on growth and reproduction in crops such as dry bean (*Phaseolus vulgaris*), soybean (*Glycine max*), broad bean (*Vicia faba*), and peanut (*Arachis hypogea*) (Schnelle and Hensley[45] and references therein). Mycorrhizal activities were affected by MCPA sprayed on (*Pisum sativum*) with ensuing decreases in growth observed.[46]

Herbicides can exert anomalous effects on plant developmental processes. They have been shown to affect seed production and seed germination. It was demonstrated as early as 1948 that applications of 2,4-D caused a delay in seed germination and growth of wheat plants sufficient to favor an increase of wireworm damage.[47] It has been known for some time that when some of these herbicides are applied to cereal crops late in their growth stage, just before seed formation, the plants produced far fewer seeds. Greenhouse experiments recently revealed that at typically used rates, dicamba and picloram reduced all or nearly all seed production while 2,4-D was much less effective.[48] Further field test experiments supported these greenhouse results.[49]

Research with glyphosate showed that depending upon the plant species, application rate, and the timing of application, effects on seed production, seed germination, and seedling development have been observed on a large number of plant species from various families.[50] Glyphosate produced an inhibitory effect on pollen germination and seed formation when applied at the flower bud stage of goldenrod (*S. canadensis*).[51] Germination, emergence, and plant establishment of native Australian plant species were impeded by the herbicide fluazifop-*p*-butyl.[52] Plants of *Stellaria media* treated with the herbicide glufosinate ammonium produced seeds with reduced germination and emergence.[53]

Herbicides can instigate unexpected effects on plants that have not been studied as part of the registration package because they are not relevant from an agronomical viewpoint. Sulfonylurea herbicides are selective herbicides that act by inhibiting the enzyme acetolactate synthase, which catalyzes the synthesis of the three branched-chain amino acids valine, leucine, and isoleucine. They are very potent herbicides applied at exceedingly low doses, in the order of a few grams per hectare. When applied at the fraction of the recommended label rate at the onset of flower formation, however, they were found to reduce the reproductive outputs of several species with significant reproductive damage occurring with only scant visible symptoms on leaves. Cherry trees sprayed at doses as low as 0.2% of the field application rate of the sulfonylurea chlorsulfuron showed a significant reduction in the production of fruits, with almost no observable damage to vegetative parts.[54] Low doses of metsulfuron methyl produced important injury on the vegetative biomass when crop and native species were sprayed at the seedling stage, but plants sprayed at later stages showed considerable reduction in the reproduction.[55] Potato bulking was greatly affected by low rates of sulfometuron methyl whereas few visible symptoms were observed on the vegetative parts.[56] Under field situations, it was found that berry production in hawthorn (*Crataegus monogyna* Jacq.) was severely affected by average spray drift concentrations higher than 2.5% of the label rate of the sulfonyl urea metsulfuron methyl (0.1 g-ai/ha) and that the effect was still observed 1 year after the spray event.[57,58]

Imidazolinones are another chemical family classified as low-dose herbicides. They also inhibit the enzyme acetolactase synthase and as such have been implicated in unforeseen effects. Potato (*Solanum tuberosum* L. from the Solanaceae family) tuber size and quality were most detrimentally affected by exposure to imidazolinone herbicides, as low as 0.1 times the recommended field rate, when exposure occurred during tuber bulking, as compared to exposure at seedling emergence or at tuber initiation.[59] While reductions in the weight and the overall yield of sensitive species were problematic, another major concern was the potential lowering of fruit quality, which could lead to significant economic losses in the case of crops.[56,59] Effects on native species have not been investigated, but it is easy to stipulate that drift of imidazolinones to non-target habitats could elicit a reduction in weight of sensitive species, including effects on underground and reproductive parts. This may lead to species of the Solanaceae family and other families to produce fruits and storage organs that are less attractive to wildlife.

Damage to or modification of plant morphology, including epinasty, inhibition of leaf expansion, and stem and root distortion, has been reported in weeds and crops, particularly for phenoxy herbicides.[60] Herbicides can cause morphological deformation with unforeseen repercussions. For instance, it was possible to demonstrate that morphological deformations of the flowering parts in *Arabidopsis thaliana* (L.) Heynh. induced by some herbicides prevented pollination and seed production.[61] Bud and flower abortion has been reported for sulfonylureas (see above). Such effects, usually undetected, can have long term adverse impacts, particularly on monocarpic species.

Population/Community Level

Herbicide use in modern agriculture alters species abundance, composition, and diversity in non-target habitats. Effects of low doses of herbicides on plants (grown in pots and placed at different distances from the spray swath) have been reported.[62,63] Marshall and Bernie[64] showed that several of the broad-leaved species found in field margins were susceptible to the six different herbicides tested separately in pots. In the field, Marrs et al.[65] assessed effects of spray drift in relation to plant damage and yield for a range of plant species of conservation interest in Britain after applying each of six herbicides with a standard agricultural hydraulic ground sprayer. They observed lethal effects at 2–6 m from field edges and damage at greater distances, although damaged plants were able to recover within the growing season. Effects on seedlings were observed up to 20 m.[66] The long-term effect of such damage to plants remains unknown. In North America, work performed by Jobin et al.[67] showed that recurrent applications of herbicides had a long-term effect on plant populations inhabiting hedgerows and woodlot edges adjacent to crop fields. In Britain, several native arable weeds are considered endangered due to destruction of their habitats and extensive use of agrochemicals.[68]

In the Netherlands, it was calculated that 9.5% of all pesticides applied was dispersed outside croplands.[18] Drift scenarios together with herbicide toxic effects investigated using bioassays and taking into account distances from spray events were used to estimate impact on biodiversity. In 2005, 41% of the linear landscape features near cropland were affected. This was an improvement since, in 1998, 59% of the area was affected. Natural areas located within farming regions were also affected by herbicide displacement in 31% and 11% of the area in 1998 and 2005, respectively. Measures in place such as unsprayed buffer zones and better equipment as well as reduced reliance on herbicides were largely responsible for this decline in unwanted effects on plant diversity. Also in the Netherlands, small plot experiments used to investigate the effects of the herbicide fluroxypyr on plants monitored for 3 years showed a decline in diversity and biomass.[69] Change in plant populations was also noticed.

Ecosystem/Trophic Level

Alterations by herbicides on primary producers through effects on morphology, physiology, phenology, species composition, diversity, and abundance can resonate considerably to other trophic levels. The best documented study implicating herbicide repercussions was conducted in Britain over several decades. The grey partridge (*Perdix perdix*) has been surveyed since 1933 in the margins of crop fields, and it was found that numbers declined by 80% between 1952 and the mid-1980s.[70] Studies conducted from the 1960s led to the conclusion that the use of herbicides and, to a lesser extent, insecticides precipitated the decline of grey partridge populations. Although partridges are largely herbivorous, newly hatched chicks feed largely on arthropods during the first 2–3 weeks of their lives. The falling number of grey partridges in agricultural land was attributed to declining chick survival early in the season due to weed removal by herbicides. A reduction in weed diversity and density on which insects feed and inhabit caused a food shortage at this very crucial period of the year.[71] Removal of field margins was also a contributing factor.

Many subsequent studies have shown that as a result of herbicidal effects, cover and diversity of flowering plant species are reduced in crop fields and field margins, thus subsequently reducing the resources available to flower- visiting insects and other arthropods.[72–74] Likewise, abundant floral diversity was found to be the prevailing factor related to high Lepidopteran diversity in farmland habitats (Boutin et al.[75] and references therein).

Factors Interacting with Herbicide Effects

Toxicity to plants and other organisms may be exacerbated by the chemical and physical properties of a compound, namely, volatility, mobility, and persistence in the environment. For example, atrazine is a widely used product in North America applied for the pre- or postemergence control of broadleaf weeds and grasses, predominately on corn crops. It is a selective compound with systemic activity resulting in chlorosis and eventual plant death. The primary mode of degradation of atrazine is through hydrolysis. It has been reported to persist in soil, water, and aquifers for years.[76] In a study conducted in Germany, it was found that the herbicide atrazine was still present 22 years after its application in soil.[77] It is regularly found in groundwater in the United States and Canada where it is widely used. Unquestionably, the persistence of atrazine explains its presence and accumulation in the environment and long-term effects.[78]

Most herbicides are highly water soluble and therefore can move away from crop fields with rainwater. Volatile compounds can elicit damage to non-target plants; clomazone applied in crop fields produced striking bleaching effects that caused the U.S. regulatory agencies to consider wild plants in their risk assessments.[79]

Persistent herbicides may cause long-lasting and unexpected contamination. Clopyralid is a growth-regulator-type herbicide used for the control of broad-leaved species. It was used on sugar beet that was later fed to cattle with the resulting manure containing sufficient residues to contaminate crops on which it was subsequently spread.[80] The persistence of clopyralid in compost and mulches was also demonstrated in the United States.[81] The same problem occurred with the related compound aminopyralid, an auxinic herbicide registered for the control of broad-leaved weeds on grassland and rangeland. Grass treated with aminopyralid persisted in the silage for more than 1 year. Cattle or horses fed with the hay produced contaminated manure toxic to receiving crops.[82]

Herbicide toxic effects may vary depending on the characteristics of the receiving environment. Residues of the sulfonylurea chlorsulfuron have been detected and caused crop damage as long as 7 years after application in Alberta, Canada.[83] Sulfonylureas are known for their increased activity at soil pH above neutrality, which is often observed in soils in western Canada.[84] Temperature may also exert effects on persistence and thus on long-term toxicity. In Canada and elsewhere where winters are long and cold, many pesticides, including chlorsulfuron, will take longer to degrade.[85]

Direct toxicity of herbicides to primary producers and other organisms is typically assessed without taking into account confounding stressors present under normal agriculture practices or occurring in natural conditions. For instance, high fertilizer application in intensive agriculture is the norm. Few studies on the misplacement of fertilizer in non-target adjacent areas have been published, but Rew et al.[86] in Britain demonstrated that off-target deposition could vary from 2% to 133.3% of the application rate in field margins, depending on the type of machinery used. Some species will thrive under high nutrient conditions, including grasses.[69] Field margins where intensive agriculture prevails tend to become simplified and dominated by grasses because both herbicides and fertilizers tend to suppress broad-leaved species.[21] Grasses are wind pollinated and, as a consequence, pollinators are reduced or eliminated.

A plethora of unforeseen events and interactions can occur when herbicides are released in the environment. Measuring phytotoxicity is a first step for assessing unwanted effects in the environment.

Measuring Phytotoxicity

Herbicides (and other pesticides) are special contaminants because they are intentionally released in the environment. Therefore, before they are registered for use, risk assessment of herbicide phytotoxicity is conducted using data submitted by registrants. In current guidelines, phytotoxicity tests designed to assess the impact to terrestrial plants are conducted under controlled greenhouse conditions, using crop plants growing individually in pots, and effects are assessed at the juvenile stage, usually before reproduction occurs.[87,88] These tests are used for risk assessment of native plants growing within communities exposed to variable outdoor conditions at various phenological stages (Figure 2). Therefore, there is a large gap in regulatory testing between actual phytotoxicity tests performed with selected non-native species under artificial conditions and risk assessment conducted for protection of native plants in natural habitats. Ecological risk assessment for effects on native plants usually takes into account both herbicide use patterns and herbicide characteristics. Elements that are typically not considered are biotic and abiotic factors, which will affect plant sensitivity to herbicides. Current phytotoxicity testing assumes similar effects at different phenological stages and does not take into account recurrent exposure of sublethal doses.

In the past, phytotoxicity testing was designed for the protection of agronomically relevant species from accidental herbicide drift; therefore, crop species were appropriately used in testing. However,

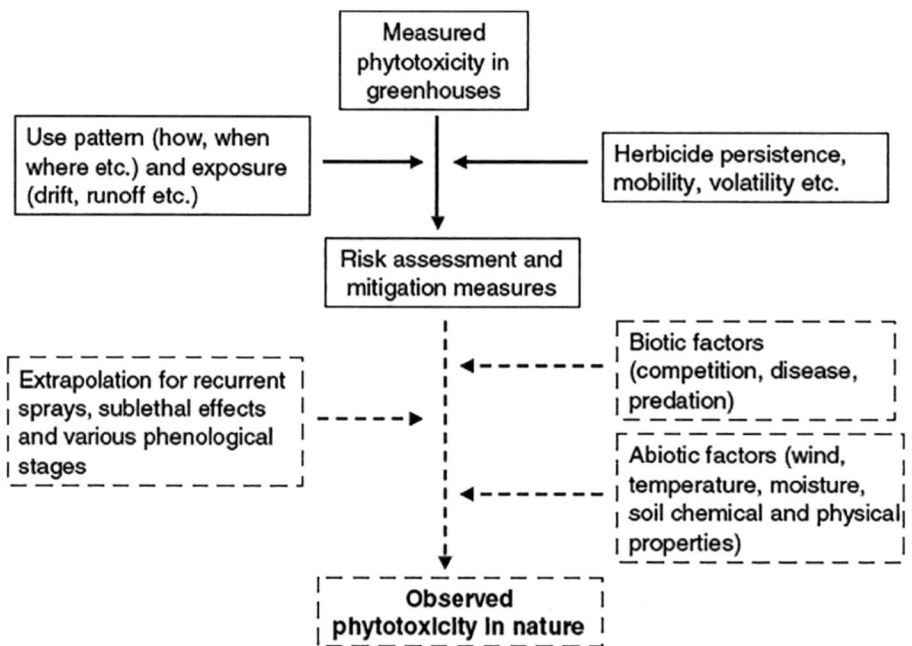

FIGURE 2 Schematic representation of measured and observed herbicide phytotoxicity testing. Phytotoxicity is measured following routine test guidelines. Subsequent ecological risk assessment for effects on native plants usually takes into account herbicide use pattern such as how (aerial or ground applications), when (time of the year and day), and where (type of terrains) herbicides are applied, which will determine to some extent exposure through drift, runoff, or overspray. Herbicide persistence, mobility, and volatility will increase exposure. Elements represented in dashed boxes are usually not considered and determined the observed phytotoxicity under natural conditions. Biotic factors include competition or interactions with other plant species, effects of diseases, and predation on plant health. Wind, low or high temperature, drought or flooding, and soil properties may affect plants and modify sensitivity to herbicides. Phytotoxicity testing assumes similar effects at different phenological stages and does not take into account recurrent exposure of sublethal doses.

it has been increasingly recognized that the vegetation bordering or in the vicinity of crop fields was important for the native plant species and for the other trophic levels relying on primary producers for food, habitat, and shelter. In studies aimed at comparing crop and noncrop species, results indicated that plant sensitivity is both herbicide and species dependent and that no obvious pattern emerged with numerous native and crop plant species tested with a number of herbicides.[89,90] For some herbicides, results with crops would mean underprotection of native species, while for other herbicides, testing with crop species would entail overprotection. However, these studies were limited to a narrow taxonomic range in line with the crop species usually tested, e.g., short-lived species. It was also found that the numerous native species selected for the studies were, for the most part, easy to grow and maintain in the greenhouse. These results suggest that testing should cover a broader range of native species in toxicity testing for regulatory purposes.

Further results indicated that a large variability in herbicide response existed among crop varieties and native plant ecotypes.[89,91,92] The number of species needed to be tested to encompass the range of sensitivity of a given herbicide has not been resolved. Boutin and Rogers[93] showed that the range of species sensitivity increases with an augmentation of numbers of species tested up to 40, suggesting that the number of species (n = 6 or 10) tested in current guidelines is insufficient. These studies and others indicated that the large emphasis placed on collecting very precise EC_{25} or EC_{50} values (effective concentration or dose causing a 25% or 50% effect on test plants) may be problematic and erroneous and that extrapolation factors should be determined in order to alleviate some of the uncertainty in testing that is currently not considered.

Tests in current guidelines are required for plants at the two- to four-leaf stage, the assumption being that the juvenile stage is more sensitive than the adult period. However, natural vegetation bordering crop fields consists of plants at various phenological stages. It was found in several studies that indeed the young stage is very sensitive to herbicides.[66,94] Conversely, plants at later stages can show higher sensitivity to some herbicides.[54,94]

For practical reasons, phytotoxicity tests are conducted under ideal greenhouse conditions using plants growing singly in pots and devoid of intra- or interspecific competition. This contrasts with plants growing within natural communities subjected to the occasional drought or flood, continuous wind, attacks from herbivores or pathogens, and other abiotic or biotic elements that interfere with herbicide effects. It is of course impossible and impractical to study the interactions of all these factors. In the past, studies comparing greenhouse and field tests elicited contradictory results with species being more or less sensitive when tested in greenhouses as compared to field tests.[53,69,95–97] A few studies have examined the direct effect of herbicide use of communities of native plants[21,64–67] (Figure 1), and although responses were extremely variable, most species showed some effects. There is considerable uncertainty in current phytotoxicity testing, and further studies are required to monitor herbicide effects on terrestrial native plants.[98]

Conclusion

Herbicides are used extensively, mostly in agriculture, with sometimes unforeseen environmental effects in non-target areas. Although measuring phytotoxicity can be straightforward when conducted on known crop plants under simple laboratory conditions (in greenhouses), determining phytotoxicity is a multifaceted affair as demonstrated by the numerous factors interacting under natural conditions (Figure 2). The linkage between measured phytotoxicity in greenhouses and observed phytotoxicity in nature implies the inclusion of environmental biotic and abiotic factors as well as considerations of test conditions, herbicide characteristics, level of exposure, and use pattern (Figure 2). It was also shown that phytotoxicity can have consequences at all trophic levels, and this is seldom taken into account.

Ecological and agronomic needs are often conflicting. Weed plants want to be suppressed in favor of crops, yet they are necessary to sustain some wildlife in otherwise desolate agrarian systems. This wildlife can in turn provide vital ecological services (e.g., pollinators for crops or biological control).

Mitigation measures to reduce phytotoxicity on desirable plants, including the elimination of aerial application, implementation of unsprayed buffer zones, more advanced equipment, and avoidance of adverse spray conditions (e.g., high wind, temperature inversion, or forecast rain), can substantially reduce herbicide impacts on the environment and are increasingly considered by the farming communities and regulators. Nevertheless, there appears to be an excessive dependence on herbicides for the control of weeds. New uses for herbicides are still being developed, including the desiccation of crops at the end of the growing season with foliar contact herbicides to facilitate harvest or the use of herbicide-resistant, genetically modified crops. Alternatives to herbicides exist and are widely used in organic farming. They comprise mechanical weed control using tillage or mowing, non-mechanical weed control through crop rotation or the planting of companion crops, and biological weed control using parasites, predators, or pathogens. Another complementary measure that can be promoted in conventional farming involves Integrated Pest Management (IPM), a management system that uses all available tactics to manage pest and weeds including crop rotation, host-resistant varieties, mechanical and physical controls, and chemical control. Weed control in agriculture should and can be made compatible with wildlife values.

Acknowledgments

The author wishes to thank D. Carpenter for useful comments on the manuscript and B. Daigneault for assistance in formatting. Funding was provided by Environment Canada (including Pesticide Science Fund) for the experimental work presented by the author and collaborators.

References

1. Zimdahl, R.L. *Fundamentals of Weed Science,* 3rd Ed.; Academic Press: San Diego, CA, 2008.
2. Matsumoto, M.; Hirose, A.; Ema, M. Review of testicular toxicity of dinitrophenolic compounds, 2-sec-butyl-4, 6-dinitrophenol, 4,6-dinitro-*o*-cresol and 2,4-dinitrophenol. Reprod. Toxicol. **2008**, *26* (3–4),185–190.
3. Duke, S.O. Overview of herbicide mechanisms of action. Environ. Health Perspect. **1990**, *87,* 263–271.
4. Walsh, M.J.; Maguire, N.; Powles, S.B. Combined effects of wheat competition and 2,4-D amine on phenoxy herbicide resistant *Raphanus raphanistrum* populations. Weed Res. **2009**, *49,* 316–325.
5. Anonymous. Common and chemical names of herbicides approved by the Weed Science Society of America. Weed Sci. **2010**, *58* (4), 511–518.
6. Brimble, S.; Bacchus, P.; Caux, P.-Y. Pesticide utilization in Canada: A compilation of current sales and use data, GoC Cat. No. En4–56/2005E; Environment Canada: Gatineau, QC, 2005.
7. Eberwine, J.W.; Hagood, E.S. Effect of johnsongrass (*Sorghum halepense*) control on the severity of virus diseases of corn (*Zea mays*). Weed Technol. **1995**, *9,* 73–79.
8. United States Environmental Protection Agency, 2001, available at http://www.epa.gov/pesticides/pestsales/ (accessed May 2011).
9. IRENA (Indicators Reporting on the Integration of Environmental Concerns into Agricultural Policy). Fact Sheet no. 9. Consumption of Pesticides, 1999.
10. Eurostat. Statistical books. The use of plant protection products in the European Union. Data 1992–2003. Statistical books. 2007 edition; 222 pp.
11. Wilson, C.; Tisdell, C. Why farmers continue to use pesticides despite environmental, health and sustainability costs. Economics, Ecology and the Environment, Working Paper no. 53; School of Economics, University of Queensland: Brisbane, Australia, 2000.
12. Statistics Canada. Census of Agriculture. Ministry of Supply and Services Canada, 1987.
13. National Research Council. Alternative Agriculture National Research Council: Washington, DC, 1989; 448 pp.

14. Olszyk, D.; Burdick, C.; Pfleeger, T.; Lee, E.H.; Watrud, L.S. Assessing the risks to non-target plants from herbicides. Jpn. J. Agric. Meteorol. **2004**, *60*, 221–242.

15. Foley, J.A.; DeFries, R.; Asner, G.P.; Barford, C.; Bonan, G.; Carpenter, S.R.; Chapin, F.S.; Coe, M.T.; Daily, G.C.; Gibbs, H.K.; Helkowski, J.H.; Holloway, T.; Howard, E.A.; Kucharik, C.J.; Monfreda, C.; Patz, J.A.; Prentice, I.C.; Ramankutty, N.; Snyder, P.K. Global consequences of land use. Science **2005**, *309* (5734), 570–574.

16. Cobham, R.; Rowe, J. Evaluating the wildlife of agricultural environments: An aid to conservation. In *Wildlife Conservation Evaluation;* Usher, M.B., Ed.; Chapman and Hall Ltd.: London, 1986; 223–246.

17. Available at http://www.fraseramerica.org/commerce.web/product_files/Environmental_Indicators_2002.pdf (accessed February 2011).

18. de Jong, F.M.W.; de Snoo, G.R.; van de Zande, J.C. Estimated nationwide effects of pesticide spray drift on terrestrial habitats in the Netherlands. J. Environ. Manage. **2008**, *86* (4), 721–730.

19. Available at http://en.wikipedia.org/wiki/EcozonesofCanada (accessed February 2011).

20. Pfleeger, T.G.; Olszyk, D.; Burdick, C.; King, G.; Kern, J.; Fletcher, J. Using a geographic information system to identify areas with potential for off target pesticide exposure. Environ. Toxicol. Chem. **2006**, *25* (8), 2250–2259.

21. Boutin, C.; Jobin, B. Intensity of agricultural practices and effects on adjacent habitats. Ecol. Appl. **1998**, *8* (2), 544–557.

22. Ware, G.W.; Cahill, W.P.; Gerhardt, P.D.; Witt, J.M. Pesticide drift, IV. On target deposits from aerial application of insecticides. J. Econ. Entomol. **1970**, *63* (6), 1982–1983.

23. Davis, B.N.K.; Williams, C.T. Buffer zone widths for honeybees from ground and aerial spraying of insecticides. Environ. Pollut. **1990**, *63* (3), 247–259.

24. Conacher, J.; Conacher, A. Herbicides in Agriculture: Minimum Tillage, Science and Society, Geowest No. 22; Department of Geography, University of Western Australia: Nedlands, WA, 1986.

25. Franzaring, J.; Kempenaar, C.; van der Eerden, L.J.M. Effects of vapours of chlorpropham and ethofumesate on wild plant species. Environ. Pollut. **2001**, *114* (1), 21–28.

26. Follak, S.; Hurle, K. Effect of airborne bromoxyniloctanoate and metribuzin on non-target plants. Environ. Pollut. **2003**, *126* (2), 139–146.

27. Yao, Y.; Harner, T.; Blanchard, P.; Tuduri, L.; Waite, D.; Poissant, L.; Murphy, C.; Belzer, W.; Aulagnier, F.; Sverko, E. Pesticides in the atmosphere across Canadian agricultural regions. Environ. Sci. Technol. **2008**, *42* (16), 5931–5937.

28. Welch, H.E.; Muir, D.C.G.; Billeck, B.N.; Lockhart, W.L.; Brunskill, G.J.; Kling, H.J.; Olson, M.P.; Lemoine, R.M. Brown snow: A long-range transport event in the Canadian Arctic. Environ. Sci. Technol. **1991**, *25* (2), 280–286.

29. Hoferkamp, L.; Hermanson, M.H.; Muir, D.C.G. Current use pesticides in Arctic media; 2000–2007. Sci. Total Environ. **2010**, *408* (15), 2985–2994.

30. Eilsler, R. *Paraquat Hazards to Fish, Wildlife, and Invertebrates: A Synoptic Review,* Biological Report 85(1.22); U.S. Fish and Wildlife Service, 1990; 1–38.

31. Andreotti, G.; Freeman, L.E.B.; Hou, L.; Coble, J.; Rusiecki, J.; Hoppin, J.A.; Silverman, D.T.; Alavanja, M.C.R. Agricultural pesticide use and pancreatic cancer risk in the agricultural health study cohort. Int. J. Cancer **2009**, *124* (10), 2495–2500.

32. Shorr, R.I.; Ray, W.A.; Daugherty, J.R.; Griffin, M.R. Incidence and risk factors for serious hypoglycemia in older persons using insulin or sulfonylureas. Arch. Intern. Med. **1997**, *157* (15), 1681–1686.

33. Anderson, R.J.; Prahlad, K.V. The deleterious effects of fungicides and herbicides on *Xenopus laevis* embryos. Arch. Environ. Contam. Toxicol. **1976**, *4* (1), 312–323.

34. Mann, R.M.; Hyne, R.V.; Choung, C.B.; Wilson, S.P. Amphibians and agricultural chemicals: Review of the risks in a complex environment. Environ. Pollut. **2009**, *157* (11), 2903–2927.

35. Kelly, D.W.; Poulin, R.; Tompkins, D.M.; Townsend, C.R. Synergistic effects of glyphosate formulation and parasite infection on fish malformations and survival. J. Appl. Ecol. **2010**, *47* (2), 498–504.

36. Feng, P.C.C.; Clark, C.; Andrade, G.C.; Balbi, M.C.; Caldwell, P. The control of Asian rust by glyphosate in glyphosate-resistant soybeans. Pest Manage. Sci. **2008**, *64* (4), 353–359.

37. Darwish, A.M.; Mitchell, A.J. Evaluation of diquat against an acute experimental infection of *Flavobacterium columnare* in channel catfish, *Ictaluruspunctatus* (Rafinesque). J. Fish Dis. **2009**, *32* (5), 401–408.

38. Thomson, W.T. *Agricultural Chemicals. Book 1 Insecticides*; Thomson Publications: Fresno, CA, 1985–1986.

39. van Lersel, M.W.; Bugbee, B. Phytotoxic effects of benzimidazole fungicides on bedding plants. J. Am. Soc. Hortic. Sci. **1996**, *121* (6), 1095–1102.

40. Heitefuss, P. Assessment of herbicide effects on interactions of weeds, crop plants, pathogens and pests. In *Field Methods for the Study of Environmental Effects of Pesticides, BCPC Monograph No. 40*; Greaves, M.P., Greig-Smith, P.W., Smith, B.D., Eds.; British Crop Protection Council: Thornton Heath, U.K., 1988; 265–274.

41. Adams, J.B.; Drew, M.E. Grain aphids in New Brunswick. III. Aphid populations in herbicide-treated oat fields. Can. J. Zool. **1965**, *43*, 789–794.

42. Ries, S.K. Subtoxic effects on plants. In *Herbicides*; Audus, L.J., Ed.; Academic Press: London, 1976; Vol. 2, 313–344.

43. Oka, I.N.; Pimentel, D. Herbicide (2,4-D) increases insect and pathogen pests on corn. Science **1976**, *193* (4249), 239–240.

44. Ishii, S.; Hirano, C. Growth responses of larvae of the rice stem borer to rice plants treated with 2,4-D. Entomol. Exp. Appl. **1963**, *6*, 257–262.

45. Schnelle, M.A.; Hensley, D.L. Effects of pesticides upon nitrogen fixation and nodulation by dry bean. Pestic. Sci. **1990**, *28* (1), 83–88.

46. Garcia-Romera, I.; Ocampo, J.A. Effect of the herbicide MCPA on VA mycorrhizal infection and growth of *Pisum sativum*. J. Plant Nutr. Soil Sci. **1988**, *151* (4), 225–278.

47. Fox, W.B. 2,4-D as a factor in increasing wireworm damage to wheat. Sci. Agric. **1948**, *28* (9), 423–424.

48. Rinella, M.J.; Haferkamp, M.R.; Masters, R.A.; Muscha, J.M.; Bellows, S.E.; Vermeire, L.T. Growth regulator herbicides prevent invasive annual grass seed production. Invasive Plant Sci. Manage. **2010**, *3* (1), 12–16.

49. Rinella, M.J.; Masters, R.A.; Bellows, S.E. Growth regulator herbicides prevent invasive annual grass seed production under field conditions. Rangeland Ecol. Manage. **2010**, *63* (4), 487–490.

50. Blackburn, L.G.; Boutin, C. Subtle effects of herbicide use in the context of genetically modified crops: A case study with glyphosate (Roundup®). Ecotoxicology **2003**, *12*, 271–285.

51. Guo, S.L.; Jiang, H.W.; Fang, F.; Chen, G.Q. Influences of herbicides, uprooting and use as cut flowers on sexual reproduction of *Solidago canadensis*. Weed Res. **2009**, *49* (3), 291–299.

52. Rokich, D.P.; Harma, J.; Turner, S.R.; Sadler, R.J.; Tan, B.H.; Fluazifop-*p*-butyl herbicide: Implications for germination, emergence and growth of Australian plant species. Biol. Conserv. **2009**, *142* (4), 850–869.

53. Riemens, M.M.; Dueck, T.; Kempenaar, C. Predicting sublethal effects of herbicides on terrestrial non-crop plant species in the field from greenhouse data. Environ. Pollut. **2008**, *155* (1), 141–149.

54. Fletcher, S.J.; Pfleeger, T.G.; Hillman, C.R. Potential environmental risks associated with the new sulfonylurea herbicides. Environ. Sci. Technol. **1993**, *27* (10), 2250–2252.

55. Boutin, C.; Elmegaard, N.; Kjaer, C. Toxicity testing of fifteen non-crop plant species with six herbicides in a greenhouse experiment: Implications for risk assessment. Ecotoxicology **2004**, *13* (4), 349–369.

56. Pfleeger, T.; Olszyk, D.; Plocher, M.; Yilma, S. Effects of low concentrations of herbicides on full-season, field-grown potatoes. J. Environ. Qual. **2008**, *37*, 2070–2082.

57. Kjær, C.; Strandberg, M.; Erlandsen, M. Metsulfuron spray drift reduces fruit yield of hawthorn (*Crataegus monogyna* L.). Sci. Total Environ. **2006**, *356*, 228–234.

58. Kjær, C.; Strandberg, M.; Erlandsen, M. Effects on hawthorn the year after simulated spray drift. Chemosphere **2006**, *63* (5), 853–859.

59. Eberlein, C.V.; Guttieri, M.J. Potato (*Solanum tuberosum*) response to simulated drift of imidazolinone herbicides. Weed Sci. **1994**, *42* (1), 70–75.

60. Tottman, D.R.; Davies, E.L.P. The effect of herbicides on the root system of wheat plants. Ann. Appl. Biol. **1978**, *90* (1), 93–99.

61. Ratsch, H.C.; Johndro, D.J.; McFarlane, J.C. Growth inhibition and morphological effects of several chemicals in *Arabidopsis thaliana* (L.) Heynh. Environ. Toxicol. Chem. **1986**, *5* (1), 55–60.

62. Breeze, V.G.; van Rensburg, E. Vapour of the free acid of the herbicide 2,4-D is toxic to tomato and lettuce plants. Environ. Pollut. **1991**, *72* (4), 259–267.

63. Marrs, R.H.; Frost, A.J. A microcosm approach to the detection of the effects of herbicide spray drift in plant communities. J. Environ. Manage. **1997**, *50* (4), 369–388.

64. Marshall, E.J.P.; Bernie, J.E. Herbicide effects on field margin flora. Proceedings of the BCPC International Congress—Weeds; British Crop Protection Council: Farnham, Surrey, U.K., 1985; 1021–1028.

65. Marrs, R.H.; Williams, C.T.; Frost, A.J.; Plant, R.A. Assessment of the effects of herbicide spray drift on a range of plant species of conservation interest. Environ. Pollut. **1989**, *59* (1), 71–86.

66. Marrs, R.H.; Frost, A.J.; Plant, R.A. Effects of herbicide spray drift on selected species of nature conservation interest: The effects of plant size and surrounding vegetation structure. Environ. Pollut. **1991**, *69* (2–3), 223–235.

67. Jobin, B.; Boutin, C.; DesGranges, J.-L. Effects of agricultural practices on the flora of hedgerows and woodland edges in Southern Québec. Can. J. Plant Sci. **1997**, *77*, 293–299.

68. Marshall, E.J.P. The impact of landscape structure and sown grass margin strips on weed assemblages in arable crops and their boundaries. Weed Res. **2009**, *49* (1), 107–115.

69. Kleijn, D.; Verbeek, M. Factors affecting the species composition of arable field boundary vegetation. J. Appl. Ecol. **1997**, *37* (2), 256–266.

70. Sotherton, N.W.; Dower, J.W.; Rands, N.R.W. The effects of pesticide exclusion strips on faunal populations in Great Britain. Ecol. Bull. **1988**, *39*, 197–199.

71. Potts, G.R. The effects of modern agriculture, nest predation and game management on the population ecology of partridges (*Perdix perdix* and *Alectoris rufa*). Adv. Ecol. Res. **1980**, *11*, 2–79.

72. Lagerlöf, J.; Stark, J.; Svensson, B. Margins of agricultural fields as habitats for pollinating insects. Agric., Ecosyst. Environ. **1992**, *40*, 117–124.

73. Longley, M.; Sotherton, N.W. Factors determining the effects of pesticides upon butterflies inhabiting arable farmland. Agric., Ecosyst. Environ. **1997**, *61* (1), 1–12.

74. Holzschuh, A.; Steffan-Dewenter, I.; Kleijn, D.; Tscharntke, T. Diversity of flower-visiting bees in cereal fields: effects of farming system, landscape composition and regional context. J. Appl. Ecol. **2007**, *44* (1), 41–49.

75. Boutin, C.; Baril, A.; McCabe, S.K.; Martin, P.A.; Guy, M. The value of woody hedgerows for moth diversity on organic and conventional farms. Environ. Entomol. **2011**, *40* (3): 560–569.

76. Lazorko-Connon, S.; Achari, G. Atrazine: Its occurrence and treatment in water. Environ. Rev. **2009**, *17*, 199–214.

77. Jablonowski, N.D.; Köppchen, S.; Hofmann, D.; Schäffer, A.; Burauel, P. Persistence of ^{14}C-labeled atrazine and its residues in a field lysimeter soil after 22 years. Environ. Pollut. **2009**, *157* (7), 2126–2131.

78. Graymore, M.; Stagnitti, F.; Allinson, G. Impacts of atrazine in aquatic ecosystems. Environ. Int. **2001**, *26*, 483–495.

79. Poster, J. Command herbicide. The rookie battles controversy. Crops Soils Mag. **1986**, *39* (1), 9–11.

80. Eagle, D.J. Agrochemical damage to U.K. crops. Pestic. Outlook **1990**, *1* (2), 14–16.

81. Nilsson, H.; Aamisepp, A. Persistence in plants and transfer of clopyralid (3,6-dichloropicolinic acid) through plant remains. In *Weeds and Weed Control,* 25th Swedish Weed Conference, Vol. 1 Reports. Uppsala, Sweden, 1984.

82. http://www.pesticides.gov.uk/garden.asp?id=2480 and http://www.guardian.co.uk/environment/2008/jun/29/food.agriculture (accessed May 2011).

83. Moyer, J.R.; Esau, R.; Kozub, G.C. Chlorsulfuron persistence and response of nine rotational crops in alkaline soils of southern Alberta. Weed Technol. **1990**, *4* (3), 543–548.

84. Blair, A.M.; Martin, T.D. A review of the activity, fate and mode of action of sulfonylurea herbicides. Pestic. Sci. **1988**, *22* (3), 195–219.

85. Walker, A. Simulation of herbicide persistence in soil. II. Simazine and linuron in long-term experiments. Pestic. Sci. **1976**, *7* (1), 50–58.

86. Rew, L.J.; Theaker, A.J.; Froud-Williams, R.J.; Boatman, N.D. Nitrogen fertilizer misplacement in field boundaries. Aspects Appl. Biol. **1992**, *30,* 203–206.

87. Organisation for Economic Co-operation and Development (OECD). Terrestrial Plant Test: Vegetative Vigour Test. New TG 227. Seedling Emergence and Seedling Growth Test. Updated TG 208, 2006.

88. United States Environmental Protection Agency (USEPA). Ecological Effects Test Guidelines: Terrestrial Plant Toxicity—Vegetative Vigor, OPPTS 850.4250, EPA 712-C-96- 163 and Early Seedling Toxicity Test, OPPTS 850.4130, EPA 712–C–96–347. Washington, DC, 1996.

89. White, A.L.; Boutin, C. Herbicidal effects on non-target vegetation: Investigating the limitations of current pesticide registration guidelines. Environ. Toxicol. Chem. **2007**, *26* (12), 2634–2643.

90. Carpenter, D.; Boutin, C. Sublethal effects of the herbicide glufosinate ammonium on crops and wild plants: Short-term effects compared to vegetative recovery and plant reproduction. Ecotoxicology **2010**, *19* (7), 1322–1336.

91. Boutin, C.; White, A.L.; Carpenter, D. Measuring variability in phytotoxicity testing using crop and wild plant species. Environ. Toxicol. Chem. **2010**, *29* (2), 237–242.

92. Bhatti, M.A.; Felsot, A.S.; Al-Khatib, K.; Kadir, S.; Parker, R. Effects of simulated chlorsulfuron drift on fruit yield and quality of sweet cherries (*Prunus avium* L.). Environ. Toxicol. Chem. **1995**, *14* (3), 537–544.

93. Boutin, C.; Rogers, C.A. Pattern of sensitivity of plant species to various herbicides—An analysis with two databases. Ecotoxicology **2000**, *9* (4), 255–272.

94. Boutin, C.; Lee, H.B.; Peart, T.E.; Batchelor, S.P.; Maguire, R.J. Effects of the sulfonylurea herbicide metsulfuron methyl on growth and reproduction of five wetland and terrestrial plant species. Environ. Toxicol. Chem. **2000**, *19* (10), 2532–2541.

95. Dalton, R.L.; Boutin, C. Comparison of the effects of glyphosate and atrazine herbicides on nontarget plants grown singly and in microcosms. Environ. Toxicol. Chem. **2010**, *29* (10), 2304–2315.

96. Damgaard, C.; Mathiassen, S.K.; Kudsk, P. Modeling effects of herbicide drift on the competitive interactions between weeds. Environ. Toxicol. Chem. **2008**, *27* (6), 1302–1308.

97. Clark, J.; Ortego, L.S.; Fairbrother, A. Sources of variability in plant toxicity testing. Chemosphere **2004**, *57* (11), 1599–1612.

98. Boutin, C.; Aya, K.L.; Carpenter, D.; Thomas, P.J.; Rowland, O. Phytotoxicity testing for herbicide regulation: shortcomings in relation to biodiversity and ecosystem services in agrarian systems. Sci. Total Environ. **2011**, *40* (3): 560–569.

5

Insecticides: Aerial Ultra-Low-Volume Application

He Zhong

Introduction

Mosquito control is necessary in order to protect public health from mosquito-borne diseases such as West Nile virus, eastern equine encephalitis, St. Louis encephalitis, malaria, and dengue. Aerial ultra-low-volume (ULV) application of mosquito insecticides is one of the most effective techniques for controlling adult mosquitoes and preventing mosquito-borne diseases.[1] During ULV application, large insecticide droplets may sometimes result in unwanted mortality to nontarget organisms.[2–8] For many years, nontarget mortality caused by mosquito adulticiding was usually accepted as a "casualty of war." With the advancement of mosquitocide residue monitoring and new spray technologies, we now recognize that "casualties" can be reduced to a minimum. The important factors that contribute to mosquito control efficacy and associated nontarget mortality are insecticide deposition,[7–9] droplet size,[1,10] spray time,[10] application dose,[1,10] topography,[10] and weather conditions (such as wind velocity, direction, and atmospheric stability).[10] Due to the complexity of aerial ULV applications, it may be difficult to achieve optimal mosquito reduction without nontarget mortality during an aerial spray mission. However, control efficacy can be increased, and nontarget mortality can be minimized, if aerial application is conducted at the right place (by increasing retention time of mosquitocide droplets in the air in order to enhance their contact with flying mosquitoes),[11,12] at the right time (dusk, or night when adult mosquitoes are actively flying),[10] and at the right dose (proper insecticide concentration in the air to kill mosquitoes but not nontarget organisms).[1,9]

Right Place

Current aerial spray technology is capable of targeting a specific zone where adult mosquitoes are actively flying. The droplet size of aerially applied insecticides governs downwind dispersal and subsequent impingement on targets.[1] Smaller droplets [5–25 pm volume median diameter (VMD)] remain aloft longer and offer a better probability for impingement on flying adult mosquitoes.[1] Larger insecticide droplets (more than 100 pm VMD) will deposit on the ground sooner after application, thereby reducing the likelihood of contact with flying mosquitoes. Insecticide that is deposited on the ground not only is wasted but also may adversely affect nontarget organisms.[7,8] Environmental contamination can be reduced by adopting application techniques that maintain droplets in the air and promote controlled downwind movement of the insecticide cloud while minimizing ground deposition (particularly in environmentally sensitive areas).[7,9] This concept is different from agricultural applications, where insecticide deposition is needed to coat the surface of crops. Moreover, agricultural applications try to reduce insecticide drift away from the target zone, such as a crop or field, rather than maximize droplet suspension in the air column.

Right Time

The best time of the day for mosquito adulticide applications (also called the "spray window") is at dusk, dawn, or nighttime when mosquitoes are actively flying, and this is when the optimal control efficacy will be achieved.[1,10] Applying an insecticide in that period will reduce impact on daytime active nontarget organisms like honeybees, dragonflies, and butterflies. These spray windows are the time periods when daytime nontargets are resting and, therefore, protected from insecticide exposure. Mosquito control operation in such spray windows will protect many daytime active nontargets from being adversely impacted by insecticide dispersal clouds.

Spraying at the right time also means spraying under optimal meteorological conditions, currently recognized as 3–10 miles/hr wind velocity without the presence of a temperature inversion. Understanding, as well as achieving, "optimal" meteorological conditions for mosquito spraying is often difficult.[1] Although aerial mosquito application technology can accurately calibrate the amount of insecticide delivered using the nozzle systems. Once released, the aerosol is in the hands of Mother Nature. The spray cloud, as it is carried by wind and influenced by gravity, starts its journey to the ground from an altitude of 100300 ft. Wind velocity, direction, and atmospheric stability greatly affect the distribution of the spray cloud.[1] Also, downwind movement and deposition of insecticide residue can vary greatly from one spray mission to another.[11,12] As wind speed increases, the impingement force of spray droplets will increase on targets and nontargets. This situation creates considerable variation in control efficacy and can often influence whether the effects on nontargets are minimal or substantial.[13]

Right Dose

Increasing the application rate may not always improve the level of control but will generally increase the risk of nontarget mortality due to escalating exposure levels. In reality, it is sometimes very difficult to apply the proper application rate to achieve adequate mosquito control without causing nontarget mortality. Nontargets' differential tolerance to insecticides may be the result of physiological as well as geographical differences within and among those organisms. The larger body size of some nontarget species may increase their tolerance levels to the insecticide. Natural topographic barriers, such as trees, bushes, and grasses, may provide refuge for them to escape lethal exposure from the insecticide aerosol.[8] Conversely, adult mosquitoes in vegetated areas may also be protected by the physical barriers,[1] and if the application rate is increased to compensate for this, adverse effects to nontargets may occur.

To determine the proper application rate or "right dose," studies are needed to determine the correlation of the insecticide concentration in the air column with adult mosquito mortality[11,12] and that of ground deposition concentrations with nontarget mortality.[7–9] The insecticide concentration at its final destination, whether at an airborne target or on the ground, is referred to as the terminal insecticide concentration (TIC), which is different from and can be influenced by the initial application dose. TIC is influenced by many environmental variables and therefore needs to be frequently monitored. If the TIC is adequate to kill the majority of adult mosquitoes and low enough that it spares nontargets of concern, the application dose will be considered appropriate. In this way, TIC critically affects control efficacy and nontarget impact. Determination of TIC during routine application of mosquitocides provides a mechanism to assess or cross-compare control efficacy and impact on nontargets during aerial mosquito control missions. This process will ensure the proper application dose to achieve the delicate balance between effective mosquito control and minimal nontarget impact.

Novel Application Technology

Mosquito control programs worldwide continue to develop novel application technologies to increase control efficacy and lessen damage to nontargets. In the late 1990s, James Robinson's group at Florida's Pasco County Mosquito Control District led an effort to develop a high-pressure nozzle system to deliver small insecticide droplets (<30 μm VMD).[11,12] The high-pressure system, with insecticide applied at two-thirds the label rate, achieved better adult mosquito control compared with a conventional flat-fan nozzle system (>80 μm VMD)[11] and dramatically reduced the mortality of caged fiddler crabs, *Uca pugilator* (Bosc), from 80% to 0 in Collier County, Florida.[7] Another field trial with honeybees (*Apis mellifera* L.) conducted during routine nighttime aerial adult mosquito control missions in Manatee County, Florida, further demonstrated the advantage of the new high-pressure system. Honeybees that clustered outside of beehive entrances were exposed to naled sprays for mosquito control. The larger insecticide droplets produced by flat-fan nozzles killed more than 90% of the bees clustered outside of the hives and resulted in an average of 35% reduction in honey yield at the end of the season.[8] In contrast, bee mortality and average honey yield were not significantly different compared with hives similarly exposed to the smaller droplets of the high-pressure nozzle system.[9] Several spray nozzles such as air-assisted nozzles, high-pressure nozzles, and Micronair nozzles— capable of delivering smaller spray droplets (<30 μm VMD) for adult mosquito control—are now available.

Insecticide Residue Monitoring

At present, bioassay techniques are widely used to measure mosquito control efficacy and nontarget organism impact. Generally, bioassays only answer "yes or no" (i.e., dead or alive) and are primarily acute toxicity tests. However, bioassays do not address critical issues such as where or how much insecticide is present following the ULV application. Insecticide residue monitoring is encouraged because it can identify TICs following mosquito control application. Research shows that the level of TIC is critically reflective of mosquito control efficacy and nontarget impacts based on the dose–response relationship. When TIC is used in conjunction with bioassays, mosquito control programs will have a powerful quantitative tool to combat mosquitoes while protecting the environment.[7–9,11,12] Insecticide residues cannot be observed by the human eye. However, they can be detected and quantified by modern analytical techniques such as gas chromatography or high-performance liquid chromatography. By monitoring insecticide residues, we can 1) determine the insecticide concentrations in the air and use these data to facilitate adjustments to the spray system to attain the right application dose; 2) identify any excessive insecticide deposited on the ground or into the water and use these data to optimize the spray system; 3) monitor the distance of aerosol dispersed downwind for an effective spray offset; 4) establish appropriate impact thresholds for nontarget mortality and effective thresholds for mosquito control efficacy; and 5) compare different application equipment or operational scenarios.

Conclusion

In summary, controlling adult mosquitoes through ULV aerial application of insecticides while minimizing nontarget impacts can be achieved by operational mosquito control programs. Instituting a residue monitoring program can be critical in protecting the environment. TIC levels identified from insecticide residue monitoring provide a powerful tool to optimize the application so that proper application of insecticides can be achieved at the right time, at the right place, and at the right dose.

Acknowledgments

I thank James Cilek and Cate Brock for their critical reviews, helpful comments, and suggestions.

References

1. Mount, G.A.; Biery, T.L.; Haile, D.G. A review of ultralow-volume aerial sprays of insecticide for mosquito control. J. Am. Mosq. Control. Assoc. **1996**, *12* (4), 601–618.
2. Hill, E.F.; Eliason, D.A.; Kilpatrick, J.W. Effects of ultralow volume application of malathion in Hale County, Texas. J. Med. Entomol. **1971**, *8* (2), 173–179.
3. McKenney, C.L.; Shirley, M.A.; Pierce, R.H. *A Two-Year Research Project in the Rookery Bay National Estuarine Research Reserve to Evaluate Populational and Physiological Responses of Crabs from Various Habitats in Relation to the Fate of a Mosquito Control Pesticide*, Contract No. NC-ND2100–3–00042; 1997; 9.
4. Pankiw, T.; Jay, S.C. Aerially applied ultra-low-volume malathion effects on caged honey bees (Hymenoptera: Apidae), caged mosquitoes (Diptera: Culicidae), and malathion residues. J. Econ. Entomol. **1992**, *85*, 687–691.
5. Pankiw, T.; Jay, S.C. Aerially applied ultra-low-volume malathion effects on colonies of honey bees (Hymenoptera: Apidae). J. Econ. Entomol. **1992**, *85*, 692–699.
6. Schoor, W.P.; Middaugh, D.P.; Marcovich, D.T. Effects of aerially applied fenthion on survival and reproduction of the panacea sand fiddler, *Uca panacea,* in laboratory habitats. Arch. Environ. Contam. Toxicol. **2000**, *38*, 327–333.
7. Zhong, H.; Dukes, J.; Greer, M.; Hester, P.; Shirley, M.; Anderson, B. Ground deposition impact of aerially applied fenthion on the fiddler crabs, *Uca pugilator.* J. Am. Mosq. Control. Assoc. **2003**, *19,* 47–52.
8. Zhong, H.; Latham, M.; Hester, P.G.; Frommer, R.L.; Brock, C. Impact of aerial applied ultra-low-volume insecticide naled with flat-fan nozzle system on honeybee survival and productivity. Arch. Environ. Contam. Toxicol. **2003**, *45,* 216–220.
9. Zhong, H.; Latham, M.; Payne, S.; Brock, C. Minimizing the impact of mosquito adulticide naled on honeybees *Apis mellifera* (Hymenoptera: Apidae): Aerial ultra-low volume application using a high-pressure nozzle system. J. Econ. Entomol. **2004**, *97,* 1–7.
10. Mount, G.A. A critical review of ultralow-volume aerosols of insecticide applied with vehicle-mounted generators for adult mosquito control. J. Am. Mosq. Control. Assoc. **1998**, *14* (3), 305–334.
11. Dukes, J.; Zhong, H.; Greer, M.; Hester, P.; Hogan, D.; Barber, J.A.S. A comparison of two spray nozzle systems used to aerially apply the ultra-low-volume adulticide fenthion. J. Am. Mosq. Control Assoc. **2004**, *20* (1), 27–35.
12. Dukes, J.; Zhong, H.; Greer, M.; Hester, P.; Hogan, D.; Barber, J.A.S. A comparison of two ultra-low-volume spray nozzle systems by using a multiple swath scenario for the aerial application of fenthion against adult mosquitoes. J. Am. Mosq. Control Assoc. **2004**, *20* (1), 36–44.
13. Zhong, H.; Hribar, L.J.; Daniels, J.C.; Feken, M.A.; Brock, C.; Trager, M.D. Aerial ultra-low-volume application of naled: Impact on nontarget imperiled butterfly larvae (*Cyclargus thomasi bethune-bakeri*) and efficacy against adult mosquitoes (*Aedes taeniorhynchus*). Environ. Entomol. **2010**, *39* (6), 1961–1972.

6

Neurotoxicants: Developmental Experimental Testing

Vera Lucia S.S.
de Castro

Introduction

Neurotoxic substances may play a role in a number of neurodevelopmental disorders. They can be released by industrial facilities and by agricultural practices, much of which ends up in the air or groundwater. Since most neuroteratogens affect multiple regions and processes, they can result in various behavioral defects.

Given the importance placed on fostering optimal cognitive development and the fact that chemical exposures can perturb the exquisite spatial and temporal choreography of brain development, it is not surprising that neurodevelopmental deficit frequently serves as the critical adverse health effect in risk assessments.[1]

Detection and characterization of chemical-induced toxic effects in the central and peripheral nervous system represent a big challenge. Prediction of neurotoxic effects is a key feature in the toxicological profile of compounds and is therefore required by many regulatory testing schemes.[2]

Despite the increasing recognition of the need to evaluate developmental neurotoxicity (DNT) in safety assessment, only very few of the commercial chemicals in current use have been examined with respect to neurodevelopmental effects. Validated rodent models exist, but they are considered expensive and are only infrequently used. The neurodevelopmental disorders include learning disabilities, attention deficit hyperactivity disorder, autism spectrum disorders, developmental delays, and emotional and behavioral problems. The causes of these disorders are unclear, and interacting genetic, environmental, and social factors are likely determinants of abnormal brain development.[3,4]

In calculations of environmental burdens of disease in children, lead neurotoxicity to the developing brain is a major contributor. Pesticide effects could well be of the same magnitude, or larger, depending on the exposure levels.[4] For example, an emerging literature provides evidence of neurobehavioral consequences resulting from exposure to relatively low levels of organochlorine and organophosphate pesticides in infants and children.

Organophosphate pesticides continue to be applied widely in agriculture and in residences throughout the world, representing about half the total annual amount of insecticides used. One of the major concerns with these agents is their propensity to elicit DNT at exposures below the threshold for any systemic symptoms, so that potentially damaging fetal or childhood exposures may go undetected until persistent functional impairments become expressed.[5]

Organophosphate-poisoned populations have shown a consistent pattern of deficits when compared to a nonexposed or non-poisoned population on measures of motor speed and coordination, sustained attention, and information processing speed.[6] In experimental models, developing animals have been shown to be more susceptible than adult animals to the acute toxicity of the organophosphate pesticide chlorpyrifos, which can cause neurobehavioral abnormalities.[7] Neonatal diazinon exposure below the threshold for appreciable cholinesterase inhibition in a non-monotonic dose–effect caused persisting neurocognitive deficits in adulthood. The organophosphorous insecticide can affect transmitter systems supporting memory function, differently, implying participation of mechanisms other than their common inhibition of cholinesterase.[5]

Future studies should examine the neurodevelopment effects in human beings associated with pesticide mixtures and other classes of pesticides (e.g., carbamates, pyrethroids), and with pesticide mixtures, because there is increasing use of these pesticides in certain communities that are replacing the organophosphate and organochlorine pesticides.[8] In this direction, the U.S. Agency for Toxic Substances and Disease Registry (ATSDR) developed a program for chemical mixtures of which an integral part is a mixtures health risk assessment. ATSDR has completed evaluations for several simple mixtures of child-specific exposure concern.[9]

Behavioral Aspects of DNT

Behavior represents an integrated response of the nervous system that can reveal functional changes important to the overall fitness and survival of the organism exposed to single pesticides or mixtures. Although some developmental neurotoxicants are structural teratogens as well, behavioral dysfunctions may be more serious than structural defects under certain circumstances.[10] The major developmental sensory systems of concern in toxicology include visual, auditory, olfactory, nociceptive (pain and other noxious stimuli), somatosensory, and vestibular. However, neurobehavioral functions are influenced by subject variables such as age, sex, education, and social and (especially in humans) cultural background.

Brain Development and Maturation

The development and maturation of the mammalian brain is an extremely complex process. Brain development involves cell division, migration and differentiation, programmed cell death (apoptosis), cell-to-cell interactions (e.g., for migration and synaptic communication), and multiple other processes under different timetables for the various brain regions. Genetic, epigenetic, and environmental factors (e.g., exposure to toxic chemicals, including certain heavy metals, industrial chemicals, and pesticides), particularly during the susceptible periods of development and aging, can result in many possible adverse central nervous system (CNS) consequences, ranging from mild to severe and involving various functions (e.g., cognition, motor, or sensory dysfunction). DNT refers to any adverse effect of perinatal exposure to a toxic substance on the normal development of nervous system structure and/ or function.[11]

The mammalian brain undergoes a period of rapid brain growth, which in humans occurs perinatally, spanning from the third trimester of pregnancy throughout the first 2 years of life. In rats and mice, the brain growth spurt occurs in the neonate, spanning the first 3–4 weeks of life and reaching its peak around postnatal day 10. This period is characterized by axonal and dendritic outgrowth and the establishment of neuronal connections, and during this period, animals acquire many new motor and sensory abilities. Neurotypic and gliotypic proteins can serve as sensitive indicators of time- and region-specific effects of chemicals on the developing nervous system. The presence of xenobiotics in the brain during this defined period of maturational processes is a critical factor for induction of persistent changes in behavior and transmitter systems.[12] For example, it has been reported that multiple neurotransmitter systems are altered following exposure to organophosphorus insecticides. Developmental chlorpyrifos exposure produces persistent deficiencies in cholinergic synaptic neurochemistry.[13] Also, there is increasing evidence that polychlorinated biphenyls (PCBs) and methyl mercury also have neurotoxic effects. An enhanced effect of these toxicants, due to either synergistic or additive effects, would be considered as a risk for fetal development. It is postulated that these neurotoxicants might interact.[14]

Epidemiological studies have demonstrated a relationship between perinatal exposure to persistent organic pollutants among others and neurological and behavioral disturbances in infants and children. Studies in animals have confirmed that contaminants like PCBs, metals, and pesticides can disrupt behavioral functioning.[15]

Evidence indicates that exposure to environmental chemicals could have an impact on children's health and development. The developing CNS of fetus and children is particularly susceptible to chemically induced damage compared with the brain of adults due to the different pharmacokinetic factors, diminished defense mechanisms, or the fact that the developing nervous system undergoes a highly complex series of ontogenetic processes that are vulnerable to chemical perturbation.[16]

Periods of Vulnerability

The developing nervous system is particularly sensitive to environmental insults during critical periods that are dependent on the temporal and regional emergence of specific and sequential developmental processes (i.e., proliferation, migration, differentiation, synaptogenesis, myelination, and apoptosis). Evidence from numerous sources demonstrates that neural development extends from the embryonic period through adolescence. In general, the sequence of events is comparable among species, although the time scales are considerably different. Developmental exposure of animals or humans to numerous agents (e.g., x-ray irradiation, ethanol, lead, methyl mercury, or chlorpyrifos) demonstrates that interference with one or more of these developmental processes can lead to DNT.[17]

For many behaviors, a critical period exists during which the animal is sensitive to these organizational effects. Functional and structural life-lasting modifications can be induced by alterations of natural conditions during these adaptive developmental stages of maturation. The critical periods have been described for some cortical circuits involved in many different sensory systems such as the auditory, somatosensory, and olfactory systems. These critical periods occur also during postnatal life. Indeed, experience-dependent plasticity during critical periods of postnatal development shapes the adult brain anatomy and function.[18,19]

Protocols for Experimental Studies

In order to reduce the risk regarding the exposure to pollutants, addressing the behavioral aspects by appropriate investigation to sustain the safe use of the compounds is suggested. Evaluation of pre- and postnatal developmental parameters can be improved by including different tests on and safety assessment of chemicals to indicate the proper functioning of the sensory, motor, emotional, and cognitive domains.

Laboratory experimental studies suggest that many currently used pesticides such as organophosphates, carbamates, pyrethroids, ethylenebisdithiocarbamates, and chlorophenoxy herbicides can cause neurodevelopmental toxicity. Adverse effects on brain development can be severe and irreversible.[4]

Emotional processes can be viewed as adaptive events or states that are likely to occur across the animal kingdom, but that may or may not have subjective components, comprising physiological, behavioral, and subjective components, depending on the species and circumstances involved. Although these different components usually act in concert, they are potentially dissociable, not always operating as a functional whole.[20]

The utilization of an experimental protocol containing indices related to reproduction and animal development can identify initial damages due to exposure to environmental pollutants. Behavioral experimental methods are used to detect and characterize developmental neurotoxic effects on sensory, cognitive, and motor system functions. Neurobehavioral evaluations are widely used to examine the potential neurotoxicity of pesticides and other chemicals, [21] since neurobehavioral performance can be a sensitive biomarker of the neurodevelopmental consequences of exposure to environmental agents.

Prevention of possible damages due to pesticides during the development of young organisms, like newborns, requires an integrated strategy capable of monitoring the standard use of these products as well as the integration of the potential effects to improve evaluation. If available, biomarkers of exposure are useful for assessing the bioavailability of toxicants to the dam and offspring in utero and after birth. The evaluation of these biomarkers needs to differentiate normal variability from changes that are adverse in response.[22]

Animal models are used to understand neurophysiological processes on the basis of human exposure to xenobiotics. They represent a basis for understanding their pathophysiological traits. There is a variety of methodologies that can be utilized to assess these processes. Cross-species comparability between human and experimental animals supports the assumption that DNT effects in animals indicate a potential to affect development in humans.[10]

The first guideline specifically designed to evaluate DNT was developed and implemented by the U.S. Environmental Protection Agency (EPA) in 1991 and has later on been updated. The Organization for Economic Cooperation and Development (OECD) initiated the development of a DNT guideline (TG 426) following the recommendations of the OECD Working Group on Reproduction and Developmental Toxicity in Copenhagen in 1995. The first draft based on the U.S. EPA DNT guideline was prepared following a 1996 Expert Consultation Meeting that addressed a number of significant issues and incorporated improvements. The draft TG 426 was distributed to National Coordinators for comments in 1998, and significant technical issues in the comments were further discussed and revised.[10,23]

Developmental toxicity may result from either prenatal or postnatal exposure, may manifest at any life stage, and may be expressed as functional deficits. The DNT study is a specialized type of developmental toxicity study designed to screen for adverse effects of pre- and postnatal exposure on the development and function of the nervous system and to provide dose–response characterizations of those outcomes. The U.S. EPA and OECD DNT guidelines recommend administration of the test substance during gestation and lactation. Cohorts of offspring (typically rat) are randomly selected from control and treated litters for evaluations of gross neurologic and behavioral abnormalities during postnatal development and adulthood. These include assessments of physical development, behavioral ontogeny, motor activity, motor and sensory function, learning and memory, and postmortem evaluation of brain weights and neuropathology.[23]

There are a number of stimulus properties shared by all sensory systems, including intensity, frequency, duration, and location in space. In this way, behavioral tests of motor dysfunction in animals include those used to detect spontaneous movement disorders such as changes in gait, tremors, and myoclonus, and those used to detect changes in induced movement such as reflexes, reactions, and movements under operant control. Tests of motor function include observation of locomotion, measurement of locomotor activity, and tests of reflexes and reactions. Also, assessment of cognitive function is a critical component of a DNT assessment to address concerns over potential long term consequences of exposures to toxicants during brain development. Cognitive function is thought to encompass learning, memory, and attention processes.[24–27]

In this way, more effort is needed to adequately evaluate the neurotoxic effects. More elaborated experimental protocols are continuously proposed. They will focus on the interpretation of data obtained in studies that link xenobiotics exposure and functional (behavioral) deficits due to specific neurotransmitter and synaptic mechanisms,[28] identifying possible chemical class-specific targets and biomarkers of effect. Above all, gene expression could be also used as a sensitive tool for the initial identification of DNT effects induced by different mechanisms of toxicity in both cell types (neuronal and glial) and at various stages of cell development and maturation.[16]

Recent literature have examined specific end points across multiple guideline DNT studies to demonstrate the value of current methods in hazard characterization and explore further opportunities for methodologic refinement, examining the interpretation of neurodevelopmental end points for human health risk assessment, data interpretation and variability, positive control data, and statistical analysis.[11,29,30]

Test method reliability, reproducibility, and relevance are attributable in part to the high level of standardization of the test methods;[23] in some cases, the variability of some end points (e.g., motor activity) is very large. Methods have been suggested to decrease such variability.[29–31] Sources of variability include factors related to environmental conditions, personnel, experimental procedures, and equipment.[21,31] The detection, measurement, and interpretation of DNT effects depend on appropriate study design and execution, using established methods with appropriate controls.

Furthermore, the nature and extent of developmental neurotoxic effects often are dependent on the timing of exposure to a toxic agent or combinations of agents and environmental conditions; i.e., organisms exhibit distinct temporal windows of susceptibility. Variations in neurotoxic outcomes across species are expected because stages of nervous system development can vary significantly between species in relation to the time of birth. Thus, the time and duration of exposure in animal models must also be selected carefully to match the window of exposure in the human situation and allow cross-species extrapolation.[32]

Detection and characterization of chemical-induced toxic effects in the central and peripheral nervous system represent a major challenge for employing newly developed technologies in the field of neurotoxicology. For example, those using specific brain cell types can produce results of general mechanism of action but not specific to the chemical tested. In addition, toxicokinetic models are to be developed in order to properly evaluate absorption, distribution, metabolism, and excretion, as well as the blood–brain barrier. Behavioral toxicologists will be needed to contribute for the experimental tests and computational models to anchor molecular initiating events to adverse outcomes. Therefore, an intensive search for the development of alternative methods using in silico models for neurotoxic hazard assessment is appropriate.[33] The following are some of the challenges that need to be overcome: predicting behavior using models of complex neurobiological pathways, standardizing study designs and dependent variables to facilitate creation of databases, and managing the cost and efficiency of behavioral assessments.[34]

Conclusion

There is growing evidence of the adverse impact of exposures to ambient and indoor air pollutants on fetal growth and both early childhood and animal neurodevelopment. The normal structure and function of the nervous system may be altered as a result of exposure to some pollutants before or after birth. Its analysis is particularly relevant in assessing the interference of a chemical pollutant with neuroendocrine maturation by behavioral methods, as it is a sensitive and broad marker of perturbation of both nervous and endocrine functions.

A number of methods can evaluate alterations in sensory, motor, and cognitive functions in laboratory animals exposed to toxicants during nervous system development. Assessment methods are being developed to examine other nervous system functions, including social behavior, autonomic processes, and biologic rhythms.[35] Fundamental issues underlying proper use and interpretation of these methods include 1) consideration of the scientific goal in experimental design; 2) selection of an appropriate animal model; 3) expertise of the investigator; 4) adequate statistical analysis; and 5) proper data interpretation.

Acknowledgments

I wish to thank Dr. Stachetti for his precious contribution to this entry, consisting of the initial text, over which the present one has been constructed. Also, I would like to thank him for the opportunity offered to contribute to this prestigious publication, by referring my name as author. Without his help and his confidence on my work, this achievement would not have been possible. I also would like to thank Dr. Stachetti for the final text comments and suggestions.

References

1. Bellinger, D.C. Interpreting epidemiologic studies of developmental neurotoxicity: Conceptual and analytic issues. Neurotoxicol. Teratol. **2009**, *31,* 267–274.
2. Crofton, K.M.; Makris, S.L.; Sette, W.F.; Mendez, E.; Raffaele, K.C. A qualitative retrospective analysis of positive control data in developmental neurotoxicity studies. Neuro-toxicol. Teratol. **2004**, *26,* 345–352.
3. Dietrich, K.N.; Eskenazi, B.; Schantz, S.; Yolton, K.; Rauh, V.A.; Johnson, C.B.; Alkon, A.; Canfield, R.L.; Pessah, I.N.; Berman, R.F. Principles and practices of neurodevelopmental assessment in children: Lessons learned from the centers for children's environmental health and disease prevention. Res. Environ. Health Perspect. **2005**, *113* (10), 1437–1446.
4. Bjørling-Poulsen, M.; Andersen, H.R.; Grandjean, P. Potential developmental neurotoxicity of pesticides used in Europe. Environ. Health **2008**, *7,* 50.
5. Timofeeva, O.A.; Roegge, C.S.; Seidler, F.J.; Slotkin, T.A.; Levin, E.D. Persistent cognitive alterations in rats after early postnatal exposure to low doses of the organophosphate pesticide, diazinon. Neurotoxicol. Teratol. **2008**, *30,* 38–45.
6. Rohlman, D.S.; Lasarev, M.; Anger, W.K.; Scherer, J.; Stupfel, J.; McCauley, L. Neurobehavioral performance of adult and adolescent agricultural workers. NeuroToxicology, **2007**, *28,* 374–380.
7. Richardson, J.; Chambers, J. Effects of repeated oral postnatal exposure to chlorpyrifos on cholinergic neurochemistry in developing rats. Toxicol. Sci. **2005**, *84,* 352–359.
8. Eskenazi, B.; Rosas, L.G.; Marks, A.R.; Bradman, A.; Harley, K.; Holland, N.; Johnson, C.; Fenster, L.; Barr, D.B. Pesticide toxicity and the developing brain. Basic Clin. Pharmacol. Toxicol. **2008**, *102,* 228–236.
9. Pohl, H.R.; Abadin, H.G. Chemical mixtures: Evaluation of risk for child-specific exposures in a multi-stressor environment. Toxicol. Appl. Pharmacol. **2008**, *233,* 116–125.
10. Hass, U. The need for developmental neurotoxicity studies in risk assessment for developmental toxicity. Reprod. Toxicol. **2006**, *22,* 148–156.
11. Tyl, R.W.; Crofton, K.; Moretto, A.; Moser, V.; Sheets, L.P.; Sobotka, T.J. Identification and interpretation of developmental neurotoxicity effects: A report from the ILSI Research Foundation/Risk Science Institute expert panel on neurodevelopmental endpoints. Neurotoxicol. Teratol. **2008**, *30* (4), 349–381.
12. Viberg, H. Exposure to polybrominated diphenyl ethers 203 and 206 during the neonatal brain growth spurt affects proteins important for normal neurodevelopment in mice. Toxicol. Sci. **2009**, *109* (2), 306–311.
13. Johnson, F.O.; Chambers, J.E.; Nail, C.A.; Givaruang-sawat, S.; Carr R.L. Developmental chlorpyrifos and methyl parathion exposure alters radial-arm maze performance in juvenile and adult rats. Toxicol. Sci. **2009**, *109* (1), 132–142.
14. Andersen, I.S.; Voie, O.A.; Fonnum, F.; Mariussen, E. Effects of methyl mercury in combination with polychlorinated biphenyls and brominated flame retardants on the uptake of glutamate in rat brain synaptosomes: A mathematical approach for the study of mixtures. Toxicol. Sci. **2009**, *112* (1), 175–184.

15. Bowers, W.J.; Nakai, J.S.; Chu, I.; Wade, M.G.; Moir, D.; Yagminas, A.; Gill, S.; Pulido, O.; Meuller, R. Early developmental neurotoxicity of a PCB/organochlorine mixture in rodents after gestational and lactational exposure. Toxicol. Sci. **2004**, *77*, 51–62.

16. Hogberg, H.T.; Kinsner-Ovaskainen, A.; Coecke, S.; Hartung, T.; Bal-Price, A.K. mRNA expression is a relevant tool to identify developmental neurotoxicants using an in vitro approach. Toxicol. Sci. **2010**, *113* (1), 95–115.

17. Rice, D.; Barone, S., Jr. Critical periods of vulnerability for the developing nervous system: Evidence from humans and animal models. Environ. Health Perspect. **2000**, *108* (3), 511–533.

18. Soiza-Reilly, M.; Azcurra, J.M. Developmental striatal critical period of activity-dependent plasticity is also a window of susceptibility for haloperidol induced adult motor alterations. Neurotoxicol. Teratol. **2009**, *31*, 191–197.

19. Crews, F.; He, J.; Hodge, C. Adolescent cortical development: A critical period of vulnerability for addiction. Pharmacol. Biochem. Behav. **2007**, *86*, 189–199.

20. Paul, E.S.; Harding, E.J.; Mendl, M. Measuring emotional processes in animals: The utility of a cognitive approach. Neurosci. Biobehav. Rev. **2005**, *29*, 469–491.

21. Slikker, W., Jr.; Acuff, K.; Boyes, W.; Chelonis, J.; Crofton, K.; Dearlove, G.; Li, A.; Moser, V.; Newland, C.; Rossi, J.; Schantz, S.; Sette, W.; Sheets, L.; Stanton, M.; Tyl, S.; So-botka, T. Behavioral test methods workshop. Neurotoxicol. Teratol. **2005**, *27*, 417–427.

22. Raffaele, K.C.; Rowland, J.; May, B.; Makris, S.L.; Schumacher, K.; Scarano, L.J. The use of developmental neurotoxicity data in pesticide risk assessments, Neurotoxicol. Teratol., **2010**, *32*(5):563–572. doi:10.1016/j.ntt.2010.04.053.

23. Makris, S.L.; Raffaele, K.; Allen, S.; Bowers, W.J.; Hass, U.; Alleva, E.; Calamandrei, G.; Sheets, L.; Amcoff, P.; Delrue, N.; Crofton, K.M. A retrospective performance assessment of the developmental neurotoxicity study in support of OECD test guideline 426. Environ. Health Perspect. **2009**, *117* (1), 17–25.

24. Després, C.; Richer, F.; Roberge, M.C.; Lamoureux, D.; Beuter, A. Standardization of quantitative tests for preclinical detection of neuromotor dysfunctions in pediatric neurotoxicology. NeuroToxicology **2005**, *26*, 385–395.

25. Luft, J.; Bode, G. Integration of safety pharmacology endpoints into toxicology studies. Fundam. Clin. Pharmacol. **2002**, *16*, 91–103.

26. Moser, V.; Phillips, P.; Levine, A.; McDaniel, K.; Sills, R.; Jortner, B.; Butt, M. Neurotoxicity produced by dibromo-acetic acid in drinking water of rats. Toxicol. Sci. **2004**, *79*, 112–122.

27. Sarter, M. Animal cognition, defining the issues. Neurosci. Biobehav. Rev. **2004**, *28*, 645–650.

28. Yanai, J.; Brick-Turin, Y.; Dotan, S.; Langford, R.; Pinkas, A.; Slotkin, T.A. A mechanism-based complementary screening approach for the amelioration and reversal of neurobehavioral teratogenicity. Neurotoxicol. Teratol., **2010**, *32* (1), 109–113.

29. Crofton, K.M.; Foss, J.A.; Hass, U.; Jensen, K.; Levin, E.D.; Parker, S.L. Undertaking positive control studies as part of developmental neurotoxicity testing. Neurotoxicol Teratol. **2008**, *30* (4), 266–287.

30. Raffaele, K.C.; Fisher, J.E.; Hancock, S.; Hazelden, K.P.; Sobrian, S.K. Determining normal variability in a developmental neurotoxicity test. Neurotoxicol. Teratol. **2008**, *30* (4), 288–325.

31. Castro, V.L.; Silva, P.A. Validation of neurobehavioral studies for evaluating the perinatal effects of single and mixture exposure to pesticides. In *Progress in Pesticides Research*; Kanzantzakis, C.M., Ed.; Nova Science Publishers: New York, 2009; 371–395.

32. Hines, R.N.; Sargent, D.; Autrup, H.; Birnbaum, L.S.; Brent, R. L.; Doerrer, N.G.; Cohen Hubal, E.A.; Juberg, D.R.; Laurent, C.; Luebke, R.; Olejniczak, K.; Portier, C.J.; Slikker, W. Approaches for assessing risks to sensitive populations: Lessons learned from evaluating risks in the pediatric population. Toxicol. Sci. **2010**, *113* (1), 4–26.

33. Coecke, S.; Eskes, C.; Gartlon, J.; Kinsner, A.; Price, A.; van Vliet, E.; Prieto, P.; Boveri, M.; Bremer, S.; Adler, S.; Pellizzer, C.; Wendel, A.; Hartung, T. The value of alternative testing for neurotoxicity in the context of regulatory needs. Environ. Toxicol. Pharmacol. **2006**, *21*, 153–167.

34. Bushnell, P.J.; Kavlock, R.J.; Crofton, K.M.; Weiss, B.; Rice, D.C. Behavioral toxicology in the 21st century: Challenges and opportunities for behavioral scientists—Summary of a symposium presented at the annual meeting of the Neurobehavioral Teratology Society, 2009. Neurotoxicol. Teratol. **2010**, *32*, 313–328.

35. Cory-Slechta, D.A.; Crofton, K.M.; Foran, J.A.; Ross, J.F.; Sheets, L.P.; Weiss, B.; Mileson, B. Methods to identify and characterize developmental neurotoxicity for human health risk assessment. I: Behavioral effects. Environ. Health Perspect. **2001**, *109* (1), 79–91.

7

Persistent Organic Pesticides

Gamini Manuweera

Introduction

Persistent organic pesticides refer to a set of synthetic organic chemical substances meant to control pests in the human environment including those concerning agriculture, veterinary health, and public health. Persistent organic pesticides are part of a larger group of chemicals known as persistent organic pollutants or POPs.[1] In addition to pesticides, POPs include industrial chemicals and unintentionally produced chemical substances or by-products of anthropogenic origin. Dioxins, for example, are unintentionally produced POPs formed during incomplete combustion processes involving organic matter and chlorine. Chemically, POPs include linear and cyclic halogenated hydrocarbons. In some POPs, functional moieties may also exist in the hydrocarbon molecule as in the case of perfluorooctane sulfonic acid and its salts.

POPs are highly toxic to living organisms. This group of chemicals does not easily undergo common environmental degradation processes including chemical, microbial, or photolytic reactions. Once released, POPs stay for a long period of time in the environment, posing higher risk of long-term exposure to human populations and ecosystems. POPs are lipophilic (has affinity to fat, lipids, etc.) due to the nonpolar organic nature of the substances. The lipophilicity allows chemical substances to readily accumulate in fatty tissues of living organisms (the process referred to as bioaccumulation). Once accumulated, the concentration of POPs in the living organisms builds up through the food chain via biomagnification processes, increasing the risk of adverse effects at the higher tropic levels.

Some of the physicochemical properties of the POPs facilitate long-range transport in the environment. POPs are found in the alpine and mountainous regions, the Arctic, Antarctica, and remote Pacific islands far away from where activities associated with POPs are taking place. In the environment, POPs undergo sorption into organic matter, intra-media, and inter-media dispersion (diffusion), as well as advection (transport mechanism of substances due to the bulk motion of the medium). Most of the POPs are semi-volatile in nature. With long residence time in transport media, POPs can travel very long distances across regions through environmental transportation processes, including atmospheric transport, making them available for human and environmental exposure on a global scale.

The toxic effects of POPs are mainly linked to long-term, low-level exposure scenarios mostly resulting in chronic health problems, while some POPs could also exert acute effects. The toxic endpoints of POPs include cancer, birth defects, reproductive problems, damages to specific organs such as the liver and kidneys, among others.

Much attention was drawn to this group of substances at the international level after it became apparent that they travel long distances across borders. As a consequence, several countries started banning POPs in the 1970s. However, actions by a limited number of countries alone were unable to control continued environmental pollution and adverse health effects from such border-crossing substances. A regional legal agreement that specifically addresses POPs was adopted in 1998 with the Aarhus Protocol on Persistent Organic Pollutants under the regional Convention on Long-Range Transboundary Air Pollution (LRTAP) of the UN Economic Commission for Europe (UNECE).[2] As an agreement at the regional level was not sufficient to ensure satisfactory protection of human health and the environment from adverse effects of POPs, negotiations of an international legally binding instrument to reduce or eliminate, where possible, releases of POPs were initiated under the auspices of the United Nations Environmental Programme (UNEP) in 1998. In May 2001, more than 100 countries agreed and adopted a global treaty, now known as the Stockholm Convention on Persistent Organic Pollutants.[3,4] Some aspects of the life cycle of POPs are considered in other international legally binding instruments. POP wastes are included in the Basel Convention on the Control of Transboundary Movements of Hazardous Wastes and Their Disposal,[5] while the international trade of most POP pesticides is addressed by the Rotterdam Convention on the Prior Informed Consent Procedure for Certain Hazardous Chemicals and Pesticides in International Trade.[6] Several other international initiatives also address POPs, notably the Global Programme of Action for the Protection of the Marine Environment from Land-Based Activities (GPA)[7] and a number of regional sea agreements.

Persistent Organic Pesticides under the Stockholm Convention

As of 2019, there were 30 POPs included in the Stockholm Convention, of which, 17 were pesticides. At the time of its entry into force in 2001, the Stockholm Convention had 9 pesticides among the total of 12 POPs listed therein.

The Convention provides provisions for Parties to make proposals to add new POPs. These proposals must contain information on the chemical relating to the screening criteria established under the Convention that include persistence, bioaccumulation, potential for long-range environmental transport, and adverse effects. The POPs Review Committee established by the Convention reviews submissions on candidate POPs, including related information from other sources, prepares a risk profile, and undertakes a risk management evaluation. This step is undertaken once the committee evaluate the risk profile and agrees that the candidate chemical is likely, as a result of its long-range environmental transport, to lead to significant adverse human health and/or environmental effects, such that global action is warranted.

Upon that it undertakes a risk management evaluation to identify specific details of control actions relevant to the chemical to accompany with its recommendation to the COPs. If the candidate chemical satisfactorily meets the POP screening criteria of the Convention, the POPs Review Committee makes recommendations to the Conference of the Parties for considering the new chemical under the Convention. The Conference of the Parties then evaluates the recommendations made by the POPs Review Committee for the inclusion of the candidate chemical.

At its fourth meeting held in 2009, the Conference of the Parties to the Convention considered and included nine new chemical substances consisting five pesticides as POPs in the Convention. During the fifth, seventh, and ninth Conferences of the Parties held in 2011, 2017 and 2019, endosulfan and pentachlorophenol and its salts and esters and dicofol were listed under the Convention, respectively, as pesticides (see Table 1).

TABLE 1   Pesticides Listed as POPs under
the Stockholm Convention

Aldrin
Alpha hexachlorocyclohexane
Beta hexachlorocyclohexane
Chlordane
Chlordecone
DDT
Dieldrin
Dicofol
Endosulfan
Endrin
Heptachlor
Hexachlorobenzene
Lindane
Mirex
Pentachlorobenzene
Pentachlorophenol and its salts and esters
Toxaphene

The Stockholm Convention requires Parties to take measures to reduce or eliminate releases from intentional production and use of POPs through two different approaches:

- Prohibit and/or take the legal and administrative measures necessary to eliminate the chemical substances listed in Annex A of the Convention.
- Restrict production and use of the chemicals listed in Annex B of the Convention in accordance with specific measures provided in that Annex.

The chemical substances for elimination are listed in Annex A of the Convention. When chemical substances with existing commercial uses are included in the Convention for elimination, some chemicals may still have certain specific use or uses for which Parties to the Convention may require a transition period to eliminate completely the reliance on the chemical substance. For such chemicals, exemptions are provided for those specific uses and related production for a limited period of time. These specific exemptions are initially available for a 5-year period. The Parties opt for the use of specific exemptions that require conformation to a set of precautionary measures relevant to its available uses established by the Convention to ensure reduced releases of the substance and effective elimination. When there are no longer any Parties registered for a particular type of specific exemption, no new registrations may be made with respect to it. Aldrin, for example, was listed in Annex A of the Convention for elimination with a specific exemption of use only as local ectoparasiticide and insecticide with no further production. At the end of the initial period, the exemptions for production and use are not available, if Parties have effectively eliminated the reliance on those POPs.

During the fourth meeting of the Conference of the Parties in 2009, it was noted that the specific exemptions provided for six POP pesticides initially listed in Annex A of the Convention were no longer needed. Accordingly, the POP pesticides aldrin, chlordane, dieldrin, heptachlor, hexachlorobenzene (also found as a by-product and industrial chemical), and mirex listed in Annex A have no specific exemptions available for Parties anymore. In addition to those six POP pesticides, Annex A also contains alpha hexachlorocyclohexane (also found as a by-product), beta hexachlorocyclohexane (also found as a by-product), chlordecone, endrin, and toxaphene, for which no specific exemptions were provided at the time of listing them in the Convention. Dicofol was listed under the Convention without exemptions. The three remaining POP pesticides listed in Annex A with specific exemptions are lindane, technical endosulfan and its related isomers and pentachlorophenol and its salts and esters.

Lindane was listed under the Convention with prohibition of its production. All uses have also been prohibited except as human health pharmaceutical for control of head lice and scabies as second-line treatment. For Parties that may wish to use lindane for the control of head lice and scabies, the specific exemptions were initially available for that purpose until 2015. When listing the chemical, the Conference of the Parties requested the Secretariat to develop a work plan in collaboration with the World Health Organization (WHO) for reporting and reviewing requirements on the use of lindane for the specific exemptions allowed by the Convention. Lindane has been listed from 1977 and 1990 in the WHO Model List with benzyl benzoate and permethrin as a cost-effective alternative to lindane for head lice and scabies. It was removed from the WHO Model List in 1992 on the basis that "it is toxic to the environment and humans, and safer alternatives are available." The specific exemptions for lindane provided under the Stockholm Convention expired for most Parties in 2015. It was noted at the ninth Conference of the Parties held in 2019 that no Party has registered for specific exemption for lindane. Hence, registration for the exemptions of lindane will no longer be available for Parties.

The specific exemptions for technical endosulfan expired for most Parties in 2017. As of 2018, two Parties, which had made declarations pursuant to the Convention, had registered exemptions that will be applicable for them until 2019.

The chemical substances listed in Annex B are those identified with specific uses for which there are no alternatives available at present or that the alternatives are not accessible or effectively available under certain settings. Such uses are recognized under the Convention as acceptable purposes with no set timeline for elimination. Parties are allowed to produce and use the chemicals for those purposes according to the recommended practices provided with respective uses. In such cases, a process is established to periodically review the continued need of the chemical in question for the acceptable purposes. The Convention has listed dichlorodiphenyltrichloroethane (DDT), for example, under Annex B with an acceptable purpose for disease vector control. It requires activities associated with DDT to be in accordance with the WHO's recommendations and guidelines on the use of DDT. Further, the use of DDT is allowed when locally safe, effective, and affordable alternatives are not available to the Party in question.

Following Article 4 of the Convention, the Parties that use POP pesticides according to the specific exemptions and acceptable purposes provided in the Convention require notifying the Secretariat to register for that purpose. The register of specific exemptions is made publicly available on the Convention's website (www.pops.int).

Major Issues concerning Persistent Organic Pesticides

Over the years, there has been an increase in both the general understanding and concern about adverse effects of POP pesticides. The latest concerns include interference of POPs with hormonal activities, acting as "endocrine disruptors," and possible interlinkages between climate change and POPs. Release, distribution, and degradation of POPs are highly dependent on environmental conditions. Climate change and increasing climate variability have the potential to affect POPs' contamination via higher releases from primary sources and environmental reservoirs, changes in transport processes and pathways, and routes of degradation. Exposure to POPs and related impacts on environmental and human health can be further exacerbated by higher atmospheric temperatures.

Most of the POP pesticides included in the Stockholm Convention are first-generation pesticides discovered in the World War II era. The subsequent advent of pesticides from the chemical families of organophosphates, carbamates, and synthetic pyrethroids challenged the continuity of favorable market for the first-generation pesticides. However, the remarkable successes achieved in the malaria eradication programs in the 1950s and 1960s using DDT for indoor residual spraying (IRS)[8] and the continued need to rely on DDT for disease vector control compelled the global community to place DDT under Annex B of the Convention, thereby allowing its continued production and use for disease vector control. The WHO recommends DDT only for IRS. Countries can use DDT as long as necessary, in the quantity needed, provided that the guidelines and recommendations of the WHO and the Stockholm Convention are all met and until locally appropriate and cost-effective alternatives are available for a sustainable transition from DDT. The continued need for DDT for disease vector control is evaluated at regular Conferences of the Parties, held every 2 years. The evaluation is undertaken in consultation with the WHO on the basis of available scientific, technical, environmental, and economic information. As a separate process, the WHO also reviews new information on adverse health effects of DDT periodically to facilitate the evaluation of DDT by the Convention. In 2010, the WHO expert consultation report on "DDT in Indoor Residual Spraying: Human Health Aspects" identified several potential hazards of DDT and its toxic metabolites. These include acute poisoning hazards for children with accidental ingestion, carcinogenicity, developmental toxicity, male reproductive effects, and concerns for women of childbearing age who live in DDT IRS-treated dwellings. However, in terms of relevant exposure scenarios for the general population in countries using IRS, the expert panel concluded that available evidence does not point to concern about levels of exposure for any of the endpoints that were assessed. The report demands further research to better evaluate risks that were suggested in the studies reviewed.[9]

The chemical substances currently added to the Convention as new POPs often have many active uses. For some of those uses, either alternatives are not currently found or cost-effective alternative products and options are not readily accessible under a certain setting, often under the conditions prevailing in developing countries. Therefore, it is not uncommon that new POPs added to the Convention consist of a relatively longer list of use exemptions. The later addition to POP pesticides, endosulfan, which is listed in Annex A, has a number of uses provided under its specific exemptions (see Table 2). Eliminating exposure to POPs is more challenging when the chemicals are included in Annex B of the Convention with many acceptable purposes, where there is no time-bound phaseout requirement.

Indiscriminate use of pesticides could result in the development of pest resistance and resurgence of new pests requiring increased dependence on pest control actions. Limitations in effective deployment

TABLE 2 Specific Exemptions of Crop–Pest Complexes of Endosulfan Available for Parties under the Stockholm Convention

Crop	Pest
Apple	Aphids
Arhar, gram	Aphids, caterpillars, pea semilooper, pod borer
Bean, cowpea	Aphids, leaf miner, whiteflies
Chilli, onion, potato	Aphids, jassids
Coffee	Berry borer, stem borers
Cotton	Aphids, cotton bollworm, jassids, leaf rollers, pink bollworm, thrips, whiteflies
Eggplant, okra	Aphids, diamondback moth, jassids, shoot, and fruit borer
Groundnut	Aphids
Jute	Bihar hairy caterpillar, yellow mite
Maize	Aphids, pink borer, stem borers
Mango	Fruit flies, hoppers
Mustard	Aphids, gall midges
Rice	Gall midges, rice hispa, stem borers, white jassid
Tea	Aphids, caterpillars, flushworm, mealybugs, scale insects, smaller green leafhopper, tea geometrid, tea mosquito bug, thrips
Tobacco	Aphids, oriental tobacco budworm
Tomato	Aphids, diamondback moth, jassids, leaf miner, shoot and fruit borer, whiteflies
Wheat	Aphids, pink borer, termites

of nonchemical pest control interventions, especially in developing countries, lead to further reliance on chemical control options. Even if the alternative pesticides are not highly toxic or do not possess POPs' characteristics, the increase of chemical load on the environment is inevitable, leading to undesirable consequences. Under certain settings where locally appropriate, cost-effective, and safer alternatives are not accessible when managing pest resistance has become a serious challenge would demand possible reintroduction or continued use of POP pesticides.

Alternative Approaches

Alternatives to POP pesticides include chemical and nonchemical products as well as control interventions that focus on avoiding pest interference or creating conditions not favorable for the prevalence of the targeted pest. Often, any one of those potential alternatives alone would not fulfill all desirable features of the POP pesticides as a successful replacement. It requires formulating approaches with combination of viable options appropriate for the given situation. The availability of a wider range of choices from different pest control options is vital for the development of such alternative approaches. New developments in organic chemistry such as increased flexibility to modify known functional insecticide chemical backbones to produce new toxophores have expanded the prospects for new chemical alternatives. Current research on nonchemical options ranges from conventional biopesticide products to gene technology and interventions on physical ecosystem management. These initiatives should present promising opportunities for a wider selection of vector control options for efficient integration.

Major issues related to the elimination of pesticides in the present list of POPs particularly concern the use of DDT for disease vector control. There are only three classes of pesticides currently available for public health vector control as alternatives to DDT: synthetic pyrethroids, organophosphates, and carbamates. They represent two different modes of actions limiting the choice of pesticides available for the management of vector resistance. The situation demands an urgent need for bringing new cost-effective public health pesticides to disease-endemic countries. All current public health pesticide active ingredients were initially developed for the agrochemical market. In the 1980s, the shift of the agrochemical target product profile, from broad-spectrum contact insecticides to target site specific selective toxicity insecticides, delivered a number of new agrochemicals that could be repurposed for public health. In efforts to eliminate DDT, governments should seek alternative approaches that are sustainable in situations prevailing in the country. It will be important to ensure that these POP pesticides are not simply replaced by other pesticides, but that the principles of integrated pest and vector management are adopted with due consideration on resistance management. Intergovernmental agencies such as the United Nations Environment Programme are promoting initiatives to demonstrate sustainable replacement of DDT in disease vector control using integrated multidisciplinary approach. It is also important to ensure that the strategies used will not be compromised by measures in other sectors. Efforts for effective management of resistance in disease vectors are hindered where the same insecticides are used in agriculture. Similarly, unplanned environmental modifications for developmental purposes could create more breeding grounds for malaria mosquitoes.

Assurance of close collaboration between sectors and key stakeholders is vital in endeavors by countries to find more sustainable solutions to POPs. These collaborations also help strengthen the base of the civil society in communities to increase other social benefits. Community participation, multisectoral initiatives on public awareness campaigns, and local social surveillance have been successfully integrated in a program implemented in Mexico and Central America on demonstrating the effectiveness of alternative methods to DDT for malaria control.[10] Such solutions must be based on the local conditions and can be best sustained through active community participation. Structures established under one sector such as Farmer Field Schools in agriculture may, for example, serve the purposes of public health and the environment. The interrelationship between the environment, agriculture, and health is, hence, a key for identifying sustainable strategies that will effectively and efficiently protect agriculture from pests, communities from diseases like malaria, and ecosystems from persistent pesticides.

Global Initiatives to Promote Alternatives to DDT

The fourth Conference of the Parties of the Stockholm Convention concluded that countries currently using DDT for disease vector control may need to continue such use until locally appropriate and cost-effective alternatives are available for a sustainable transition away from DDT. To support the countries still using DDT to reduce their reliance, it also endorsed the establishment of the Global Alliance for the development and deployment of products, methods, and strategies as alternatives to DDT for disease vector control.

The Global Alliance for alternatives to DDT provides an instrument for partnership and collaboration among all stakeholders at the global and national level to increase momentum on achieving the common goals and to catalyze new initiatives for the development and deployment of alternatives towards the elimination of reliance on DDT. The work of the Global Alliance is organized in a manner that respects its noninvolvement in funding and executing programs on the ground, yet addresses expectations that it will trigger significant actions in support of the development and deployment of alternatives to DDT.

The instrument is expected to stimulate the research community and chemical industry to accelerate the release of safer chemical alternatives to DDT. It draws global development initiatives to strengthen in-country capacity and knowledge base for efficient integration of vector control options into cost-effective and sustainable programs in disease-endemic countries. It involves global authorities and experts to review and develop support tools for countries to implement related activities and support efficient and effective networks and communication to promote indigenous knowledge and innovative concepts on nonchemical approaches.

The platform has been established with partners from parties to the Convention with due consideration on malaria disease-endemic countries and WHO, including research and academic institutions, the donor community, civil society organizations, and the pesticide industry.[11]

The UNEP in response to an invitation by the sixth Conference of the Parties to the Stockholm Convention held in 2013 prepared a road map for the development of locally safe, effective, affordable, and environmentally sound alternatives to DDT in consultation with WHO, the DDT expert group, established by the Conference of the Parties to assess continued need for DDT for disease vector control, and the Secretariat of the Convention. The basis of the Conference of the Parties to develop a road map was the conclusion that countries relying on DDT for disease vector control may need to continue such use until locally safe, effective, affordable, and environmentally sound alternatives are available for a sustainable transition away from DDT. The purpose of the road map is to provide a thematic guide and logical steps that are needed to make locally safe, effective, affordable, and environmentally sound alternatives to DDT.[12]

Conclusion

The purpose of pesticides as a tool to control and, including in many cases, to kill living organisms presents the greatest drawback for its own existence and use. While the adverse effects of pesticides are many and diverse, some pesticides pose unique risks. Irrespective of the point of release, POP pesticides exert serious adverse health and environmental effects on the global scale, leaving little or no options except avoiding the reliance on them towards total elimination.

In spite of constant efforts, challenges for the development and deployment of sustainable solutions to avert continued reliance on some POP pesticides remain. The socioeconomic and ecological dimensions in specific settings may impede bringing in straightforward global solutions without compromising the benefits associated with certain uses of POP pesticides. The conventional approach of finding a chemical replacement with same pesticidal properties of the POP pesticide is not always viable. When long residual effects on the control of targeted pest, a property link to POPs' characteristics, are among the reasons for continued need for a POP pesticide, finding a chemical replacement with similar properties becomes even more challenging. The solution should, therefore, respect a multidisciplinary approach targeting interventions on a broader scope encompassing the life cycle of the pest consisting of a series of control options. Any individual element of such multidisciplinary control approach shall not produce a complete control over the targeted pest. The implementation of properly formulated complementary control interventions offers a successful and sustainable outcome. Such an approach requires the collaboration of different sectors of the society in implementing respective control actions with proper coordination in a strategic framework. It also includes action by stakeholders at global, regional, and local levels to ensure enhanced sustainability of the initiatives. Benefits to the global community on such integrated approach are not limited to the protection of human health and the environment from adverse effects of POP pesticides. It also helps to achieve enhanced coordination and collaboration within the civil society for sustainable development especially under resource-limited settings.

References

1. United Nations Environment Programme. Stockholm Convention on Persistent Organic Pollutants (POPs), 2009.
2. United Nations Economic Commission for Europe: Environment, available at www.unece.org/leginstr/env.html (accessed November 2011).
3. Persistent Organic Pollutant; United Nations Environmental Programme, available at www.chem.unep.ch/pops (accessed August 2011).
4. About the Convention. Stockholm Convention—Protecting Human Health and the Environment from Persistent Organic Pollutants, available at http://chm.pops.int/Conven-tion/tabid/54/Default.aspx (accessed October 2011).
5. Basel Convention on the Control of Transboundary Movements of Hazardous Wastes and their Disposal, available at http://archive.basel.int/ (accessed March 2011).
6. Rotterdam Convention: Shared Responsibility, available at www.pic.int/ (accessed November 2011).
7. Global Programme of Action for the Protection of the Marine Environment for Land-based Activities (GPA); United Nations Environment Programme, available at www.gpa.unep.org/ (accessed November 2011).
8. Mörner, J.; Bos, R.; Fredrix, M. *Reducing and/or Eliminating Persistent Organic Pesticides—Guidance on Strategies for Sustainable Pest and Vector Management*; United Nations Environment Programme: Geneva, 2002.
9. World Health Organization. *DDT in Indoor Residual Spraying: Human Health Aspects, Environmental Health Criteria*; WHO Press: Geneva, 2011, p. 241.
10. Regional Program of Action and Demonstration of Sustainable Alternative to DDT for Malaria Vector Control in Mexico and Central America (project DDT/UNEP/GEF/PAHP), available at www.paho.org/english/ad/sde/ddt-regionals.htm (accessed October 2011).
11. Global Alliance for Alternatives to DDT. Stockholm Convention: Protecting Human Health and the Environment from Persistent Organic Pollutants, available at www.pops.int/ga (accessed March 2011).
12. Road map for the development of alternatives to DDT, UNEP/DTIE, Chemicals Branch, February 2015, available at www.unenvironment.org/explore-topics/chemicals-waste/what-we-do/persistent-organic-pollutants/roadmap-alternatives-ddt (accessed July 2019).

8

Pollutants: Organic and Inorganic

A. Paul Schwab

Introduction

A soil pollutant can be broadly defined as any chemical or other substance which either is not normally found in soil or is present at high enough concentrations to be harmful to any living organisms.[1] This very general definition could be applied to human-derived (anthropogenic) as well as naturally occurring constituents. This entry will focus primarily on those pollutants that have anthropogenic origins and will address carbon-based (organic) and inorganic chemicals. Organic contaminants can include pesticides, solvents, preservatives, petroleum products, hormones, and antibiotics. Inorganic pollutants are comprised of heavy metals (e.g., lead and mercury), nonmetals (selenium), metalloids (arsenic and antimony), radionuclides, and simple soluble salts (sodium chloride). Whether the source of the contaminants is natural or anthropogenic, understanding their chemistry, toxicity, and bioavailability is crucial to responding to soil pollution.

Organic Pollutants

Thousands of organic chemicals exist in nature, and many are acutely toxic as well as carcinogenic, mutagenic, or teratogenic. When discussing toxic organic chemicals in soils, the focus normally is on those pollutants resulting from human activities because their release has the potential to be controlled. Typical synthetic organic pollutants include fuels, lubricants, herbicides, fungicides, insecticides, solvents, and propellants.

Soil Contamination by Organic Chemicals

Organic chemicals can find their way into the soil accidentally through spills, leaking storage tanks, and unintentional discharges. However, not all soil pollution is accidental. More than one-half million metric tons of pesticides are used annually, the majority of which are used on agricultural fields.[1]

Approximately 20,000 metric tons of pesticides are used for non-agricultural applications, including railroad right of way weed control, turf, and horticulture. Although many pesticides do not persist in soils, others are highly persistent and have been studied extensively because of their negative impacts. Chlordane (termite control), dichlorodiphenyltrichloroethane (DDT) (mosquitoes), and atrazine (weeds) are excellent examples of organic pesticides that have been determined to have human health effects and severe ecological impacts; chlordane and DDT have been banned in the United States, and the banning of atrazine has been debated.

Thousands of organic chemicals are in use today, and listing all the specific compounds that contaminate soils is beyond the scope of this entry. However, contaminants can be sorted into categories of chemicals that are used frequently and are commonly found in soils (Table 1). Insecticides, herbicides, fungicides, and nematicides may be the most frequently encountered, and some pesticides are quite toxic. Atrazine has been found to be somewhat persistent and is mobile such that atrazine applied to soil can migrate to groundwater and surface water. The insecticide DDT and its metabolites continue to be found in ecosystems despite being banned for decades. Methyl bromide (1,2-dibromomethane) is used as a nematicide, but it is a gas that is known to be ozone depleting. Trichloroethylene (TCE) is a useful solvent that, when disposed on soil, moves rapidly downward and contaminates groundwater. Petroleum-based fuels such as diesel and gasoline are problematic when spilled or originating from leaking underground storage tanks. Many of the other classes of compounds listed are of ecological concern when found in soil, again due to their potential toxicity to a wide range of organisms.

On December 11, 1980, the United States Congress enacted the Comprehensive Environmental Response, Compensation, and Liability Act (CERCLA), commonly known as Superfund, to provide wide-ranging federal authority to respond to contamination or threats of contamination by hazardous substances. Among other actions, CERCLA established a list of priority pollutants with known or suspected

TABLE 1 Some Classes of Organic Chemicals Found in Soils

Contaminant Class	Example(s)	Sources, Impacts
Insecticides	DDT, chlordane, diazinon	Used for insect control; DDT and chlordane have serious health effects; diazinon has been found in water supplies.
Herbicides	Atrazine, 2,4-D, 2,4,5-T	Atrazine used on corn; 2,4-D widely used in lawns and agriculture; defoliant 2,4,5-T implicated in health effects in Agent Orange.
Fungicides	Benomyl, propiconazole, chlorothalonil	Agricultural fungicides that are directly applied to soil.
Nematicides	Methyl bromide (1,2-dibromomethane)	Used in fumigation of soil to remove nematodes; neurotoxin and ozone depleting.
Solvents	Trichloroethylene, trichloroethane	Widely used solvents and degreasers; chronic health effects not clear.
Fuels	Diesel, kerosene	From spills, leaking tanks; can lead to groundwater contamination. Some constituents of fuels are carcinogenic.
Polyaromatic hydrocarbons	Chrysene, benzo[a]pyrene	Components of petroleum, particularly after combustion; many of these are carcinogenic.
Polychlorinated biphenyls	Aroclors (1260, 1016, 1242); coplanar congeners (3,4,3', 4'- tetrachlorobiphenyl, 3,4,5,3', 4'- pentachlorobiphenyl)	Dielectric fluids in transformers, capacitors, and coolants. Can lead to skin conditions, ocular lesions, teratogenic effects in animals, endocrine disruption.
Explosives, propellants	TNT, RDX	Residuals from manufacture are the largest source of contamination; acute and chronic toxicities documented.

Source: Adapted from Schwarzenbach, Gschwend, and Imboden[2] and Evangelou.[3]

ecological or health impacts. The list is periodically updated, and the pollutants are prioritized. Thirty-two of these compounds are given in Table 2 along with the CERCLA priority ranking[4] and soil screening levels (SSLs).[5] This list includes compounds that have been banned in the United States (e.g., DDT) as well as those that are part of our everyday lives (e.g., BTEX—benzene, toluene, ethylbenzene, xylenes—found in gasoline). Most of the SSLs are in the range of 1 to 50 mg/kg. Some compounds have significantly higher SSL values (5000 mg/kg for toluene), indicating that these compounds are far less toxic. Others have very low SSL values (5.0×10^{-4} mg/kg for benzidine), suggesting that these compounds are a threat at very low concentrations. Soil concentrations of these compounds required to prevent threats to groundwater are approximately 100 times lower than the residential SSLs.

TABLE 2 Organic Compounds on the CERCLA Priorities List and Their Ranking, Health Impacts, and SSLs

| | | Soil Screening Level | |
| | | Residential Soil | Protect Groundwater |
Constituent	CERCLA Rank	mg/kg	
Vinyl chloride	4	6.0×10^{-2}	5.6×10^{-6}
PCBsa	5	1.4×10^{-1}	5.2×10^{-3}
Benzene	6	1.1×10^{-0}	2.1×10^{-4}
Polycyclic aromatic hydrocarbons	8	1.5×10^{-2}	2.7×10^{-4}
Chloroform	11	2.9×10^{-1}	5.3×10^{-5}
DDT, p, p'	12	1.7×10^{0}	6.7×10^{-2}
Trichloroethylene	16	2.8×10^{0}	7.2×10^{-4}
Dieldrin	17	3.0×10^{-2}	1.7×10^{-4}
Chlordane	20	1.6×10^{0}	1.3×10^{-2}
DDE, p, p'	21	1.4×10^{0}	4.7×10^{-2}
Hexachlorobutadiene	22	6.2×10^{0}	1.7×10^{-3}
Aldrin	24	2.9×10^{-2}	6.5×10^{-4}
DDD, p, p'	25	1.4×10^{0}	6.6×10^{-2}
Benzidine	26	5.0×10^{-4}	2.4×10^{-7}
Toxaphene	31	4.4×10^{-1}	9.4×10^{-3}
Hexachlorocyclohexane, γ(lindane)	32	5.2×10^{-1}	3.6×10^{-4}
Tetrachlorethylene	33	5.5×10^{-1}	4.9×10^{-5}
Heptachlor	34	1.1×10^{-1}	1.2×10^{-3}
1,2-Dibromomethane	35	2.5×10^{1}	2.5×10^{-3}
Hexachlorocyclohexane, ß	36	2.7×10^{-1}	2.2×10^{-4}
Acrolein	37	1.5×10^{-1}	8.4×10^{-6}
Disulfoton	38	2.4×10^{0}	2.7×10^{-3}
3,3'-Dichlorobenzidine	40	1.1×10^{0}	9.4×10^{-4}
Endrin	41	1.8×10^{1}	4.4×10^{-1}
Pentachlorophenol	45	3.0×10^{0}	5.7×10^{-3}
Heptachlor epoxide	46	5.3×10^{-2}	1.5×10^{-4}
Carbon tetrachloride	47	6.1×10^{-1}	1.7×10^{-4}
Diazinon	56	4.3×10^{1}	1.6×10^{-1}
Xylenes	58	6.3×10^{2}	2.0×10^{-1}
Toluene	71	5.0×10^{3}	1.6×10^{0}
Ethylbenzene	99	5.4×10^{0}	1.7×10^{-3}

All screening levels are for residential soils unless stated otherwise. Radionuclides were excluded from this list.

a This is a general class of contaminants, and individual members have unique SSLs. The most restrictive value was chosen for this table.

The legacy of high use of persistent, potentially toxic compounds became clear in a study published in 2010 in which the presence of organochlorine pesticides, polychlorinated biphenyls, and perfluorinated compounds were determined in food samples purchased in supermarkets in the United States.[6] The tracked compounds were detected in nearly all the food samples: DDT metabolite *p*, *p′*-dichlorodiphenyldichloroethylene was found in milk products; polychlorinated biphenyls (PCBs) were found in fish; and perfluorinated compounds were found in over half of all samples. Results such as these add to already enhanced sensitivities concerning organic contaminants in soils.

Potential Impacts of Organic Contaminants

After an organism is exposed to an organic pollutant, a number of antagonistic effects are possible if concentrations are high enough. The most dramatic impact is acute toxicity, in which symptoms are quickly apparent and readily identified. Consuming large quantities of the pure contaminant is not necessarily a requirement for acute toxicity. A case of acute parathion poisoning was reported when a child consumed contaminated soil.[7] Although reports of such cases are rare, the possibility for acute poisoning through soil consumption by children with pica is realistic for other compounds, such as phenol.[8] The estimated lethal dose for phenol is estimated to range from 10 to 50 mg/kg body mass, but the ingestion of only 5 g of soil contaminated with an SSL of 47,000 mg/kg would result in a dose of 18 mg/kg.

At lower concentrations, the impacts of organic contaminants become less obvious and take longer to be expressed. For many carcinogens, for example, decades are required to develop cancerous tumors. Pathway of exposure, concentrations, and duration of the exposure all dictate the resulting health effects and are essential components of risk analysis. Exposure to contaminated soil can result in chronic toxicity if the soil repeatedly comes in contact with the skin, if particulates are frequently inhaled, or if volatile compounds migrate into closed living spaces. Most SSLs have been developed based on risk associated with chronic exposure.

Bioremediation of Organic Contaminants

Although contamination of soil by organic compounds is an important environmental problem, many of these pollutants can be removed from the soil through bioremediation. Soil microorganisms have a remarkable capacity to degrade organic contaminants. Degradation can be direct, using the organics as a source of energy and carbon, or indirect, in which the compounds are cometabolized by organisms seeking similar compounds. End products can be $CO_2(g)$ after total mineralization; alteration and humification; or incorporation into the microbial biomass. Polyaromatic hydrocarbons (PAHs) are readily degraded in the soil, as are many pesticides and components of BTEX. However, highly chlorinated compounds such as PCBs, some solvents, and the explosive cyclotrimethylenetrinitramine (RDX) are far more difficult to degrade, and some of the initial degradation products are as toxic as or more toxic than the parent compound. Reviews have been published for the bioremediation of pesticides,[9] a wide range of organic contaminants by fungi,[10] bacterial degradation of aromatic compounds,[11] PAHs,[12] aliphatic hydrocarbons,[13] explosives,[14] petroleum hydrocarbons by mycorrhizae,[15] earthworm-assisted bioremediation,[16] and composting as a general bioremediation approach.[17] Bioremediation of specific compounds also has been reviewed: pyridine, indole, and quinoline[18]; dieldrin and endrin[19]; and catechols.[20] This subject is treated in depth in another section of this entry.[21]

Phytoremediation is another approach taken to enhance the dissipation of organic contaminants in soil. The mechanism of removal of the contaminant from the soil depends upon the properties of the organic compound, the soil, and the chemistry of the roots. In some instances, organic contaminants are assimilated by the plants and either degraded or volatilized as part of the transpiration stream. Trichloroethylene has been observed to be effectively remediated in the root zone of poplar trees, but the mechanism has been the subject of debate. In some instances, uptake of TCE and eventual volatilization have been observed,[22] but other studies detected no volatilization of TCE and complete degradation in the soil.[23] In nearly all cases for PAHs and PCBs, plant uptake is negligible, and phytoremediation of these compounds is accomplished by microbial degradation in the rhizosphere.

Uptake of organic contaminants during phytoremediation is an important consideration for many reasons. From an ecological standpoint, accumulation of contaminants in the aboveground portions of plants is undesirable because of the potential for introduction into the food chain or dispersal of the contaminants. From the remediation perspective, uptake may be desirable to help remove the compounds from the soil and allow degradation within plant tissues. Uptake of volatile compounds followed by release to the atmosphere would be prohibited in many regulatory environments. Therefore, efforts have been taken to predict the transfer from soil to roots to the transpiration stream of higher plants. The most useful parameter in this analysis is the transpiration stream concentration factor (TSCF):

$$TSCF = \frac{Concentration\ in\ xylem\ sap}{Concentration\ in\ external\ solution}$$

Concentrations in the xylem are difficult to quantify and are estimated as the amount of a compound assimilated over a given period of time that has been corrected for degradation with the plant. If the compound of interest is neither actively accumulated nor excluded from the plant (i.e., passive uptake), then TSCF = 1.0. Non-ionized organic compounds are of particular interest in phytoremediation because target contaminants are in this class, including many pesticides, PAHs, PCBs, etc. Trends in TSCF as a function of the octanol-water coefficient (K_{OW} or log K_{OW}) have been investigated.[24,25] Compounds that are soluble in water have small or even negative values of log K_{OW}; hydrophobic compounds have high values of log K_{OW}. Briggs et al.[24] investigated substituted phenylureas and o-methylcarbamoyloximes, and Burken and Schnoor[25] investigated a suite of compounds including RDX, phenol, benzene, atrazine, TCE, pentachlorophenol (PCP), and others. The combined data from the two studies are shown in Figure 1 and follow the relationship

$$TSCF = 0.76 \exp\left(-\left(\log K_{OW} - 2.45\right)^2 / 4.38\right)$$

This relationship predicts that compounds with a log K_{OW} of 2.45 will have the maximum movement into the transpiration stream. The shape of the curve is viewed as reflecting the balance of the various tendencies for the compounds to desorb from soil surfaces and pass through the hydrophobic

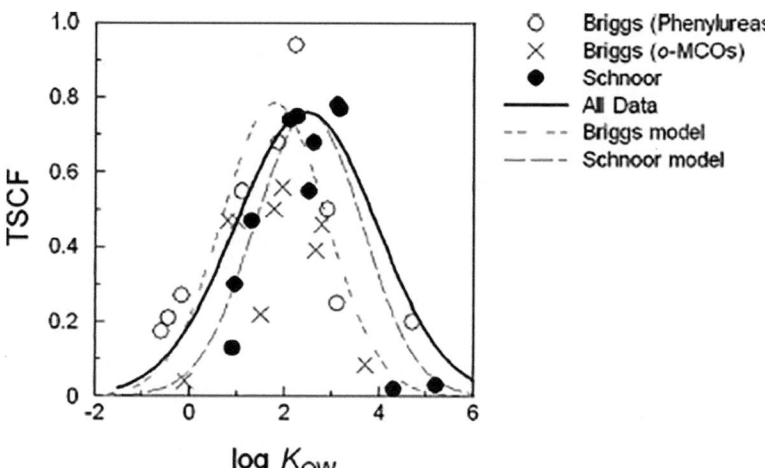

FIGURE 1 Transpiration stream concentration factor (TSCF) as a function of log K_{OW} (octanol-water partition coefficient). Data come from Briggs et al.[24] and Burken and Schnoor.[25] The models take the following form: TSCF = $a \exp(-(\log K_{OW} - b)^2 c)$, in which a, b, and c are variables.

cellular membranes. Compounds with a low log K_{OW} will desorb readily from the soil but will not pass through the hydrophobic plasmalemma. Organic contaminants with high log K_{OW} values are predicted to have the capability to penetrate the root membranes, but they either will not desorb from soil components or will irreversibly adsorb to the lipophilic root membranes.[26]

Inorganic Pollutants

Inorganic pollutants are those contaminant compounds that do not contain carbon (with a few exceptions, including cyanide and carbonate). These contaminants include acutely toxic heavy metals, metalloids, oxyanions, and radioactive elements and the far-less-toxic soluble salts. Although all elements are naturally occurring, thousands of inorganic compounds are formed only through human activities. Some are highly toxic in trace quantities, but all have a threshold concentration above which they can be harmful.

Classes of Inorganic Pollutants and Concentration Ranges

As with organic pollutants, thousands of inorganic contaminants exist and are found over a wide range of concentrations. It is convenient to group them into classes (Table 3) consistent with the periodic table. Metals include Pb, Zn, Ni, and Cd, and their typical soil concentrations vary over a wide range, as do the concentrations considered anomalous. Most metals in soils occur as the cations in solution and in the solid phase. Nearly all metals can have toxic effects on humans, although some metals (such as Zn) require high concentrations for toxicity. Metalloids, including As and Sb, generally occur in soil solution and solid phases as oxyanions (such as AsO_4^{3-}), resulting in significantly different chemical behavior than the metals. Metalloids can be highly toxic. Soluble salts are typical inorganic species that, in normal concentrations, generally do not constitute a health or environmental threat. Typical salts are sulfates, chlorides, and nitrates of calcium, magnesium, potassium, and sodium. These components are always present in soil and do not pose an environmental problem until concentrations become excessive (Table 3).

Similar to organic contaminants, inorganic compounds have been listed and prioritized by CERCLA,[4] and most have SSLs established by the U.S. Environmental Protection Agency.[5] Radionuclides are not included in these listings because radioactive elements are considered under different legislative initiatives. Table 4 lists 27 inorganic species, their CERCLA priority ranking, and SSLs. The inorganic constituents have a prominent position in the top 10 on the CERCLA priority list, with arsenic being number one and lead, mercury, and cadmium also in the top ten. However, the priority ranking for the remaining inorganic contaminants is generally lower than those for the

TABLE 3 Inorganic Contaminants in Soil, Their Typical Ranges Found in Uncontaminated Soils, and Those Concentrations Considered to Be Unusual

Contaminant	Sources, Impacts	Typical Range in Soil	Extreme Concentrations
Metals Pb, Zn, Cd, Cu, Ni, Ba, Sr, Mn, Cr	Mining, smelting, electroplating	0.1 to 1000 mg/kg	25 to >10,000 mg/kg (metal dependent)
Metalloids As, B, Ge, Sb	Industrial activities, smelting	<0.1 to 100 mg/kg	>50 mg/kg
Nonmetals Se, I, P, S	Manufacturing, processing	<0.1 to 1000 mg/kg	up to 2000 mg/kg (element dependent)
Soluble salts Halides, alkalis, alkaline earths, sulfates, nitrates	Mining, manufacturing, petroleum extraction, natural sources	<500 mg/kg	>1000 mg/kg

TABLE 4 Inorganic Compounds on the CERCLA Priorities List and Their Ranking, Health Impacts, and SSLs

| | | Soil Screening Level | |
| | | Residential Soil | Protect Groundwater |
Constituent	CERCLA Rank	mg/kg	
Antimony	219	4.0×10^1	6.6×10^{-1}
Arsenic	1	3.9×10^{-1}	1.3×10^{-3}
Barium	109	1.5×10^4	3.0×10^2
Beryllium	42	1.6×10^2	2.5×10^1
Cadmium	8	7.0×10^1	1.4×10^0
Chlorine	91	7.5×10^2	1.6×10^0
Chromium(VI)	18	2.9×10^{-1}	8.3×10^{-4}
Cobalt	49	2.3×10^1	4.9×10^{-1}
Copper	128	3.1×10^3	5.1×10^1
Cyanide	28	1.6×10^3	7.4×10^0
Fluoride	211	3.1×10^3	NL[a]
Hydrazine	84	2.1×10^{-1}	NL[a]
Iodine	NL[a]	7.8×10^2	NL[a]
Lead	2	4.0×10^2	NL[a]
Manganese	117	1.8×10^3	5.7×10^1
Mercury	3	2.3×10^1	5.7×10^{-1}
Molybdenum	NL[b]	3.9×10^2	3.7×10^0
Nickel	53	1.5×10^3	4.8×10^1
Nitrate	216	1.3×10^5	NL[a]
Nitrite	212	7.8×10^3	NL[a]
Perchlorates	NL[b]	5.5×10^1	NL[a]
Phosphorus (white)	19	1.6×10^0	2.7×10^{-3}
Selenium	147	2.9×10^2	9.5×10^{-1}
Silver	214	3.9×10^2	1.6×10^0
Uranium	98	2.3×10^2	4.9×10^1
Vanadium	198	5.5×10^1	2.6×10^1
Zinc	74	2.3×10^4	6.8×10^2

All screening levels are for residential soils unless stated otherwise. Radionuclides were excluded from this list.
[a] Not listed.
[b] Not listed because radionuclides are treated separately.

organics (Table 2). The residential SSLs for the inorganics tend to be far higher than for the organics and typically have a range of 50 to 3000 mg/kg. Arsenic (0.39), Cr(VI) (0.29), and hydrazine (0.21) are the only inorganics with SSL values less than 1 mg/kg. SSLs for protection of groundwater are roughly 100 times lower than residential SSLs.

Potential Health Impacts of Inorganic Contaminants in Soils

Many of the metals, metalloids, and oxyanions have negative effects on human health, and these impacts have been studied for decades. The most widely distributed and best-known toxic metal in soils is Pb. If Pb-contaminated soil is consumed, the risk increases for lead-associated nervous system, brain, and blood disorders. The impacts of Pb in soil from car exhaust[27] and smelter sites[28] are among the many studies examining this well-known metal. The SSL for Pb is 400 mg/kg.

Cadmium exposure for humans occurs mostly through the consumption of food grown in Cd-contaminated soils. Health impacts of Cd include kidney function, bone strength, and central nervous system disorder. Cadmium contamination in soils can be the result of mining, refining, smelting, battery production, battery disposal, and other industrial operations. Generally, cadmium uptake by plants increases with decreasing pH and increasing total Cd in the soil.[29] The mechanism of cadmium uptake by plants is similar to that of zinc and calcium.

Chromium in soils is viewed to be problematic only if the Cr is in the Cr(VI) (chromate) oxidation state. Reduced chromium, Cr(III), is less mobile and far less toxic. High chromate can result in renal failure, DNA damage, and cancer. Because Cr(VI) is present in the soil as the anion CrO_4^{2-}, it is not strongly held by the soil, tends to be mobile, and is readily assimilated by plants. Fugitive dust from contaminated soil also can lead to increased incidence of cancer.[30] If Cr(VI) can be reduced to Cr(III) in the soil, the health and environmental threats are greatly diminished.[31,32] Labile organic matter, microorganisms, and reduced inorganic species such as Fe^{2+} can readily convert Cr(III) to Cr(VI). Although surface soils in equilibrium with atmospheric O2(g) are predicted to maintain chromium as Cr(VI), even small amounts of organic matter in moist, shallow subsoils can reduce Cr(VI) to Cr(III).

Mercury contamination in humans can result in skin rashes, hypertension, rapid pulse, kidney dysfunction, and memory failure. The most common pathway of exposure to Hg poisoning is through fish and meat, but high concentrations in soils can result in increased exposure through plant uptake and inhalation. The SSL for Hg is 23 mg/kg for residential soils and 0.57 mg/kg to protect groundwater (Table 4). Mercury contamination in soil is fairly uncommon, but in some mining areas, concentrations in the soil as high as 2700 mg/kg have been observed and are associated with plant concentrations as high as 1100 mg/kg. A threshold concentration in the soil is required before plant concentrations increase significantly above background.[33]

In most natural systems, arsenic occurs as one of two redox species: arsenate, AsO_3^{3-}, or arsenite, AsO_4^{3-}, Although both forms appear to induce toxicity in humans, the arsenite is more readily absorbed by human tissues.[34] The residential soil SSL for arsenic is 0.39 mg/kg and 0.013 mg/kg to prevent impacts on groundwater. Most of the attention to arsenic poisoning in recent years has been paid to As in drinking water, particularly as a result of extreme concentrations of total As in wells in Bangladesh. However, plant uptake of As and the exposure of foods to high-As irrigation water also can result in arsenic intake in excess of the World Health Organization's recommended limits.[35]

Remediation of Inorganic Contaminants

The best approach to keeping soil free from contamination is prevention. When this approach fails, removing the contaminants may be necessary, and many options are available.[36] With few exceptions, biological approaches to decontamination of soils with high concentrations of pollutant metals are limited to phytoremediation. Some inorganic contaminants, such as nitrate, chromate, perchlorate, and other redox-active species in which one redox state is less of a threat than another, can be bioremediated through processes in which the chemical state of the contaminant is altered. Nitrate can be denitrified through the production of gaseous nitrogen compounds.[37] Perchlorate can be reduced to benign chloride,[38] and Cr(VI) can be reduced to Cr(III).[39,40] For metals or metalloids in which only one chemical state is generally found in soils, a biological approach not involving phytoremediation is rare.

During the past two decades, phytoremediation of metals in soils has been vigorously explored. For some metals, phytoremediation has been highly successful and has found commercial applications: hyperaccumulation of Zn, Cd, and Ni by *Thlaspi caerulescens*;[41] brake ferns for removal of As from soil;[42] and phytovolatilization of mercury.[43,44] One of the most common metal contaminants in soils in Pb, but effective means of phytoremediation of Pb-contaminated soils have been elusive.[45]

Phytoremediation is not the only remediation alternative for metal-contaminated soils. Traditional and engineering approaches have been used with reasonable success. Although basic excavation methods are effective at cleaning up sites, the soil remains contaminated and is merely transported to a containment facility. Excavation is the least expensive means of remediation but damages the site and does not improve the contaminated soil resource. Soils may be incinerated to burn off organic pollutants and to volatilize metals. Other means of containment include capping the site, encasing the soil in cement, or in situ vitrification in which the soil is melted followed by solidification of the soil into a single mass of glass. These methods are effective in isolating the soil, but considerable waste is generated, the contaminants remain in place, and the soil is usually lost as a viable resource.[46]

Less destructive approaches often are preferred because the soil is not seriously altered or destroyed, and the techniques are often less expensive. Such methods are typified by soil washing,[47] electrokinetics and electromigration processes,[39] and chemical stabilization.[39]

Pathways of Exposure

Once in the soil, organic and inorganic pollutants can follow many pathways to potentially impacted organisms.[48] For all contaminants, direct consumption of contaminated soil can be an important mechanism of exposure for soil-borne organisms, grazing mammals that consume the soil, and humans. For some pollutants, direct soil consumption is the only means of exposure if the contaminants have low mobility and low bioavailability; Pb and five-or six-ring PAHs are examples.

Eight pathways exist by which contaminants in soil can reach humans.[48] The pathways include direct soil consumption; uptake of contaminants from soil into food crops and subsequent consumption by humans; consumption of animals that have been contaminated by tainted soil; inhalation of metals in air or airborne dust; and drinking contaminated groundwater or surface water. Half of the mechanisms involve plant uptake and subsequent consumption of the plants. Most anionic inorganic species are readily assimilated by plants, as are many cationic species. However, some inorganic species form highly insoluble solid phases or are otherwise non-bioavailable, and they are not translocated within the plant. For nonionic organic species, plant uptake is a function of water solubility and hydrophobicity/lipophilicity. Compounds with intermediate lipophilicity are most suitable for plant uptake. Once the pollutants are in the plant, organisms that consume the vegetation may be exposed to the pollutant, and some of the contaminants are passed along the food chain.

Other pathways of exposure include the contaminant moving from the soil to the water or air or becoming airborne on particulates. All of these mechanisms have been shown to contribute to negative health effects. Drinking contaminated water is one of the better-known pathways. If the drinking water originates from groundwater, then the pollutants must be soluble in water and percolate readily through the soil and into the groundwater. When surface water is used for drinking, the transport mechanism can be movement of dissolved contaminant in the water, or contaminants may be adsorbed onto the suspended sediment in the soil runoff. Virtually every soil pollutant can be transported by this mechanism, and only careful control of runoff water and sediments can avoid this problem.

Direct volatilization from the soil into the air is important for a limited number of contaminants, but when volatilization of a potentially toxic compound or element occurs, exposure is directly through the lungs. Mercury (Hg) is a well-known volatile metal, and direct inhalation of high concentrations of gaseous Hg or dimethyl mercury can be poisonous. Inhalation of airborne, contaminated particles is a more frequent occurrence but less insidious. After inhalation, contaminants must be released from the dust before they can be assimilated. Inhalation of contaminated particulates can be reduced through standard soil erosion control measures, such as establishing windbreaks and adequate plant cover on contaminated soils.

Conclusions

Hundreds of contaminants are found in soils, and many reach concentrations that require action. Regulated organic contaminants are usually anthropogenic in their origins and, as a result, their impact can be minimized by controlling their manufacture and use. Even after strict regulations, including outright banning of manufacture or use, detectable or even dangerous concentrations can be found for decades. Residual DDT and its degradation products are examples of highly persistent organic compounds that remain dangerous long after the termination of their use. Metals and other inorganic contaminants can be as problematic as persistent organics because they are not easily transformed to benign forms and they will persist in soils for extended periods. Although most metals do not have a pathway of exposure that could result in human toxicity, some metals (e.g., Se, Mo, Cd, Pb, As) do have such a pathway and must be monitored carefully.

References

1. Pierzynski, G.M.; Sims, J.T.; Vance, G.F. Classification of pollutants. In *Soils and Environmental Quality,* 2nd Ed.; Lewis Publishers: Boca Raton, Florida, 1993; 47–50, 167183, 185–210.
2. Schwarzenbach, R.P.; Gschwend, P.M.; Imboden, D.M. *Environmental Organic Chemistry;* Wiley-Interscience: New York, 1992; 8–41, 255–344.
3. Evangelou, V.P. Water quality. In *Environmental Soil and Water Chemistry;* Wiley Interscience: New York, 1998; 476–499.
4. Agency for Toxic Substances Registry. 2007 CERCLA Priority List of Hazardous Substances, available at http://www.agriculturedefensecoalition.org/sites/default/files/pdfs/4H_2007_CDC_ATSDR_CERCLA_Priority_List_of_Hazard ous_Substances_2007.pdf, (accessed May 2012).
5. United States Environmental Protection Agency. Regional Screening Levels (RSL) Summary Table May 2010, available at http://www.epa.gov/reg3hwmd/risk/human/rb-concentration_table/Generic_Tables/pdf/master_sl_table_run_MAY2012.pdf, (accessed May 2012).
6. Schecter, A.; Colacino, J.; Haffner, D.; Patel, K.; Opel, M. Perfluorinated compounds, polychlorinated biphenyls, and organochlorine pesticide contamination in composite food samples from Dallas, Texas, USA. Environ. Health Per-spect. **2010,** *118* (6), doi:10.1289/ehp.0901347.
7. Quinby, G.E.; Clappison, G.B. Parathion poisoning—A nearfatal pediatric case treated with 2-pyridine aldoxime methio- dide (2-PAM). Arch. Environ. Health **1961,** *3,* 438–447.
8. Calabrese, E.J.; Stanek, E.J.; James, R.C.; Roberts, S.M. Soil ingestion—A concern for acute toxicity in children. J. Environ. Health **1999,** *61,* 18–23.
9. Castelo-Grande, T.; Augusto, P.A.; Monteiro, P.; Estevez, A.M.; Barbosa, D. Remediation of soils contaminated with pesticides: A review. Internat. J. Environ. Anal. Chem. **2010,** *90,* 438–467.
10. Pinedo-Rivello, C.; Aleu, J.; Collado, I.G. Pollutants biodegradation by fungi. Curr. Org. Chem. **2009,** *13,* 11941214.
11. Seo, J.-S.; Keum, Y.S.; Li, Q.X. Bacterial degradation of aromatic compounds. Int. J. Environ. Res. Public Health **2009,** 6, 278–309.
12. Bamforth, S.M.; Singleton, I. Bioremediation of polycyclic aromatic hydrocarbons: Current knowledge and future directions. J. Chem. Technol. Biotechnol. **2005,** *80,* 723736.
13. Stroud, J.L.; Paton, G.I.; Semple, K.T. Microbe-aliphatic hydrocarbon interactions in soil: Implications for biodegradation and bioremediation. J. Appl. Microbiol. **2007,** *102,* 1239–1253.
14. Lewis, T.A.; Newcombe, D.A.; Crawford, R.L. Bioremediation of soils contaminated with explosives. J. Environ. Manag. **2004,** *70,* 291–307.
15. Roberston, S.J.; McGill, W.B.; Massicotte, H.B.; Rutherford, P.M. Petroleum hydrocarbon contamination in boreal forest soils: A mycorrhizal ecosystems perspective. Biol. Rev. **2007,** *82,* 213–240.
16. Hickman, Z.A.; Reid, B.J. Earthworm assisted bioremediation of organic contaminants. Environ. Internat. **2008,** 34, 1072–1081.
17. Semple, K.T.; Reid, B.J.; Fermor, T.R. Impact of composting strategies on the treatment of soils contaminated with organic pollutants. Environ. Pollut. **2001,** *112,* 269–283.
18. Fetzner, S. Bacterial degradation of pyridine, indole, quinoline, and their derivatives under different redox conditions. Appl. Microbiol. Biotechnol. **1998,** *49,* 237–250.

19. Bhatt, P.; Kumar, M.S.; Chakrabarti, T. Fate and degradation of POP-hexachlorocyclohexane. Crit. Rev. Environ. Sci. Technol. **2009**, *39*, 655–695.

20. Zeyaullah, M.; Abdelkafe, A.S.; Ben Zabye, W.; Ali, A. Biodegradation of catechols by micro-organisms—A short review. African J. Biotechnol. **2009**, *13*, 2916–2922.

21. Radosevich, M.; Rhine, E.D. Pollutants (Ch. 281). In *Encyclopedia of Soil Science,* 2005; 1338–1343.

22. Ma, X.; Burken, J.G. TCE diffusion to the atmosphere in phytoremediation applications. Environ. Sci. Technol. **2003**, *37*, 2534–2539.

23. Strand, S.E.; Dossett, M.; Harris, C.; Wang, X. Doty, S.L. Mass balance studies of volatile chlorinated hydrocarbon phytoremediation. Z. Naturforsch. **2005**, *60*, 325–330.

24. Briggs, G.G.; Bromilow, R.H.; Evans, A.A.. Relationship between lipophilicity and root uptake and translocation of nonionized chemicals by barley. Pestic. Sci. **1982**, *13*, 495–504.

25. Burken, J.G.; Schnoor, J.L. Predictive relationships for uptake of organic contaminants by hybrid poplar trees. Environ. Sci. Technol. **1998**, *32*, 3379–3385.

26. Bromilow, R.H.; Chamberlain, K. Principles governing uptake and transport of chemicals. In *Plant Contamination: Modeling and Simulation of Organic Chemical Processes*; Trapp, S, McFarlane, C., Eds.; Lewis Publishers: Boca Raton, Florida, 1995; 37–68.

27. Rhue, R.D.; Mansell, R.S.; Ou, L.T.; Tang, S.R.; Ouyang, Y. The fate and behavior of lead alkyls in the environment—A review. Crit. Rev. Environ. Control **1992**, *22*, 169–193.

28. Aelion, C.M.; Davis, H.T.; McDermott, S.; Lawson, A.B. Soil metal concentrations and toxicity: Associations with distances to industrial facilities and implications for human health. Sci. Total Environ. **2009**, *407*, 2216–2223.

29. Nawrot, T.S.; Staessen, J.A.; Roels, J.A.; Roels, H.A.; Munters, E.; Cuyupers, A.; Richart, T.; Ruttens, A.; Smeets, K.; Clijsters, H.; Vangronsveld, J. Cadmium exposure in the population: From health risks to strategies of prevention. Biometals. **2010**, *23*, 769–782.

30. Le Bot, B.; Gilles, E.; Durand, S.; Glorrenc, P. Bioaccessible and quasi-total metals in soil and indoor dust. Euro. J. Mineral **2010**, *22*, 651–657.

31. Stern, A.H. A quantitative assessment of the carcinogenicity of hexavalent chromium by the oral route and its relevance to human exposure. Environ. Res. **2010**, *110*, 798–807.

32. Dhal, B.; Thatoi, H; Das, N; Pandey, B.D. Reduction of hexavalent chromium by *Bacillus* sp. isolated from chromite mine spoils and characterization of reduced product. J. Chem. Technol. Biotechnol. **2010**, *85*, 1471–1479.

33. Molina, J.A.; Oyarzun, R.; Esbri, J.M., Higuereas, P. Mercury accumulation in soils and plants in the Almaden mining district, Spain: One of the most contaminated sites on Earth. Environ. Geochem. Health **2006**, *28*, 487–498.

34. Saha, J.C.; Dikshit, A.K.; Bandyopadhyay, M.; Saha, K.C. A review of arsenic poisoning and its effects on human health. Crit. Rev. Environ. Sci. Technol. **1999**, *29*, 281–313.

35. Roychowdhry, T. The impact of sedimentary arsenic through irrigated groundwater on soil, plant, crops and human continuum from Bengal delta: Special reference to raw and cooked rice. Environ. Eng. Sci. **2003**, *20*, 405–422.

36. LaGrega, M.D.; Buckingham, P.L.; Evans, J.C. In *Hazardous Waste Management;* McGraw-Hill: New York, 1994; 447–831.

37. Nie, S.W.; Gao, W.S.; Chen, Y.Q.; Sui, P.; Eneji, A.E. Review of current status and research approaches to nitrogen pollution in farmlands. Agric. Sci. China **2009**, *7*, 843–849.

38. Xu, J.L.; Song, Y.U.; Min, B.K.; Steinberg, L.; Logan, B.E. Microbial degradation of perchlorate: Principles and application. Environ. Eng. Sci. **2003**, *20*, 405–422.

39. Jeyasingh, J.; Somasundaram, V.; Philip, L.; Bhallamudi, S.M. Bioremediation of Cr(VI) contaminated soil/sludge: Experimental studies and development of a management model. Chem. Eng. J. **2010**, *2*, 556–564.

40. Mulligan, C.N.; Yong, R.N.; Gibbs, B.F. Remediation technologies for metal-contaminated soils and groundwater: An evaluation. Eng. Geol. **2001**, *1-4*, 197–203.
41. Milner, M.J.; Kochian, L.V. Investigating heavy-metal hyperaccumulation using *Thlaspi caerulescens* as a model system. Ann. Bot. **2008**, *102*, 3–13.
42. Gonzaga, M.I.S.; Santos, J.A.G.; Ma, L.Q. Phytoextraction by arsenic hyperaccumulator *Pteris vittata* L from six arsenic-contaminated soils: Repeated harvests and arsenic redistribution. Environ. Pollut. **2008**, *2*, 212–218.
43. Nagata, T.; Morita, H.; Akizawa, T.; Pan-Hou, H. Development of a transgenic tobacco plant for phytoremediation of methylmercury pollution. Appl. Microbiol. Biotechnol. **2010**, *87*, 781–786.
44. Lomonte, C.; Doronila, A.I.; Gregory, D.; Baker, A.J.M.; Spas, D.K. Phytotoxicity of biosolids and screening of selected plants with potential for mercury phytoextraction. J. Haz. Mater. **2010**, *173*, 494–501.
45. Hadi, F.; Bano, A.; Fuller, M.P. The improved phytoextraction of lead (Pb) and the growth of maize (*Zea mays* L.); the role of plant growth regulators (GA(3) and IAA) and EDTA alone and in combination. Chemosphere **2010**, *80*, 457–462.
46. Fleri, M.A.; Whetstone, G.T. In situ stabilisation/solidification: Project lifecycle. J. Haz. Mater. **2007**, *Special Issue S1*, 141, 441–456.
47. Lestan, D.; Luo, C.L.; Li, X.D. The use of chelating agents in the remediation of metal-contaminated soils: A review. Environ. Pollut. **2008**, *153*, 3–13.
48. Chaney, R.L; Ryan, J.A.; Brown, S. Environmentally acceptable endpoints for soil metals. In *Environmental Availability of Chlorinated Organics, Explosives, and Metals in Soils;* Anderson, W.C., Loehr, R.C., Reible, D., Eds.; American Academy of Environmental Engineers: Annapolis, Maryland, 1999; 111–155.

9

Pollution: Genotoxicity of Agrotoxic Compounds

Vera Lucia S.S.
de Castro and
Paola Poli

Introduction

Thousands of chemicals are in common use, but only a portion of them has undergone significant toxicological evaluation, leading to the need to prioritize the remainder for targeted testing. Various assays were set up and performed to identify the potential hazard (genotoxicity, carcinogenicity, reproductive toxicity, etc.) of pollutants for human health. Increasing need for effective tools to assess the risks derived by the large number of both natural and anthropic-origin organic and inorganic noxious substances, both of natural origin and released in the environment by human activities, leads to the development of very sensitive detectors of harmful substances. It is an accepted assumption that the simple measurement of chemical concentration, with reference to established regulatory rules, will not give an accurate account of the environmental hazard. The measurement of environmental physical–chemical parameters is the first step in monitoring environmental quality; however, attention and alarm thresholds of these parameters only concern the toxic effects of the studied polluting substance but do not take into consideration the issue of chronic exposure at low doses of physical agents and chemicals, frequently present in complex mixtures. On the other hand, the monitoring by biological assays can effectively define the risks for the environment and humans. These analyses may be able to assess the complexity of natural environment, in terms of both different organisms and differences in physiological status, to establish cause/effect relationships between presence and concentration of pollutants and consequent environmental damages and to detect the possible synergistic effect of complex mixtures of chemicals. Therefore, much attention was paid to biological sensors, markers, or detectors able to provide information on the effects of exposure to and/or susceptibility against a variety of environmental contaminants through the knowledge of action mechanisms and the identification of possible endpoints.

Due to their role as bioactive chemicals, pesticides tend to form electrophilic metabolites capable of reacting and combining with biological macromolecules. Preferred sites of action include nucleophilic oxygen as well as nitrogen atoms, both of which are abundant in DNA, predisposing the genetic material to mutagenic covalent binding. One main deleterious effect resulting from exposure to environmental mutagens is the possible initiation of cancer. Other manifestations of genotoxic environmental pollutants such as pesticides include heritable genetic diseases, reproductive dysfunction, and birth defects.[1] Epidemiological studies provide evidence that several types of tumors and other carcinogenic manifestations are in excess among farmers and other occupational groups associated with pesticide handling.[2–4] In addition to epidemiological evidence, laboratory and field monitoring data indicate that increases in chromosome aberration, recombination, sister chromatid exchange (SCE), and other genotoxic events are in excess for pesticide-exposed groups, pointing out a generic genetic activity of pesticides in humans.[5] Recently, a number of laboratory investigations and field studies have documented a correlation between genotoxic pollutants and heritable reproduction effects on individuals as well as a potential link to the declines of fish populations.[6]

Reproductive Toxicity

It is estimated that close to 30% of all pregnancies end in spontaneous abortion. Although about 60% of spontaneous abortions are thought to be due to genetic, infectious, hormonal, and immunological factors, the role of the environment remains poorly understood. Pregnancy involves a delicate balance of hormonal and immunological functions, which may be affected by environmental substances. Many toxic substances that are persistent in the environment and accumulate in the fatty tissues may disrupt this equilibrium.[7]

Several evidence link paternal exposure to genotoxic agents with an increased risk of pregnancy loss, developmental and morphological defects, infant mortality, infertility, and genetic diseases in the offspring, including cancer.[8] Toxicogenomic analysis of environmental chemicals may be performed to investigate the ability of genomics to predict toxicity, categorize chemicals, and elucidate mechanisms of toxicity. The concordance of in vivo observations and gene expression findings demonstrated the ability of genomics to accurately categorize chemicals, identify toxic mechanisms of action, and predict subsequent pathological responses.[9]

The central nervous system appears to be especially susceptible to toxic insults during development and there is evidence that functional changes can be induced at a lower exposure level than those resulting in toxicity in adults. As an example, fetal exposure to some environmental contaminants such as organophosphorus pesticides at apparently non-toxic doses may alter some important behaviors at adulthood in mice.[10] Neurobehavioral performance can be a sensitive biomarker of the neurodevelopmental consequences of exposure of environmental agents.[11,12]

Environmental contaminants as pesticides can come from a variety of sources, including diet, drinking water, and both indoor and outdoor residential use. These chemicals may alter gene expression profiles.[13] Also, some metals can disturb the reproduction process. A number of mechanisms of cadmium toxicity have been suggested, including ionic and molecular mimicry, interference with cell adhesion and signaling, oxidative stress, apoptosis, genotoxicity, and cell cycle disturbance.[14] Some mercury compounds are known as teratogenic agents, especially affecting the normal development of the central nervous system. Since the 1990s, a genotoxic effect has been demonstrated in human populations exposed to mercury through diet,[15] occupation,[16] or carrying dental fillings.[17]

In fact, concentrations of methylmercury causing significant genotoxic alterations in vitro below both safety limit and concentration were associated with delayed psychomotor development with minimal signs of methylmercury poisoning. Based on mercury's known ability to bind sulfhydryl groups, several hypotheses were raised about potential molecular mechanisms for the metal genotoxicity. Mercury may be involved in four main processes that lead to genotoxicity: generation of free radicals and oxidative stress, action on microtubules, influence on DNA repair mechanisms, and direct interaction with

DNA molecules.[18] DNA damage can also be a sensitive bioindicator of mercury contamination in other organisms as in fish.[19]

Some pesticides may act as endocrine disruptors. As an example, the fungicide fenarimol acts both as an estrogen agonist and as an androgen antagonist.[11] In addition, fenarimol affects rat aromatase activity in human tissues.[20] This compound also affects other enzymes of the cytochrome P450 gene family that are involved in the metabolism of steroids.[21] Studies of reproductive toxicity in rats have shown that, as a result of fenarimol maternal exposure, some neuromuscular and behavioral deficits in nursing pups may occur principally during the last gestational period and lactation.[12]

Pesticides may also operate through hormonal or genotoxic pathways to affect male reproduction. They may penetrate the blood–testis barrier to potentially affect spermatogenesis, by affecting either genetic integrity or hormone production. Effects may be at different stages of the cell cycle such as during meiotic disjunction, and such abnormalities can have deleterious effects on reproduction and offspring.[22]

Human and Environmental Genotoxicity

Mutation is a manifestation of change of the structure of DNA. Agents that cause DNA damage and mutation, as well as being potentially capable of causing hereditary disorders in the offspring and succeeding generations of exposed populations, are also likely to be carcinogenic. For these reasons, testing for the induction of DNA damage and for mutagenicity, using a variety of short-term tests, has become an accepted part of the toxicological evaluation of drugs, industrial intermediates, cosmetics, food and feed additives, pesticides, biocides, etc. Regulatory agencies and international authorities recommend a test scheme consisting of *in vitro* and *in vivo* methods to identify genotoxic/mutagenic substances as those described on Organization for Economic Cooperation and Development (OECD) and United States Environmental Protection Agency (USEPA) guidelines, as recent examples.[23,24]

The current EPA mutagenicity testing battery is required for pesticides and toxic substances that are regulated under the Federal Insecticide, Fungicide and Rodenticide Act (FIFRA) and the Toxic Substances Control Act (TSCA), respectively. The battery is a three-tiered system of various mutagenicity tests (including the *Salmonella,* mouse lymphoma, and mouse micronucleus assays in the first tier). Guidelines for the conduct of the tests employed in this battery were issued and were ultimately harmonized within EPA and with the OECD.[25]

In addition to the risks to human health posed by mutagenic pollutants such as pesticides, there are important ecological risks as well, such as the threat posed by pesticides to the stability of the ecosystems through the cumulative introduction of deleterious mutations into the genetic pool of populations. Indeed, it has been demonstrated that pesticides and other environmental genotoxicants are capable of altering the genetic makeup of some natural populations.[26,27]

The occurrence of genotoxic pollutants in the aquatic environment is of increasing concern. Pollution with anthropogenic toxicants may create pronounced environmental gradients that impose strong local selection pressures. Toxic contaminants may also directly impact genetic structure in natural populations by exhibiting genotoxicity. The genetic variation within natural populations and, hence, the potential for local adaptation can itself be impacted by anthropogenic pollution.[26]

Reservoirs are complex aquatic systems mediating between rivers and lakes; they usually reflect multiple impacts generated by a variety of anthropogenic activities. The sediment compartment is the intermediate or final receptor of insoluble (or slightly water soluble) pollutants and can act as a sink for various substances. Sediments accumulate chemicals up to concentrations many times higher than in the free water column. As pollutants may be made available under certain environmental conditions (such as dredging or flood events), sediments can also become a source of diffuse contamination to the free water space. Sediment pollutants are linked not only to organisms in aquatic ecosystem dynamics but also to human health via water and fish consumption. Due to their ability to metabolize xenobiotics and accumulate pollutants, fish represent important monitoring systems within aquatic genotoxicity assessment.[6]

In general, chemical mixtures may influence local ecosystems in a site-specific way defined by all aspects along the source and availability of the mixture and the nontarget receptor. Also, humans are concurrently exposed to a number of chemicals via food and environment. These chemicals may have a combined action that causes a lower or higher toxic effect than would be expected from knowledge about the single compounds. Consequently, combined actions need to be addressed in the risk assessment process. In addition, developments in the area of toxicogenomics have also been suggested as a way of increasing our knowledge of mechanism of toxicity in order to better understand and improve the approaches for risk assessment of combined actions of chemicals.[28]

Both chronic and acute contamination of watershed by pesticides dissolved or adsorbed to soil particles can affect aquatic organisms. Pesticide toxicity may occur in a broad range of non-target aquatic organisms, both in plants from microalgae to macrophytes and in animals from microinvertebrates to fish predators. The possible effects of pesticide exposure on fish are of interest because of the position of fish in the food chain and because early life stages of fish have been shown to be highly sensitive to pollutants. Genotoxicity assessment in fish has been highlighted since the implications of the genotoxic effects are impacting on fitness traits such as reproductive success, genetic patterns, and subsequent population dynamics.[29]

Pesticide Genotoxicity Assessment

Traditionally, the impact of pollutants discharged into the environment by human activities has been assessed using chemical assays or by evaluating physical parameters. High-performance liquid chromatography or gasmass spectrometry techniques have been widely used. However, the information is limited to the concentration of the pollutants, not on their toxicity. To overcome these limitations, biological analyses have been introduced. The use of effect-based screening tools has the advantage of indicating the real impact of all chemicals present in a given sample or ecosystem. Rapid biological tests are playing their major role in hazard and risk assessment, especially at the screening level. Bioassays measure changes in physiology or behavior of living organisms resulting from stresses induced by biological or chemical toxic compounds, which can cause disruption of the metabolism.[30]

Toxicological animal-based assays on organisms from different trophic levels (algae and plants, worms and crustaceans, fishes, etc.), employed in the identification of hazardous chemicals are sometimes expensive and time consuming, require large sample volumes, and raise ethical problems. In vitro methods, also commonly used for screening and ranking chemicals, must be included in battery tests for risk assessment purposes. Their major promise is to supply mechanism-derived information, considered pivotal for adequate risk assessment. Mutagenesis short-term assays can directly detect the genetic effect of chemical and physical agents on the tested cells/organisms, able to assess the DNA damage resulting from exposure to both single chemicals and heterogeneous mixtures.

Fundamental research in the mechanisms of induced mutation and carcinogenesis induction has benefited from the development of highly refined short-term tests for genotoxicity. The knowledge of toxicity pathways that is derived from genomic data would highlight the potential mechanisms. The challenge is to comprehensively integrate the disparate chemical, biological, toxicological, and toxicogenomic data in order to elucidate the mechanisms and networks involved in toxicity and to develop quantitative models capable of accurately predicting thresholds.[31]

However, despite the progress on newer methodologies, quantitative data on toxicological effects of some widely used pesticides in agricultural practices are not well established even at the single-organism level. Among the pesticides in common use, attention is drawn towards the use of triazole fungicides that are used in the control of several fungal and plant diseases. These fungicides have demonstrated some adverse effects in mammalian species derived from genomic data.[32] Also, genotoxicity studies conducted in microalgae showed that the exposure to epoxiconazole can result in an increase in the extent of DNA strand breaks depending on the concentration.[33]

Nevertheless, results from this kind of assays are prone to the criticisms concerning the differences between the real conditions in the ecosystems and those of the in vitro assays. In fact, since various factors (chemical, physical, and biological) affect environmental conditions, the transfer of results obtained by in vitro techniques to the field is a complex task, and the establishment of extrapolation parameters is a crucial issue. In this context, a combination of in vitro and in situ (in vivo) bioassays represents a promising approach to better understand both the real exposure situation in the environment and the action mechanisms studied by the in vitro bioassays.[34]

Also, although the genotoxicity testing strategies employed prior to product registration are designed to identify potential in vivo genotoxins, concerns that exposure to pesticides may result in long-term adverse effects still exist.[35] A positive association between occupational exposure to complex pesticide mixtures and the presence of chromosomal aberrations (CAs), SCEs, and micronuclei has been detected in a number of studies, although some of these failed to detect cytogenetic damage. Chromosomal damage induced by pesticides appears to have been transient in acute or discontinuous exposure, but cumulative in continuous exposure to complex agrochemical mixtures.[36]

Assessment Methodology

Human and wildlife risks associated with pesticide mutagenic capability can be assessed by using screening systems that are sensitive and able to detect the whole mutagenic spectrum. There are more than 100 short-term bioassays for evaluating the potential genotoxic effects of pesticides, but since no single system encompasses the whole spectrum of possible genetic toxic effects, a combination of evaluation procedures is recommended for the assessment of pesticide mutagenesis. Genotoxicity testing batteries were specifically established for hazard identification, the first step in risk assessment. Studies of sensitivity and correlation among test systems are fundamental for a more accurate evaluation of the environmental risks, as well as extrapolation of data to other target organisms.

As a consequence of the very large number of genetic toxicity evaluation techniques described in the specialized literature, most assays can be considered ancillary and will be employed only for specific ends.[37] The assays accepted for routine evaluation of pesticides fall in one of six testing categories: 1) microbial assays: a) prokaryotic (bacterial, such as the *Salmonella typhimurium*, *Escherichia coli*, *Bacillus subtilis*) or b) eukaryotic assays (fungi, such as *Saccharomyces cerevisiae*, *Neurospora crassa*); 2) in vitro isolated eukaryotic cell lines; 3) host-mediated assays; 4) in vivo animal; and 5) in vivo plant bioassays.

In order to approximate the studies to higher-organism subjects, several assays involve in vitro exposure of mammalian cell lines. Mouse bone marrow cells, erythrocytes, and white blood cells; hamster ovary cells; and human lung fibroblast are some examples, and several genetic endpoints can be evaluated, such as DNA damage, unscheduled DNA synthesis, chromatid exchanges, and micronucleus frequency. Aside from the practical advantages offered by the microbial and isolated cell line bioassays (ease of manipulation, asepticism, small space and large population assayed, low cost), the tests often depend upon an external metabolic activation complement, since the microbial and isolated cell lines may be incapable of responding to pro-mutagenic compounds if these are not partially metabolized.[38] The host-mediated assays are devised to circumvent this deficiency. Mammalian subjects are exposed to the pesticide; the microbial cell line is injected into this treated subject and later recovered and evaluated for mutation induction. Alternatively, for plant metabolic activation, whole plants are treated with the pesticide and their extracts are applied directly into the microbial assay.

Finally, plant bioassays involve a variety of endpoints, from DNA damage in leaf cells of *Impatiens balsamina* to micronuclei in pollen mother cells (*Tradescantia*) and root tip meristematic cells (*Allium*, *Vicia*), reversion and crossing over in chlorophyll-deficient lines (*Hordeum*, *Glycine*), sugar-specific starch production in pollen (*Zea*), flower pigmentation alteration (*Tradescantia*), and many other endpoints in many different species.[39–41]

Some recent studies[42–49] mainly use bioassays such as Comet, micronucleus, CAs, and SCE to detect pesticide genotoxic potential both in vitro and in vivo.

Comet Assay

The single-cell gel electrophoresis assay (or Comet assay) is a simple, rapid, and sensitive technique for analyzing and measuring DNA damage in individual mammalian (and to some extent prokaryotic) cells. The method published by Sing et al.[50] makes the Comet assay versatile and forms the basis for all developments that have taken place in this field. In the Comet assay, the cells are embedded in a thin agarose gel on a microscope slide. The cells are lysed to remove all cellular proteins and the DNA subsequently allowed unwinding under alkaline/neutral conditions. Following unwinding, the DNA is electrophoresed and stained with a fluorescent dye. During electrophoresis, broken DNA fragments (damaged DNA) or relaxed chromatin migrates away from the nucleus. The Comet assay essentially measures the sizes of DNA fragments within the cell. It is therefore necessary to convert DNA damage to DNA fragments by introducing breaks at the sites of DNA damage before they can be detected on Comet assay. The simplest types of DNA damage detected by Comet assay are the double-strand breaks (DSBs). DSBs within the DNA result in DNA fragments and can be detected by merely subjecting them to electrophoretic mobility at the neutral pH. Single-strand breaks (SSBs) do not produce DNA fragments unless the two strands of the DNA are separated/denatured. This is accomplished by unwinding the DNA at pH 12.1. Other types of DNA damage broadly termed as alkali labile sites are expressed when the DNA is treated with alkali at pH greater than 13.

The in vivo Comet assay is a well-established genotoxicity test.[51–55] It is currently mainly performed with somatic cells from different organs to detect a genotoxic activity of potential carcinogens. It is regarded as a useful test for follow-up testing of positive or equivocal in vitro test results and for the evaluation of local genotoxicity. In current test strategies, the in vivo Comet assay is mainly performed with somatic cells from different organs to detect the genotoxic activity of potential carcinogens and is regarded as a useful test for follow-up testing of positive in vitro findings. Furthermore, the comet assay also has the potential to be a useful tool for investigating genotoxicity in germ cells.[56]

Micronucleus Assay

The purpose of the micronucleus assay is to detect chromosome structure modifying agents: segregation is the way to lead the induction of micronuclei in interphase cells. These micronuclei may originate from acentric fragments (chromosome fragments lacking a centromere) or whole chromosomes unable to migrate with the remainder of the chromosomes during the anaphase of cell division.[57–59] The in vitro micronucleus assay is a mutagenic test system for the detection of chemicals that induce the formation of small membrane-bound DNA fragments, i.e., micronuclei in the cytoplasm of interphase cells.

After exposure to a test substance and addition of cytochalasin B for blocking cytokinesis, cell cultures are grown for a period sufficient to allow chromosomal damage to lead to the formation of micronuclei in bi- or multinucleated interphase cells. Harvested and stained interphase cells are then analyzed microscopically for the presence of micronuclei. Micronuclei are scored in those cells that complete nuclear division following exposure to the test item. Additionally, the cells are classified as mononucleates, binucleates, or multinucleates to estimate the proliferation index as a measure of toxicity.

The micronucleus test is also an efficient biological assay for monitoring population exposure to mixtures of agrochemicals as shown in a study that was performed in farm workers directly exposed to large amounts of agrochemicals (fungicides, insecticides, and herbicides) in an area of grain farming (wheat and soybeans).[36,60,61] High MN frequency was detected in people at higher cancer risk due to occupational/environmental exposure to a wide variety of carcinogens.[62–66]

CA Test

Chromosome mutations and related events are the cause of many human genetic diseases, and there is substantial evidence that chromosome mutations and related events causing alterations in oncogenes and tumor suppressor genes of somatic cells are involved in cancer induction in humans and experimental animals.

The purpose of the in vitro CA test is to identify agents that cause structural chromosome aberrations in cultured cells.[67,68] With the majority of chemical mutagens, induced aberrations are of the chromatid type, but chromosome-type aberrations also occur.

Structural CAs may be induced by direct DNA breakage, by replication on a damaged DNA template, by inhibition of DNA synthesis, and by other mechanisms (e.g., topoisomerase II inhibitors).[69] Based on morphological criteria, structural CAs can be divided into two main classes: chromosome-type aberrations (CSAs), involving both chromatids of one or multiple chromosomes, and chromatid-type aberrations (CTAs) involving only one of the two chromatids of a chromosome or several chromosomes.[69,70] An increase in polyploidy may indicate that a chemical has the potential to induce numerical aberrations.[70] CTAs (e.g., chromatid type breaks and exchanges) arise predominantly in vitro during S-phase of cultured lymphocytes, in response to base modifications and SSBs induced in vivo by S-phase-dependent clastogens.[69,70]

Proliferating cells are treated with the test substance in the presence and absence of a metabolic activation system for 3–6 hr and sampled at a time equivalent to about 1.5 normal cell cycle length after the beginning of treatment.[71] If this protocol gives negative results both with and without activation, an additional experiment without activation should be done, with continuous treatment until sampling at a time equivalent to about 1.5 normal cell cycle lengths. Certain chemicals may be more readily detected by treatment/sampling times longer than 1.5 cycle lengths. At predetermined intervals after exposure of cell cultures to the test substance, they are treated with a metaphase arresting substance (e.g., colchicine), harvested, and stained, and metaphase cells are analyzed microscopically for the presence of chromosome aberrations. Though the purpose of the test is to detect structural chromosome aberrations, it is important to record polyploidy and endoreduplication when these events are seen.

Structural CAs in peripheral blood lymphocytes as assessed by the chromosome aberration assay in vivo have been used for more than 30 years in occupational and environmental settings as a biomarker of early effects of genotoxic carcinogens. A high frequency of structural CAs in lymphocytes (reporter tissue) is predictive of increased cancer risk.[36,60,72]

Sister Chromatid Exchange

SCE is the exchange of homologous stretches of DNA sequence between sister chromatids and occurs normally in cells during mitosis. In the presence of genotoxic agents that provoke DNA damage, the rate of SCE increases. Equal SCE has been thought to be an important mechanism of DSB repair in eukaryotes, but this has never been proven due to the difficulty of distinguishing SCE products from parental molecules.[73]

The SCE analysis was adopted as an indicator of genotoxicity. SCEs represent the interchange of DNA replication products at apparently homologous loci. The exchange process presumably involves DNA breakage and reunion, although little is known about its molecular basis.[73–80]

Some studies revealed that the nucleotide pool imbalance can have severe consequences on DNA metabolism and it is critical in SCE formation. The modulation of SCE by DNA precursors raises the possibility that DNA changes are responsible for the induction of SCE in mammalian cells.[81,82]

Detection of SCEs in in vitro test requires some means of differentially labeling sister chromatids, and this can be achieved by incorporation of bromodeoxyuridine (BrdU) into chromosomal DNA for two cell cycles.[83,84] Cells in an exponential stage of growth are exposed to the test substance for a suitable period of time; in most cases, 1–2 hr may be effective, but the treatment time may be extended up to two complete cell cycles in certain cases. Cells without sufficient intrinsic metabolic activity should be exposed to the test chemical in the presence and absence of an appropriate metabolic activation system. At the end of the exposure period, cells are washed free of test substance and cultured for two rounds of replication in the presence of BrdU. As an alternative procedure, cells may be exposed simultaneously to the test chemical and BrdU for the complete culture time of two replication cycles.[85] Cells are analyzed in their second posttreatment division, ensuring that the most sensitive cell cycle stages have been exposed to the chemical. Cell cultures are treated with a spindle inhibitor (e.g., colchicine) 1–4 hours prior to harvesting. Chromosome preparations are made by standard cytogenetic techniques. Staining of slides to show SCEs can be performed by several techniques (e.g., the fluorescence plus Giemsa method).[85,86]

While increased levels of CA have been associated with increased cancer risk,[87,88] a similar conclusion has not been reached for SCE. However, high levels of SCE frequency have been observed in persons at higher cancer risk due to occupational or environmental exposure to a wide variety of carcinogens.[62–66]

Biomonitoring

Plants

Higher plants are recognized as excellent genetic models to detect environmental mutagens and are frequently used in monitoring studies.[89–94] Plant systems represent a complex multicellular environment where the efficiency of different protection or repair mechanisms can be modulated by cellular homeostasis. Higher plants, even showing low concentrations of oxidase enzymes and a limitation in the substrate specification in relation to other organism groups,[40] present consistent results that may serve as a warning to other biological systems, since the target is DNA, common to all organisms.[95]

They represent a stable sensor in an ecosystem and hence allow following the evolution of the genotoxic impact. Well-defined higher plants represent an excellent basis for cytogenetic evaluations after exposure to genotoxic pollutants, especially since the maturation of their gametes (meiosis) follows the same patterns as in animals and humans,[96] although sometimes plant and animal assays are differentially responsive to some pesticides as pendimethalin.[97]

Among the plant species, *Allium cepa* has been widely used to evaluate DNA damage, such as chromosome aberrations and disturbances in the mitotic cycle. Employing *A. cepa* as a test system to detect mutagens dates back to the 1940s.[40,98,99] It has been used to this day to assess a great number of chemical agents, which contributes to its increasing application in environmental monitoring. *A. cepa* is characterized as a low-cost test. It is easily handled and has advantages over other short-term tests that require previous preparations of the samples, as well as the addition of exogenous metabolic system. *A. cepa* test also enables the evaluation of different endpoints (mitotic index, chromosome aberrations, nuclear abnormalities, and micronucleus).[40,99,100] Among the endpoints, chromosome aberrations have been the most used one to detect genotoxicity along the years. The mitotic index and some nuclear abnormalities are used to evaluate both cytotoxicity and mutagenicity of different chemicals. Moreover, *A. cepa* test system provides important information to evaluate action mechanisms of an agent about its effects on the genetic material (clastogenic and/or aneugenic effects). Transgenic/transformed plants could be used as a model to better check whether, in response to environmental stress, DNA damage might be regulated by alterations in single genes and to understand the mode and mechanisms of DNA damage.[101]

Animals and Human

The response to a genotoxic agent can range considerably in the exposed populations. It is currently accepted that susceptibility to genotoxic exposure varies interindividually and that this could be the result of hereditary or acquired characteristics. For this reason, attention has been focused on genetic polymorphisms, which are able to modulate human response to genotoxic environmental agents. This individual variation suggests the importance of individual susceptibility factors. Many studies[102–108] showed the influence of metabolic and DNA repair polymorphisms in the individual response to exposure to carcinogens. Genetic polymorphisms of metabolizing enzymes widely influence xenobiotic effective dose.[109] On the other hand, the role of polymorphisms linked to the DNA repair genes appears to be essential in the modulation of genotoxic risk linked to carcinogen exposure.[105,107,110]

Tests for Some Novel Materials: The Case of Nanomaterials

Nanomaterials are generally defined as having one or more external dimensions or an internal or surface structure on the nanoscale (about 1–100 nm). Nanomaterials display novel properties (small size, particular shape, large surface area, and large surface activity) that make them attractive in many applications. However, these properties may contribute to their toxicological profile and may also affect nanomaterials' possible direct or indirect interaction with the DNA.[111]

Although it has been shown that cationic functionalized carbon nanotubes can condense with DNA and that gold nanoparticles binding to the major groove of DNA is associated with killing of cancer cells, for most nanomaterials, it is even unknown whether they directly interact with DNA or whether they promote indirect effects. The identification of the different ways by which various nanomaterials interact with DNA will improve the extrapolations of genotoxicity test results to human risk evaluation. Application of standard methods to nanomaterials demands, however, several adaptations, and the interpretation of results from the genotoxicity tests may need additional considerations. The use of a battery of standard genotoxicity testing methods covering a wide range of mechanisms towards establishing methodological adaptations and better test conditions is a practical and pragmatic approach for this genotoxic evaluation.[111]

Recent reviews have concluded that information on the genotoxicity of nanomaterials is still inadequate for general conclusions, e.g., on its characteristics critical for genotoxicity. It is presently unclear how well standard genotoxicity tests, designed for soluble chemicals, can be used to assess the genotoxicity of nanomaterials. It appears that more genotoxicity studies on nanomaterials are urgently required, and the need for novel methods to assess nanomaterials–genome interactions is imminent. Current evidence indicates that various nanomaterials may carry genotoxic potential although the mechanisms of which remain to be identified.[112]

Above all, some nanomaterials can generate great concern about their possible adverse effects. As an example, although titanium dioxide is frequently used for industrial and cosmetics purposes because of its low toxicity, nanosized or ultrafine TiO_2 (UF-TiO_2) (<100 nm in diameter) can generate pulmonary fibrosis and lung tumor in rats and cytotoxicity in rat lung alveolar macrophages.[113] These authors also observed that it can cause genotoxicity and cytotoxicity in cultured human cells in experimental models as well.

TiO_2 nanoparticles in the absence of photoactivation are potentially genotoxic to fish cells under in vitro conditions. This effect becomes more pronounced in the presence of UVA, along with cytotoxic effects, which only occurred during combined exposure to TiO_2 and UVA.[114]

Although the biological effects of some nanomaterials have already been assessed, information on toxicity and possible mechanisms of various particle types are insufficient. The comparative analysis demonstrated that particle composition probably played a primary role in the cytotoxic effects of different nanoparticles, whereas the genotoxicity potential might be mostly attributed to particle shape.[115] Furthermore, the genotoxic effectiveness could be mediated through lipid peroxidation and oxidative stress as in the case of ZnO nanoparticles on human epidermal cells, even at low concentrations.[116]

Future Concerns and New Genomic Techniques

The evidence of metabolic activation of pesticides into direct-acting mutagens and the complex interactions that may occur as a result of the synergistic action of the multitude of chemicals presently under common usage warrant devoted attention to the evaluation of the potential genetic consequences of pesticide mixtures and derived metabolites accumulated through food chains. This complexity is further deepened by the unpredictability of mutational effects, as a result of pleiotropism and other unforeseeable events. Studies of this nature should attempt to 1) point out the impossibility of predicting all genetic effects of pesticides and 2) recognize that small, improbable effects may have important consequences when imposed onto the very large populations exposed to pesticides in the environment.

With the advent of new technologies (e.g., genomics, automated analyses, and in vivo monitoring), new regulations (e.g., the reduction of animal tests by the European Registration, Evaluation, and Authorization of Chemicals), and new approaches to toxicology (e.g., Toxicity Testing in the 21st Century, National Research Council), the field of regulatory genetic toxicology is undergoing a serious re-examination. However, it is appropriate to apply a prudent approach to risk assessment, maintaining current testing standards that are working properly until others have proven superior by rigorous scientific evidence and widespread agreement.[117]

Human cytogenetics has proven to be effective over a 50 years life span. It is in reality a collection of techniques that, while common, are cheap, fast, and wide ranging. Therefore, in genotoxicology, they continue to be useful to identify mutagenic agents as well as to evaluate and analyze exposed populations.[118]

The introduction of advanced molecular techniques leads to improved risk assessment and also provides an alternative to the massive use of animal testing. Transcriptional profiling and DNA chips are highly informative and are among the most promising novel techniques for environmental risk assessment. Moreover, information discerned from these chips enables the identification of new discriminative biomarker genes. Based on these biomarker genes, cellular reporters can be constructed. These can be used in a high-throughput setup and can significantly facilitate ecotoxicological risk assessment.[119] However, some important technical and interpretative hurdles still need to be overcome before a full implementation of ecotoxicogenomics in regulatory settings can occur. Toxicogenomics uses molecular biology's entire arsenal to analyze the changes in DNA induced by genotoxic agents.[120,121] These changes include those induced in certain genes not related to the mechanisms of genotoxicity, like early response genes. These genes can be used as indicators for which there is a stress response to the genotoxic agent. Quantitative real-time polymerase chain reaction (RT-PCR) and microarrays are the most important procedures to analyze these genes, which can be numerous since current technology allows it.

RT-PCR uses fluorescent probes to measure the exact amount of a nucleic acid. In genotoxicology, it can be used to quantify gene expression to detect genetic polymorphisms and to quantify chromosome deletions. These probes are used to verify the expressed gene or to selectively analyze its expression over time or dose parameters. It will also become more important to analyze expression in specific cell populations in order to profile the global alterations in gene expression involved in chronic chemical exposure that may lead to tumor development.[122]

The results of the microarray studies in toxicogenomics show that genotoxic agents are able to induce changes in gene expression profiles. These studies demonstrate that the gene expression pattern is able to generate the information necessary to determine the agent's type of action and even to predict it. Chemically induced changes in gene expression can result in the identification of simple, sensitive, and relevant biomarkers of effect that can be used in dose–response studies to more readily identify precursor effects in the low-dose range.[123] Therefore, the ability to analyze the effect of mutagenic agents on a large number of genes in a single experiment using gene expression profiling analysis has been used to demonstrate that certain genes may or may not become activated when exposed to a toxic agent, depending on the type of cells and the type of genotoxic agent they have been exposed to.[124]

Furthermore, the DNA damage itself can activate genes codifying proteins involved in DNA repair and/or induce the apoptosis process, when the genes associated with this process are activated. Although broad consensus is still lacking concerning the guidelines for the reproducibility of experiments, toxicogenomic studies could lead to development of early biomarkers of toxic injury and may also help to resolve issues related to interspecies extrapolation and susceptibility variation among individuals.[122] As an application of this toxicogenomic approach, a study[32] reports data that linked genomic data to specific toxicological endpoints of the antifungal triazole, which causes varying degrees of hepatic toxicity and disrupts steroid hormone homeostasis in rodent in vivo models. Overall, these analyses revealed functional categories of chemical response genes that indicate mechanisms and provide direction for further research on triazole mechanisms of action.

Conclusion

The occurrence of genotoxic pollutants in the environment is of increasing concern. Genotoxicity studies can be applied to elucidate potential mechanisms of physiological or molecular alterations to minimize exposure of contaminants to levels that maintain sustainability of the environment. There is a large number of genetic toxicity evaluation assays described in the specialized literature that are able to identify action modes and biological targets both in the environment and in the organisms. Microorganisms as well as cell lines are widely used as in vitro models for risk assessment. Animal and plant models may be valuable for a better comprehension of the metabolic pathways of genotoxic chemicals. Various methods can be applied both in vitro and in epidemiological studies. New approaches through molecular techniques are also proposed for testing the genotoxic load of environmental pollutants. Moreover, in order to better establish biological hazards, it is also necessary to study the effects induced by chemicals on organism development, physiology, and behavior to evaluate their consequences. In relation to this, few attempts to link the genotoxic effects of contaminants with effects at the physiological or behavioral level have been made, which would lead to a detrimental impact on vital life processes including the reproductive potential of the organisms.

Acknowledgments

We wish to thank Dr. Stachetti for his precious contribution to this entry, consisting of the initial text, over which the present one has been constructed. Also, we would like to thank him for the opportunity offered to contribute to this prestigious publication, by referring our names as authors. Without his help and his confidence on our work, this achievement would not have been possible. We also would like to thank Dr. Stachetti for the final text comments and suggestions.

References

1. Waters, M.; Sandhu, S.S.; Simmon, V.F.; Mortelmans, K.E.; Mitchell, A.D.; Jorgenson, T.A.; Jones, D.C.L.; Valencia, R.; Garrett, N.E. Study of pesticide genotoxicity. In Genetic Toxicology: An Agricultural Perspective; Fleck, R.A., Hollaender, A., Eds.; Basic Life Sciences, Plenum Press: New York, 1982; Vol. 21, 275–326.
2. Maele-Fabry, G.; Duhayon, S.; Mertens, C.; Lison, D. Risk of leukaemia among pesticide manufacturing workers: A review and meta-analysis of cohort studies. Environ. Res. 2008, *106*, 121–137.
3. Rull, R.P.; Gunier, R.; Behren, J.; Hertz, A.; Crouse, V.; Buffler, P.A.; Reynolds, P. Residential proximity to agricultural pesticide applications and childhood acute lymphoblastic leukemia. Environ. Res. 2009, *109*, 891–899.
4. Alavanja, M.C.R.; Hoppin, J.A.; Kamel, F. Health effects of chronic pesticide exposure: Cancer and neurotoxicity. Annu. Rev. Public Health 2004, *25*, 155–197.
5. Maroni, M.; Fait, A. Health effects in man from long-term exposure to pesticides—A review of the 1975–1991 literature. Toxicology 1993, *78* (1–3), 1–180.
6. Rocha, P.S.; Luvizotto, G.L.; Kosmehl, T.; Bottcher, M.; Storch, V.; Braunbeck, T.; Hollert, H. Sediment genotoxicity in the Tiete River (São Paulo, Brazil): In vitro comet assay versus in situ micronucleus assay studies. Ecotoxicol. Environ. Saf. 2009, *72*, 1842–1848.
7. Weselak, M.; Arbuckle, T.; Walker, M.; Krewski, D. The influence of the environment and other exogenous agents on spontaneous abortion risk. J. Toxicol. Environ. Health, Part B 2008, *11* (3–4), 221–241.

8. Aitken, R.J.; Koopman, P.; Lewis, S.E. Seeds of concern. Nature **2004**, *432*, 48–52.

9. Martin, M.T.; Brennan, R.J.; Hu, W.; Ayanoglu, E.; Lau, C.; Ren, H.; Wood, C.R.; Corton, J.C.; Kavlock, R.J.; Dix, D.J. Toxicogenomic study of triazole fungicides and perflu-oroalkyl acids in rat livers predicts toxicity and categorizes chemicals based on mechanisms of toxicity. Toxicol. Sci. **2007**, *97* (2), 595–613.

10. Ricceri, L.; Venerosi, A.; Capone, F.; Cometa, M.; Lorenzini, P.; Fortuna, S.; Calamandrei, G. Developmental neurotoxicity of organophosphorous pesticides: Fetal and neonatal exposure to chlorpyrifos alters sex-specific behaviors at adulthood in mice. Toxicol. Sci. **2006**, *93* (1), 105–113.

11. Andersen, H.; Vinggaard, A.; Rasmussen, T.; Gjermandsen, I.; Bonefeld-Jørgensen, E. Effects of currently used pesticides in assays for estrogenicity, androgenicity, and aromatase activity in vitro. Toxicol. Appl. Pharmacol. **2002**, *179* (1), 1–12.

12. Castro, V.L.; Mello, M.A.; Diniz, C.; Morita, L.; Zucchi, T.; Poli, P. Neurodevelopmental effects of perinatal fenarimol exposure on rats. Reprod. Toxicol. **2007**, *23*, 98–105.

13. Tully, D.B.; Bao, W.; Goetz, A.K.; Blystone, C.R.; Ren, H.; Schmid, J.E.; Strader, L.F.; Wood, C.R.; Best, D.S.; Narotsky, M.G.; Wolf, D.C.; Rockett, J.C.; Dix, D.J. Gene expression profiling in liver and testis of rats to characterize the toxicity of triazole fungicides. Toxicol. Appl. Pharmacol. **2006**, *215*, 260–273.

14. Thompson, J.; Bannigan, J. Cadmium: Toxic effects on the reproductive system and the embryo. Reprod. Toxicol. **2008**, *25*, 304–315.

15. Pinheiro, M.C.N.; Crespo-Lopez, M.E.; Vieira, J.L.F.; Oikawa, T.; Guimarães, G.A.; Araujo, C.C.; Amoras, W.W.; Ribeiro, D.R.; Herculano, A.M.; do Nascimento, J.L.M.; Silveira, L.C.L. Mercury pollution and childhood in Amazon riverside villages. Environ. Int. **2007**, *33*, 56–61.

16. Zachi, E.C.; Ventura, D.F.; Faria, M.A.; Taub, A. Neuropsychological dysfunction related to earlier occupational exposure to mercury vapor. Braz. J. Med. Biol. Res. **2007**, *40* (3), 425–433.

17. Di Pietro, A.; Visalli, G.; La Maestra, S.; Micale, R.; Baluce, B.; Matarese, G.; Cingano, L.; Scoglio, M.E. Biomonitoring of DNA damage in peripheral blood lymphocytes of subjects with dental restorative fillings. Mutat. Res **2008**, *650* (2), 115–22.

18. Crespo-López, M.E.; Macêdo, G.L.; Pereira, S.I.D.; Arrifano, G.P.F.; Picanço-Diniz, D.L.W.; do Nascimento, J.L.M.; Herculano, A.M. Mercury and human genotoxicity: Critical considerations and possible molecular mechanisms. Pharmacol. Res. **2009**, *60* (4), 212–220.

19. Della Torre, C.; Petochi, T.; Corsi, I.; Dinardo, M.M.; Baroni, D.; Alcaro, L.; Focardi, S.; Tursi, A.; Marino, G.; Frigeri, A.; Amato, E. DNA damage, severe organ lesions and high muscle lev-els of As and Hg in two benthic fish species from a chemical warfare agent dumping site in the Mediterranean Sea. Sci. Total Environ. **2010**, *408* (9), 2136–2145.

20. Vinggaard, A.; Jacobsen, H.; Metzdorff, S.; Andersen, H.; Nellemann, C. Antiandrogenic effects in short-term in vivo studies of the fungicide fenarimol. Toxicology **2005**, *207*, 21–34.

21. Paolini, M.; Mesirca, R.; Pozzetti, L.; Sapone, A.; Cantelli-Forti, G. Molecular non-genetic biomarkers related to fenarimol cocarcinogenesis, organ- and sex-specific CYP induction in rat. Cancer Lett. **1986**, *101*, 171–178.

22. Perry, M.J. Effects of environmental and occupational pesticide exposure on human sperm: A systematic review. Hum. Reprod. Update **2008**, *14* (3), 233–242.

23. OECD. *OECD Guideline for the Testing of Chemicals. 414 Prenatal Developmental Toxicity Study;* Organisation for Economic Co-operation and Development: Paris, 2009.

24. Rudén, C.; Hansson, S.O. Registration, Evaluation, and Authorization of Chemicals (REACH) is but the first step— How far will it take us? Six further steps to improve the European chemicals legislation. Environ. Health Perspect. **2010**, *118* (1), 6–10.

25. Dearfield, K.L.; Cimino, M.C.; McCarroll, N.E.; Mauer, I.; Valcovic, L.R. Genotoxicity risk assess-ment: A proposed classification strategy. Mutat. Res. **2002**, *521*, 121–135.

26. Coors, A.; Vanoverbeke, J.; De Bie, T.; De Meester, L. Land use, genetic diversity and toxicant toler-ance in natural populations of Daphnia magna. Aquat. Toxicol. **2009**, *95*, 71–79.

27. Bickham, J.W.; Sandhu, S.; Hebert, P.D.; Chikhi, L.; Athwal, R. Effects of chemical contaminants on genetic diversity in natural populations: Implications for biomonitoring and ecotoxicology. Mutat. Res. **2000,** *463,* 33–51.

28. Andersen, M.E.; Krewski, D. Toxicity testing in the 21st century: Bringing the vision to life. Toxicol. Sci. **2009,** *107,* 324–330.

29. Bony, S.; Gillet, C.; Bouchez, A.; Margoum, C.; Devaux, A. Genotoxic pressure of vineyard pesticides in fish: Field and mesocosm surveys. Aquat. Toxicol. **2008,** *89,* 197–203.

30. Girotti, S.; Ferri, E.N.; Fumo, M.G.; Maiolini, E. Monitoring of environmental pollutants by bioluminescent bacteria. Anal. Chim. Acta **2008,** *608,* 2–29.

31. Boverhof, D.R.; Zacharewski, T.R. Toxicogenomics in risk assessment: Applications and needs. Toxicol. Sci. **2006,** *89,* 352–360.

32. Goetz, A.K.; Dix, D.J. Toxicogenomic effects common to triazole antifungals and conserved between rats and humans. Toxicol. Appl. Pharmacol. **2009,** *238,* 80–89.

33. Akcha, F.; Arzul, G.; Rousseau, S.; Bardouil, M. Comet assay in phytoplankton as biomarker of genotoxic effects of environmental pollution. Mar. Environ. Res. **2008,** *66,* 59–61.

34. Pellacani, C.; Buschini, A.; Furlini, M.; Poli, P.; Rossi, C. A battery of in vivo and in vitro tests useful for genotoxic pollutant detection in surface waters. Aquat. Toxicol. **2006,** *77,* 1–10.

35. Bull, S.; Fletcher, K.; Boobis, A.R.; Battershill, J.M. Evidence for genotoxicity of pesticides in pesticide applicators: A review. Mutagenesis **2006,** *21* (2), 93–103.

36. Bolognesi, C. Genotoxicity of pesticides: A review of human biomonitoring studies. Mutat. Res., Rev. Mutat. Res. **2003,** *543* (3), 251–272.

37. Epstein, S.S.; Legator, M.S. The Mutagenicity of Pesticides: Concepts and Evaluation; The MIT Press: Cambridge, 1971; 220 pp.

38. Hrelia, P.; Vigagni, F.; Maffei, F.; Morotti, M.; Colacci, A.; Perocco, P.; Grilli, S.; Cantelli-Forti, G. Genetic safety evaluation of pesticides in different short-term tests. Mutat. Res. **1994,** *321* (4), 219–228.

39. Poli, P.; de Mello, M.A.; Buschini, A.; de Castro, V.L.S.S.; Restivo, F.M.; Rossi, C.; Zucchi, T.M.A.D. Evaluation of the genotoxicity induced by the fungicide fenarimol in mammalian and plant cells by the use of single-cell gel electrophoresis assay. Mutat. Res. **2003,** *540,* 57–66.

40. Leme, D.M.; Marin-Morales, M.A. Allium cepa test in environmental monitoring: A review on its application. Mutat. Res. **2009,** *682,* 71–81.

41. Rodrigues, G.S.; Ma, T.H.; Pimentel, D.; Weinstein, L.H. Tradescantia bioassays as monitoring systems for environmental mutagenesis—A review. Crit. Rev. Plant Sci. **1997,** *16* (4), 325–359.

42. Li, X.; Li, S.; Liu, S.; Zhu, G. Lethal effect and in vivo genotoxicity of profenofos to Chinese native amphibian (Rana spinosa) tadpoles. Arch. Environ. Contam. Toxicol. 2010, *59,* 478–483. DOI 10.1007/s00244-010-9495-4.

43. Soloneski, S.; Larramendy, M.L. Sister chromatid exchanges and chromosomal aberrations in Chinese hamster ovary (CHO-K1) cells treated with the insecticide pirimicarb. J. Hazard. Mater. **2010,** *174,* 410–415.

44. Kumar, R.; Nagpure, N.S.; Kushwaha, B.; Srivastava, S.K.; Lakra, W.S. Investigation of the genotoxicity of malathion to freshwater teleost fish Channa punctatus (Bloch) using the micronucleus test and comet assay. Arch. Environ. Contam. Toxicol. **2010,** *58,* 123–130.

45. Candioti, J.V.; Natale, G.S.; Soloneski, S.; Ronco, A.E.; Larramendy, M.L. Sublethal and lethal effects on Rhinella arenarum (Anura, Bufonidae) tadpoles exerted by the pirimicarb-containing technical formulation insecticide Aficida. Chemosphere **2010,** *78,* 249–255.

46. Yin, X.; Zhu, G.; Li, X.B.; Liu, S. Genotoxicity evaluation of chlorpyrifos to amphibian Chinese toad (Amphibian: Anura) by comet assay and micronucleus test. Mutat. Res. **2009,** *680,* 2–6.

47. Vega, L.; Valverde, M.; Elizondo, G.; Leyva, J.F.; Rojas, E. Diethylthiophosphate and diethyldithiophosphate induce genotoxicity in hepatic cell lines when activated by further biotransformation via cytochrome P450. Mutat. Res. **2009,** *679,* 39–43.

48. Mladinic, M.; Berend, S.; Vrdoljak, A.L.; Kopjar, N.; Radic, B.; Zeljezic, D. Evaluation of genome damage and its relation to oxidative stress induced by glyphosate in human lymphocytes in vitro. Environ. Mol. Mutagen. **2009,** *50,* 800–807.

49. Moore, P.D.; Yedjou, C.G.; Tchounwou, P.B. Malathion-induced oxidative stress, cytotoxicity, and genotoxicity in human liver carcinoma (HepG2) cells. Environ. Toxicol. **2009,** *25,* 221–226. DOI 10.1002/tox.20492.

50. Sing, N.P.; McCoy, M.T.; Tice, R.R.; Schneider, E.L. A simple technique for quantitation of low levels of DNA damage in individual cells. Exp. Cell Res. **1988,** *175,* 184–191.

51. Tice, R.R.; Agurell, E.; Anderson, D.; Burlinson, B.; Hartmann, A.; Kobayashi, H.; Miyamae, Y.; Rojas, E.; Ryu, J-C.; Sasaki, Y.F. Single cell gel/comet assay: Guideline for in vitro and in vivo genetic toxicology testing. Environ. Mol. Mutagen. **2000,** *35,* 206–221.

52. Tsuda, S.; Matsusaka, N.; Madarame, H.; Miyamae, Y.; Ishida, K.; Satoh, M.; Sekihashi, K.; Sasaki, Y.F. The alkaline single cell electrophoresis assay with eight mouse-organs: Results with 22 monofunctional alkylating agents (including 9 dialkyl N-nitrosoamines) and 10 DNA crosslinkers. Mutat. Res. **2000,** *467,* 83–98.

53. Sekihashi, K.; Yamamoto, A.; Matsumura, Y.; Ueno, S.; Watanabe-Akanuma, M.; Kassie, F.; Knasmüller, S.; Tsuda, S.; Sasaki, Y.F. Comparative investigation of multiple organs of mice and rats in the comet assay. Mutat. Res. **2002,** *517,* 53–74.

54. Castro, V.L.; Mello, M.A.; Poli, P.; Zucchi, T.M. Prenatal and perinatal fenarimol-induced genotoxicity in leukocytes of in vivo treated rats. Mutat. Res. **2005,** *583,* 95–104.

55. Tewari, A.; Dhawan, A.; Gupta, S.K. DNA damage in bone marrow and blood cells of mice exposed to municipal sludge leachates. Environ. Mol. Mutagen. **2006,** *47,* 271–276.

56. Speit, G.; Vasquez, M.; Hartmann, A. The comet assay as an indicator test for germ cell genotoxicity. Mutat. Res. **2009,** *681,* 3–12.

57. Fenech, M. The cytokinesis-block micronucleus technique. In Technologies for Detection of DNA Damage and Mutations; Pfeifer, G.P., Ed.; Plenum Press: New York, 1996; 25–36.

58. Fenech, M. The in vitro micronucleus technique. Mutat. Res. **2000,** *455,* 81–95.

59. Parry, J.M.; Parry, E.M. The use of the in vitro micronucleus assay to detect and assess the aneugenic activity of chemicals. Mutat. Res. **2006,** *607* (1), 5–8.

60. Joksic, G.; Vidakovic, A.; Spasojevic-Tisma, V. Cytogenetic monitoring of pesticide sprayers. Environ. Res. **1997,** *75,* 113–118.

61. Pacheco, A.O.; Hackel, C. Chromosome instability induced by agrochemicals among farm workers in Passo Fundo, Rio Grande do Sul, Brazil. Cad. Saude Publica **2002,** *18* (6), 1675–1683.

62. Fucic, A.; Markucic, D.; Mijic, A.; Jazbec, A.M. Estimation of genome damage after exposure to ionising radiation and ultrasound used in industry. Environ. Mol. Mutagen. **2000,** *36,* 47–51.

63. Vaglenov, A.; Nosko, M.; Georgieva, R.; Carbonell, E.; Creus, A.; Marcos, R. Genotoxicity and radioresistance in electroplating workers exposed to chromium. Mutat. Res. **1999,** *446,* 23–34.

64. Somorovska, M.; Szabova, E.; Vodicka, P.; Tulinska, J.; Barancokova, M.; Fabri, R.; Liskova, A.; Riegerova, Z.; Petrovska, H.; Kubova, J.; Rausova, K.; Dusinska, M.; Collins, A. Biomonitoring of genotoxic risk in workers in a rubber factory, comparison of the Comet assay with cytogenetic methods and immunology. Mutat. Res. **1999,** *445,* 181–192.

65. Fenech, M.; Perepetskaya, G.; Mikhalevic, L. A more comprehensive application of the micronucleus technique for biomonitoring of genetic damage rates in human populations. Experiences from the Chernobyl catastrophe. Environ. Mol. Mutagen. **1997,** *30,* 112–118.

66. Sinues, B.; Sanz, A.; Bernal, M.L.; Tres, A.; Alcala, A.; Lanuza, J.; Ceballos, C.; Saenz, M.A. Sister chromatid exchanges, proliferating rate index, and micronuclei biomonitoring of internal exposure to vinyl chloride monomer in plastic industry workers. Toxicol. Appl. Pharmacol. **1991,** 108, 37–45.

67. Evans, H.J. Cytological methods for detecting chemical mutagens. In *Chemical Mutagens, Principles and Methods for their Detection*; Hollaender, A., Ed.; Plenum Press: New York and London, 1976; Vol. 4, 1–29.

68. Ishidate, M., Jr.; Sofuni, T. The in vitro chromosomal aberration test using Chinese hamster lung (chl) fibroblast cells in culture. In *Progress in Mutation* Research; Ashby, J., et al., Eds.; Elsevier Science Publishers: Amsterdam–New York–Oxford, 1985; Vol. 5, 427–432.

69. Albertini, R.J.; Anderson, D.; Douglas, G.R.; Hagmar, L.; Hemminki, K.; Merlo, F.; Natarajan, A.T.; Norppa, H.; Shuker, D.E.; Tice, R.; Waters, M.D.; Aitio, A. IPCS guidelines for the monitoring of genotoxic effects of carcinogens in humans, International Programme on Chemical Safety. Mutat. Res. **2000**, *463*, 111–172.

70. Hagmar, L.; Stromberg, U.; Bonassi, S.; Hansteen, I.L.; Knudsen, L.E.; Lindholm, C.; Norppa, H. Impact of types of lymphocyte chromosomal aberrations on human cancer risk: Results from Nordic and Italian cohorts. Cancer Res. **2004**, *64*, 2258–2263.

71. Galloway, S.M.; Aardema, M.J.; Ishidate, M., Jr.; Ivett, J.L.; Kirkland, D.J.; Morita, T.; Mosesso, P.; Sofuni, T. Report from working group on in in vitro tests for chromosomal aberrations. Mutat. Res. **1994**, *312*, 241–261.

72. Mateuca, R.; Lombaert, N.; Aka, P.V.; Decordier, I.; Kirsch-Volders, M. Chromosomal changes: Induction, detection methods and applicability in human biomonitoring. Biochimie **2006**, *88*, 1515–1531.

73. González-Barrera, S.; Cortés-Ledesma, F.; Wellinger, R.E.; Aguilera, A. Equal sister chromatid exchange is a major mechanism of double-strand break repair in yeast. Mol. Cell Biol. **2003**, *11* (6), 1661–1671.

74. Domínguez, I.; Pastor, N.; Mateos, S.; Cortés, F. Testing the SCE mechanism with non-poisoning topoisomerase II inhibitors. Mutat. Res. **2001**, *497*, 71–79.

75. Speit, G.; Hochsattel, R.; Vogel, W. The contribution of DNA single-strand breaks to the formation of chromosome aberrations and SCEs. Basic Life Sci. **1984**, *29*, 229–244.

76. Das, B.C. Factors that influence formation of sister chromatid exchanges in human blood lymphocytes. Crit. Rev. Toxicol. **1988**, *19* (1), 43–86.

77. Shiraishi, Y. Nature and role of high sister chromatid exchanges in Bloom syndrome cells. Some cytogenetic and immunological aspects. Cancer Genet. Cytogenet. **1990**, *50* (2), 175–187.

78. Tucker, J.D.; Auletta, A.; Cimino, M.C.; Dearfield, K.L.; Jacobson-Kram, D.; Tice, R.R.; Carrano, A.V. Sister-chromatid exchange: Second report of the Gene-Tox Program. Mutat. Res. **1993**, *297* (2), 101–180.

79. Tilman, G.; Loriot, A.; Van Beneden, A.; Arnoult, N.; Londoño-Vallejo, J.A.; De Smet, C.; Decottignies, A. Subtelomeric DNA hypomethylation is not required for telomeric sister chromatid exchanges in ALT cells. Oncogene **2009**, *28* (14), 1682–1693.

80. White, J.S.; Choi, S.; Bakkenist, C.J. Transient ATM kinase inhibition disrupts DNA damage-induced sister chromatid exchange. Sci. Signal. **2010**, *3*, ra44. DOI: 10.1126/scisig nal.2000758.

81. Popescu, N.C. Sister chromatid exchange formation in mammalian cells is modulated by deoxyribonucleotide pool imbalance. Somatic Cell Mol. Genet. **1999**, *25*, 101–108.

82. Ashman, C.R.; Davidson, R.L. Bromodeoxyuridine mutagenesis in mammalian cells is related to deoxyribonucleotide pool imbalance. Mol. Cell. Biol. **1981**, *1*, 254–260.

83. Latt, S.A. Sister chromatid exchanges, indices of human chromosome damage and repair: Detection by fluorescence and induction by mitomycin-C. Proc. Natl. Acad. Sci. U. S. A. **1974**, *71*, 3162–3166.

84. Latt, S.A.; Schreck, R.R. Sister chromatid exchange analysis. Am. J. Hum. Genet. **1980**, *32*, 297–313.

85. EPA. In vitro sister chromatid exchange assay. Test methods and guidelines/OPPTS Harmonized test guidelines OPPTS 870.5900. U.S. Environmental Protection Agency, 1998, available at http://www.epa.gov/epahome/research.htm (accessed May 2010).

86. Morgan, W.F.; Schwartz, J.L.; Murnane, J.P.; Wolff, S. Effect of 3-aminobenzamide on sister chromatid exchange frequency in X-irradiated cells. Radiat. Res. **1983**, *93*, 567–571.

87. Hagmar, L.; Brogger, A.; Hansteen, I.L.; Heim, S.; Hogstedt, B.; Knudsen, L.; Lambert, B.; Linnainmaa, K.; Mitelman, F.; Nordenson, I.; Reuterwall, C.; Salomaa, S.I.; Skerfving, S.; Sorsa, M. Cancer risk in human predicted by increased levels of chromosomal aberrations in lymphocytes: Nordic Study Group on the Health Risk of Chromosome Damage. Cancer Res. **1994,** *54,* 2919–2922.
88. Hagmar, L.; Bonassi, S.; Stromberg, U.; Brogger, A.; Knudsen, L.; Norppa, H.; Reuterwall, C. The European Study Group on Cytogenetic Biomarkers and Health. Chromosomal aberrations in lymphocytes predict human cancer. A report from the European Study Group on Cytogenetic Biomarkers and Health (ESCH). Cancer Res. **1998,** *58,* 4117–4121.
89. Gichner, T.; Menke, M.; Stavreva, D.A.; Schubert, I. Maleic hydrazide induces genotoxic effects but no DNA damage detectable by the comet assay in tobacco and field beans. Mutagenesis **2000,** *15,* 385–389.
90. Citterio, S.; Aina, R.; Labra, M.; Ghiani, A.; Fumagalli, P.; Sgorbati, S.; Santagostino A. Soil genotoxicity assessment: A new strategy based on biomolecular tools and plant bioindicators. Environ. Sci. Technol. **2002,** *36* (12), 2748–2753.
91. Restivo, F.M.; Laccone, M.C.; Buschini, A.; Rossi, C.; Poli, P. Indoor and outdoor genotoxic load detected by the Comet assay in leaves of *Nicotiana tabacum* cultivars Bel B and Bel W3. Mutagenesis **2002,** *17* (2), 127–134.
92. Klumpp, A.; Ansel, W.; Klumpp, G.; Calatayud, V.; Garrec, J.P.; He, S.; Penuelas, J.; Ribas, A.; Ro-Poulsen, H.; Rasmussen, S.; Sanz, M.J.; Vergne, P. Tradescantia micronucleus test indicates genotoxic potential of traffic emissions in European cities. Environ. Pollut. **2006,** *139* (3), 515–522.
93. Liu, W.; Zhu, L.S.; Wang, J.; Wang, J.H.; Xie, H.; Song, Y. Assessment of the genotoxicity of endosulfan in earthworm and white clover plants using the comet assay. Arch. Environ. Contam. Toxicol. **2009,** *56* (4), 742–746.
94. Villarini, M.; Fatigoni, C.; Dominici, L.; Maestri, S.; Ederli, L.; Pasqualini, S.; Monarca, S.; Moretti, M. Assessing the genotoxicity of urban air pollutants using two in situ plant bioassays. Environ. Pollut. **2009,** *157* (12), 3354–3356.
95. Rodrigues, F.P.; Angeli, J.P.; Mantovani, M.S.; Guedes, C.L.; Jordão, B.Q. Genotoxic evaluation of an industrial effluent from an oil refinery using plant and animal bioassays. Genet. Mol. Biol. **2010,** *33* (1), 169–175.
96. Sadowska, A.; Pluygers, E.; Niklinska, W.; Maria, M.R.; Obidoska, G. Use of higher plants in the biomonitoring of environmental genotoxic pollution. Folia Histochem. Cytobiol. **2001,** *39,* 52–53.
97. Dimitrov, B.D.; Gadeva, P.G.; Benova, D.K.; Bineva, M.V. Comparative genotoxicity of the herbicides Roundup, Stomp and Reglone in plant and mammalian test systems. Mutagenesis **2006,** *21* (6), 375–382.
98. Grant, W.F. The present status of higher plant bioassays for the detection of environmental mutagens. Mutat. Res. **1994,** *310* (2), 175–185.
99. Ma, T.H.; Cabrera, G.L.; Owens, E. Genotoxic agents detected by plant bioassays. Rev. Environ. Health **2005,** *20* (1), 1–13.
100. Feretti, D.; Zerbini, I.; Zani, C.; Ceretti, E.; Moretti, M.; Monarca, S. Allium cepa chromosome aberration and micronucleus tests applied to study genotoxicity of extracts from pesticide-treated vegetables and grapes. Food Addit. Contam. **2007,** *24* (6), 561–572.
101. Mancini, A.; Buschini, A.; Restivo, F.M.; Rossi, C.; Poli, P. Oxidative stress as DNA damage in different transgenic tobacco plants. Plant Sci. **2006,** *170,* 845–852.
102. Norppa, H. Cytogenetic biomarkers and genetic polymorphisms. Toxicol. Lett. **2004,** *149* (1–3), 309–334.
103. Vodicka, P.; Koskinen, M.; Naccarati, A.; Oesch-Bartlomowicz, B.; Vodickova, L.; Hemminki, K.; Oesch, F. Styrene metabolism, genotoxicity, and potential carcinogenicity. Drug Metab. Rev. **2006,** *38* (4), 805–853.
104. Lampe, J.W. Diet, genetic polymorphisms, detoxification, and health risks. Altern. Ther. Health Med. **2007,** *13* (2), S108–S111.

105. Lin, J.; Swan, G.E.; Shields, P.G.; Benowitz, N.L.; Gu, J.; Amos, C.I.; de Andrade, M.; Spitz, M.R.; Wu, X. Mutagen sensitivity and genetic variants in nucleotide excision repair pathway: Genotype–phenotype correlation. Cancer Epidemiol. Biomarkers Prev. **2007**, *16* (10), 2065–2071.

106. Rueff, J.; Teixeira, J.P.; Santos, L.S.; Gaspar, J.F. Genetic effects and biotoxicity monitoring of occupational styrene exposure. Clin. Chim. Acta **2009**, *399* (1–2), 8–23.

107. Rohr, P.; da Silva, J.; Erdtmann, B.; Saffi, J.; Guecheva, T.N.; Henriques, J.A.; Kvitko, K. BER gene polymorphisms (OGG1 Ser326Cys and XRCC1 Arg194Trp) and modulation of DNA damage due to pesticides exposure. Environ. Mol. Mutagen. **2010** [Epub ahead of print].

108. Goode, E.L.; Ulrich, C.M.; Potter J.D. Polymorphisms in DNA repair genes and associations with cancer risk. Cancer Epidemiol. Biomarkers Prev. **2002**, *11*, 1513–1530.

109. Buschini, A.; De Palma, G.; Poli, P.; Martino, A.; Rossi, C.; Mozzoni, P.; Scotti, E.; Buzio, L.; Bergamaschi, E.; Mutti, A. Genetic polymorphism of drug-metabolizing enzymes and styrene-induced DNA damage. Environ. Mol. Mutagen. **2003**, *41* (4), 243–252.

110. Vodicka, P.; Kumar, R.; Stetina, R.; Sanyal, S.; Soucek, P.; Haufroid, V.; Dusinska, M.; Kuricova, M.; Zamecnikova, M.; Musak, L.; Buchancova, J.; Norppa, H.; Hirvonen, A.; Vodickova, L.; Naccarati, A.; Matousu, Z.; Hemminki, K. Genetic polymorphisms in DNA repair genes and possible links with DNA repair rates, chromosomal aberrations and single-strand breaks in DNA. Carcinogenesis **2004**, *25*, 757–763.

111. Landsiedel, R.; Kapp, M.D.; Schulz, M.; Wiench, K.; Oesch, F. Genotoxicity investigations on nanomaterials: Methods, preparation and characterization of test material, potential artefacts and limitations—Many questions, some answers. Mutat. Res. **2009**, *681*, 241–258.

112. Savolainen, K.; Alenius, H.; Norppa, H.; Pylkkänen, L.; Tuomi, T.; Kasper, G. Risk assessment of engineered nanomaterials and nanotechnologies—A review. Toxicology **2010**, *269*, 92–104.

113. Wang, J.J.; Sanderson, B.J.S.; Wang, H. Cyto- and genotoxicity of ultrafine TiO_2 particles in cultured human lymphoblastoid cells. Mutat. Res. **2007**, *628*, 99–106.

114. Reeves, J.F.; Davies, S.J.; Dodd, N.J.F.; Jha, A.N. Hydroxyl radicals (•OH) are associated with titanium dioxide (TiO_2) nanoparticle-induced cytotoxicity and oxidative DNA damage in fish cells. Mutat. Res. **2008**, *640*, 113–122.

115. Yang, H.; Liu, C.; Yang, D.; Zhanga, H.; Xi, Z. Comparative study of cytotoxicity, oxidative stress and genotoxicity induced by four typical nanomaterials: The role of particle size, shape and composition. J. Appl. Toxicol. **2009**, *29*, 69–78.

116. Sharma, V.; Shukla, R.K.; Saxena, N.; Parmar, D.; Das, M.; Dhawan, A. DNA damaging potential of zinc oxide nanoparticles in human epidermal cells. Toxicol. Lett. **2009**, *185*, 211–218.

117. Elespuru, R.K.; Agarwal, R.; Atrakchi, A.H.; Bigger, C.A.H.; Heflich, R.H.; Jagannath, D.R.; Levy, D.D.; Moore, M.M.; Ouyang, Y.; Robison, T.W.; Sotomayor, R.E.; Cimino, M.C.; Dearfield, K.L. Current and future application of genetic toxicity assays: The role and value of in vitro mammalian assays. Toxicol. Sci. **2009**, *109* (2), 172–179.

118. Garcia-Sagredo, J.M. Fifty years of cytogenetics: A parallel view of the evolution of cytogenetics and genotoxicology. Biochim. Biophys. Acta **2008**, *1779*, 363–375.

119. Robbens, J.; van der Ven, K.; Maras, M.; Blus, R.; De Coen, W. Ecotoxicological risk assessment using DNA chips and cellular reporters. Trends Biotechnol. **2007**, *25* (10), 460–466.

120. Gant, T.W.; Zhang, S.D. In pursuit of effective toxicogenomics. Mutat. Res. **2005**, *575*, 4–16.

121. Ellinger-Ziegelbauerb, H.; Aubrechta, J.; Kleinjans, J.C.; Ahr, H.-J. Application of toxicogenomics to study mechanisms of genotoxicity and carcinogenicity. Toxicol. Lett. **2009**, *186*, 36–44.

122. Waters, M.D.; Olden, K.; Tennant, R.W. Toxicogenomic approach for assessing toxicant-related disease. Mutat. Res. **2003**, *544*, 415–424.

123. Sen, B.; Mahadevan, B.; DeMarini, D.M. Transcriptional responses to complex mixtures: A review. Mutat. Res. **2007**, *636*, 144–177.

124. Lettieri, T. Recent applications of DNA microarray technology to toxicology and ecotoxicology. Environ. Health Perspect. **2006**, *114*, 4–9.

10

Pollution: Pesticides in Agro-Horticultural Ecosystems

J.K. Dubey and
Meena Thakur

Introduction

Agro-horticultural ecosystems are man-made ecosystems that are greatly influenced by human activities. In the last few decades, agricultural intensification and diversification in most of the developing and developed countries have resulted in high-input farming, i.e., excessive use of pesticides (herbicides, insecticides, fungicides) and fertilizers, which has helped to meet the rising food demand. Agricultural and horticultural ecosystems are now dominated by monoculture and hybrid varieties to enhance crop yields. Hybrid varieties are high yielding and more demanding. The excessive use of inorganic fertilizers results in rapid multiplication and subsequent outbreaks of many pests, simultaneously resulting in enhanced pesticide use. Rice and cotton are important crops grown worldwide, and because of monoculture, they are attacked by hundreds of pests and receive a disproportionately high share of pesticides (17% and 24%) worldwide.[1] It has been observed that chemicals that have been banned for most of the food crops are still being used on cotton as it is not consumed directly. For example, nearly one-quarter of all pesticides used in the United States are applied to cotton, and the overall amount and intensity per acre are increasing every year. Worldwide, more and extremely toxic pesticides are sprayed on cotton than on any other crop,[2] which find an entry in the human body in the form of salad dressings, baked goods, and snacks like Fritos and Goldfish.

Based on the trend of herbicide use in the United States, agriculture accounts for three quarters of total chemical pesticides used. According to the data published by the US Environmental Protection Agency (EPA), in 2001, 675 million lb of chemical pesticides were used in agriculture.[3] Thus, horticultural and agricultural ecosystems are highly dependent on pesticides. It is estimated that if pesticides are not used, food supplies would fall to 30%–40% due to the ravages of pests.[4] Worldwide, research data show that without effective pest management, preharvest losses in crops would average about 40%

and the world's food and fiber production as well as environmental and human health would be seriously threatened.[5]

According to market research report, pesticide consumption has been found to be highest in China followed by the United States and Brazil and India ranked 11th after Spain, and as per Tata Strategic Management Group[6] (2016), the world agrochemical market is expected to be worth $60 billion in 2016 of which Indian market will account for $6.8 billion of the total revenues and is expected to hike at $8 billion by 2025 in comparison to world market share of $8.0 billion.[6]

No doubt, pesticides play an important role in enhancing agricultural productivity, but intensive use of pesticides in agricultural and horticultural ecosystems has resulted in degradation of environment, whether it is in the form of development of resistant pests, secondary pest outbreak, or pest resurgence, which leads to changes in ecosystem biodiversity or adverse effects on pollinators, natural enemies, and many incurable human diseases. Several pesticides are known to persist for a longer period in the environment or on the substrate to which they are applied and have long-term side effects on human health and the natural environment. Persistent organic pesticides, especially DDT (dichloro diphenyl trichloroethane) and HCH (hexachloro cyclohexane), have been detected in various systems, even in human blood, fat, and milk samples. Even the soft drinks that are water-based flavored drinks are known to contain pesticides. The hazards of pesticide pollution have been much realized presently, and the demands for their reduction are desired globally in various pesticide monitoring programs.

Pesticide Consumption Trend

By using nonchemical techniques such as sanitation, cultivation, crop rotation, resistant cultivars, and biological control (including introduction of transgenic) for pest control, many pests cannot be controlled adequately. Therefore, there is a continuous need for application of substantial quantities of chemical pesticides, as a result of which crop protection in many developing countries is still dominated by an increasing use of pesticides.

The world pesticide market now amounts $60 billion and is expected to increase to $80 billion in 2025, whereas, Indian market of pesticides is recently reported to be $6.8 billion out of which $2.9 billion is by domestic use and $3.9 billion is earned by export of pesticides to various countries. Both the domestic and the export markets are expected to enhance to $4.0 billion each in 2025, thereby increasing the expected Indian pesticide market to $8.0 billion.[6]

Globally, pesticide consumption has been found to be highest in China 18,07,000 and 17,72,449 metric tonnes in 2014 and 2015, respectively. In the United States, 4,07,779 metric tonnes of pesticide consumption in 2012 was observed. After China and the United States, next in the list are Brazil (3,52,336 and 3,95,646 metric tonnes in 2014 and 2015), Argentina (2,07,706 metric tonnes in 2014), Mexico (98,814 metric tonnes in 2014), Ukraine (78,201 metric tonnes in 2014), Canada (76,314 metric tonnes in 2014), and France (75,339 and 63,818 metric tonnes in 2014 and 2015).[7]

According to the data given by FAOSTAT 2018, the pesticide consumption reported in Europe is 4,36,361, 4,28,506, 4,49,392, and 3,56,279 metric tonnes in 2012, 2013, 2014, and 2015, respectively.[7] Out of all European countries, the top five consumers have been France, Italy, Spain, Germany, and Poland with the highest consumption in France, i.e., 63,844, 66,676, 75,339, and 63,818 metric tonnes in 2012, 2013, 2014, and 2015, respectively.

The per hectare consumption of pesticides in India is among the lowest in the world and currently stands at 0.6 kg/ha against 5 kg/ha in the UK and at almost 20 times approximately 13 kg/ha in China. In order to increase yield and ensure food security for its enormous population, agrochemical penetration in India is bound to go up.[6]

Further, government policies on subsidies, establishment of market for agrochemicals, availability of technology packet, etc., encourage the farmers to use pesticides excessively. Today, more than 80% of worldwide pesticide sales fall to the share of only six companies. In 2004, three agrochemical companies, each with sales of more than $4 billion, together controlled the global market for pesticides,[8] by controlling such a large stake of the market, these companies have a considerable influence on the way in which plant protection is practiced in farming ecosystems.

Pesticides Use Pattern in India

In many cases, farmers go for prophylactic applications, whether they are required or not. India was one of the first countries in the third world to start large-scale use of pesticides for the control of insect pests of public health as well as of agricultural importance, initially DDT and HCH were imported soon after independence. A moderate beginning in indigenous manufacture of pesticides was made with the setting of the plant to produce HCH in 1952 at Rishra (West Bengal). This was followed by the indigenous production of DDT.[9]

The export of various pesticides from India has been higher than their import. In 2010, pesticides of $1.14 billion were exported which gradually increased to $1.38, $1.66, and $2.13 billion in 2011, 2012, and 2013, respectively, whereas the export of pesticides from India was decreased after 2013 and was nearly of $1.94 and $1.93 billion in 2014 and 2015, respectively. The amount of pesticides imported into India increased from $0.63 (in 2010) to $0.99 billion (in 2014) and somewhat decreased to $0.89 billion in 2015.[7]

Considering the pesticide consumption and area under cultivation in 2016–2017 in different states, the consumption of pesticides (kg per hectare) has been found to be highest in Jammu and Kashmir (1.89) followed by Tripura (0.78), Punjab (0.74), Haryana (0.62), Telangana (0.61), and Maharashtra (0.58). The pesticide consumption in Kerala, Uttar Pradesh, Himachal Pradesh, Tamil Nadu, Jharkhand, and West Bengal was found to be 0.41, 0.39, 0.36, 0.34, 0.32, and 0.27 kg/ha, respectively.[10]

In India, Kerala is one of the leading agricultural states of India, and currently, India is the leading manufacturer of basic pesticides in Asia and ranks 12th globally. Compiling the data on the consumption level of pesticides in agriculture in Kerala (1995–1996 to 2007–2008), the total quantity is estimated at 462.05 metric tonnes (2007–2008). Pesticide application in the state is prophylactic and is one of the most important risk management strategies; e.g., pesticide application on bitter gourd starts from the time of transplanting. The prophylactic application of the pesticides is resorted to at an interval of 2 weeks initially, which gets reduced to 2 days as the crop nears flowering and fruit set. There is a tendency among farmers to change the chemicals in each spray. Thus, on an average, acetamiprid is sprayed 6 times, phorate and dimethoate 5 times each, quinalphos and indoxacarb 4 times each, and the rest 3–4 times each. During a crop cycle of 90 days in bitter gourd, farmers apply pesticides as many as 50 times.[11]

The pesticides that are used in the state include chemicals that are banned for sale in Kerala (endosulfan), banned for use in fruits/vegetables (monocrotophos), and those permitted for restricted use only (methyl parathion, lindane, and methoxy ethyl mercury chloride). The farmers are investing a large portion of their income on pesticides, and pesticide consumption is reported to be the primary method of suicide in Kerala. Of the 900–1000 suicides per year, 60% are by consuming poisons. The commonly used poisons are furadan, malathion, and rat poison.[12] Moreover, farmers also go for suicide because of indebtedness due to purchase of chemicals and sometimes due to complete failure of crops in spite of heavy investment.[12]

Impact of Pesticide Use on the Pollinators

Wild bees, bumblebees, honeybees, and solitary bees are well known and valued as important pollinators of crops/plants and are in commercial use for pollination. About 33% of all crops require pollination. Intensive cultivation and excessive use of pesticides lead to a sharp decline in the population of these pollinators, which is one of the major causes of low productivity of agricultural and horticultural crops. Pesticides, such as DDT, benzene hexachloride (BHC), cyclodiene, and most of the organophosphorus and carbamate compounds are highly toxic to bees. Although endosulfan is listed as a persistent organic pollutant, it is the only available pesticide known to be safe for honeybees and other beneficial insects and is still extensively used in many countries. Health and environmental causalities related to excessive use of this pesticide are reported from many parts of the world and European countries withdrew its registration in 2005. Still, there are many instances where most of the pesticides fail to be effective and endosulfan is recommended. For example, in 2008, there was heavy weevil infestation on hazelnut crops in Italy, and the Italian government had to prescribe the use of endosulfan for 120 days although it endangered the health of its citizens. Pesticides like neonicotinoids used as a substitute for endosulfan in agricultural ecosystems in countries such as Germany, France, UK, and the United States have resulted in mass bee killing and colony collapse disorders. In 1959, carbaryl was used against certain orchard pests and later registered for many other crops. In 1967, it caused the destruction of an estimated 70,000 colonies of honeybees in California from use in cotton and an estimated 33,000 colonies in Washington from use in corn. The estimated national loss from all pesticide poisoning from the same year was 500,000 colonies. Carbaryl is still one of the most destructive bee-killing chemical.[13]

A horticultural ecosystem, particularly the apple ecosystem, is heavily polluted with pesticides. Apple is one of the important commercial horticultural crops grown in temperate regions of the world. Honeybees are important pollinators of this crop, and quality beehives are placed in apple orchards for enhancing apple productivity. Since the crop is important from an economic point of view, much attention is paid to it. In India, the temperate northern regions of the country (Himachal Pradesh, Jammu, and Kashmir) are known for quality apple production. To obtain quality fruits in Himachal Pradesh, a number of pesticides, which affect the pollinators, are applied on the crop right from fruit set to harvest. The situation has reached an alarming level in the state, and if required measures are not initiated to conserve and rear the population of these pollinators, it could impinge on the total agricultural and fruit production in the years to come. Currently, about 943,000 ha of land is under fruit cultivation in the state, and it requires at least 5 lakh colonies of honeybees alone for pollination to enhance the production. There is a great need to encourage organic farming to enhance the population of natural pollinators and avoid pesticide applications when crops, cover crops, weeds, and wildflowers are in bloom in the treatment area or nearby.

Impact of Pesticide Use on Natural Enemies

Pesticides including insecticides and miticides are primarily used to regulate insect and mite populations in agricultural and horticultural crop production systems. However, continuous dependence on pesticides may eventually result in a number of potential ecological problems including resistance, secondary pest outbreaks, and/or target pest resurgence.[14,15] Therefore, implementation of alternative management strategies is justified in order to conserve existing pesticides and produce crops with minimal damage from insect pests. One option that has gained interest by producers is integrating pesticides with biological control agents or natural enemies including parasitoids and predators.[16] This is often referred to as "compatibility," which is the ability to integrate or combine natural enemies with pesticides so as to regulate arthropod pest populations without directly or indirectly affecting the life history parameters or population. This may also refer to pesticides being effective against targeted insect pests but relatively non-harmful to natural enemies.[17,18] Pesticides vary in their activity, which not only impacts how they kill arthropod pests but also how they may indirectly influence natural enemy populations. Pesticides may be classified as contact, stomach poison, systemic, and/or translaminar.[19,20] In addition, the application method—foliar versus drench or granular—may determine the extent of any indirect effects on natural enemies[21] as well as the pesticide mode of action. The type of natural enemy-parasitoid or predator may be influenced differently based on the factors mentioned above. Furthermore, the type of pesticide may substantially contribute to any indirect effects on natural enemies. For example, broad-spectrum, nerve toxin pesticides such as most of the older pesticides in the chemical classes, organophosphate (acephate and chlorpyrifos), carbamate (carbaryl and methiocarb), and pyrethroid (bifenthrin and cyfluthrin) may be both directly and indirectly more harmful to natural enemies than non-nerve toxin-type pesticides (often referred to a "selective pesticides") including insect growth regulators (kinoprene and pyriproxyfen), insecticidal soaps (potassium salts of fatty acids), horticultural oils (petroleum or neem-based), selective feeding blockers (flonicamid and pymetrozine), and microbials (entomopathogenic fungi and bacteria and other microorganisms).[22] The non-nerve toxin pesticides are generally more specific or selective in regards to arthropod pest activity with broader modes of action than nerve toxin pesticides.[16] The effects of pesticides on natural enemies are typically associated with determining direct effects such as mortality or survival over a given time period (24–96 hours).[23] While evaluations associated with the direct effects of pesticides on natural enemies are important, what are actually more relevant are the indirect or delayed effects of pesticides because this provides information on the long-term stability and overall success of a biological control program when attempting to integrate the use of pesticides with natural enemies.[24–27] Any indirect effects, which are sometimes referred to as sublethal, latent, or cumulative adverse effects, may be associated with interfering with the physiology and behavior of natural enemies by inhibiting longevity, fecundity, reproduction (based on the number of progeny produced or eggs laid by females), development time, mobility, searching (foraging) and feeding behavior, predation and/or parasitism, prey consumption, emergence rates, and/or sex ratio.[15,25,28,29–30]

In apple ecosystems, many host species of phytophagous arthropods among which red spider mite and two-spotted spider mite are substantial worldwide. Predatory mites play an important role in checking the population of these mites. In India, the red spider mite *Panonychus ulmi* was a minor pest up to 1990, but the commercialization of apple led to excessive and repeated use of pesticides for quality apple production, as a result of which the natural mite predators were destroyed and it emerged as a serious pest of apples. Most of the spray schedules are now focused to this pest, further deteriorating the condition. Pyrethroids and carbamates, i.e., carbaryl (Sevin), are highly toxic to predatory mites, e.g., *Typhlodromus occidentalis* and use of Sevin for thinning causes mite flare-ups. Another well-known example is the resurgence of brown plant hopper (BPH) in rice ecosystem. If no pesticides are used, BPH is kept under control by its natural enemies (mirid bugs, ladybird beetles, spiders, and various pathogens). Since rice is a heavily sprayed crop, pesticides kill the natural enemies and create a situation where BPH can multiply rapidly. Thus, similar to *P. ulmi*, it has also become a serious man-made pest. Synthetic pyrethroids result in spider mite resurgence.[31] In a study conducted by Beers,[32] pyrethroids, carbamates, organophosphates, Assail, Calypso, and Actara are toxic to *T. occidentalis* and *Zetzellia mali*, which are mostly found associated with *P. ulmi* and other phytophagous mites. Organophosphates and carbamates are reported to cause high levels of mortality to coccinellids and lacewings.

Pesticide Use and Bt Transgenics

A common pest management technology used in agroecosystems is the use of *Bacillus thuringiensis* (Bt) transgenic. Bt crops, particularly cotton, are grown all over the world. Bt crops were mainly introduced with an aim to reduce pesticide use, but growing secondary pest populations and efforts to control them have further increased the use of pesticides The major cotton-growing countries, i.e., the United States, China, India, and Argentina have quickly adopted this technique for cotton seeds. For example, before the commercialization of Bt cotton, the Chinese farmers applied an average of 20 pesticide treatments in a season to control bollworm infestations. With the adoption of Bt, the average number of treatments has fallen to only 6.6 in the early stages of Bt adoption.[33] As a result, the pesticide use decreased by 43.3 kg/ha in 1999, i.e., a 71% decrease in pesticide use. For the years 2000 and 2001, Bt cotton was associated with an average reduction of 35.7 kg/ha of pesticide or a percentage deduction of 55%.[34] Similar results have been found in other major cotton-growing countries, and Indian farmers save 39% of expenditures by planting Bt crops.[35] Argentine farmers save 47% of expenditures,[36] Mexican farmers can save 77%,[37] and South African farmers can save 58% by planting Bt.[38] Evidence shows that, though Bt seed costs 2 to 3 times more than a conventional seed, savings on pesticide expenditures guarantee a much higher net return for Bt adopters. Using a household survey from 2004, 7 years after the initial commercialization of Bt cotton in China, we show that total pesticide expenditure for Bt cotton farmers in China is nearly equal to that of their conventional counterparts, about $101/ha. Bt farmers in 2004, on the average, have to spray pesticide 18.22 times, which are more than 3 times higher compared with 6 times the pesticide spray in 1999. Detailed information on pesticide expenditures reveals that, though Bt farmers saved 46% of bollworm pesticides relative to non-Bt farmers, they spend 40% more on pesticides targeted to kill an emerging secondary pest. These secondary pests, e.g., mirid bugs, were rarely found in the field prior to the adoption of Bt cotton, presumably kept in check by bollworm populations and regular pesticide spraying.

Cotton is attacked by more than 165 pests, and farmers repeatedly spray pesticides, which increase the chances of resurgence of secondary pests. In Andhra Pradesh, the number of attacks of aphids, thrips, and jassids has increased since the introduction of Bt cotton in 2002. Many diseases and pests such as tobacco leaf streak virus and tobacco caterpillars have newly emerged in Bt cotton ecosystems in this state.[11]

Pesticide Use in Weeds

Herbicide should be applied at the time when their impact on weeds is highest. If preemergence weed control is optimized, the need for pest emergence measures may be reduced. The cultivation of genetically modified herbicide-tolerant crops has the potential to reduce herbicide inputs. The world sales of agrochemicals is dominated by herbicides (46%) followed by insecticides (26%), fungicides (23%), and others (5%).[39] On the other hand, the Indian market trend indicates domination of insecticides (61.39%) followed by fungicides (19.06%), herbicides (16.75%), and others (2.80%).[40]

The large-scale adoption of dwarf high-yielding varieties (HYV) and hybrids and the increased use of irrigation, fertilizers, and monocropping have increased weed problem in agro-horticultural ecosystems, simultaneously leading to increased herbicide use. Herbicides, such as isoproturon, atrazine, alachlor, butachlor, and oxyfluorfen, are applied on agro-horticultural ecosystems for control of weeds. Globally, herbicides constitute 52% of the total pesticide sales, and in some countries such as the United States, Germany, and Australia, the figure is as high as 60%–70%.[41] According to USDA–NASS (US Department of Agriculture–National Agricultural Statistics Service) report, the use of genetically modified crops is the main reason for the rise in herbicide use.[42] For example, widespread introduction of genetically modified soybeans, cotton, and corn by Monsanto resulted in a 15-fold increase in the use of glyphosate (Roundup) from 1994 to 2005 on these three crops in the United States. The excessive use of glyphosate has resulted in resistant weeds, as a result of which the application of glyphosate, atrazine, 2,4-D, and other leading weed-killing chemicals has further increased since 2002. 2,4-D, the second most heavily used herbicide on soybeans (after glyphosate) in the United States, is associated with a number of adverse health impacts on agricultural workers. These herbicides have increased the risk of cancer, have increased the rate of birth defects in children of men who apply the herbicide, and are also a suspected endocrine disruptor. Similarly, atrazine, the most heavily used herbicide on corn, has been linked to endocrine disruption, neuropathy, and cancer (particularly breast and prostate cancer). It is regularly detected in drinking water supplies in the United States and has been associated with low sperm counts in men. Exposure to extremely low levels of atrazine can cause sex change and/or deformities in frogs, fish, and other organisms. Based on this evidence, and the widespread presence of atrazine in drinking water supplies, the European Union announced a ban on atrazine in 2006. However, the US EPA reregistered atrazine in 2003 despite objections from scientists and environmental groups.[43] Cheaper formulations of herbicides containing 2,4-D and 2-methyl-4-chlorophenoxyacetic acid (MCPA) are still used in many countries, and weeds have developed resistance, e.g., in Bulgaria, 47% of wheat and barley crops were affected by 2,4-D-resistant weeds in 2000.[44]

Pesticide Residues in Agro-Horticultural Ecosystems

Agro-horticultural produce constitutes an essential part of human diet, and as per the recommendation of World Health Organization (WHO), there should be at least 30% fruits and vegetables in our daily diet, depending on the body weight of the person. Vegetables are the major source of vital nutrients.

But it is not heartening to know that instead of fulfilling the nutritional requirements, these fruits and vegetables carry pesticide residue harmful to the health of the consumers. Main reasons of finding the pesticide residues in these food stuffs may be because of providing shiny and fresh appearance and good color to the crop. This leads to usage of synthetic chemicals beyond the safe limits. Due to the persistent nature of some of the highly toxic pesticides, or maybe due to illegal use of prohibited/banned pesticides, these compounds have been detected in the environment worldwide.[45] So, usage of restricted pesticides is a matter of concern especially in case of vegetables. There is a general belief that these food items are much harmful if impregnated with pesticides in comparison to other food stuffs, because they are generally consumed raw or semi-cooked. Many studies support the presence of pesticide residues in vegetables, on an average, this percentage is 50%–70% in India as mentioned by [46-48]. In India, 51% of the food commodities have been detected with pesticide residues.[49] A study conducted by Charan et al. in 2010[47] revealed that 67% of total contaminated samples exceeded the maximum residue limit (MRL) values recommended by the Food and Agriculture Organization (FAO)/WHO. Another study revealed the presence of monocrotophos, chlorpyrifos, cypermethrin, and endosulfan, etc. in the vegetables.[48] According to a report, over 98% of sprayed insecticides and 95% of herbicides reached nontarget destinations such as other species, air, water, and soil.

Most of the pesticides used on crops are persistent, especially organochlorines, which persist for a longer period in the environment (substrate). The organochlorine insecticides (such as DDT and BHC) that were banned still persist in soil and contaminate both organic and conventional crop produce. Baker and co-workers observed pesticide residues in organic fruit samples. The reasons for residues in organic fruit samples were in violation of organic methods of cultivation, pesticide-contaminated water used for irrigation, or pesticide residues left in the soil, if previously used to grow conventional crops.[50]

Apple fruit crop is attacked by a number of insect pests and diseases such as apple scab, San Jose scale, wooly apple aphid, fruit scrapper, defoliating beetles, and tent caterpillar. Pesticides such as chlorpyrifos, endosulfan, carbendazim, propineb, and mancozeb are applied to control these pests. After spraying/treatment, pesticide residues get deposited on the fruits and dissipate slowly depending upon the number of factors such as physiochemical characteristics of pesticide, weather conditions, and time after treatment. Preharvest or postharvest interval or waiting period between spray and harvest is required for safe consumption of fruits. Sometimes, the produce is sent to the market immediately after spraying, and consumers unknowingly consume the product and may be badly affected. Similarly, under Indian conditions, a number of synthetic pesticides, such as deltamethrin, cypermethrin, dimethoate, quinalphos, oxydemeton methyl and carbaryl, are used to control mango crop pests such as mango hopper, mango mealy bug, and fruit fly, as well as powdery mildew and malformations. Deltamethrin at 0.002% does not require any waiting period, but cypermethrin requires 11 days of waiting period. Mango is eaten after removing the peel, but the residues on its peel also find their way into the consumer by contact. Residues of mancozeb and lindane though within the permissible limit were detected in mango fruit samples.[51] The repeated spray of bifenthrin on mango from flowering to 1 month before harvest resulted in residues that persisted on the peel for more than a month, and rate of degradation was very low.[52]

The consumption of pesticide in India is low as compared to other countries, in spite of this, there is widespread contamination of food commodities with pesticide residues due to non-judicious use of pesticides. An earlier survey carried out by the Indian Council of Medical Research, New Delhi, revealed that 51% of food commodities contained pesticide residues, and out of these, 20% had pesticide residues above the MRL values, as compared to 21% contamination with only 2% of samples above the MRL on a worldwide basis.[52]

Now, the scenario in India has started changing very rapidly as new pesticide molecules, whose application rate (as well as persistence in the environment) is very low, are being introduced every year. Heavy-duty pesticides have been either banned or put under restricted use. The pesticide load on the agro-horticultural ecosystem has declined as compared to the last decades. Maximum pesticides in India are used on cotton and rice. The Malwa area of Punjab, which is famous for cotton growing, has been named as the cancer belt of Punjab because pesticides have contaminated the whole environment, including groundwater, and caused cancer among its people. Out of the total pesticides used in India, only 13%–14% is used on fruits and vegetables; despite this, half of the fruits and vegetables were found contaminated with pesticide residues.[52] Pesticide residues in 10% of the samples were above the MRL value. Residues of methyl parathion, endosulfan, chlorpyrifos, Dimethyl 2,2-Dichlorovinyl Phosphate (DDVP), dimethoate, fenitrothion, monocrotophos, cypermethrin, deltamethrin, copper, etc., were above the MRL in fruits and vegetables.[52]

Approaches for Pesticide Use Reduction

Since the excessive and indiscriminate use of pesticides has polluted every component of the environment, people all over the world have realized the need for pesticide reduction so as to prevent the environment from further deterioration. The use of pesticides in fruits and vegetable production has been developed in many countries and the range of pesticides is very large, most of them are chemicals used since the 1950s, quite often pesticides not licensed for use on food crops (typically cotton pesticides and consumers health).

However, it is not an easy task, as agrochemical market and crop protection knowledge are increasingly controlled by few multinationals. Today, more than 80% of worldwide pesticide sales fall to the share of only six companies. Presently, efforts are being made to reduce pesticide applications worldwide by organic and integrated pest management (IPM) approaches. IPM system relies on biological, cultural, and other less chemically intensive approaches to pest management and are best options to minimize residues in several horticultural commodities requiring cooperation between farming community and pesticide dealers while providing high quality and pest-free produce in developing countries. IPM strategies need to be concentrated on improving the cultural practices and reducing pesticide use, substituting less costly ecofriendly pesticide, continuous monitoring of the pest menace, and finally, training the extension officers/farming community to make IPM a successful event.

However, in countries like India, this alternative pest management approach to reduce pesticide use could not find much success due to poor farmer participation. An all-India survey confirmed that 34% of the respondents have no idea about IPM and only less than 5% of them follow complete IPM technology.[53] Nowadays in many states of India, government authorities are promoting natural farming and urging the farming community to adopt it in order to discourage the use of chemical pesticides.

However, IPM techniques are still characterized by a large amount of pesticide use and by the application of many different pesticides, e.g. organic apple production does not use any herbicides and applies only biological control, but fungal diseases like apple scab demand for the intensive use of sulfur and copper in organic apple orchards and copper has a negative impact on the environment. IPM techniques have not been widely implemented on many crops, e.g., wheat. There is a great need to modify "good agricultural practice" and change it to "pesticide avoidance practice" and to improve the education of farmers so as to promote organic farming or natural farming and IPM as the best alternatives for pesticides.

References

1. Dhaliwal, G.S., and Arora, R. *Integrated Pest Management Concepts and Approaches*. Kalyani Publishers, New Delhi, 2001, 427pp.
2. Imhoff, D. *King Cotton—Pesticide Residue is Commonin Cotton Byproducts Used in Agriculture—Brief Article*. Sierra, 1999.
3. Pesticides Industry Sales and Usage: 2000 and 2001 Market Estimates; U.S. Environmental Protection Agency, May 2004, Table 3.4, available at www.epa.gov/oppbead1/pestsales/01pestsales/market_estimates2001.pdf (accessed March 2, 2019).
4. Anonymous. *Scientific Agriculture Prevents Mass Starvation*. Herxter. J. Agvet Div., Hoechst, Australia Inc., 1992.

5. Kennedy, I.R. Pesticides in Perspective: Balancing their benefits with the need for environmental protection and remediation of their residues in seeking agricultural produce free of pesticide residues. In Kennedy, I.R., Skerritt, J.H., Johnson, G.I., and Highley, E. (eds.) *Proceedings of an International Workshop held in Yogyakarta, Indonesia, 17–19 February, 1998*; The Australian Centre for International Research (ACIAR), 1998.

6. Tata Strategic Management Group: Next generation Indian agriculture-Role of crop protection solutions. A report on Indian Agrochemical Industry, 2016.

7. FAOSTAT. 2018, available at http//www.fao.org/faostat/en/#data/EP

8. Pesticides and the Agrochemical Industry. Pesticide use reduction in Germany Pesticide Action Network Germany Pesticide Action Network Europe, available at www.pan-germany.org (accessed February 27, 2019).

9. Dhaliwal, G.S., and Koul, O. *Quest for Pest Management from Green Revolution to Gene Revolution*. Kalyani Publishers, New Delhi, 2010.

10. Anonymous. *Ministry of Statistics and Programme Implementation*. Government of India, 2018. http://www.mospi.gov.in/ Website (assessed on May 15, 2020).

11. Indira, D.P. Pesticides in agriculture—A boon or a curse? A case study of Kerala. *Economic and Political Weekly*, 2010, xlv (26 and 27) EPW, 50–52.

12. Jayakrishnan, T. Health impacts of pesticides used in agriculture. In Devi, P.L. (ed.) *Pesticide Use and Environmental Health*, compendium of papers presented in the workshop, KAU/SANDEE; Kerala Agricultural University Publishers, 2006.

13. Johansen, C.A. Pesticides and pollinators. *Annual Review of Entomology,* 1977, 22: 177–192.

14. Hardin, M.R., Benrey, B., Coll, M., Lamp, W.O., Roderick, G.K., and Barbosa, P. Arthropod pest resurgence: An overview of potential mechanisms. *Crop Protection*, 1995, 14: 3–18.

15. Ruberson, J.R., Nemotom, H., and Hirose, Y. Pesticides and conservation of natural enemies in pest management. In Barbosa, P. (ed.) *Conservation Biological Control*. Academic Press, San Diego, CA, 1998, pp. 207–220.

16. Croft, B.A. *Arthropod Biological Control Agents and Pesticides*. John Wiley & Sons, New York, 1990.

17. De Clercq, P., De Cock, A., Tirry, L., Vinuela, E., and Degheele, D. Toxicity of Diflubezuron and Pyriproxyfen to the predatory bug *Podisus maculiventris. Entomologia Experimentalis et Applicata*, 1995, 74: 17–22.

18. Charleston, D.S., Kfir, R., Dicke, M., and Vet, L.E.M. Impact of botanical pesticides derived from *Melia azedarach* and *Azadirachta indica* on the biology of two parasitoid species of the Diamond back moth. *Biological Control*, 2005, 33: 131–142.

19. Ware, G.W., and Whitacre, D.M. *The Pesticide Book*. Meister Pro Information Resources, Willoughby, OH, 2005.

20. Cloyd, R.A. Managing insect and mite pests. In Nau, J. (ed.) *Ball Red Book* (Vol. 2, 18th Edition). Ball Publishing, West Chicago, IL, 2011, pp. 107–119.

21. Cloyd, R.A., and Bethke, J.A. Impact of neonicotinoid insecticides on natural enemies in greenhouse and interior environments. *Pest Management Science*, 2011, 67: 3–9.

22. Cloyd, R.A. Compatibility of insecticides with natural enemies to control pests of greenhouses and conservatories. *Journal of Entomological Science*, 2006, 41: 189–197.

23. Stapel, J.O., Cortesero, A.M., and Lewis, W.J. Disruptive sublethal effects of insecticides on biological control: Altered foraging ability and life span of a parasitoid after feeding on extra floral nectar of cotton treated with systemic insecticides. *Biological Control*, 2000, 17: 243–249.

24. Jacobs, R.J. Kouskolekas, C.A., and Gross, H.R. Jr. Responses of *Trichogramma pretiosum* (Hymenoptera: Trichogrammatidae) to residues of permethrin and endosulfan. *Environmental Entomology*, 1984, 13: 355–358.

25. Elzen, G.W., O'Brien, P.J., and Powell, J.E. Toxic and behavioral effects of selected insecticides on the *Heliothis* parasitoid *Micropliti scroceipes. Entomophaga*, 1989, 34: 87–94.

26. Elzen, G.W. Sublethal effects of pesticides on beneficial parasitoids. In Jepson, P.C. (ed.) *Pesticides and Non-Target Invertebrates*. Intercept, Wimborne, 1990, pp. 129–150.

27. Roger, C., Vincent, C., and Coderre, D. Mortality and predation efficiency of *Coleomegilla maculata lengi* Timb. (Col., Coccinellidae) following application of neem extracts (*Azadirachta indica* A. Juss., Meliaceae). *Journal of Applied Entomology*, 1995, 119: 439–443.

28. Desneux, N., Decourtye, A., and Delpuech, J.M. The sublethal effects of pesticides on beneficial arthropods. *Annual Review of Entomology*, 2007, 52: 81–106.

29. Moriarty, F. The Sublethal effects of synthetic insecticides on insects. *Biological Reviews*, 1969, 44: 321–357.

30. Wright, D.J., and Verkerk, R.H.J. Integration of chemical and biological control systems for arthropods: Evaluation in a multitrophic context. *Pesticide Science*, 1995, 44: 207–218.

31. Gerson, U., and Cohen, E. Resurgence of spider mites (Acari: Tetranychidae) induced by synthetic pyrethroids. *Experimental and Applied Acarology*, 1989, 6(1): 29–46.

32. Beers, E.H. Integrated Mite Control: Nontarget effects on predator and prey. *84th orchard Pest and Disease Management Conference*, 13–15 January, 2010, Portland Hilton, Portland, OR, 2010.

33. Huang, J.K., Ruifa, H., Fan, C., Pray, C.E., and Rozelle, S. Bt cotton benefits, costs and impacts in China. *AgBioForum*, 2002, 5(4): 153–166.

34. Pray, C.E., Huang, J., Hu, R., and Rozelle, S. Five years of Bt cotton in China—The benefits continue. *Plant Journal*, 2002, 31(4): 423–430.

35. Qaim, M., and Zilberman, D. Yield effects of genetically modified crops in developing countries. *Science*, 2003, 299: 900–902.

36. Qaim, M., and Dejanvry, A. Genetically modified crops, corporate pricing strategies, and farmers' adoption: The case of Bt cotton in Argentina. *American Journal of Agricultural Economics*, 2003, 85(4), 814–828.

37. Traxler, G., Godoy-Avila, S., Falck-Zepeda, J., and Espinoza-Arellano, J. Transgenic cotton in Mexico: A case study of the Comarca Lagunera. In Kalaitzandonakes, N. (ed.) *The Economic and Environmental Impacts of Agribiotech*. Kluwer, New York, 2003, pp. 183–202.

38. James, C. *Global Status of Commercialized Transgenic Crops: 2002*. ISAAA Briefs 27. Ithaca, NY, 2002.

39. Pimental, D. Area-wide pest management: Environmental, Economic and food issues. In Vreysen, M.J.B., Robinson, A.S., and Hendrichs, J. (eds.) *Area-wide Control of Insect Pests: From Research to Field Implementation*. Springer, Dordrecht, 2007, pp. 35–47.

40. Sheety, P. K., Murugan, M., and Sreeja, K. G. Crop protection stewardship in India: Wanted or Unwanted. *Current Sciences*, 2008, 95: 457–464.

41. Dixit, A. Herbicide recommendation in different crops. *Crop Care*, 2009, 35(2): 33–38.

42. Pispini, M., Schimpf, M., Lopez, J., and Chandrasekaran, K. Who Benefits from GM Crops? The Rise in Pesticide Use. Friends of the Earth International and Center for Food Safety, 2008, see especially pp. 8–12, available at www.centerforfoodsafety.org/WhoBenefitsPR2_13_08.cfm (accessed December 14, 2010).

43. Anonymous. 2011, available at www.beyondpesticides.org/pesticides/factsheets/Atrazine.pdf (accessed February 18, 2011).

44. Nikolova, S. Pesticide use, issues and how to promote sustainable agriculture in Bulgaria. Published by Pesticide Action Network Germany (PAN Germany) in co-operation with Association, Agrolink, 2004.

45. Rajendran, R.B., and Subramanian, A.N. Chlorinated pesticide residues in surface sediments from the river Kaveri, South India. *Journal of Environmental Science and Health*, 1999, 34: 269–288.

46. Karanth, N.G.K. Challenges of limiting pesticide residues in fresh/vegetables: The Indian experience. In Hanak, E.E., Boutrif, P., and Fabre, M.P. (eds.) *Food Safety Management in Developing Countries*. CIRAD-FAO, Montpellier, 2002, pp. 11–13.

47. Charan, P.D., Ali, S.F., Kachhawa, Y., and Sharma, K.C. Monitoring of pesticide residues in farm gate vegetables of central Aravalli region of Western India. American-Eurasian Journal *of* Agricultural & Environmental *Sciences*, 2010, 7: 255–258.

48. Ranga Rao, G.V., Sahrawat, K.L., Srinivasa, R.C., Binitha, D., Reddy, K.K., and Bharath, B.S. Insecticide residues in vegetable crops grown in Kothapalli Watershed, Andhra Pradesh, India: a case Study. *Indian Journal of Dryland Agricultural Research and Development*, 2009, 24: 21–27.

49. Gupta, P.K. Pesticide exposure. Indian Science. *Toxicology*, 2004, 198: 83–90.

50. Baker, B.P., Benbrook, C., Groth, E., III, and Benbrook, K.L. Pesticide residues in conventional, integrated pest management (IPM)—Grown and organic foods: Insights from three US data sets. *Food Additives & Contaminants*, 2002, 19: 427–446.

51. Bhandari, R. Pesticide residues in vegetables and fruits. *International Journal of Scientific Research in Chemical Sciences*, 2015, 2(1): 11–17.

52. Soudamini, M., and Rekha, A. Persistence of dicofol residues in/on acid lime. *Pesticide Research Journal*, 2005, 17(1): 64–65.

53. Shetty, P.K., Murugan, M., and Sreeja, K.G. Crop protection stewardship in India: Wanted or unwanted. *Current Science*, 2008, 95(4): 457.

11

Pollution: Pesticides in Natural Ecosystems

J.K. Dubey and
Meena Thakur

Introduction

Pesticides play an important role in boosting the economy of the agricultural industry by providing effective pest control, and their continued use is essential for enhancing the productivity.[1] It is estimated that food supplies would immediately fall to 30%–40% due to the ravages of pests if pesticides are not used.[2] A United Nations report stated that population growth is a major problem facing our planet. In 1900, there were 1.6 billion people on the planet. In 1992, this has risen to 5.25 billion, and by the year 2050, it will reach 10 billion. Developing countries are more affected by this explosive increase in world population. Presently, our dependence on pesticides has increased up to the extent that if modern agriculture was operated without chemical control, the crop production will probably decline in many areas, food price will soar far higher, and food shortage will become more severe. Although pesticides have played an important role in enhancing crop yields, they have also come up with various environmental problems. When present above permissible limits, they act as pollutants, creating pesticide pollution. Many pesticides are present today in different concentrations in various components of our environment such as air, water, and soil. More than 5,00,000 people are either killed or incapacitated every year by poisoning, and most of these casualties occur in developing nations.[3]

Ecologically, however, pesticides have created two major problems that were not previously anticipated. As pollutants, they contaminate numerous natural ecosystems [terrestrial: forest, grassland, desert, etc.; aquatic: fresh water (running water such as spring, stream, or rivers or standing water such as lake, pond, pools, puddles, ditch, and swamp); and marine (deep water bodies such as ocean or shallow ones such as a sea and estuary)] not intended to be targets. Second, most of them have directly/indirectly affected human health. The objective of this entry is to provide basic knowledge on pesticide exposure and to understand issues on residues in the natural ecosystem.

History of Pesticides and Pesticide Problems

The term *pesticide* covers a wide range of compounds including insecticides, fungicides, herbicides, rodenticides, molluscicides, nematicides, plant growth regulators, and others. In the 1940s, dichlorodiphenyltrichloethane (DDT) became the first widely available synthetic insecticide. It was highly effective, but it showed signs of becoming less effective as insects became resistant to it. It accumulated in the bodies of animals and high up the food chain by biomagnifications and bioconcentrations, causing problems with reproduction. Rachel Carson's *Silent Spring* in 1962 drew the attention of environmentalists to the disaster that was gathering pace across the globe. Public awareness of problems with pesticides grew by the 1970s when DDT was banned in many countries. It is still used in some places for malaria control, and it is still present in the bodies of many animals, even hundreds of miles away from where it has not been used. The introduction of other synthetic insecticides—organophosphate (OP) insecticides in the 1960s, carbamates in the 1970s, and pyrethroids in the 1980s, as well as herbicides and fungicides in 1970s–1980s, contributed to a great extent in pest control and agricultural output. The consequences of pesticide use have resulted in serious health implications to man and his environment. There is now overwhelming evidence that some of these chemicals pose potential risk to humans and other forms of life and unwanted side effects to the environment.[4] The worldwide deaths and chronic illnesses due to pesticide poisoning numbered about 1 million per year.[5]

The problem is more serious when pesticides that are banned are used indiscriminately. Banned pesticides are still used on crops that are not consumed directly, e.g., cotton. Few people think of cotton as food, but once the fiber is removed, two-thirds of the cotton crop winds up in the food we eat. Every year in the United States, half a million tons of cottonseed oil goes into processed salad dressings, baked goods, and snacks like Fritos and Goldfish. Another 3 million tons of cottonseed is fed to beef and dairy cattle, which also eat vast amounts of the cotton by-products known as "gin trash."[6]

How Do Pesticides Sprayed on Agro-/Horti-Ecosystems Enter Natural Ecosystems?

Almost less than 1% of the total pesticides applied actually hit the target organisms.[7] Most reach nontarget sectors of agro-ecosystems and/or spread to surrounding ecosystems as chemical pollutants. The pesticide somehow "leaks" into another ecosystem via movement of water from one body to another via outflow streams or seepage into the water table. Some pesticides might evaporate into the atmosphere and be carried elsewhere by winds. Regardless of how the leak occurs, the pesticide could affect accidental targets; e.g., a volatile insecticide used to control mosquitoes evaporates and kills bees; thus, a wide variety of plants do not get pollinated, thereby affecting their yield. The pesticide may also be taken in by migratory animals (birds in particular) and carried elsewhere; the toxin may affect the birds' reproduction in some way, or those birds might be eaten up by a higher order of predators and the toxin may inflict some injury to them. Either way, this would affect the balance of predation in some land-based ecosystem. The movement of systemic insecticides' active ingredient into floral parts may indirectly impact natural enemies that feed on plant pollen or nectar as a nutritional food source.[8,9]

Pesticide Pollution and Natural Ecosystems

Effect on the Soil Environment

Many pesticides contain chemicals that are persistent soil contaminants; their effects may last for years. Pesticides move with water in soil to groundwater and on soil to surface water. They decrease biodiversity in the soil by killing soil organisms; when life in the soil is killed off, the soil quality deteriorates and has a knock-on effect upon the retention of water. This is a problem for farmers particularly in times of drought.[10] At such times, organic farms have been found to have yields 20%–40% higher than conventional farms. Soil fertility is affected in other ways, too. When pesticides kill off most of the active soil organisms, the complex interactions that result in good fertility break down. Risk assessment of pesticide impact on human health is not an easy and particularly accurate process because of difference in the periods and the levels of exposure types of pesticides mixtures in the field and geographic and meteorological characteristics of the agricultural area of pesticide applied.[11,12] The data is usually collected from various tests conducted on metabolism patterns, acute toxicity, subchronic, chronic, carcinogenecity, genotoxicity, teratogenecity, and generation study using rat as a model mammal or dogs and rabbits.[13]

Application of systemic pesticides as drenches or granules may exhibit indirect effects on natural enemies via several mechanisms including elimination of floral parts by consumption of active ingredient while ingesting plant fluids and contamination of prey ingesting either lethal or sublethal concentration.[14–16]

Plants depend on millions of bacteria and fungi to bring nutrients to their rootlets. When these cycles are disrupted, plants become more dependent upon exact doses of chemical fertilizers at regular intervals. Even so, the incredibly rich interactions in healthy soil cannot be fully replicated by the farmer with chemicals. Hence, the soil and our nutritionist compromised. We get large but watery vegetables and fruits, which often lack natural taste and nutrients and may even contain harmful toxic pesticide residues. Studies of pesticide effects on the soil fauna have reported increased numbers of collembolan, because chemicals reduced populations of natural enemies, especially of predatory mites.[17]

Effect on the Aquatic Environment

Pesticides enter the freshwater ecological systems either from direct application of pesticides for the control of harmful aquatic fauna or as runoff from the treated areas, drift during aerial spraying, and industrial effluents from washing and spraying of equipment and containers. Several groups of organochlorine pesticides such as DDT, endosulfan, and chlorinated phenoxy acetic acid used as herbicides and fungicides such as hexachlorobenzene and pentachlorophenol are of interest in water pollution. Because of their solubility in water and tendency to be absorbed on solid surfaces, only traces of these chemicals are found in solid surfaces and treated water.[18]

Microorganisms form a vital part of the freshwater environment. Bishop[19] measured the effects of DDT on Mastigophora, Infusoria, and Sarcodina in ponds near Savannah, Georgia, and found little change in population numbers after treatment at relatively low rates. Hoffman and Olive[20] found that the growth of populations in Colorado lakes was inhibited after the addition of rotenone and toxaphene. Phytoplankton (beneficial/detrimental) can be seriously affected by agricultural chemicals. DDT sprays have caused serious reductions of bottom-dwelling invertebrates, the reductions in some cases amounting to 95% of the population. Malathion has also caused destruction of stream invertebrates. Cushing and Olive[21] found that toxaphene and rotenone reduced numbers of midge larvae in Colorado reservoirs and algae. On higher plants in the freshwater environment, adsorption of pesticides in/on the vegetation resulted in phytotoxicity, which either retarded the growth of or killed aquatic plants. Kolleru Lake is the largest natural freshwater body of Andhra Pradesh in India where agriculture and aquaculture are some of the primary activities at the lake basin. The increased use of pesticides in agriculture and aquaculture had a negative impact on the quality of water in the lake.[22,23]

A major environmental impact has been the widespread mortality of fish and marine invertebrates due to the contamination of aquatic systems by pesticides. Most of the fish in Europe's Rhine River were killed by the discharge of pesticides, and at one time, fish populations in the Great Lakes became very low due to pesticide contamination. In addition, many of the organisms that provide food for fish are extremely susceptible to pesticides, so the indirect effects of pesticides on the fish food supply may have an even greater effect on fish populations. Some pesticides, such as pyrethroids, are extremely toxic to most aquatic organisms. It is evident that pesticides cause major losses in global fish production.

Effect on the Terrestrial Environment

A wide variety of pesticides are applied on horticultural and agricultural crops. Some of them are highly specific and others are broad spectrum; both types can affect terrestrial wildlife, soil, water systems, and humans. The misuse of pesticides can cause valuable pollinators such as bees and hoverflies to be killed, and this in turn can badly affect food crops. Bees are extremely important in the pollination of crops and wild plants; about 33% of all crops require pollination. Although pesticides are screened for toxicity to bees, and the use of pesticides toxic to bees is permitted only under stringent conditions, many bees are killed by pesticides, resulting in the considerably reduced yield of crops dependent on bee pollination. Bee population has been suffering a serious decline in recent years. Without bees, many food crops would simply fail to grow.

Neonicotinoid pesticides developed in 1980 are under fire for risks to pollinators, and European Union Commission had already imposed this class of pesticides as researchers have become concerned about their potential to harmless birds, earthworms, aquatic insects, and pollinators.[24] Three neonicotinoids—thiamethoxam, clothianidin, and imidacloprid—posed an unacceptable risk to bees and were banned for use for some period on flowering crops such as corn, oilseed rape, and sunflower upon which the bees feed.[25]

It has been observed that through natural selection, some pests eventually become quite resistant to pesticides and farmers may need increasing amounts of pesticides, making the problem worse. Orchards are complex ecosystems easily perturbed by the extensive use of pesticides, and there are many instances of increased pest attacks in orchards after the use of pesticides, e.g., outbreaks of codling moth, leafrollers, aphids, scales, and tetranychid mites.[26] When pesticides were first used on tropical cotton crops, they controlled two or three important pests of the crops and greatly increased yields. Within a few seasons, however, the chemicals reduced the population of natural enemies and a number of other arthropod species became pests.[27]

Amphibians such as frogs are particularly vulnerable to concentrations of pesticides in their habitat. Atrazine, the most heavily used herbicide, is regularly detected in drinking water supplies in the Midwest, the United States, and exposure to extremely low levels of atrazine can cause sex change and/or deformities in frogs, fish, and other organisms.[28] Based on this evidence, and the widespread presence of atrazine in drinking water supplies, the European Union (EU) announced a ban on atrazine in 2006. The US Environmental Protection Agency re-registered atrazine in 2003 despite objections from scientists and environmental groups.

Pesticides have had some of their most striking effects on birds, particularly those in the higher trophic levels of food chains, such as bald eagles, hawks, and owls. These birds are often rare, endangered, and susceptible to pesticide residues such as those occurring from the bioconcentration of organochlorine insecticides through terrestrial food chains. Pesticides may kill grain- and plant-feeding birds, and the elimination of many rare species of ducks and geese has been reported. Populations of insect-eating birds such as partridges, grouse, and pheasants have decreased due to the loss of their insect food in agricultural fields through the use of insecticides. Pesticides can affect animal reproduction directly, as evident by the deleterious effect of the persistent organochlorine insecticides on reproduction in receptors and other birds. The US National Academy of Sciences stated that the DDT metabolite, dichlorodiphenyldichloroethane (DDE), causes eggshell thinning and that the bald eagle population in the United States declined primarily because of exposure to DDT and its metabolites.[29] Fish-eating birds are more severely affected than terrestrial predatory birds, because the former acquire more pesticides via their food chain.[30] Pesticides can also affect reproduction in invertebrates, e.g., sublethal doses of DDT, dieldrin, and parathion increased egg production of Colorado potato beetle after 2 weeks by 50%, 33%, and 65%, respectively.[31] Aquatic ecosystems with flowing water can usually recover their structure and function more quickly from pesticide effects than ponds with standing water.

Effects on Humans

There are a number of ways in which humans can be exposed to pesticides through the environmental route. Man's primary exposure to pesticides is probably via those used domestically in wood preservation or as household insecticides. Pesticides can endanger workers during production and transportation or during and after use. Bystanders may also be affected at times, e.g., walkers using public rights-of-way on adjacent land or families whose homes are close to crop-spraying activities. One of the main hazards of pesticide use is to farm workers and gardeners. A recent study by the Harvard School of Public Health in Boston discovered a 70% increase in the risk of developing Parkinson's disease for people exposed to even low levels of pesticides.[32]

The effects of pesticide residues in food and water probably cause a great public concern, although reports of clinical poisoning due to residues are extremely rare. Their residual population in food commodities is alarming. Leafy vegetables, cereals, fruits, rice, meat, milk, fish, and even human milk have been contaminated by various pesticides in a range of 0.1–25.7 mg/kg. The herbicide 2,4-D is identified as a carcinogen in humans and dogs. Acephate is a mutagen, carcinogen, fetotoxic, feminizes rats, and kills birds. In a multicountry study (Belgium, China, Federal Republic of Germany, India, Israel, Japan, Mexico, Sweden, the United States, and Yugoslavia) on the assessment of human exposure to selected organochlorine compounds, the residue levels for pp'-DDE and β-HCH were found to be higher in the human milk samples collected from developing countries such as China, India, and Mexico than in the participating developed countries. A higher level of these chemicals in mother's milk is a clear-cut reflection of their increased burden through their translocation passage.[33]

A number of bottles of wine were tested for pesticide residues, and 100% of conventional wines included in the analysis were found to contain pesticides, with one bottle containing ten different pesticides. On an average, each wine sample contained more than four pesticides. The analysis revealed 24 different pesticide contaminants, including five classified as being carcinogenic, mutagenic, neurotoxic, or endocrine disrupting by the EU. Human health is at risk when chemical residues are present in so much of our food supplies.

Children are particularly vulnerable to the toxic effects of pesticides. Studies have found higher rates of brain cancer, leukemia, and birth defects in children who suffered early exposure to pesticides. A survey of baby foods in 2000 showed detectable pesticide residues in nearly 50% of foods sampled. Fourteen percent of foods tested showed residues of more than one pesticide at levels 30 times the proposed limit of 0.01 mg/kg.[34] A UK government report in 2003 showed that more than 70%, 90%, 61%, 54%, and 35% of apples, lemons, bread, rice, and potatoes analyzed had pesticide residues, respectively. The main source of exposure to pesticides for most people is through diet. A study in 2006 measured organophosphorus levels in 23 school children before and after changing their diet to organic food. The levels of organophosphorus exposure dropped immediately and dramatically when the children began the organic diet.[34]

Over the last few decades, agricultural pesticides have become common household items in rural areas of the developing world. In 2014, the Natural Crime Records Bureau of India reported 5650 farmer suicides.[35] In India, the first report of poisoning due to pesticides was from Kerala in 1958, where more than 100 people died after consuming wheat flour contaminated with parathion.[36] This prompted the Special Committee on Harmful Effects of Pesticides constituted by the Indian Council of Agricultural Research to pay more attention on the problem.[37] Exposure to accidental emissions of methyl isocyanate from a pesticide factory in Bhopal, India, killed more than 5000 people, leaving more than 50,000 with permanent damage.

Alternatives for Pesticide Problems

The toxic effects of pesticides on our foods, land, and their effects on the health of human beings and their progeny make it an issue that is becoming more and more crucial. Integrated pest management (IPM; pest surveillance, use of crop varieties resistant to pest, sound cultural practices, biological control, and use of ecofriendly pesticides) emphasizes the need for simpler and ecologically safer measures for pest control to reduce environmental pollution and other problems caused by excessive and indiscriminate use of pesticides. Preference should be given to organic foods that are grown without toxic pesticides by organic methods. There are now many biological control tactics available where benign species are used to manage less benign ones. Ladybirds (ladybugs) are often introduced to control aphids (greenfly and others). Organic nontoxic sprays are used to stimulate the soil. They work by stimulating fungi in the soil that help to feed the plants and help them in developing resistance to disease and insect attack. There are also many successful barrier methods that help to deter insect attacks (the use of nets to ward off birds and larger insects). Companion planting is also used; garlic, for example, helps some plants resist insect attacks. There are a small number of organic pesticides that are legitimate to use in organic food production system. Some of these can be made at home using simple ingredients such as soap and alcohol.

It is evident that misuse, overuse, and abuse of pesticides lead to many environmental problems as discussed. Pesticides must be used as part of a planned systematic pest management program utilizing as many control techniques as applicable (IPM).

Novel pesticides' modes of action, improved safety profiles, and the implementation of alternative cropping systems which are less dependent on pesticides could minimize exposure to pesticides and undesirable exposure on human health. Moreover, the use of appropriate and well-maintained spraying equipment along with taking all the precautions required in all the stages of pesticide handling could also reduce exposure to pesticides. The overall pesticide handling according to regulations reducing the public concerns in food and drinking water minimize the effects of pesticides in human health and environment.[38]

Emphasis should be placed on using all the techniques of organic farming and supplementing these with the use of pesticides, i.e., using pesticides as part of an organic farming system. The IPM approach will help to minimize the effects of pesticide pollutants on the environment and natural ecosystems and will also help in economic and ecological sustainable food production.

References

1. Kent, J. Education and training in farm chemical management. *Proc. Conf. Agriculture, Education and Information Transfer*; Murrumbidgee College of Agriculture, Yanco, 1991.
2. Anonymous. *Scientific Agriculture Prevents Mass Starvation*; The Herxter. J. of the Agvet Division, Hoechst, Australia Inc., 1992.
3. Anonymous. Water pollution—Case study, 2010, www.environmentandpeople.org. All Rights Reserved © Environment and People. Web site designed and maintained by HS visual FX. (accessed October 8, 2010).
4. Igbedioh, S.O. Effects of agricultural pesticides on humans, animals and higher plants in developing countries. *Archives of Environmental Health* **1991**, *46*, 218.
5. Environews Forum. Killer environment. *Environmental Health Perspectives* **1999**, *107*, A62.
6. Imhoff, D. *King Cotton 1999—Pesticide Residue is Commonin Cotton Byproducts Used in Agriculture; Brief Article*; Sierra, May 1999.
7. Pimentel, D.; Levitan, L. Pesticides: Amounts applied and amounts reaching pests. *BioScience* **1986**, *36*, 86–91.
8. Kiman, Z.B.; Yeargan, K.V. Development and the reproduction of the predator *Orius insidious* (Hemiptera: Anthocoridae) reared on the diets of the selected plant material and arthropod prey. *Annals of the Entomological Society of America* **1985**, *78*, 464–467.

9. Hagen, K.S. Ecosystem analysis: Plant cultivars (HPR), Entomophagus species and food supplements. In: Boethal, D.J.; Eikenbary, R.D. (eds.) *Interactions of Plant Resistance and Parasitiods and Predators of Insects*; John Wiley & Sons, Inc., New York, 1986, pp. 151–197.

10. Roger, A.; Simpson, I.; Oficialc, R.; Ardales, S.; Jimen, R. Effects of pesticides on soil and water microflora and mesofauna in wetland ricefields: A summary of current knowledge and extrapolation to temperate environments. *Australian Journal of Experimental Agriculture* **1994**, *34*, 1057–1068.

11. Bolognesi, C. Genotoxicity of pesticides: A review of human biomonitoring studies. *Mutation Research* **2003**, *543*, 251–272.

12. Pastor, S.; Creus, A.; Parron, T.; Cebulska-Wasilewska, A.; Siffel, C.; Piperakis, S.; Marcos, R. Biomonitoring of four European populations occupationally exposed to pesticides: Use of micronuclei as biomarkers. *Mutagenesis* **2003**, *18*, 249–258.

13. Matthews, G.A. *Pesticides: Health, Safety and the Environmen*; Blackwell Publishing, Oxford, 2006.

14. Tillman, P.G.; Mullinix, B.G. Comparison of susceptibility of pest *Euschistusservus* and predator *podisus maculiventris* (Heteroptera: Pentatomidae) to selected Insecticides. *Journal of Economic Entomology* **2004**, *97*, 800–806.

15. Krischiik, V.A.; Landmark, A.L.; Heimpel, G.E. Soil applied imidacloprid is translocated to nectar and kills nectar feeding *Anagyrus pseudococci* (Girault) (Hymenoptera: Encyrtidae). *Environonmental Entomology* **2007**, *36*, 1238–1245.

16. Szczepaniec, A.; Creaky, S.F.; Laslowski, K.L.; Nyrop, J.P.; Raupp, M.J. Neonicotinoid insecticide imidacloprid causes outbreaks of spider mites on elm trees in urban landscapes. *PLoS ONE* **2011**, *6*(5), e2018.

17. Edwards, C.A.; Thompson, A.R. Pesticides and the soil fauna. *Residue Reviews* **1973**, *45*, 1–79.

18. Beitz, H.; Schmidt, H.; Herzel, F. Occurrence, toxicological and ecotoxicological significance of pesticides in ground-water and surface water. *Chemistry of Plant Protection* **1994**, *8*, 3–53.

19. Bishop, E.L. Effects of DDT mosquito larviciding on wildlife; the effects on the plankton population of routine larviciding with DDT. *Public Health Reports* **1947**, *62*(35), 1263–1268.

20. Hoffman, D.A.; Olive, J.R. The effects of rotenone and toxaphene upon plankton of two Colorado reservoirs. *Limnology and Oceanography* **1961**, *6*(2), 219–222.

21. Cushing, C.E. Jr.; Olive, J.R. Effects of toxaphene and rotenone upon the macroscopic bottom fauna of two northern Colorado reservoirs. *Transactions of the American fisheries Society* **1957**, *86*, 294–301.

22. Victor, L.L. Fleisher J.H. and B.J. Jandorf (ed.). Pesticides in sea water and the possibilities of their use in mariculture. In: *Research in Pesticides*; Chichester Academic Press, New York and London, 1965.

23. Edwards, C.A. U.S. Environmental Protection Agency Website. Pesticides, www.epa.gov/pesticides (accessed October 8, 2010).

24. Campwell, P.J. Declining European bee health: Banning the neonicotinoids is not the answer. *Outlooks on Pest Management* **2013**, *24*(2), 52–57.

25. Cressey, D. Europe debates risk to bees. *Nature* **2013**, *496* (7446), 408.

26. Brown, A.W.A. *Ecology of Pesticides*; John Wiley & Sons, New York, 1978.

27. ICAITI (Instituto Centro Americano de Investigacion y Technología Industrial). *An Environmental and Economic Study of the Consequences of Pesticide Use in Central American Cotton Production: Final Report*; Central American Research Institute for Industry (Guatemala), United Nations Environment Programme, Nairobi, Kenya, 1977.

28. Anonymous. 2011, www.beyondpesticides.org/pesticides/factsheets/Atrazine.pdf (accessed October 8, 2010).

29. Liroff, R.A. Balancing risks of DDT and malaria in the global POPs treaty. *Pesticide Safety News* **2000**, *4*, 3.

30. Pimentel, D. *Ecological Effects of Pesticides on Non-Tar-get Species*; Executive Office of the President, Office of Science and Technology, Washington, DC, 1971, 220 pp.

31. Abdallah, M.D. The effect of sublethal dosages of DDT, parathion, and dieldrin onoviposition of the Colorado potato beetle (*Leptinotarsa decemlineata* Say) (Coleoptera: Chrysomelidae). *Bulletin of the Entomological Society of Egypt, Economic Series* **1968**, *2*, 211–217.

32. Ascherio, A.; Chen, H.; Weisskopf, M.G.; O'Reilly, E.; McCullough, M.L.; Calle, E.E.; Schwarzschild, M.A.; Thun, M.J. Pesticide exposure and risk for Parkinson's disease. *Annals of Neurology* **2006**, *60*(2), 197–203.

33. UNEP/WHO. *Assessment of Human Exposure to Selected Organochlorine Compounds through Biological Monitoring*; Slorach, S.A.; Vaz, R., Eds.; Swedish National Food Administration, Uppsala, 1983, p. 49.

34. Anonymous. *Pesticide Problems are a Growing Concern*, 2011. Copyright © Greenfootsteps.com 2006–2010.

35. National Crime Reports Bureau, ADSI Report Annual-2014. Government of India, 2014, p. 242.

36. Karunakaran, C.O. The Kerala food poisoning. *Journal of Indian Medical Association* **1958**, *31*, 204.

37. ICAR. *Harmful Effects of Pesticides. Report of the Special Committee of ICAR*; Wadhwani, A.M.; Lall, I.J., Eds.; Indian Council of Agricultural Research, New Delhi, 1972, p. 44.

38. Damalas, C.A.; Eleftherohorinos, I.G. Pesticide exposure, safety and risk assessment indicators. *International Journal of Environment Research and Public Health* **2011**, *8*, 1402–1419.

12

Polychlorinated Biphenyls (PCBs)

Marek Biziuk and
Angelika Beyer

Introduction

There is growing concern about the trace quantities of highly chlorinated organic compounds (e.g., dioxins, PCBs, and certain pesticides) that exist in diverse environmental media (air, water, soils, sediments, and biota), enter the trophic chain, and reach humans and wildlife. Consequently, there is growing scientific, regulatory, and social interest in measuring the levels of chlorinated chemicals in environmental media, and in determining the environmental effects of such contamination.

Representative of these synthetic organic chlorine compounds are polychlorinated biphenyls commonly known as PCBs. PCBs are man-made chemicals that never existed in nature until the 1900s when they started to be released into the environment by manufacturing companies and consumers. Although production of PCBs was banned, when their ability to accumulate in the environment and to cause harmful effects became apparent (in 1970, Sweden; in 1972, Japan; in 1977, the United States), these chemicals still are found in the environment.[1]

PCBs make up a group of 209 individual chlorinated biphenyl rings—congeners. They were typically manufactured as mixtures of 60 to 90 different congeners and were usually contaminated with small amounts of very toxic chemicals such as polychlorinated dibenzofurans (furans) or polychlorinated dibenzodioxins (dioxins). The trade names of some commercial PCB mixtures are Aroclor (United States), Clophen (Germany), Fenclor (Italy), Kanechlor (Japan), and Phenoclor (France).[2]

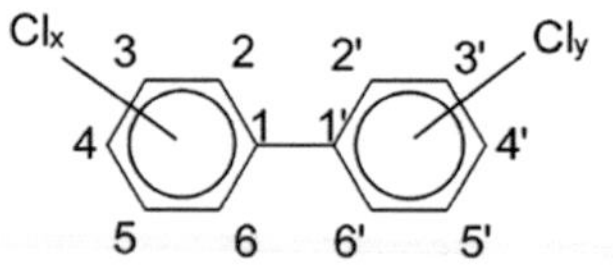

FIGURE 1 Structure of PCBs.

Chemical Identity

PCBs are a mixture of individual chemicals. The general chemical structure of chlorinated biphenyl (two benzene rings with a carbon–carbon bond between carbon 1 on one ring and carbon 1' on the second ring) with a varying number of chlorines is shown in Figure 1. It can be seen from the structure that a large number of chlorinated compounds are possible (209 possible congeners) in which 2–10 chlorine atoms are attached to the biphenyl molecule. Chlorines can be attached to any of the carbons by removing the hydrogen from that carbon and substituting the chlorine in its place. The common nomenclature used for identifying the location of chlorine atoms on the biphenyl rings is as shown in Figure 1.

The congeners are arranged in ascending numerical order using a numbering system developed by Ballschmiter and Zell[3] that follows the International Union of Pure and Applied Chemistry (IUPAC) rules of substituent characterization in biphenyls. The resulting PCB numbers, also referred to as congener, IUPAC, or BZ numbers, are widely used for identifying individual congeners.

Properties of PCBs

Because of physical and chemical properties of PCBs (summarized in Figure 2[4–6]), these chemicals were quickly acclaimed as an industrial breakthrough. PCBs are either oily liquids or solids that are colorless to light yellow without known smell or taste. In general, PCBs are relatively insoluble in water (solubility decreases with increased chlorination) but freely soluble in nonpolar organic solvents and biological lipids.[7] Because of their thermal stability, they do not easily burn, hence their past popular use as coolants, as insulating materials, and for electrical applications. The properties of PCBs vary from one congener to the next, e.g., color of PCB mixture darkens, viscosity increases, the flash point rises, and the substance becomes less combustible with rising chlorine content. Also, as the number of chlorines in a PCB mixture increases, the mixture is more stable and thus resistant to biodegradation. The congeners with large numbers of chlorines are also proving to be the ones that present the greatest environmental and health risks.

The properties that make PCB mixtures so desirable and applicable in industry (general inertness, thermal stability) are the ones that make the mixtures so hazardous to the environment. The toxicity of a PCB congener is dependent on the number of chlorines present on the biphenyl structures and the positions of the chlorines. The congeners in which there is a coplanar confirmation with chlorine substituents on the *meta* and *para* positions of the phenyl rings are the most toxic and bioaccumulative ones. For instance, congeners with chlorines in both *para* positions (4 and 4') and at least 2 chlorines at the *meta* positions (3, 5, 3', 5') are considered to be "dioxin like" and are particularly toxic.[8]

The high thermal and chemical resistance of PCB congeners means that they do not readily break down when exposed to heat or chemical treatment. However, since PCBs do not break down, they remain in the environment. Due to their persistence in the environment and the fact that they are poorly biodegraded, PCBs accumulate in the environment.

Physicochemical properties of PCBs

- A low degree of reactivity (very stable even when exposed to heat and pressure).
- Good insulating properties.
- Nonflammable.
- Good solubility in nonpolar solvents, oils, and fats.
- Virtual insolubility in water.
- Low vapor pressure (nonvolatile).
- Low electrical conductivity.
- High thermal conductivity.
- High ignition temperature.
- Very high resistance to chemical factors—do not undergo oxidation, reduction, addition, elimination or electrophilic substitution reactions except under extreme conditions.

FIGURE 2   Physicochemical properties of PCBs.

Global Distribution and Sources

PCBs were first produced in 1929 for a wide variety of uses because of their unique physical properties that made them attractive compounds for industries (see Table 1[9–12]). As more uses were found for PCBs, their production increased exponentially. In Table 1, there are identified PCB use areas based on their presence in closed, partially closed, and open systems. These designations refer to how easily the PCBs contained within a product can escape to the surrounding environment. In closed applications, PCBs are held completely within the equipment. Under ordinary circumstances, no PCBs would be available for exposure to the user or the environment. However, PCB emissions may occur during equipment servicing/repairing and decommissioning, or as a result of damaged equipment. Partially closed PCB applications, in which the PCB oil is not directly exposed to the environment, but may become so periodically during typical use, lead also to PCB emissions, through air or water discharge, whereas in open systems, PCBs are in direct contact with the environment and thereby may be easily transferred to the environment. Generally, closed and partially closed systems contain PCB oils or fluids. The PCBs in open systems take on the form (type of media) of the product they have been used in as an ingredient. Therefore, PCBs in open applications may be found in forms ranging from paint to plastic or rubber.[9]

The first indication that PCBs may be damaging to human health occurred four decades after PCBs were first introduced into the environment. Preliminary studies suggested that PCB s may pose a serious health threat to humans, and at the same time, there were indications of widespread distribution and longevity throughout the environment. As more attention was turned towards PCBs, it became clearer that PCBs were having a negative impact on many biological systems. Eventually, all production and importation of PCBs was banned in the 1970s.[13] Today, the production of PCBs has been ceased in many countries with the exception of small quantities manufactured strictly for research purposes. However, the ecotoxicological problems created by PCB contamination will be evident for many years to come, despite the restrictions of PCB utilization.

TABLE 1 Examples of Applications of PCBs

Closed Applications	Partially Closed Applications	Open Applications

• Electrical transformers
• Electrical capacitors
  • Power factor capacitors in electrical distribution systems
  • Lighting ballasts
  • Motor start capacitors in refrigerators, heating systems, air conditioners, hair dryers, water well motors
  • Capacitors in electronic equipment including television sets and microwave ovens
• Electrical motors in some specialized fluid-cooled motors
• Electrical magnets in some fluid-cooled separating magnets
	• Heat transfer fluids
	• Hydraulic fluids
	• Vacuum pumps
	• Switches
	• Voltage regulators
	• Liquid-filled electrical cables
	• Liquid-filled circuit breakers
		• Lubricants
		  • Immersion oil for microscopes (mounting media)
		  • Brake linings
		  • Cutting oils
		  • Lubricating oils
		• Casting waxes
		  • Pattern waxes for investment castings
		• Surface coatings
		  • Paints
		  • Surface treatment for textiles
		  • Carbonless copy paper
		  • Flame retardants
		  • Dust control
		• Adhesives
		  • Special adhesives
		  • Adhesives for waterproof wall coatings
		• Plasticizers
		  • Gasket sealers
		  • Filling material in joints of concrete
		  • Polyvinyl chloride plastics
		  • Rubber seals
		• Inks
		  • Dyes
		  • Printing inks
		• Other uses
		  • Insulting materials
		  • Pesticides

Although the manufacture, processing, distribution, and use of PCBs are widely prohibited, they have been released to the environment solely by human activity and still are redistributed from one environmental compartment to another.[1] There still exist a lot of different activities that generate PCB wastes. PCBs entered air, water, and soil during their manufacture, use, and disposal, mainly by leakage of supposedly closed systems, from landfill sites, incineration of waste, agricultural lands, industrial discharges, and sewage effluents. For more details, see Table 2.[14–23]

Nowadays, PCBs are present in all environmental media because of global circulation. The most important mechanism for global dispersion of these contaminations is atmospheric transport, which depends on the number of chlorines present on the biphenyl molecule:

- Biphenyls with one chlorine atom remain in the atmosphere.
- Those with one to four chlorines gradually migrate toward polar latitudes in a series of volatilization/de- position cycles between the air and the water and/or soil.
- Those with four to eight chlorines remain in midlatitudes.
- Those with eight to nine chlorines remain close to the source of contamination.[24]

There are two classic approaches to model the distribution of PCBs (and other persistent organic pollutants [POPs]).[25] Multicompartment models use just limited meteorological data but include detailed descriptions of the partitioning of the species within and between the different environmental media,[26] while chemistry transport models have a detailed treatment of transport and chemistry in

TABLE 2   Sources and Transport of PCBs within the Environment

Part of the Environment	Sources of PCBs in the Environmental System	Factors Influencing the Pattern and Rates of PCB Movement in the Media
Aquatic system— higher concentrations in the sediments of aquatic systems	• Accidental spills of PCB-containing hydraulic fluids • Improper disposal • Combined sewer overflows, or storm water runoff • Runoff and leaching from PCB-contaminated sewage sludge applied to farmland	• Properties of PCB congeners—desorption of PCBs from particulate is more likely to occur from lower-chlorinated, more water-soluble PCB congeners • Sorption reactions—with the chlorine content of PCB congener, surface area increases; with the organic content of the sediment, sorption increases • Sudden hydrographic activity like flooding or dredging causes sediments to be resuspended and redistributed and can cause the release of PCBs from sediments to overlying waters
Air system	• Volatilization from soil and water • Escape from uncontrolled landfills and hazardous waste sites • Incineration of PCB-containing wastes • Leakage from older electrical equipment • Improper waste disposal or spills • Leakage from supposedly closed systems • Incineration of waste • Industrial discharges • Sewage effluents	• Air temperature • Wind speed • Storm frequency • Rainfall rates • Volatility of individual PCB congeners
Soil system	• Accidental leaks and spills • Release from contaminated soils in landfills and hazardous waste sites • Deposition of vehicular emissions near roadway soil • Land application of sewage sludges containing PCBs	• Sorption reactions—i.e., highly chlorinated congeners are sorbed by soils and remain significantly immobile against leaching • Vapor phase transport—PCB congeners have a moderate vapor pressure, so vapor phase transport may allow for redistribution or migration through the saturated soil pores

the atmosphere but a rather simple description of the compartments other than the atmosphere.[27,28] The multicompartment models have been successful in describing the global distribution of POPs and their long-term environmental fate in the various compartments.[29,30] Unfortunately, not many direct comparisons between model results and measurement data have been made so far.[31]

Environmental Fate

PCBs, as it was indicated before, can partition between environmental media such as atmosphere, oceans, rivers, or soils. Differences in partitioning behavior among PCBs reflect differences in their physicochemical properties and persistence in the various media.[32]

PCBs do not readily break down in the environment and thus may remain there for a very long time. PCBs can travel long distances in the air and be deposited in areas far away from where they were released. In water, a small amount of PCBs may remain dissolved, but most stick to organic particles and bottom sediments. These toxic compounds can also bind strongly to soil. The degradation and transformation of PCBs entail difficult mechanisms of chemical, biochemical, or thermal destruction.[33] These substance may be (and are) accumulated through the trophic chain and reach aquatic organisms, fish, and humans. Consequently, PCBs accumulate in fish and marine mammals, reaching levels that may be many thousands of times higher than in water. That is why there is great interest in different pathways for PCB loss, such as volatilization, adsorption on organic matter, and biodegradation, which can reduce PCB bioavailability.

Volatilization

PCBs enter the atmosphere from volatilization from both soil and water surfaces.[34] It was reported over 35 years ago by Haque,[35] who found minimal PCB loss at ambient temperature. Heat, airflow (hood storage), coarse grain size, high water content, and enrichment in lower *ortho*-chlorinated congeners all were expected to increase the rate and extent of PCB volatilization. As indicated by their higher vapor pressures, the lower-chlorinated homologs in particular are subject to volatilization.[36] This can result in both a loss and a source of lighter homologs—a source because upon volatilization, the atmosphere is enriched with these homologs, which are then subject to atmospheric deposition. Once in the atmosphere, PCBs are both present in the vapor phase and sorbed to particles. PCBs in the vapor phase appear to be more mobile and are transported further than particle-bound PCBs.[24]

Because of their persistence and semi-volatility, PCBs have a great potential for long-range atmospheric transport, which enables them to migrate from the mid-latitudes to the Arctic regions, for instance.[31,37] Atmospheric transport may occur in many mechanisms, one of them is a mechanism known as cold condensation,[37] by which some PCBs are preferably removed from the atmosphere in cold regions and by which they can reach surprisingly high concentrations in the Arctic environment where they can bioaccumulate in animals and humans. Less volatile compounds that sorb strongly to atmospheric particles or that dissolve easily in rain droplets tend to have a more limited potential for long-range atmospheric transport, whereas semivolatile species can be transported over long distances in one or more steps towards the Arctic region.[37,38]

Both wet deposition and dry deposition remove PCBs from the atmosphere.[39]

Adsorption and Desorption on Organic Matter and Bioavailability

Once released into the environment, PCBs adsorb strongly to soil and sediment. As a result, these compounds tend to persist in the environment, with half-lives for most congeners ranging from months to years. Over time, contaminated sediments can be a source of hydrophobic organic contaminants (such as PCBs) and a significant health risk to aquatic food webs.[40] Leaching of PCBs from sediment and soil is slow, particularly for the more highly chlorinated congeners, and translocation to plants via soil is insignificant. Cycling of PCBs through the environment involves volatilization from land and water surfaces into the atmosphere, with subsequent removal from the atmosphere by wet or dry deposition, then revolatilization.[41]

Sorption properties of PCBs play a significant role in their mobility, ultimate fate in the sediments, and availability for degradation. The literature suggests that PCBs preferentially adsorb onto organic matter over adsorbing onto clay. Moreover, PCBs can sorb to dissolved organic matter (DOM) or particulate organic matter (POM). When associated with DOM, PCB contaminants are unavailable for uptake by organisms and, hence, become less bioavailable. In contrast, although PCBs sorbed to POM prevents or constrains direct uptake of PCBs, these contaminants are still available to the detrital food web, which is an important pathway in rivers. Planar PCBs bind strongly to POM and are less bioavailable.[42] Highly chlorinated homologs sorb strongly to POM and are not assimilated easily by detritus feeders.[43]

Bioavailability of sedimentary PCBs is traditionally assessed by measuring PCB uptake into benthic organisms over a standard exposure time. More recently, passive samplers have been used experimentally to estimate bio-availability.[44–46]

It is known that a combination of binding processes (sorption) and mass-transport processes (diffusion) is responsible for the partitioning of PCBs between aqueous and solid phases, and for their transport between these phases. These processes are also directly involved in and affect the environmental fate of PCBs. Precise quantitative predictions of phase speciation may allow an a priori estimate of the directly bioavailable, dissolved fractions of pollutants, as well as their tendency for long-term dispersion in the environment. Such predictions are critical in assessing the environmental risk from PCB contamination.[47]

The data generated during bench-scale adsorption studies and molecular-level study of the mechanism of adsorption of PCBs on substrates in the environment can help in effective desorption and the destruction of the persistent PCBs.[48]

Biodegradation and Transformation

The environmental persistence of PCBs results primarily from the inability of natural aquatic and soil biota to metabolize and/or degrade the compound at a significant rate. Studies on the biodegradation (degradation by bacteria or other microorganisms) of PCBs show that there are two biologically mediated processes for the degradation of PCBs: anaerobic and aerobic.[49] Microorganisms participate in the biodegradation by producing enzymes, which modify the organic pollutant into simpler compounds in such a way that the negative effects may be minimized. Biodegradation is of two forms:

- Mineralization—competent organisms use the organic pollutant as a source of carbon and energy resulting in the reduction of the pollutant to its constituent elements.
- Cometabolism—it requires a second substance as a source of carbon and energy for the microorganisms, but the target pollutant is transformed at the same time.[50,51]

If the products of cometabolism are amenable to further degradation, they can be mineralized; otherwise, incomplete degradation occurs. This may result in the formation and accumulation of metabolites that are more toxic than the parent molecule requiring a consortium of microorganisms, which can utilize the new substance as source of nutrients.[49]

The effectiveness of biodegradation depends on many factors, which are summarized in Table 3.[51-53]

Biodegradation is the only process known to degrade PCBs in soil or aquatic systems. Theoretically, the biological degradation of PCBs should result in CO_2, chlorine, and water. This process involves the removal of chlorine from the biphenyl ring (anaerobic reductive dechlorination) followed by cleavage and oxidation of the resulting compound (aerobic oxidative degradation).[54] The anaerobic process removes chlorine atoms of highly chlorinated PCBs, those with five or more chlorine atoms, which are then mineralized under aerobic condition.[49]

Under anaerobic condition, reductive dechlorination of PCBs occurs in soils and sediments. Different microorganisms with distinct dehalogenating enzymes, each exhibiting a unique pattern of congener selectivity resulting in various patterns of PCB dechlorination, exist in PCB- contaminated sites,[55] including the following isolated bacteria: *Desulfomonile tiedjei*,[56] *Desulfitobacterium, Dehalobacter restrictus, Dehalospirillum multivorans, Desulforomonas chloroethenica, Dehalococcoides ethenogenes*, and the facultative anaerobes *Enterobacter* strain MS-1 and *Enterobacter agglomeram*.[57] The rate, extent, and route of dechlorination is dependent on the composition of the active microbial community, which in turn are influenced by environmental factors such as availability of carbon sources, hydrogen or other electron donors, the presence or absence of electron acceptors other than PCBs, temperature, and pH.[58] However, a similarity between degradation patterns exists. The position of chlorine atoms on the rings affects the rate of biodegradation. Not only are PCBs with *para*- and *meta*-substituted rings more easily degraded than the *ortho*-substituted compounds, as shown in Figure 3,[59] but PCBs containing all chlorines on one ring are biodegraded faster than those that contain chlorines throughout both rings.

Persistence of PCBs in the environment increases with the degree of chlorination of the congener, i.e., compounds with a high degree of chlorination are resistant to biodegradation and degrade slowly in the environment. Anaerobic PCB dechlorination reduces the potential risk and potential exposure to PCBs because it significantly reduces the bioconcentration potential of the PCB mixture through conversion to

TABLE 3   Environmental Factors That Affect the Biodegradation of PCBs

Factor	How It Affects Biodegradation
Structure of the compound, i.e., the presence of substituents and their position in the molecule	• A high degree of halogenation requires high energy from the microorganisms to break the stable carbon-halogen bonds • Chlorine as the substituent alters the resonant properties of the aromatic substance as well as the electron density of specific sites; it may result in the deactivation of the primary oxidation of the compound by microorganisms • The positions occupied by substituted chlorines have stereochemical effects on the affinity between enzymes and their substrate molecule
Solubility of the compound	• Microorganisms easily access compounds with high aqueous solubility • Highly chlorinated congeners that are very insoluble in water are also very resistant to biodegradation
Concentration of the pollutant	• At a low concentration range, degradation increases linearly with increase in concentration until such time that the rate essentially becomes constant regardless of further increase in pollutant concentration • In general, a low pollutant concentration may be insufficient for the induction of degradative enzymes or to sustain growth of competent organisms • A very high concentration may render the compound toxic to the organisms
Temperature pH Presence of toxic or inhibitory substance and competing substances Availability of suitable electron acceptors Interactions among microorganisms	• The conditions should be optimal for the microorganism

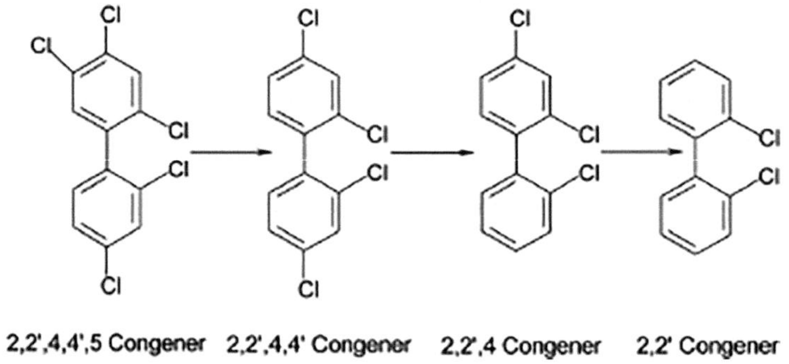

2,2',4,4',5 Congener   2,2',4,4' Congener   2,2',4 Congener   2,2' Congener

FIGURE 3   A potential pathway for anaerobic degradation of highly chlorinated PCB congeners to less chlorinated ones.

congeners that do not significantly bioaccumulate in the trophic chain.[60] Moreover, lightly chlorinated congeners produced by dechlorination can be readily degraded by indigenous bacteria.[61,62]

Aerobic biodegradation involving biphenyl ring cleavage is restricted to the lightly chlorinated PCB congeners, those with four or less chlorine atoms, resulting from the dechlorination of highly chlorinated congeners.[63,64] Aerobic oxidative destruction involves two clusters of genes. The first one is responsible for the transformation of PCB congeners to chlorobenzoic acid, and the second cluster is responsible for the degradation of the chlorobenzoic acid. A common growth substrate for

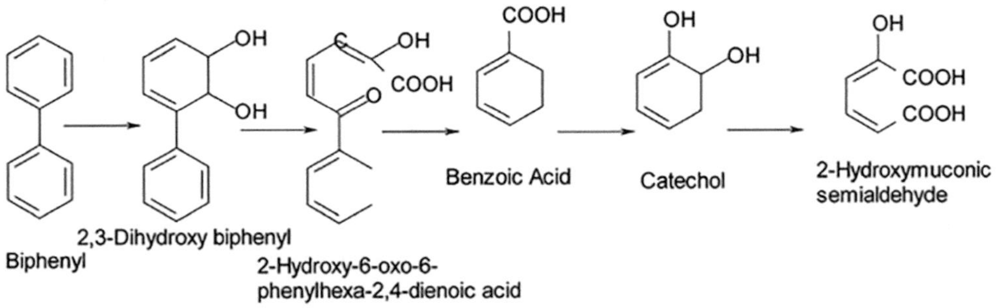

FIGURE 4   A possible pathway for the aerobic oxidative dehalogenation of PCBs.

PCB-degrading bacteria is biphenyl or monochlorobiphenyl. When biphenyl is utilized by bacteria, *meta-ring* cleavage product is produced. This has been observed in most bacteria studied especially in *Pseudomonas* sp.,[54] as well as in *Micrococcus* sp.[65] The metabolic pathway used by this family of bacteria is illustrated in Figure 4.[65]

Both anaerobic and aerobic metabolism modes transform PCBs. Different microorganisms show preferential attack on the PCB molecule, resulting in different patterns of biodegradation. The degree of chlorination of the congener and environmental factors influence the degradation potential of the compound.[49] Higher-chlorinated biphenyls therefore are potentially fully biodegradable in a sequence of anaerobic reductive dechlorination followed by aerobic mineralization of the lower-chlorinated products.[66]

Bioaccumulation, Bioconcentration, and Biomagnification

Organisms can accumulate high concentrations of PCBs relative to concentrations of these substances in non-biotic portions of the environment. This phenomenon is variously referred to as bioconcentration, bioaccumulation, and biomagnification.[67] Because some confusion exists in the literature about these definitions, we try to explain these more precisely following the terms set out by Gobas and Morrison.[68]

Bioaccumulation is a selective process that causes an increased chemical concentration in an organism compared to that in the surrounding medium and results from uptake by all exposure routes including transport across respiratory surfaces, dermal absorption (bioconcentration), and dietary absorption (biomagnification). Bioaccumulation can thus be viewed as a combination of bioconcentration and biomagnification.

The bioaccumulation factor (BAF) in fish is the ratio of the concentration of the chemical in the organism C_B to that in water, similar to that of the bioconcentration factor (BCF).

$$BAF = C_B/C_{WT} \text{ or } C_B/C_{WD},$$

where BAF is the bioaccumulation factor and $C_B/C_{Wt(WD)}$ is the concentration of the chemical in the organism/in water.

The most common approach for evaluating levels of bioaccumulation is to compare the levels retained by the organism with levels in the contaminated medium in which they live.[69]

Bioconcentration results from uptake of chemicals from water (usually under laboratory conditions). Uptake occurs via the respiratory surface and/or skin and results in the chemical concentration in an organism being greater than that in the surrounding medium.

BCF is defined as the ratio of the chemical concentration in an organism, C_B, to the total chemical concentration in the water, C_{WT}, or to the freely dissolved chemical concentration in water, C_{WD} (it only takes into account the fraction of the chemical in the water that is biologically available for uptake). The BCF is expressed as follows:

$$\text{BCF} = C_B / C_{WT} \text{ or } C_B / C_{WD},$$

where BCF is the bioconcentration factor and C_B/C_{WT}(WD) is the concentration of the chemical in the organism/in water.

Although sometimes applied to other aquatic species, the principal target organism for BCF assessment tends to be fish, primarily because of their importance as food for many species, including humans.

Biomagnification, on the other hand, is the bioaccumulation of a substance up the trophic chain when residues are transferred from consumption of smaller organisms by larger ones in the chain. It generally refers to the sequence of processes that produces higher concentrations in organisms at higher levels in the food chain (at higher trophic levels). These processes always results in an organism having higher concentrations of a substance than is present in the organism's food. Biomagnification also results in higher concentrations of the substance than would be expected if water were the only exposure mechanism.[70] A biomagnification factor (BMF) can be defined as the ratio of the concentration of chemical in the organism (C_B) to that in the organism's diet (C_A), and can be expressed as

$$\text{BMF} = C_B / C_A,$$

where BMF is the biomagnification factor, C_B is the concentration of chemical in the organism, and C_A is the concentration of chemical in the organism's diet.

This is the simplest definition of a BMF. It can also be described as the ratio of the observed lipid-normalized BCF to K_{ow}, which is the theoretical lipid-normalized BCF. This is equivalent to the multiplication factor above the equilibrium concentration. If this ratio is equal to or less than one, then the compound has not been biomagnified. If the ratio exceeds one, then the chemical is biomagnified by that factor.

The mechanism of biomagnification is not completely understood. Achieving a concentration of a chemical greater than its equilibrium value indicates that the elimination rate is slower than for chemicals that reach equilibrium. Transfer efficiencies of the chemical would affect the relative ratio of uptake and elimination. There are many factors that control the uptake and elimination of a chemical after contaminated food is consumed; these include factors specific to the chemical (solubility, K_{ow}, molecular weight and volume, and diffusion rates between organism gut, blood, and lipid pools), as well as factors specific to the organism (the feeding rate, diet preferences, assimilation rate into the gut, rate of chemical's metabolism, rate of egestion, and rate of organism growth). Because humans occupy a very high trophic level, we are particularly vulnerable to adverse health effects from exposure to chemicals that biomagnify.[71]

Chemicals that bioaccumulate do not necessarily biomagnify, although many papers report that PCB congeners do in fact biomagnify.[72,73] Some early bioaccumulation models used the concept of a food-chain multiplier, which is now considered excessively simplistic.[74] Exposure of PCBs solely from one source only occurs in laboratory experiments.[75] In nature, organisms are always exposed to different sources of contaminants, and therefore, what happens in the field is more complex than reflected in laboratory studies and cannot easily be emulated by laboratory studies. Mass balance models are simple tools that allow evaluation of various uptake and loss processes.[67] A variety of mass balance models have been developed to address water quality issues in lakes, estuaries, and slow-flowing water bodies.[76] The simpler models only consider advection and an overall loss due to the combined processes of volatilization, net transfer to sediment, and degradation. The rate constant for the overall loss is derived from fugacity calculations for a single segment system. The more rigorous models perform fugacity calculations for each segment and explicitly include the processes of advection, evaporation, water–sediment exchange, and degradation in both water and sediment. In this way, chemical exposure in all compartments (including equilibrium concentrations in biota) can be estimated.[77,78] In general, these models consider the organism to be a single "box."[67,79] These models require information about the chemicals, the organism, and associated environmental parameters.[79]

PCB congeners with less *ortho*-substitution are accumulated up the trophic chain at a greater rate than other congeners in their homolog group.[74] Non-*ortho*- substituted congeners, especially those that lack adjacent unsubstituted *meta* and *para* sites and unsubstituted *ortho* and *meta* sites, are undoubtedly metabolically recalcitrant in invertebrate and vertebrate tissues.[80] Changes in distributions of congeners are mainly caused by transfers among biotic compartments. There is no enrichment in higher trophic levels of mono- and non-*ortho*-substituted congeners. However, many coplanar congeners, especially very toxic PCB 77, are depleted with increasing trophic levels; PCB 77 is therefore almost certainly metabolized.[74]

Exposure of PCBs solely from one source only occurs in laboratory experiments. In nature, there are always multiple sources of contaminants, and therefore, field results must be studied carefully. Moreover, the properties of individual PCB congeners substantially affect accumulation or degradation pathways. Empirical models only reflect one of several possible mechanisms.[81]

Health Effects

As PCBs persist in the environment, the general population is potentially exposed to a variety of PCBs via food (especially fish caught in contaminated lakes or rivers, meat, and dairy products), air, surface soils, drinking water, and groundwater. In the workplace, people might be exposed to PCBs during repair and maintenance of PCB transformers and other old electrical devices, and disposal of PCB materials. Although mixtures used in industry are not identical to the combinations of PCBs present in the environment (or in breast milk), these mixtures have been found to have similar harmful effects.

The health effects of PCBs have been very widely studied in people (studies of industrial workers exposed to PCB-containing mixtures in the course of their work, as well as studies of adults and children exposed to PCBs as a result of consuming contaminated fish), laboratory animals, and wildlife in contaminated areas. These studies indicate that people who are regularly exposed to PCBs are at greater risk for a variety of health problems.[82] Moreover, evidence on the health effects of exposure to PCBs has been obtained from two episodes of mass poisoning that occurred in Japan (the 1968 Yusho incident) and Taiwan (the 1979 Yu-Cheng incident). Some of the most important findings are summarized below.

Research shows that PCBs cause a variety of adverse health effects depending on the route of exposure, age, sex, and area of the body where PCBs are concentrated. Studies on animals show conclusive evidence that PCBs are carcinogenic. Animals that ate food containing large amount of PCBs for short periods of time had mild liver damage and some died. PCBs have also been implicated as a cause of mass mortalities in seabirds.[49]

Moreover, a number of epidemiological studies of workers exposed to PCBs have been performed. The Department of Health and Human Services (DHHS) has concluded that PCBs may reasonably be anticipated to be carcinogens. Also, the Environmental Protection Agency (EPA) and the International Agency for Research on Cancer (IARC) have determined that PCBs are probably carcinogenic to humans.[83,84] Research also shows that exposure to PCBs in high concentration can have various acute effects including a skin disease known as chloracne (skin lesions), liver damage, other non-cancer short-term effects like body weight loss, impaired immune function, and clinically diagnosable damage to the central nervous system, causing headaches, dizziness, depression, nervousness, and fatigue. Other adverse health effects of PCBs are liver, stomach, and thyroid gland injuries; behavioral alterations; and impaired reproduction.[82]

The EPA has set a limit of 0.0005 milligrams of PCBs per liter of drinking water (0.0005 mg/L). Moreover, the Food and Drug Administration (FDA) requires that infant foods; eggs, milk, and other dairy products; fish and shellfish; poultry; and red meat contain no more than 0.2–3 parts of a PCB per million (0.2–3 ppm).[84]

Regulations

After the impact of PCBs on the environment was recognized, in 1976, the U.S. Congress charged the EPA with regulating the issue of PCBs. The ban on the manufacturing, processing, distribution in commerce, and use of PCBs, as well as the PCB disposal and marking regulations, was enclosed in the Toxic Substances Control Act (TSCA) of 1976.[85] In 1979, after subsequent amendments, the regulations stipulate that the production of PCBs in the United States is generally banned, the use of PCB-containing materials still in service is restricted, the discharge of PCB- containing effluents is prohibited, the disposal of materials contaminated by PCBs is regulated, and the import or export of PCBs is only permitted through an exemption granted from EPA.

In the European community, the use of PCBs in open applications such as printing inks and adhesives was banned in 1976 (Directive 76/403/EEC[86]). Use of PCBs as a raw material or chemical intermediate has been banned in the European Union (EU) since 1985.[87] In 1996, the 1976 directive was replaced by Directive 96/59/EC,[88] which set a deadline of 2010 for complete phase out or decontamination of equipment containing PCBs. However, the United Nations Environment Programme (UNEP) global treaty adopted at the Stockholm Convention on Persistent Organic Pollutants (May 2001) stipulates that the use in equipment shall be eliminated by 2025.[89] This date is a minimum requirement and does not prevent individual governments, or groups of governments, from maintaining earlier phase-out dates. However, the most important regulations—Council Directive 96/59/EC[88] and HELCOM Recommendation 6/1[90]—concerning total banning of PCBs have been fully implemented only by EU countries. Furthermore, the Commission has adopted community strategy for dioxins, furans, and PCBs aimed at reducing as far as possible the release of these substances in the environment and their introduction in the trophic chains.[91]

Despite the existing regulations, there is still a substantial amount of PCBs in use, because exemption has been given in many countries for contained use in existing equipment with long lifetimes, at least for an initial period after a production ban was decided. There are also quantities in storage awaiting disposal. The chemical industry is currently making a proposal to solve the disposal problem.[92]

Disposal of PCBs from the Environment

PCBs and PCB-contaminated equipment and oil are required to be properly disposed of in a manner similar to that of hazardous waste.

Much effort has been directed towards the selection of technology options for the disposal of PCBs from the environment. Although the baseline remediation technology for PCBs is incineration, other options do exist. Destruction of PCBs requires the breaking of molecular bonds by an input of thermal or chemical energy. The main features of combustion and non-combustion processes are summarized in Table 4.[93]

While destruction is to be preferred, some PCBs from a range of consumer goods are likely to enter landfills accepting municipal waste. PCBs deposited in landfills may therefore contaminate groundwater and surface water following migration into leachate. The behavior of PCBs in landfills is far from fully understood and a precautionary approach is recommended.

TABLE 4   Destruction Processes for PCB Wastes

Process		Waste Types Accepted	Advantages	Disadvantages
Incineration		• Liquids and dilute slurries • PCB-containing waste equipment (may require preprocessing)	• High destruction efficiencies (99.9999% or more) • Meeting legal requirements • Facilities can treat a range of wastes, both chlorinated and non-chlorinated	• Risk of emission of harmful substances if inadequately controlled • Careful process control is required to maintain important parameters (residence time, temperature, turbulence, and oxygen concentration) at the desired level and to ensure the effectiveness of the gas cleaning system • Costly, especially if wastes have to be shipped off-site • Some equipment may require preprocessing by mechanical alteration—e.g., shredding to expose contents of capacitors, draining and disassembly of transformers, cutting large transformers to size, or packing solids and sludges in drums and feeding via a chute
Dechlorination processes	Gas phase chemical reduction (GPCR)	• PCB-contaminated liquids, soils, sediments, equipment, and material	• High destruction efficiencies (99.9999%) • Modular, transportable, or fixed configurations • The expected throughput of the main reactor system is 1000 to 3000 tons/mo	• Need to establish treatment conditions for individual components
	Base catalyzed decomposition (BCD)	• – Liquids • – Soil and building rubble contaminated by POPs	• High destruction efficiencies (99.9999%) • Modular, transportable, or fixed plants • The process can tolerate inorganic and organic debris provided this material is smaller than 50 mm or can be shredded down to this size	• The first step of BC, designed to treat solid matrices, requires mechanical pretreatment • Need to establish treatment conditions for individual components

(*Continued*)

TABLE 4 (*Continued*)   Destruction Processes for PCB Wastes

Process	Waste Types Accepted	Advantages	Disadvantages
Sodium reduction	Oils with a PCB content of up to 10,000 pg/dm³	Transportable and fixed plants Widely used for in situ removal of PCBs from active transformers	Need to establish treatment conditions for individual components
Supercritical water oxidation (SCWO)	Liquid wastes or solids less than 200 microns in diameter, and an organic content of less than 20%	A compact, totally enclosed system All emissions and residues may be captured for assay and reprocessing if needed	Need to establish treatment conditions for individual components
Plasma arc	Liquid waste streams of any concentration (the most cost-effective method is to treat concentrated wastes) Solids in the form of a pumpable fine slurry	Various plasma reactors developed for the thermal destruction of hazardous waste Transportable and fixed units The system can treat its own fly ash plus filtration media, minimizing secondary wastes	Contaminated soil, very viscous liquids or sludges, other equipment (capacitors and transformers) can be treated after pretreatment
Pyrolysis	Solid, liquid and gaseous wastes	Transportable and fixed configurations Off gases can be reused as synthesis gas	Need to establish treatment conditions for individual components
Molten salt oxidation	Liquids Solids—only if reduced to small particle sizes for pneumatic conveying	The reaction takes place within the salt bath, virtually eliminating the fugitive inventories found in incineration	Need to establish treatment conditions for individual components
Solvated electron technology	Liquids Solid materials (up to 45 cm diameter)	POP wastes are reduced to metal salts and simple hydrocarbon compounds; PCBs are reduced to petroleum hydrocarbons, sodium chloride, and sodium amide	Material with a high water content (>40% w/w) must be dewatered prior to treatment

Conclusion

For decades, PCBs have been recognized as important and potentially harmful environmental contaminants. The intrinsic properties of PCBs, such as high environmental persistence, resistance to metabolism in organisms, and tendency to accumulate in lipids have contributed to their ubiquity in environmental media and have induced concern for their toxic effects after prolonged exposure.

PCBs are bioaccumulated mainly by aquatic and terrestrial organisms and thus enter the food web. Humans and wildlife that consume contaminated organisms can also accumulate PCBs in their tissues. Such accumulation is of concern, because it may lead to body burdens of PCBs that could have adverse health effects in humans and wildlife.

Moreover, PCBs are slower to biodegrade in the environment than are many other organic chemicals. The low water solubility and the low vapor pressure of PCBs, coupled with air, water, and sediment transport processes, mean that they are readily transported from local or regional sites of contamination to remote areas.

PCBs are transformed mainly through microbial degradation, and particularly reductive dechlorination via organisms that take them up. Metabolism by microorganisms and other animals can cause relative proportions of some congeners to increase while others decrease. Because the susceptibility of PCBs to degradation and bioaccumulation is congener specific, the composition of PCB congener mixtures that occur in the environment differs substantially from that of the original industrial mixtures released into the environment. Generally, the less-chlorinated congeners are more water soluble, more volatile, and more likely to biodegrade. On the other hand, higher-chlorinated PCBs are often more resistant to degradation and volatilization and sorb more strongly to particulate matter. Some higher-chlorinated PCBs tend to bioaccumulate to greater concentrations in tissues of animals than do lower- molecular-weight ones. The higher-chlorinated PCBs can also biomagnify in food webs.

There is still much to be learned about the chemistry of PCBs. Current research focuses on finding ways to break the molecules down into harmless compounds. The biodegradation of PCBs utilizing microorganisms presently appears to be our best hope for removing PCBs from the environment.

Glossary

BAF: Bioaccumulation factor
BCF: Bioconcentration factor
BMF: Biomagnification factor
BZ number: A system of sequential numbers for the 209 PCB congeners introduced in 1980 by Ballschmiter and Zell that identifies a given congener simply and precisely. Also referred to as congener, IUPAC, or PCB number.
DHHS: Department of Health and Human Services
DOM: Dissolved organic matter
EPA: Environmental Protection Agency
FDA: Food and Drug Administration
IARC: International Agency for Research on Cancer
PCBs: Polychlorinated biphenyls
POM: Particulate organic matter
POPs: Persistent organic pollutants
TSCA: Toxic Substances Control Act
UNEP: United Nations Environment Programme

Acknowledgments

This research was financially supported by the Polish Ministry of Science and Higher Education (grant no. N N312 300535). A. Wilkowska is grateful for financial support from the Human Capital Programme (POKL.04.01.01-00-368/09).

References

1. US Department of Health and Human Services. Toxicological profile for polychlorinated biphenyls (update). Agency for Toxic Substances and Disease Registry: Atlanta, 2000, available at http://www.atsdr.cdc.gov/toxprofiles/tp17.pdf (accessed January 2010).

2. De Voogt, P.; Brinkman, U.A. Production, properties and usage of polychlorinated biphenyls. In *Halogenated Biphenyls, Terphenyls, Naphthalenes, Dibenzodioxins and Related Products,* 2nd Ed.; Kimbrough, R.D., Jensen, A.A., Eds.; Elsevier Science Publishers: Amsterdam, 1989; 3–45.

3. Ballschmiter, K.; Zell, M. Analysis of polychlorinated biphenyls (PCB) by glass capillary gas chromatography: Composition of technical Aroclor and Clophen-PCB mixtures. Fresenius Z. Anal. Chem. **1980**, *302,* 20–31.

4. Dunnivant, F.M.; Elzerman, A.W. Aqueous solubility and Henry's law constant data for PCB congeners for evaluation of quantitative structure–property relationships (QSPRs). Chemosphere **1988**, *17,* 525–541.

5. Hutzinger, O.; Safe, S.; Zitko, V. Photochemical degradation of chlorobiphenyls (PCBs). Environ. Health Perspect. **1972**, *1,* 15–20.

6. Beyer, A.; Biziuk, M. Environmental fate and global distribution of polychlorinated biphenyls. Rev. Environ. Contam. Toxicol. **2009**, *201,* 137–158.

7. Hazard waste generation and commercial hazardous waste management capacity: An assessment, SW-894. Prepared by Booz-Allen and Hamilton, Inc. and Putnam, Hayes and Barlett, Inc. for the Office of Planning and Evaluation and the Office of Solid Waste. U.S. Environmental Protection Agency: Washington, DC, 1980; D-4.

8. Barbalace, R.C. The Chemistry of Polychlorinated Biphenyls. EnvironmentalChemistry.com. 2003, available at http://EnvironmentalChemistry.com/yogi/chemistry/pcb.html (accessed January 2010).

9. Guidelines for the Identification of PCBs and Materials Containing PCBs. 1999, available at http://www.chem.unep.ch/pops/pdf/PCBident/pcbid1.pdf (accessed January 2010).

10. Durfee, R.L. Production and usage of PCB's in the United States. In Proceedings of the National Conference on Polychlorinated Biphenyls, EPA-560/6–75–004.U.S. Environmental Protection Agency: Washington, DC, 1976; 103–107.

11. Orris, P.; Kominsky, J.R.; Hryhorczyk, D.; Melius, J. Exposure to polychlorinated biphenyls from an overheated transformer. Chemosphere **1986**, *15,* 1305–1311.

12. Welsh, M.S. Extraction and gas chromatography/electron capture analysis of polychlorinated biphenyls in railcar paint scrapings. Appl. Occup. Environ. Hyg. **1995**, *10* (3), 175–181.

13. Boate, A.; Deleersnyder, G.; Howarth, J.; Mirabelli, A.; Peck L. Chemistry of PCBs, available at http://wvlc.uwaterloo.ca/biology447/modules/intro/assignments/Introduction2a.htm (accessed January 2010).

14. Eisenreich, S.J.; Baker, J.E.; Franz, T.; Swanson, M.; Rapaport, R.A.; Strachan, W.M.J.; Hites, R.A. Atmospheric deposition of hydrophobic organic contaminants to the Laurentian Great Lakes. In *Fate of Pesticides and Chemicals in the Environment;* Schnoor, J.L., Ed.; John Wiley and Sons, Inc.: New York, 1992; 51–78.

15. Pham, T.T.; Proulx, S. PCBs and PAHs in the Montreal urban community (Quebec, Canada) wastewater treatment plant and in the effluent plume in the St. Lawrence River. Water Res. **1997**, *31*(8), 1887–1896.

16. Hansen, L.G.; O'Keefe, P.W. Polychlorinated dibenzofurans and dibenzo-*p*-dioxins in subsurface soil, superficial dust, and air extracts from a contaminated landfill. Arch. Environ. Contam. Toxicol. **1996**, *31*(2), 271–276.

17. Hansen, L.G.; Green, D.; Cochran, J.; Vermette, S.; Bush, B. Chlorobiphenyl (PCB) composition of extracts of subsurface soil, superficial dust and air from a contaminated landfill. Fresenius J. Anal. Chem. **1997**, *357,* 442–448.

18. Swackhamer, D.L.; Armstrong, D.E. Estimation of the atmospheric and nonatmospheric contributions and losses of polychlorinated biphenyls for Lake Michigan on the basis of sediment records of remote lakes. Environ. Sci. Technol. **1986**, *20,* 879–883.

19. Wallace, J.C.; Basu, I.; Hites, R.A. Sampling and analysis artifacts caused by elevated indoor air polychlorinated biphenyl concentrations. Environ. Sci. Technol. **1996**, *30*(9), 2730–2734.

20. Ohsaki, Y.; Matsueda, T. Levels, features and a source of non-ortho coplanar polychlorinated biphenyl in soil. Chemosphere **1994**, *28*(1), 47–56.

21. Blumbach, J.; Nethe, L.P. Organic components reduction (PCDD/PCDF/PCB) in flue-gases and residual materials from waste incinerators by use of carbonaceous adsorbents. Chemosphere **1996**, *32*(1), 119–131.

22. Gunkel, G.; Mast, P.G.; Nolte, C. Pollution of aquatic ecosystems by polychlorinated biphenyls (PCB). Limnologica **1995**, *25*(3/4), 321–331.

23. Alcock, R.E.; Bacon, J.; Bardget, R.D.; Beck, A.J.; Hay-garth, P.M.; Lee, R.G.M.; Parker, C.A.; Jones, K.C. Persistence and fate of polychlorinated biphenyls (PCBs) in sewage sludge-amended agricultural soils. Environ. Pollut. **1996**, *93* (1), 83–92.

24. Wania, F.; Mackay, D. Tracking the distribution of persistent organic pollutants. Environ. Sci. Technol. **1996**, *30*, 390A–396A.

25. Hansen, K.M.; Prevedouros, K.; Sweetman, A.J.; Jones, K. C.; Christensen, J.H. A process-oriented intercomparison of a box model and an atmospheric chemistry transport model: Insights into model structure using alpha-HCH as the modelled substance. Atmos. Environ. **2006**, *40*, 2089–2104.

26. Mackay, D. *Multimedia Environmental Models. The Fugacity Approach*, 2nd Ed.; Lewis Publishers: Boca Raton, FL, 2001.

27. Gong, S.L.; Huang, P.; Zhao, T.L.; Sahsuvar, L.; Barrie, L. A.; Kaminski, J.W.; Li, Y.F.; Niu, T. GEM/POPs: A global 3-D dynamic model for semi-volatile persistent organic pollutants—Part 1: Model description and evaluations of air concentrations. Atmos. Chem. Phys. **2007**, *7*, 4001–4013.

28. Huang, P.; Gong, S.L.; Zhao, T.L.; Neary, L.; Barrie, L.A. GEM/POPs: A global 3-D dynamic model for semi-volatile persistent organic pollutants—Part 2: Global transports and budgets of PCBs. Atmos. Chem. Phys. **2007**, *7*, 4015–4025.

29. Wania, F.; Daly, G.L. Estimating the contribution of degradation in air and deposition to the deep sea to the global loss of PCBs. Atmos. Environ. **2002**, *36*, 5581–5593.

30. Wania, F.; Su, Y.S. Quantifying the global fractionation of polychlorinated biphenyls. Ambio **2004**, *33*, 161–168.

31. Eckhardt, S.; Breivik, K.; Li, Y.F.; Mano, S.; Stohl, A. Source regions of some persistent organic pollutants measured in the atmosphere at Birkenes, Norway. Atmos. Chem. Phys. **2009**, *9*, 6597–6610.

32. Li, N.Q.; Wania, F.; Lei, Y.D.; Daly, G.L. A comprehensive and critical compilation, evaluation, and selection of physical chemical property data for selected polychlorinated biphenyls. J. Phys. Chem. Ref. Data **2003**, *32*, 1545–1590.

33. Erickson, M.D. *Analytical Chemistry of PCBs*; Butterworth Publishers: Stoneham, Massachusettes, 1986.

34. Hansen, L.G. *The Ortho Side of PCBs: Occurrence and Disposition*; Kluwer Academic Publishers: Boston, 1999.

35. Haque, R.; Schmedding, D.; Freed, V. Aqueous solubility, adsorption and vapor behavior of polychlorinated biphenyl Aroclor 1254. Environ. Sci. Technol. **1974**, *8*, 139–142.

36. Chiarenzelli, J.; Scrudato, R.; Arnold, G.; Wunderlich, M.; Rafferty, D. Volatilization of polychlorinated biphenyls during drying at ambient conditions. Chemosphere **1996**, *33*, 899–911.

37. Wania, F.; Mackay, D. Global fractionation and cold condensation of low volatility organochlorine compounds in polar regions. Ambio **1993**, *22*, 10–18.

38. Wania, F. Potential of degradable organic chemicals for absolute and relative enrichment in the Arctic. Environ. Sci. Technol. **2006**, *40*, 569–577.

39. Nelson, E.D.; McConnell, L.L.; Baker, J.E. Diffusive exchange of gaseous polycyclic aromatic hydrocarbons and polychlorinated biphenyls across the air-water interface of the Chesapeake Bay. Environ. Sci. Technol. **1998**, *32* (7), 912–919.

40. Luthy, R.G.; Aiken, G.R.; Brusseau, M.L.; Cunningham, S.D.; Gschwend, P.M.; Pignatello, J.J.; Reinhard, M.; Traina, S.J.; Weber, W.J.; Westall, J.C. Sequestration of hydrophobic organic contaminants by geosorbents. Environ. Sci. Technol. **1997**, *31* (12), 3341–3347.

41. ATSDR—Agency for Toxic Substances and Disease Registry. Toxicological profile for polychlorinated biphenyls (update). Atlanta: US Department of Health and Human Services, 2000.

42. Van Bavel, B.; Andersson, P.; Wingfors, H.; Ahgren, J.; Bergqvist, P.A.; Norrgren, L.; Rappe, C.; Tysklind, M. Multivariate modeling of PCB bioaccumulation in three-spined stickleback (*Gasterosteus aculeatus*). Environ. Toxicol. Chem. **1996**, *15*, 947–954.

43. Boese, B.L.; Winsor, M.; Lee, II, H.; Echols, S.; Pelletier, J.; Randall, R. PCB Congeners and hexa-chlorobenzene biota sediment accumulation factor for *Macoma nasuta* exposed to sediments with different total organic carbon contents. Environ. Toxicol. Chem. **1995**, *14,* 303–310.

44. Friedman, C.L.; Burgess, R.M.; Perron, M.M.; Cantwell, M.G.; Ho, K.T.; Lohmann R. Comparing polychaete and polyethylene uptake to assess sediment resuspension effects on PCB bioavailabil-ity. Environ. Sci. Technol. **2009**, *43*(8), 2865–2870.

45. Trimble, T.A.; You, J.; Lydy, M.J. Bioavailability of PCBs from field-collected sediments: Application of Tenax extraction and matrix-SPME techniques. Chemosphere **2008**, *71*, 337–344.

46. Verweij, F.; Booij, K.; Satumalay, K.; van der Molen, N.; van der Oost, R. Assessment of bioavailable PAH, PCB and OCP concentrations in water, using semipermeable membrane devices (SPMDs), sediments and caged carp. Chemosphere **2004**, *54* (11), 1675–1689.

47. Gdaniec-Pietryka, M.; Wolska, L.; Namiesnik, J. Physical speciation of polychlorinated biphenyls in the aquatic environment. Trends Anal. Chem. **2007**, *26*, 1005–1012.

48. Adsorption and Desorption of Organic Contaminants to Predict Fate, Transport, and Propensity to Electrochemically Degrade, available at http://www.epa.gov/nrmrl/lrpcd/wm/projects/135925. htm (accessed February 2010).

49. Borja, J.; Taleon, D.M.; Auresenia, J.; Gallardo, S. Polychlorinated biphenyls and their biodegradation. Process Biochem. **2005**, *40*, 1999–2013.

50. Dobbins, D.C. Biodegradation of pollutants. In *Encyclopedia of Environmental Biology*; Academic Press Inc.: New York, 1995.

51. McEldowney, S.; Hardman, D.J.; Wait, S. *Pollution: Ecology and Biotreatment*; Longman Scientific and Technical: New York, 1993.

52. Furukawa, K. Modification of PCBs by bacteria and other microorganisms. In *PCBs and the Environment;* Waid, J.S., Ed.; CRC Press: Florida, 1986; 89–100.

53. Sylvestre, M.; Sandossi, M. Selection of enhanced PCB- degrading bacterial strains for bioremediation: Consideration of branching pathways. In *Biological Degradation and Remediation of Toxic Chemicals;* Chaudhry, G.R., Ed.; Chapman and Hall: New York, 1994.

54. Boyle, A.W.; Silvin, C.J.; Hassett, J.P.; Nakas, J.P.; Tanenbaum, S.W. Bacterial PCB biodegradation. Biodegradation **1992**, *3*, 285–298.

55. Alder, A.C.; Haggblom, M.M.; Oppenheimer, S.R.; Young, L.Y. Reductive dechlorination of polychlorinated biphenyls in anaerobic sediments. Environ. Sci. Technol. **1993**, *27*, 530–538.

56. Mohn, W.W.; Tiedje, J.M. Microbial reductive dechlorination. Microbiol. Rev. **1992**, *56*, 482–507.

57. Holliger, C.; Wohlfarth, G.; Diekert, G. Reductive dechlorination in the energy metabolism of anaerobic bacteria. FEMS Microbiol. Rev. **1998**, *22*, 383–398.

58. Wiegel, J.; Wu, Q. Microbial reductive dehalogenation of polychlorinated biphenyls. FEMS Microbiol. Ecol. **2000**, *32*, 1–15.

59. Fish, K.M.; Principe, J.M. Biotransformations of Arochlor 1242 in Hudson River test tube microcosms. Appl. Environ. Microbiol. **1994**, *60*, 4289–4296.

60. Ye, D.; Quensen, III, J.F.; Tiedje, J.M.; Boyd, S.A. Anaerobic dechlorination of polychlorinated biphenyls (Arochlor 1242) by pasteurized and ethanol-treated microorganisms from sediments. Appl. Environ. Microbiol. **1992**, *58* (4), 1110–1114.

61. Moore, J.A. *Reassessment of Liver Findings in PCB Studies for Rats*; Institute of Evaluating Health Risks: Washington, 1991.

62. Safe, S. Toxicology, structure-function relationship, and human and environmental health impacts of polychlorinated biphenyls: progress and problems. Environ. Health. Perspect. **1992**, *100*, 259–268.

63. Cookson, Jr., J.T. *Bioremediation Engineering: Design and Application;* McGraw Hill: New York, 1995.

64. Kuipers, B.; Cullen, W.R.; Mohn, W.W. Reductive dechlorination of nonachloro biphenyls and selected octachloro biphenyls by microbial enrichment cultures. Environ. Sci. Technol. **1999**, *33*, 3579–3585.

65. Benvinakatti, B.G.; Ninnekar, H.Z. Degradation of biphenyl by a *Micrococcus* species. Appl. Microbiol. Biotechnol. **1992**, *38*, 273–275.

66. Field, J.A.; Sierra-Alvarez, R. Microbial transformation and degradation of polychlorinated biphenyls. Environ. Pollut. **2008**, *155*, 1–12.

67. Mackay, D.; Fraser, A. Bioaccumulation of persistent organic chemicals: Mechanism and models. Environ. Pollut. **2000**, *110*, 375–391.

68. Gobas, F.A.P.C.; Morrison, H.A. Bioconcentration and biomagnification in the aquatic environment. In *Handbook of Property Estimation Methods for Chemicals;* Boethling, R.S., Mackay, D., Eds.; CRC Press: Boca Raton, FL, 2000; 189–231.

69. Kucklick, J.; Harvey, H.R.; Ostrom, P.; Ostrom, N.; Baker, J. Organochlorine dynamics in the pelagic food web of Lake Baikal. Environ. Toxicol. Chem. **1996**, *15*, 1388–1400.

70. Glossary for Chemists of Terms Used in Toxicology: Pure and Applied Chemistry, V. 65, no. 9; International Union of Pure And Applied Chemistry (IUPAC): Bethesda, Maryland, 1993; 2003–2122, available at (http://sis.nlm.nih.gov/enviro/glossarymain.html—online version posted by the U.S. National Library of Medicine).

71. Bierman, Jr., V.J. Equilibrium partitioning and biomagnification of organic chemicals in benthic animals. Environ. Sci. Technol. **1990**, *24*, 1407–1412.

72. Burreau, S.; Zebuhr, Y.; Broman, D.; Ishaq, R. Biomagnification of PBDEs and PCBs in food webs from the Baltic Sea and the northern Atlantic Ocean. Sci. Total Environ. **2006**, *366*, 659–672.

73. Nfon, E.; Cousins, I.T.; Broman, D. Biomagnification of organic pollutants in benthic and pelagic marine food chains from the Baltic Sea. Sci. Total Environ. **2008**, *397*, 190–204.

74. Campfens, J.; Mackay, D. Fugacity-based model of PCB bioaccumulation in complex aquatic food webs. Environ. Sci. Technol. **1997**, *31*, 577–583.

75. Pelka, A. Bioaccumulation models and applications: Setting sediment cleanup goals in the Great Lakes. National Sediment Bioaccumulation Conference Proceedings, EPA 823-R-98–002U.S.; Environmental Protection Agency Office of Water: Washington, DC, 1998; 5-9-5–30.

76. Chapra, S.C.; Reckhow, K.H. *Engineering Approaches for Lake Management. Mechanistic Modeling;* Butterworth Publishers/Ann Arbor Science: Woburn, MA, 1983; Vol. 2.

77. Warren, C.; Mackay, D.; Whelan, M.; Fox, K. Mass balance modelling of contaminants in river basins: A flexible matrix approach. Chemosphere **2005**, *61*, 1458–1467.

78. Warren, C.; Mackay, D.; Whelan, M.; Fox, K. Mass balance modelling of contaminants in river basins: Application of the flexible matrix approach. Chemosphere **2007**, *68*, 1232–1244.

79. Arnot, J.A.; Gobas, F. A food web bioaccumulation model for organic chemicals in aquatic ecosystems. Environ. Toxicol. Chem. **2004**, *23*, 2343–2355.

80. Bright, D.A.; Grundy, S.L.; Reimer, K.J. Differential bioaccumulation of non-ortho-substituted and other PCB congeners in coastal arctic invertebrates and fish. Environ. Sci. Technol. **1995**, *29*, 2504–2512.

81. Antunes, P.; Gil, O.; Reis-Henriques, M.A. Evidence for higher biomagnification factors of lower chlorinated PCBs in cultivated seabass. Sci. Total Environ. **2007**, *377*, 36–44.

82. Available at http://www.atsdr.cdc.gov/DT/pcb007.html (accessed February 2010).

83. U.S. Environmental protection agency. PCBs: A cancer dose-response assessment and applications to environmental mixtures, EPA/600/P-96/001F; 1996, available at http://cfpub.epa.gov/ncea/CFM/recordisplay.cfm?deid=12486 (accessed February 2010).

84. Available at http://www.atsdr.cdc.gov/tfacts17.pdf (accessed February 2010).

85. Online version of TSCA, from the Government Printing Office, available at http://frwebgate.access.gpo.gov/cgi-bin/usc.cgi?ACTION=BROWSE&TITLE=15USCC53 (accessed January 2010).

86. Council Directive 76/403/EEC of 6 April 1976 on the disposal of polychlorinated biphenyls and polychlorinated terphenyls.

87. Council Directive 85/467/EEC of 1 October 1985 amending for the sixth time Directive 76/769/EEC on the approximation of the laws, regulations and administrative provisions of the Member States relating to restrictions on the marketing and use of certain dangerous substances and preparations (PCBs/PCTs).

88. Council Directive 96/59/EC of 16 September 1996 on the disposal of polychlorinated biphenyls and polychlorinated terphenyls (PCBs/PCTs).

89. Manila Bulletin. Global chemical treaty (opinion/editorial). Manila Bulletin Publishing Corp.: Manila, 2001.

90. Helcom Recommendation 6/1, Recommendation regarding the elimination of use of PCBs and PCTs—adopted 13 March 1995, having regard to article 13, Paragraph b of the Helsinki Convention.

91. Communication from the Commission to the Council, the European Parliament and the Economic and Social Committee. Community strategy for dioxins, furans and polychlorinated biphenyls (2001/C 322/02), available at http://ec.europa.eu/environment/waste/pcbs/pdf/en.pdf (accessed January 2010).

92. ICCA/WCC Position Paper (1999): Best Available Techniques for Destruction of PCBs-Incineration Technology, presented at the UNEP-INC 3 meeting on POPs in Geneva, September 6–10, 1999.

93. Inventory of World-wide PCB Destruction Capacity, Second Issue; United Nations Environment Programme: Switzerland, 2004; Prepared by UNEP Chemicals, available at http://www.chem.unep.ch/pops/pcb_activities/pcb_dest/PCB_Dest_Cap_SHORT.pdf (accessed January 2010).

13

Toxic Substances

Sven Erik Jørgensen

Introduction

This entry gives a short overview of the toxic contaminants that we most often detect in the environment and consider the most threatening compounds. As we are using about 100,000 different chemicals in modern society, we can expect to find 100,000 different compounds plus intermediates of these compounds resulting from a wide spectrum of decomposition processes that take place in the environment. To be able to give an overview of many chemical compounds, their fates, effects, and associated risks, it is necessary to make a classification of the compounds. Below is a brief treatment of the properties, effects, characteristic processes, and particular risks of the following classes of contaminants that represent the most toxic compounds that we are using and the most harmful compounds for the environment.

1. Petroleum hydrocarbons
2. PCBs and dioxins
3. Pesticides
4. Polycyclic aromatic hydrocarbons (PAHs)
5. Heavy metals
6. Detergents
7. Synthetic polymers and xenobiotics applied in the plastics industry

Only the organometallic compounds are considered in this overview, while several other entries cover contamination by heavy metals in general. The seven classes are presented in the succeeding sections.

For more comprehensive coverage of toxic chemical compounds in the environment, refer to Loganathan and Lam,[1] Newman and Unger,[2] Hoffman et al.,[3] and Schuurmann and Markert.[4]

Petroleum Hydrocarbons

Petroleum hydrocarbons include a variety of organic compounds. Hydrocarbons (compounds composed of carbon and hydrogen) constitute only 50%–90% of petroleum. They are n-alkanes, branched alkanes, cycloalkanes, and aromatics. Cycloalkanes comprise usually the largest portion of hydrocarbons in

139

petroleum, while aromatics are usually present to the extent of 20% or less. Characteristic compounds are benzene, alkyl-substituted benzenes, and fused-ring PAHs (see also group D). Some of these aromatics are carcinogenic, and this is probably the group that is of greatest environmental concern. In addition to hydrocarbons, petroleum contains sulfur and nitrogen compounds, such as thiophene, ethanethiol, and pyridine derivatives.

Petroleum compounds are emitted or discharged into all spheres.

Evaporation removes the lower molecular weight of the more volatile components of the petroleum mixture. Hydrocarbons with vapor pressures equal to that of *n*-octane (0.019 atm at room temperature) or greater will be lost quickly via evaporation.

The lower-molecular-weight hydrocarbons tend also to be the most water soluble, but for the same molecular weight, aromatic hydrocarbons are more soluble than cycloalkanes, which are more soluble than branched alkanes, with the *n*-alkanes being the most insoluble in water. Petroleum products are discharged directly to the sea by accidents or by violation of international regulations. In a massive discharge of petroleum products, which has been recorded in several ship accidents, most of the petroleum will initially float on the surface of marine waters as a slick. Eventually, most slicks are dispersed widely and form a 0.1 mm thick layer on the water. If drift results in landfall, or contact with coral or mangrove communities, a disastrous environmental impact may occur.

Fortunately, chemical transformation and degradation processes act on petroleum compounds in the environment. Microbial transformation and photo oxidation are of particular importance.

The aromatic compounds of petroleum are the most toxic substances present. They are lethal to crustaceans and fish with LC_{50} in the range of 0.1–10 mg/L.

PCBs and Dioxins

PCBs and dioxins are characterized as aromatic compounds with a high content of chlorine. As the names cover polychlorinated biphenyls and polychlorinated dibenzo(1,4)dioxins, there are many different individual compounds under these labels. For instance, 209 different PCB compounds are known, although only about 130 are found in commercial mixtures. Figures 1 and 2 show the molecular structures and names of some of the most common PCBs and dioxins.

The applications of PCBs have been quite diverse (capacitor oil, plasticizers, printer's ink, etc.), but due to investigations in the 1960s and 1970s, in which it was found that PCBs occurred widely in the environment and significant bioaccumulation took place, voluntary restrictions were introduced and all "open" applications were banned.

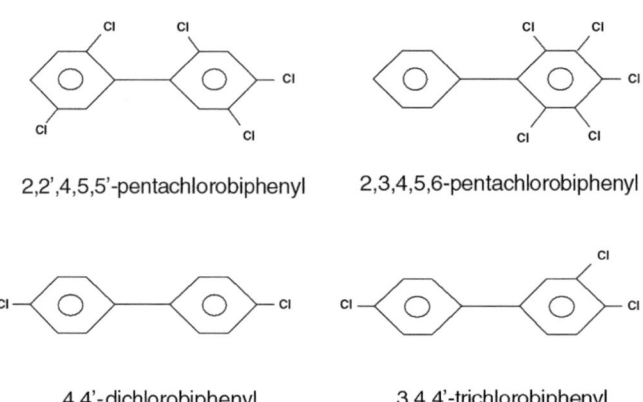

2,2',4,5,5'-pentachlorobiphenyl   2,3,4,5,6-pentachlorobiphenyl

4,4'-dichlorobiphenyl   3,4,4'-trichlorobiphenyl

FIGURE 1   The molecular structure and names of four common PCBs.

Dioxins

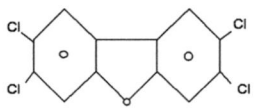

1,2-dichloridbenzo (1,4) dioxin   1,2,6-trichloridbenzo (1,4) dioxin

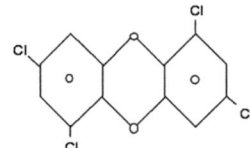

2,3,7,8-tetrachloridbenzofuran   1,2,3,4,7,8-hezachlorodibenzo (1,4) dioxin

1,3,6,8-tetrachlorobenzo (1,4) dioxin

FIGURE 2   Molecular structure and names of five common dioxins.

Dioxins are not deliberately produced, but are by-products of chemical processes involving chlorine, for instance, production of various organochlorine and bleaching of pulp and of combustion processes if chlorine-containing compounds are present.

Both PCBs and dioxins are characterized by a low water solubility and a high K_{ow} (most components have log K_{ow} > 5). Both groups of compounds are very persistent to decomposition processes, which explains why they are strong bioaccumulators, although dioxins have a UV–VIS absorption spectrum that results in significant absorption from solar radiation. Some dioxins have a half-life time in the troposphere of a few days.

Pesticides

Pesticides are used to remove pests and, due to their direct use in nature, have probably been the most criticized environmental contaminants. Usage of dichlorodiphenyltrichloroethane (DDT) and relatedinsecticides accelerated during the 1940s and the subsequent decades until environmental doubt occurred in the mid-1960s. Since 1970, DDT has been banned in most industrialized countries, but it is still used in developing countries, for instance, India, where it has resulted in very high body concentrations in the Indian population. All chlorinated hydrocarbon insecticides are banned in most industrialized countries due to their persistence and ability to bioaccumulate (K_{ow} is high).

Pesticides can be divided into the following classes depending on their use and their chemical structure:

- Herbicides comprise carbamates, phenoxyacetic acids, triazines, and phenylureas.
- Insecticides encompass organophosphates, carbamates, organochlorines, pyrethrins, and pyrethroids.
- Fungicides are dithiocarbamates, copper, and mercury compounds. See also the entries covering these two metals.

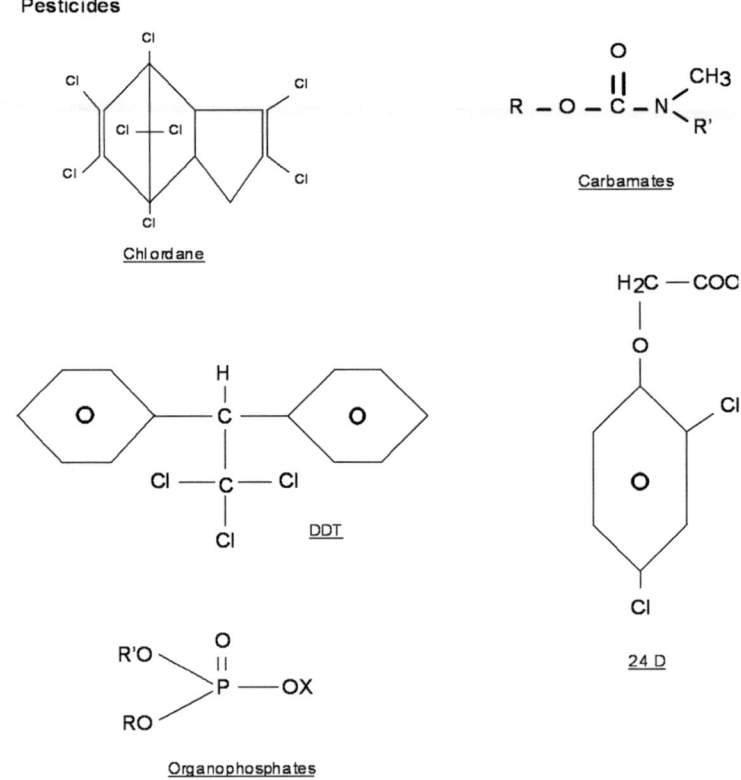

FIGURE 3   Molecular structure and names of five common pesticides.

The pesticides are an extremely chemically diverse group of substances, as they only have in common their toxicity to pests. A few of the most important molecules are shown in Figure 3. They are mostly produced synthetically although the natural pesticide pyrethrin has achieved commercial success.

Chlorohydrocarbons are strongly bioconcentrated, as already emphasized. In addition, they are very toxic to a wide range of biota, particularly to aquatic biota.

Organophosphates are almost equally toxic to biota, but due to these compounds' lack of persistence, higher solubility in water, and bioaccumulation capacity, they are still in use.

Carbamates are relatively water soluble and have limited persistence. They are, however, toxic to a wide range of biota. They act by inhibiting cholinesterase.

The pyrethins have a complex chemical structure and a high molecular weight. Thus, they are poorly soluble in water and tend to be lipophilic. They are, however, readily degraded by hydrolysis. They are more attractive to use than most of the other pesticides due to their very low mammalian toxicity.

Phenoxyaceticacidisaveryeffectiveherbicidebutcontainstraceamountsoftetrachloro-dibenzo-dioxin.

Pesticides are banned in organic agriculture where they are replaced by other methods, for instance, mechanical and biological methods (use of predator insects).

Polycyclic Aromatic Hydrocarbons

PAHs are molecules containing two or more fused 6C aromatic rings. They are ubiquitous contaminants of the natural environment, but growing industrialization has increased environmental concern about these components. Two common members are naphthalene and benzo(*a*)pyrene (see Figure 4). PAHs are usually solids, with naphthalene (lowest molecular weight) having a melting point of 81°C.

PAHs

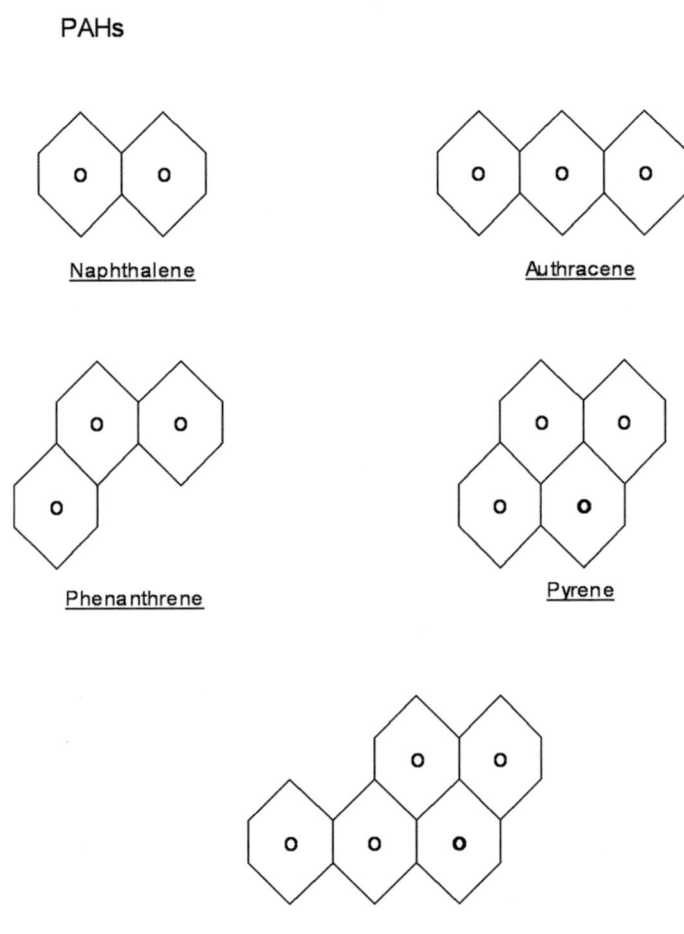

FIGURE 4   Molecular structure and names of five common PAHs.

The natural sources of PAHs in the environment are forest fires and volcanic activity. The anthropogenic sources are coal-fired power plants, incinerators, open burning and motor vehicle exhaust. As a result of these sources, PAHs commonly occur in air, soil, and biota. They are lipophilic compounds able to bioaccumulate. The low-molecular- weight compounds are moderately persistent, while for example benzo(*a*)pyrene, with a higher molecular weight, persists in aquatic systems for up to about 300 weeks.

They are relatively toxic to aquatic organisms and have LC_{50} values for fish in the range of 0.1–10 mg/L. The major environmental concern of PAHs is that many PAHs are carcinogenic. It is has been shown[5] that benzo(*a*)pyrene is an endocrine disrupter and that many more PAHs have the environmental adverse effect of disturbing the hormone balance of nature.

Human exposure to PAHs occurs through tobacco smoking as well as through compounds in food and the atmosphere.

Organometallic Compounds

Organometallic compounds are compounds having metal carbon bonds, where the carbon atoms are part of an organic group. The best known example is probably tetraethyl lead, which is used as an additive to gasoline. It has now been phased out of use in many countries—all industrialized countries—due to its environmental consequences. Organometallic compounds can be formed in nature from metal or metal ions, for instance, di- methylmercury, or are produced for various purposes, as catalysts (e.g., organo-aluminum), as pesticides (e.g., organoarsenic and organotin compounds), as stabilizers in polymers (e.g., organotin compounds), and as gasoline additives (e.g., organolead compounds). Organometallic compounds exhibit properties that are different from those of the metal itself and inorganic derivatives of the metal, for instance, a relatively higher toxicity than the metals.

Most organometallic compounds are relatively unstable and undergo hydrolysis and photolysis easily. Most organometallic compounds have weakly polar carbon–metal bonds and are often hydrophobic. They therefore only dissolve in water to a small extent and are readily sorbed onto particulates and sediments.

The most harmful organometallic compounds from an environmental point of view are organomercury, organotin, organolead, and organoarsenic, which are all very toxic to mammals.

Detergents (and Soaps)

Detergents and soaps contain surface active agents (surfactants) that are classified according to the charged nature of the hydrophilic part of the molecule:

- Anionic: negatively charged
- Cationic: positively charged
- Nonionic: neutral, but polar
- Amphoteric: a zwitterion containing positive and negative charges

They are produced and consumed in large quantities and are mostly discharged into the sewage system and end up in wastewater plants. The early surfactants contained highly branched alkyl hydrophobes that were resistant to biodegradation. These surfactants are largely obsolete today, having been replaced by linear alkyl benzene sulfonates (LAS) and other biodegradable surfactants.

The toxicity to mammals is generally low for all surfactants, while the toxicity to aquatic organisms is relatively high (LC_{50} from about 0.1 to about 77 mg/L). The toxicity will generally increase with the carbon chain length (see Figure 5).

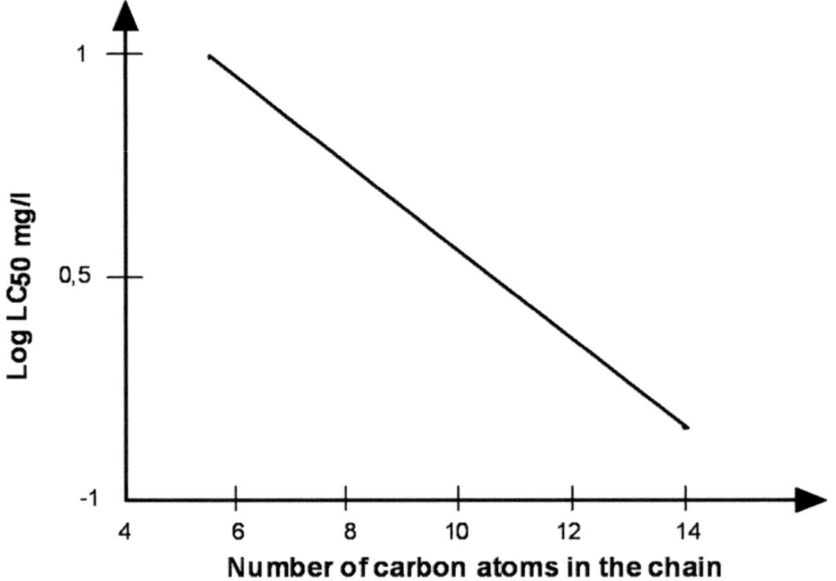

FIGURE 5   Log LC_{50} is plotted versus the number of carbon atoms in the chain for LASs. As seen, increased chain length implies increased toxicity.

Many surfactants bind strongly to soils and sediments, which implies that, to the extent that they are not biodegraded in a biological treatment plant, they will mainly be found in the sludge phase.

Synthetic Polymers and Xenobiotics Applied in the Plastics Industry

Synthetic polymers and xenobiotics applied in the plastics industry form a very diverse group of compounds from a chemical viewpoint. Synthetic polymers are useful (plumbing, textiles, paint, floor, covering, and as the basic material for a wide spectrum of products) because they are resistant to biotic and abiotic processes of transformation and degradation. These properties, however, also cause environmental management problems associated with the use of these components. In addition, several xenobiotic compounds are used as additives, softeners, stabilizers, and so on in synthetic polymer to improve their properties. Some of these additives are very toxic and may cause other and additional environmental problems; for instance, phthalates are widely used in the plastics industry and it has been demonstrated that phthalates have effects as endocrine disrupters.

After use, synthetic polymers are usually incinerated together with industrial and household garbage (solid waste). The presence of poly-vinyl-plastic (PVC) will imply that hydrochloric acid and, to a certain extent, dioxins are formed, but this is strongly dependent on the incineration conditions. As it is difficult to separate different types of plastics, it has been discussed to phase out the use of PVC, but due to the unique properties of PVC, this has not yet been decided.

References

1. Loganathan, B.G.; Lam, P.K.S. *Global Contamination Trends of Persistent Organic Chemicals;* CRC Press: Boca Raton, London, and New York, 2012; 639 pp.
2. Newman, M.C.; Unger, M.A. *Fundamentals of Ecotoxicology,* 2nd Ed.; CRC Press: Boca Raton, London, and New York, 2003.
3. Hoffman, D.J.; Rattner, B.A.; Burton, G.A., Jr.; Cairns, J., Jr. *Handbook of Ecotoxicology;* Lewis Publ./CRC Press: Boca Raton, London, and New York, 1994.
4. Schuurmann, G.; Markert, B. *Ecotoxicology;* John Wiley: New York, 1998; 902 pp.
5. Andersen, H.R. Examination of Endocrine Disruptors. Thesis, DFU, Copenhagen University, 1998.

II

Natural Elements and Chemicals

14

Allelochemics

John Borden

Terminology

To clarify in part the emerging maze of newly discovered message-bearing chemicals, or *semiochemicals* (Gk. *semeion,* sign or signal), the term *allelochemic* (Gk. *allelon,* of each other) was coined in 1970 to embrace any semiochemical with interspecific activity.[1] Thus allelochemics are distinguished from *pheromones* (Gk. *pherum,* to carry; *horman,* to excite) that convey a message between organisms of the same species. Three categories of allelochemics are commonly recognized.

Kairomones (Gk. *kairos,* opportunistic) are allelochemics that provide an adaptive advantage to the perceiver. In most cases there is no benefit, or even harm, to the emitter, for example, the attraction of predators to the odor of their prey. The evolution of such chemicals as true biological signals would be disadaptive, and therefore unlikely. Therefore some semantic purists remove all evolutionary implications pertaining to kairomones by referring to them as infochemicals.

Allomones (Gk. *allos,* other) are allelochemics that convey an adaptive advantage to the emitter. The repellent odor of an alarmed skunk is often used as an example. However, in an evolutionary sense it may also be adaptive for the receiver to be able to detect and avoid the skunk's odor, for example, for a predator not to be "tagged" with an aroma that warns potential prey of its presence. Therefore, a skunk's odor is more aptly termed a *synomone* (Gk. *syn,* with), an allelochemic that conveys a mutual advantage to both the emitter and the receiver.

Most allelochemics have a *releaser* effect, in which behavioral responses are evoked. However, they may also have a *primer* effect, in which a physiological or biochemical function is stimulated or inhibited.

Natural Occurrence

Table 1 provides a small window on the thousands of allelochemic interactions that occur in nature. The compounds that mediate these interactions are equally diverse (Figure 1). Very few interactions are mediated by a single compound; most involve relatively simple blends; and some, for example, floral fragrances, comprise dozens of compounds in a single blend. Although Table 1 provides examples of allelochemic interactions among terrestrial plants, arthropods, and vertebrates, many are also found among aquatic organisms, and examples occur in all Kingdoms and Phyla.

Kairomonal interactions include the attraction of many species of phytophagous insects to their host plants, entomophagous insects to their prey or to insects that they parasitize, and blood-feeding diptera to their vertebrate hosts.[1-3] They also include the avoidance by prey species of odors associated

TABLE 1 Examples of Natural Occurrence and Function of Allelochemics[a]

Type of Allelochemic and Source	Example
Kairomone:	
Plants	Attraction of Colorado potato beetles, Leptinotarsa decemlineata, to 6-carbon leaf volatiles, e.g., (E)-2-hexen-1-ol-(2), from solanaceous plants
	Attraction of ambrosia beetles, Trypodendron lineatum, Gnathotrichus sulcatus and G. retusus, to ethanol (1) from moribund coniferous trees, logs, and stumps
	Attraction and stimulation of oviposition by onion maggots, Delia antiqua, in response to mono- and disulfides, e.g., dipropyl disulfide (3) from onions, Allium cepa
	Employment of host tree kairomones, e.g., a-pinene (8) from conifers and a-cubene (10) from elms as synergists of aggregation pheromones that mediate mass attack of trees by bark beetles e.g., the Douglas-fir beetle, Dendroctonus pseudotsugae, and the smaller European elm bark beetle, Scolytus multistriatus, respectively
Insects	Stimulation in a parasitic chalcidoid wasp, Trichogramma evanescens, of searching for and oviposition in corn earworm eggs, Helicoverpa zea, by tricosane (15) in moth scales adhering to newly laid eggs
	Attraction of predaceous clerid beetles, Thanasimus and Enoclerus spp. to aggregation pheromones, e.g., ipsenol (19), ipsdienol (20), and frontalin (22) of their bark beetle hosts
Vertebrates	Attraction of blood-feeding mosquitoes to CO_2 (23) exhaled by mammals
	Attraction of tsetse flies, Glossina spp., to volatiles, e.g., acetone (24) and 1-octen-3-ol (25) from bovine animals on which they feed
	Avoidance by voles, Microtus spp., of volatile chemicals in the urine of mustellid predators, e.g., 2-propylthiotane (26) and 3-propyl-1,2-dithiolane (27) from short-tailed weasels, Mustella erminea
Allomone:	
Plants	Inhibition of germination of growth of one species of plant by allelopathic chemicals produced in the leaves, roots, or other tissues of another plant, e.g., by juglone (7) leaching from the leaves of black walnut trees, Juglans nigra
	Disruption of growth, metamorphosis, or reproduction of insect herbivores by producing insect hormones or hormone analogues, e.g., juvabione (11), a juvenile hormone mimic in true firs, Abies spp.
Spiders	Emission of moth sex pheromones, e.g., (Z)-9-tetradecenyl acetate (18) by bolas spiders, Mastophora spp., to attract male moths as prey
Insects	Mimicking the cuticular recognition compounds, e.g., 11-methyl pentacosane (16) of larval ants by caterpillars of lycenid butterflies, syrphid fly larvae, and scarab beetle larvae, thereby gaining entry into and acceptance within ant nests, where they prey on ant brood
Synomone:	
Plants	Avoidance of nonhost plants by insects in response to volatiles emitted by the plants, e.g., repellency of coniferophagous bark beetles, e.g., the mountain pine beetle, Dendroctonus ponderosae, to conophthorin (13) (also a repellent pheromone of cone and twig beetles) in the bark of birches, Betula spp.
	Repellency of black bean aphids, Aphis fabae, to methyl salicylate (4) in the volatiles of nonhost plants
	Tritophic interaction in which corn plants, Zea mays, respond to volicitin (14) (a kairomone) in the saliva of beet armyworm caterpillars, Spodoptera exigua, feeding on them by producing specific blends of volatiles (synomones), e.g., (E)-4, 8-dimethyl-1,3,7-nonatriene (9) that attract females of a parasitic wasp, Cotesia marginiventris, which in turn oviposit on the feeding caterpillars
	Attraction of honey bees, Apis mellifera, to multicomponent blends of floral volatiles, e.g., geraniol (12), of many species of angiosperm plants, with mutual benefit of pollination to the plant and a pollen and nectar source to the bees
Insects	Antagonists in the blends of moth sex pheromones that repel males of related species, ensuring reproductive isolation even though the major pheromone components are attractive to males of both species, e.g., (Z)-9-tetradecanal (17), a pheromone component of threelined leafrollers, Pandemis limitata, that inhibits response of oblique banded leafroller males, Choristoneura rosaceana, to threelined leafroller females
	Mutual repellency between aggregation pheromones of two species of bark beetles, e.g., (R)-(−)-ipsdienol (20) produced by pine engravers, Ips pini, and (S)-(−)-ipsenol (19) produced by California fivespined ips, I. paraconfusus, reserving the host phloem resource for the first-arriving species, and avoiding interspecific exploitative competition

[a] *Numbers in parentheses correspond with numbered compounds in Figure 1.*

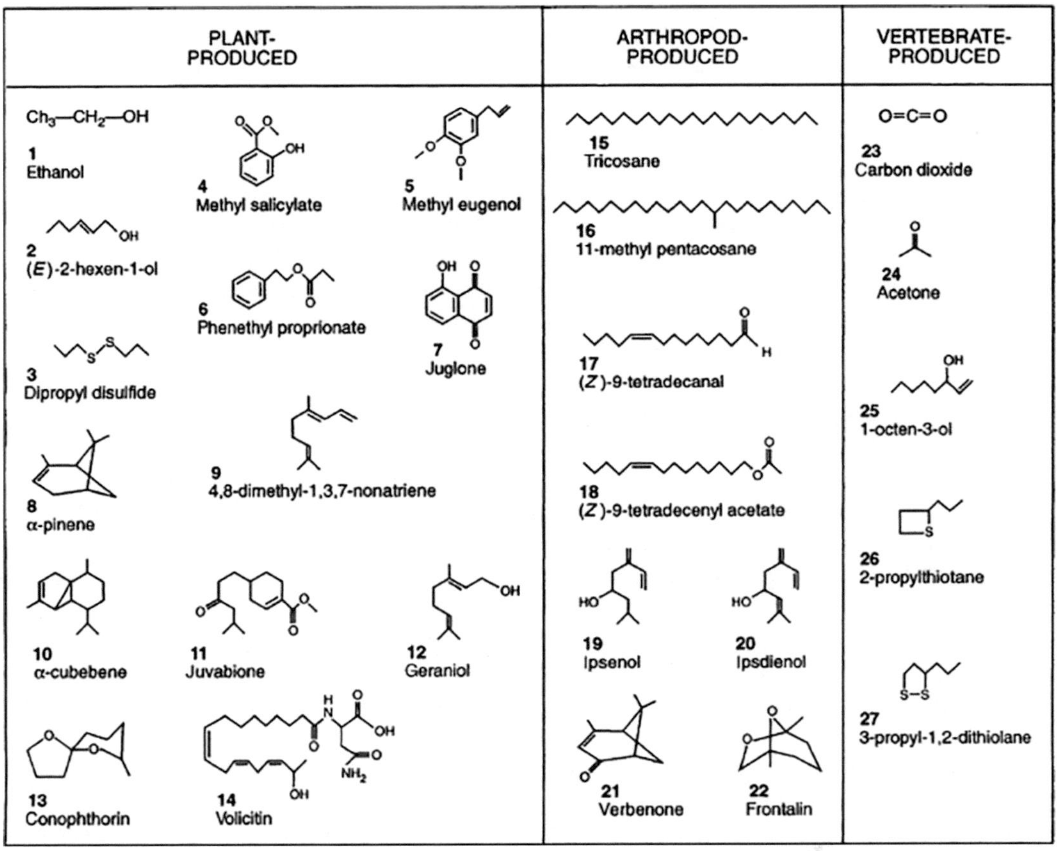

FIGURE 1 Structural formulae of compounds given in Table 1 exemplifying some of the chemical diversity among allelochemics as follows: primary alcohol (**1, 2**), secondary alcohol (**25**), disulfide (**3**), aromatic ester (**6**), unsaturated ester (**18**) monoterpene (**8**), sesquiterpene (**9, 10**), sesquiterpenoid (**11**), terpene alcohol (**12, 19, 20**), terpene ketone (**21**), spiroacetal (**13**), fatty acid derivative conjugated to an amino acid (**14**), straight chain hydrocarbon (**15**), branched hydrocarbon (**16**), unsaturated aldehyde (**17**), bicyclic ketal (**22**), atmospheric gas (**23**), ketone (**24**), thiotane (**26**), and thiolane (**27**).

with predators, a phenomenon found in five animal phyla, but curiously not yet among terrestrial insects.[4] Sometimes more than one type of semiochemical may be involved; for example, the aggregation of bark beetles necessary to mass attack and kill a tree is mediated by a blend of aggregation pheromones synergized by host tree kairomones.

Allomonal interactions may employ "trickery",[5] for example, bolas spiders that emit moth sex pheromones that lure mate-seeking male moths to their death, and many species of myrmecophiles (ant lovers) that gain access to ant nests by chemically mimicking the cuticular recognition compounds of ants on which they prey. The most well-known allomones are released by or are contained in plants, and have a primer effect. Allelopathic allomones are often leached from the leaves (or other parts) of plants of one species, and inhibit the germination of seeds or growth of plants in other species that could be potential competitors.[6] Some plants may also produce defensive allomones against insect herbivores, for example, hormones or analogues of hormones that disrupt the growth and metamorphosis of their insect enemies.

Among the many examples of synomones are repellents that ensure reproductive isolation between closely related species of insects, or mitigate against the occurrence of interspecific exploitative competition for a limited host resource. Often synomones are the same compounds as one or more of the components that convey a pheromonal message, for example, to attract mates or to aggregate on or near a food source. There is increasing evidence that host-seeking phytophagous insects not only use kairomones to find their host plants, but also use synomones to avoid nonhost plants on which their fitness would be greatly reduced. Synomonal floral scents provide a mutual benefit to flowering plants that gain from pollination by insects that in turn are attracted to a nutritious nectar or pollen source.

Another type of allelochemic interaction involves three trophic levels and the action of both kairomonal and synomonal stimuli.[3] In one remarkable example of this type of tritrophic interaction, corn plants being fed on by beet armyworm caterpillars are exposed to minute amounts of a kairomone called volicitin in the insect's saliva. Volicitin has a primer effect, eliciting the plant to synthesize a specific blend of volatile synomonal compounds that attract females of a parasitic wasp. The wasp oviposits in the beet armyworm larvae, benefiting by finding its host, and in turn providing an advantage to the plant by parasitizing and killing the caterpillar.

Practical Applications

Knowledge about the natural occurrence and role of allelochemics opens up a huge, but relatively untapped, potential for exploiting them as pest management tools.[2] In some cases the knowledge itself is important. Plant breeders may seek varieties of plants that contain or release chemicals that deter feeding or development by phytophagous insects. Plants with allelopathic characteristics, for example, *Eucalyptus* spp., may be useful in landscaping to reduce weed problems. Species or varieties rich in attractive kairomones may be used as trap crops for various insect pests. In the production of transgenic agricultural crops it is critical not to lose the capacity for tritrophic interaction that will ensure parasitism of herbivorous insects, lest the genetically modified plants be more vulnerable to insect pests than unmodified plants.

In other cases, the capacity to use allelochemics as pest management tools may be demonstrated, but technological, economic, or social limitations may prevent their use. Allelopathic allomones from plants are under consideration for development as a new class of biodegradable herbicides.[7] But to date none can compete with conventional chemical herbicides with regard to ease of synthesis, capacity for formulation, efficacy, and/or safety. If used widely, some allelo-pathogens may pose an unacceptable threat to environmental or human health. Recent investigations show considerable promise for using nonhost volatiles to "disguise" herbivorous host plants or trees as nonhosts, but commercial formulations have not yet appeared on the market, in part because of the challenge and expense of registering those allelochemic products as pesticides. Similar problems beset the development and use of predator volatiles to protect plants from damage by herbivorous vertebrates such as deer and voles.

Despite the above limitations, a few kairomones have found widespread commercial use.[2,8] Among them is methyl engenol, which is used worldwide for capturing tephritid fruit flies, both for detection of unwanted introductions and for direct suppression of populations in a lure and kill tactic employing an insecticide-laced substrate baited with methyl engenol. A lure and kill tactic is also used effectively for control of tsetse flies that are drawn by acetone and 1-octen-3-ol baits to insecticide-treated "target" traps that simulate the silhouette of a large vertebrate. Other applications combine kairomones with pheromones. A combination of phenethyl proprionate and methyl engenol with the sex pheromone of the Japanese beetle is used in many thousands of traps in the United States each year. Similarly the kairomones ethanol and α-pinene have been used since 1981 in combination with aggregation pheromones in commercial mass trapping programs for three species of ambrosia beetles in British Columbia.

Allelochemics may also find use in the future in the application of "push-pull" tactics, in which one repellent volatile treatment is used to protect a plant or group of plants from attack by insects (push), and another attractive treatment is used in baited traps or trap plants to pull the insects away. One outstanding example of a successful push-pull application saved a rare stand of endangered Torrey pines in California from being killed by the California five spined ips. Two repellent synomones, verbenone, produced by western pine beetles, and (–)-ipsdienol, produced by pine engravers, were deployed inside the uninfested portion of the stand, and traps baited with attractive aggregation pheromone were arrayed in an adjacent area of beetle-killed pines. Over 86 weeks beginning in May 1999, 330,717 beetles were captured.

Conclusion

Despite many studies, most natural allelochemic interactions are yet to be discovered. The adoption of allelochemics as pest management tools has been limited. However, there is great potential for judicious selection and commercial development of allelochemics, particularly in integrated pest management programs that will combine a number of alternative ecologically based tactics with the reduced use of conventional chemical pesticides.

References

1. Whittaker, R.; Feeney, P. Allelochemics: chemical interactions between species. Science **1971**, *171*, 757–770.
2. Metcalf, R.; Metcalf, E. *Plant Kairomones in Insect Ecology and Control*; Chapman and Hall: New York, 1992; 168.
3. Vet, L.; Dicke, M. Ecology of infochemical use by natural enemies in a tritrophic context. Annu. Rev. Entomol. **1992**, *37*, 141–172.
4. Kats, L.; Dill, L. The scent of death: chemosensory assessment of predation risk by prey animals. Ecoscience **1998**, *5*, 361–394.
5. Stowe, M.; Turlings, T.; Loughrin, J.; Lewis, W.; Tumlinson, J. The Chemistry of Evesdropping, Alarm, and Deceit. In *Chemical Ecology: The Chemistry of Biotic Interaction*; Eisner, T., Meinwald, J., Eds.; National Academy Press: Washington, DC, 1995; 51–65.
6. Zindahl, R. *Fundamentals of Weed Science*; Academic Press: New York, 1993; 135–146.
7. Cutler, G. Allelopathy for Weed Suppression. In *Pest Management: Biologically Based Technologies*; Lumsden, R., Vaughn, J., Eds.; American Chemical Society: Washington, DC, 1993; 290–302.
8. Borden, J. Disruption of Semiochemical-Mediated Aggregation in Bark Beetles. In *Insect Pheromone Research: New Directions*; Cardé, R., Minks, A., Eds.; Chapman and Hall: New York, 1997; 421–438.

15

Aluminum

Johannes Bernhard
Wehr, Frederick
Paxton Cardell
Blamey, Peter
Martin Kopittke,
and Neal William
Menzies

Introduction

Aluminum (Al) is a metallic element characterized by its low density (2.7 g/cm^3) and resistance to corrosion due to the formation of a protective oxide layer on the surface. It is the most common (8%) metallic element of the earth's crust and the third most common element (after oxygen and silicon) on earth (Wehr, Blamey, and Menzies 2007). The element reacts strongly with oxygen-containing ligands and never occurs naturally as pure metal. It forms both octahedral and tetrahedral coordination compounds and occurs only in the trivalent oxidation state. In soil, Al-containing minerals dissolve at low pH, especially below pH 4.5, and release Al into the soil solution and aquatic environment. Al can also occur in the atmosphere as aeolian dust or ash.

Soluble forms of Al can have toxic effects on plants, humans, as well as soil and aquatic organisms, but alleviation of Al toxicity can be achieved by increasing the pH to above 4.5. Aluminum present in water and food is not readily taken up by humans. Nano-sized Al particles in the environment may result in some effects on plants and animals but more research is needed.

The chemistry and uses of Al are discussed, as well as the sources and alleviation of Al toxicity in soils, water, and atmosphere. Finally, the environmental toxicology of Al to plants, microorganisms, and animals is briefly discussed.

Chemistry of Al

Dissolution of Al minerals at low pH releases octahedral Al^{3+}, which is hexa-coordinated with water molecules (Martin 1996). Soluble Al^{3+} is classified as a hard acid due to its small ionic radius of 0.053 nm and reacts strongly with "hard" ligands such as oxygen and fluoride (Martin 1996, Martell et al. 1996). It also forms stable complexes with didentate and multidentate ligands (chelate effect) (Martell et al. 1996, Salifoglou 2002). As the pH of an acidic Al solution is increased, especially >4, hydrolysis of Al occurs, giving rise to a series of Al-hydroxy species [$AlOH^{2+}$, $AlOH_2^+$, and $Al(OH)_3$] (Martin 1996), which may undergo aggregation (polymerization) via OH bridges if the Al concentration is sufficiently high (>0.3 mg/L) (>10 μM) (Furrer, Trusch, and Muller 1992). As the pH is raised above pH 6, Al changes its coordination number to 4 and yields tetrahedral aluminate ($AlOH_4^-$) (Martin 1996). Depending on the pH and Al concentration in solution, hydrolyzed Al species may aggregate and form polymeric (polycationic) Al species such as $Al_2(OH)_2^{4+}$, $Al_3(OH)_4^{5+}$, and $Al_8(OH)_{20}(H_2O)x^{4+}$, the "gibbsite fragment" model forms, $Al_6(OH)_{12}(H_2O)_{12}^{6+}$ through $Al_{54}(OH)_{144}(H_2O)_{36}^{18+}$, $Al_{13}O_4(OH)_{24}(H_2O)_{12}^{7+}$ (Al_{13}), and $Al_2O_8Al_{28}(OH)_{56}(H_2O)_{26}^{18+}$ species (Bertsch and Parker 1996, Brown et al. 1985, Orvig 1993, Sarpola et al. 2006, Wang and Muhammed 1999). The Al_{13} polycation in solution is highly toxic to root elongation (Bertsch and Parker 1996). While many more Al polymers have been proposed, the experimental evidence in support of these species is limited.

Al Uses and Production

Due to its corrosion resistance, light weight, and excellent thermal and electrical conductivity, the metal is extensively used in the building and construction industries (window frames, doors, external cladding, A/C ducts, thermal insulation), automotive (engine blocks, car bodies), shipping (hulls) and aerospace industries (aircraft bodies), power lines, and food packaging (cans and other containers, foil). Aluminum salts and compounds are used for water purification, in which alum ($KAl(SO_4)_2.12H_2O$) is important, as catalysts in the chemical industry, and as ingredients in cosmetics (antiperspirants), pharmaceuticals (antacids, vaccine adjuvant), and foods (baking powder, spreading agent) (Table 1).

The main producer of bauxite ore is Australia (30% of world production, 88 Mt in 2017), with China, Brazil, Guinea, Jamaica and India producing lesser quantities (Resources-and-Energy-Quarterly 2018) (Table 2). The main producers of alumina in 2017 were Australia and China (World-Aluminum.org 2019), whereas smelting (refining of bauxite) is occurring in regions with cheap electrical energy (Table 2).

TABLE 1 Global Uses of Al (2018)

Use	(%)
Transport and manufacturing	41
Packaging	20
Construction	14
Electrical	8
Consumer durables	7
Machinery	7
Other	3

Source: https://archive.industry.gov.au/Office-of-the-Chief-Economist/
Publications/ResourcesandEnergyQuarterlyMarch2018/documents/
Resources-and-Energy-Quarterly-March-2018-Aluminium-alumina-
and-bauxite.pdf (accessed 29 March 2019).

TABLE 2 Worldwide Production of Alumina and Primary Al Metal in 2017

Geographic Region	Alumina (kt)	Primary Al Metal (kt)
Africa and Asia (excl China)	8,382	10,776
North America	3,033	3,950
South and Central America	12,713	1,378
East and Central Europe	4,499	3,999
Western Europe	5,890	3,776
China	70,699	35,905
Oceania	20,783	1,817
Rest of World	6,391	1,800
Total	132,400	63,404

Source: www.world-aluminium.org/statistics/primary-aluminium-production (accessed 29 March 2019).

Sources of Al in Soil

Al Minerals

The main Al-containing primary minerals are feldspars and micas. Several gemstones (ruby, sapphire, tourmaline) also contain Al. The primary Al-containing minerals weather initially to 2:1 layer aluminosilicate clay minerals (e.g., vermiculite, montmorillonite, smectite), which upon further weathering form 1:1 layer aluminosilicates such as kaolinite. Further weathering of clay minerals leads to leaching of silica and base cations (calcium and magnesium), leaving behind hydrous aluminum oxide (e.g., gibbsite, boehmite, and diaspore) in the form of bauxite, which is an important aluminum ore (Wehr, Blamey, and Menzies 2007).

Forms of Soluble Al in Soil

In acidic soils (pH < 4.5), aluminosilicate clay minerals are unstable and dissolve, releasing the trivalent Al^{3+} cation. The solubility of Al increases three orders of magnitude for every unit decrease in solution pH: at pH 4.5 the soluble Al concentration is around 0.09 mg/L (3.5 μM), decreasing to 95 ng/L (3.5 nM) at pH 5.5 (Menzies, Bell, and Edwards 1994). The trivalent Al^{3+} ion is toxic to plant roots at concentrations of 0.1–1 mg/L (5–50 μM) (Brown et al. 2008, Horst, Wang, and Eticha 2010, Poschenrieder et al. 2008). The Al^{3+} ion is readily complexed by soil organic matter and the soil organic matter controls the Al availability in most soils containing sufficient organic matter (Adams et al. 2000, Brown et al. 2008, Lofts et al. 2001, Simonsson 2000, Skyllberg 1999, Guo et al. 2007). In mineral soils low in organic matter, the availability of Al is determined by the cation exchange capacity, ionic strength, and pH (Guo et al. 2007). The polymeric Al species (e.g., Al_{13}) are metastable with a half-life for Al_{13} of several hundred hours (Etou et al. 2009, Furrer, Gfeller, and Wehrli 1999) which implies that the species undergo depolymerization or crystallization with time. The Al_{13} species is also rhizo-toxic to plant roots, with concentrations between 0.1 and 2 mg/L (0.1–2 μM) inhibiting root growth (Kopittke, Menzies, and Blamey 2004, Bertsch and Parker 1996). It is generally accepted that sulfate, phosphate, fluoride, silicate, and organic acids prevent or reverse Al_{13} formation (Bertsch and Parker 1996, Casey 2005, Masion et al. 1994, Yamaguchi et al. 2003, Kerven, Larsen, and Blamey 1995) and crystalline gibbsite may induce crystallization and depolymerization of Al_{13} (Sanjuan and Michard 1987). Since sulfate and silicate ions are prevalent even in acid soils, the natural occurrence of Al_{13} in soil solution is considered unlikely (Bertsch and Parker 1996, Gerard, Boudot, and Ranger 2001, Hiradate, Taniguchi, and Sakurai 1998). However, there is evidence that Al_{13} may form in the cell wall of plant roots (Kopittke, Menzies, and Blamey 2004, Masion and Bertsch 1997, Xia and Rayson 1998) even if the conditions in bulk solution are not favorable for Al_{13} formation. A report by Hunter and Ross (Hunter and Ross 1991) claiming that up to 30% of acid forest soils contained Al_{13} could not be replicated by other research groups. In alkaline soils, the aluminate anion $[Al(OH)_4^-]$ can form but this species is not very toxic to plants (Kinraide 1991, Kopittke, Menzies, and Blamey 2004) but may be toxic to aquatic organisms (Griffitt et al. 2008, Sjostedt et al. 2009).

Mobilization of Al in Soil

Aluminosilicate and aluminum sesquioxide clay minerals slowly dissolve at pH < 4.5 and release Al into the soil solution. Currently, more than 30% of potentially arable soils are acidic and affected by Al toxicity (von Uexküll and Mutert 1995). Soil acidification and consequential Al toxicity can be exacerbated by anthropogenic acid deposition (acid rain), use of ammonium-containing fertilizers, use of legumes in crop rotations, and the removal of crops from agricultural land (von Uexküll and Mutert 1995, Lesturgez et al. 2006). In unfarmed soils, atmospheric acid inputs are the main causes of acidification (Lapenis et al. 2004, Bergholm, Berggren, and Alavi 2003), whereas acidification in fertilized soils is mainly caused by nitrification of ammonium N-fertilizer, especially when the N is not taken up by the plant and is leached from the soil (Bergholm, Berggren, and Alavi 2003, Guo et al. 2010).

Aluminum in soil solution decreases turn-over of soil organic matter by limiting microbial degradation of the stable Al-organic matter complexes formed in soil, leading to an increase in soil organic matter (Scheel, Dorfler, and Kalbitz 2007, Scheel et al. 2008, Rasmussen, Southard, and Horwath 2006, Takahashi and Dahlgren 2016, Miyazawa et al. 2013).

Aluminum can also be released from soil minerals in strongly acidic conditions found in acid sulfate soils (Dent and Pons 1995, Faltmarsch, Astrom, and Vuori 2008), monosulfidic black ooze (Bush, Fyfe, and Sullivan 2004) and acid mine drainage (Liang and Thomson 2009, Pu et al. 2010). Acid sulfate soils are limited in occurrence to permanently or temporarily waterlogged low-lying areas (up to 5 m above sea level) (Dent and Pons 1995), whereas organic matter in irrigation channels can lead to the formation of monosulfidic black ooze (Bush, Fyfe, and Sullivan 2004). Exposure of these sulfide-containing materials to air leads to the formation of sulfuric acid and release of Al (Dsa et al. 2008, Liang and Thomson 2009, Pu et al. 2010, Soucek, Cherry, and Zipper 2003).

Fluoride ions (F^-), present as an impurity in phosphate fertilizers, can form strong complexes with Al and the resultant lesser charged complexes, AlF^{2+}, AlF_2^+, and AlF_3, are mobile in soil solution (Martin 1996, Martinent-Catalot et al. 2002) and can be either removed from soil by drainage or seepage water, or taken up by plants (Manoharan et al. 2007) and alleviate Al toxicity (Yang et al. 2016).

Organic acids form complexes with Al and can mobilize Al in the soil (Takahashi et al. 2008, Lange, Solberg, and Clarke 2006a). In the field, movement of organic acid–Al complexes gives rise to podzolization of soils and downward movement of Al until conditions are favorable for dissociation of Al complexes deeper in the profile or export to waterways (Bardy et al. 2007).

Alleviation of Al in Soils

Since the solubility of Al in soil solution is lowest around pH 6–7 (McBride 1994), controlling the pH is the most effective way in reducing potential Al toxicity in ecosystems. The pH value to which acid soils need to be limed depends upon the plant species, but is often near 5.5. Compounds to increase the pH of acid soils include calcitic ($CaCO_3$) or dolomitic ($CaMgCO_3$) limestone (Machacha 2004), slaked lime [$Ca(OH)_2$] (Sun et al. 2000), wood ash (Materechera and Mkhabela 2002), alkaline poultry manure (Mokolobate and Haynes 2002, Tang et al. 2007), charcoal (Steiner et al. 2007), biochar (Dai et al. 2017), fly ash (Morikawa and Saigusa 2002, Tarkalson et al. 2005), ground basalt (Panhwar et al. 2016), alkaline slag (Shi et al. 2017, Li et al. 2015) or waste cement (Morikawa and Saigusa 2002). Any waste product with high ash alkalinity can be used to ameliorate acid soil by increasing the pH (Naramabuye and Haynes 2006). Surface application of liming agents with subsequent incorporation into the soil (ploughing, harrowing, etc.,) is commonly used. The mobility of limestone in the soil profile is very low (Conyers et al. 2003, Godsey et al. 2007, Scott et al. 2007). Therefore, rapid alleviation of subsoil acidity can only be achieved by deep placement of limestone (deep ripping). This, however, is currently not economically possible for most low-value broadacre crops. Depending on the severity of soil acidity, and the buffer capacity of the soil as determined by the cation exchange capacity of the soil, between 0.1 and 1 ton lime-stone/ha/yr are needed to maintain soil pH in agricultural production systems (Scott et al. 2007, Machacha 2004). Numerous methods have been developed to determine the required liming rate based on laboratory soil analyses. The liming of soil needs to be repeated every few years depending on the land management system. To overcome the low mobility of limestone in soils, higher rates of gypsum can be used instead. The Ca from gypsum moves more readily down the soil profile (Liu and Hue 2001) and the increase in Ca and ionic strength can lower the Al toxicity, despite increasing the overall concentration of Al (as $AlSO_4^+$, which is less toxic than Al^{3+}). Liming has greater beneficial effect on soil bacterial communities than on soil fungal communities (Mota et al. 2008, Nelson and Mele 2006).

Application of organic matter (green manure, farm yard manure) can lower Al toxicity by complexing Al (Qin and Chen 2005, Vieira et al. 2008). However, application of organic matter requires higher application rates (>10 tons/ha) than limestone (1 ton/ha) and is, therefore, not as effective as limestone (Raboin et al. 2016). Humic acid has been shown to either precipitate or bind Al_{13} (Yamaguchi et al. 2004) and Al^{3+}, (Matthias, Maurer, and Parlar 2003, Shoba and Chudnenko 2014), thereby reducing root growth inhibition. Biochar can also adsorb Al-OH complexes and reduce rhizotoxicity in crops (Qian, Chen, and Hu 2013). Sewage sludge can also be alkaline and the presence of organic acids in sludge can bind and immobilize Al (Lopez-Diaz, Mosquera-Losada, and Rigueiro-Rodriguez 2007).

Alkaline bauxite refinery waste (red mud) poses a challenge for plant growth due to the high Fe oxide concentration leading to phosphate immobilization, absence of diverse microbial communities, and lack of organic matter (Wehr, Menzies, and Fulton 2006). However, the presence of aluminate in these waste materials is not limiting to plant root growth (Kopittke, Menzies, and Blamey 2004).

Sources of Al in Water

Acid Deposition

Atmospheric gases such as NO_2, SO_2, and CO_2 dissolve in rainwater and form acid rain (Lapenis et al. 2004, Lawrence 2002). The atmospheric acid inputs in Europe are considered to be around 0.2–4 kmol/ha/yr (de Vries, Reinds, and Vel 2003). The acids can either dissolve Al minerals in soil (see above) or dissolve sediments in water bodies receiving acidic water. Furthermore, acid deposition can release Al complexed to organic matter. Introduction of gaseous emission standards in the 1980s in Europe, and later in the United States, has lowered acid rain, with a consequential decrease in Al in surface water (Lange, Solberg, and Clarke 2006b, Skjelkvale et al. 2001, Vuorenmaa et al. 2018). Podzolization of lateritic soils was found to export vast quantities of Al to waterways (Bardy et al. 2007). The concentration of Al in drainage water from forest watersheds is highly variable, however (Lange, Solberg, and Clarke 2006b). In coastal areas subjected to atmospheric salt deposition, the salt can displace Al from cation exchange sites, leading to an increase in Al in drainage water (Lange, Solberg, and Clarke 2006b). Concentrations of Al in groundwater are governed by pH since Al solubility is pH dependent: less than 0.01 mg/L Al is observed at pH 7, which increases to 51 mg/L at pH < 4 (Fest et al. 2007).

Acid Mine Drainage and Acid Sulfate Soils

Sulfidic mine wastes and acid sulfate soils release acid when exposed to air, resulting in leaching of acid drainage water and soluble Al into water bodies. While there are indications that the high Fe concentration in acid mine drainage can protect aluminosilicate clay mineral from dissolution and minimize Al release (Dubikova et al. 2002), no such mechanism has been reported for acid sulfate soils. At Trinity inlet in North Queensland, Australia, Al concentrations in drainage water from acid sulfate soils were measured in the range 4.2–10 mg/L (Hicks, Bowman, and Fitzpatrick 1999).

Aluminum from Water Purification

Water clarification involving the settling of fines (suspended colloids and microorganisms) may be accomplished with alum (potassium aluminum sulfate). Since drinking water will be near neutral (pH 6.5–8.5 (NHMRC 2004)), solubility of Al is low. The European, Australian, New Zealand, and WHO regulations for drinking water propose a threshold concentration of Al of 0.1-0.2 mg/L (Schaefer and Seifert 2006). Traces of Al are found in raw and alum-treated potable water (Cech and Montera 2000, Srinivasan and Viraraghavan 2002). The amount of Al taken up from drinking water has been estimated at 1%–2% of total Al uptake per day (Yokel, Hicks, and Florence 2008, Stauber et al. 1999, Willhite, Ball, and McLellan 2012), and less than 2% of Al in drinking water is bioavailable (Willhite, Ball, and McLellan 2012, Yokel and McNamara 2001).

Water treatment residues (alum sludges) have been used to immobilize phosphate and heavy metals in soils and wastes (Mahdy et al. 2009, Oladeji, Sartain, and O'Connor 2009). Land application of these sludges has not resulted in elevated Al concentrations in plants grown in these soils (Oladeji, Sartain, and O'Connor 2009), especially when the sludges were aged to reduce Al availability (Agyin-Birikorang and O'Connor 2009).

Alleviation of Al in Water

Raising the pH of water to pH 6–7 will precipitate Al and establish an equilibrium concentration of Al determined by the solubility product of Al-hydroxide in water. At pH < 4.5, solubility of Al minerals increases rapidly, releasing Al^{3+} into the waterbody and at pH > 8, aluminate anions are formed (Wehr, Blamey, and Menzies 2007). Therefore, pH adjustment of water to near neutrality is the most effective method to decrease Al concentrations in water (Fest et al. 2007). The Al remaining in potable water after purification can only be removed by ion exchange resins (Othman, Abdullah, and Abd Aziz 2010).

Sources of Al in the Atmosphere

Atmospheric dust derived from raised soil particles is the main source of Al in the atmosphere, but volcanic activity can be a source also. In the atmosphere, Al occurs generally as silicate, sulfate, or oxide compounds. Flue ash and flue gases can also contribute Al to the atmosphere (Kabata-Pendias and

TABLE 3   Concentrations of Al in the Atmosphere

Region	Al Concentration ($\mu g/m^3$)
South Pole	0.0003–0.0008
Greenland	0.24–0.38
Shetland Islands	0.06
Norway	0.03
Germany	0.16–2.90
Japan	0.04–10.60
China	2.00–6.00
North America	0.60–2.33
Central America	0.76–0.88
South America	0.46–15.00

Source: Adapted from Kabata-Pendias and Kabata 2001 and Wang et al., 2007.

Kabata 2001). Dust contains around 2.6% Al (Schussler, Balzer, and Deeken 2005). Concentrations of Al in the atmosphere vary between geographic locations (Table 3). Inhalation of Al from the atmosphere has been estimated as 4–20 µg Al/day (Yokel, Hicks, and Florence 2008).

Sources of Al in Foodstuff, Cosmetics, Pharmaceuticals, and Workplaces

The concentration of Al in foodstuff is generally very low since little Al is taken up by plants and translocated to the aboveground parts (Table 4) (Chen et al. 2008, Fung et al. 2009, Mueller, Anke, and Illing-Guenther 1998, Ertl and Goessler 2018, Schaefer and Seifert 2006). However, the concentration of Al in black tea is high (Table 4), but most of the Al in infused tea is complexed with phenolics and organic acids, though addition of lemon or lime juice may increase Al availability (Fung et al. 2009, Krewski et al. 2007). The bioavailability of Al from black tea is around 0.4% (Yokel and Florence 2008).

Preparation of acidic foodstuff in Al containers or covering with Al foil may increase Al in foodstuff (Ertl and Goessler 2018, Turhan 2006, Ranau, Oehlenschlager, and Steinhart 2001). The addition of sodium aluminum phosphate as a spreading agent to processed foodstuff (e.g., cheese spread) can also contribute to the dietary Al intake (Yokel, Hicks, and Florence 2008, Schaefer and Seifert 2006). Salts of Al are also used as a food additive for clarification of beverages, as anticaking agent, as baking powder, and to enhance color stability (Stahl et al. 2018, Aguilar et al. 2008, Schaefer and Seifert 2006). The Al in food contributes around 95% of daily oral Al uptake and the daily dietary Al intake has been estimated as 4–16 mg, of which 0.1%–0.3% is bioavailable (Krewski et al. 2007, Yokel, Hicks,

TABLE 4 Concentration of Al in Foodstuff

Foodstuff	Al Concentration (mg/kg DW)
Grains	1–135
Pulses	3–16
Fruit	0.5–20
Vegetables	1–270
Herbs	8–280
Spices	6–695
Meat	0.5–16
Fish	0.5–10
Dairy	1–16
Tea	900–1000

Source: Adapted from Kabata-Pendias and Kabata 2001, Yi and Cao 2008, Fung et al. 2009, Ertl and Goessler 2018, Mueller, Anke, and Illing-Guenther 1998, Schaefer and Seifert 2006, Chen et al. 2008.

TABLE 5 Estimated Sources and Daily Exposure to Al, its Bioavailability, and Estimated Daily Intake

Environmental Exposure	Exposure (µg/day)	Availability (%)	Intake (µg/day)
Air	2–200	2	0.04–4
Industrial air	25,000	2	
Water	200–1,000	<2	2–10
Food	8,000–16,000	<0.3	8–48
Cosmetics	<70,000	0.2	140
Vaccines	1–8	95	1–8
Antacids	<5,000,000	<2	100,000

Source: Adapted from Yokel, Hicks, and Florence 2008, Krewski et al. 2007, Gourier-Frery and Frery 2004.

and Florence 2008, Stahl et al. 2018) (Table 5). The tolerable weekly intake of Al has been set by the European Food Safety Authority to 1 mg Al per kilogram of bodyweight per week (Stahl et al. 2018, Aguilar et al. 2008).

Cosmetics, especially antiperspirants, contain 25% Al as Al-chlorohydrate (Shen and Nardello-Rataj 2009). Daily antiperspirant use can contribute around 70 mg Al per day, yet dermal absorption of Al is only 0.02%–0.2% (Yokel, Hicks, and Florence 2008, Flarend et al. 2001). Buffered aspirin and antacids often contain Al hydroxide suspensions and can contribute up to 5 g Al/day (Yokel, Hicks, and Florence 2008).

Al-phosphate and Al-hydroxide are used as adjuvants to enhance vaccine effectiveness (Exley, Siesjo, and Eriksson 2010). It has been estimated that the amount of Al taken up from Al-containing adjuvant in vaccine is in the range 1–8 µg Al/day (Yokel, Hicks, and Florence 2008), with a lifetime intake of 15 mg, of which most is likely bioavailable (Yokel and McNamara 2001).

Environmental Toxicology of Al

Al Toxicity in Plants

The leading cause of poor fertility of acidic soils is Al toxicity (Asher, Grundon, and Menzies 2002, Horst, Wang, and Eticha 2010, Poschenrieder et al. 2008). Toxicity of Al in plants is mainly manifested by an inhibition of root growth, with Al^{3+} and Al_{13} considered the most toxic species (Kinraide 1991, 1997, Poschenrieder et al. 2008, Singh et al. 2017). The action of Al on plant cells is both in the apoplast (binding to cell wall components) (Kopittke et al. 2015) and in the cytoplasma (formation of reactive oxygen species and oxidative stress responses (Yamamoto et al. 2003, Singh et al. 2017)). The correlation between Al concentration and root grow inhibition is sometimes poor. This is due to the several Al-hydroxy species that coexist within a narrow pH band and cannot be investigated in isolation. Furthermore, the activities of individual species must be calculated from equilibrium data that may be uncertain (Boudot et al. 1996). The critical Al concentration at which root elongation is inhibited is as low as 0.1–0.5 mg/L (5–20 µM) in solutions of low ionic strength, representative of acid soils. Exact critical values depend on the plant species and the conditions in which the plants are grown. Generally, root hairs are more sensitive to Al toxicity than roots (Brady et al. 1993). Root growth inhibition results in short stubby roots and absence of root hairs, leading to poor water and nutrient utilization by plants. The main site affected by Al is near the elongation zone of roots (1–4 mm behind the root tip) (Blamey, Nishizawa, and Yoshimura 2004, Sivaguru and Horst 1998, Kopittke et al. 2015). The reaction of aluminum with phosphate anions in the soil may result in the precipitation of Al-phosphate minerals (e.g., variscite), lowering phosphorus availability to plants. This effect is accentuated by the lower phosphorus uptake capacity of Al-damaged roots.

In plants, detoxification of Al is generally achieved by organic acid exudation (Chen, Wang, and Yeh 2017, Ryan, Delhaize, and Randall 1995), but co-precipitation of Al with Si in the cell wall to form insoluble phytolith has also been reported (Hodson and Sangster 1999, Sangster and Hodson 2001).

Many *Rhizobium* strains necessary for nodulation in legumes are inhibited by soil Al, with 0.03 mg/L Al at pH 5 being toxic (Kinraide and Sweeney 2003). Therefore, nodulation and nitrogen fixation are decreased in acidic soils (Jaiswal, Naamala, and Dakora 2018, Ramirez et al. 2018). Soil bacteria are considered to be more sensitive to low pH and Al than soil fungi (Shirokikh et al. 2004). Swarming of some bacteria can also be inhibited by Al (Illmer and Schinner 1997).

Nano-sized aluminum oxide particles are more toxic than micro-sized Al oxide particles but soluble Al^{3+} is most toxic (Shabnam and Kim 2018). In whole plants, nanoparticles are adsorbed onto roots and not translocated, thus cause no toxicity (Shabnam and Kim 2018), but are toxic when applied to callus cultures (Poborilova, Opatrilova, and Babula 2013). The effect of nanoparticles on soil microbial process is still unclear (Doshi et al. 2008, Mishra and Kumar 2009, Jiang, Mashayekhi, and Xing 2009) and requires more research.

Al Toxicity in Aquatic Organisms

In solution, Al is highly toxic to aquatic organisms in fresh waters (Gensemer and Playle 1999) with >0.4 mg/L Al in freshwater streams decreasing macroinvertebrates (Waters and Webster-Brown 2013) and >0.02 mg/L causing mortality in brown trout (Andren and Rydin 2012). In seawater, awhich is alkaline, presence of aluminate anions can affect marine organisms (Golding et al. 2015).

In freshwater fish and crayfish, Al impairs gill function by inducing secretion of mucus and causes oxygen stress (Ward, McCrohan, and White 2006, Griffitt et al. 2011). No biomagnification of Al was observed in aquatic food webs (Winterbourn, McDiffett, and Eppley 2000).

Maximal Al accumulation occurs on gills at pH 6–8 (Ward, McCrohan, and White 2006, Gensemer and Playle 1999) but complexation with organic matter can prevent Al binding to gills and decrease toxicity (Trenfield et al. 2012).

Nanoparticulate Al is less toxic than Al^{3+} in aquatic ecosystems and AlNP did not cause gill modification and no mortality up to 12 mg/L in moderately hard water (Griffitt et al. 2011). Freshwater algae showed toxicity at 40–45 mg/L after 3 day exposure to nano Al (Sadiq et al. 2011) The freshwater Daphnids are sensitive to nano-sized Al (Strigul et al. 2009, Pakrashi et al. 2013) Toxicity of nanoparticles appears to be caused both by triggering oxidative stress responses in the target and by dissolution of Al^{3+} (Pakrashi et al. 2013).

Al Toxicity in Humans

Aluminum has been implicated as a possible risk factor in Alzheimer's disease and dialysis dementia in humans (Nayak 2002, Yokel 2000, Krewski et al. 2007). Alzheimer's disease is a multifactorial disease, with environmental and genetic factors playing a role in the pathogenesis (Mutter et al. 2007, Frisardi et al. 2010, Liang 2018).

It has been shown that Al exerts its cytotoxic effects by disrupting lipid membrane fluidity, and disturbing iron (Fe), magnesium, and calcium homeostasis, and causing oxidative stress (Tönnies and Trushina 2017, Mailloux, Lemire, and Appanna 2011, Yokel 2000, Michalke, Halbach, and Nieschwitz 2009). The oxidative stress response triggered by Al may affect mitochondria (Kumar and Gill 2009) and triggering neurodegenerative disorders (Campbell and Bondy 2000).

Parenterally and intramuscularly applied Al is 100% bioavailable (Schafer and Jahreis 2009). By contrast, availability of Al via oral, inhalation, or dermal routes is less than 2% (Schafer and Jahreis 2009). In plasma, Al is bound to transferrin (Schafer and Jahreis 2009) and citric acid (Yokel and McNamara 2001). Most of the Al taken up by humans is excreted in the urine (Yokel and McNamara 2001) and feces (Yokel and McNamara 2001, Schafer and Jahreis 2009). Ionic Al^{3+} may form in lysosomes from Al adjuvants (Mold, Shardlow, and Exley 2016).

Bauxite mining in open-cut mines does not appear to affect the respiratory health of mine workers (Wesdock and Arnold 2014). Bauxite mine and Al refinery workers have no increased risk of cancer (Fritschi et al. 2008, Wesdock and Arnold 2014), but some bauxite ores may contain radioactive elements (radium and thorium), which may pose a long-term radiation exposure risk. In Al smelter workers, exposure to fluoride may affect lung function more than Al itself (Wesdock and Arnold 2014) and no congenital abnormalities have been observed in Al smelter workers (Kiesswetter et al. 2009, Wesdock and Arnold 2014). Likewise, Al welders are not showing increased incidence of neurobehavioral problems (Kiesswetter et al. 2009).

Nanoparticles of Al oxide have been shown to cause neurotoxicity, and they can cross the blood–brain barrier (Jiang, Mashayekhi, and Xing 2009, Sharma et al. 2009). Nano-sized Al is more biotoxic than micro-sized Al (Jiang, Mashayekhi, and Xing 2009, Park et al. 2015, Ates et al. 2015). However, more research is needed into the environmental threats posed by nanoparticles of Al and to determine the threlold levels for various shapes and sizes of Al nanoparticles in humans. Limited studies with human cell cultures in-vitro showed detrimental effects when applied at 50 mg/L (Song et al. 2017).

Conclusion

Aluminum has no known biological benefit to living organisms and is toxic in micromolar concentrations. Since solubility of Al is minimal at circumneutral pH, alleviation of Al toxicity can be achieved by pH adjustment. Complexation of Al by organic acids and humic substances also lowers biotoxicity of Al. While Al toxicity in soil can be easily overcome by incorporating lime into the soil by tilling, new management strategies need to be developed to counteract soil acidification and consequent Al toxicity in zero-till farming systems. The decrease in atmospheric acid inputs (acid rain) over the last two decades has resulted in lower Al solubility and less Al damage to aquatic ecosystems. Uptake of Al in humans through foodstuff and water poses an uncertain risk, but pollution with nano-particulate Al may result in human health problems. The effect of nano-sized Al particles on the functioning of soil and water ecosystems, as well as human health, needs more research.

Acknowledgments

This research was supported under the Australian Research Council's Discovery Projects Funding scheme (project number DP0665467).

References

Adams, M. L., D. J. Hawke, N. H. S. Nilsson, and K. J. Powell. 2000. The relationship between soil solution pH and Al^{3+} concentrations in a range of South Island (New Zealand) soils. *Australian Journal of Soil Research* 38 (1):141–153.

Aguilar, F., H. Autrup, S. Barlow, L. Castle, R. Crebelli, W. Dekant, K. H. Engel, N. Gontard, D. Gott, S. Grilli, R. Gurtler, J. C. Larsen, C. Leclercq, J. C. Leblanc, F. X. Malcata, W. Mennes, M. R. Milana, I. Pratt, I. Rietjens, P. Tobback, F. Toldra, and Members Panel. 2008. Safety of aluminium from dietary intake. *EFSA Journal* 6 (7). doi:10.2903/j.efsa.2008.754.

Agyin-Birikorang, S., and G. A. O'Connor. 2009. Aging effects on reactivity of an aluminum-based drinking-water treatment residual as a soil amendment. *Science of the Total Environment* 407 (2):826–834.

Andren, C. M., and E. Rydin. 2012. Toxicity of inorganic aluminium at spring snowmelt--in-stream bioassays with brown trout (Salmo trutta L.). *Science of the Total Environment* 437:422–32. doi:10.1016/j. scitotenv.2012.08.006.

Asher, C. J., N. J. Grundon, and N. W. Menzies. 2002. *How to Unravel and Solve Soil Fertility Problems, ACIAR Monograph No. 83.* Canberra: Australian Centre for International Agricultural Research.

Ates, M., V. Demir, Z. Arslan, J. Daniels, I. O. Farah, and C. Bogatu. 2015. Evaluation of alpha and gamma aluminum oxide nanoparticle accumulation, toxicity, and depuration in *Artemia salina* larvae. *Environmental Toxicology* 30 (1):109–118. doi:10.1002/tox.21917.

Bardy, M., C. Bonhomme, E. Fritsch, J. Maquet, R. Hajjar, T. Allard, S. Derenne, and G. Calas. 2007. Al speciation in tropical podzols of the upper Amazon basin: a solid-state Al-27 MAS and MQMAS NMR study. *Geochimica et Cosmochimica Acta* 71 (13):3211–3222.

Bergholm, J., D. Berggren, and G. Alavi. 2003. Soil acidification induced by ammonium sulphate addition in a Norway spruce forest in Southwest Sweden. *Water Air and Soil Pollution* 148:87–109.

Bertsch, P. M., and D. R. Parker. 1996. Aqueous polynuclear aluminum species. In *The Environmental Chemistry of Aluminum*, edited by G. Sposito, 117–168. Boca Raton: CRC Lewis Publishers.

Blamey, F. P. C., N. K. Nishizawa, and E. Yoshimura. 2004. Timing, magnitude, and location of initial soluble aluminum injuries to mungbean roots. *Soil Science and Plant Nutrition* 50:67–76.

Boudot, J. P., O. Maitat, D. Merlet, and J. Rouiller. 1996. Occurrence of non-monomeric species of aluminium in undersaturated soil and surface waters: consequences for the determination of mineral saturation indices. *Journal of Hydrology* 177 (1–2):47–63.

Brady, D. J., D. G. Edwards, C. J. Asher, and F. P. C. Blamey. 1993. Calcium amelioration of aluminium toxicity effects on root hair development in soybean *Glycine max*. *New Phytologist* 123:531–538.

Brown, T. T., R. T. Koenig, D. R. Huggins, J. B. Harsh, and R. E. Rossi. 2008. Lime effects on soil acidity, crop yield, and aluminum chemistry in direct-seeded cropping systems. *Soil Science Society of America Journal* 72 (3):634–640.

Brown, P. L., R. N. Sylva, G. E. Batley, and J. Ellis. 1985. The hydrolysis of metal ions. Part 8. Aluminium (III). *Journal of the Chemical Society, Dalton Transactions* 1985:1967–1970.

Bush, R. T., D. Fyfe, and L. A. Sullivan. 2004. Occurrence and abundance of monosulfidic black ooze in coastal acid sulfate soil landscapes. *Australian Journal of Soil Research* 42 (5–6):609–616.

Campbell, A., and S. C. Bondy. 2000. Aluminum induced oxidative events and its relation to inflammation: a role for the metal in Alzheimer's disease. *Cellular and Molecular Biology* 46:721–730.

Casey, W. H. 2005. Large aqueous aluminum hydroxide molecules. *Chemical Reviews* 106:1–16.

Cech, I., and J. Montera. 2000. Spatial variations in total aluminum concentrations in drinking water supplies studied by geographic information system (GIS) methods. *Water Research* 34 (10):2703–2712. doi:10.1016/s0043-1354(00)00026-9.

Chen, R. F., R. F. Shen, P. Gu, H. Y. Wang, and X. H. Xu. 2008. Investigation of aluminum-tolerant species in acid soils of South China. *Communications in Soil Science and Plant Analysis* 39 (9–10):1493–1506.

Chen, Y.-T., Y. Wang, and K.-C. Yeh. 2017. Role of root exudates in metal acquisition and tolerance. *Current Opinion in Plant Biology* 39:66–72. doi:10.1016/j.pbi.2017.06.004.

Conyers, M. K., D. P. Heenan, W. J. McGhie, and G. P. Poile. 2003. Amelioration of acidity with time by limestone under contrasting tillage. *Soil & Tillage Research* 72 (1):85–94. doi:10.1016/s0167-1987(03)00064-3.

Dai, Z. M., X. J. Zhang, C. Tang, N. Muhammad, J. J. Wu, P. C. Brookes, and J. M. Xu. 2017. Potential role of biochars in decreasing soil acidification – a critical review. *Science of the Total Environment* 581:601–611. doi:10.1016/j.scitotenv2016.12.169.

de Vries, W., G. J. Reinds, and E. Vel. 2003. Intensive monitoring of forest ecosystems in Europe 2: atmospheric deposition and its impacts on soil solution chemistry. *Forest Ecology and Management* 174 (1–3):97–115. doi:10.1016/s0378-1127(02)00030-0.

Dent, D. L., and L. J. Pons. 1995. A world perspective on acid sulfate soils. *Geoderma* 67 (3–4):263–276.

Doshi, R., W. Braida, C. Christodoulatos, M. Wazne, and G. O'Connor. 2008. Nano-aluminum: transport through sand columns and environmental effects on plants and soil communities. *Environmental Research* 106 (3):296–303.

Dsa, J. V., K. S. Johnson, D. Lopez, C. Kanuckel, and J. Tumlinson. 2008. Residual toxicity of acid mine drainage-contaminated sediment to stream macroinvertebrates: relative contribution of acidity vs. metals. *Water Air and Soil Pollution* 194 (1–4):185–197.

Dubikova, M., P. Cambier, V. Sucha, and M. Caplovicova. 2002. Experimental soil acidification. *Applied Geochemistry* 17 (3):245–257. doi:10.1016/s0883-2927(01)00081-6.

Ertl, K., and W. Goessler. 2018. Aluminium in foodstuff and the influence of aluminium foil used for food preparation or short term storage. *Food Additives and Contaminants: Part B* 11:153–159.

Etou, A., S. Q. Bai, T. Saito, H. Noma, Y. Okaue, and T. Yokoyama. 2009. Formation conditions and stability of a toxic tridecameric Al polymer under a soil environment. *Journal of Colloid and Interface Science* 337 (2):606–609.

Exley, C., P. Siesjo, and H. Eriksson. 2010. The immunobiology of aluminium adjuvants: how do they really work? *Trends in Immunology* 31:103–109.

Faltmarsch, R. M., M. E. Astrom, and K. M. Vuori. 2008. Environmental risks of metals mobilised from acid sulphate soils in Finland: a literature review. *Boreal Environment Research* 13 (5):444–456.

Fest, E. P. M. J., E. J. M. Temminghoff, J. Griffioen, B. Van Der Grift, and W. H. Van Riemsdijk. 2007. Groundwater chemistry of Al under Dutch sandy soils: effects of land use and depth. *Applied Geochemistry* 22 (7):1427–1438.

Flarend, R., T. Bin, S. D. Elmore, and S. L. Hem. 2001. A preliminary study of the dermal absorption of aluminium from antiperspirants using aluminium-26. *Food and Chemical Toxicology* 39:163–168.

Frisardi, V., V. Solfrizzi, C. Capurso, P. G. Kehoe, B. P. Imbimbo, A. Santamato, F. Dellegrazie, D. Seripa, A. Pilotto, A. Capurso, and F. Panza. 2010. Aluminum in the diet and Alzheimer's disease: from current epidemiology to possible disease-modifying treatment. *Journal of Alzheimers Disease* 20:17–30.

Fritschi, L., J. L. Hoving, M. R. Sim, A. Del Monac, E. MacFarlane, D. McKenzie, G. Benke, and N. de Klerk. 2008. All cause mortality and incidence of cancer in workers in bauxite mines and alumina refineries. *International Journal of Cancer* 123 (4):882–887.

Fung, K. F., H. P. Carr, B. H. T. Poon, and M. H. Wong. 2009. A comparison of aluminum levels in tea products from Hong Kong markets and in varieties of tea plants from Hong Kong and India. *Chemosphere* 75 (7):955–962.

Furrer, G., M. Gfeller, and B. Wehrli. 1999. On the chemistry of the Keggin Al_{13} polymer: kinetics of proton-promoted decomposition. *Geochimica et Cosmochimica Acta* 63:3069–3076.

Furrer, G., B. Trusch, and C. Muller. 1992. The formation of polynuclear aluminum under simulated natural conditions. *Geochimica et Cosmochimica Acta* 56:3831–3838.

Gensemer, R. W., and R. C. Playle. 1999. The bioavailability and toxicity of aluminum in aquatic environments. *Critical Reviews in Environmental Science and Technology* 29 (4):315–450.

Gerard, F., J. P. Boudot, and J. Ranger. 2001. Consideration on the occurrence of the Al-13 polycation in natural soil solutions and surface waters. *Applied Geochemistry* 16 (5):513–529.

Godsey, C. B., G. M. Pierzynski, D. B. Mengel, and R. E. Lamond. 2007. Management of soil acidity in no-till production systems through surface application of lime. *Agronomy Journal* 99 (3):764–772. doi:10.2134/agronj2006.0078.

Golding, Lisa A., Brad M. Angel, Graeme E. Batley, Simon C. Apte, Rick Krassoi, and Chris J. Doyle. 2015. Derivation of a water quality guideline for aluminium in marine waters. *Environmental Toxicology and Chemistry* 34 (1):141–151. doi:10.1002/etc.2771.

Gourier-Frery, C., and N. Frery. 2004. Aluminium. *EMC-Toxicologie Pathologie* 1:79–95.

Griffitt, R. J., A. Feswick, R. Weil, K. Hyndman, P. Carpinone, K. Powers, N. D. Denslow, and D. S. Barber. 2011. Investigation of acute nanoparticulate aluminum toxicity in zebrafish. *Environmental Toxicology* 26 (5):541–551. doi:10.1002/tox.20669.

Griffitt, R. J., J. Luo, J. Gao, J. C. Bonzongo, and D. S. Barber. 2008. Effects of particle composition and species on toxicity of metallic nanomaterials in aquatic organisms. *Environmental Toxicology and Chemistry* 27 (9):1972–1978.

Guo, J. H., X. J. Liu, Y. Zhang, J. L. Shen, W. X. Han, W. F. Zhang, P. Christie, K. W. T. Goulding, P. M. Vitousek, and F. S. Zhang. 2010. Significant acidification in major Chinese croplands. *Science* 327:1008–1010.

Guo, J. H., X. S. Zhang, R. D. Vogt, J. S. Xiao, D. W. Zhao, R. J. Xiang, and J. H. Luo. 2007. Evaluating main factors controlling aluminum solubility in acid forest soils, southern and southwestern China. *Applied Geochemistry* 22:388–396.

Hicks, W. S., G. M. Bowman, and R. W. Fitzpatrick. 1999. Environmental impact of acid sulfate soils near Cairns, QLD. In *CSIRO Land and Water Technical Report 15/99*. Canberra: Commonwealth Scientific and Industrial Research Organisation.

Hiradate, S., S. Taniguchi, and K. Sakurai. 1998. Aluminum speciation in aluminum-silica solutions and potassium chloride extracts of acidic soils. *Soil Science Society of America Journal* 62 (3):630–636.

Hodson, M. J., and A. G. Sangster. 1999. Aluminium/silicon interactions in conifers. *Journal of Inorganic Biochemistry* 76 (2):89–98. doi:10.1016/s0162-0134(99)00119-1.

Horst, W. J., Y. Wang, and D. Eticha. 2010. The role of the root apoplast in aluminium-induced inhibition of root elongation and in aluminium resistance of plants: a review. *Annals of Botany* 106:185–197.

Hunter, D., and D. S. Ross. 1991. Evidence for a phytotoxic hydroxy-aluminum polymer in organic soil horizons. *Science* 251 (4997):1056–1058.

Illmer, P., and F. Schinner. 1997. Influence of aluminum on motility and swarming of *Pseudomonas* sp. and *Arthrobacter* sp. *FEMS Microbiology Letters* 155 (1):121–124.

Jaiswal, S. K., J. Naamala, and F. D. Dakora. 2018. Nature and mechanisms of aluminium toxicity, tolerance and amelioration in symbiotic legumes and rhizobia. *Biology and Fertility of Soils* 54 (3):309–318. doi:10.1007/s00374-018-1262-0.

Jiang, W., H. Mashayekhi, and B. S. Xing. 2009. Bacterial toxicity comparison between nano- and micro-scaled oxide particles. *Environmental Pollution* 157 (5):1619–1625.

Kabata-Pendias, A., and H. Kabata. 2001. *Trace elements in soils and plants*. 3rd ed. Boca Raton: CRC Press.

Kerven, G. L., P. L. Larsen, and F. P. C. Blamey. 1995. Detrimental sulfate effects on formation of Al-13 tridecameric polycation in synthetic soil solutions. *Soil Science Society of America, Journal* 59:765–771.

Kiesswetter, E., M. Schaper, M. Buchta, K. H. Schaller, B. Rossbach, T. Kraus, and S. Letzel. 2009. Longitudinal study on potential neurotoxic effects of aluminium: II. Assessment of exposure and neurobehavioral performance of Al welders in the automobile industry over 4 years. *International Archives of Occupational and Environmental Health* 82 (10):1191–1210.

Kinraide, T. B. 1991. Identity of the rhizotoxic aluminium species. In *Plant-soil interactions at low pH*, edited by R. J. Wright, V. C. Baligar and R. P. Murrmann, 717–728. Dordrecht: Kluwer Academic Publishers.

Kinraide, T. B. 1997. Reconsidering the rhizotoxicity of hydroxyl, sulphate, and fluoride complexes of aluminium. *Journal of Experimental Botany* 48 (310):1115–1124.

Kinraide, T. B., and B. K. Sweeney. 2003. Proton alleviation of growth inhibition by toxic metals (Al, La, Cu) in rhizobia. *Soil Biology & Biochemistry* 35 (2):199–205.

Kopittke, P. M., N. W. Menzies, and F. P. C. Blamey. 2004. Rhizotoxicity of aluminate and polycationic aluminium at high pH. *Plant and Soil* 266 (1–2):177–186.

Kopittke, P. M., K. L. Moore, E. Lombi, A. Gianoncelli, B. J. Ferguson, F. P. C. Blamey, N. W. Menzies, T. M. Nicholson, B. A. McKenna, P. Wang, P. M. Gresshoff, G. Kourousias, R. I. Webb, K. Green, and A. Tollenaere. 2015. Identification of the primary lesion of toxic aluminum in plant roots. *Plant Physiology* 167 (4):1402–1411. doi:10.1104/pp.114.253229.

Krewski, D., R. A. Yokel, E. Nieboer, D. Borchelt, J. Cohen, J. Harry, S. Kacew, J. Lindsay, A. M. Mahfouz, and V. Rondeau. 2007. Human health risk assessment for aluminium, aluminium oxide, and aluminium hydroxide. *Journal of Toxicology and Environmental Health-Part B-Critical Reviews* 10:1–269.

Kumar, V., and K. D. Gill. 2009. Aluminium neurotoxicity: neurobehavioural and oxidative aspects. *Archives of Toxicology* 83 (11):965–978.

Lange, H., S. Solberg, and N. Clarke. 2006a. Aluminum dynamics in forest soil waters in Norway. *Science of the Total Environment* 367:942–957.

Lange, H., S. Solberg, and N. Clarke. 2006b. Aluminum dynamics in forest soil waters in Norway. *Science of the Total Environment* 367 (2–3):942–957. doi:10.1016/j.scitotenv.2006.01.033.

Lapenis, A. G., et al. 2004. Acidification of forest soil in Russia: From 1893 to present. *Global Biogeochemical Cycles* 18(1): GB1037.

Lawrence, G. B. 2002. Persistent episodic acidification of streams linked to acid rain effects on soil. *Atmospheric Environment* 36 (10):1589–1598. doi:10.1016/s1352-2310(02)00081-x.

Lesturgez, G., R. Poss, A. Noble, O. Grunberger, W. Chintachao, and D. Tessier. 2006. Soil acidification without pH drop under intensive cropping systems in Northeast Thailand. *Agriculture Ecosystems & Environment* 114 (2–4):239–248.

Li, J. Y., Z. D. Liu, W. Z. Zhao, M. M. Masud, and R. K. Xu. 2015. Alkaline slag is more effective than phosphogypsum in the amelioration of subsoil acidity in an Ultisol profile. *Soil & Tillage Research* 149:21–32. doi:10.1016/j.still.2014.12.017.

Liang, R. 2018. Cross talk between aluminum and genetic susceptibility and epigenetic modification in Alzheimer's disease. In *Neurotoxicity of Aluminum*, edited by Qiao Niu, 173–191. Singapore: Springer Singapore.

Liang, H. C., and B. M. Thomson. 2009. Minerals and mine drainage. *Water Environment Research* 81 (10):1615–1663.

Liu, J., and N. V. Hue. 2001. Amending subsoil acidity by surface applications of gypsum, lime, and composts. *Communications in Soil Science and Plant Analysis* 32 (13–14):2117–2132. doi:10.1081/css-120000273.

Lofts, S., C. Woof, E. Tipping, N. Clarke, and J. Mulder. 2001. Modelling pH buffering and aluminium solubility in European forest soils. *European Journal of Soil Science* 52 (2):189–204.

Lopez-Diaz, M. L., M. R. Mosquera-Losada, and A. Rigueiro-Rodriguez. 2007. Lime, sewage sludge and mineral fertilization in a silvopastoral system developed in very acid soils. *Agroforestry Systems* 70 (1):91–101. doi:10.1007/s10457-007-9046-9.

Machacha, S. 2004. Comparison of laboratory pH buffer methods for predicting lime requirement for acidic soils of eastern Botswana. *Communications in Soil Science and Plant Analysis* 35 (17–18):2675–2687. doi:10.1081/lcss-200030441.

Mahdy, A. M., E. A. Elkhatib, N. O. Fathi, and Z. Q. Lin. 2009. Effects of co-application of biosolids and water treatment residuals on corn growth and bioavailable phosphorus and aluminum in alkaline soils in Egypt. *Journal of Environmental Quality* 38 (4):1501–1510.

Mailloux, R. J., J. Lemire, and V. D. Appanna. 2011. Hepatic response to aluminum toxicity: dyslipidemia and liver diseases. *Experimental Cell Research* 317 (16):2231–2238. doi:10.1016/j.yexcr.2011.07.009.

Manoharan, V., P. Loganathan, R. W. Tillman, and R. L. Parfitt. 2007. Interactive effects of soil acidity and fluoride on soil solution aluminium chemistry and barley (*Hordeum vulgare* L.) root growth. *Environmental Pollution* 145:778–786.

Martell, A. E., R. D. Hancock, R. M. Smith, and R. J. Motekaitis. 1996. Coordination of Al(III) in the environment and in biological systems. *Coordination Chemistry Reviews* 149:311–328.

Martin, R. B. 1996. Ternary complexes of Al^{3+} and F^- with a third ligand. *Coordination Chemistry Reviews* 141:23–32.

Martinent-Catalot, V., J.-M. Lamerant, G. Tilmant, M.-S. Bacou, and J. P. Ambrosi. 2002. Bauxaline: a new product for various applications of Bayer process red mud. *Light Metals* 2002:125–131.

Masion, A., and P. M. Bertsch. 1997. Aluminium speciation in the presence of wheat root cell walls: a wet chemical study. *Plant, Cell and Environment* 20:504–512.

Masion, A., F. Thomas, D. Tchoubar, J. Y. Bottero, and P. Tekely. 1994. Chemistry and structure of Al(OH)/organic precipitates. A small-angle X-ray scattering study. 3. Depolymerization of the Al_{13} polycation by organic ligands. *Langmuir* 10:4353–4356.

Materechera, S. A., and T. S. Mkhabela. 2002. The effectiveness of lime, chicken manure and leaf litter ash in ameliorating acidity in a soil previously under black wattle (Acacia mearnsii) plantation. *Bioresource Technology* 85 (1):9–16. doi:10.1016/s0960-8524(02)00065-2.

Matthias, A., M. Maurer, and H. Parlar. 2003. Comparative aluminium speciation and quantification in soil solutions of two different forest ecosystems by Al-27-NMR. *Fresenius Environmental Bulletin* 12 (10):1263–1275.

McBride, M. B. 1994. *Environmental chemistry of soils*. New York: Oxford University Press.

Menzies, N. W., L. C. Bell, and D. G. Edwards. 1994. Exchange and solution phase chemistry of acid, highly weathered soils. 2. Investigations of mechanisms controlling Al release into solution. *Australian Journal of Soil Research* 32:269–283.

Michalke, B., S. Halbach, and V. Nieschwitz. 2009. JEM Spotlight: metal speciation related to neurotoxicity in humans. *Journal of Environmental Monitoring* 11:939–954.

Mishra, V. K., and A. Kumar. 2009. Impact of metal nanoparticles on the plant growth promoting *Rhizobacteria*. *Digest Journal of Nanomaterials and Biostructures* 4 (3):587–592.

Miyazawa, M., T. Takahashi, T. Sato, H. Kanno, and M. Nanzyo. 2013. Factors controlling accumulation and decomposition of organic carbon in humus horizons of Andosols. *Biology and Fertility of Soils* 49:929–938.

Mokolobate, M. S., and R. J. Haynes. 2002. Comparative liming effect of four organic residues applied to an acid soil. *Biology and Fertility of Soils* 35 (2):79–85. doi:10.1007/s00374-001-0439-z.

Mold, M., E. Shardlow, and C. Exley. 2016. Insight into the cellular fate and toxicity of aluminium adjuvants used in clinically approved human vaccinations. *Scientific Reports* 6:31578. doi:10.1038/srep31578.

Morikawa, C. K., and M. Saigusa. 2002. Si amelioration of Al toxicity in barley (*Hordeum vulgare* L.) growing in two Andosols. *Plant and Soil* 240 (1):161–168. doi:10.1023/a:1015804401190.

Mota, D., F. Faria, E. A. Gomes, I. E. Marriel, E. Paiva, and L. Seldin. 2008. Bacterial and fungal communities in bulk soil and rhizospheres of aluminum-tolerant and aluminum-sensitive maize (*Zea mays* L.) lines cultivated in unlimed and limed Cerrado soil. *Journal of Microbiology and Biotechnology* 18 (5):805–814.

Mueller, M., M. Anke, and H. Illing-Guenther. 1998. Aluminium in foodstuffs. *Food Chemistry* 61:419–428.

Mutter, J., J. Naumann, R. Schneider, and H. Walach. 2007. Mercury and Alzheimer's disease. *Fortschritte der Neurologie und Psychiatrie* 75:528–540.

Naramabuye, F. X., and R. J. Haynes. 2006. Effect of organic amendments on soil pH and Al solubility and use of laboratory indices to predict their liming effect. *Soil Science* 171 (10):754–763. doi:10.1097/01.ss.0000228366.17459.19.

Nayak, P. 2002. Aluminum: impacts and disease. *Environmental Research* 89 (2):101–115.

Nelson, D. R., and P. M. Mele. 2006. The impact of crop residue amendments and lime on microbial community structure and nitrogen-fixing bacteria in the wheat rhizosphere. *Australian Journal of Soil Research* 44 (4):319–329. doi:10.1071/sr06022.

NHMRC. 2004. Australian drinking water guidelines 6. National water quality management strategy. Canberra, ACT, Australian Government - National Health and Medical Research Council. 615p.

Oladeji, O. O., J. B. Sartain, and G. A. O'Connor. 2009. Land application of aluminum water treatment residual: aluminum phytoavailability and forage yield. *Communications in Soil Science and Plant Analysis* 40 (9–10):1483–1498.

Orvig, C. 1993. The aqueous coordination chemistry of aluminum. In *Coordination chemistry of aluminum*, edited by G. H. Robinson, 85–121. New York: VCH.

Othman, M. N., M. P. Abdullah, and Y. F. Abd Aziz. 2010. Removal of aluminium from drinking water. *Sains Malaysiana* 39 (1):51–55.

Pakrashi, S., S. Dalai, A. Humayun, S. Chakravarty, N. Chandrasekaran, and A. Mukherjee. 2013. *Ceriodaphnia dubia* as a potential bio-indicator for assessing acute aluminum oxide nanoparticle toxicity in fresh water environment. *PLoS ONE* 8 (9). doi:10.1371/journal.pone.0074003.

Panhwar, Q. A., U. A. Naher, J. Shamshuddin, R. Othman, and M. R. Ismail. 2016. Applying limestone or basalt in combination with bio-fertilizer to sustain rice production on an acid sulfate soil in Malaysia. *Sustainability* 8 (7). doi:10.3390/su8070700.

Park, E. J., G. H. Lee, J. H. Shim, M. H. Cho, B. S. Lee, Y. B. Kim, J. H. Kim, Y. Kim, and D. W. Kim. 2015. Comparison of the toxicity of aluminum oxide nanorods with different aspect ratio. *Archives of Toxicology* 89 (10):1771–1782. doi:10.1007/s00204-014-1332-5.

Poborilova, Z., R. Opatrilova, and P. Babula. 2013. Toxicity of aluminium oxide nanoparticles demonstrated using a BY-2 plant cell suspension culture model. *Environmental and Experimental Botany* 91:1–11. doi:10.1016/j.envexpbot.2013.03.002.

Poschenrieder, C., B. Gunse, I. Corrales, and J. Barcelo. 2008. A glance into aluminum toxicity and resistance in plants. *Science of the Total Environment* 400 (1–3):356–368.

Pu, X. X., O. Vazquez, J. D. Monnell, and R. D. Neufeld. 2010. Speciation of aluminum precipitates from acid rock discharges in Central Pennsylvania. *Environmental Engineering Science* 27 (2):169–180.

Qian, L. B., B. L. Chen, and D. F. Hu. 2013. Effective alleviation of aluminum phytotoxicity by manure-derived biochar. *Environmental Science & Technology* 47 (6):2737–2745. doi:10.1021/es3047872.

Qin, R. J., and F. X. Chen. 2005. Amelioration of aluminum toxicity in red soil through use of barnyard and green manure. *Communications in Soil Science and Plant Analysis* 36 (13–14):1875–1889. doi:10.1081/css-200062480.

Raboin, L. M., A. H. D. Razafimahafaly, M. B. Rabenjarisoa, B. Rabary, J. Dusserre, and T. Becquer. 2016. Improving the fertility of tropical acid soils: Liming versus biochar application? A long term comparison in the highlands of Madagascar. *Field Crops Research* 199:99–108. doi:10.1016/j.fcr.2016.09.005.

Ramirez, M. D. A., J. D. Silva, N. Ohkama-Ohtsu, and T. Yokoyama. 2018. In vitro rhizobia response and symbiosis process under aluminum stress. *Canadian Journal of Microbiology* 64:511–526.

Ranau, R., J. Oehlenschlager, and H. Steinhart. 2001. Aluminium levels of fish fillets baked and grilled in aluminium foil. *Food Chemistry* 73 (1):1–6. doi:10.1016/s0308-8146(00)00318-6.

Rasmussen, C., R. J. Southard, and W. R. Horwath. 2006. Mineral control of organic carbon mineralization in a range of temperate conifer forest soils. *Global Change Biology* 12:834–847.

Resources-and-Energy-Quarterly. 2018. https://archive.industry.gov.au/Office-of-the-Chief-Economist/Publications/ResourcesandEnergyQuarterlyMarch2018/documents/Resources-and-Energy-Quarterly-March-2018-Aluminium-alumina-and-bauxite.pdf.

Ryan, P. R., E. Delhaize, and P. J. Randall. 1995. Malate efflux from root apices and tolerance to aluminium are highly correlated in wheat. *Australian Journal of Plant Physiology* 22:531–536.

Sadiq, I. M., S. Pakrashi, N. Chandrasekaran, and A. Mukherjee. 2011. Studies on toxicity of aluminum oxide (Al2O3) nanoparticles to microalgae species: Scenedesmus sp and Chlorella sp. *Journal of Nanoparticle Research* 13 (8):3287–3299. doi:10.1007/s11051-011-0243-0.

Salifoglou, A. 2002. Synthetic and structural carboxylate chemistry of neurotoxic aluminum in relevance to human diseases. *Coordination Chemistry Reviews* 228:297–317.

Sangster, A.G., and Hodson, M.J. 2001. Silicon and aluminium codeposition in the cell wall phytoliths of gymnosperm leaves. *In Phytoliths - Applications in earth science and human history* (eds. Meunier, J.D., Colin F.). A.A. Balkema, Lisse, The Netherlands. pp. 343–355.

Sanjuan, B., and G. Michard. 1987. Aluminum hydroxide solubility in aqueous solutions containing fluoride ions at 50°C. *Geochimica et Cosmochimica Acta* 51:1823–1831.

Sarpola, A. T., V. K. Hietapelto, J. E. Jalonen, J. Jokela, and J. H. Ramo. 2006. Comparison of hydrolysis products of AlCl3.6H2O in different concentrations by electrospray ionization time of flight mass spectrometer (ESI TOF MS). *International Journal for Environmental Analytical Chemistry* 86:1007–1018.

Schaefer, U., and M. Seifert. 2006. Oral intake of aluminum from foodstuffs, food additives, food packaging, cookware and pharmaceutical preparations with respect to dietary regulations. *Trace Elements and Electrolytes* 23:150–161.

Schafer, U., and G. Jahreis. 2009. Update on regulations of aluminium intake - biochemical and toxicological assessment. *Trace Elements and Electrolytes* 26:95–99.

Scheel, T., C. Dorfler, and K. Kalbitz. 2007. Precipitation of dissolved organic matter by aluminum stabilizes carbon in acidic forest soils. *Soil Science Society of America Journal* 71:64–74.

Scheel, T., K. Pritsch, M. Schlater, and K. Kalbitz. 2008. Precipitation of enzymes and organic matter by aluminum-impacts on carbon mineralization. *Journal of Plant Nutrition and Soil Science-Zeitschrift für Pflanzenernährung und Bodenkunde* 171 (6):900–907.

Schussler, U., W. Balzer, and A. Deeken. 2005. Dissolved Al distribution, particulate Al fluxes and coupling to atmospheric Al and dust deposition in the Arabian Sea. *Deep-Sea Research Part Ii-Topical Studies in Oceanography* 52 (14–15):1862–1878. doi:10.1016/j.dsr2.2005.06.005.

Scott, B. J., I. G. Fenton, A. G. Fanning, W. G. Schumann, and L. J. C. Castleman. 2007. Surface soil acidity and fertility in the eastern Riverina and western slopes of southern New South Wales. *Australian Journal of Experimental Agriculture* 47 (8):949–964.

Shabnam, N., and H. Kim. 2018. Non-toxicity of nano alumina: a case on mung bean seedlings. *Ecotoxicology and Environmental Safety* 165:423–433. doi:10.1016/j.ecoenv.2018.09.033.

Sharma, H. S., S. F. Ali, S. M. Hussain, J. J. Schlager, and A. Sharma. 2009. Influence of engineered nanoparticles from metals on the blood-brain barrier permeability, cerebral blood flow, brain edema and neurotoxicity. An experimental study in the rat and mice using biochemical and morphological approaches. *Journal of Nanoscience and Nanotechnology* 9 (8):5055–5072.

Shen, J. S., and V. Nardello-Rataj. 2009. Deodorants and antiperspirants: chemistry under arms. *Actualite Chimique* (331):8–18.

Shi, R. Y., J. Y. Li, N. Ni, K. Mehmood, R. K. Xu, and W. Qian. 2017. Effects of biomass ash, bone meal, and alkaline slag applied alone and combined on soil acidity and wheat growth. *Journal of Soils and Sediments* 17 (8):2116–2126. doi:10.1007/s11368-017-1673-9.

Shirokikh, I. G., A. A. Shirokikh, N. A. Rodina, L. M. Polyanskaya, and O. A. Burkanova. 2004. Effects of soil acidity and aluminum on the structure of microbial biomass in the rhizosphere of barley. *Eurasian Soil Science* 37 (8):839–843.

Shoba, V. N., and K. V. Chudnenko. 2014. Ion exchange properties of humus acids. *Eurasian Soil Science* 47 (8):761–771. doi:10.1134/s1064229314080110.

Simonsson, M. 2000. Interactions of aluminium and fulvic acid in moderately acid solutions: stoichiometry of the H$^+$/Al^{3+} exchange. *European Journal of Soil Science* 51:655–666.

Singh, S., D. K. Tripathi, S. Singh, S. Sharma, N. K. Dubey, D. K. Chauhan, and M. Vaculík. 2017. Toxicity of aluminium on various levels of plant cells and organism: A review. *Environmental and Experimental Botany* 137:177–193. doi:10.1016/j.envexpbot.2017.01.005.

Sivaguru, M., and W. J. Horst. 1998. The distal part of the transition zone is the most aluminum-sensitive apical root zone of maize. *Plant Physiology* 116:155–163.

Sjostedt, C., T. Wallstedt, J. P. Gustafsson, and H. Borg. 2009. Speciation of aluminium, arsenic and molybdenum in excessively limed lakes. *Science of the Total Environment* 407 (18):5119–5127.

Skjelkvale, B. L., K. Torseth, W. Aas, and T. Andersen. 2001. Decrease in acid deposition – Recovery in Norwegian waters. *Water Air and Soil Pollution* 130 (1–4):1433–1438. doi:10.1023/a:1013956829092.

Skyllberg, U. 1999. pH and solubility of aluminium in acidic forest soils: a consequence of reactions between organic acidity and aluminium alkalinity. *European Journal of Soil Science* 50:95–106.

Song, Z. M., H. Tang, X. Y. Deng, K. Xiang, A. N. Cao, Y. F. Liu, and H. F. Wang. 2017. Comparing toxicity of alumina and zinc oxide nanoparticles on the human intestinal epithelium in vitro model. *Journal of Nanoscience and Nanotechnology* 17 (5):2881–2891. doi:10.1166/jnn.2017.13056.

Soucek, D. J., D. S. Cherry, and C. E. Zipper. 2003. Impacts of mine drainage and other nonpoint source pollutants on aquatic biota in the upper Powell River system, Virginia. *Human and Ecological Risk Assessment* 9 (4):1059–1073.

Srinivasan, P. T., and T. Viraraghavan. 2002. Characterisation and concentration profile of aluminium during drinking-water treatment. *Water SA* 28 (1):99–106.

Stahl, T., S. Falk, H. Taschan, B. Boschek, and H. Brunn. 2018. Evaluation of human exposure to aluminum from food and food contact materials. *European Food Research and Technology* 244 (12):2077–2084. doi:10.1007/s00217-018-3124-2.

Stauber, J. L., T. M. Florence, C. M. Davies, M. S. Adams, and S. J. Buchanan. 1999. Bioavailability of Al in alum-treated drinking water. *Journal American Water Works Association* 91 (11):84–93.

Steiner, C., W. G. Teixeira, J. Lehmann, T. Nehls, J. L. V. de Macedo, W. E. H. Blum, and W. Zech. 2007. Long term effects of manure, charcoal and mineral fertilization on crop production and fertility on a highly weathered Central Amazonian upland soil. *Plant and Soil* 291 (1–2):275–290. doi:10.1007/s11104-007-9193-9.

Strigul, N., L. Vaccari, C. Galdun, M. Wazne, X. Liu, C. Christodoulatos, and K. Jasinkiewicz. 2009. Acute toxicity of boron, titanium dioxide, and aluminum nanoparticles to *Daphnia magna* and *Vibrio fischeri*. *Desalination* 248 (1–3):771–782.

Sun, B., R. Poss, R. Moreau, A. Aventurier, and P. Fallavier. 2000. Effect of slaked lime and gypsum on acidity alleviation and nutrient leaching in an acid soil from Southern China. *Nutrient Cycling in Agroecosystems* 57 (3):215–223. doi:10.1023/a:1009870308097.

Takahashi, T., and R. A. Dahlgren. 2016. Nature, properties and function of aluminum-humus complexes in volcanic soils. *Geoderma* 263:110–121.

Takahashi, T., A. Mitamura, T. Ito, K. Ito, M. Nanzyo, and M. Saigusa. 2008. Aluminum solubility of strongly acidified allophanic Andosols from Kagoshima prefecture, southern Japan. *Soil Science and Plant Nutrition* 54 (3):362–368.

Tang, Y., H. Zhang, J. L. Schroder, M. E. Payton, and D. Zhou. 2007. Animal manure reduces aluminum toxicity in an acid soil. *Soil Science Society of America Journal* 71 (6):1699–1707. doi:10.2136/sssaj2007.0008.

Tarkalson, D. D., G. W. Hergert, W. B. Stevens, D. L. McCallister, and S. D. Kackman. 2005. Fly, ash as a liming material for corn production. *Soil Science* 170 (5):386–398. doi:10.1097/01.ss.0000169910.25356.3a.

Tönnies, E., and E. Trushina. 2017. Oxidative stress, synaptic dysfunction, and Alzheimer's disease. *Journal of Alzheimer's Disease* 57 (4):1105–1121.

Trenfield, Melanie A., Scott J. Markich, Jack C. Ng, Barry Noller, and Rick A. van Dam. 2012. Dissolved organic carbon reduces the toxicity of aluminum to three tropical freshwater organisms. *Environmental Toxicology and Chemistry* 31 (2):427–436. doi:10.1002/etc.1704.

Turhan, S. 2006. Aluminium contents in baked meats wrapped in aluminium foil. *Meat Science* 74 (4):644–647. doi:10.1016/j.meatsci.2006.03.031.

Vieira, F. C. B., Z. L. He, C. Bayer, P. J. Stoffella, and V. C. Baligar. 2008. Organic amendment effects on the transformation and fractionation of aluminum in acidic sandy soil. *Communications in Soil Science and Plant Analysis* 39 (17–18):2678–2694. doi:10.1080/00103620802358813.

von Uexküll, H. R., and E. Mutert. 1995. Global extent, development and economic impact of acid soils. In *Plant soil interactions at low pH*, edited by R. A. Date, N. J. Grundon, G. E. Rayment and M. E. Probert, 5–19. Dordrecht: Kluwer Academic.

Vuorenmaa, J., A. Augustaitis, B. Beudert, W. Bochenek, N. Clarke, H. A. de Wit, T. Dirnbock, J. Frey, H. Hakola, S. Kleemola, J. Kobler, P. Kram, A. J. Lindroos, L. Lundin, S. Lofgren, A. Marchettom, T. Pecka, H. Schulte-Bisping, K. Skotak, A. Srybny, J. Szpikowski, L. Ukonmaanaho, M. Vana, S. Akerblom, and M. Forsius. 2018. Long-term changes (1990–2015) in the atmospheric deposition and runoff water chemistry of sulphate, inorganic nitrogen and acidity for forested catchments in Europe in relation to changes in emissions and hydrometeorological conditions. *Science of the Total Environment* 625:1129–1145. doi:10.1016/j.scitotenv.2017.12.245.

Wang, M., and M. Muhammed. 1999. Novel synthesis of Al_{13}-cluster based alumina materials. *NanoStructured Materials* 11:1219–1229.

Wang, P., et al. 2007. Study of aluminium distribution and speciation in atmospheric particles of different diameters in Nanjing, China. *Atmospheric Environment* 41(27): 5788–5796.

Ward, R. J. S., C. R. McCrohan, and K. N. White. 2006. Influence of aqueous aluminium on the immune system of the freshwater crayfish *Pacifasticus leniusculus*. *Aquatic Toxicology* 77 (2):222–228.

Waters, A. S., and J. G. Webster-Brown. 2013. Assessing aluminium toxicity in streams affected by acid mine drainage. *Water Sci Technol* 67 (8):1764–72. doi:10.2166/wst.2013.051.

Wehr, J. B., F. P. C. Blamey, and N. W. Menzies. 2007. Aluminum. In *Encyclopedia of Soil Science*, edited by R. Lal, 1–6. New York: Taylor and Francis.

Wehr, J. B., N. W. Menzies, and I. Fulton. 2006. Revegetation strategies for bauxite refinery residue: a case study of Alcan Gove in Northern Territory, Australia. *Environmental Management* 37:297–306.

Wesdock, James C., and Ian M. F. Arnold. 2014. Occupational and environmental health in the aluminum industry: key points for health practitioners. *Journal of occupational and environmental medicine* 56 (5 Suppl):S5–S11. doi:10.1097/JOM.0000000000000071.

Willhite, C. C., G. L. Ball, and C. J. McLellan. 2012. Total allowable concentrations of monomeric inorganic aluminum and hydrated aluminum silicates in drinking water. *Critical Reviews in Toxicology* 42 (5):358–442. doi:10.3109/10408444.2012.674101.

Winterbourn, M. J., W. F. McDiffett, and S. J. Eppley. 2000. Aluminium and iron burdens of aquatic biota in New Zealand streams contaminated by acid mine drainage: effects of trophic level. *Science of the Total Environment* 254 (1):45–54.

World-Aluminum.org. 2019. www.world-aluminium.org/statistics/primary-aluminium-production.

Xia, H., and G. D. Rayson. 1998. Investigation of aluminum binding to a *Datura innoxia* material using[27]Al NMR. *Environmental Science and Technology* 32:2688–2692.

Yamaguchi, N. U., S. Hiradate, M. Mizoguchi, and T. Miyazaki. 2003. Formation and disappearance of Al tridecamer in the presence of low molecular weight organic ligands. *Soil Science and Plant Nutrition* 49:551–556.

Yamaguchi, N., S. Hiradate, M. Mizoguchi, and T. Miyazaki. 2004. Disappearance of aluminum tridecamer from hydroxyaluminum solution in the presence of humic acid. *Soil Science Society of America Journal* 68 (6):1838–1843. doi:10.2136/sssaj2004.1838.

Yamamoto, Y, Y. Kobayashi, S. R. Devi, S Rikiishi, and H Matsumoto. 2003. Oxidative stress triggered by aluminum in plant roots. *Plant and Soil* 255:239–243.

Yang, Y., Y. Liu, C. F. Huang, J. de Silva, and F. J. Zhao. 2016. Aluminium alleviates fluoride toxicity in tea (Camellia sinensis). *Plant and Soil* 402 (1–2):179–190. doi:10.1007/s11104-015-2787-8.

Yi, J., and J. Cao. 2008. Tea and fluorosis. *Journal of Fluorine Chemistry* 129 (2):76–81.

Yokel, R. A. 2000. The toxicology of aluminum in the brain: a review. *Neurotoxicology* 21 (5):813–828.

Yokel, R. A., and R. L. Florence. 2008. Aluminum bioavailability from tea infusion. *Food and Chemical Toxicology* 46 (12):3659–3663. doi:10.1016/j.fct.2008.09.041.

Yokel, R. A., C. L. Hicks, and R. L. Florence. 2008. Aluminum bioavailability from basic sodium aluminum phosphate, an approved food additive emulsifying agent, incorporated in cheese. *Food and Chemical Toxicology* 46:2261–2266.

Yokel, R. A., and P. J. McNamara. 2001. Aluminium toxicokinetics: an updated mini review. *Pharmacology and Toxicology* 88:159–167.

16

Boron: Soil Contaminant

Rami Keren

Introduction

Boric acid is moderately soluble in water. Its solubility increases markedly with temperature due to the large negative heat of dissolution. Boron is considered as a typical metalloid having properties intermediate between the metals and the electronegative non-metals. Boron has a tendency to form anionic rather than cationic complexes. Boron chemistry is of covalent B compounds and not of B^{3+} ions because of its very high ionization potentials. Boron has five electrons, two in the inner spherical shell (1s^2), two in the outer spherical shell (2s^2), and one in the dumbbell shaped shell $\left(2p_x^1\right)$.[1] In the hybrid orbital state, the three electrons in the 2s and 2p orbitals form a hybrid orbital state $\left(2s^1 2p_x^1 2p_y^1\right)$, where each electron is alone in an orbit whose shape has both spherical and dumbbell characteristics. Each of these three orbits can hold one electron from another element to form a covalent bond between the element and B (BX$_3$). This leaves one 2p electron orbit that can hold two electrons, which if filled would completely fill the eight electron positions (octet) associated with the second electron shell around B. BX$_3$ compounds behave as acceptor Lewis acids toward many Lewis bases such as amines and phosphines. The acceptance of two electrons from a Lewis base completes the octet of electrons around B. Boron also completes its octet by forming both anionic and cationic complexes.[1] Therefore, tri-coordinate B compounds have strong electron-acceptor properties and may form tetra- coordinate B structures. The charge in tetra-coordinate derivatives may range from negative to neutral and positive, depending upon the nature of the ligands.

For the unshared oxygen atoms bound to B, they are, probably, always OH groups. Thus, in accordance with the electron configuration of B, boric acid acts as a weak Lewis acid:

$$B(OH)_3 + 2H_2O = B(OH)_4^- + H_3O^+ \tag{1}$$

The formation of borate ion is spontaneous. The first hydrolysis constant of B(OH)$_3$, K_{h1}, is 5.8×10^{-10} at 20°C,[2] and the other K_{h2} and K_{h3} values are 5.0×10^{-13} and 5.0×10^{-14}, respectively.[3] A dissociation beyond $B(OH)_4^-$ is not necessary to explain the experimental data, at least below pH 13.[4,5] Boron species other than B(OH)$_3$ and $B(OH)_4^-$, however, can be ignored in soils for most practical purposes. The first hydrolysis constant of B(OH)$_3$ varies with temperature from 3.646×10^{-10} at 178 K to 7.865×10^{-10} at 318 K.[6]

Both B(OH)$_3$ and $B(OH)_4^-$ ion species are essentially monomeric in aqueous media at low B concentration ($\leq 0.025\,mol L^{-1}$). However, at high B concentration, polyborate ions exist in appreciable amount.[7] The equilibria between boric acid, monoborate ions, and polyborate ions in aqueous solution are rapidly reversible. In aqueous solution, most of the polyanions are unstable relative to their monomeric forms B(OH)$_3$

and $B(OH)_4^-$.[8] Results of nuclear magnetic resonance[9] and Raman spectrometry[10] lead to the conclusion that B(OH)3 has a trigonal-planar structure, whereas the $B(OH)_4^-$ ion in aqueous solution has a tetrahedral structure. This difference in structure can lead to differences in the affinity of clay for these two B species.

Boron–Soil Interaction

The elemental form of boron (B) is unstable in nature and found combined with oxygen in a wide variety of hydrated alkali and alkaline earth-borate salts and borosilicates as tourmaline. The total B content in soils, however, has little bearing on the status of available B to plants.

Boron can be specifically adsorbed by different clay minerals, hydroxy oxides of Al, Fe, and Mg, and organic matter.[11] Boron is adsorbed mainly on the particle edges of the clay minerals rather than the planar surfaces. The most reactive surface functional group on the edge surface is the hydroxyl exposed on the outer periphery of the clay mineral. This functional group is associated with two types of sites that are available for adsorption: Al(III) and Si(IV), which are located on the octahedral and tetrahedral sheets, respectively. The hydroxyl group associated with this site can form an inner sphere surface complex with a proton at low pH values or with a hydroxyl at high pH values. The B adsorption process can be explained by the surface complexation approach, in which the surface is considered as a ligand.[12] Such specific adsorption, which occurs irrespective of the sign of the net surface charge, can occur theoretically for any species capable of coordination with the surface metal ions. However, because oxygen is the ligand commonly coordinated to the metal ions in clay minerals, the B species $B(OH)_3$ and $B(OH)_4^-$ are particularly involved in such reactions. Possible surface complex configurations for B—broken edges of clay minerals—were suggested by Keren, Grossl, and Sparks.[12]

Keren and Bingham[11] reviewed the factors that affect the adsorption and desorption of B by soil constituents and the mechanisms of adsorption. Soil pH is one of the most important factors affecting B adsorption. Increasing pH enhances B adsorption on clay minerals, hydroxy-Al and soils, showing a maximum in the alkaline pH range (Figure 1).

The response of B adsorption on clays to variations in pH can be explained as follows. Below pH 7, $B(OH)_3$ predominates and since the affinity of the clay for this species is relatively low, the amount of

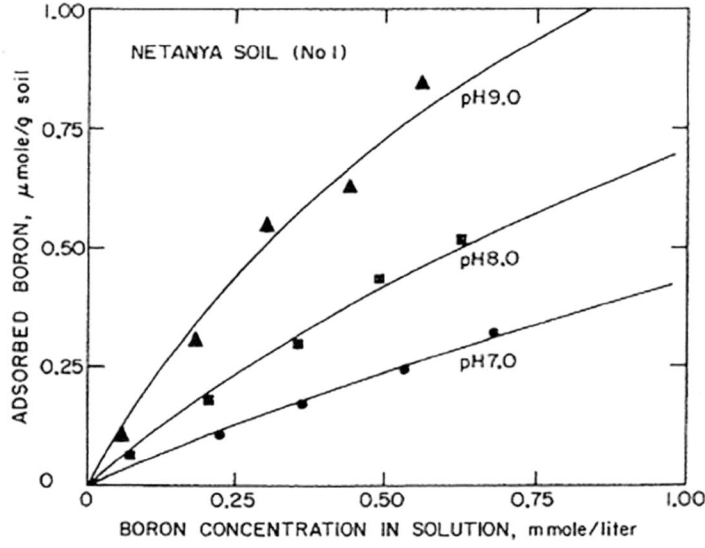

FIGURE 1 Boron adsorption isotherms for a soil as a function of solution B concentration and pH. Bold lines—calculated values.
Source: Mezuman and Keren.[28]

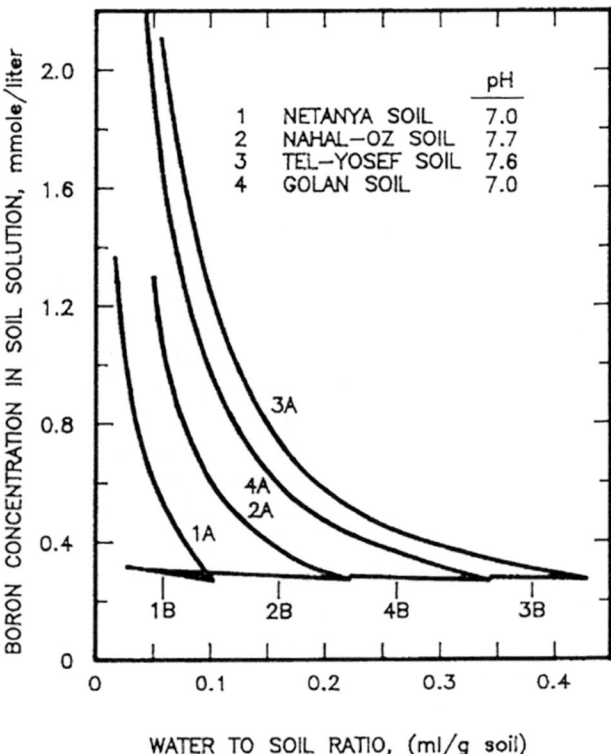

FIGURE 2   Boron concentration in soil solution as a function of solution-to-soil ratio for a given total amount of B. (a) No interaction between B and soil, (b) Boron adsorption account for.
Source: Mezuman and Keren.[28]

adsorption is small. Both $B(OH)_4^-$ and OH^- concentrations are low at this pH; thus, their contribution to total B adsorption is small despite their relatively strong affinity for the clay. As the pH is increased to about 9, the $B(OH)_4^-$ concentration increases rapidly. Since the OH^- concentration is still low relative to the B concentration, the amount of adsorbed B increases rapidly. Further increases in pH result in an enhanced OH^- concentration relative to $B(OH)_4^-$, and B adsorption decreases rapidly due to the competition by OH^- at the adsorption sites. Adsorption models for soils, clays, aluminum oxide, and iron oxide minerals have been derived by various workers.[13–17]

In assessing B concentration in irrigation water, however, the physicochemical characteristics of the soil must be taken into consideration because of the interaction between B and soil. Boron sorption and desorption from soil adsorption sites regulate the B concentration in soil solution depending on the changes in solution B concentration and the affinity of soil for B. Thus, adsorbed B may buffer fluctuations in solution B concentration, and B concentration in soil solution may change insignificantly by changing the soil-water content (Figure 2). When irrigation with water high in B is planned, special attention should be paid to this interaction because of the narrow difference between levels causing deficiency and toxicity symptoms in plants.

Boron–Plant Interaction

Boron is an essential micronutrient element required for growth and development of plants.

Many of the experimental data suggest that B uptake in plants is probably a passive process. There are clear evidences, however, that B uptake differs among species.[18] Several mechanisms have been postulated to explain this apparent paradox.[18–20] Boron deficiency in plants initially affects meristematic tissues, reducing or terminating growth of root and shoot apices, sugar transport, cell-wall synthesis and structure, carbohydrate metabolism and many biochemical reactions.[21,22] Tissue B concentrations associated with the appearance of vegetative deficiency symptoms have been identified in many crop species. It is essential to remember that for B, as for phosphorus and several other plant nutrient elements, deficiency may be present long before visual deficiency symptoms occur.

Excess and toxicity of boron in soils of semi-arid and arid areas are more of a problem than deficiency. Boron toxicity occurs in these areas either due to high levels of B in soils or due to additions of B in irrigation water. A summary of B tolerance data based upon plant response to soluble B is given by Maas.[23] Bingham et al.[24] showed that yield decrease of some crops (wheat, barley, and sorghum) due to B toxicity could be estimated by using a model for salinity response, suggested by Maas and Hoffman.[25]

There is a relatively small difference between the B concentration in soil solution causing deficiency and that resulting in toxicity symptoms in plants.[11] A consequence of this narrow difference is the difficulty posed in management of appropriate B levels in soil solution.

The suitability of irrigation water has been evaluated on the basis of criteria that determine the potential of the water to cause plant injury and yield reduction. In assessing the B in irrigation water, however, the physicochemical characteristics of the soil must be taken into consideration because the uptake by plants is dependent only on B activity in soil solution.[26,27] Boron uptake by plants grown in a soil of low-clay content is significantly greater than that of plants grown in a soil of high-clay content at the same given level of added B (Figure 3). This knowledge may improve the efficacy of using water of different qualities, whereby water with relatively high B levels could be used to irrigate B-sensitive crops in soils that show a high affinity to B. Such water can be used for irrigation as long as the equilibrium B concentration in soil solution is below the toxic concentration threshold (the maximum permissible concentration for a given

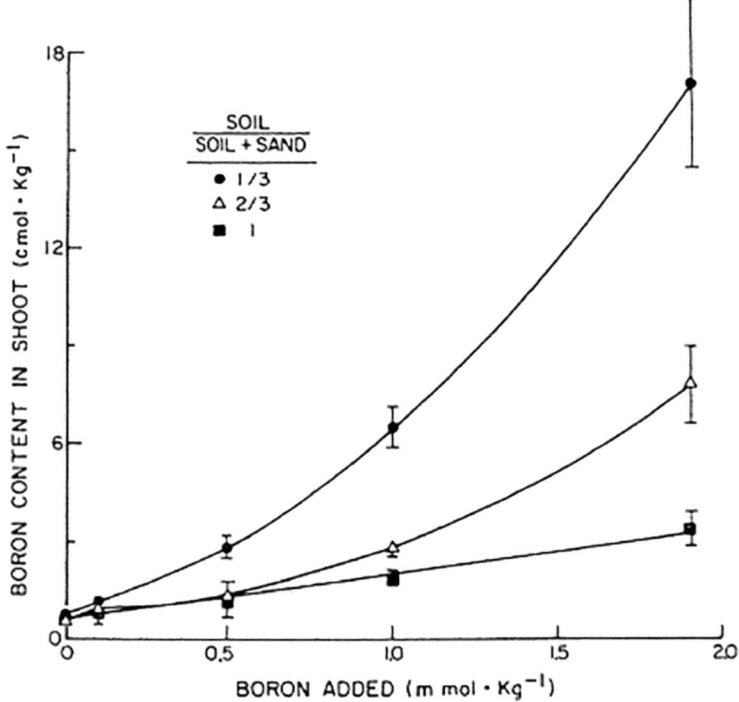

FIGURE 3   Relationship between B content in wheat shoot and the amount of B added to soil, for three ratios of soil-sand mixtures.
Source: Keren et al.[26]

crop species that does not reduce yield or lead to injury symptoms) for the irrigated crop. The existing criteria for irrigation water, however, make no reference to differences in soil type.

References

1. Cotton, F.A.; Wilkinson, G. *Advanced Inorganic Chemistry,* 5th Ed.; Wiley and Sons: New York, NY, 1988.
2. Owen, B.B. The dissociation constant of boric acid from 10 to 50°. J. Am. Chem. Soc. **1934**, *56,* 1695–1697.
3. Konopik, N.; Leberl, O. Colorimetric determination of PH in the range of 10 to 15. Monatsh **1949**, *80,* 420–429.
4. Ingri, N. Equilibrium studies of the polyanions containing B^{III}, Si^{IV}, Ge^{IV} and V^{V}. Svensk. Kem. Tidskr. **1963**, *75,* 199–230.
5. Mesmer, R.E.; Baes, C.F., Jr.; Sweeton, F.H. Acidity measurements at elevated temperature. VI. Boric acid equilibria. Inorg. Chem. **1972**, *11,* 537–543.
6. Owen, B.B.; King, E.J. The effect of sodium chloride upon the ionozation of boric acid at various temperatures. J. Am. Chem. Soc. **1943**, *65,* 1612–1620.
7. Adams, R.M. *Boron, Metallo-Boron Compounds and Bo- ranes*; John Wiley and Sons: New York, 1964.
8. Onak, T.P.; Landesman, H.; Williams, R.E.; Shapiro, I. The B^{II} nuclear magnetic resonance chemical shifts and spin coupling values for various compounds. J. Phys. Chem. **1959**, *63,* 1533–1535.
9. Good, C.D.; Ritter, D.M. Alkenylboranes: II. Improved preperative methods and new observations on methylvinylboranes. J. Am. Chem. Soc. **1962**, *84,* 1162–1166.
10. Servoss, R.R.; Clark, H.M. Vibrational spectra of normal and isotopically labeled boric acid. J. Chem. Phys. **1957**, *26,* 1175–1178.
11. Keren, R.; Bingham, F.T. Boron in water, soil and plants. Adv. Soil Sci. **1985**, *1,* 229–276.
12. Keren, R.; Grossl, P.R.; Sparks, D.L. Equilibrium and kinetics of borate adsorption-desorption on pyrophyllite in aqueous suspensions. Soil Sci. Soc. Am. J. **1994**, *58,* 1116–1122.
13. Keren, R.; Gast, R.G.; Bar-Yosef, B. pH-dependent boron adsorption by na-montmorillonite. Soil Sci. Soc. Am. J. **1981**, *45,* 45–48.
14. Keren, R.; Gast, R.G. pH dependent boron adsorption by montmorillonite hydroxy-aluminum complexes. Soil Sci. Soc. Am. J. **1983**, *47,* 1116–1121.
15. Goldberg, S.; Glaubig, R.A. Boron adsorption on aluminum and iron oxide minerals. Soil Sci. Soc. Am. J. **1985**, *49,* 1374–1379.
16. Goldberg, S.; Glaubig, R.A. Boron adsorption on California soils. Soil Sci. Soc. Am. J. **1986**, *50,* 1173–1176.
17. Goldberg, S.; Forster, H.S.; Heick, E.L. Boron adsorption mechanisms on oxides, clay minerals and soils inferred from ionic strength effects. Soil Sci. Soc. Am. J. **1993**, *57,* 704–708.
18. Nable, R.O. Effects of B toxicity amongst several barley wheat cultivars: a preliminary examination of the resistance mechanism. Plant Soil **1988**, *112,* 45–52.
19. Nable, R.O.; Lance, R.C.M.; Cartwright, B. Uptake of boron and silicon by barley genotypes with differing susceptibilities to boron toxicity. Ann. Bot. **1990**, *66,* 83–90.
20. Brown, P.H.; Hu, H. Boron uptake by sunflower, squash and cultured tobacco cells. Physiol. Plant **1994**, *91,* 435–441.
21. Loomis, W.D.; Durst, R.W. Chemistry and biology of boron. BioFactors **1992**, *3,* 229–239.
22. Marschner, H. *Mineral Nutrition of Higher Plants,* 2nd Ed.; Academic Press: London, 1995.
23. Maas, E.V. Salt tolerance of plants. In *Handbook of Plant Science in Agriculture;* Christie, B.R., Ed.; CRC Press, Inc.: Cleveland, Ohio, 1984.
24. Bingham, F.T.; Strong, J.E.; Rhoades, J.D.; Keren, R. An application of the Maas-Hoffman salinity response model for boron toxicity. Soil Sci. Soc. Am. J. **1985**, *49,* 672–674.

25. Maas, E.V.; Hoffman, G.J. Crop salt tolerance—current assessment. ASCE J. Irrig. Drainage Div. **1977**, *103*, 115–134.
26. Keren, R.; Bingham, F.T.; Rhoades, J.D. Effect of clay content on soil boron uptake and yield of wheat. Soil Sci. Soc. Am. J. **1985**, *49*, 1466–1470.
27. Keren, R.; Bingham, F.T.; Rhoades, J.D. Plant uptake of boron as affected by boron distribution between liquid and solid phases in soil. Soil Sci. Soc. Am. J. **1985**, *49*, 297–302.
28. Mezuman, U.; Keren, R. Boron adsorption by soils using a phenomenological adsorption equation. Soil Sci. Soc. Am. J. **1981**, *45*, 722–726.

17

Cadmium: Toxicology

Sven Erik Jørgensen

Introduction: Dispersion and Application

From 1940 to 1960, Japanese in the Toyama Prefecture were poisoned by cadmium in their rice, because the river water used for irrigation was contaminated by cadmium from a cadmium mine.[1] Cadmium replaces calcium in the bones, which makes them soft. It causes extreme pain and the disease was named *itai-itai,* which means "ouch- ouch." When it was discovered that cadmium was causing the disease, the mine waste was controlled, but by then, several thousands (mainly farmers) were already suffering from the very painful disease.

Cadmium is used for metal surface treatment, as a stabilizing agent in plastic and many alloys. Phosphorus fertilizers have a relatively high cadmium concentration—in the order of 10–80 mg/kg. Together with the cadmium emitted from coal-fired power plants, it is the most important source of global cadmium dispersion. More than 1000 tons of cadmium is globally dispersed from coal-fired power plants. Cadmium was previously applied in ceramic, but this use of the very toxic metal is now banned in most industrialized countries.

Like other heavy metals, cadmium shows biomagnification and bioaccumulation and is accumulated in the sediment of aquatic ecosystems. It entails that the dispersion of cadmium in aquatic ecosystems can, in most cases, best be determined by analyses of the sediment where relatively higher concentrations are found.[2] Similar to lead (*Lead: Ecotoxicology*, p. 1651), it is possible by analyses of sediment cores to determine the history of cadmium emission. The phosphorus fertilizer industry has, for several decades, discharged cadmium-containing production waste to the Little Belt in Denmark. An environmental impact assessment was carried out for Little Belt by analyses of a large number of sediment cores, whereby the history of the contamination was determined. Fortunately, cadmium can form complexes with chloride, which are less toxic than the cadmium ions.[3,4] The complexes are at the same time more soluble, which means that less cadmium is transferred to the sediment and more to the open sea where it is diluted significantly. The discharge of cadmium waste was stopped as a result of the investigation, but it was also concluded that the contamination of the Little Belt was less than expected, probably due to the formation of cadmium–chloride complexes.

Due to the high cadmium concentration in phosphorus fertilizers, there is a risk for cadmium contamination of agricultural land, particularly by the application of intensive agriculture. Cadmium is taken up by plants (see also *Bioremediation*, p. 408). A cadmium model has been developed to relate the cadmium concentration in crops as a function of the cadmium contamination by the use of fertilizers.[5]

It is beneficial for all toxic substances and particularly for the most toxic heavy metals to make a regional mass balance to identify the sources and the dispersion to assess whether a specified toxic substance would reach an unacceptably high concentration and thereby do the most harm. Jørgensen and Fath[5] have exemplified regional mass balances. Figure 1 shows one example, namely, a cadmium balance for Danish agricultural land, but similar balances can also be found for lead, mercury, and copper, as well as in other regions. The cadmium pollution of the agricultural land comes from the use of fertilizers (1.7 g/ha/yr), from the use of sludge (0.18 g/ha/yr), and from air pollution, dry deposition, and rainwater (1.7 g/ha/yr). A total of 1.7 g/ha/yr is taken up from the soil by the plants and 0.2 g/ha/yr contaminates vegetable products, while only 0.01 g of cadmium per year contaminates the animal food, even when the cadmium in imported fodder is considered. By far, most of the cadmium-contaminating domestic animals go back to the agricultural land by animal waste. In total, about 3.8 g of cadmium is added to the agricultural land per year, but 0.7 g of cadmium per hectare per year is transported by drainage water and groundwater to other sites (ecosystems) and 0.2 g/ha/yr will be removed by the harvest of vegetables. The net accumulation in the agricultural soil is therefore 3.1 g/ha/yr. The mass balance can be used to calculate the effect of using fertilizers with less cadmium or the result of a reduction of cadmium discharged to the atmosphere. If more complex pollution abatement strategies are applied, it is necessary to apply an ecotoxicological model with the state variables, processes, and transfers as shown in Figure 1 as the core of the model.

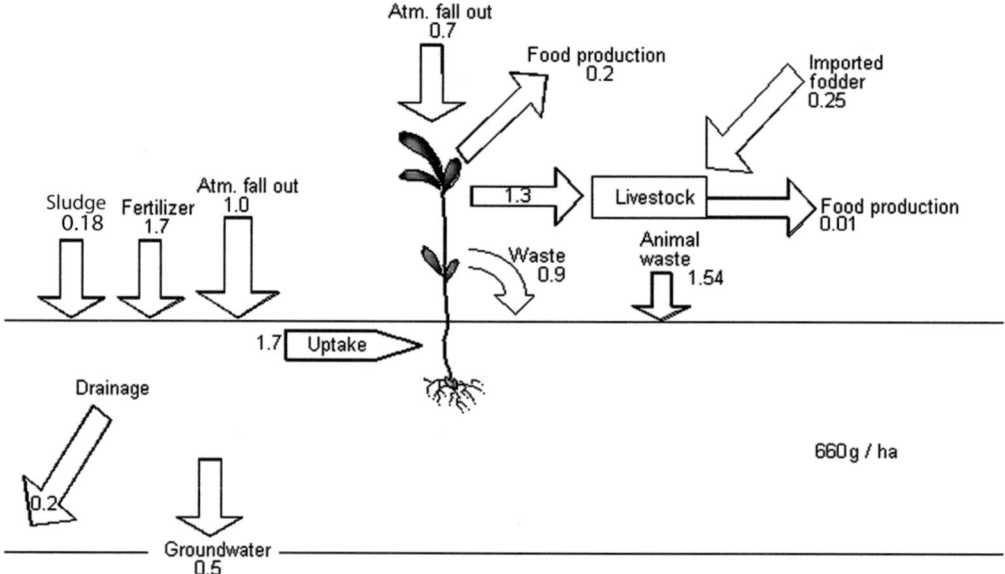

FIGURE 1 Cadmium balance for Danish agricultural land. The accumulated amount is about 660 g/ha, while the annual input by fertilizers is 1.7 g/ha and by air pollution another 1.7 g/ha, of which 1 g is accumulated directly on the bare soil and the 0.7 g on the plants. The amount from fertilizers corresponds to the amount taken up by plants. 1.3 g/ha of the cadmium contamination of plants and 0.25 g/ha of cadmium in imported fodder give an accumulation of 1.55 g of cadmium per hectare per year in domestic animals, but only the 0.01 g/ha will end up as cadmium contamination of animal products.

Toxicity and Ecotoxicity

Cadmium has carcinogenic and teratogenic effects and is also highly toxic, as indicated by the LD_{50} value for rats—70–90 mg/kg. LD_{50} for cadmium in smoke (as cadmium oxide) by inhalation is 500 min/mg/m^3. The cadmium concentration in rice that caused the itai-itai disease in Japan was about 1 mg/kg.

Uptake from food of cadmium is about the same as for other heavy metals—7%–10%. Cadmium is accumulated mainly in the kidneys. The total cadmium concentration in the body increases with age, as the uptake is bigger than the excretion, mainly by the urine. The excretion follows a first-order reaction with a coefficient of 0.0001 1/24h. At an age of 50, the cadmium content in the body will be about 25 mg, of which the 8–9 mg is accumulated in the kidney and the 3 mg is accumulated in the liver.[6] The daily accumulation is about 2–3 µg.

It is possible to express cadmium accumulation in the body as a function of time by the following differential equation:

$$dCd/dt = \text{daily uptake} - \text{excretion coefficient} * Cd$$

Cd is the amount of cadmium in the body and the daily uptake is found as about 10% of the cadmium content in the food and 50% of the cadmium in the 20 m^3 of air used for respiration per 24 hr. The intake by food is about 25 µg/24 hr and the intake by respiration is very minor. Therefore, the equation can be written as:

$$dCd/dt = 0.025*10/100 - 0.0001*Cd \text{ mg/24hr}$$

At steady state, the cadmium concentration becomes 0.0025/0.0001 = 25 mg, which (of course) is close to the average value at the age of 50 years.

References

1. Newman, M.C.; Unger, M.A. *Fundamentals of Ecotoxicology*, 2nd Ed.; CRC Press: Boca Raton, London, and New York, 2003.
2. Jørgensen, S.E. *Principles of Pollution Abatement;* Elsevier: Amsterdam, 2000; 520 pp.
3. Jørgensen, S.E.; Halling-Sørensen, B.; Mahler, H. *Handbook of Estimation Methods in Ecotoxicology and Environmental Chemistry;* Taylor and Francis Publ.: Boca Raton, FL; 230 pp.
4. Jørgensen, S.E.; Jørgensen, L.A.; Nors Nielsen, S. *Handbook of Ecological and Ecotoxicological Parameters;* Elsevier: Amsterdam, 1991; 1380 pp.
5. Jørgensen, S.E.; Fath, B. *Fundamentals of Ecological Modelling*, 4th Ed.; Elsevier: Amsterdam, Oxford, 2011; 396 pp.
6. *Om Metaller*, 2nd Ed.; Statens Naturvørdsverk: Sweden, 1980.

18

Carbon: Soil Inorganic

Donald L. Suarez

Introduction

Concerns related to global climate change and greenhouse gas warming have led to extensive interest in changes in carbon storage and release of carbon dioxide into the atmosphere. While the burning of fossil fuels was a major contributor to the increase in atmospheric carbon dioxide over the last 200 years, there has been extensive research on the impact of other processes, such as changes in the organic carbon (OC) stocks in soils. Inorganic carbon (IC) storage and changes associated with land use have received relatively little attention, despite the fact that IC is a major carbon pool in the near-surface environment. At the global scale, total soil OC is estimated at 1550 Pg and soil IC at 950 Pg.[1] In addition, in arid regions, IC can comprise more than 90% of the total C in the soil. The IC in the soil is present in the minerals calcite and dolomite. Both minerals are relatively insoluble; however, dolomite dissolution is much slower than calcite dissolution at the intermediate pH values relevant to most soils. Also, dolomite does not generally precipitate under earth surface conditions. As a result, it can be considered that over the time frames of current interest related to climate change, the dolomite component in soils remains constant or decreases due to dissolution, while calcium carbonate content may increase or decrease.

This entry considers the soil processes and land-use changes that impact changes in soil IC and its implication for the atmospheric carbon dioxide budget. Among the land-use changes considered are land clearing, tillage, cropping, fertilization, irrigation in both humid and arid environments, irrigation management (leaching), and composition and source of the irrigation water. The impacts of these changes on atmospheric carbon dioxide concentrations are more difficult to analyze and differ with regard to long-term and short-term effects.

Soil Processes

In terms of OC at the land surface and in the soil, a decrease in OC is directly related to either a net removal of OC from the site such as timber or crop harvesting or input of carbon dioxide to the atmospheric in the case of oxidation of soil organic matter. In contrast, for dissolution of the carbonate minerals, a decrease in IC in the soil can result in either a release or an uptake of carbon dioxide from the atmosphere. The impact of these processes is entirely dependent on the local environmental and hydrologic conditions.

In the instance of dissolution under non-acidic conditions, the net reaction for dissolution of calcium carbonate is:

$$C_aCO_3 + H_2O + CO_2 \rightarrow C_a^{2+} + 2HCO_3^-. \tag{1}$$

Thus, dissolution of calcium carbonate results in a sink for atmospheric carbon. Increased soil carbon dioxide concentrations also lead to a net increase in carbonic acid in solution (H_2CO_3). This carbonic acid is partially recharged to the groundwater and is thus a sink for atmospheric CO_2 that is greater than that indicated by Eq. 1 and generally not considered when evaluating organic matter decomposition.

Dissolution of calcium carbonate under acidic conditions results in the following net reaction:

$$C_aCO_3 + 2H^+ \rightarrow C_a^{2+} + H_2O + CO_2, \tag{2}$$

thus leading to a net release of carbon dioxide to the atmosphere. In the case of dissolution of IC, the residence time in the groundwater will determine the net impact on atmospheric carbon dioxide in the time scale of most concern—tens to hundreds of years.

The majority of the alkalinity in the earth's near-surface environment is derived from dissolution of carbonate rocks. The net long-term impact is uptake of carbon dioxide from the atmosphere, dissolution of IC, movement of alkalinity in the surface and subsurface, and discharge to the oceans via rivers.

In terms of the overall C cycle, most of the carbon dioxide released from precipitation of carbonates in the soil (an important process in arid regions) results in a return of C to the atmosphere, as seen by the reverse reaction of Eq. 1. Precipitation of IC results in a net release of 1 mol of carbon dioxide for each mole of IC produced. Sequestration of C in the form of soil IC is carbon storage but the net result of the process is a net release of C to the atmosphere. The quantity released is less than indicated by Eq. 1 since a small amount of the carbon dioxide generated may be recharged with the groundwater, a process that occurs in the absence of precipitation of soil IC.

An important factor to consider when evaluating the implication of an increase in carbon storage is the origin of the calcium and bicarbonate ions. The above instance considered dissolution of IC from the soil. Weathering of silicate minerals is important over long time scales, resulting in release of calcium, sodium, magnesium, and potassium ions as well as bicarbonate. In this instance, the net reaction for dissolution is consumption of 2 mol of carbon dioxide to produce 2 mol of bicarbonate ions, in addition to the release of an equivalent quantity of cations into solution. Thus, silicate dissolution in the soil results in sequestering atmospheric carbon in the form of dissolved bicarbonate. It may be possible in arid regions that the resultant soil solution becomes concentrated, resulting in precipitation of IC. In this process, 1 mol of C is released during IC precipitation and 2 mol are sequestered from silicate dissolution. In this instance, the net result of the silicate weathering is still carbon sequestration from the atmosphere since not all of the alkalinity generated by silicate dissolution is reprecipitated. In terms of IC, the result is an increase in soil IC. Silicate dissolution has a relatively small role in terms of changes related to human activity.

Dissolved IC Cycling

The net impact of soil IC changes on atmospheric carbon dioxide is critically related to the fate of alkalinity (primarily bicarbonate ions) transported into the subsurface below the root zone. If we consider that most of the recharge is ultimately discharged to surface systems, then the potential exists for reprecipitation in these systems, which are at lower carbon dioxide concentrations than the subsurface. It is estimated[2] that inland waters receive approximately 1.9 Pg C/yr, of which 0.8 is returned to the atmosphere as carbon dioxide while 0.9 Pg C/yr is discharged to the oceans, of which 0.45 Pg C/yr is IC. The direct degassing of groundwater is estimated at 0.01 Pg/yr with an uncertainty of 0.003–0.03 Pg/yr.[2] Combining soil liming as well as irrigation practices and other land-use changes, it is likely that the IC flux to groundwater has increased due to human activities. It is estimated that the dissolved IC (mostly bicarbonate) flux into groundwater is on the order of 0.2 Gg/yr.[3] The impact of changes in this subsurface flux, however, is likely relatively long term as the groundwater residence times are mostly on the order of hundreds to thousands of years.[3]

Land Use

Land use, among other environmental factors, affects changes in soil IC with the full impact on atmospheric carbon dioxide expressed only over the long term. Both increases and decreases in IC storage are possible as a result of various management practices. In irrigated lands, there are various factors that cause either a decrease or an increase in soil IC. In most instances, multiple land-use changes have occurred as a result of human activity, and it is difficult to isolate the specific process that is dominant in explaining changes in soil IC. It is calculated that China has a total of 53 Pg of soil IC and that human activity has resulted in a net loss of 1.6 Pg of IC.[4]

Soil respiration processes include plant respiration and release of carbon dioxide via the roots, as well as microbial decomposition of soil organic matter. These processes result in an elevated carbon dioxide concentration in the soil relative to the atmospheric condition. The soil CO_2 concentration is typically in the range of 0.1%–5%, orders of magnitude higher than the atmospheric value. Increased plant biomass productivity due to crop selection irrigation and/or fertilization results in increased soil carbon dioxide production and concentration, thus contributing to increased weathering of both silicates and carbonates. Inadvertent or deliberate introduction of non-native species also impacts biomass productivity, soil CO_2 concentrations, water use, recharge, and erosion with subsequent impact on soil IC.

Land Clearing

Land clearing generally results in increased water runoff and soil erosion. This process or any other process such as tillage that increases erosion serves to remove the surface soil horizons. Since these horizons are generally depleted in IC relative to less weathered, deeper horizons, erosion causes an apparent increase in the IC content of the top 1–2 m of carbonate-containing soils. This may result in erroneous estimates of changes in soil IC when comparing the top 1 m of disturbed vs. native vegetation sites. In terms of carbonate dissolution, the impact of land clearing is not certain. After clearing, there is increased runoff, thus decreased surface infiltration, favoring less dissolution of carbonates from the surface horizons. This effect may be compensated by the decreased water consumption (lower evapotranspiration) after clearing, resulting in increased deep recharge and possibly greater carbonate dissolution or less reprecipitation at depth, depending on the volumes of water. Depending on how much biomass remains after clearing, there is likely a short-term increase in soil CO_2 followed by a longer-term reduction, favoring less carbonate dissolution in the soil.

Cropping and Tillage

The impact of cropping and tillage on IC storage is not certain and there are limited studies, but the sum of the studies suggests net IC accumulation. The impact of tillage and cropping on IC in the Northern Great Plains of the United States has been evaluated by comparing long-term cultivation (likely >80 years) with uncultivated grassland.[5] There was a statistically significant increase in IC with cropping for one soil type (at the 90% confidence level but not at the 95% confidence level) but no statistically significant differences for the other two soil types examined. The lack of a clear impact is likely related to the fact that these soils did not have significant quantities of IC in the upper portion of the profile where most of the water transport and root water uptake would occur.

In a study of Russian Chernozem soils, the soil IC was greater on a continuously cropped field and on a fallow field relative to a native grassland, with most of the approximately 100 Mg/ha increase in IC being in the upper portion of the profile.[6] A grassland field where hay was harvested showed no significant changes over the native vegetation (grassland) field. The authors attributed the increase in the cropped and fallow fields to cultivation, irrigation, and fertilization, with the calcium being supplied by fertilizer and manure additions. The manure also supplied alkalinity necessary for carbonate formation. The effects in this study may be more related to fertilizer application than changes due to cultivation per se.

When comparing IC in cropped soils to native grassland, it needs to be considered that grassland and native vegetative soils can also have net accumulation or loss of IC with net increase in pedogenic and recently formed IC. We cannot assume that increases in pedogenic carbonate necessarily correspond to increases in IC storage. For example, it is reported that for grasslands and forested soils in Saskatchewan, Canada, there is an increase in pedogenic IC of 1.2–1.8 kg/ha/yr,[7] and this has been cited as an increase in IC. However, the soils contain dolomite and re-cystallization of carbonates is listed by the authors as a major process producing the pedogenic carbonate.[7] Only one of the soils (Black Chernozem) had net IC accumulation based on comparison of the pedogenic carbonate accumulated and the depletion of the lithogenic carbonate.[7]

Land clearing, tillage, and overgrazing in arid lands serve to increase wind erosion and to redistribute soil in the landscape. In this manner, non-calcareous soils receive inputs of carbonates. This process likely increases net dissolution of carbonates, as it spreads carbonates across the landscape and into areas with non-calcareous soils.

In humid environments, IC is leached from the soil. The elevated CO_2 concentrations in the soil enhance IC solubility relative to earth surface conditions. In humid environments, carbonates are successively leached from the upper portions of the soil profile. Agricultural practices may serve to enhance or reduce the net removal of carbonates. Removal of vegetation from a site with cropping practices such as tree harvesting or crop or forage harvesting serves to remove base cations and causes net acidification of the upper portions of the soil profile. If carbonates are present deeper in the soil, this acidification increases IC dissolution. The impact of removal of vegetation in humid environments with carbonates in the subsoil can be calculated by assuming that the net harvested alkalinity is compensated by an equal increase in carbonate dissolution in the subsurface.

Fertilization and Liming

Since optimum plant growth is generally at a pH lower than that observed in untreated calcareous soils, acid fertilizers are commonly applied in arid regions. Use of sulfur with subsequent oxidation to sulfate results in acid release to the soil (2 mol of protons per mole of sulfur). Application of ammonia salts with subsequent fixation into organic matter or oxidation to nitrous oxide or elemental nitrogen also releases protons (2 and 1 mol of protons per mole of ammonia ions, respectively). This acidification will increase carbonate dissolution proportionately and has a significant effect since ammonia salts are widely applied as fertilizers.

Application of fertilizer as urea or ammonia gas should have no net effect on carbonate dissolution (upon oxidation to nitrous oxide or elemental nitrogen) other than the indirect impact on soil CO_2. In contrast, use of nitrate fertilizers serves to increase pH and thus reduce carbonate dissolution. Generally, nitrate is not utilized on calcareous soils, so the impact on IC storage in soil is slight. The quantitative impact of fertilization on changes in IC is not easily calculated, as it depends on the extent of N incorporation into organic matter, mineralization, the extent to which the harvested biomass is removed from the site, and the occurrence of carbonates in the subsurface.

The practice of liming soils is very prevalent for crop production and urban landscaping in humid environments and represents a major anthropogenic input. The liming is done by application of calcium carbonate or dolomite to the soil. Addition of liming products, primarily calcite, are reported as 3.7 Tg C/yr in the United States for 1978.[8] This is a significant but temporary addition to the soil IC pool. Since liming is not generally needed in arid regions, it is reasonably assumed that the majority of the material is applied to acid soils and thus it is readily dissolved. The liming serves two functions, to increase the pH of acid soils and as an inexpensive calcium source in regions with leached, calcium-deficient soils. The net impact of liming is neutralization of most of the carbonate in the mineral and release of the generated carbon dioxide to the atmosphere, as represented in Eq. 2. The amount of carbon dioxide release to the atmosphere is less than indicted by Eq. 2 for all but the highly acidic soils, as some of the carbon is sequestered in the form of dissolved bicarbonate. In humid environments, most of the bicarbonate produced will likely not reprecipitate, even in the long term; instead, it will be discharged into the ocean via river drainage systems.

Humid Region Irrigation

Large quantities of water applied in excess of plant transpiration needs result in leaching and maintenance of elevated water content at or near the soil surface. The additional water applied as irrigation enhances dissolution of IC as does the increased soil carbon dioxide concentrations due to increased water content as in the soil. Irrigation in humid environments thus serves to increase the net recharge through the soils and thus removal of carbonates. These changes may be relatively difficult to detect in view of the limited amount of irrigation water added and the fact that irrigation in humid environments, although increasing rapidly, was very limited in the past. Field studies are needed to determine the impact of irrigation on changes in IC in humid environments.

Arid and Semiarid Region Irrigation

Arid zone soils usually contain at least minor amounts of carbonates, even if they are not classified as calcareous. In the absence of irrigation, there may be redistribution of carbonates within the soil but likely little net precipitation. The majority of the pedogenic calcite is reprecipitated calcite with relatively small amounts added as a result of silicate mineral weathering. This process is of geological significance and central to explaining soil formation but of less importance to the time frames of interest with regard to recent changes in atmospheric CO_2. Significant amounts of carbonates are added to the surface of arid land soils as dust. IC is leached from the upper part of the soil profile by dissolution into the infiltrating rain and is mostly reprecipitated at depth after plant extraction of the available water.

Irrigation in arid and semiarid environments may result in a net increase or decrease in soil IC, depending on the water source and fraction of water applied that is leached (leaching fraction). There are various opposing effects. First, elevated CO_2 concentrations in the root zone relative to the atmospheric condition results in enhanced calcite and dolomite solubility and dissolution and, hence, depletion of soil IC. Second, plant water extraction and evaporation concentrate the soluble salts into a smaller volume of water and enhance calcite precipitation, thus increasing IC. Plant roots extract relatively dilute water from the soil, leaving most of the salts behind. At low leaching fractions (where the quantity of water applied plus rain is only 2%–10% greater than the amount of water transpired), the effect of concentration of salts due to plant water extraction and evaporation is greater than the enhanced CO_2 effect and there is net precipitation of calcite.

For a calcite-supersaturated surface water such as the Colorado River, it is estimated[9] that at a leaching fraction of 0.1, there is net precipitation of 125 kg/ha/yr of C, based on water consumption of 1.2 m/yr and an average soil CO_2 partial pressure of 3 kPa (approximately 3% of the soil atmosphere). Model simulations indicate that net precipitation of calcite occurs in the soil profile, with loss of IC in the upper portion of the root zone and precipitation of IC in the lower portion. At high leaching fractions, there is net dissolution of carbonates. Using a predictive calcite precipitation model that accounts for 3-fold calcite supersaturation in precipitating environments, it is predicted that at a leaching fraction of 0.4, there will be a net dissolution of IC of 70 kg/ha/yr of C. Again, this calculation is based on water consumption of 1.2 m/yr and soil CO_2 partial pressure of 3 kPa.[10] In all instances, there is a prediction of net dissolution in the upper portion of the soil root zone and net precipitation in the lower portions of the profile. Using average leaching fractions for the western United States, it is estimated that irrigation with surface waters on 12 million ha[11] results in an increase in soil IC of 1 Tg/yr,[10] or 80 kg/ha/yr. Considering the reverse reaction in Eq. 1, this increase in soil IC corresponds to a net release of almost 1 Tg/yr of C to the atmosphere.

Irrigation with groundwater saturated with respect to calcite will result in precipitation of carbonates at almost all leaching fractions, since the irrigation water is equilibrated at the groundwater CO_2 partial pressure and is highly supersaturated with respect to calcite upon degassing and application to the soil surface. Calcite-saturated groundwater is used for irrigation on an estimated 3.12 million ha in the United States.[11] It is estimated that irrigation on these soils results in a net IC precipitation of 1.3 Tg/yr or 420 kg/ha/yr.[10] This corresponds to slightly less than 1.3 Tg/yr of carbon dioxide release to the atmosphere.

The model calculations of IC precipitation are dependent on several assumptions regarding calcite precipitation and soil CO_2 concentrations and do not consider the effect of fertilizers. Generally, acidifying fertilizers are applied in arid regions, thus reducing the extent of expected increases in IC with irrigation. A partial validation of the modeling is available by comparing modeling results with data from Palo Verde, California. Using measured CO_2 partial pressure in the groundwater in Palo Verde Valley, there is prediction of no net change in IC at a leaching fraction of 0.5, obtained from the measured water diversions and drainage volumes. Consistent with these predictions, the groundwater composition draining back to the Colorado River from Palo Verde Valley shows no evidence for net precipitation or dissolution of carbonates in the soil.

There is some direct field evidence for the influence of irrigation on soil IC, but as expected from the above discussion, the results are site specific depending on water source, water composition, and leaching, as well as quantity and type of fertilizer used. It is also difficult to be certain that differences among sites are only related to changes in management. Researchers[12] observed a net decrease in the calcium carbonate content of three pairs of soil profiles taken from sites in Israel irrigated for approximately 40 years as compared to non-irrigated sites. The estimated input of 4.40 m of water per year at those sites is contrasted with the yearly potential evapotranspiration of 1.93 m. The observed trend is qualitatively consistent with model predictions if we account for the input of rain and the estimated high leaching fraction of 0.56. Isotopic evidence indicated that there was precipitation of pedogenic carbonate at depth despite a net decrease in IC content at depth.[12] This suggests solubilization and reprecipitation, but the measured net impact was still loss of IC in the soil profile, as expected from modeling this water budget data using surface water for irrigation.

In a study in the San Joaquin Valley in California, researchers compared samples of a soil taken from irrigated and native vegetation sites.[13] They also measured a net loss of carbonates attributed to 8 years of irrigation. Net carbonate loss was estimated as 7×10^3 kg/ha/yr (800 kg C/ha/yr). Leaching fractions at the site were not reported, but this value corresponds to approximately 10 times greater dissolution than expected based on model simulations. However, another study by the same author[14] found no change in total carbonate when comparing pedons with native vegetation and those irrigated for 5–25 years.

In this instance, both gypsum and sulfur were applied as amendments for reclamation. Gypsum would tend to greatly increase precipitation of carbonates and thus increase IC, while sulfur would acidify the soil and cause net dissolution of carbonates.

In other studies, the results are also mixed as expected if we consider variations in irrigation water composition and leaching fractions. In a semiarid northwestern U.S. study, irrigated crops showed a net increase in IC while irrigated pasture showed a net decrease relative to native sagebrush vegetation.[15] In the San Joaquin Valley, California, there was a net gain in IC of 1.8 kg/m² after 30 years of irrigation and a net loss of IC after 55 years of irrigation at another site, while in Imperial Valley, there was a net gain of IC of 4 kg/m² or 40 Mg/ha after 85–90 years of irrigation.[16] Another, more extensive study in Imperial Valley[17] on paired soil cores (irrigated and adjacent non-irrigated sites) observed no changes in IC storage after 90 years of irrigation and no isotopic C shifts indicative of recrystallization. Accurately modeling the net global impact of irrigation on soil IC based on water types, water application quantities, and evapotranspiration will first require the ability to accurately simulate the impacts on specific locations.

Sodic Soil Reclamation

Reclamation of sodic soils can result in either an increase or a decrease in soil IC. Gypsum application to a sodic and alkaline soil will increase the soil carbonate content, as the increased Ca will precipitate most of the soluble bicarbonate and carbonate. This will increase IC sequestration in the soil but have a net effect of also increasing the carbon dioxide flux to the atmosphere (and decreasing the net flux of C and alkalinity to the groundwater). Gypsum is also very commonly used in arid regions to improve water infiltration. At least some of the applied calcium from the gypsum application is reprecipitated as calcium carbonate, increasing the soil IC. Application of sulfuric acid, sulfur, or green manuring all serve to dissolve soil carbonates, thus depleting IC stocks in the soil. Addition of acid will result in an increase in atmospheric carbon dioxide equal to the IC removed.

Green manuring is a management practice of adding fresh plant organic matter to a calcareous soil, enhancing CO_2 concentrations in the soil and thus enhancing carbonate dissolution, providing a calcium source to replace the soil exchangeable sodium. This process will decrease soil IC and result in net release of CO_2 to the atmosphere. The dissolution process will deplete atmospheric carbon, but the effect is negated by the larger increase in carbon dioxide release related to the oxidation of the OC. It is estimated that this process can dissolve on the order of 400 to 800 kg/ha during a year of reclamation. Use of acid is currently a widespread and generally recommended practice to prevent emitter clogging in drip irrigation systems. This practice may result in total removal of carbonates within 10–20 years, for soils with less than 3% carbonate content.

Impact of IC on Atmospheric Carbon Dioxide

Dissolution of carbonates in neutral to alkaline environments results in consumption of CO_2 gas and formation of aqueous bicarbonate $\left(HCO_3^-\right)$, while precipitation of carbonates results in release of CO_2. The net effect of dissolution or precipitation of soil IC on atmospheric CO_2 depends on the solution flow path. In regions irrigated with surface water, the dissolution of carbonates results in a net C sink. The high alkalinity drainage water usually flows back to the river, but even in shallow groundwater systems, this would take on the order of tens to several hundred years. The resultant degassing of carbonic acid in arid environments will result in reprecipitation of carbonate in the river or reservoir and releases CO_2 back into the atmosphere. If the water is recharged into deep aquifers, the net soil flux of dissolved IC is preserved as an IC sink. In acid environments, liming of soils results in CO_2 release to the atmosphere as there is little or no net alkalinity produced. Examination of records of volumes of water and concentrations of alkalinity (mostly bicarbonate) revealed that within the past 50 years, there has been a dramatic increase in net alkalinity export from the Mississippi River to the Gulf of Mexico.[18] The 17.7 Tg/yr increase in IC is attributed to increased rainfall in the basin and increased proton delivery to the land due to enhanced carbon dioxide production in the soil, resulting in increased weathering. It seems reasonable to consider that management changes for increased crop productivity may have increased soil liming and dissolution of IC. Such a drastic change does not likely reflect a response to a changing groundwater composition but is likely related to increased alkalinity in surface overland flow. The impact of a drier climate on the net export of alkalinity was not evaluated. However, it is reasonable to assume that decreased runoff would result in decreased mass of alkalinity transported to surface waters.

Conclusion

Land-use practices have a long-term impact on soil IC. Due to the large C pools in the soil, these impacts are not generally observed in short-term studies. Use of acidifying fertilizers such as ammonia and sulfur reduces soil IC. Practices that favor dissolution of carbonates include irrigation with surface waters and irrigation with water in large excess of plant transpiration. Practices that favor accumulation of IC in the soil include lower water applications relative to transpiration when irrigating in arid and semiarid regions (leaching less than 30% of the applied water), irrigation with groundwaters at elevated CO_2 concentrations, application of gypsum to alkaline soils, and use of nitrate fertilizer. Other factors that affect soil carbonate content include land clearing, cropping practices, and erosion.

In humid environments, the major anthropogenic impacts on inorganic C are liming of surface soils, use of fertilizers, removal of vegetation, and soil erosion. In semiarid and arid environments, increased IC is favored by use of groundwater vs. surface water for irrigation and application of gypsum amendments used to improve water infiltration. Decreased IC is favored by inefficient irrigation with surface water and application of NH_4 fertilizer. The net effect of irrigation on a global scale, neglecting the effects of fertilizer addition, is an estimated increase in soil inorganic C by 30 Tg/yr as well as a release of an almost equal amount of C to the atmosphere. Liming practices in humid regions throughout the world are estimated to have no net effect on inorganic soil C but release up to 85 Tg C/yr to the atmosphere. Of course this is the net effect from IC processes and does not include sequestration of OC by crop production.

References

1. Batjes, N.H. Total carbon and nitrogen in soils of the world. Eur. J. Soil Sci. **1966**, *47*, 151–163.
2. Cole, J.J.; Prairie, Y.T.; Caraco, N.F.; McDowell, W.H.; Tranvik, L.J.; Striegl, R.G.; Duarte, C.M.; Kortelainen, P.; Downing, J.A.; Middelburg, J.J.; Melack, J. Plumbing the global carbon cycle: Integrating inland waters into the terrestrial carbon budget. Ecosystems 2007, *10*, 171–184.
3. Kessler, T.J.; Harvey, C.F. The global flux of carbon dioxide into groundwater. Geophys. Res. Lett. **2009**, *28*, 279–282.
4. Wu, H.; Guo, Z.; Gao, Q.; Peng, C. Distribution of soil inorganic carbon storage and its changes due to agricultural land use activity in China. Agric. Ecosyst. Environ. **2009**, *129*, 413–421.
5. Cihacek, L.J.; Ulmer, M.G. Effects of tillage on inorganic carbon storage in soils of the Northern Great Plains of the U.S. In *Agricultural Practices and Policies for Carbon Sequestration in Soil*; Kimble, J.M., Lal, R., Follett, R.F., Eds.; CRC Press: Boca Raton, 2002; 63–69.
6. Mikhailov, E.; Post, C.A. Effects of land use on soil inorganic carbon stocks in the Russian Chernozem. J. Environ. Qual. **2006**, *35*, 1384–1388.
7. Landi, A.; Mermut, A.R.; Anderson, D.W. Origin and rate of pedogenic accumulation in Saskatchewan soils, Canada. Geoderma **2003**, *117*, 143–156.
8. Voss, R.D. What constitutes an effective liming material. In *National Conference on Agricultural Limestone*; National Fertilizer Development Center: Muscle Shoals, AL, 1980; 52–61.
9. Suarez, D.; Rhoades, J. Effect of leaching fraction on river salinity. J. Irrig. Drain. Div., Am. Soc. Civ. Eng. **1977**, *103* (2), 245–257.

10. Suarez, D. Impact of agriculture on CO_2 as affected by changes in inorganic carbon. In *Global Climate Change and Pedogenic Carbonates;* Lal, R., Kimble, J., Eswaran, H., Stewart, B., Eds.; Lewis: Boca Raton, 1999; 257–272.

11. Solley, W.B.; Pierce, R.R.; Perlman, H.A. *Estimated Use of Water in the United States in 1990;* U.S. Geol. Surv. Circular 1004; U.S. Gov. Printing Office: Washington, DC, 1993.

12. Magaritz, M.; Amiel, A. Influence of intensive cultivation and irrigation on soil properties in the Jordan Valley, Israel: Recrystallization of carbonate minerals. Soil Sci. Soc. Am. J. **1981**, *45*, 1201–1205.

13. Amundson, R.G.; Smith, V.S. Effects of irrigation on the chemical properties of a soil in the Western San Joaquin Valley, California. Arid Soil Res. Rehabil. **1988**, *2*, 1–17.

14. Amundson, R.D.; Lund, L. The stable isotope chemistry of a native and irrigated Typic Natrargid in the San Joaquin Valley of California. Soil Sci. Soc. Am. J. **1987**, *51*, 761–767.

15. Entry, J.A.; Sojka, R.E.; Shewmaker, G.E. Irrigation increases inorganic carbon in agricultural soils. Environ. Manage. **2004**, *33*, s309–s317.

16. Wu, L.; Wood, Y.; Jiang, P.; Li, L.; Pan, G.; Lu, J.; Chang, A.C.; Enloe, H.A. Carbon sequestration and dynamics of two irrigated agricultural soils in California. Soil Sci. Soc. Am. J. **2007**, *72*, 808–814.

17. Suarez, D.L. Impact of agriculture on soil inorganic carbon. Annual Meetings, Baltimore, MD, Oct. 18–23, 1998; Agronomy Abstracts; Soil Science Society of America: Madison, WI, 1998; 258–259.

18. Raymond, P.A.; Cole, J.J. Increase in the export of alkalinity from North America's largest river. Science **2003**, *301*, 88–91.

19

Chromium

Bruce R. James and
Dominic A. Brose

Introduction

Chromium is a heavy metal that is essential for human health as a cofactor of insulin in its trivalent form [(Cr(III)] but may cause lung cancer if inhaled in the hexavalent oxidation state [Cr(VI)]. Trivalent Cr is only sparingly soluble in neutral to alkaline natural waters as $Cr(OH)_3$ or Cr_2O_3, but it can be oxidized to Cr(VI) by naturally occurring manganese (III,IV) (hydr)oxides and by hydrogen peroxide, ozone, chlorine gas, hypochlorite, and other electron acceptors used in the environmental remediation of water, sediments, and soils. Hexavalent Cr can be reduced to Cr(III) by elemental Fe and Fe(II), sulfides (H_2S and HS^-), easily oxidized organic C compounds, and other electron donors. The rates and extent of oxidation and reduction reactions of Cr are governed by the redox potential (Eh) and acidity (pH) of natural soil–water systems and of engineered environments, such as anthropogenic wetlands and wastewater treatment facilities.

Health Effects

Concerns surrounding the presence of chromium (Cr) in natural waters and drinking water supplies must address the contrasting solubilities and toxicities of its common oxidation states in natural environments: Cr(III) and Cr(VI). Currently, Cr regulation principally focuses on "total Cr," without distinguishing the oxidation states in a solid or water sample. The U.S. Environmental Protection Agency's (EPA) national standard for total Cr in drinking water is 100 μg/L (100 ppb), except that California has set its current drinking water standard at 50 μg/L.[1,2] In 1999, California set a Public Health Goal of 2.5 μg/L, based on a 1968 study in Germany that found stomach tumors in animals that drank chromium-enriched water. The U.S. EPA rejected that study as flawed and determined that there was no evidence it was carcinogenic in water, which resulted in California rescinding its goal in 2001 and reverting back to the 50 μg/L standard.[2] The point of contention regarding the 1968 study was whether Cr(VI) is reduced to Cr(III) in the stomach by gastric acids.

Chromium(VI) is genotoxic in a number of in vitro and in vivo toxicity tests.[3] Because the mechanisms of geno-toxicity and carcinogenicity are not fully understood, the National Toxicology Program (NTP) conducted animal tests to assess the potential for cancer due to ingestion of Cr(VI).[4] Reduction of Cr(VI) to Cr(III) is hypothesized to occur primarily in the stomach, as a mechanism of detoxification. In this 2 years NTP study, no neoplasms or non-neoplastic lesions were observed in the forestomach or glandular stomach of rats or mice. However, observed increases in neoplasms, or abnormal growths of tissue, in the small intestine of mice, toxicity to red and white blood cells and bone marrow, and uptake of Cr(VI) into tissues of rats and mice suggest that at least a portion of the administered Cr(VI) was not reduced in the stomach.[4] The stepwise reduction of Cr(VI) to Cr(III) through Cr(V) and Cr(IV) creates short-lived oxidation states that may be the forms of Cr that are the actual carcinogens.

This finding, in addition to the absence of increases in neoplasms or non-neoplastic lesions of the small intestine in rats or mice exposed to chromium picolinate monohydrate (CPM), an organically bound form of Cr(III),[5] provides evidence that Cr(VI) is not completely reduced in the stomach and is responsible for these carcinogenic effects. Additionally, it should be noted that Cr(III), like that found in CPM, is essential for human health in trace amounts as an activator of insulin,[6] but exists predominantly in nature in cationic forms that are typically only sparingly soluble in near-neutral pH soils, plants, cells, and natural waters.[7]

Sources of Chromium and Occurrence in Natural Waters and Water Supplies

When soluble Cr is detected in natural waters, especially at high concentrations, it is usually Cr(VI) derived from industrial wastes containing Cr(VI) or possibly resulting from the oxidation of certain forms of Cr(III) in soils or sediments.[8,9] Chromium is the seventh most abundant metal on earth with an average content of 100 mg/kg in the earth's crust and 3700 mg/kg for the earth as a whole, principally as Cr(III) in unreactive, insoluble miner-als, such as chromite ($FeO.Cr_2O_3$).[10] Roasting chromite ore under alkaline, high-temperature conditions oxidizes Cr_2O_3 to soluble Cr(VI), a widely used starting material for production of stainless steel, pressure-treated lumber, chrome-tanned leather, pigments, chrome-plated metals, and other common products used in modern societies.[11] As a result, Cr(VI) remaining in industrial by-products, such as chromite ore processing residue, chrome plating bath waste, paint aerosols, and other industrial wastes, may enrich soils and contaminate surface waters and groundwater that are supplies for domestic uses, irrigation, and industrial processes.

In contrast to these concentrated, anthropogenic sources of Cr(VI); naturally occurring sources of Cr are predominantly Cr(III) and occur at low concentrations. Ultramafic and basaltic rocks (and soils developed from these parent materials), however, may contain up to 2400 mg Cr/kg, and can release small fractions of the Cr contained in them as Cr(VI), either through dissolution of Cr(VI) minerals or possibly via oxidation of Cr(III) by Mn(III,IV) (hydr)oxides. As a result, Cr(VI) has been detected in groundwater (0.05–0.5 mg/L) in arid regions dominated by these alkaline, Cr-rich rocks and soils. A concentration of Cr(VI) of 7.5 mg/L in pH 12.5 groundwater from Jordan is the highest known level that is not due to human influence. Naturally occurring Cr in alkaline, aerobic ocean water exists principally as Cr(VI) at concentrations in the range of 3–7.3 nM (0.16–0.38 µg/L).[12]

The balance of the different forms and the solubilities of Cr(III) and Cr(VI) in natural waters is governed by pH, aeration status, and other environmental conditions (Table 1). Understanding and predicting the oxidation state, solubility, mobility, and bioavailability of Cr in water are further complicated by the fact that Cr(III) can be oxidized (lose three electrons) to form Cr(VI), whereas Cr(VI) can gain three electrons and be reduced to Cr(III).[13,14] Natural variation and human-induced changes in pH and the oxidation–reduction status of soil and water can control the solubility of Cr. As a result, purification of drinking water supplies and treatment of wastewaters contaminated with Cr are possible through chemical and microbiological processes that modify the acidity and the relative abundance of oxidizing and reducing agents for Cr.[15,16]

Solubility Controls of Chromium Concentrations in Water

Most inorganic compounds of Cr(III) are less soluble in water than are those of Cr(VI) because Cr(III) cations have high ionic potentials (charge-to-size ratio) and hydrolyze to form covalent bonds with OH^- ions (Table 1). When three OH^- anions surround the Cr^{3+} cation, it is particularly stable in water as the sparingly soluble compound, $Cr(OH)_3$ (Table 2). Upon aging and dehydration, $Cr(OH)_3$ slowly converts to the more crystalline, less soluble Cr_2O_3.[12] Incorporation of Fe(III) or Fe(II) into solid phases and precipitates containing Cr(III) renders the Cr(III) less soluble, often by a factor of 10^3 in the

TABLE 1   Oxidation States and Forms of Chromium in Natural Waters

Oxidation State	Form	Name	Chemical Conditions of Water under Which It Is Found and Pertinent Reaction in Natural Waters
Chromium (III) (trivalent chromium)	$Cr(H_2O)_6^{3+}$	Hexaquochromium(III)	pH < 3.5; strong affinity for negatively charged ions (e.g., phosphate) and colloid surfaces (e.g., living cells and phyllosilicate clays or fulvic and humic acids); green color
	$Cr(H_2O)_5OH^{2+}$	Monohydroxychromium(III)	First hydrolysis product formed at pH > 3.5 upon dilution of or addition of base to solution of Cr(III); green
	$Cr(H_2O)_4(OH)_2^+$	Dihydroxychromium(III)	Second hydrolysis product of Cr(III); may dimerize and polymerize to form high molecular weight cations in planes of octahedron; green
	$Cr(H_2O)_3(OH)_3^0$	Chromium hydroxide	Metastable, uncharged hydrolysis product that precipitates as the sparingly soluble $Cr(OH)_3$
	$Cr(H_2O)_2(OH)_4^-$	Hydroxochromate	Fourth hydrolysis product of Cr(III) that may form at pH > 11; may oxidize to Cr(VI) by O_2
	Cr(III)–organic acid complexes and chelates	For example: chromium citrate, chromium picolinate, chromium fulvate	Soluble complexes and chelates in which water molecules of hydration surrounding $Cr(H_2O)_6^{3+}$ are displaced by carboxylic acid and N-containing ligands; formation is pH and concentration dependent; blue–green–purple colors, depending on ligand binding Cr(III)
Chromium (VI) (hexavalent chromium)	H_2CrO_4	Chromic acid	Fully protonated form of Cr(VI) formed at pH < 1; see Figure 2 for key Eh values for redox
	$HCrO_4^-$	Bichromate	Form of Cr(VI) that predominates at 1 < pH < 6.4; yellow; see Figure 2 for key Eh values for redox
	CrO_4^{2-}	Chromate	Form of Cr(VI) that predominates at pH > 6.4; yellow; see Figure 2 for key Eh values for redox
	$Cr_2O_7^{2-}$	Dichromate	Form of Cr(VI) that predominates at pH < 3 and in concentrated solutions (>1.0 mM); rapidly reverts to $HCrO_4^-$ or CrO_4^{2-} upon dilution or pH change

TABLE 2   Solubility in Water at pH 7 of Selected Chromium Compounds

Oxidation State	Compound Name	Formula	Approximate Solubility (mol Cr/L)
Chromium (III) (trivalent chromium)	Chromium(III) hydroxide	$Cr(OH)_3$ (am)	10^{-12}
	Chromium(III) oxide	Cr_2O_3 (Cr)	10^{-17}
	Chromite	$FeO \cdot Cr_2O_3$ (Cr)	10^{-20}
	Chromium chloride	$CrCl_3$	Highly soluble
	Chromium sulfate	$Cr_2(SO_4)_3$	Highly soluble
	Chromium phosphate	$CrPO_4$	10^{-10}
	Chromium fluoride	CrF_3	1.2×10^{-3}
	Chromium arsenate	$CrAsO_4$	10^{-10}
Chromium (VI) (hexavalent chromium)	Potassium chromate	K_2CrO_4	3.2
	Sodium chromate	Na_2CrO_4	5.4
	Calcium chromate	$CaCrO_4$	0.14
	Barium chromate	$BaCrO_4$	1.7×10^{-3}
	"Zinc yellow" pigment	$3ZnCrO_4 \cdot K_2CrO_4 \cdot Zn(OH)_2H_2O$	8.2×10^{-3}
	Strontium chromate	$SrCrO_4$	5.9×10^{-3}
	Lead chromate	$PbCrO_4$	1.8×10^{-6}
	Chromium(VI) jarosite	$KFe_3(CrO_4)_2(OH)_6$ (Cr)	10^{-30}

solubility product (Ksp).[17,18] In the pH range of 5.5–8.0, Cr(III) reaches minimum solubility in water due to this hydrolysis and precipitation reaction, an important process that controls the movement of Cr(III) in soils enriched with industrial wastewaters and solid materials. Under strongly acidic conditions (pH < 4), unhydrolyzed $Cr(H_2O)_6^{3+}$ cations exist in solution, while $Cr(OH)4^-$ forms under strongly alkaline conditions (pH > 11), particularly in response to adding base to solutions of soluble salts of Cr(III), e.g., $CrCl_3$, $Cr(NO_3)_3$, or $Cr_2(SO_4)_3$.

Other anions besides OH^- coordinate with $Cr(H_2O)_6^{3+}$ and displace water molecules of hydration to form sparingly soluble compounds and soluble chelates (Table 2). In water treatment facilities and in natural waters, phosphate ($H_2PO_4^-$, HPO_4^{2-}, PO_4^{3-}), arsenate ($H_2AsO_4^-$, $HAsO_4^{2-}$) and fluoride (F^-) may form low solubility compounds with Cr(III). Organic complexes of Cr(III) with carboxylic acids (e.g., citric, oxalic, tartaric, fulvic) remain soluble at pH values above which $Cr(OH)_3$ forms. By increasing the solubility of Cr(III) in neutral and alkaline waters, such organic complexes enhance the potential for absorption of Cr(III) by cells. Stable, insoluble complexes of Cr(III) also form with humic acids and other high molecular aggregate weight organic moieties in soils, sediments, wastes, and natural waters.[19]

With the exception of chromium(VI) jarosite (Table 2), Cr(VI) compounds are more soluble over the pH range of natural waters than are those of Cr(III), thereby leading to the greater concern about the potential mobility and bioavailability of Cr(VI) than Cr(III) in natural waters. The alkali salts of Cr(VI) are highly soluble, $CaCrO_4$ is moderately soluble, and $PbCrO_4$ and $BaCrO_4$ are only sparingly soluble. In colloidal environments containing aluminosilicate clays and (hydr)oxides of Al(III), Fe(II,III), and Mn(III,IV) (e.g., in soils and sediments), Cr(VI) anions may be adsorbed similarly to SO_4^{2-}. Low pH and high ionic strength promote retention of $HCrO_4^-$ and CrO_4^{2-} on positively charged sites, especially those associated with colloidal surfaces dominated by pH-dependent charge. Such electrostatic adsorption may be reversible, or the sorbed Cr(VI) species may gradually become incorporated into the structure of the mineral surface (chemisorption). Recently precipitated $Cr(OH)_3$ can adsorb Cr(VI) or incorporate Cr(VI) within its structure as it forms, thereby forming a Cr(III)–Cr(VI) compound.[20]

Oxidation-Reduction Chemistry of Chromium in Natural Waters

The paradox of the contrasting solubilities and toxicities of Cr(III) and Cr(VI) in natural waters and living systems is complicated by two reduction–oxidation (electron transfer) reactions: Cr(III) can oxidize to Cr(VI) in soils and natural waters, and Cr(VI) can reduce to Cr(III) in the same systems, and at the same time. Understanding the key electron transfer processes (redox) and predicting environmental conditions governing them are central to treatment of drinking water, wastewaters, and contaminated soils, and to predicting the hazard of Cr in natural systems.[21] The metaphor of a seesaw (Figure 1) is useful in picturing the undulating nature of the changes in Cr speciation in water due to oxidation of Cr(III) and reduction of Cr(VI). A balance for the two redox reactions is achieved in accordance with the quantities and reactivities of reductants and oxidants in the system [e.g., organic matter and Mn(III,IV) (hydr)oxides, as modulated by pH and pe as a master variables].[15,22]

The thermodynamics (energetics predicting the relative stability of reactants and products of a chemical reaction) of interconversions of Cr(III) and Cr(VI) compared to other redox couples can be used to predict the predominance of Cr(III) or Cr(VI) in water supplies (Figure 2). Certain electron-poor species may act as oxidants (electron acceptors) for Cr(III), especially soluble forms of Cr(III), in the treatment of water supplies or in soils enriched with Cr(III) (Nieboer and Jusys, Figure 3).[19] Examples are those above the bold line for Cr(VI)–Cr(III) on the Eh–pH diagram: Cl_2, OCl^-, H_2O_2, O_3, and MnOOH. In contrast, electron-rich species may donate electrons to electron-poor Cr(VI) and reduce it to Cr(III): Fe^{2+} [or Fe(0)], H_2S, H_2, ascorbic acid (and organic compounds, generally), and SO_2. Sunlight may affect the kinetics of both oxidation and reduction reactions for Cr, a relevant fact for natural processes in lakes and streams and for treatment technologies for drinking water purification. Depending on pH, temperature, and the concentrations of oxidants and reductants, Cr(VI)-to-Cr(III) ratio in natural waters may be predicted.

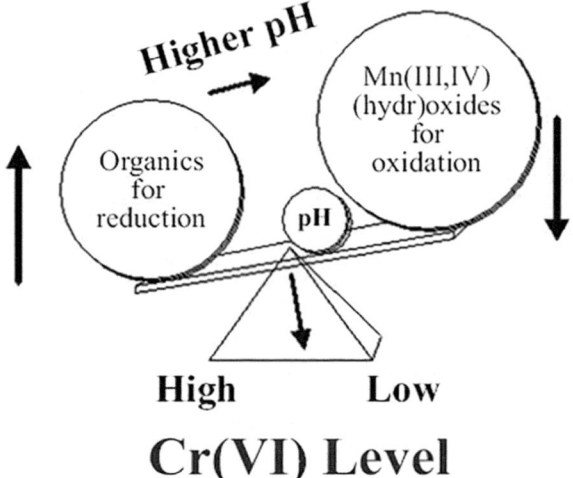

FIGURE 1 Seesaw model depicting a balance of the oxidation of Cr(III) by Mn(III,IV) (hydr)oxides and the reduction of Cr(VI) by organic compounds, with the pH acting as a sliding control (master variable) on the seesaw to set the redox balance for given quantities and reactivities of oxidants and reductants. The equilibrium quantity of Cr(VI) in the water is indicated by the pointing arrow from the fulcrum.

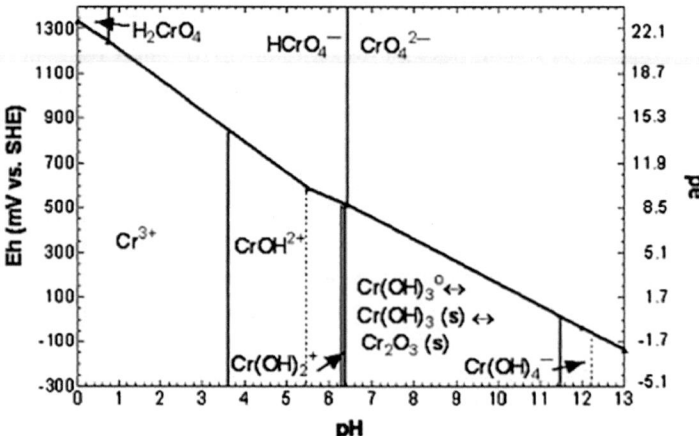

FIGURE 2   Eh–pH diagram illustrating the stability field defined by Eh (redox potential relative to the standard hydrogen electrode, SHE) and pH for Cr(VI) and Cr(III) at 10^{-4} M total Cr. The vertical dashed lines indicate semiquantitatively the pH range in which $Cr(OH)_3$ is expected to control Cr(III) cation activities in the absence of other ligands besides OH^-.

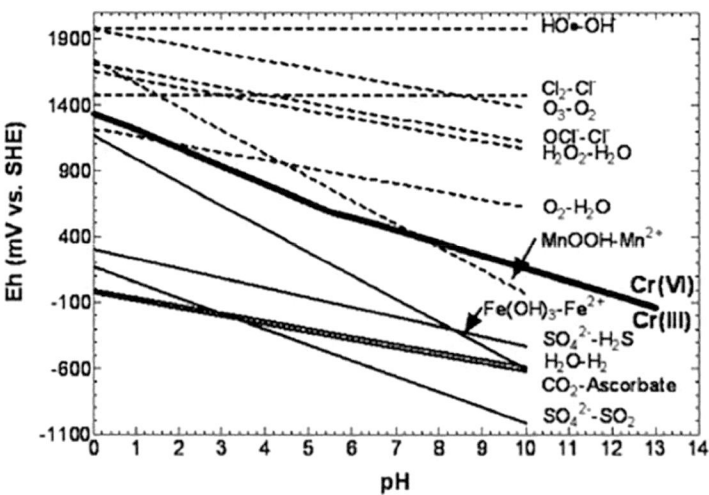

FIGURE 3   Eh–pH diagram showing potential oxidants for Cr(III) in natural waters as dashed lines above the bold Cr(VI)–Cr(III) line and potential reductants for Cr(VI) below the line. Each line for an oxidant (first species of the pair) and reductant (second species) combination represents the reduction potential (in mV) at a given pH established by that oxidant–reductant pair (e.g., O_3-O_2). The oxidant member of a pair for a higher line is expected to oxidize the reductant member of the lower line, thereby establishing the area and species between the lines as thermodynamically favored to exist at chemical equilibrium.[23]

Conclusion

Hexavalent chromium remaining in industrial by-products may contaminate soils, surface waters, and groundwater that are supplies for domestic uses, irrigation, and industrial processes. Prediction of the likelihood of Cr(III) oxidation and Cr(VI) reduction occurring is important for water treatment and for establishing health-based regulations and allowable limits for Cr(VI) and Cr(III) in water supplies. In agricultural soil–plant–water systems, Cr(VI) added in irrigation water or formed via oxidation of Cr(III) will reduce to Cr(VI) if electron donors (e.g., Fe^{2+}, H_2S, and organic matter) and Eh–pH conditions are sufficiently reducing (Bartlett and James[8] and James and Bartlett[20]; Figure 2). If not reduced, Cr(VI) may leach from surface soils to subsoils and groundwater. Therefore, prediction of Cr bioavailability and mobility in natural waters must consider redox reactions of this heavy metal.

References

1. US Environmental Protection Agency (EPA). Basic Information about Chromium in Drinking Water, 2010. Available at http://www.epa.gov/safewater/contaminants/basicinformation/chromium.html#four. (Accessed on May 13, 2012).
2. California Department of Public Health (DPH). *Chromium-6: Timeline for Drinking Water Regulations,* 2009. Available at: http://www.cdph.ca.gov/certlic/drinkingwater/Pages/Chromium6 timeline.aspx. (Accessed May 13, 2012).
3. International Agency for Research on Cancer (IARC). Chromium and chromium compounds. IARC Monogr. Eval. Carcinog. Risks Hum. **1990**, *49,* 49–256.
4. Stout, M.D.; Herbert, R.A.; Kissling, G.E.; Collins, B.J.; Travlos, G.S.; Witt, K.L.; Melnick, R.L.; Abdo, K.M.; Malarkey, D.E.; Hooth, M.J. Hexavalent chromium is carcinogenic to F344/N rats and B6C3F1 mice after chronic oral exposure. Environ. Health Perspect. **2009**, *117* (5), 716–722.
5. National Toxicology Program (NTP). Toxicology and carcinogenesis studies of chromium picolinate monohydrate (CAS No. 27882-76-4) in F344/N rats and B6C3F1 mice (feed studies). TR556. 2010. Research Triangle Park, N.C.
6. Anderson, R.A. Essentiality of chromium in humans. Sci. Total Environ. **1989**, *86,* 75–81.
7. Kimbrough, D.E.; Cohen, Y.; Winer, A.M.; Creelman, L.; Mabuni, C.A. Critical Assessment of Chromium in the Environment. Crit. Rev. Environ. Sci. Technol. **1999**, *29* (1), 1–46.
8. Bartlett, R.; James, B. Behavior of chromium in soils: III. Oxidation. J. Environ. Qual. **1979**, 8, 31–35.
9. James, B.R.; Petura, J.C.; Vitale, R.J.; Mussoline, G.R. Oxidation–reduction chemistry of chromium: Relevance to the regulation and remediation of chromate-contaminated soils. J. Soil Contam. **1997**, *6* (6), 569–580.
10. Nriagu, J.O. Production and uses of chromium. In *Chromium in the Natural and Human Environments;* Nriagu, J.O., Nieboer, E., Eds.; Wiley-Interscience: New York, **1988**; 81–103.
11. Barnhart, J. Chromium chemistry and implications for environmental fate and toxicity. J. Soil Contam. **1997**, *6* (6), 561–568.
12. Ball, J.W.; Nordstrom, D.K. Critical evaluation and selection of standard state thermodynamic properties for chromium metal and its aqueous ions, hydrolysis species, oxides, and hydroxides. J. Chem. Eng. Data **1998**, *43,* 895–918.
13. Bartlett, R.J. Chromium redox mechanisms in soils: Should we worry about Cr(VI)? In *Chromium Environmental Issues;* FrancoAngeli: Milan, 1997; 1–20.
14. James, B.R. Remediation-by-reduction strategies for chromate-contaminated soils. Environ. Geochem. Health **2001**, *23* (3), 175–179.
15. James, B.R. The challenge of remediating chromium-contaminated soils. Environ. Sci. Technol. **1996**, *30,* 248A–251A.

16. Fendorf, S.; Wielenga, B.W.; Hansel, C.M. Chromium transformations in natural environments: The role of biological and abiological processes in chromium(VI) reduction. Int. Geol. Rev. **2000**, *42*, 691–701.

17. Rai, D.; Saas, B.M.; Moore, D.A. Chromium(III) hydrolysis constants and solubility of chromium(III) hydroxide. Inorg. Chem. **1987**, *26*, 345–369.

18. Sass, B.M.; Rai, D. Solubility of amorphous chromium(III)–Iron(III) hydroxide solid solutions. Inorg. Chem. **1987**, *26*, 2228–2232.

19. Nieboer, E.; Jusys, A.A. Biologic chemistry of chromium. In *Chromium in the Natural and Human Environments;* Nriagu, J.O., Nieboer, E., Eds.; Wiley-Interscience: New York, 1988; 21–80.

20. James, B.R.; Bartlett, R.J. Behavior of chromium in soils: VII. Adsorption and reduction of hexavalent forms. J. Environ. Qual. **1983**, *12*, 177–181.

21. James, B.R. Redox phenomena. In *Encyclopedia of Soil Science,* 1st Ed.; Lal, R., Ed.; Marcel Dekker, Inc.: New York, 2002; 1098–1100.

22. James, B.R.; Brose, D.A. Oxidation - Reduction Phenomena. In *Handbook of Soil Sciences: Properties and Processes.* 2nd edn.; Huang, P.M., Li, Y., Summer, M.E., Ed.; CRC Press: Boca Raton, 2012; 14-1–14-24.

23. James, B.R. Chromium. In *Encyclopedia of Water Science;* Stewart, B.A.; Howell, T., Ed., Marcel Dekker: New York, 2003; 75–79.

20

Cobalt and Iodine

Ronald G. McLaren

Soil Cobalt

The Co concentration in soils depends primarily on the parent materials (rocks) from which they were formed and on the degree of weathering undergone during soil development.[2] Cobalt tends to be most abundant as a substituent ion in ferromagnesian minerals, and therefore has relatively high concentrations in mafic and ultramafic rocks (rocks containing high or extremely high proportions of ferromagnesian minerals). Conversely, Co concentrations are relatively low in felsic rocks (rocks containing large amounts of silica-rich minerals) such as granite, and in coarse-textured quartz-rich sedimentary rocks (sandstones). Higher concentrations of Co may be associated with finer textured sediments (shales) in which Co has become surface adsorbed by, or incorporated into, secondary layer silicates by isomorphous substitution.[3] Typical concentrations of Co reported in different rock types are shown in Table 1. As a result of the large range in Co concentrations of soil parent materials, and variation in the degree of weathering, total soil Co concentrations also vary widely. However, the mean values reported for agricultural soils from many countries appear to have a somewhat restricted range of between approximately 2 and 20 mg/kg (Table 1).

Forms of Cobalt in Soils

Cobalt in soils, whether released from parent materials during soil development, or derived from anthropogenic contaminant sources, occurs in several different forms or associations. Cobalt may be present as 1) the simple Co^{2+} ion, or as complexes with various organic or inorganic ligands in the soil solution; 2) exchangeable Co^{2+} ions; 3) specifically adsorbed Co, bound to the surfaces of inorganic soil colloids (clays and oxides/hydrous oxides of Al, Fe, and Mn); 4) Co complexed by soil organic colloids; 5) Co occluded by soil oxide materials; and 6) Co present within the crystal structures of primary and secondary silicate minerals.[6] In some soils, there appears to be a particularly strong association between Co and manganese (Mn) oxides, especially in soils where Mn oxides occur as distinct nodules or coatings.[7,8]

TABLE 1 Cobalt and Iodine Concentrations in Rocks and Soils

	Co Concentration (mg/kg)	I Concentration (mg/kg)
Rock type		
Ultramafic (e.g., serpentinite)	100–300	0.01–0.5
Mafic (e.g., basalt, gabbro)	30–100	0.08–50
Intermediate (e.g., diorites)	1–30	0.3–0.5
Felsic (e.g., granites, gneiss)	<1–10	0.2–0.5
Sandstones	0.3–10	0.5–1.5
Shales/argillites	11–40	2–6
Limestones	0.1–3.0	0.5–3.0
Soils		
Complete range	0.1–300	<0.1–25.4
Range of mean values	2–21.5	1.1–13.1

Source: Kabata-Pendias and Pendias,[1] Aubert and Pinta,[4] and Swaine.[5]

Plant Availability of Soil Cobalt

The immediate source of Co for plant uptake is the soil solution. However, Co concentrations in the soil solution are extremely low, generally much less than $0.1\,\mu g/mL$.[1] Soil solution Co appears to be in equilibrium with Co adsorbed at the surfaces of soil colloids, and the distribution between solution and surface phases is strongly influenced by soil pH. As pH increases, soluble Co decreases.[6] Similarly, soils with high capacities to adsorb Co, particularly those soils with high Mn oxide contents, also have low solution Co concentrations.[9] Thus, in addition to soils with low total Co concentrations, the plant availability of Co may be low in soils with high pH or high Co adsorption capacities. Cobalt availability is also influenced by soil moisture status, availability increasing under waterlogged conditions.[10]

Determination of soil Co dissolved by various extractants is the most common way of assessing the plant availability of soil Co. Ideally, the forms of Co extracted should include any soluble and exchangeable Co, together with any solid-phase forms of Co that are able to move readily into the soil solution in response to changes in solution Co concentrations.[6] The two extractants used most commonly for this purpose are 2.5% acetic acid and solutions of ethylenediaminetetraacetic acid (EDTA). However, the ability of these extractants to accurately assess Co availability appears to be somewhat limited.[11,12]

Cobalt Deficiency

Cobalt deficiency in grazing sheep and cattle was first diagnosed in the 1930s, initially in New Zealand, Australia, and Scotland.[13–15] The condition causes animals to lose their appetite, become weak and emaciated, suffer severe anemia, and eventually die. Subsequently, it was shown that Co is a constituent of both vitamin B_{12} and a closely related coenzyme, and that Co deficiency is in effect a deficiency of vitamin B_{12}.[16] It has been concluded from field studies that pasture containing Co below 0.08 mg/kg for sheep or below 04 mg/kg for cattle is unlikely to meet nutritional requirements in terms of maintaining adequate serum and liver vitamin B_{12} concentrations, and healthy growth rates.[17,18]

Cobalt deficiency is commonly treated by applying Co-containing fertilizers to pastures, usually at very low rates, e.g., 350 g/ha/yr of $CoSO_4 \cdot 7H_2O$. However, on some soils such treatments appear to be ineffective.[11] Alternative treatments for Co deficiency involve injecting the animal with vitamin B_{12}, the use of Co drenches, or the insertion of slowly dissolving Co "bullets" in the animal's rumen.

Soil Iodine

The range of I concentrations found in rocks and soils is shown in Table 1. Iodine occurs as a minor constituent in various minerals, where it can replace anions such as OH^-, SiO_4^{4-}, and CO_3^{2-}, and has relatively low concentrations in most types of rock.[19] Highest I concentrations are generally found in fine-grained sedimentary rocks (Table 1). However, soil I concentrations are generally higher than in the rocks from which they have been derived, a fact attributed predominantly to atmospheric accessions of I.[20] Iodine is known to be present in the atmosphere in vapor form and associated with particles of dust. In coastal areas, accession of I may also be related to sea spray.[1]

Forms of Iodine in Soils

Information of the forms of I in soils is limited, most published analyses of soils have determined total I concentrations only. Of the three most common forms of elemental (I_2), iodide (I^-), and iodate (IO_3^-), it seems likely that most I in soils is present as iodide or possibly as elemental I. The presence of iodate in soils has also been postulated, but would require high oxidation conditions in neutral or alkaline soils.[20] Indeed, there is some evidence that when iodate is added to soils it is rapidly reduced to elemental I or iodide by soil organic matter.[21] There is also evidence that, under some conditions, elemental I can be volatilized from soils.[22]

Most I in soils appear to be associated with soil organic matter and iron (Fe) and aluminium (Al) oxides, materials by which both iodide and elemental I are known to be strongly adsorbed.[21,23] Indeed the distribution of I in soil profiles appears to follow the distribution of these soil components.[24] The atmospheric accessions of I and the affinity between I and organic matter often result in maximum concentrations of I in the surface horizons of soils.[20,24]

Plant Availability of Soil Iodine

Interest in soil analysis as a means of assessing plant availability of I has been minimal,[20] and relationships between the I-status of soils and the concentrations of I in plants appear to be poor.[1] Even soil extractants designed to determine the most soluble forms of I in soils do not provide a good indication of I availability to plants.[25] Plants are capable of absorbing I directly from the atmosphere,[1] and plant species or varietal differences appear to have a greater influence on plant I concentration than soil I status.[20,26] Dicotyledonous pasture species (e.g., clovers) generally have higher I contents than do grasses.[26] Concentrations of I in plants may be reduced by liming,[25] by the application of N fertilizer,[26] and by the application of farmyard manure.[25] Seasonal effects on pasture I concentrations have also been observed, with decreases in the summer, and slight increases in the autumn.[26]

Iodine Deficiency

Low concentrations of I in food and water have been associated with the occurrence of endemic goitre (enlargement of the thyroid gland) in humans and farm livestock.[27] Early work suggested a close relationship between goitre incidence and low soil levels of I, however, it is now recognized that other factors are also involved. In particular, the presence of a group of substances known as goitrogens, which occur in various plant species, has been shown to reduce thyroid hormone synthesis and metabolism.[20] In the absence of goitrogens, it is considered that diets containing 0.5 mg I/kg DM will more than adequately meet the I requirements of all classes of animals, while levels as low as 0.15 mg/kg might be sufficient to meet the requirements for growing animals.[28] In the presence of goitrogens, I requirements may be as high as 2 mg I/kg DM.[28] Attempts to increase pasture I concentrations with I-containing fertilizers have been generally ineffective, with very low recoveries of the applied I.[25,26] Direct supplementation of livestock is normally the preferred way to increase I intakes.

Conclusions

Cobalt and Iodine deficiencies in livestock result from low soil concentrations and/or low availability of these trace elements for uptake by pasture plants. Plant availabilities of Co and I are determined by several factors including soil forms, soil sorption properties, soil pH, soil moisture status, season, plant species, and fertilizer applications. Deficiencies of Co and I can be prevented by application of fertilizers (Co) to the soil, or by direct treatment of livestock (Co and I).

References

1. Kabata-Pendias, A.; Pendias, H. *Trace Elements in Soils and Plants*; CRC Press: Boca Raton, 1984; 238–246.
2. Mitchell, R.L. Cobalt in soil and its uptake by plants. In *Atti del IX Simposio Internazionale di Agrochimica su La Fitonutrizione Oligominerale*; Punta Ala: Italy, 1972; 521–532.
3. Hodgson, J.F. Chemistry of micronutrient elements in soils. Adv. Agron. **1963**, *15*, 119–159.
4. Aubert, H.; Pinta, M. *Trace Elements in Soils, Development in Soil Science 7*; Elsevier Scientific Publishing Co.: Amsterdam, 1977; 395 pp.
5. Swaine, D.J. *The Trace Element Content of Soils*; CAB: Harpenden, England, 1955; 157 pp.
6. McLaren, R.G.; Lawson, D.M.; Swift, R.S. The forms of cobalt in some Scottish soils as determined by extraction and isotopic exchange. J. Soil Sci. **1986**, *37*, 223–234.
7. Taylor, R.M.; McKenzie, R.M. The association of trace elements with manganese minerals in Australian soils. Aust. J. Soil Res. **1966**, *4*, 29–39.
8. Jarvis, S.C. The association of cobalt with easily reducible manganese in some acidic permanent grassland soils. J. Soil Sci. **1984**, *35*, 431–438.
9. Tiller, K.G.; Honeysett, J.L.; Hallsworth, E.G. The isotopically exchangeable form of native and applied cobalt in soils. Aust. J. Soil Res. **1969**, 7, 43–56.
10. Adams, S.N.; Honeysett, J.L. Some effects of soil waterlogging on Co and Cu status of pasture plants grown in pots. Aust. J. Agric. Res. **1964**, *15*, 357–367.
11. McLaren, R.G.; Lawson, D.M.; Swift, R.S.; Purves, D. The effects of cobalt additions on soil and herbage cobalt concentrations in some S.E. Scotland pastures. J. Agric. Sci. Camb. **1985**, *105*, 347–363.
12. McLaren, R.G.; Lawson, D.M.; Swift, R.S. The availability to pasture plants of native and applied soil cobalt in relation to extractable soil cobalt and other soil properties. J. Sci. Food Agric. **1987**, *39*, 101–112.
13. Grange, L.I.; Taylor, N.H. Bush sickness. Part IIA. The distribution and field characteristics of bush-sickness soils. Bull. N. Z. Dept. Sci. Ind. Res. **1932**, *32*, 21.
14. Underwood, E.J.; Filmer, J.F. The determination of the biologically potent element (cobalt) in limonite. Aust. Vet. J. **1935**, *11*, 84–92.
15. Corner, H.H.; Smith, A.M. The influence of cobalt on pine disease in sheep. Biochem. J. **1938**, *32*, 1800–1805.
16. Smith, R.M.; Gawthorne, J.M. The biochemical basis of deficiencies of zinc, manganese, copper and cobalt in animals. In *Trace Elements in the Soil-Plant-Animal System*; Nicholas, D.J.D., Egan, A.R., Eds.; Academic Press: New York, 1975; 243–258.
17. Andrews, E.D. Observations on the thrift of young sheep on a marginally cobalt deficient area. N. Z. J. Agric. Res. **1965**, *8*, 788–817.
18. Gardner, M.R. *Cobalt in Ruminant Nutrition: A Review*; Department of Agriculture: Western Australia, 1977; 602–620.
19. Goldschmidt, V.M. *Geochemistry*; Oxford University Press: Oxford, 1954.
20. Fleming, G.A. Essential micronutrients. II. Iodine and selenium. In *Applied Soil Trace Elements*; Davies, B.E., Ed.; John Wiley and Sons, Ltd.: Chichester, 1980; 199–234.

21. Whitehead, D.C. The influence of organic matter, chalk, and sesquioxides on the solubility of iodide, elemental iodine, and iodate incubated with soil. J. Soil Sci. **1974**, *25,* 461–470.

22. Whitehead, D.C. The volatilisation, from soils and mixtures of soil components, of iodine added as potassium iodide. J. Soil Sci. **1981**, *32,* 97–102.

23. Whitehead, D.C. The sorption of iodide by soils as influenced by equilibrium conditions and soil properties. J. Sci. Food. Agric. **1973**, *24,* 547–556.

24. Whitehead, D.C. Iodine in soil profiles in relation to iron and aluminium oxides and organic matter. J. Soil Sci. **1978**, *29,* 88–94.

25. Whitehead, D.C. Uptake by perennial ryegrass of iodide, elemental iodine and iodate added to soil as influenced by various amendments. J. Sci. Food. Agric. **1975**, *26,* 361–367.

26. Hartmans, J. Factors affecting the herbage iodine content. Neth. J. Agric. Sci. **1974**, *22,* 195–206.

27. Underwood, E.J. *Trace Elements in Human and Animal Nutrition,* 3rd Ed.; Academic Press: New York, 1971; 543 pp.

28. Agricultural Research Council. In *The Nutrient Requirements of Ruminant Livestock;* Commonwealth Agricultural Bureau: Slough, U.K., 1980; 351 pp.

21

Copper

David R. Parker and
Judith F. Pedler

Copper in Soils

The average total concentration of Cu in the Earth's crust is estimated to be 70 mg/kg, although levels of 20–30 mg/kg are prevalent in average soils.[1] Common primary minerals include Cu sulfides, with Cu largely in the +I oxidation state, which dissolve by weathering processes. Secondary minerals of Cu(II) include oxides, carbonates (malachite), silicates, sulfates, and chlorides, most of which are relatively soluble. Copper(II) may substitute for Fe, Mg, and Mn in an assortment of minerals, especially silicates and carbonates.[1]

The Cu^{2+} ion can form strong inner-sphere complexes, and is thus immobilized by carboxylic, carbonyl, or phenolic functional groups, even at low pH. Exchangeable and weak acid-extractable Cu represent a small percentage of total Cu in most soils. The bulk of the Cu is complexed by organic matter, occluded in oxides, and substituted in primary and secondary minerals. Organic matter and Mn oxides are the most likely materials to retain Cu in a nonexchangeable form in soils. Alkali extraction techniques that remove organic matter from soils usually release large fractions of the total soil Cu.[2] Addition of organic matter to soils, and biological exudation of organic acids may both increase dissolved organic carbon, thus solubilizing Cu from mineral forms, increasing the total dissolved Cu in soil solution,[3] but predictive models of humic acid binding of Cu in soil solution are generally inadequate.[4] Overall, Cu is one of the least mobile of the trace elements, maintained in a form sufficiently available to plants but relatively resistant to movement by leaching.[1]

Free Cu^{2+} in soil solution decreases with increasing pH, reaching a minimum above pH 10. In the absence of organic ligands, Cu speciation is dominated by free Cu^{2+}, and increasingly by carbonate and hydroxy complexes as pH rises above 6.5.[1] The dissolved organic carbon found in most surface soils has a strong affinity for Cu, but estimates of the percentage of soluble Cu that is organically complexed can vary widely.

Copper and Plants

Plant uptake of Cu appears to be directly related to the concentration of the free ion, Cu^{2+}, but may also be influenced by the total concentration in soil solution, including organic complexes.[5] As with most trace metals, it is not known whether Cu is passively absorbed or actively taken up across the root-cell membrane.[6] Rates of absorption are generally low, on the order of 1 nmole h^{-1} (g root dry weight)$^{-1}$.[7] The activity of free Cu^{2+} required in nutrient solution for optimal plant growth is just 10^{-14}–10^{-16} M.[5] Copper absorption is generally halted by metabolic inhibitors and uncoupling agents which disrupt the normal transmembrane potential.[7]

Uptake of Cu is strongly affected by pH: increasing concentration of hydrogen ions decreases the absorption of Cu ions by plant roots.[8] Uptake is also affected by the presence of Ca, and to a lesser extent by Mg, both of which compete with Cu for binding sites at the root plasmalemma.[9] The effects of other trace metals on Cu uptake have frequently been seen as inhibitory (Zn), or stimulatory (Mn), but not necessarily under well-defined conditions.[2]

Copper as an Essential Element

Copper is an essential element for plant growth and is a component of many enzymes, including plasto-cyanin, and thus is an indispensable prosthetic group in Photosystem 2. Cu-containing proteins are also important in respiration (cytochrome c oxidase is the terminal oxidase of the mitochondrial electron chain), in detoxification of superoxide radicals (superoxide dismutase), and in lignification (polyphenol oxidase). Ascorbate oxidase, which contains eight Cu^{2+} ions, has been proposed as an indicator of plant Cu status, although its relevant biological function has yet to be determined.[10]

The critical concentration of Cu in shoot tissue for optimal growth does not vary greatly between plant species, ranging from 1 to 6 µg/g dry weight of young leaf tissue.[11] Most crops are recorded as having a requirement of 3–5 µg/g.[11] The average concentration of Cu in plant parts varies with age and with the level of Cu and N supply. Translocation of Cu to plant shoots increases with an increasing supply of N. In the xylem and phloem saps, Cu is probably complexed by amino acids.[2] Copper is usu-ally described as having "variable" phloem mobility in plants,[2] as the retranslocation of Cu from older tissues is regulated by both Cu-supply and by N-status. Lack of a sufficiently long-lived radioisotope makes study of Cu transport and translocation problematic.

Plant Growth on Copper-Deficient Soils

Copper deficiency most often occurs on organic soils where excessive leaching has occurred, or on calcareous sands. In general, crops grown on mineral soils with Cu contents less than 4 to 6 µg/g, or on organic soils with less than 20 to 30 µg/g are the most likely to suffer Cu deficiency,[3] although this varies with specific soil type and the crop grown.

Symptoms of Cu-deficiency include chlorosis, necrosis, leaf distortion and terminal dieback, and are most evident in new leaf growth. Wilting can also occur, indicating structural weaknesses due to reduced lignification of the xylem elements. These symptoms are not entirely specific to Cu deficiency, and can be observed in plants under a variety of stresses. The most profound symptoms of Cu deficiency are those seen in the reproductive cycles of many sensitive species: delay of flowering, and/or reductions in seed and fruit yield as a consequence of sterile pollen or reduced floret numbers. Because these latter symptoms are not observed until maturity or harvest, Cu deficiency is often termed a "hidden hunger." Rice, citrus and cassava are sometimes referred to as indicator species that are more sensitive to Cu deficiency, but are still not reported to require more than 5–6 µg/g Cu to avoid Cu deficiency. Cereal rye and canola are crops more tolerant of Cu deficiency, requiring only 1–2 µg/g Cu.[11]

Copper deficiency in legumes depresses nodulation and N2 fixation, leading to N deficiencies. Unlike Mo and Co, there seems to be no specific Cu requirement for N2 fixation in nodules beyond that required for plant growth and the production of carbohydrates.[12]

Copper deficiency decreases polyphenol oxidase activity and thus lignification.[14] Susceptibility of Cu-deficient plants to pathogenic attack may be increased due to reduced lignification of xylem elements, or due to impaired lignification in response to pathogenic invasion (wounding response). Application of Cu to soil, at rates too low to directly affect the pathogen, has controlled powdery mildew in wheat.[15]

It has also been suggested that, where Cu in soils is more than sufficient, the accumulation of additional Cu in roots provides a fungistatic defense against pathogens.[16] Conversely, where Cu is deficient, roots are more vulnerable to pathogenic invasion.

There are genetic differences in the absorption of Cu by plant roots. Rye is able to take up significantly more Cu from soil than wheat, and is thus viewed as being more Cu-efficient. Tritcale, the wheat-rye hybrid, inherits the efficiency factor.[7] Copper efficiency could be a useful trait in breeding crops for regions where soils are commonly Cu-deficient.[17] However, the mechanism of the efficiency factor is not clear.

Plant Growth on High-Copper Soils

As some plant species have adapted to soils of low Cu status, others have evolved tolerance to Cu-toxic conditions. The 16th century author, Agricola, wrote of indicator plants that grow on naturally Cu-rich soils. Natural revegetation of mine spoils has been shown to reflect rapid genetic evolution of Cu tolerance by grasses and other plants.[14] There seem to be several possible mechanisms of Cu tolerance, although exclusion from shoots is a common feature. The exceptions are a few Cu-accumulator species which may contain in excess of 1000 µg/g Cu in shoot tissue.[12] In other Cu-tolerant species, root compartmentation or immobilization of Cu may be achieved by immobilization in cell walls, by complexation with intracellular proteins, or by removal of Cu to the vacuole.[14]

With nontolerant taxa, plant growth is likely to be depressed when Cu concentrations in the whole shoots exceed ~20 µg Cu g^{-1}. Symptoms of Cu toxicity include poorly developed and discolored root systems, reduced shoot vigor, and leaf chlorosis. Toxicity thresholds (e.g., for a 10% yield reduction) seem to vary widely, probably because of the low translocation of Cu from roots to shoots. Only when roots are overwhelmed by Cu rhizotoxicity does sufficient Cu reach shoots to affect growth and function. Other syndromes, such as Fe deficiency, can readily occur as secondary consequences of excess Cu.[14]

Exclusion of Cu from the shoot protects photosynthetic activity, which is highly sensitive to excess Cu. Photosynthetic electron transport is blocked by high levels of free Cu, at the oxidizing side of Photosystem 2, and at the reducing side of Photosystem 1. Excess Cu supply results in reduced lipid content and noticeable changes in the fatty-acid composition of tomato roots and primary leaves, indicating enhanced activity of enzymes which catalyze lipid peroxidation.[18]

Concentrated Cu sprays have historically been used to control foliar pathogens, especially in vineyard and orchard crops. These fungicidal sprays often included limestone to reduce their phytotoxicity, and to make them more rainfast. The accumulation of Cu in the soils under these crops has not regularly caused Cu toxicity, which indicates the remarkable ability of most plant roots to accumulate Cu, while regulating its flow to the shoots. Both Cu deficiency and toxicity may result in non-specific symptoms of plant stress. Assessment of the Cu status of a soil, or of crop plants, is most accurate when soil type, soil history, and soil and plant analyses for Cu are all considered.

References

1. McBride, M.B. Forms and distribution of copper in solid and solution phases of soil. In *Copper in Soils and Plants*; Loneragan, J.F., Robson, A.D., Graham, R.D., Eds.; Academic Press: Sydney, Australia, 1981; 25–45.
2. Loneragan, J.F. Distribution and movement of copper in plants. In *Copper in Soils and Plants*; Loneragan, J.F., Robson, A.D., Graham, R.D., Eds.; Academic Press: Sydney, Australia, 1981; 165–188.
3. Sanders, J.R.; McGrath, S.P. Experimental measurements and computer predictions of copper complex formation by soluble soil organic matter. Environ. Pollut. **1988**, *49*, 63–79.
4. Robertson, A.P.; Leckie, J.O. Acid/base copper binding, and Cu^{2+}/H^+ exchange properties of a soil humic acid, an experimental and modelling study. Environ. Sci. Technol. **1999**, *33*, 786–795.

5. Bell, P.F.; Chaney, R.L.; Angle, J.S. Free metal activity and total metal concentrations as indices of micronutrient availability to barley *(Hordeum vulgare* L. Cv Klages). Plant Soil **1991**, *130,* 51–62.

6. Strange, J.; Macnair, M.R. Evidence for a role for the cell membrane in copper tolerance of Mimulus guttatus fischer Ex DC. New Phytol. **1991**, *119,* 383–388.

7. Graham, R.D. Absorption of copper by plant roots. In *Copper in Soils and Plants*; Loneragan, J.F., Robson, A.D., Graham, R.D., Eds.; Academic Press: Sydney, Australia, 1981; 141–163.

8. Lexmond, Th.M.; van der Vorm, P.D.J. The effect of pH on copper toxicity to hydroponically grown maize. Neth. J. Agric. Sci. **1981**, *29,* 217–238.

9. Parker, D.R.; Pedler, J.F.; Thomason, D.T.; Li, H. Alleviation of copper rhizotoxicity by calcium and magnesium at defined free metal-ion activities. Soil Sci. Soc. Am. J. **1998**, *62,* 965–972.

10. Maksymiec, W. Effect of copper on cellular processes in higher plants. Photosynthetica **1997**, *34,* 321–342.

11. Reuter, D.J.; Robinson, J.B. *Plant Analysis: an Interpretation Manual*; CSIRO Press: Melbourne, Australia, 1997; 83–566.

12. Römheld, V.; Marschner, H. Function of micronutrients in plants. In *Micronutrients in Agriculture*, 2nd Ed.; Mortvedt, J.J., Cox, F.R., Shuman, L.M., Welch, R.M., Eds.; SSSA: Madison, WI, 1991; 297–328.

13. Jarvis, S.C. Copper concentrations in plants and their relationship to soil properties. In *Copper in Soils and Plants*; Loneragan, J.F., Robson, A.D., Graham, R.D., Eds.; Academic Press: Sydney, Australia, 1981; 265–285.

14. Woolhouse, H.W.; Walker, S. The physiological basis of copper toxicity and copper tolerance in higher plants. In *Copper in Soils and Plants*; Loneragan, J.F., Robson, A.D., Graham, R.D., Eds.; Academic Press: Sydney, Australia, 1981; 235–262.

15. Graham, R.D. Susceptibility to powdery mildew *(Erisiphe graminis)* of wheat plants deficient in copper. Plant Soil **1980**, *56,* 181–185.

16. Graham, R.D.; Webb, M.J. Micronutrients and disease resistance and tolerance in plants. In *Micronutrients in Agriculture*; Mortvedt, J.J., Cox, F.R., Shuman, L.M., Welch, R.M., Eds.; Soil Science Society of America: Madison, WI, 1991; 329–370.

17. Owuoche, J.O.; Briggs, K.G.; Taylor, G.J. The efficiency of copper use by 5A/5RL wheat-rye translocation lines and wheat *(Triticum aestivum* L.) cultivars. Plant Soil **1996**, *180,* 113–120.

18. Ouariti, O.; Boussama, N.; Zarrouk, M.; Cherif, A.; Ghorbal, M.H. Cadmium-and copper-induced changes in tomato membrane lipids. Phytochemistry **1997**, *45,* 1343–1350.

22

Globalization

Alexandru V. Roman

Introduction

Within the last two decades, the world has witnessed an extraordinary increase in the flow of resources, goods, services, knowledge, and ideas among countries. Technological developments and competition imperatives have led to higher levels of interconnectedness between nations and their economies. Today, it would be impossible to discuss major innovations, products, services, or companies only within the contexts of state boundaries. Globalization has also paved the way for the evolution of a global community—one that acts and functions in a coordinated manner in regard to common dilemmas. However, for all the benefits that can be directly or indirectly associated with economic growth and globalization, many argue that it imposes great burdens on the environment.

This entry discusses the challenges associated with understanding environmental impacts and tracing policy priorities within the framework and dynamics of globalization. The discussion revolves around four dominant themes: the globalization debate, the concepts of sustainable economic growth, global governance, and the expectations about the future developments within environmental management and globalization. The entry makes two holistic arguments: 1) the understanding of the environmental and governance impacts of globalization is evolving, but it is still seriously hindered by the conflict and competing discourses; and 2) an adequate response to the environmental consequences of current trends has to be built on collaborative efforts that are comprehensive in nature and do not identify the environmental impacts as mere externalities.

Understanding Globalization: The Debate, Perspectives, and the Complexity of the Environmental Challenges

The term globalization became an essential part of governance discourse in the 1990s. Academics have not fared well in their attempts to reach a single definition of globalization. The phenomenon is very complex, and its ramifications are still not fully understood and induce a great deal of debate.

How globalization is defined will depend on the assumptions that one makes about its nature and dynamics. The term is commonly used to describe the increased flow of knowledge, labor, resources, goods, services, and policy considerations, but also the standardization of values across societies, within a broader context of global awareness, and social transformational influences that lead to higher degrees of interconnectedness and interdependence. In essence, globalization is a description of the social, financial, economic, and political conditions that became pronounced in the second part of the 1980s and intensified after the end of the Cold War.

To a certain extent, the roots of globalization can be traced back to the 19th century,[1] while the underlining trade patterns possibly as far back as several thousand years. The actual concept entered the academic vocabulary in the 1960s but at the time it failed to attract any serious attention from the literature or the social context. With time, the idea of globalization started to symbolize the rise of consciousness of the world society in regard to normative developments.[2] As patterns of interconnectedness evolved, they led to manifold pressure but also opportunities to restructure or to regiment societies through a possible homogenization of beliefs and perspectives.[2]

Currently, globalization is primarily associated with the increased levels of financial and economic interdependencies. These dynamics are expected to remain an important part of the future economic landscape.[3] The Organisation for Economic Co-operation and Development identified three leading forces behind the phenomenon: 1) information and communication technologies; 2) liberalization of capital movements and deregulation; and 3) increased access to markets for trade and investments.[4] Others discern only two broad categories of motivators: on the one hand, the technological changes led to reduction in transport and communication costs, and on the other hand, policy decisions prodded regional and global integration initiatives.[5] While economic dimensions are important drivers behind globalization, it would be misleading to pick them as the only consequential aspects. It is probably best to think of globalization as a broad set of social processes or as an interpretative framework that aside from economic interdependencies also acknowledges associations within the world's political systems, beliefs, cultures, and individuals.

The concept of globalization and the consequences associated with it are believed to be doomed to controversy and inimical debate.[6,7] The dispute over whether globalization is "good" or "bad" can ignite anger, which can easily spill into violent street manifestations, as was the case in Seattle in 1999. Challenges and issues caused by integration trends will probably continue to define the future of global political and economic discourses. Historically, there always existed a significant conceptual gap between the advocates and opponents of globalization. The fissure in the perspectives of the two camps appears to be highly dependent on and correlated with the economic and political states of the world. At the same time, it has also been argued that the two opposing views are nothing more than reflections of other historical factious debates such as pro/anti-capitalism or pro/anti-American.[6]

The supporters believe that amidst economic growth, globalization also improves social equality and standards of living, helps the environment, and provides a platform for developing sensitivity for global issues. The globalization advocates construct their support on three main grounds. Foremost, it is argued that technological and knowledge spillovers that result from globalization lead to improvements in the socioeconomic status of individuals around the globe.[7] In addition, it is thought that globalization might provide the solutions to wicked problems that otherwise would not have been possible to envision or develop. Finally, based on the assumption of the rationality of the individual behavior within the market frameworks, it is expected that globalization will create the conditions for improved democratic outcomes within the design of domestic policies but also in international politics.

The critics of globalization condemn it on the basis of two overarching ideas. First, its opponents claim that the concept is an evolutionary form of a domination discourse promoted by the powerful, both within the national and the international arenas. This discourse demonstrates limited sensitivity to or acceptance of the competing views of marginalized groups, and that inexorably worsens the economic, environmental, and human conditions of the global community. Second, many point out that the economic rewards and the environmental burdens are not fairly distributed among the established economies and the less developed ones, with the former receiving the greater proportions of the benefits.[8] It is commonly suggested that globalization leads to harmful economic and social behaviors, such as "race to the bottom"—corporations seeking lax regulatory environments in order to achieve lower production costs at the expense of the environment. In the same vein, some are concerned that even developed countries might bid down their regulatory standards with the intent to attract investments.[9,10] Most importantly, it appears that the ecological consequences have been systematically understated within economic analyses and when discussing long-term strategic developments.[11]

Both camps, as well as those caught somewhere in between, are able to present well-constructed arguments. At the moment neither of the two extremes can claim an overwhelming advantage in the debate nor that the globalization dynamics have been overly responsive to any single perspective. Using the KOF index of globalization, which measures globalizations along political, social, and economic components, scholars have shown that the degree to which countries become interdependent and interconnected has been increasing constantly, with developing Asia experiencing the most rapid rates of globalization.[5] While social and political globalization patterns are similar to those of economic globalization, the former are less stable and follow a more volatile path than the latter.[5] As a result it is almost impossible to identify a community, agency, network, country, or society that is immune to the effects of globalization. One simply cannot avoid being part of the globality—the social condition.[12] Even the antiglobalizers are an essential part of globalization.[6,12]

Therefore, one could easily argue that the nature of globalization is far more complex than it is habitually considered within the contexts of partisan rhetoric. Adding environmental impacts to the picture complicates matters even further. First, as has already been pointed out, we have yet to reach an adequate understanding of globalization on its own. Attempting to account for its long-term environmental implications will most likely reduce any rational analysis to idealized guesstimations. Second, globalization is routinely associated with global climate change. The latter concept has generated cult-like reactions ever since its rise to agenda frontline status. The increased saliency of climate change is primarily due to awareness-building efforts, such as Al Gore's *An Inconvenient Truth* (2006) documentary. The belief that there is a direct causal relationship between globalization and global warming leaves little time for authentic and reasonable dialogue among pro- and antiglobalizers. Finally, the majority of current academic efforts in the area are fragmented, and there are not many credible and coherent interdisciplinary efforts to adequately understand the nature of globalization. Researchers frequently address only specific aspects of globalization. More often than not, these academic accounts are biased in some way and rarely allow for the possibility of differing interpretations.

Globalization, Sustainable Economic Growth, and Environmental Impacts

To globalize or not to globalize is not the question. What is of import is "how" to globalize.[13] Quintessentially, the matter is that of sustainable economic growth—maximizing economic development, but only within the framework imposed by environmental constraints.[14] While at times, globalization is too closely associated or even believed to be synonymous with international trade, thus underestimating other facets of the phenomenon, for the purpose of this entry, the economic hyperactivity of the last few decades warrants principal attention since it impacts the environment the most.

The globalization-driven changes have created real and perceived shifts in terms of the ecological, sociopolitical, and cultural systems. In many ways, the understanding of the impacts of global economic development and the inertia dynamics that come with it is lagging the development itself. From a broad perspective, the environmental challenges associated with globalization can be grouped within two categories: 1) direct and interdependent issues; and 2) transformational and wicked issues.

Direct and interdependent challenges are the ones that arise from the scale of human activities or from the incremental evolution of the existing conditions. For example, much of the growth in demand for trade will be fueled from the demand side of the equation rather than the supply side of it. Existing consumption habits will stimulate additional demand. Technological changes and competition will cause short life cycles for products and increased interdependency among producers and consumers. Once certain consumption dynamics are set in place, they become part of the larger culture and are eventually institutionalized. Population growth will lead to higher demand for food, and the global community will have to face the pollution and ecological trade-offs associated with meeting that demand.[15]

The transformational and wicked challenges represent the environmental impacts caused by the lack of a complete understanding of the nature of the ecological system but also as a direct consequence of the efforts to mitigate current environmental pressures. The ecological dilemma associated with the production of nuclear power is one challenge that can be categorized as wicked. Such a form of energy production is a highly needed solution for alleviating the pressures caused by high demand for energy. However, the events of Chernobyl in 1986 and Fukushima in 2011 serve as painful reminders of the vulnerability of this form of energy production. At the same time, once an industry or a belief is established within a policy system, it becomes very problematic to transform it or terminate it, even if a more appropriate alternative is available. Essentially, "older" responses to the environmental dilemmas, once institutionalized, could stymie the development of improved solutions when the "new" solutions endanger their own survival.

Other wicked challenges are those associated with the transformational impacts of the efforts to meet the global demand for resource and mainly alimentary demands.[15] According to some estimates, it would take 1.5 "planets earth" to sustain the current demand for resources.[16] The extra 50% in resource demand has to be covered through increased efficiency, genetic manipulations, and gains in productivity. The consequences of many of these transformations are yet to be fully understood, and on many occasions, they are either speeding up the rates of ecological damages (e.g., farming) or generating new environmental challenges that have yet to make it on the policy agenda. Solutions for some of the issues within the latter category are beyond our present-day technological or cognitive capacity (e.g., accurately tracing the influences of genetic changes in food).

The liberalization of the markets of the recent years has led to an increase in trade, which stimulated economic growth[5,17–19] and possibly led to convergence in terms of economic development and income levels.[5,20] Higher levels of trade will constitute a significant part of the future global developments. With the exception of 2009, for the last two decades, international trade experienced average annual growth rates above 6%, and at the same time, international travel and transport grew, on average, 14% a year.[20,21] Research by the Hamburg Institute of International Economics suggests that the values of trade among countries will continue to see larger annual growths until 2030—among East Asia and Pacific countries, 12.6%; South Asia, 10.9%; Latin America, 8.5%; and industrialized countries, 5.7%.[22] These economic patterns are not expected to change,[23] and they will continue to place unprecedented levels of pressure on the environment.[4] Here it is important to note that it has not yet been conclusively proven whether the degradation of the environment would have been considerably different under scenarios of zero or minimal international trade. Globalization might have simply increased the speed of proliferation of otherwise inevitable scale consequences of growth.[6]

The majority of the reports on environmental impacts of trade liberalization is not positive. A study that looked at the data from 63 countries for the period of 1960 to 1999 finds that increased trade has harmful effects on the environment; for every percent increase in trade openness, there is a 0.579% increase in carbon dioxide.[24] Global transportation is responsible for 14% of the total greenhouse emissions; of this, trucks account for 3.2%, ships 1.4%, and aviation 0.98%.[3] It is estimated that in the United States, the environmental damages from transportation using trucks are $0.0023/mi,[25] while in Europe, these costs are between 0.0209 and 0.0746 EUR/km.[26] Damages from air transport are believed to be between 0.01 and 0.05 EUR per passenger/km.[27] Greenhouse gases emitted by air transports leave deeper pollution footprints since their emissions happen at high altitudes. Unfortunately, this will also be the type of emission that will grow the fastest among the group.[3]

In addition to the direct harms to the environment from transportation of goods, there are also significant concerns about oil spills. In the period of 2000 to 2010, there has been an average of 3.3 major oil spills (700 tons or more) per year.[28] In spite of constant growth in oil trade, the above numbers actually represent a significant improvement from the average of 7.8 spills/year in the 1990s, 9.3 spills/year in the 1980s, and 25.3 spills/year in the 1970s.[28]

It is important to note that many of the highest-polluting and most environmentally unfriendly industries, such as ocean fisheries and industries using coal or water, are currently overproducing as a result of the governmental subsidies that they receive in order to meet inflated market demands.[14,15] It is believed that the aggregate and continuous pressures from global economic activities will lead to the fact that by 2100, the global temperature will rise anywhere from 1.1°C to 6.3°C.[29] By 2050, the outcomes or higher temperatures could displace up to 200 million people (e.g., due to droughts, floods, rising sea levels).[3]

Higher levels of trade and economic growth as a result of globalization are not perceived by everyone as being first and foremost negative for environment. At the moment, the available empirical data are not sufficient to reach a decisive conclusion on whether economic development harms or benefits the environment,[30] as many empirical studies reach mixed results.[20,31,32] Some scholars suggest that free trade benefits the environment—more trade leads to higher individual incomes, which motivate the demand for a cleaner environment and lower levels of sulfur dioxide pollution concentrations.[6,14,33–35] It is expected that the world will grow cleaner as it grows richer.[6,33,35] The latter belief, which assumes the existence of the environmental Kuznets curve (EKC), is highly dependent on institutional arrangements.[34] Countries that are able to institutionally enforce environmental concerns will be able to devise the contexts necessary for economic growth and ecological needs to coexist.

The possibility of the existence of EKC was met by a lot of enthusiasm from the academic community when the idea was introduced in the early 1990s. However, the promises associated with the EKC research have not materialized. Recent studies have questioned the validity of the earlier empirical results, which were highly dependent on the specifications of the econometric models and on the quality of the data.[20,30] While one can still make a case for an EKC effect being present in regard to certain pollutants, there is strong evidence that there is no EKC relationship present in terms of CO_2 emissions.[20]

The large range of disagreements within the academic research is primarily due to the failures to emphasize the role of the institutional contexts. Institutions appear to be much more important in determining the degree of environmental effects of globalization than was previously thought. At least in the case of poor countries, it is doubtful that, at this point in time, they would have the strong institutional frameworks necessary for the growing-cleaner- while-growing-richer effect to become a reality. Corruption and unstable regulative and enforcement structures within these nations would probably preclude, at least in the short run, any significant environmental improvements expected from market liberalization or from growing richer.

Global Governance: Organizations, Policy Efforts, and Environmental Justice

It is often the case that when discussing globalization, one unavoidably encounters politically laden narratives that at times can reduce otherwise constructive debates to rhetoric- driven exchanges. In order to permit constructive progress within the global environmental management efforts, it is vital to acknowledge and understand the forces behind the dominant globalization metaphors. Only those initiatives that will be identified as legitimate and will rely on citizens' participation and input in the design process will be successful in the long run.

In a globalized world, almost all policy actors are global actors. Not so long ago, one could have argued that national governments, the International Monetary Fund, the World Bank, the United Nations, the World Trade Organization, and several of the larger corporations were the principal players one should consider in the global policy arena. This is not the case any longer. Information communication technologies have reduced transaction costs and barriers to entry and increased the possible impact and saliency of the actions by smaller policy actors. These developments extended the set of participants to thousands of nongovernmental organizations (NGOs) and networks. According to United Nations' estimates, there are more than 40,000 internationally active NGOs, while at the national levels, these numbers are in the millions. These institutions have become regular actors in defining, designing, and implementing environmental and global governance policies. The power of the NGOs cannot be easily dismissed, as many of them have levels of influence and public acceptance that can frequently challenge the validity of governmental actions.[36] A large number of actors in the global governance arena is welcomed in terms of offering a diversification of perspectives, but at the same time, this makes coordinated efforts less feasible because of the following: 1) NGOs hold competing views and interests that preclude effective action when a collaborative framework is not established; and 2) the presence of a multitude of actors with quasi-veto powers increases the instability and unpredictability of the policy systems. The latter enforces a short-term strategic outlook, as actors choose to emphasize short-range progress rather than facing the complexity of the long-run issues.

Desire for fast economic growth at times entices countries to choose economic benefits over environmental considerations.[37] Within this context, developing a unified global environmental policy framework becomes critical but problematic. The inability to generate an encompassing global policy approach is associated with the fragmentation of interests of the global actors and with the convoluted nature of pollution and its narratives. Greenhouse gas (carbon dioxide, nitrous oxide, and methane) emissions from increased trade and economic activity do not stay within borders, and climate change does not discriminate in its effects.[14] The location of their emission is irrelevant since their effects are global. Meanwhile, the positive environmental impacts from increased trade, such as changes in demand patterns due to higher incomes and technological spillovers, might be only localized in nature.

While economic theory can suggest many viable regulatory solutions that have been tested in practice, more often than not, these solutions are not employed.[14,15] The failure to do so is primarily due to the difficulties in parsing an acceptable understanding of the existing condition within the competing discourses but also due to the inability to break down the politics of financial interests and certain undesired consequential habits of the markets. The economic and social activities that cause the highest damages to the environment are still receiving the bulk of all governmental subsidies, and governments appear unwilling to promote strict environmental standards and technology.[15]

Some of the leading social concerns associated with globalization are in regard to the ethic of pollution distribution and cost. There is a significant unease in regard to the migration of polluting industries toward poor countries (pollution haven hypothesis) with weak institutional and social structures. Is it appropriate to concentrate polluting industries in certain areas? Should countries that do not significantly benefit from trade be recompensed for the increased pollution from transportation emissions? These are just a few of the questions that need to be answered.

As is the case with most globalization dimensions, the implications of such dynamics are not necessarily clear. Researchers suggest that given that high-polluting industries are generally capital intensive,[31,32] they would in fact concentrate in countries with strong financial and economic systems.[14,34] It is believed that the industrial composition of United States has shifted to relatively higher-polluting industries as a result of globalization.[38] Also, at least in the short run, the economic costs of stricter environmental standards are disproportionally borne by the low-income individuals[39]—that is, the poor will have to pay a higher relative price for a cleaner environment.

Future of the Global Environment: Research, Controversies, Current Successes, and Future Expectations

The complexity of the globalization dilemmas requires coordinated efforts on the part of the global community. The task of reaching a common denominator on policy actions is by no means a trivial one. Some suggest that the efforts to generate constructive solutions and progress in the globalization debate are frustrated by many concerns, with the inability to define, assess, and understand globalization acting as the primary stymieing barriers.[5,40] The fact that the study of globalization has not settled into a disciplinary field can be both beneficial and detrimental toward generating an adequate understanding of globalization. It is valuable because such a multifaceted matter would require interdisciplinary collaboration, but it is disadvantageous for at least two reasons. First, the research fragmentation reduces the probability of coherent and focused knowledge creation. Second, globalization research efforts might have a difficult time generating the prestige and respectability that comes with a recognized field of study. Nevertheless, the progress made within the academic community and in terms of political awareness of the environmental implications of globalization warrants hope that viable solutions to environmental dilemmas can be generated.

Currently, concerns regarding climate change are high on the political agendas of most countries, as it becomes evident that environmental degradation is a negative-sum game—if adequate actions are not undertaken, everyone will lose. Questions such as the availability of clean water, health impacts of changing environment, and loss of biodiversity, directly or peripherally linked to the globalization dynamics, are also becoming more salient. Even with an incomplete understanding of the environmental ramifications of economic growth and globalization, there is a silent confidence that there are possible ways to remedy the condition.[6,15] The necessary policies and solutions are available, realistic, and at a reasonable cost.[40] It is very important that corrective actions be taken immediately[41]- since the costs of the same policies might become prohibitive in 10 to 20 years.[3]

Policies that are restricted jurisdictionally to state or regional delineations will not be effective in addressing the nature of the environment impacts or in dealing with transportation fleets that operate under flags of convenience, thus not subject or responsive to operational regulations. In order to make a difference, environmental considerations need to be global in character and intertwined with all financial and economic policies domestically and internationally. Environmental policies should not be limited to the efforts of environmental agencies. Every contract, every school curriculum, every professional meeting, and every legislative act should have a subsection acknowledging or addressing environmental considerations. Research suggests that stricter regulations on emissions of nitrogen and sulfur oxides would lower such discharges by one- third as soon as 2030.[41] The trade-off of achieving this milestone in 2030 would be a global gross domestic product of 97% higher than today, rather than 99% higher. By employing an optimal pollution fee (marginal social cost equals marginal benefits), the greenhouse gas emissions can be reduced by 25% by 2050 and up to 45% by 2100.[42] Pollution-curtailing policies will be more effective if they tax the "bad" (e.g., pollution emissions, farming runoffs), which can be measured relatively more easily, rather than if they subsidize the "good," which is harder to identify and measure.[41]

In order to successfully manage the environmental impacts of globalization, changes at the cultural levels are unavoidable. The impacts that consumerist-type financial and economic systems have on the environment through increased pressures from the demand side are not minor. Efficiency gains in terms of energy consumptions obtained through the production of more environmentally friendly products and through ecologically conscious designs or eco-labeling are a start. Current products from lightbulbs to cars can be made to be more efficient to the point where the energy savings can reach 90% of the amount some of the existing products use.[15] Interestingly, it has been identified that "going green" is good for business—as such, industries prefer and lobby for stricter environmental regulations once trade is liberalized.[43] Foreign industries are generally more environmentally conscious than similar domestic firms operating in the same market.[44] Still, it is doubtful that such efficiency gains can provide sufficiently fast ecological relief without being supplemented by significant changes in social behavior and dominant environmental management discourse.

Based on what has been discussed thus far in the entry it becomes clear that there is no simple solution to managing the environmental changes caused by economic activity. Climate change is considered to be the greatest market failure in history.[3] An effective response would have to account for a multitude of dimensions:

1. Environmental ownership—environmental impacts need to be directly accounted for within the production costs and also clearly delineated within educational efforts. Environmental consequences, positive or negative, should be reinterpreted outside the traditional language of the "externality" narrative, and emissions accounting should become *a core consideration* of policy design and implementation.
2. "Environmentalization" of perspectives—redefining the environment as a local, domestic, and global priority in terms of both social norms and economic and security demands. The designs of environmental regulations should move past the traditional command- and-control policies and encourage more innovation.
3. Sustained collaboration—holistic, continuous, and coordinated global cooperative efforts.

Embracing these approaches does not necessarily mean an acceptance of lower levels of economic growth. By some accounts, stricter environmental regulations might not even hinder economic activity; on the contrary, such approaches lead to more innovative and productive economies.[45]

Conclusions

It is hard to foresee how globalization dynamics and clash of national values will affect democracy and the administrative approaches toward the understanding and the design of environmental policy, specifically in the long run.[46] It is certain that the answers cannot be identified within traditional and simplistic dichotomies like unique vs. universal or us vs. the other. Ironically, even antiglobalizers, in order to oppose globalization, need mechanisms, tools, and communication channels that are part of the globalization-driven transformative patterns. This is just one of many paradoxical complications found in the debate over the nature and interconnectedness of the environmental impacts of globalization. What has become clear is that in order for any policy to effectively address environmental challenges, it would have to incorporate both micro (individual) and macro (societal) imperatives, since managing the impacts of globalization calls for fundamental changes in interpretations and regulations.

Any viable and practical solution would have to attend to the complexity of globalization as well as the unpredictable character of the dynamics of ecological shifts. The 2009 and 2010 United Nations Conventions on Climate Change (UNCC) held in Copenhagen and Cancun, respectively, have not been as successful as one might have wished in legally binding governments in terms of serious actions. The Kyoto Protocol of 1997 that went into effect in 2005, despite its rather modest character, is still probably the main regulatory success at a global level. The scope and outcomes of the protocol would be significantly more consequential should the United States decide to ratify. The participation of United States is crucial since it will provide an incentive for nonparticipating developing countries to follow. Nevertheless, the dialogue during the 2009 and 2010 conventions suggests that participants understand the urgency of the situation and are ready to engage in global long-term environmental measures. Climate change and sustainable growth are top priorities for the United Nations.[47] The 2011 UNCC in Durban, South Africa, should succeed in having the parties move past nonbinding dialogue and into real action above what was achieved by the Kyoto Ptorocol.

Lester Brown[15] argues for the need for two major redefinitions of traditional policies and interpretations in order to manage the ecological challenges and avoid the "perfect storm" or the "ultimate recession." First, taxes on income should be decreased, while those on carbon emissions should be raised. Second, there is a need for redefining the idea of security as environmental security and to allocate resources toward addressing these challenges. According to Brown, the global community needs to reinterpret climate change, population growth, water shortages, poverty, rising food prices, and failing states rather than military aggression as the main threats to global well-being.[15] In an encompassing sense, global governance is in dire need of what some have called "clumsy solutions"— policies that creatively integrate opposing or competing understandings and solutions to address the common problem of climate change.[48]

Economists love to cite Milton Friedman stating that there is no such a thing as a free lunch. The deterioration of the global environment is the price that the global community is paying for its habits and economic expectations. Overlooking the environmental costs, either on purpose or due to the inability to quantify them, does not make those costs any less serious or real. Removing or reversing globalization trends will not solve the environmental challenges. The scale effects from economic and population growth would have resulted in ecological challenges regardless of international trade. The latter has intensified the criticality of the condition, but it is doubtful that stopping trade or the exchange of knowledge, services, or cultures would reduce the strain on the environment. Paradoxically, globalization is exactly what is needed to manage the environmental impacts of globalization.

References

1. Masson, P. Globalization: Facts and figures, IMF Policy Discussion Paper 2001, 01/04.
2. Meyer, J. Globalization. Int. J. Comp. Sociol. **2007**, *48* (4), 261–273.
3. Stern, N. *The Economics of Climate Change: The Stern Review*; Cambridge University Press: Cambridge, 2007.
4. OECD. *Measuring Globalisation: OECD Handbook on Economic Globalisation. Indicators 2005*; OECD Publishing, 2005.
5. Villaverde, J.; Maza, A. Globalisation, growth and convergence. World Econ. **2011**, *34* (6), 952–971.
6. Bhagwati, J. *In defense of globalization*; Oxford University Press: New York, 2007.
7. Kwong, J. Globalization's effects on the environment. Society **2005**, *42* (2), 21–28.
8. Stiglitz, J. *Globalization and its discontents*; Norton: New York, 2002.
9. Oates, W.; Schwab, R. Economic completion among jurisdictions: Efficiency enhancing or distortion inducing? J. Pub. Econ. **1998**, *35* (3), 333–354.
10. Levison, A. Environmental regulatory competition: A status report and some new evidence. Natl. Tax J. **2003**, *56* (1), 91–106.
11. Weitzman, M. On modeling and interpreting the economics of catastrophic climate change. Rev. Econ. Stat. **2009**, *91* (1), 1–19.
12. Steger, D.M. *Globalization: A Very Short Introduction*; New York: Oxford University Press, 2009.
13. Rodrick, D. Globalisation, social conflict and economic growth. World Econ. **1998**, *21* (2), 143–158.
14. Frankel, J. The environment and globalization, Working Paper 10090; NBER Working Paper Series, available at http://www.nber.org/papers/w10090 (accessed August 2011).
15. Brown, L.B. *World on the Edge: How to Prevent Environmental and Economic Collapse*; New York: W. W. Norton and Company, 2011.
16. Wackernagel, M.; Schulz, N.B.; Deumling, D.; Linares, A.C.; Jenkins, M.; Kapos, V.; Chad Monfreda, C.; Loh, J.; Norman Myers, N.; Norgaard, R.; Randers, J. *Tracking the Ecological Overshoot of the Human Economy*, Proceedings of the National Academy of Science, 99 (14), July 2002; 9266–9271.
17. Crafts, N. Globalisation and economic growth: A historical perspective. World Econ. **2004**, *27* (1), 45–58.
18. Frankel, J.; Romer, D. Does trade cause growth? Am. Econ. Rev. **1999**, *89* (3), 379–399.
19. Singh, T. Does international trade cause economic growth? A survey. World Econ. **2010**, *33* (11), 1517–1164.
20. Frankel, J; Rose, A. Is trade good or bad for the environment? Sorting out the causality. Rev. Econ. Stat. **2005**, *87* (1), 85–91.
21. WTO. *World Trade Report 2008—Trade in a Globalizing World*; WTO: Geneva, 2008.
22. Berenberg Bank and Hamburg Institute for International Economics. *Strategy 2030—Maritime Trade and Transport Logistics*. Berenger Bank: Hamburg, 2006.
23. WTO. *World Trade Report 2011—The WTO and Preferential Trade Agreements: From Co-existence to Coherence*; WTO: Geneva, 2011.

24. Magani, S. Trade liberalization and the environment: Carbon dioxide for 1960–1999. Econ. Bull. **2004**, *17* (1), 1–5.

25. Forkenbrock, D. Comparison of external costs of rail and truck freight transportation. Transp. Res. Part A **2001**, *35* (4), 321–337.

26. Bickel, P.; Schmid, S.; Friedrich, R. Environmental costs. In *Measuring the Marginal Social Cost of Transport Research in Transportation Economics*; Nash, C., Ed.; Elsevier: Oxford, 2005; Vol. 14, 185–210.

27. Dings, J.; Wit, R.; Leurs, B.; Davidson, M. *External Costs of Aviation*, Research Report 299 96 1006; Federal Ministry of the Environment, Nature Conservation and Nuclear Safety, 2003.

28. Huijer, K. Trends in oil spill from tanker ships. The International Tanker Owners Pollution Federation Limited, 2011, available at http://www.itopf.com/information-services/data-and-statistics/statistics/ (accessed August 2011).

29. Intergovernmental Panel on Climate Change. *Climate Change 2007: Synthesis Report. Contribution of Working Groups I, II and III to the Fourth Assessment Report of the Intergovernmental Panel on Climate Change*; IPCC: Geneva, 2007.

30. William, H.T.; Levinson, A.; Wilson, D.M. Reexamining the empirical evidence for an environmental Kuznets curve. Rev. Econ. Stat. **2002**, *84* (3): 541–551.

31. Cole, M.; Elliott, R. FDI and the capital intensity of "dirty" sectors: A missing piece of the pollution haven puzzle. Rev. Dev. Econ. **2005**, *9* (4), 530–548.

32. Shen, J. Trade liberalization and environmental degradation in China. Appl. Econ. **2008**, *40* (8), 997–1004.

33. Grossman, G; Kruger, A. Economic growth and the environment. Q. J. Econ. **1995**, *110* (2), 353–77.

34. Antweiler, W.; Copeland, B.; Taylor, S. Is free trade good for the environment? Am. Econ. Rev. **2000**, *91* (4), 877–908.

35. Bhagwati, J. *Free Trade Today*; Princeton University Press: Princeton, 2002.

36. Koppell, J. *World Rule: Accountability, Legitimacy and the Design of Global Governance*; The University of Chicago Press: Chicago, 2010.

37. Thai, K.; Rahm, D.; Coggburn. J. *Handbook of Globalization and Environment*; Taylor and Francis: Boca Raton, 2007.

38. Ederington, J.; Levinson, A; Miner, J. Footloose and pollution-free. The Review of Economics and Statistics **2005**, *87* (1), 92–99.

39. Hasset, K.; Mathur, A.; Metcalf, G. The incidence of a US carbon tax: A lifetime and regional analysis. Energy J. **2009**, *30* (2), 155–177.

40. Scholte, J. A. Defining globalization. World Econ. **2008**, *31* (11), 1299–1313.

41. OECD. *Globalisation in OECD,* OECD Environmental Outlook to 2030; OECD Publishing, 2008.

42. Nordhaus, W. *A Question of Balance: Weighing the Options on Global Warming Polices;*Yale University Press: New Haven, 2008.

43. McAusland, C. Trade, politics, and the environment: Tailpipe vs. smokestack. J. Environ. Econ. Manage. **2008**, *55* (1), 52–71.

44. Eskeland, G.; Harrison, A. Moving to greener pastures? Multinationals and the pollution haven hypothesis. J. Dev. Econ. **2003**, *70* (1), 1–23.

45. Porter, M; van der Linde, C. Toward a new conception of the environment-competitiveness relationship. J. Econ. Perspect. **1995**, *9* (4), 97–118.

46. Box, R.; Marshall, G.; Reed, B.; Reed, M. New public management and substantive democracy. Pub. Admin. Rev. **2001**, *61* (5): 608–619.

47. United Nations. UN chief sees sustainable development as top priority in his second term, 2011, available at http://www.un.org/apps/news/story.asp?NewsID=39288 (accessed September 2011).

48. Verweij, M.; Douglas, M.; Ellis, R.; Engel, C.; Hendriks, F.; Lohmann, S.; Ney, S.; Rayner, S.; Thompson, M. Clumsy solutions for a complex world: The case of climate change. Pub. Admin. **2006**, *84* (4), 817–843.

23

Heavy Metals

Mike J. McLaughlin

Introduction

It is generally accepted that metals having a specific gravity (weight per unit volume) greater than 5 Mgm^{-3} are termed heavy metals. In soils, these elements include cadmium (Cd), cobalt (Co), chromium (Cr), copper (Cu), iron (Fe), mercury (Hg), manganese (Mn), molybdenum (Mo), nickel (Ni), lead (Pb) and zinc (Zn)[1] (http://www.soils. org/sssagloss; accessed February 2001). The term "heavy metals" is often used synonymously with the term "trace elements," but this is incorrect as trace elements are generally defined as those elements normally occurring in soil at concentrations less than 100 mg kg^{-1},[2] which precludes several heavy metals, e.g., Cr, Fe, and Mn. Arsenic (As) is often included in the group "heavy metals," but is more correctly classified as a metalloid.

General Chemistry

Some heavy metals are essential for either plant or animal survival on land (Co, Cr, Cu, Fe, Mn, Mo, Ni, and Zn) while others are nonessential and toxic at low concentrations (Cd, Hg, and Pb). All the heavy metals, except Pb, are transition elements, belonging to the d-block in the periodic table. Many of these elements differ from the alkaline earth metals (e.g., Na, Ca, and Mg) in that they can exist in several valence states in soil (Table 1). In particular, Cr, Fe, and Mn are markedly affected by soil redox potential and undergo both oxidation and reduction depending on soil conditions. This has important implications for the availability and toxicity of many heavy metals. As both Fe and Mn are major structural metals in soil minerals, reduction of the Fe^{3+} and Mn^{4+} ions may lead to a change in soil mineral surfaces important for retention of many elements, including other metals. For Cr, oxidation converts the nontoxic Cr^{3+} ion to the toxic and carcinogenic Cr^{6+} ion. This reaction has even more significance in soils as a strongly sorbed or precipitated cation (Cr^{3+}) is converted into a poorly sorbed or soluble anion (CrO_4^{2-}). However, even in aerobic soils, Cr^{3+} is the thermodynamically stable state, so added Cr^{6+} ion is rapidly converted to Cr^{3+} in most soils.

Solubility and availability/toxicity to organisms of heavy metal cations (Cd^{2+}, Cr^{3+}, Fe^{n+}, Pb^{2+}, Mn^{n+}, Hg^{2+}, Ni^{2+}, and Zn^{2+}) decrease as soil pH increases. This is due to the increase in negative charge on variable charge surfaces in soil[4] and the propensity for these metals to precipitate as sparingly soluble compounds (phosphates, carbonates, and hydroxides) as soil pH increases.[5] On the other hand, solubility and availability/toxicity to organisms of anionic heavy metals (CrO_4^{2-}, MoO_4^{2-}) may increase as soil pH increases, again due to increases in surface negative charge on soil particles affecting sorption.

TABLE 1 Physician and Chemical Properties of the Heavy Metals

Element	Symbol	Atomic No.	Molecular Weight	Valence	Natural Isotopes	Density (MgnT-3)	Melting Point (°C)	First Ionization Potential (eV)	Dominant Species in Soil[a]	Dominant Species in Soil Solution[b]	
										pH 3.5–6.0	pH 6.0–8.5
Cadmium	Cd	48	112.41	2	8	8.65	321	8.96	Cd^{2+}	Cd^{2+}, $CdCl^+$, $CdSO_4^0$	Cd^{2+}, $CdCl^+$, $CdSO_4^0$
Chromium	Cr	24	52.01	2, 3, 6	4	7.19	1875	6.76	Cr^{3+}, CrO_4^{2-}	Cr^{3+}, $CrOH^{2+}$	$Cr(OH)_4-$
Cobalt	Co	27	58.94	2, 3	1	8.90	1493	7.86	Co^{2+}	—	—
Copper	Cu	29	63.54	1, 2	2	8.94	1083	7.73	Cu^{2+}	Cu^{2+}, Cu-org.	Cu-hydroxy species. $CuCO_3^0$, Cu-org.
Iron	Fe	26	55.85	2, 3	4	7.87	1536	7.87	Fe^{2+}, Fe^{3+}	Fe-hydroxyspecies, Fe-org.	Fe-hydroxy species. Fe-org.
Lead	Pb	82	207.19	2, 4	4	11.35	327	7.42	Pb^{2+}	Pb^{2+}, $PbSO_4^0$, Pb-org.	Pb-hydroxy and carbonate species, Pb-org.
Manganese	Mn	25	54.94	2, 3, 4, 7	1	7.44	1244	7.44	Mn^{2+}, Mn^{4+}	Mn^{2+}, $MnSO_4^0$, Mn-org.	Mn^{2+}, $MnSO_4^0$, $MnCO_3^0$
Mercury	Hg	80	200.61	1, 2	7	13.54	−39	10.44	Hg^{2+}, $(CH_3)_2Hg$	—	—
Molybdenum	Mo	42	95.94	6	7	10.22	2610	7.10	MoO_4^{2-}	—	—
Nickel	Ni	28	58.71	2, 3	5	8.91	1453	7.64	Ni^{2+}	Ni^{2+}, $NiSO_4^0$, Ni-org.	Ni^{2+}, $NiHCO_3^+$, $NiCO_3$
Zinc	Zn	30	65.37	2	5	7.14	420	9.39	Zn^{2+}	Zn^{2+}, $ZnSO_4^0$, Zn-org.	Zn^{2+}, Zn-hydroxy and carbonate species, Zn-org.

[a] From Logan.[2]
[b] From Ritchie and Sposito.[3]

Under reducing conditions in soil, many of the metals form insoluble metal sulfides, therefore reducing availability to plants and animals.

Heavy metals are also subject to complexation reactions in soil with both inorganic and organic ligands, which may markedly increase mobility in soil. This may be used to good effect by plants in scavenging essential metals from soil, e.g., phytochelatins (see below). However, for nonessential metals, complexation and increased mobility also may lead to increased environmental risks through leaching and plant uptake, e.g., chloride-induced uptake of Cd by food crops in saline soils.[6]

Abundance in Rocks and Soils

Heavy metals become incorporated into primary minerals in igneous rocks through isomorphous substitution. In sedimentary rocks, heavy metals are incorporated as constituents of minerals, or are removed from the water column and trapped in sediments by adsorption or precipitation processes.[7] Apart from Cr, Fe, and Mn, heavy metals are generally present at trace concentrations (<100 mg kg^{-1}) in most soils, with the exception of soils developed over mineralized parent materials or soils developed through biological enrichment.[8] Background concentrations in soil usually reflect the composition of the parent rock material. Background concentrations of heavy metals are usually determined where no known anthropogenic inputs have occurred, but this assessment is often problematic due to the global spread of anthropogenic emissions through atmospheric transport processes.[9] The typical range in abundances of heavy metals in unpolluted soils is shown in Figure 1.

Concentrations of heavy metals in soil can be significantly increased through human activity. A number of primary and secondary sources have been identified as contributing to enhanced concentrations of heavy metals in soil[12] (Table 2). Many countries have introduced legislation to minimize the amount of metals accumulating in soil, and set ceiling concentrations above which further metal additions must stop.

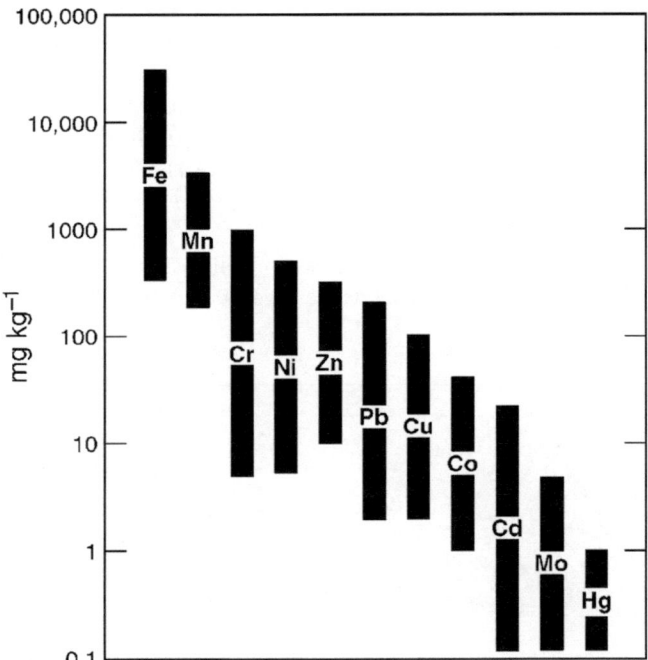

FIGURE 1 Range of heavy metal concentrations of soil. Bars represent commonly found values.
Source: Logan,[2] Alloway,[7] Barry and Rayment,[8] Lantzy and Mackenzie,[9] Kabata-Pendias and Pendias,[10] Swaine,[11] and Ferguson.[12]

TABLE 2 Sources of Heavy Metal Contamination in Soils

Source	Main Heavy Metals
Primary sources	
Fertilizers	Cd, Cu, Mo, Pb, Zn
Irrigation water	Cd, Fe
Manures and composts	Cd, Cr, Cu, Fe, Hg, Mn, Mo,Ni, Pb, Zn
Pesticides	Cu, Hg, Pb, Zn
Sewage biosolids (sludges)	Cd, Cr, Cu, Fe, Hg, Mn, Mo, Ni, Pb, Zn
Soil amendments	Cu, Mn, Pb, Zn
(lime, gypsum, etc.)	
Secondary sources	
Automobile aerosols	Pb
Coal combustion	Pb
Mine waste and effluents	Cd, Cu, Fe, Hg, Mn, Ni, Pb, Zn
Nonferrous smelter waste	Cd, Cu, Hg, Mn, Ni,
	Pb, Zn
Paint dispersal	Cd, Pb
Tire wear	Cd, Zn
Waste combustion	Cd, Pb

Source: Barry and Rayment.[8]

Biological Effects

Co, Cr, Cu, Fe, Mn, Mo, Ni, and Zn are all essential for either healthy plant or animal functioning in soil and all these metals, except Ni, are used in fertilizers or animal stock supplements to ensure efficient agricultural production on soils deficient in essential metals. Due to the low solubility of essential heavy metals in neutral and alkaline soils, many plants have developed strategies to mobilize solid phase forms to facilitate uptake by roots (e.g., Fe, Mn, and Zn). Nonproteinogenic amino acids, or phytochelatins, secreted by actively growing roots are important in acquisition of Cu, Fe, Zn, and possibly Mn from deficient soils.[13] At high concentrations in soil, all the essential elements may pose risks to microorganisms, plants, animals, or humans (Figure 2). Nonessential heavy metals in soil (Cd, Hg, and Pb) have no beneficial effects at low concentrations (Figure 2) and may be toxic at even trace concentrations (e.g., Cd, Hg).

Critical exposure pathways for expression of heavy metal toxicity in soil have been examined as part of regulation governing re-use of biosolids (sewage sludge) on soil,[14] but act as a general risk

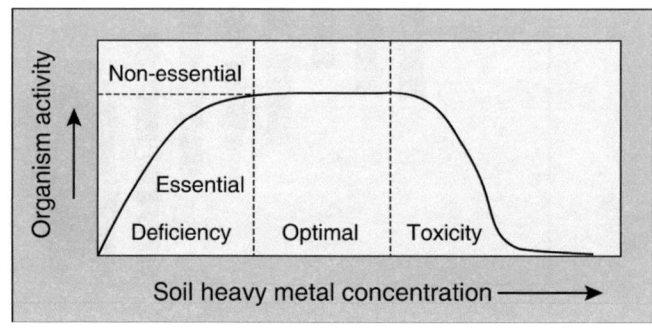

FIGURE 2 Typical concentration–response relationship for essential and nonessential heavy metals in soil.

TABLE 3 Critical Risk Pathway Assessment for Heavy Metal Pollution of Soils

Metal	Dominant Risk Pathway	Secondary Risk Pathway	Most Important Predictor Required
Cd	Food chain transfer	Phyto- and ecotoxicity	Soil–plant uptake
Co	Food chain transfer	Phyto- and ecotoxicity	Soil–plant uptake
Cr	Phyto- and ecotoxicity	Metal leaching	Toxic threshold definition, speciation
Cu	Phyto- and ecotoxicity	Soil ingestion by animals/humans	Toxic threshold definition
Fe	Phyto- and ecotoxicity[a]	None	Toxic threshold definition
Hg	Soil ingestion by animals/humans	Metal leaching	Toxic threshold definition, speciation
Mn	Phyto- and ecotoxicity	Soil ingestion by animals/humans	Toxic threshold definition
Ni	Phyto- and ecotoxicity	Soil ingestion by animals/humans	Toxic threshold definition
Pb	Soil ingestion by animals/humans	Phyto- and ecotoxicity	Oral bioavailability assessment
Zn	Phyto- and ecotoxicity	Food chain transfer	Soil-plant uptake

[a] Only in acid soils under reducing condition.
Source: McLaughlin.[15]

analysis template for all types of pollution of soil by heavy metals.[15] For some elements, e.g., Cd and Co, food chain transfer is the main risk pathway as these elements are easily accumulated by plants in edible tissues. For other elements, sorption in soil is strong, bioaccumulation by microorganisms low, and plant uptake and translocation so low that the dominant risk pathway is (for higher animals and humans) through direct ingestion of soil, e.g., Hg and Pb. For the other heavy metals, behavior in soil and bioaccumulation characteristics result in toxicity to plants and microorganisms (phyto- and ecotoxicity) being the dominant risk pathway at high concentrations, e.g., Cr, Cu, Mn, Ni, and Zn. Thus, prediction of risks from heavy metal pollution of soils through soil testing requires a different emphasis depending on the metal considered (Table 3).

References

1. *Internet Glossary of Soil Science Terms*; Soil Science Society of America: Madison, WI.
2. Logan, T.J. Soils and environmental quality. In *Handbook of Soil Science*; Sumner, M.E., Ed.; CRC Press: Boca Raton, FL, 2000; G155–G163.
3. Ritchie, G.S.P.; Sposito, G. Speciation in soils. In Chemical *Speciation in the Environment*; Ure, A.M.E., Davidson, C.M., Eds.; Blackie Academic and Professional: London, 1995; 201–233.
4. Barrow, N.J. *Reactions with Variable Charge Soils;* Marti- nus Nijhoff: Dordrecht, The Netherlands, 1997; 191 pp.
5. Lindsay, W.L. *Chemical Equilibria in Soils;* Wiley: New York, 1979; 449 pp.
6. McLaughlin, M.J.; Tiller, K.G.; Smart, M.K. Speciation of cadmium in soil solution of saline/sodic soils and relationship with cadmium concentrations in potato tubers. Aust. J. Soil Res. **1997**, *35,* 1–17.
7. Alloway, B.J. The origins of heavy metals in soils. In *Heavy Metals in Soils,* 2nd Ed.; Alloway, B.J., Ed.; Blackie Academic and Professional: London, 1995; 38–57.
8. Barry, G.; Rayment, G.E. Heavy metals and nutrients in soils and sediments of Raine island, great barrier reef. Land Contam. Reclam. **1997**, *5* (4), 281–285.
9. Lantzy, R.J.; Mackenzie, F.T. Atmospheric tracemetals: global cycles and assessment of man's impact. Geochim. Cosmochim. Acta **1979**, *43,* 511–525.
10. Kabata-Pendias, A.; Pendias, H. *Trace Elements in Soils and Plants,* 2nd Ed.; CRC Press: Boca Raton, FL, 1992; 365 pp.
11. Swaine, D.J. *The Trace Element Content of Soils;* Commonwealth Bureau of Soil Science Technical Communication No. 48; Commonwealth Agricultural Bureaux: Farnham Royal, U.K., 1955; 157 pp.

12. Ferguson, J.E. The Heavy Elements: Chemistry, Environmental Impact and Health Effects; Pergamon Press: Oxford, 1990; 614 pp.
13. Römheld, V. Different strategies for iron acquisition in plants. Physiol. Plant. **1987,** *70,* 231–234.
14. USEPA—United States environment protection agency. *Part 503—Standards for the Use and Disposal of Sewage Sludge;* Federal Register 58, 1993; 9387–9404.
15. McLaughlin, M.J.; Zarcinas, B.A.; Stevens, D.P.; Cook, N. Soil testing for heavy metals. Commun. Soil Sci. Plant Anal. **2000,** *31* (11–14), 1661–1700.

Inorganic Carbon: Composition and Formation

Larry P. Wilding and
H. Curtis Monger

Introduction

It has become increasingly common for "soil inorganic carbon" to mean soil carbonate mineral carbon, mainly $CaCO_3$. In the strict sense, however, inorganic carbon not only encompasses carbon in carbonate minerals, but also carbon in the carbonic acid system.[1] The carbonic acid system includes gaseous carbon dioxide ($CO_{2(g)}$), aqueous carbon dioxide ($CO_{2(aq)}$), carbonic acid ($H_2CO_{3(aq)}$), bicarbonate ion ($HCO_{3(aq)}^-$), and carbonate ion ($CO_{3(aq)}^{2-}$).

Composition

In the soil solution, as with other solutions, the interaction of these species can be represented by the following reaction:[1]

$$CO_{2(g)} + H_2O_{(l)} = H_2CO_{3(aq)} = HCO_{3(aq)}^- + H_{(aq)}^+$$

$$= CO_{3(aq)}^{2-} + 2H_{(aq)}^+ \tag{1}$$

Cations, such as Ca^{2+}, Mg^{2+}, Fe^{2+}, Mn^{2+}, and Na^+, precipitate with HCO_3^- (which is the dominant anion between pH 6.5 and 10.5) and CO_3^{2-} (which is the dominant anion above pH 10.5) to form a variety of carbonate minerals. The reaction of $Ca_{(aq)}^{2+}$ with $HCO_{3(aq)}^-$ to form calcite is illustrated below:

$$Ca_{(aq)}^{2+} + 2HCO_{3(aq)}^- = CaCO_{3(s)} + CO_{2(g)} + H_2O_{(l)} \tag{2}$$

There are about 60 carbonate minerals, which in addition to calcite, include aragonite ($CaCO_3$), dolomite [$CaMg(CO_3)_2$], siderite ($FeCO_3$), magnesite ($MgCO_3$), rhodocrosite ($MnCO_3$), cerussite ($PbCO_3$), and malachite [$CuCO_3Cu(OH)_2$]. In soil, the overwhelmingly abundant carbonate mineral is calcite.[2] In unique soil environments, however, other carbonate minerals have been found, such as pedogenic siderite and dolomite.[3]

Individual carbonate crystals of pedogenic origin are generally too small to be seen with the unaided eye. Yet when concentrated together, their combined presence takes on a white color with a variety of macroscopic forms. These forms include carbonate filaments (also called my- celia, pseudomycelia, and threads), films, coatings, soft spheroidal segregations (white eyes), nodules, cylindroids, concretions, glaebules, and veins. Soil fabric which is impregnated with carbonate to the point that it occurs as an essentially continuous medium has been termed "k-fabric."[4] Stages of morphogenetic carbonate accumulation, in which progressively greater amounts of carbonate occur in progressively older soils, are important chronologic indicators. The calcic and petrocalcic horizons are diagnostic horizons in soil taxonomy.[5] Calcic horizons generally contain greater than 15% carbonate by weight. Petrocalcic horizons are indurated forms of calcic horizons. Examples of these horizons are shown in Figures 1A and B. Examples of carbonate crystals as viewed with optical microscopy and scanning electron microscopy are shown in Figures 1C and D.

Dissolution of carbonates in soil systems can be represented by the following reaction (Eq. 3). In humid regions, soluble products of this weathering reaction flux through the vadose zone into groundwater, or precipitate as pedogenic carbonates deep in the soil or geologic system. In arid

(a)　　(b)

(c)　　(d)

FIGURE 1 Examples of inorganic carbon as it exists in the field and under the microscope. The white horizon is a calcic horizon in (a), and petrocalcic horizon in (b). The small golden crystals in (c) are calcite crystals coating sand grains; the black region is a pore space as it appears in cross-polarized light. A calcified fungal filament (cf) viewed with scanning electron microscopy is shown in (d).

regions, soluble products precipitate at relatively shallow depths as a result of sparse rainfall and insufficient leaching.

$$HCO_{3(aq)}^- + H_{(aq)}^+ + CaCO_{3(s)} = Ca(HCO_3)_{2(aq)} \tag{3}$$

Formation

Being located in an arid, semiarid, or subhumid climate is the primary factor that controls carbonate formation. In many areas, the boundary between carbonate-accumulating soil and noncarbonate-accumulating soil is about 500 mm (20 in.) mean annual rainfall.[6] This relationship is confounded, however, by the effects of soil temperature, soil drainage, nature and properties of the parent material (e.g., soil texture, carbonate content, carbonate mineralogy, and porosity), soil drainage, landform position, geomorphic stability, and effectiveness of precipitation (rainfall intensity and duration). Hence, there are many examples in humid and subhumid environments where soil carbonate persists in the soil system at depths inconsistent with regional models. In humid regions, for example, inorganic carbon persists as calcite or dolomite detritus in soils derived from certain parent materials (e.g., calcareous loess, till, outwash, alluvial deposits, sedimentary and metamorphic rocks). In seasonally wet soils, carbonate can accumulate in upper subsoils from capillary rise of bicarbonates via evaporative pumping from shallow groundwater.[7] In addition, carbonate minerals can occur in wetland soils which commonly contain soluble carbonates, bicarbonates or carbonic acid depending on the pH of the local environment

Pedogenic vs. Geogenie Carbonate

Many soils develop in calcareous parent materials. For these soils, it has been a challenge quantifying carbonate that formed in the soil profile vs. carbonate mechanically inherited from parent material. Carbonate formed in the soil profile has been termed "secondary," "authigenic," or "pedogenic."[4,8] On the other hand, carbonate mechanically inherited from parent material has been termed "primary," "geogenic," or "lithogenic."[2,9] Criteria for distinguishing pedogenic from geogenic carbonates involve the scrutiny of both field and laboratory evidence. Field evidence, for example, includes differences in the presence of marine fossils, carbonate morphology (such as nodules, pendants, and laminar caps which indicate pedogenic), and distribution patterns with depth, where, for example, a carbonate horizon of pedogenic origin is overlain and underlain by noncalcareous soil. Laboratory evidence includes comparing mineralogy, particle size, microfabric, and $^{13}C/^{12}C$ ratios of carbonate with unknown origin to those of carbonate with known geogenie origin.[10–12]

Models of Carbonate Formation

There are several processes that cause carbonates to form in soil. Excluding geologic processes, such as lacustrine and deep groundwater cementation that preserves the original sedimentary structure, the formation pedogenic carbonate can broadly be placed into four models—per descensum, per ascensum, in situ, and biogenic models.

Per Descensum Model

The per descensum model accounts for carbonate formation resulting from downward moving meteoric water and can be subdivided into three types. First is the dissolution of pre-existing carbonates in the upper profile, their vertical translocation, and their precipitation in the subsoil. This model was invoked to explain why progressively shallower carbonates occur in progressively drier climates.[13] Later, this per descensum model was used as the basis for calculating the number of wetting-fronts required to leach carbonates to a particular depth.[14] In both cases, it was assumed that carbonate was uniformly distributed in parent material at the beginning of pedogenesis.

Second is the case in which pedogenic carbonate forms in soils with noncalcareous parent materials. Unlike the model described before, noncalcareous parent material does not have carbonate uniformly distributed throughout the profile at the beginning of pedogenesis. In southern New Mexico, for example, prominent calcic and petrocalcic horizons occur in soils with rhyolite alluvium as parent material. This alluvium would yield low amounts of calcium if the rhyolite particles were thoroughly decomposed, which they were not.[15] Therefore, atmospheric additions, another per descensum model, was judged to be the source of carbonates.[15] Initially calcareous dust was measured and considered to be the source of carbonate. Later it was realized that Ca^{2+} in rain was an additional, and even larger source of Ca^{2+} for reaction with soil HCO_3^- to form carbonates.[15] Building on these per descensum concepts, compartmental models have been constructed that compute the depth, amount, and distribution of pedogenic carbonate as a function of climate and time.[16,17]

Third, in addition to vertical illuviation within a soil profile, lateral, downslope migration of the soil solution containing soluble products of carbonate is another per descensum model. In this case, carbonate is thought to precipitate after carbonate-charged waters migrate from upslope positions to lower landscape positions.[18]

Per Ascensum Model

The per ascensum model accounts for carbonate formation resulting from bottom–up movement. A primary example is the capillary rise of Ca^{2+} and bicarbonate from shallow water tables by evaporative pumping, which leads to the precipitation of carbonates in the upper subsoil.[7] Moreover, chemical studies have shown that in some environments plants promote carbonate formation by transporting Ca^{2+} upward to the land surface from subsoil, rock, and groundwater.[19]

In Situ Model

Third, in the in situ model, pedogenic carbonate is the result of in-place dissolution and reprecipitation of bedrock composed of marine carbonate.[20] Limestone, for instance, is progressively transformed into pedogenic carbonate as a result of short-range carbonate dissolution and reprecipitation proximal to the depth of the upper contact with limestone. This is a rather unique method to form pedogenic carbonates where the total carbonate content of the zone of enriched pedogenic products is less than the carbonate content of the limestone originally. These pedogenic zones have a much higher macro- and micro-porosity than the limestone.

In addition to marine carbonates, the in situ model also includes carbonate formation resulting from in-place chemical weathering of Ca-bearing igneous rock. Upon release into the soil solution by weathering, Ca^{2+} precipitates with bicarbonate formed from the reaction of water with CO_2 generated by root and microbial respiration. In many cases, however, igneous parent material has been considered as an insufficient source of Ca^{2+} and hence external sources, such as atmospheric additions of Ca^{2+}, have been sought.[6]

Biogenic Model

Fourth, some plants, microorganisms, and termites produce calcium carbonate. Evidence that various plants play a direct role in carbonate formation comes from the presence of euhedral calcite crystals on plant roots.[21] Moreover, several references in the Russian literature note carbonate formation by plant tissue and the downward translocation of these carbonates with wetting fronts.[22] Evidence that some microorganisms precipitate carbonates is based on observations of calcified bacteria and fungal hyphae with electron microscopy and in vitro laboratory experiments.[23,24] Evidence that termites precipitate carbonate in certain environments is based on the studies of termite mounds in Africa and southeast Asia.[25] Such mounds can be calcareous even though surrounding soils are noncalcareous, making the mounds attractive to native farmers who spread them over their agricultural fields.[25]

Conclusions

The formation of pedogenic carbonate may be dominated by one of the models listed before or may involve several of the models working together in different magnitudes. Understanding pedogenic carbonate formation has been extremely useful for understanding relative ages of geomorphic surfaces and landscape evolution.[15] A knowledge of pedogenic carbonate formation has also been useful for soil classification. Marbut,[26] for instance, used the presence of carbonate as a criterion for the highest category of his soil classification—Pedocal (soils with carbonate accumulation) and Pedalfers (soil with Al and Fe accumulation). Today, studies of pedogenic carbonate have expanded to include questions about paleoclimate, paleoecology, paleoatmospheric composition, global carbon cycles and the greenhouse effect.

References

1. Morse, J.W.; Mackenzie, F.T. *Geochemistry of Sedimentary Carbonates. Developments in Sedimentology* 48; Elsevier: Amsterdam, 1990.
2. Doner, H.E.; Lynn, W.C. Carbonate, halide, sulfate, and sulfide minerals. In *Minerals in Soil Environments*, 2nd Ed.; Dixon, J.B., Weed, S.B., Eds.; Soil Science Society of America: Madison, WI, 1989; 279–330.
3. Capo, R.C.; Whipkey, C.E.; Blachère, J.R.; Chadwick, O.A. Pedogenic origin of dolomite in a basaltic weathering profile, Kohala Peninsula, Hawaii. Geology **2000**, *28*, 271–274.
4. Gile, L.H.; Peterson, F.F.; Grossman, R.B. The K horizon—a master soil horizon of carbonate accumulation. Soil Sci. **1965**, *99*, 74–82.
5. Soil Survey Staff. *Soil Taxonomy—A Basic System of Soil Classification for Making and Interpreting Soil Surveys*, 2nd Ed.; USDA Agriculture Handbook Number 436; U.S. Govt. Printing Office: Washington, DC, 1999.
6. Birkeland, P.W. *Soils and Geomorphology*, 3rd Ed.; Oxford University Press: New York, 1999; 430.
7. Sobecki, T.M.; Wilding, L.P. Formation of calcic and argil-lic horizons in selected soils of the texas coast prairie. Soil Sci. Soc. Am. J. **1983**, *47*, 707–715.
8. Pal, D.K.; Dasog, G.S.; Vadivelu, S.; Ahuja, R.L.; Bhat- tacharyya, T. Secondary calcium carbonate in soils of arid and semiarid regions of India. In *Global Climate Change and Pedogenic Carbonates*; Lal, R., Kimble, J.M., Eswaran, H., Stewart, B.A., Eds.; Lewis Publishers: London, 2000; 149–185.
9. West, L.T.; Wilding, L.R.; Rabenhorst, M.C. Differentiation of pedogenic and lithogenic carbonate forms in texas. Geoderma **1987**, *43*, 271–287.
10. Rabenhorst, M.C.; Wilding, L.P.; West, L.T. Identification of pedogenic carbonates using stable carbon isotopes and microfabric analysis. Soil Sci. Soc. Am. J. **1984**, *48*, 125–132.
11. Drees, L.R.; Wilding, L.P. Micromorphic record and interpretations of carbonate forms in the rolling plains of texas. Geoderma **1987**, *40*, 157–175.
12. Nordt, L.C.; Hallmark, C.T.; Wilding, L.P.; Boutton, T.W. Quantifying pedogenic carbonate accumulations using stable carbon isotopes. Geoderma **1998**, *82*, 115–136.
13. Jenny, H.; Leonard, C.D. Functional relationships between soil properties and rainfall. Soil Sci. **1934**, *38*, 363–381.
14. Arkley, R.J. Calculation of carbonate and water movement in soil fromclimatic data. Soil Sci. **1963**, *96*, 239–248.

15. Gile, L.H.; Hawley, J.W.; Grossman, R.B. *Soils and Geomorphology in the Basin and Range Area of Southern New Mexico—Guidebook to the Desert Project*; New Mexico Bureau of Mines and Mineral Resources, Memoir 39: Socorro, New Mexico, 1981; 222 pp.

16. McFadden, L.D.; Tinsley, J.C. Rate and depth of pedogenic- carbonate accumulation in soils: formation and testing of a compartment model. In *Soils and Quaternary Geology of the Southwestern United States;* Weide, D.L., Ed.; Special Paper 203; Geological Society of America: Boulder, CO, 1985; 23–42.

17. Marion, G.M.; Schlesinger, W.H. Quantitative modeling of soil forming processes in deserts: the CALDEP and CALGYP models. In *Quantitative Modeling of Soil-Forming Processes;* Bryant, R.B., Arnold, R.W., Eds.; Soil Sci. Soc. Am. Spec. Publ. 39: Madison, WI, 1994; 129–145.

18. Scharpenseel, H.W.; Mtimet, A.; Freytag, J. Soil inorganic carbon and global change. In *Global Climate Change and Pedogenic Carbonates;* Lal, R., Kimble, J.M., Eswaran, H., Stewart, B.A., Eds.; Lewis Publishers: London, 2000; 27–42.

19. Goudie, A. *Duricrusts in Tropical and Subtropical Landscapes;* Oxford Univ. Press: London, 1973; 174.

20. Rabenhorst, M.C.; Wilding, L.P. Pedogenesis on the Edwards plateau, Texas: III. A new model for the formation of petrocalcic horizons. Soil Sci. Soc. Am. J. **1986**, *50,* 693–699.

21. Monger, H.C.; Gallegos, R.A. Biotic and abiotic processes and rates of pedogenic carbonate accumulation in the southwestern United States—relationship to atmospheric CO_2 sequestration. In *Global Climate Change and Pedogenic Carbonates;* Lal, R., Kimble, J.M., Eswaran, H., Stewart, B.A., Eds.; Lewis Publishers: London, 2000; 273–290.

22. Labova, E. *Soils of the Desert Zone of the USSR;* Israel Program for Scientific Translation: Jerusalem, Israel, 1967.

23. Phillips, S.E.; Milnes, A.R.; Foster, R.C. Calcified filaments: an example of geological influences in the formation of calcretes in South Australia. Aust. J. Soil Res. **1987**, *25,* 405–428.

24. Monger, H.C.; Daugherty, L.A.; Lindemann, W.C.; Liddell, C.M. Microbial precipitation of pedogenic calcite. Geology **1991**, *19,* 997–1000.

25. Thorp, J. Effects of certain animals that live in soils. Sci. Monthly **1949**, *68,* 180–191.

26. Marbut, C.F. Subcommission II. Classification, nomenclature, and mapping of soils in the United States. Soil Sci. **1928**, *25,* 61–71.

25

Lead: Ecotoxicology

Sven Erik Jørgensen

Introduction: Dispersion and Application

Lead is found in all environmental components—both living and non-living. The use of lead has dispersed the metal worldwide due to its long-term use in gasoline, batteries, solders, pigments, ammunition, paint, ceramic, and even piping. It is found for instance on the glacial ice and snow of Greenland, one of the most uncontaminated places on the Earth. Many toxic substances are generally widely dispersed and a global increase in the concentration of heavy metals and pesticides has been recorded, as exemplified for lead in Figure 1. The concentration of lead is shown in micrograms per ton (μg/t) as a function of time, which can be found by the use of analyses of ice cores. As it can be seen, the lead concentration has, since the mid-eighth century, increased 10 times—from about 10 μg/t to about 200 μg/t of snow. The dispersion of lead is caused by the many uses of this metal: in mining and smelting, in batteries, in lead-based paints, in electronic devices, in leaded gasoline, and in shots applied in hunting

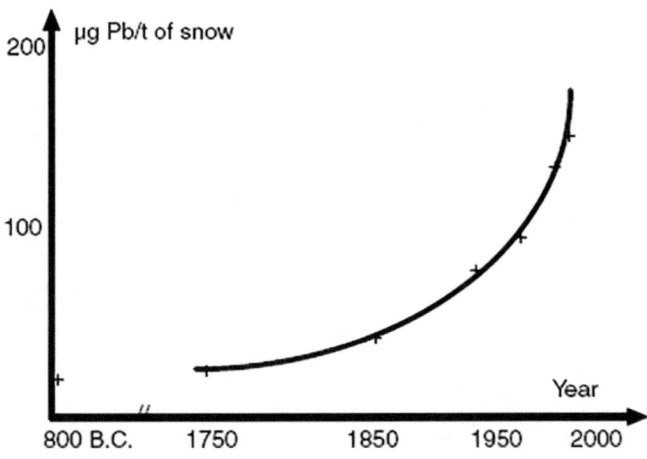

FIGURE 1 The lead concentration in the glacial ice, Greenland, as a function of time, from 800 B.C. to year 2000. The lead concentration has increased tenfold during the last 250 years, mainly due to combustion of leaded gasoline. **Source:** After Jørgensen[2] and Chemistry.[3]

and target shooting, The last two applications were phased out in most industrialized countries more than 25 years ago, but lead gasoline is still in use in many developing countries. In the United States, the manufacture of batteries is the dominant use of lead today.[1] The global dispersion of lead is particularly caused by the combustion of leaded gasoline, which today is inconceivable because the organic lead chemicals were applied in the gasoline to obtain a sufficiently high octane number at the lowest cost, and the less harmful alternatives could only have increased the gasoline price by about a quarter of an American cent per liter.

The concentrations of lead in food items are shown in Table 1 to illustrate the presence of lead in our food—an-other illustration of the consequences of the global dispersion of lead. The concentrations in the table are taken from the mid-1980s, i.e., before the introduction of lead-free gasoline had shown any significant effect. The differences between the three countries are explained by the differences in the traffic (and population) densities.

The concentration of lead in completely uncontaminated water is about 1 ng/L, while concentrations of 20 ng/L are often found when only minor discharge of lead has taken place. In contaminated and very contaminated water, a lead concentration of 100–200 ng/L is often found.

Heavy metals are dispersed globally, but the regional concentrations of most heavy metals may of course be much higher regionally than globally. The relationship between a global and a regional pollution problem and the role of dilution for this relationship are illustrated in Table 2, where the ratios of heavy metal concentrations in the River Rhine and the North Sea are shown. Notice that the amount of nickel and lead used in the region of the River Rhine is the same but the ratio is 70 times higher for

TABLE 1   Lead in Food (see Jørgensen[2])

Food Items	Typical Lead Concentration (mg/kg fresh weight)		
	England	Holland	Denmark
Milk	0.03	0.02	0.005
Cheese	0.10	0.12	0.05
Meat	0.05	<0.10	<0.10
Fish	0.27	0.18	0.10
Eggs	0.11	0.12	0.06
Butter	0.06	0.02	0.02
Oil	0.10	–	–
Corn	0.16	0.045	0.05
Potatoes	0.03	0.1	0.05
Vegetables	0.24	0.065	0.15
Fruits	0.12	0.085	0.05
Sugar	–	0.01	0.01
Soft drinks	0.12	0.13	–

TABLE 2   Heavy Metal Pollution in the River Rhine (from 1985) (see Jørgensen[2])

	River Rhine (t/yr)	Ratio: Conc. in the Rhine/Conc. in the North Sea
Cr	1,000	20
Ni	2,000	10
Zn	20,000	40
Cu	200	40
Hg	100	20
Pb	2,000	700

lead than for nickel due to the application of leaded gasoline, which disperses the lead uncontrolled, while nickel has more closed applications, which allow recycling of the metal. Notice also that lead is transported in the atmosphere primarily in the particulate phase.[4]

Ecotoxicity and Environmental Problems of Lead

The toxicity of Pb is mainly associated with the free ions Pb^{2+} (the +2 oxidation state), but lead can form complexes with hydroxide ions, carbonate, chloride ions, and many organic compounds, for instance, humic acid and amino acids. The toxicity of the complexes is generally lower than the toxicity of the free ions, because the uptake of the complexes is slower than the uptake of the free ions. It implies that it is necessary in every case study to determine by analytical methods or chemical calculation the concentrations of lead as free ions and in the form of the various complexes to determine the toxicity. The toxicity of lead declines with higher concentrations of the hardness ions, calcium, and magnesium. Lead shows, as other heavy metals, bioaccumulation and biomagnification, which is more pronounced at lower pH, because the solubility and the relative concentration of the free ions are increasing with decreasing pH.

Contaminated aquatic ecosystems have a significantly elevated concentration of lead in the sediment. Generally, the sediment has higher concentrations of heavy metals than the water, and as it is possible to analyze sediment core,[5,6] it is possible to find the contamination of heavy metals as functions of the time, provided the settling rate is known or can be estimated. This is particularly informative in the case of lead, because the use of leaded gasoline started shortly after the Second World War and was banned in industrialized countries before or around the mid-1980s. It entails that the sediment, from the approximately 40 years when leaded gasoline was used, will show a particularly high lead concentration, which of course facilitates the dating of the sediment. The lead concentration in sediment is usually 10–200 mg/kg dry weight, but as much as 3000–10,000 mg/kg dry weight can be found in contaminated areas.

A filter feeding bivalve mollusc shows a contamination of lead (and other heavy metals) that is proportional to the concentration in the sediment. The proportional constant is dependent on the composition of the sediment, but it is frequently between 0.01 and 0.05—the highest values for sediment with a high concentration of organic matter.[7]

A major source of lead exposure and toxicity for wild birds is the ingestion of lead-based ammunition. For birds, concentration of lead is in the order of 0.2 mg/kg dry weight, while a toxic effect would correspond to 100 times as much and death to 250 times as much.[5] LD_{50} for rats is 130 mg/kg dry weight (see *Inorganic Compounds: Eco-Toxicity,* p. 1479). There is a primary lead shot poisoning from direct ingestion of lead-based ammunition and a secondary lead shot poisoning when birds (and of course also other animals) ingest lead shotgun pellets and bullet fragment embedded in the flesh of dead or wounded animals shot with lead-based ammunition.

Lead is bound to SH- groups in the proteins and can therefore generally be taken up by plants with high concentrations of proteins more effectively than by plants with low protein concentrations; see also the entry about "Bioremediation." Removal of lead from areas that have been applied for target shooting with ammunition containing lead is possible by bioremediation.[8]

With other heavy metals, the uptake of lead from food is relatively low—about 7%–10%,[9,10] while lead is taken up from the atmosphere by the lungs with a higher efficiency. Figure 2 shows a steady-state model of the uptake for an average European in the mid-1980s, just before the use of leaded gasoline was banned for all new cars. The uptake from food is as seen 10%, namely, 30 μg/day out of the 300 mg/day that is in the food (compare with Table 1). The amount of lead in the $20 m^3$ of air that is daily used for respiration is about 25 mg, but as it is taken up with an efficiency of 50%, as much as 12.5 μg lead per day was accumulated in the body of an average European 25 years ago due to direct atmospheric pollution. The 30 mg/day is excreted mainly through the urine and therefore the 12.5 mg lead per day is accumulated in the body—mainly in the bones, where their effect fortunately is very minor. Due to the reduced use of leaded gasoline, the average European will today have less lead in the body. Food contains roughly half as much lead today than 25 years ago and the atmospheric pollution is also one half of the level

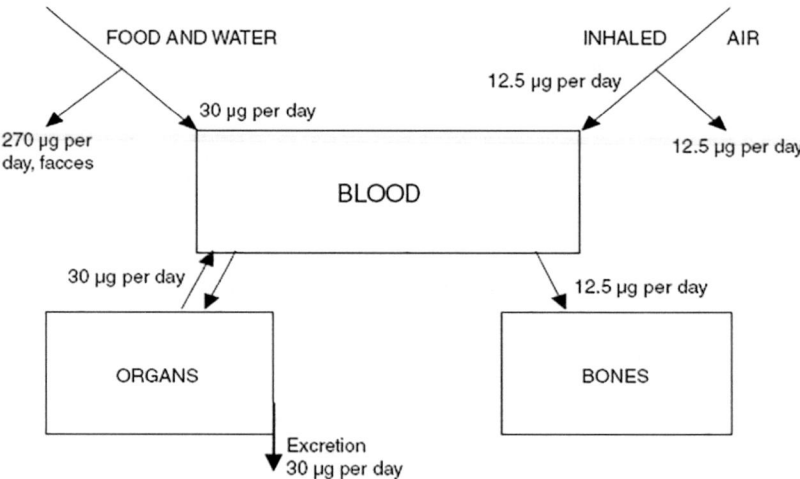

FIGURE 2   A steady-state model of an average European in 1985. The uptake from food was 30 µg/day out of 300 µg/day (10% efficiency) and 12.5 µg lead per day by respiration (uptake efficiency, 50%). The excretion balances the uptake from the food, and the lead taken up by respiration is accumulated in the bones.

today compared with 1985. It means that the amount of lead from food today is 15 mg lead per day and that from respiration is 6.25 µg lead per day. Excretion is also reduced to a level about 15–20 µg lead per day, which implies that less lead is accumulated in the bones today. The indicated amounts of lead for an average European today are found by use of a model that was calibrated and validated by the use of the amounts from 1985—it means the values that are shown in Figure 2.

A conceptual diagram for a lead model of a food chain in an aquatic ecosystem is shown in Figure 3. The boxes represent state variables, which in this case are lead in water, lead in phytoplankton, lead in zooplankton, lead in planktivorous fish (fish I), lead in carnivorous fish (fish II), and lead in the sediment. The two latter concentrations are the highest in most cases. At each level in the food chain, lead is taken up from the water and from the food. The process rates are dependent on the organisms, the temperature, the pH, and the concentrations of the free lead ions and lead complexes. Quantification of the all the processes will reveal that the lead concentration is increasing at a factor of 10–100 through the food chain. This is in contrast to a factor of about 10,000 for DDT. The difference is due to the less effective uptake of lead from the food—as mentioned, the uptake efficiency for heavy metals in food items is only about 7%–10%. DDT is taken up from food with an efficiency of 90% due to the low solubility in water and high solubility in fat tissue.

The spectrum of toxicological and ecotoxicological effects of lead is very wide. Acute toxicity of lead can cause headaches, irritability, and loss of appetite. Chronic toxicity can cause brain damage, reduced memory, anemia, liver and kidney damage, and possibilities of cancerous tumors of the kidney. It has been shown that if children are exposed to high lead concentrations, it will have a pronounced effect on their learning ability.[11] Elevated lead concentrations have also teratogenic effects,[12] and prenatal lead exposure has been demonstrated to be associated with an increased risk of malformations.[13] Genetic effects on animals have also been observed.[12] Epidemiological studies have shown that lead is related to the risk of elevated blood pressure.[14]

Significantly different lead concentrations have been found in the blood of people living far from towns and those in urban areas. Indians in the Amazons have about 8 ng lead per gram of blood. Farmers living in the European countryside have about 50 ng/g of blood and people living in very big cities (New York for instance) have about 150 ng/g of blood.[15] The highest concentrations of lead have been found in policemen who regulated traffic in industrialized countries before lead was phased out: 150–700 ng/g blood. Lead in clean air is as low as 0.0005 ng/L, while in urban areas, it is mostly between 2 ng/L and 25 ng/L.[16] In very polluted urban areas, the concentration may reach 50 ng/L, which could

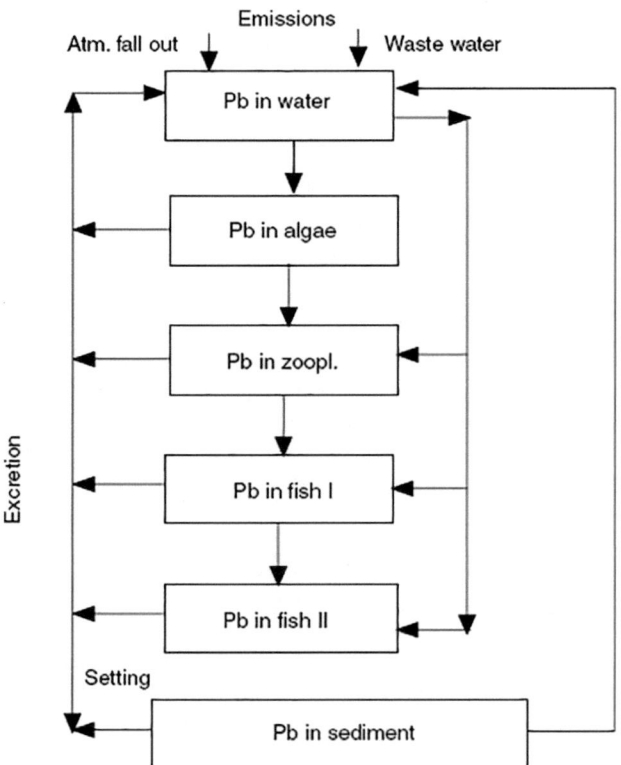

FIGURE 3 Conceptual diagram of biomagnifications of lead through the food chain in an aquatic ecosystem. Notice that each level in the food chain takes lead up from the water and from the food, but each level also excretes lead to the water.

give a daily uptake of 0.345 mg per person and a lead concentration in blood about 720 ng/g.[16] Figure 2 shows a model that could be used to find these values in the blood, when the air pollution level is known.

Lead in soil shows a similar wide range of values. Generally, a lead concentration of 0.2–5 mg/kg dry matter in agricultural areas is found,[17] but as a concentration as high as 100 mg/kg dry matter is found in particularly contaminated soil,[18] and in some extreme cases, concentrations as high as 10,000 mg/kg dry matter have been found.[19] A typical average soil in an industrialized country contains about 20 mg/kg dry matter of lead, and 95% of soil samples randomly sampled will show concentrations between 10 and 60 mg/kg dry matter. Lead concentrations above 1000 mg/kg dry weight in soil can kill earthworms and springtails.

The half-life of lead in human is about 6 years (whole body) and about 3 times as much for the skeletal system. It has been shown that skeletal burdens of lead increase linearly with age, while the non-skeletal burden is eliminated faster (as indicated, whole-body lead has a half-life of 6 years, which means that the non-skeletal lead is exchanged faster than once every 6 years). It is possible to reduce the body burden for patients clinically affected by lead by ethylene-diamine-tetra-acetate (EDTA).[20]

Abatement Methods for Reduction of Lead Pollution

Two major sources of lead pollution, namely, the combustion of leaded gasoline and the use of lead ammunition, have been eliminated in most industrialized countries by environmental legislation. The use of lead in ceramic and paints has similarly been reduced significantly by legislation or by agreement between environmental agencies and the industry.

Lead pollution has been reduced considerably during the last 25–30 years due to environmental legislation and the treatment of wastewater and contaminated smoke and air. Environmental technological solutions have been increasingly applied for the treatment of water and air. A description of the environmental technological methods can be found in the following entries of this encyclopedia:

- *Ion Exchange Application for Treatment of Water and Wastewater*
- *Wastewater Treatment: Overview of Conventional Methods*
- *Municipal Wastewater and Its Treatment*
- *Air Pollution and Environmental Technology: An Overview*

Contaminated land has also been treated, both by environmental technological methods and by bioremediation (see *Bioremediation*, p. 408).

A few cases of successful application of cleaner technology (replacing lead by less harmful components) have also been reported in the journal *Cleaner Technology*. The results of these efforts are encouraging for the use of a consequent environmental management. When environmental legislation, environmental technology, cleaner technology, and ecotechnology are working hand in hand, it is possible to achieve good pollution abatement results, as the lead pollution problem has demonstrated.

Use of Integrated Environmental Management in the Case of Lead Contamination

As with all pollution problems, lead contamination is complex. This section briefly discusses how to go around this complexity to propose an integrated and holistic environmental management and thereby solve the problems properly. The discussion in this section is in principle valid for all pollution problems (compare to the Topical Table of Contents and the How to Use This Encyclopedia in the frontmatter) although particularly for all heavy metal contaminations. The following crucial questions require answers in the case of lead contamination:

1. Which forms have the lead-free ions, or which complexes in which concentrations?
2. What are the sources to the problem? Quantitatively?
3. What possibilities do we have to eliminate which sources? Will that be sufficient to solve the problem?
4. How can we best combine the methods to solve the problem?

It is recommended to consider the following points to be able to arrive at an answer for the crucial questions:

1. The forms of lead can be found either analytically or by chemical calculations;[21] where straightforward chemical calculations are clearly shown for heavy metals with illustrative examples.
2. It is advantageous to set up a mass balance. Figure 4 shows a mass balance for the lead contamination of 1 ha Danish agricultural land. The mass balance clearly reveals the important sources. Atmospheric fallout is the dominant source, although it may also be beneficial to reduce the lead contamination coming from sludge and fertilizers. Particularly the first one of these two could be eliminated.
3. It is possible to eliminate the air pollution of lead as it has been discussed by phasing out the use of lead in gasoline. The second most important air pollution source, in many countries, is coal-fired power plants, which can be eliminated by changing to other forms of fossil fuel or to alternative energy sources. This change of the energy policy will, however, often be prohibitively expensive, and it is therefore as indicated a political question. The phasing out of lead in gasoline may be sufficient. The core question is this: how much will the lead contamination in food be reduced if we reduce the atmospheric fallout so and so much? The answer requires calculations and sometimes the use of models. Lead models have been developed by Jørgensen.[7,9] Lead models considering

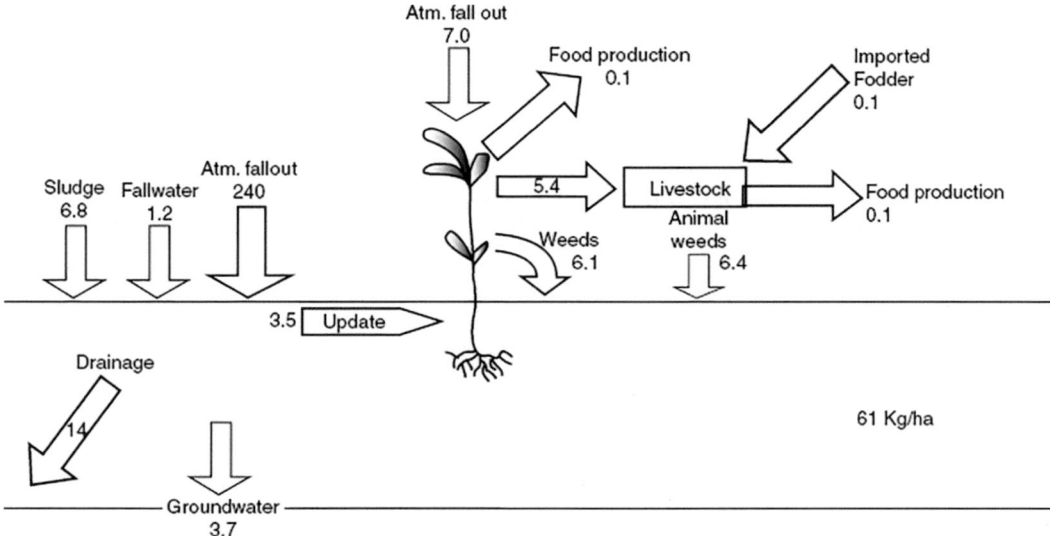

FIGURE 4   Lead balance of average Danish agriculture land. All rates are expressed as grams of Pb per hectare per year.

hydrodynamics, bioconcentration, bioaccumulation, excretion, and sedimentation can furthermore be found in Lam and Simons[22] and Aoyama et al.[23] Another core question is: if we are able to reduce the concentration in food, sediment, or soil to desired levels, would this reduction be sufficient to reduce or even eliminate the effect? This question will require a comprehensive overview of the toxicological and ecotoxicological literature about lead and its effect. If information about the effect is not available, it is necessary to perform bioassay.[24,25] These references describe in detail how to perform bioassay for heavy metals including lead.

4. A discussion of the answer to the fourth question is covered in "The Contents of the Encyclopedia and How to Use the Encyclopedia for Integrated Ecological and Environmental Management." Other examples are given in Jørgensen et al.[21]

References

1. Laws, E.A. *Aquatic Pollution: An Introductory Text,* 2nd Ed.; John Wiley and Sons: New York, 1993; 380 pp.
2. Jørgensen, S.E. *Principles of Pollution Abatement;* Elsevier: Amsterdam, 2000; 520 pp.
3. Chemistry. The lead we breathe. Chemistry **1968**, *41*, 7–12.
4. Haygarth, P.M; Jones, K.C. Atmospheric deposition of metals to agricultural surfaces. In *Biochemistry of Trace Metals;* Adriano, D.C., Ed.;. Lewis Publ.: Boca Raton, 1992; 423 pp.
5. Jørgensen, S.E.; Fath, B. *Ecotoxicology;* Elsevier: Amsterdam, Oxford, 2010; 390 pp.
6. Forstner, U.; Wittmann, G.T.W. *Metal Pollution in the Aquatic Environment;* Springer Verlag: Heidelberg, Berlin, New York, 1979; 490 pp.
7. Jørgensen, S.E.; Fath, B. *Fundamentals of Ecological Modelling,* 4th Ed.; Elsevier: Amsterdam, Oxford, 2011; 396 pp.
8. Mitsch, W.J.; Jørgensen, S.E. *Ecological Engineering and Ecosystem Restoration;* John Wiley: New York, 2004; 410 pp.
9. Jørgensen, S.E.; Bendoricchio, G. *Fundamentals of Ecological Modelling,* 3rd Ed.; Elsevier: Amsterdam, Oxford, 2001; 530 pp.

10. Newman, M.C.; Unger, M.A. *Fundamentals of Ecotoxicology*; 2nd Ed.; CRC Press: Boca Raton, London and New York, 2003.

11. Bellinger, D.; Leviton, A.; Needleman, H.L.; Waternaux, C.; Rabinowitz, H.B. Low-level lead exposure and infant development in the first year. Neurobehav. Toxicol. Teratol. **1986**, *8*, 151–161.

12. Swedish EPA. *Om Metaller*, 2nd Ed.; Statens Naturvårdsverk: Sweden, 1980.

13. Needleman, H.L.; Rabinowitz, M.; Leviton, A.; Linn, S.; Schoenbaum, S. The relationship between prenatal exposure to lead and congenital anomalies. J. Am. Med. Assoc. **1984**, *251*, 2956.

14. Schwartz, J.; Angle, C.; Pitcher, H. Relationship between childhood blood lead level and stature. Pediatrics **1986**, *77*, 281–288.

15. Jørgensen, S.E.; Jorgensen, L.A.; Nors Nielsen, S. *Handbook of Ecological and Ecotoxicological Parameters*; Elsevier: Amsterdam, 1991; 1380 pp.

16. Francis, B.M. *Toxic Substances in the Environment*; John Wiley and Sons: New York, 1994; 362 pp.

17. Prost, R. *Contaminated Soil*; Institut National de la Recherche Agronomique: Ann Arbor, MI, 1995; 525 pp. +CD.

18. Hutschinson, T.C.; Gordon, C.A.; Meema, K.M. *Global Perspectives on Lead Mercury and Cadmium Cycling in the Environment*; Wiley Eastern Limited: Ann Arbor, MI, 1994; 412 pp.

19. Kabata-Pendias, A. *Effects of Trace Metals Excess in Soils and Plants*; CRC: Boca Raton, 1986; 330 pp.

20. Nigel, B. *Environmental Chemistry*; 2nd and 3rd Eds.; Wuerz Publ. Ltd.: Winnepeg, 1994 and 2000; 378 pp. and 420 pp.

21. Jørgensen, S.E.; Tundisi, J.G.; Tundisi, T. *Handbook of Inland Aquatic Ecosystem Management*; CRC: Boca Raton, 2012; 480 pp.

22. Lam, D.C.L.; Simons, T.J. Computer model for toxicant spills in Lake Ontario. In *Metals Transfer and Ecological Mass Balances, Environmental Biochemistry*; Nriago, J.O., Ed.; Ann Arbor Science; Ann Arbor, MI, 1976; Vol. 2, 537–549.

23. Aoyama, I.; Inoue, Yos; Inoue, Yor. Simulation analysis of the concentration process of trace heavy metals by aquatic organisms from the viewpoint of nutrition ecology. Water Res. **1978**, *12*, 837–842.

24. Newman, M.C.; Jagoe, C.H. *Ecotoxicology. A Hierarchical Treatment*; CRC: Boca Raton, 1996; 412 pp.

25. Landis, W.G.; Yu, M.H. *Introduction to Environmental Toxicology*; Lewis Publ.: Boca Raton, 1995; 330 pp.

26

Lead: Regulations

Lisa A. Robinson

Introduction

Phasing out lead in gasoline is often identified as one of the U.S. Environmental Protection Agency's (USEPA's) greatest achievements.[1] As part of the regulatory process, USEPA economists developed innovative approaches for expressing the benefits of the resulting risk reductions in monetary terms, so that they could be directly compared to the costs of different control options.[2] As evidence mounts that lead may pose hazards even at relatively low exposure levels,[3,4] the methods used to estimate these monetary benefit values are in need of continual improvement and updating for application in future regulatory assessments. This entry reviews how these benefit values have been developed and identifies challenges for subsequent work. While it focuses on the approaches used in U.S. regulatory analyses, these approaches can be adapted for use in other countries and have global relevance.

The entry first reviews previous assessments, summarizing the framework for regulatory benefit-cost analysis and providing examples of how this framework has been implemented to assess the benefits of lead regulations. Because these analyses indicate that the most significant quantifiable benefits include the relationship between lead-related neurological damage and future earnings (for children) and between lead exposure and cardiovascular disease (for adults), it next discusses these benefit categories in more detail. It then describes benefits that have not been fully incorporated into previous regulatory analyses, but that may be desirable to explore in the future. The concluding section summarizes the discussion and highlights related challenges.

Review of Previous Analyses

The USEPA's pathbreaking work on valuing lead benefits was initiated in the late 1970s and early 1980s, as the United States moved to phase out lead in gasoline.[2,5] Lead was completely banned from gasoline in 1996. The valuation approaches evolved as the USEPA subsequently considered regulating lead in drinking water, sewage sludge, and paint, as well as ambient air. Typically, the quantified benefits are dominated by the effect of lead on earnings (due to its effect on IQ) and, in some cases, by its effects on cardiovascular disease, particularly the risks of premature mortality. However, analysts have been unable to quantify many other potential impacts, due largely to gaps or inconsistencies in the research on the association between lead exposure and additional health outcomes.

General Framework

When the USEPA began assessing the benefits of reducing lead exposure, requirements for conducting regulatory benefit–cost analysis were in their infancy.[6] The first comprehensive Executive Order mandating such analysis was issued by President Reagan in 1991.[7] The requirements for these analyses have since been refined, updated, and expanded, although the basic principles remain the same. Thus, the early work on lead was both innovative and pioneering, evolving into approaches used in numerous subsequent analyses.

Currently, benefit–cost analysis is required for major rules under President Clinton's Executive Order 12866,[8] as supplemented by President Obama's Executive Order 13563.[9] These executive orders mandate that Federal agencies assess alternative policies for actions that may be economically significant, i.e., that may lead to a rulemaking that has an annual economic effect of $100 million or more or has important adverse effects. To support implementation of these requirements, the U.S. Office of Management and Budget (OMB) issued Circular A-4, *Regulatory Analysis*,[10] in 2003 and summarized and clarified related requirements in a 2010 checklist for agencies.[11]

Typically, the resulting analyses contain five major components, in addition to sections discussing the rationale for the rulemaking and the regulatory options considered, as summarized in Table 1. These analyses also include both quantitative and qualitative assessment of related uncertainties, as well as information on effects that could not be quantified or valued.

This entry is primarily concerned with the economic valuation of benefits, under step 4b in Table 1. As discussed later, the ability to carry out such valuation is dependent on the extent to which the changes in risks can be assessed under step 4a. Building a strong research base that allows analysts to estimate the changes in risks of different types associated with various changes in lead exposures has been a major challenge.

Ideally, economic values reflect individuals' willingness to pay (WTP) for the risk reductions or other benefits they would receive from a regulation or policy. This approach is consistent with the theoretical framework underlying benefit–cost analysis, which is based on respect for individual preferences (often referred to as "consumer sovereignty"), assuming that each individual is the best judge of his or her own welfare. The consideration of WTP also reflects the types of trade-offs being considered in policy decisions. Given constrained resources, pollution abatement policies inevitably require trading off increased expenditures on risk reductions against decreased expenditures on other desired goods

TABLE 1 Overview of Analytic Components

1. Estimate current and potential future *baseline conditions* in the absence of government intervention.
2. Predict *responses* to each policy option under consideration.
3. Estimate the *national costs* associated with each option, summing the costs of the predicted compliance actions across those subject to the provisions and accounting for resulting market impacts (e.g., changes in consumption due to price increases).
4. Estimate the *national benefits* associated with each option, including the effects on human health and the natural and built environment. For environmental hazards, this generally consists of two steps:
    a. A *risk assessment,* which considers the link between changes in pollution levels and each health or environmental outcome of concern.
    b. An *economic analysis,* which includes monetary valuation of each outcome to the extent possible.
5. Assess the distribution of the impacts across subpopulations of concern. Such subpopulations typically include small businesses as well as sensitive and/or vulnerable subgroups (such as children or low-income individuals) whom policymakers wish to protect against disproportionate adverse effects.

and services on a society-wide level. WTP, the maximum amount of income (or wealth) an individual is willing to exchange for a beneficial outcome, represents this type of exchange. Willingness to accept compensation (WTA), or the smallest amount an individual would accept to forego the improvement, is also consistent with this framework. WTA is used infrequently in regulatory analysis, however, both because it can be difficult to measure and because regulations often involve paying for improvements rather than compensating for harms.

In some cases, individual WTP can be estimated based on consumer demand for market goods. For outcomes not directly bought and sold in the marketplace (such as health risk reductions), WTP is instead estimated from revealed or stated preference studies. Revealed preference studies use data from market transactions or observed behaviors to estimate the value of related non marketed goods or outcomes. For example, the value of mortality risk reductions is often estimated from the relationship between earnings and job-related risks, controlling for other influencing factors. Alternatively, stated preference studies rely on responses to survey questions or similar approaches. For example, a survey respondent may be asked whether he or she would be willing to pay "$X" for a 1-in-10,000 reduction in the risk of death associated with decreased exposure to an air pollutant. Each approach has advantages and limitations. In particular, revealed preference studies rely on data from actual markets but often address scenarios that differ from those of concern in policy analysis. Stated preference studies enable researchers to better tailor the scenario to the risks of concern, but the responses are hypothetical and must be carefully elicited.

For many outcomes, estimates of WTP are lacking, and analysts often rely instead on averted costs as rough proxies. In particular, health risk reductions may be valued using cost of illness estimates, including expenditures on medical treatment and often lost productivity.[12,13] The latter is typically valued using the human capital approach, which assumes that workers are paid the value of their marginal product.[12–16] Thus, compensation data are used to value illness-related lost work time. The human capital approach also can be used to assess the value of changes in unpaid work time (i.e., household production and volunteer work), based on estimates of either the wages foregone when an individual chooses to engage in unpaid rather than paid work or the cost of replacing the unpaid worker with one who is compensated.

Another, less commonly used approach is the friction cost method, which assumes that productivity will decrease temporarily while the employer implements measures to replace the absent individual rather than over the full course of the illness.[14,16,17] The "friction period" is defined as the time it takes to find and train a new employee or reallocate duties among existing employees. However, this approach is likely to understate the loss particularly during periods of full employment, because it does not take into account the additional loss that accrues if the new employee was previously working in a different job or involved in nonmarket production.

WTP to reduce the risk of illness may differ from these averted costs. Costs reflect incurred cases, not expected risk reductions, and being ill and treated is typically worse than not being ill at all. Medical costs can be difficult to estimate accurately due to the distorting effects of insurance and other third-party payments, and analysts often rely on average per case costs, which may differ from the marginal costs of small changes in the risk of illness.

Approaches for valuing time losses also often rely on a number of simplifying assumptions regarding the functioning of the labor market and individual choices between paid and unpaid work and leisure. In reality, productivity losses at a societal level will depend on unemployment rates and on the extent to which coworkers compensate for the loss of the ill worker's time. At an individual level, sick leave and disability insurance may reduce the impact of illness on earnings. In addition, inflexible work hours and other factors limit workers' ability to choose jobs that reflect their willingness to trade off paid and unpaid work and leisure time; hence, wages may over- or understate their WTP for changes in time use.

For these and other reasons, estimates of averted costs may be less than, equal to, or greater than WTP to avoid the illness. For example, medical costs may frequently understate WTP because they do not reflect the value of avoiding pain and suffering, but may overstate WTP in cases where the availability of insurance leads individuals to seek treatment that they would not be willing to pay for themselves. In addition, although average costs per case are often reported on an yearly basis, few studies track individuals longitudinally and provide data on the lifetime costs associated with particular illnesses. Thus, while the concept of averted costs seems straightforward, the direct and indirect costs of illness can be difficult to estimate. However, they provide a reasonable and widely-used proxy when WTP estimates are not available.

As discussed below, the valuation approaches used in lead regulatory analyses rely largely on averted costs, due to the lack of suitable WTP estimates. Very few WTP studies directly address lead exposures[18] or IQ-related decrements;[19] more generally, WTP for morbidity risks has not been well-studied.[20] In particular, the approach typically used to value lead-related IQ decreases is an averted cost approach, focusing on lost earnings; some analyses also consider related educational costs. The logic is that individual WTP to avoid the IQ loss would be at least equal to the income foregone. Medical costs are typically not included, because no fully effective treatment now exists. While chelation therapy has been used to address high blood lead levels, its effectiveness has been questioned in recent years.[21]

Averted costs are often also used to estimate the value of morbidity risk reductions, including reductions in nonfatal cardiovascular disease associated with decreased lead exposure. WTP estimates are available for mortality risk reductions, however.

Types of Benefits Assessed

The approaches used to value lead-related lost earnings and cardiovascular effects in regulatory analysis have evolved over the past 30 years, reflecting new data as well as increasingly sophisticated methods. One of the first comprehensive benefit analyses of reduced lead exposures was published in 1985 for USEPA regulations developed under the Clean Air Act, addressing the phase down of the amounts of lead allowed in gasoline.[5] The outcomes assessed include reductions in the costs of lead screening and treatment; compensatory education; the risks of hypertension, myocardial infarction, stroke, and premature mortality; and damages to pollution control equipment, vehicle maintenance, and fuel economy; as well as health and welfare effects associated with other pollutants. While the authors indicate that the majority of the benefits (75%) stem from the reduced risks of cardiovascular disease (with mortality risk reductions dominating the results), these effects were not considered in the USEPA's decision making because the underlying epidemiological studies had not yet undergone widespread review. However, even if the benefits of cardiovascular risk reductions are not counted, the costs of the regulations are less than the benefits.

Soon after, the USEPA's 1986 analysis of a drinking water rule added an important new benefit category, the effects of changes in IQ on earnings, while also assessing many of the other benefit categories included in the 1985 analysis of lead in gasoline.[22] Rather than summing the estimates of lost earnings and educational costs, this analysis compares them as alternative measures of cognitive damage. For this rulemaking, the majority of the benefits result from reduced water system corrosion. Of the health-related benefits, lost earnings and total cardiovascular risk reductions are similar in magnitude.

These approaches were refined in subsequent analyses. A particularly significant example is the USEPA's 1997 *Retrospective Analysis of the Clean Air* Act.[23] That analysis covers the effects of the full set of regulations implemented under the Clean Air Act over the years 1970 to 1990. For lead, the benefits assessed include compensatory education and other educational costs, future earnings, neonatal mortality, and cardiovascular disease (hypertension, coronary heart disease, stroke, and premature mortality). The results in this case are dominated by the value of reducing the mortality risks associated with cardiovascular disease.

More recent USEPA analyses focus on the effects of lead on IQ and earnings, including its 2008 assessment of the Lead Renovation, Repair, and Painting Program rule developed under the Toxic Substances Control Act (TSCA) and its 2008 analysis of the lead National Ambient Air Quality Standards (NAAQS) under the Clean Air Act.[24,25] The benefits associated with reducing cardiovascular risks were not quantified in these analyses, because the USEPA found that data on adult blood lead levels were outdated and that more information on the association between these levels and related risks was needed. For IQ and earnings, the two analyses use somewhat different assumptions. For the renovation rule, USEPA relies solely on 1995 estimates of the relationship of IQ to earnings developed by Salkever, while for the NAAQS rule, it also considers the effects of a lower 1994 estimate developed by Schwartz, because recent research suggests that the Salkever estimates may be overstated. These estimates are discussed in more detail in the next section.

In general, these regulatory analyses indicate that the monetized benefits of reductions in lead exposure often far exceed the costs. The most significant benefit categories are the impact of lead on future earnings (for children) and its impact on mortality risks from cardiovascular disease (for adults). Regulations that focus on lead in ambient air or drinking water will affect both adults and children, while those that primarily address the ingestion of lead paint or contaminated soil will largely affect children.

The benefit categories addressed are similar across analyses in part because of the epidemiological research available. The benefit assessments focus on those areas where the risk assessments have found the most consistently strong associations, and where the outcomes are amenable to monetary valuation. The number of lead epidemiological studies is large, and they consider a much wider range of endpoints than reflected in these benefit analyses.[4] Some outcomes are excluded from the benefit-cost analysis because the risk data are weak or inconsistent; for others, the risk assessment shows strong associations but addresses endpoints that are difficult to value in monetary terms. For example, small changes at the cellular level cannot be easily monetized unless information is also available on the likelihood that such changes will lead to noticeable health impairments. As a result, these analyses are likely to understate benefits, but the degree of understatement is uncertain.

Valuing IQ-Related Benefits

The benefit category most frequently assessed in lead-related regulatory analyses is the effect of IQ on earnings. This assessment generally includes two components: determining the percent change in lifetime earnings associated with each one-point change in IQ and estimating the dollar value of lifetime earnings for each cohort affected. Because the first component is far more difficult to address, the discussion that follows focuses on how the percent change in earnings is estimated. The second component, lifetime earnings, can be estimated based on readily accessible data on earnings and survival probabilities, as demonstrated by Grosse et al.,[15] who provide estimates by year of age as of 2007.

Describing these benefits as the effect of lead-associated IQ decrements on earnings is somewhat misleading, however. The underlying epidemiological studies report associations between lead and varying measures of cognitive abilities, as well as measures of behavioral and other problems that may affect both school- and work-related achievement. Thus, this category may include a number of interrelated neurological effects that impact educational attainment, the likelihood of employment, the amount of earned income, and household production.

This entry does not discuss the effects of lead or IQ on school-related costs in detail, because they tend to be a much smaller proportion of total benefits. Reductions in lead exposure may have counterbalancing impacts on these costs. To the extent that reductions improve IQ and other aspects of neurological functioning, the years of education may increase and the need for compensatory education may decrease. These changes may not be completely offsetting, however, due to differences in per-student costs and in the number of students affected. The change in the need for compensatory education is likely to affect a relatively small number of very low IQ children, while the increase in regular education may affect a larger number of children throughout the IQ spectrum. In addition, to the extent that IQ gains increase the number of years of schooling, they will also defer full employment while increasing compensation once employed.

Previous Regulatory Analyses

In simple terms, expected earnings are the product of the likelihood of employment (i.e., labor force participation) and the wages earned if employed. Cognitive ability can affect both wage levels and the probability of employment directly. In addition, cognitive ability can affect the number of years of schooling, which in turn also affects wages and participation rates. Again, the effects of lead may be greater than the effects of cognitive ability alone, because of associated behavioral problems, attentional difficulties, and other neurological effects that can influence functioning in school and at work. These relationships are illustrated conceptually in Figure 1.

Assessing the pathways that lead from blood lead levels to changes in earnings is a complicated task. The foundation for the approach used in USEPA regulatory analyses is described in a 1994 article by Schwartz.[26] These relationships were then re-estimated in 1995 by Salkever,[27] whose estimates have been used in many, if not most, subsequent analyses. However, more recent work by Grosse[28,29] suggests that Salkever may overstate these effects. Disentangling the direct and indirect effects on earnings, to avoid double counting while capturing lead's impacts on earnings to the fullest extent possible, has been a major challenge.

Schwartz focuses on the relationships marked as solid lines (a) through (g) in Figure 1: the direct effect of lead on IQ and earnings (a → b → e), the indirect effect of lead on earnings through schooling and wages (c → d → e), and the indirect effect of lead on earnings through schooling and labor force participation (c → f → g). While these effects are interrelated, they can be summed as long as the underlying studies control for each effect separately. In total, Schwartz estimates that these factors lead to a 1.76% change in earnings for each one-point change in IQ. However, because his analysis includes estimates of the direct effects of lead on schooling, it likely captures some of the effects of lead on behavior (e.g., on attention span) rather than purely measuring the effects of IQ. While the percent change may appear small on an individual basis, the large number of individuals affected by many regulations leads to high total benefits.

Salkever then revisits the approach developed by Schwartz, using data from the National Longitudinal Study of Youth (NLSY) to extend and re-estimate many of the components of the analysis. Salkever adds the two dashed arrows marked (x) and (y) in Figure 1, estimating the effect of IQ on schooling and on labor force participation. He develops estimates separately for women and men.

When weighted to reflect the relative contribution of women and men to total earnings, the Salkever estimates average about a 2.4% change in earnings per IQ point, significantly higher than the Schwartz estimate (1.76%). For example, Schwartz calculated that a 1 μg/dl reduction in1984 childhood blood lead concentrations would yield increased earnings of $5.1 billion annually (1989 dollars). If we instead apply the factors from Salkever, this amount increases by more than 30%, to close to $7 billion. While lead

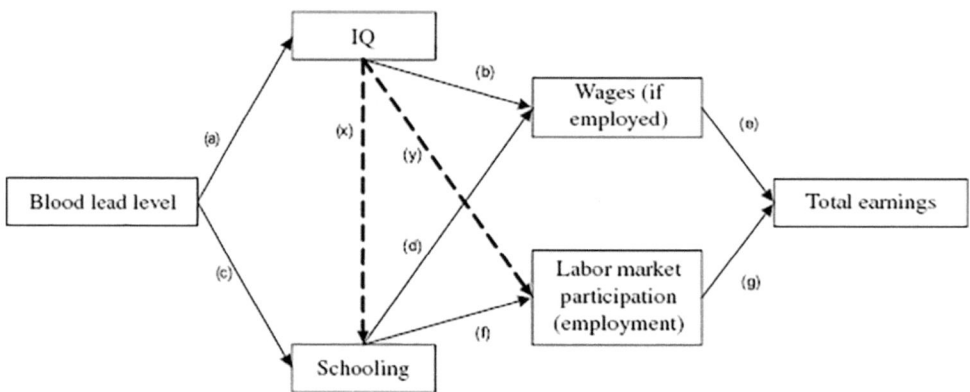

FIGURE 1   Conceptual model of relationship between lead and earnings.

exposure has dropped in recent years, newer epidemiological studies suggest that the effects of lead on IQ may be larger than suggested by this older research, which would further increase these 1984 benefits estimates.

Recent Research Findings

Linking lead-related IQ changes to changes in earnings is difficult for two reasons. First, the longitudinal databases used in the econometric analyses often rely on tests of cognitive ability that differ from those used in epidemiological studies of the associations between lead and cognitive ability. The epidemiological studies generally rely on the Wechsler Intelligence Scale for Children (WISC), which is a standard and widely used test of IQ administered in childhood. Many of the econometric studies (including the Salkever analysis) rely on data from the NLSY, which instead includes results from the Armed Forces Qualification Test (AFQT) administered in 1980 when NLSY participants were between ages 15 and 23. The AFQT measures achievement as well as ability. Thus, while IQ tends to be relatively stable after early childhood, the inclusion of achievement means that AFQT scores can change as a result of subsequent education and work experience.

Second, the causal relationships are complex and not necessarily well understood, posing challenges in terms of both measurement and statistical analysis. A wide range of personal, family, and community factors (including education system and job market characteristics) may affect earnings, and also may be affected by IQ. The causal relationships can be somewhat circular. Green and Riddell[30] provide a simple example: if we are trying to estimate the independent effects of IQ and years of schooling on earnings, but higher-IQ individuals stay in school longer because their psychic costs of schooling (e.g., the difficulty of studying) are lower, then IQ is in part causing the in-creased schooling and the contributions of the two factors to earnings are difficult to separate.

More recent evidence on the relationship between IQ and earnings is reviewed by Grosse.[28,29] Some of these studies use the same NLSY data set as the Salkever study but implement various types of adjustments. The NLSY began tracking individuals in 1979 when they were between ages 14 and 22; Salkever and the more recent studies discussed below use 1990 NLSY data, which include information on educational attainment, hourly wages, and hours worked as of ages 25 to 33. As noted earlier, the measure of cognitive ability in the NLSY is the AFQT, administered in 1980, when many respondents were already working or pursuing postsecondary education.

Recognizing that postsecondary job experience and education are likely to increase AFQT scores, in 1996, Neal and Johnson[31] re-estimated the relationship of IQ to earnings as part of their study of earnings differentials between blacks and whites. They restrict their NLSY sample to those who were age 18 or younger when they took the AFQT, so that their estimates of the relationship between ability and earnings are less inflated by the effects of further education and experience than the estimates used by Salkever.

In 2002, Grosse, Schwartz, and colleagues[28] used the results from the Neal and Johnson study to assess the economic gains associated with the U.S. reduction in blood lead levels that occurred from 1976 to 1999. According to Grosse et al., the Neal and Johnson study indicates that a one-point IQ loss results in a 1.15% difference in earnings for men and a 1.52% difference for women. These results are from models that do not control for schooling; hence, Grosse et al. do not add separate estimates of the effects of schooling on earnings to avoid double counting. However, because Neal and Johnson do not estimate the impact of ability on labor force participation, Grosse et al. add the estimate of this effect from Salkever. Once weighted to reflect the relative share of earnings by men and women in the labor force, Grosse et al. find that a one-point IQ loss is associated with a 1.66% change in earnings, slightly lower than the Schwartz estimate of 1.76% and substantially less than the Salkever estimate of 2.4%.

In his 2007 review, Grosse also discusses a newer study by Heckman et al.[32] These authors rely on the same NLSY data as discussed previously, considering the factors that influence wages at age 30. The modeling is substantially more sophisticated than prior studies, considering a wide array of personal, family, and community characteristics, including standard demographic variables as well as behavioral choices (such as smoking and participation in crime). It takes into account latent (i.e., unobserved) factors representing cognitive and non-cognitive ability and considers how these factors influence schooling choices as well as labor market choices and earnings.

Comparing the Heckman et al. findings to the findings from other studies requires some conversion. Heckman et al. multiply their coefficients by the standard deviation of the associated variable in reporting their results. Because IQ is standardized so that one standard deviation equals 15 IQ points, this entry follows Grosse[29] and divides the Heckman et al. coefficients on cognitive ability by 15 to determine the effect per IQ point. However, it uses results from Heckman et al.'s model 2 (with corrected AFQT scores) in Table 5 of Heckman[32] rather than their model 3 (as in Grosse, which relies on latent measures of ability), because the latter are more difficult to interpret. Depending on the controls included in the statistical modeling, the results suggest that a one-unit change in cognitive ability raises wages for males by 0.44% to 0.95% per IQ point.

This result is lower than the results from Neal and Johnson (1.15% for men and 1.52% for women), which do not control for schooling. Adding the pathways not included in the Heckman et al. analysis, i.e., the effect of ability on labor force participation, would increase the Heckman et al. estimates somewhat, as would including women (for whom returns to ability appear higher than for men). Perhaps more importantly, Heckman et al. find that the effects of non-cognitive factors are significant. Lead has many behavioral effects not captured in the epidemiological studies of the association between lead exposure and IQ; considering the effect of these factors could noticeably increase the impact of lead exposure on earnings.

Grosse also identifies another study, by Zax and Rees,[33] that uses an alternative data source: the Wisconsin Longitudinal Study (WLS) of Social and Psychological Factors in Aspiration and Attainment. This study includes IQ scores based on the Henmon–Nelson Test of Mental Ability administered in the 11th grade. The authors explore the effects of factors measured at age 17 (in 1957) on male earnings at ages 35 and 53. Zax and Rees find that household, community, and peer characteristics all affect earnings at both ages. Depending on the extent to which the model controls for these factors, the effect of a one-IQ-point change on earnings ranges from 0.36% to 0.74% at age 35, rising to 0.90%–1.39% at age 53, for the subsample who had earnings at both ages (Tables 5 and 6 of Zax and Rees[33]). The effect of IQ on earnings for the full sample at age 35 was almost identical to the effect for this subsample at that age.

These findings have four important implications for considering the effects of lead on earnings. First, the socioeconomic status of those affected by reduced lead exposures may affect the relationship between IQ and earnings. The Zax and Rees results for age 35 are somewhat lower than the factors found by Heckman et al. at age 30. However, the Zax and Rees sample is from a relatively affluent state, which may not be representative of the national population, and they rely on data for high school graduates. In contrast, the NLSY data used by Heckman et al. represent a wider population. Different relationships between IQ and earnings may hold if lead exposures disproportionately affect lower-income individuals, who may be less likely to complete high school and for whom the returns to IQ may be influenced dissimilarly by household and community characteristics.

Second, these findings suggest that using a constant factor, based on the IQ-to-earnings relationship for ages around 30 to 35, will understate the effect of IQ on earnings at older ages. The effect of this changing relationship on the present value of lifetime earnings is unclear, however, because it depends on three factors: the extent to which the IQ-and-earnings relationship varies across all years of age, the extent to which earnings change with age, and the discount rate used to estimate the present value of future earnings.

Third, the Zax and Rees analysis further emphasizes some of the difficulties inherent in sorting out causal links. Their model with the most controls (which leads to the lowest proportional changes; i.e., the 0.36% and 0.90% above) includes variables such as whether parents encouraged the individual to attend college. Parents may be more encouraging if their child has a higher IQ, or this encouragement may be related to other factors (such as whether the parents attended college or believe that college is affordable). To the extent that IQ leads to such encouragement, then controlling for this encouragement will understate the effects of IQ.

Finally, applying a longitudinal study to the present time increases uncertainty. To track individuals from childhood to their peak earning years requires starting at least 30 years in the past. Whether these past relationships apply to children exposed to lead in the current year is uncertain, given changes in the educational system, labor market, and other factors over time.

In sum, newer research suggests that both the Salkever estimate of a 2.4% change in earnings per one-point change in IQ and the older Schwartz estimate of 1.76% are overstated. The degree of overstatement is uncertain, however, due to the difficulties inherent in disentangling the effects represented in Figure 1. In the near term, analysts may want to test the sensitivity of their results to a range of factors, as does the USEPA in its 2008 NAAQS analysis.[25]

In the longer term, more detailed review of recent research on the relationship between IQ and earnings would be desirable, given both the complexity of the relationships and their importance in assessing the benefits of lead abatement. This review could include determining whether it is possible to use the data that underlie the reported results to estimate factors that are more directly applicable to the populations most affected by changes in lead exposures. New research may be needed that directly addresses these populations and focuses on the data needed to estimate earnings-related benefits. The research cited above suggests that the impacts of lead on non-cognitive functioning may also have important effects on education and earnings that should be further investigated. In addition, more work is needed to estimate individual WTP to avoid these risks, given that WTP is a more conceptually correct indicator of the value of related benefits.

Valuing Cardiovascular Risk Reductions

For cardiovascular risks, suitable WTP estimates that incorporate both morbidity and mortality are not currently available; thus, these two components are typically valued separately in regulatory analysis and then summed. Well-established approaches exist for valuing mortality risk reductions, based on estimates of WTP, but few studies address WTP for morbidity risks.[20] As a result, cost-of-illness estimates are generally used to value reductions in nonfatal cardiovascular risks.

The suite of lead-related cardiovascular conditions assessed varies across USEPA's benefit analyses but has at times included hypertension, stroke, and myocardial infarction or coronary heart disease more generally. The approaches for valuing these conditions have evolved over time, reflecting changes in the available data.

For mortality risk reductions, the USEPA currently follows a valuation approach based on a review conducted in the early 1990s and described in both its 2000 and 2010 *Guidelines for Preparing Economic Analyses*.[34,35] The estimates are typically expressed as the "value per statistical life" or VSL, to parallel how risk reductions are typically expressed in policy analyses.[36,37] Most regulations lead to small risk changes at the individual level, which are often presented as "statistical" cases aggregated across the affected population. For example, a 1-in-10,000 risk reduction affecting 10,000 individuals can be expressed as a statistical case (1/10,000 risk reduction × 10,000 individuals = 1 statistical case), as can a 1-in-100,000 risk reduction affecting 100,000 individuals (1/100,000 risk reduction × 100,000 individuals = 1 statistical case). For most regulations, the specific individuals who would avoid illness or whose lives would be extended by the policy cannot be identified in advance. A regulation that is expected to "save" a statistical life is one that is predicted to result in one less death in the affected population during a particular time period. "Saving" a statistical life is not the same as saving an identifiable individual from certain death.

The VSL is typically calculated by dividing individual WTP for a small risk change (in a defined time period) by the size of the risk change. For example, if an individual is willing to pay $700 for a 1-in-10,000 reduction in his or her risk of dying in the current year, the VSL is $7 million ($700 ÷ 1/10,000 = $7 million). Alternatively, individual WTP for small risk reductions could be aggregated across a population. A $7 million VSL also results if each member of a population of 10,000 is willing to pay an average of $700 for a 1-in-10,000 annual risk reduction ($700 × 10,000 = $7 million).

The USEPA has devoted considerable attention to research on the value of mortality risk reductions. At present, it recommends applying a mean VSL of 7.4 million dollars with a standard deviation of $4.7 million (in 2006 dollars), updated to the appropriate dollar year.[35] Analysts typically adjust this value to reflect changes in real income over time as well as any significant delays between changes in exposure and changes in mortality incidence (i.e., latency or cessation lag).

This approach is likely to evolve in the near future, as a result of new research and expert review. The USEPA asked its Science Advisory Board to review a 2010 White Paper it drafted on changes in its approach.[38] In 2011, the Science Advisory Board recommended that EPA 1) change the VSL terminology to avoid confusion between the value of small risk reductions and the value of saving an individual's life; 2) recognize that values vary across contexts and, to the extent possible, use values tailored to the particular risk to be regulated; 3) apply enhanced criteria to determine which valuation studies are of sufficient quality for application in policy analysis; and 4) conduct additional research on topics such as whether individuals are willing to pay differing amounts for risk reductions with differing characteristics.[39] In the near term, the extent to which these recommendations will significantly change the values used in USEPA regulatory analyses is unclear; over the longer term, an ambitious research agenda will be needed to fully meet the challenges raised by the report.

For reductions in morbidity risks, the USEPA has long relied on cost-of-illness estimates due to the lack of suitable estimates of WTP. These estimates have changed over time, but typically include the direct costs of a wide range of inpatient and outpatient services, supplies, and medications. For indirect costs, the estimates include lost market work time and, at times, lost household production.

The most recent (2008) USEPA analyses do not include cardiovascular risks in the monetized benefits estimates, because of concerns about the data available on adult blood lead levels and on the effects of lead on these risks. Previous lead analyses, such as a 2006 assessment of the USEPA's proposed renovations rule,[40] relied on relatively old cost data to value nonfatal risks, collected in the 1970s and 1980s. Given that medical treatment and labor market characteristics have changed substantially since that time, more review of research is needed to determine the appropriate values for these impacts, regardless of whether WTP or cost-of-illness methods are used.

Other Challenges and Opportunities

While there are many opportunities to improve the approaches used previously to assess the benefits of increased regulation of lead exposure, perhaps the largest challenge is expanding the types of outcomes considered. When comparing benefits to costs, policy analysts and decision makers understand that monetized benefits are likely to understate total benefits, because they exclude the value of many potential outcomes. However, without more data and analysis, it is difficult to determine whether the excluded outcomes would have a relatively small or relatively large impact on the benefit estimates, and hence whether they suggest that a substantial increase in spending on these regulations might be worthwhile.

Expanding the types of benefits that are quantified poses challenges in terms of both the risk assessment and the valuation approach. As illustrated in Box 1, estimating the benefits of lead regulations first requires data on the associated risk reductions (step 4a). The discussion of previous regulatory analyses indicates how concerns about the quality of the risk data can limit the inclusion of some benefit categories (e.g., cardiovascular risk reductions), despite the availability of a valuation method. In addition, double counting may be a concern when various epidemiological studies cover overlapping outcomes, as illustrated by the discussion of the relationship between IQ, schooling, labor force participation, and earnings, as well as the relationship to other types of neurological damage associated with lead.

The USEPA periodically reviews the scientific evidence linking lead exposures to different outcomes. The most recent draft of its *Integrated Science Assessment for Lead*,[4] which is now being reviewed, covers seven health-related categories: neurological effects, cardiovascular effects, renal effects, immune system effects, effects on heme synthesis and red blood cell function, reproductive effects and birth outcomes, and cancer. It concludes that there is a causal relationship for all of these outcomes except cancer, for which the causal relationship is described as "likely." In addition, the report covers a number of ecological effects and determines that lead affects growth, mortality, and other factors related to the health of terrestrial and aquatic systems. Careful review of these data is needed to develop estimates of the relationships between exposure reductions and risk reductions that are suitable for use in regulatory analysis.

Another challenge is determining which outcomes will be noticeably affected by a particular regulation. For example, while complete elimination or major decreases in lead exposure might avert the need for periodic testing of young children, an individual rulemaking may not have widespread or large enough effects to change the testing requirements, in which case related savings would not be expected. Regulations that address childhood lead exposures may affect their cardiovascular risks when they become adults; cardiovascular risks will be more immediately affected by a rule that addresses adult exposures. Careful comparison of baseline and postregulatory effects (steps 1 and 2 in Box 1) is needed to identify those outcomes that may be most important within the context of a particular analysis.

In addition, developing values for some outcomes included in the risk assessment may be difficult. Valuation requires information on discernable impacts, the effects of which can be weighed by individuals in determining how much they would be willing to pay to reduce related risks. For example, to value changes in red blood cell function, information on how these changes might ultimately affect an individual's activities and quality of life is needed. Similarly, valuing changes in blood pressure requires linking these changes to the need for treatment of hypertension and to the likelihood of stroke or other serious illnesses. Generally, in these cases, analysts must develop a model that estimates how these types of changes affect the likelihood of more significant health impairments as well as premature death.

Once data on the risk reductions attributable to a change in lead exposure are available, the effects can be valued using a number of different approaches. Scholars have assessed many lead-related effects that are not typically addressed in the USEPA's regulatory analyses, due in part to differences in the nature of the risk change and in part to concerns about the quality of the data.

Many of these analyses focus on elimination of lead-related hazards, rather than the smaller marginal changes likely to be associated with a federal regulation. For example, Landrigan et al.[41] assess the effects of eliminating all childhood asthma, cancer, and developmental disabilities (as well as lead poisoning) associated with environmental pollutants, using an averted cost approach that could be updated and adapted for use in other contexts. Another example is an analysis by Gould,[42] based in part on earlier work by Korfmacher, which explores the costs and benefits of eliminating the hazards that household lead paint poses to young children. She explores lead's effects on crime and attention deficit-hyperactivity disorder (ADHD) as well as other outcomes, using an averted cost approach. For example, for ADHD, Gould notes that the effects are lifelong and can include a range of types of misconduct and antisocial behavior, including drug use and crime. Considering only medical treatment and parental lost work time, she estimates that lead-associated ADHD costs $267 million annually (1996 dollars). She also estimates the effects of reducing average preschool blood lead levels on the incidence of burglaries, robberies, aggravated assaults, rapes, and murders. She concludes that the total direct cost of these crimes is approximately $1.8 billion, including victim costs, costs of legal proceedings and incarceration, and lost earnings for both the criminal and victim. Finally, she estimates that an additional $11.6 billion is lost in indirect costs, including psychological and physical damage that requires treatment and measures taken to prevent criminal actions. In addition to suggesting that the benefits associated with these large changes in exposure are significant, these and other studies indicate that data are available that can be used to estimate the value of additional benefit categories.

Conclusions

The types and magnitude of benefits from lead regulations are determined by the complex interactions of a number of factors, such as the characteristics and size of the exposed population, the magnitude of the averted exposures, the types and severities of the health effects, and the monetary values of the impacts. However, this entry indicates that the effects of lead on future earnings, and on premature mortality from cardiovascular disease, are the most significant benefit categories monetized in previous regulatory analyses. It further indicates that these analyses could be expanded to include numerous other potentially significant benefit categories.

For childhood exposures, the impact of lead on earnings (due to its effects on IQ and schooling) dominates the results. On a per-person basis, the changes in earnings are small. However, because USEPA regulations often affect a substantial number of individuals, the earnings-related benefits become large when aggregated nationally. More work is needed, however, to update and refine the estimates of the effect of a change in IQ on earnings, as well as to better understand the relationship between lost earnings and WTP for IQ changes.

For adult cardiovascular effects, premature mortality dominates the results, due in part to the fact that the value of mortality risk reductions is generally measured in millions of dollars while morbidity values are often at least an order of magnitude less. While the value of mortality risk reductions is well studied and current practices are evolving as a result of recent research and expert review, WTP estimates for many types of morbidity risks are not currently available.[20] Work currently underway (e.g., by Hammitt and Haninger[43] and Cameron and DeShazo.[44]) may eventually provide useful estimates of WTP for cardiovascular risks. More research on the lifetime costs of cardiovascular disease would also be informative.

Regulatory analyses currently do not include quantified and monetized estimates for several other benefit categories. The expansion of the epidemiological research base, represented by USEPA's recent *Integrated Science Assessment,* provides opportunities for better characterizing the relationships between lead and those outcomes traditionally included in the monetary valuation of benefits, as well as opportunities for expanding the analysis to include important additional outcomes. Some areas worthy of additional exploration include lead's non-cognitive effects on educational attainment and earnings and its effects on delinquent behavior and crime. Analysts will need to pay careful attention to the interrelationships between outcomes as the number of endpoints considered expands. Tracing these relationships is important both to avoid double counting and to ensure that the most significant benefit categories are included in the analysis to the greatest extent possible.

Acknowledgments

I would like to thank Scott D. Grosse (Centers for Disease Control and Prevention), James E. Neumann (Industrial Economics, Incorporated), and two anonymous reviewers for their helpful comments on earlier drafts of this entry.

References

1. Aspen Institute. *40 Years: EPA 40th Anniversary*; Aspen Institute: Washington, DC, 2010.
2. Robinson, L.A. *Benefits of Reduced Lead Exposure: A Review of Previous Studies.* Prepared for the U.S. Environmental Protection Agency under subcontract to Industrial Economics, Incorporated, 2007.
3. Lanphear, B.P.; Hornung, R.; Khoury, J.; Yolton, K.; Baghurst, P.; Bellinger, D.C.; Canfield, R.L.; Dietrich, K.N.; Bornschein, R.; Greene, T.; Rothenberg, S.J.; Needleman, H.L.; Schnaas, L.; Wasserman, G.; Graziano, J.; Roberts, R. Low-level environmental lead exposure and children's intellectual function: An international pooled analysis. Environ. Health Perspect **2005**, *113* (7), 894–899.

4. U.S. Environmental Protection Agency. Integrated Science Assessment for Lead. U.S. Environmental Protection Agency: Washington, DC, forthcoming. Available at http://www.epa.gov/ncea/isa/index.htm.

5. U.S. Environmental Protection Agency. *Costs and Benefits of Reducing Lead in Gasoline*; EPA 230–05–85–006. U.S. Environmental Protection Agency: Washington, DC, 1985.

6. *U.S. Office of Management and Budget. Report to Congress on the Costs and Benefits of Federal Regulations, 1997.*

7. Reagan, R. Executive Order 12291: Federal regulation. Fed. Regist. **1981**, *46* (190), 13193–13198.

8. Clinton, W.J. Executive Order 12866: Regulatory planning and review. Fed. Regist. **1993**, *58* (190), 51735–51744.

9. Obama, B. Executive Order 13563: Improving regulation and regulatory review. Fed. Regist. **2011**, *76* (14), 3821–3823.

10. U.S. Office of Management and Budget. *Circular A-4: Regulatory Analysis*; U.S. Office of Management and Budget: Washington, DC, 2003.

11. U.S. Office of Management and Budget. *Agency Checklist: Regulatory Impact Analysis*; U.S. Office of Management and Budget: Washington, DC, 2010.

12. Rice, D.P. Estimating the cost of illness. Am. J. Public Health **1967**, 57, 424–440.

13. Yabroff, K.R. et al., Eds. Health care costing: Data, methods, future directions. Med. Care **2009**, *47* (7), Sup 1.

14. Grosse, S.D.; Krueger, K.V. The income-based human capital valuation methods in public health economics used by forensic economics. J. Forensic Econ. **2011**, *22* (1), 43–57.

15. Grosse, S.D.; Krueger, K.V.; Mvundura, M. Economic productivity by age and sex: 2007 estimates for the United States. Med. Care **2009**, *47* (7), Sup. 1.

16. Zhang, W.; Bansback, N.; Anis, A.H. Measuring and valuing productivity loss due to poor health: A critical review. Soc. Sci. Med. **2011**, 72, 185–192.

17. Koopmanschap, M.A.; Rutten, F.F.H.; van Ineveld, B.M.; van Roijen, L. The friction cost method for measuring indirect costs of disease. J. Health Econ. **1995**, 4, 171–189.

18. Agee, M.D.; Crocker, T.D. Parental altruism and child lead exposure: Inferences from demand for chelation therapy. J. Hum. Resour. **1996**, *31* (3), 677–691.

19. von Stackelberg, K.; Hammitt, J.K. Use of contingent valuation to elicit willingness-to-pay for the benefits of developmental health risk reductions. Environ. Resour. Econ. **2009**, *43*, 45–61.

20. Robinson, L.A.; Hammitt, J.K. Valuing health and longevity in regulatory analysis: Current issues and challenges (Ch. 30). In *Handbook on the Politics of Regulation*; Levi-Faur, D., Ed.; Edward Elgar: Cheltenham, U.K. and Northampton, 2011.

21. Rischitelli, G.; Nygren, P.; Bougatsos, C.; Freeman, M.; Helfand, M. Screening for elevated lead levels in childhood and pregnancy: An updated summary of evidence for the U.S. Preventive Services Task Force. Pediatrics **2006**, *118*, 1867–1895.

22. U.S. Environmental Protection Agency. *Reducing Lead in Drinking Water: A Benefit Analysis;* EPA 230–09–86-019. U.S. Environmental Protection Agency: Washington, DC, 1986 (revised 1987).

23. U.S. Environmental Protection Agency. *The Benefits and Costs of the Clean Air Act: 1970–1990*; EPA 410-R-97-002. U.S. Environmental Protection Agency: Washington, DC, 1997.

24. U.S. Environmental Protection Agency. *Economic Analysis for the TSCA Lead Renovation, Repair, and Painting Program Final Rule for Target Housing and Child-Occupied Facilities;* U.S. Environmental Protection Agency: Washington, DC, 2008.

25. U.S. Environmental Protection Agency. *Regulatory Impact Analysis of the Proposed Revisions to the National Ambient Air Quality Standards for Lead;* U.S. Environmental Protection Agency: Washington, DC, 2008.

26. Schwartz, J. Societal benefits of reducing lead exposure. Environ. Res. **1994**, *66*, 105–124.

27. Salkever, D.S. Updated estimates of earnings benefits from reduced exposure of children to environmental lead. Environ. Res. **1995**, *70*, 1–6.

28. Grosse, S.D.; Matte, T.D.; Schwartz, J.; Jackson, R.J. Economic gains resulting from reduction in children's exposure to lead in the United States. Environ. Health Perspect. 2002, *110* (6), 563–569.

29. Grosse, S.D. How much does IQ raise earnings? Implications for regulatory impact analyses. AERE Newsl. **2007**, *27* (2), 17–21.

30. Green, D.A.; Riddell, W.C. Literacy skills, non-cognitive skills, and earnings: An economist's perspective. In *Towards Evidence-Based Policy for Canadian Education*; de Broucker, P., Sweetman, A., Eds.; McGill-Queens University Press: Montreal and Kingston, 2002.

31. Neal, D.A.; Johnson, W.R. The role of premarket factors in black-white wage differences. J. Political Econ. **1996**, *104*, 869–895.

32. Heckman, J.J.; Stixrud, J.; Urzua, S. The effects of cognitive and noncognitive abilities on labor market outcomes and social behavior. J. Labor Econ. **2006**, *24*, 411–482.

33. Zax, J.S.; Rees, D.I. IQ, academic performance, environment, and earnings. Rev. Econ. Stat **2002**, *84*, 600–616.

34. U.S. Environmental Protection Agency. *Guidelines for Preparing Economic Analysis*; EPA 240-R-00–003. U.S. Environmental Protection Agency: Washington, DC, 2000.

35. U.S. Environmental Protection Agency (EPA). *Guidelines for Preparing Economic Analysis*; EPA 240-R-10–001.U.S. Environmental Protection Agency: Washington, DC, 2010.

36. Hammitt, J.K. Valuing mortality risk: Theory and practice. Environ. Sci. Technol. **2000**, *34*, 1396–1400.

37. Robinson, L.A. How US government agencies value mortality risk reductions. Rev. Environ. Econ. Policy. **2007**, *1* (2), 283–299.

38. U.S. Environmental Protection Agency. *Valuing Mortality Risk Reductions for Environmental Policy: A White Paper (Review Draft)*. Prepared by the National Center for Environmental Economics for consultation with the Science Advisory Board-Environmental Economics Advisory Committee, U.S. Environmental Protection Agency: Washington, DC, 2010.

39. Kling, C. et al. *Review of Valuing Mortality Risk Reductions for Environmental Policy: A White Paper (December 10, 2010)*; EPA-SAB-11-011. U.S. Environmental Protection Agency: Washington, DC, 2011.

40. U.S. Environmental Protection Agency. *Economic Analysis for the Renovation, Repair, and Painting Program Proposed Rule*; U.S. Environmental Protection Agency: Washington, DC, 2006.

41. Landrigan, P.J.; Schechter, C.B.; Lipton, J.M.; Fahs, M.C.; Schwartz, J. Environmental pollutants and disease in American children: Estimates of morbidity, mortality, and costs for lead poisoning, asthma, cancer, and developmental disabilities. Environ. Health Perspect. **2002**, *110* (7), 721–728.

42. Gould, E. Childhood lead poisoning: Conservative estimates of the social and economic benefits of lead hazard control. Environ. Health Perspect. **2009**, *117* (17), 1162–1167.

43. Hammitt, J.K.; Haninger, K. Valuing morbidity risk: Willingness to pay per quality-adjusted life year. Toulouse School of Economics, LERNA Working Paper Series, 2011.

44. Cameron, T.A.; DeShazo, J.R. Demand for health risk reductions. Unpublished manuscript, 2010.

27

Mercury

Sven Erik Jørgensen

Introduction

Mercury is an extremely toxic element. It has no role as biological element beyond its toxicity and is not essential for any organism. It is element number 80 in the periodic table and has an atomic weight of 200.59. Mercury caused one of the most significant environmental catastrophes in history. Mercury was discharged with the wastewater from a chemical factory in Minamata Bay, Japan, in the 1950s.[1] Mercury accumulated in the sediment, where it could react microbiologically with the organic matter and form methyl mercury ions and dimethyl mercury, which has a low boiling point and can be transferred from the sediment to the water phase or even to the atmosphere. Organic mercury compounds can be taken up by fish, where concentrations 3000 times higher than in water can be recorded. These processes explain why the fish caught in the Minamata Bay had a very high mercury concentration. As a result, hundreds of people died, and over a thousand became invalids due to mercury contamination. In the 1950s, the victims were considered to have a new, previously unknown disease, the Minamata disease. Later in the 1960s, it was detected that the disease was caused by high mercury concentration in fishermen's families. The accumulation in the sediment, the biomagnifications through the food chain (water–phytoplankton–zooplankton–fish–fisherman), and the ability of mercury to form organic compounds by microbiological reactions explain the emergence of Minamata disease and mercury poisoning.

Sources of Mercury Pollution

The use of mercury has declined during the last decades due to its extreme toxicity. Mercury compounds were previously applied as fungicides, as dyestuff (cinnabar red), and for the production of chlorine, but these applications are now banned in all industrialized countries. Chlorine and sodium hydroxide are unfortunately still produced in some developing countries by a method based on a dripping mercury electrode that causes mercury to be discharged with wastewater to the environment. Mercury still is applied by dentists and is used for some electrical instruments. Today's major mercury contamination is, however, caused by coal-fired power plants due to a small concentration of mercury in all coal. Each year, 3300 tons of mercury is discharged as air pollution to the atmosphere from coal-fired power plants and the incineration of solid waste.[2] The natural emission of mercury is about 1500 tons/yr, mainly from volcanic activity.

Pollution Effects

The most important acute effects of mercury are on the lungs and the central nervous system. The chronic effect is complete damage of the central nervous system, which was the main symptom of the Minamata fishermen. In addition to central nervous system symptoms, the victims also lose weight and appetite. The LD50 value for mice is 5 mg/kg, and the World Health Organization recommends a maximum concentration of 1 ppm in food items. (Mercury has further teratogenic and genetic effects. For details about all effects, see Jørgensen et al.[3–5])

Important Mercury Processes

Evaluation of mercury pollution requires that environmental management consider the environmental processes in which mercury can participate. It is clear from Section 1 that the formation of organic mercury compounds mainly in the form of methyl and dimethyl mercury is extremely important, and these processes should therefore be included in the environmental considerations, included when an ecotoxicological model for the distribution and effect of mercury is developed.

Figure 1 shows a conceptual diagram of mercury distribution and effects in Mex Bay, close to Alexandria. The details of the model can be found in Jørgensen and Bendoricchio.[6] From the figures, it is clear that the methylation processes are included and that the uptake from water and biomagnifications through the food chain are included to account for the most important state variable of the model: the mercury concentration in the top carnivorous fish (tuna fish), which is an important consumer fish in the Mediterranean region. Figure 2 shows the input/output processes that are included in the model.

The uptake of mercury by fish is highly dependent on the pH, due to the pH dependence of mercury solubility. This dependence is of course of interest mainly when the mercury pollution is inorganic. Figure 3 shows the pH dependence for the mercury in fish when exposed to 1.5 ppm inorganic mercury.[7]

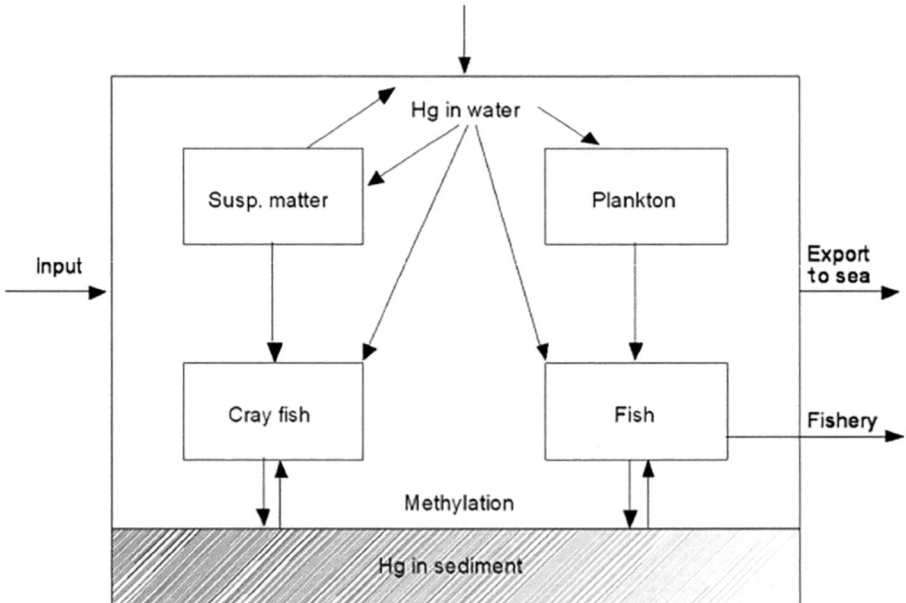

FIGURE 1 Conceptual diagram of the mercury model for Mex Bay, Egypt. Notice that methylation is considered for the release of mercury from the sediment. Notice also that the fish mercury is determined by uptake directly from the water (including the concentration of organic mercury) and through the food chain—here indicated only as water–phytoplankton–fish, but one to two additional steps could be considered.

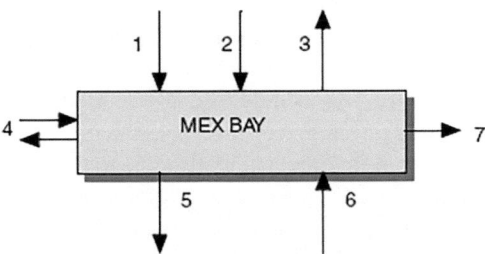

FIGURE 2 A model of the distribution of mercury in an aquatic ecosystem is a bio-geo-chemical model that must include all inputs and outputs. The numbers indicate the following: 1) discharge from waste and tributary to the bay; 2) deposition of mercury from the atmosphere (for instance, from coal-fired power plants); 3) evaporation of mercury; 4) input and output from the open sea; 5) sedimentation; 6) release from the sediment; and 7) fishery.

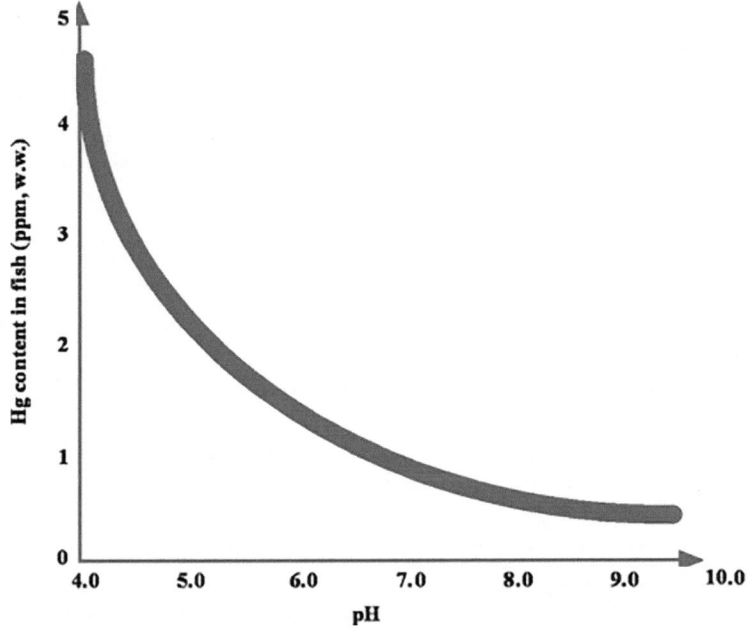

FIGURE 3 The concentration of mercury in fish expressed as ppm wet weight versus pH. The concentrations are measured at steady state and for a concentration of mercury chloride in the water of 1.5 ppm (mg/L). The values are based on many measurements for a wide spectrum of different fish species, and the concentrations in the fish are therefore indicated as a range.

The uptake of mercury from food is very different for organic mercury and inorganic mercury, namely, 90% and 20%, respectively.[7] This difference entails that a mercury model should follow and include the processes for both inorganic and organic mercury to be able to account for the final concentration, for instance, in fish as a result of a mercury contamination, whether the discharge is inorganic or organic mercury. The transfer between the two forms should of course be included, too.

Abatement of Mercury Pollution

The mercury pollution caused by coal-fired power plants is reduced for the part of the mercury adsorbed to particulate matter by the methods used for treatment of air pollutants to the extent that they are removing particulate matter. However, oil and natural gases are depleted within 60–80 years, which implies that a coal-fired power plant will still emit mercury if the power plant is using coal as fossil fuel. A significant part of mercury emission is not adsorbed to the particulate matter. The question is if we, due to emission of carbon dioxide, can continue to use coal-fired power plants after years 2050–2080. Today's emission of mercury is, as indicated above, 3300 tons/yr, and it is therefore a question of whether we can use coal-fired power plants after oil and natural gas have been depleted, not only due to the continuous emission of carbon dioxide but also due to the continuous emission of mercury, which is a consequence of the use of coal. This problem has not yet been examined in sufficient detail to give a clear answer—so this question is still open and requires further attention. It will probably be necessary, due to the greenhouse effects of carbon dioxide, to shift partially or completely to alternative energy sources before year 2080.

The mercury pollution causing wastewater problems can be solved by ion exchange—and ion exchangers with a particularly high efficiency of mercury removal are available. Furthermore, adsorption on activated carbon (see *Wastewater Treatment: Conventional Methods,* p. 2677) removes mercury, even organic mercury compounds, rather effectively. In this context, it should be mentioned that bioremediation methods are also able to reduce mercury concentration in contaminated soil (see *Bioremediation,* p. 408).

In conclusion, mercury pollution problems can be solved by today's technology, except for the emission of mercury from coal-fired power plants, which requires particular attention if the use of coal as an energy source continues to grow, unless the problem is solved by shifts to alternative energy. It is furthermore expected that general restrictions in the use of mercury compounds will be tightened in the years to come.

References

1. Newman, M.C.; Unger, M.A. *Fundamentals of Ecotoxicology,* 2nd Ed.; CRC Press: Boca Raton, London and New York, 2003.
2. *NESCAUM. Mercury Emissions from Coal-Fired Power Plants, The Case for Regulatory Action; 2003.*
3. Jørgensen, S.E.; Halling-Sprensen, B.; Mahler, H. *Handbook of Estimation Methods in Ecotoxicology and Environmental Chemistry*; Taylor and Francis Publ: Boca Raton, Boston, London, New York, Washington, D.C.; 1998. 230 pp.
4. Jørgensen, S.E.; Jørgensen L.A.; Nors Nielsen, S. *Handbook of Ecological and Ecotoxicological Parameters.* Elsevier: Amsterdam, 1991; 1380 pp.
5. Jørgensen, L.A.; Jørgensen, S.E.; Nors Nielsen, S. *Ecotox,* CD; Elsevier: Amsterdam, 2000; 4000 pp.
6. Jørgensen, S.E.; Bendoricchio, G. *Fundamentals of Ecological Modelling,* 3rd Ed.; Elsevier: Amsterdam; 530 pp.
7. Jørgensen, S.E. *Principles of Pollution Abatement*; Elsevier: Amsterdam, 2000; 520 pp.

Mycotoxins

J. David Miller

Introduction

The agriculturally important mycotoxins represent a major challenge for food and feed safety. They are unavoidable and no technology has been developed to make them disappear from our major food crops and none are on the horizon. This briefly reviews the major toxins and provides an insight into their major effects on human and animal health in fully developed market economies, the impact of fungicides and, in one case, pesticides on their reduction, and the impact on occupational health. This is one of the most intensively studied of any area of science because of the economic impact on western economies.[1,2] In addition, there is a massive toll in human suffering in parts of Africa, Latin America and East Asia.[3,4] At the time of writing, a PubMed search of the term "mycotoxins" resulted in 47,100 entries, which underscores the importance of the issue.

Five Important Mycotoxins

Although there are hundreds of fungal metabolites that are toxic in experimental systems, there are only five that are of major agricultural importance: deoxynivalenol, aflatoxin, fumonisin, zearalenone, and ochratoxin.[5,6] A number of other mycotoxins that can be important in some areas & in some years including ergot alkaloids, T-2/HT-2 toxins and some *Penicillium* toxins that can occur in animal feed.[7,8]

Deoxynivalenol occurs when wheat, barley, maize, and sometimes oats and rye are infected by *Fusarium graminearum, F. asiaticum* or *F. culmorum*. These species are the primary cause of Fusarium head blight in small grains. This disease is a major agricultural problem worldwide, as well as a similar disease in maize called Gibberella ear rot. Disease incidence is most affected by moisture at flowering, and many cultivars and hybrids used lack genetic tolerance to the disease.[9–11] *F. graminearum* is common in wheat from North America and China, where *F. asiaticum* is also important. Thirty years ago, *F. culmorum* had been the dominant species in cooler wheat-growing areas, such as Finland, France, Poland, and the Netherlands, but this trend has changed in recent years as European summers have reached record warm temperatures such that *F. graminearum* largely dominates.[9,12,13] Fusarium head blight is caused by *F. asiaticum* in warmer regions.[14]

Within these taxa, there are a number of chemotypes. Most strains fall into one of two chemoyptes, one producing deoxynivalenol by a pathway involving 3-acetyl deoxynivalenol and the other by

15-acetyl deoxynivalenol. Some strains of *F. graminearum* and *F. asiaticum* produce nivalenol.[15] Some strains of *F. graminearum* produce a trichothecene reported in 2015, NX.[16] Finally, some strains of *F. graminearum* produce both deoxynivalenol and NX.[17]

Humans are not less and are probably more sensitive to deoxynivalenol than swine.[7] Swine are the most sensitive domestic animal species to the effects of deoxynivalenol. Several mechanisms account for the toxicity of deoxynivalenol including affecting different aspects of the appetite suppression, modulation of the insulin-like growth factor acid-labile subunit and alters brain chemistry.[15,18] In addition, deoxynivalenol causes changes in immune system function in male mice, occurring at dietary concentrations often encountered by humans. As with other trichothecenes, high exposures increase susceptibility to facultative pathogens such as *Listeria*.[19] Cattle, cows, and poultry species are tolerant to deoxynivalenol and milk production is not affected at typical field concentrations.[20] The Provisional Maximum Tolerable Daily Intake (PMTDI) of the Joint Expert Committee on Food Additives and Contaminants (JECFA) of the World Health Organization (WHO) and Food and Agriculture Organization (FAO) for this toxin was revised in 2010 by making a group PMTDI for deoxynivalenol and the two acetylated precursors. These can co-occur with the parent compound in grains.[21]

Crops that are contaminated by deoxynivalenol can often contain zearalenone albeit at a lower frequency. Zearalenone is more common in maize than in small grains. Zearalenone is an estrogen analogue and causes hyperestrocism in female pigs at low levels; the dietary no-effect level is less than 1 mg/kg. Cows and sheep are also sensitive to the estrogenic effects of this toxin with depressed ovulation and lower lambing percentages.[20] Non-human primates and possibly humans are also sensitive to the estrogenic effects of zearalenone.[22,23]

Aflatoxin is mainly produced by *Aspergillus flavus,* which is a problem in many commodities, but most human exposure comes from contaminated maize, groundnuts, and rice. *Aspergillus parasiticus* is uncommon outside North and South America, and they are more associated with peanuts. There are also some less common species that produce aflatoxin. *A. flavus* contamination of maize or peanuts occurs in two basic ways. Either airborne or insect-transmitted conidia contaminate the silks and grow into the ear when the maize is under high-temperature stress or (more commonly) insect- or bird-damaged kernels become colonized with the fungus and accumulate aflatoxin. In maize and peanut plants, drought-, nutrient-, or temperature-stressed plants are more susceptible to colonization by *A. flavus* or *A. parasiticus*.[5,24]

In poultry, aflatoxin exposure results in liver damage, impaired productivity and reproductive efficiency, decreased egg production in hens, inferior egg-shell quality, inferior carcass quality, and increased susceptibility to disease. The effects of acute and chronic exposure in swine are largely attributable to liver damage. In cattle, the primary symptoms are reduced weight gain, liver and kidney damage, and reduced milk production.[5] Aflatoxin is also immunotoxic in domestic and laboratory animals with oral exposures in the microgram-per-gram range. Cell-mediated immunity (lymphocytes, phagocytes, mast cells, and basophils) is more affected than humoral immunity (antibodies and complement).[19] Naturally-occurring mixtures of aflatoxins including aflatoxin B1 are human liver carcinogens. In much of Africa as well as parts of Latin America and East Asia, aflatoxin exposure can be multiples, sometimes orders of magnitude above tolerable levels. Aside from increased rates of liver cancer, aflatoxin is associated with child stunting.[3,24-26]

Fumonisins are produced by *Fusarium verticillioides, Fusarium proliferatum,* and a number of less common fusaria. Fumonisins are a very common contaminant of maize-based food and feed in the United States, China, Europe, southern Africa, South America, and Southeast Asia.[27,28] *F. verticillioides* and *F. proliferatum* can be recovered from virtually all maize kernels including those that are healthy, which suggests that it may be an endophyte, i.e., a mutualistic relationship.[5,28,29] *F. verticillioides* and *F. proliferatum* cause a "disease" called Fusarium kernel rot. In parts of the United States and lowland tropics, this is one of the most important ear diseases and is associated with warm, dry years and insect damage. Maize plant disease stress also promotes the growth of *F. verticillioides* and fumonisin formation.[28]

Fumonisin causes equine leukoencephalomalacia (ELEM), which involves a liquefactive necrosis of the cerebral hemisphere. Clinical manifestations include abnormal movements, aimless circling, lameness, etc., followed by death. In swine, high exposures of fumonisin results in porcine pulmonary edema (PPE) caused by fumonisin-induced heart failure.[20] At lower exposures, both liver damage and kidney damage have been reported in swine. Fumonisin causes feed refusal and changes in carcass quality at dietary concentrations in the low milligram-per-kilogram range.[20,30] Fumonisin exposure results in impaired growth rates in a variety of domestic and laboratory animals.[31] Fumonisin exposure results in neural tube birth (NTD) defects in rodents.[32,33] The unifying mechanism for these toxicities is that fumonisin interferes with sphingolipid metabolism unleashing a cascade of harmful cellular effects.[34]

Exposure to *F. verticillioides*-contaminated maize has been linked to the elevated rates of esophageal cancer in the Transkei and this has since been linked to fumonisin exposure.[30] Fumonisin caused tumors in mice and rats in the U.S. National Toxicology Program study and is an IARC class 2B carcinogen (possible human carcinogen).[27] People exposed to dietary fumonisin above the Tolerable Daily Intake demonstrate fumonisin-induced sphigolipid alteration.[32,33] As with aflatoxin, exposure to fumonisin in Africa has been associated with child stunting.[27,31]

Human exposure to ochratoxin A in temperate zones mainly comes from eating foods made from contaminated small grains.[35] Ochratoxin production in small grains results from the growth of *Penicillium verucosum* typically in small pockets of grain that get wet or go in the bin wet.[36] A small number of highly contaminated kernels in a portion of cereal (~1/5000) can put the lot over the regulatory limit in the EU.[37] Ochratoxin exposure can arise from *Aspergillus carbonarius* and some strains of *A. niger*, and several related species also produce ochratoxin on grapes and coffee.[38] In Europe, minor exposures occur from meat, especially pork, from animals fed contaminated grain and from the consumption of moldy sausage, etc. resulting from the growth of *P. nordicum*.[39]

Ochratoxin is a potent nephrotoxin in swine and causes kidney cancer in male Fisher 344 rats. Pigs appear to be the most sensitive species with respect to the nephrotoxic effects.[20,40] Low exposures result in kidney damage in swine but, typically there are no overt signs. At higher concentrations (>2 μg/g), decreased weight gains occur. Poultry are affected, showing reduced growth rate and egg production at low ochratoxin concentrations >2 μg/g. Cattle are resistant to ochratoxin concentrations found in naturally contaminated grain.[20]

At the expense of competing hypotheses, ochratoxin was suspected as the cause of Balkan Endemic Nephropathy.[6] This a syndrome manifesting as late life urinary tract cancers and kidney damage in areas where ochratoxin exposure was believed to be high in parts of Eastern Europe.[5,6,40] In recent years, a compelling case for human disease seen has been made for exposure to the human carcinogen aristolochic acid. Dietary exposure occurs from *Aristolochia* seeds contaminating harvested wheat and from crops that take up aristolochic acids from the soil[41–43] As noted by JECFA,[40] evidence for the carcinogenicity of ochratoxin remains confined to laboratory strains of mice and rats. Mechanisms that explain the rodent carcinogenicity have been proposed unrelated to the production of ochratoxin-DNA adducts.[40,44]

Mycotoxins and Pesticides

Insect damage is known to promote accumulation of deoxynivalenol, fumonisin, aflatoxin, and probably zearalenone. Steps taken to reduce insect herbivory or drought stress on crops have the general effect of reducing mycotoxin contamination. Bt corn has lowered fumonisin content compared to similar non-Bt hybrids.[45–47] The situation with respect to lower aflatoxin and Bt is not clear[48] Potential reductions in deoxynivalenol in Bt hybrids depends on the degree of insect damage.[46,49]

Modeling of mycotoxin accumulations in small grains based on weather at anthesis is a part of the risk management strategy for Fusarium head blight. If the weather is conducive to disease development, farmers are able to make informed decisions about fungicide application[11,50] This has the effect of restricting the prophylactic use of fungicides when not warranted. This is an important area of research for food protection and food security.[51]

Occupational Exposures

Grain dusts represent an occupational hazard for a number of reasons including the possibility of the allergic disease hypersensitivity pneumonitis and organic dust toxic syndrome, a poorly understood disease. Inhalation of mycotoxins contained in airborne dusts is also a potential health risk. Inhalation of mycotoxins in spores and dust affects macrophage function and other aspects of lung biology. Inhalation of mycotoxins is a more potent route of exposure for systemic toxicity for aflatoxin, trichothecenes, and ochratoxin. Workplace exposure to airborne aflatoxin results in increased relative risk of liver cancer. However, the practical concern is the impact that fungal glucan and the aflatoxin has on the function of lung macrophages.[52-54]

Risk characterizations of inhalation exposure to grains containing high concentrations of deoxynivalenol exposures indicate that aside from the allergic and inflammatory diseases resulting from exposure to grains, the presence of dexoynivalenol is a concern.[54] Urinary biomarkers for deoxynivalenol (DON, DOM-1) were higher in active farmers in France, particularly from larger farms, than from retired farmers where exposure was from the diet.[55] Workplace inhalation exposure-outcome data for other toxins including zearalenone, fumonisin and ochratoxin are limited.[54] In modern grain handling facilities, dust control is a priority not only for workplace safety but also to prevent grain dust explosions. Thus, inhalation exposure to mycotoxins is likely now a minor problem but remains a potential concern in informal grain handing systems.

Conclusion

As noted, mycotoxin contamination of field crops is a problem that cannot be "solved" in the normal use of that term, results in important losses of human life, and has major economic and social consequences. Recognition that fungi produced toxins in crops began with ergot in Europe during the Middle Ages and has been recognized as low-molecular-weight chemicals since the 1880s and solving their structures and understanding appropriate management strategies took agonizingly long periods of study.[5,6] The current strategies for managing mycotoxins in the diet require modeling, crop breeding, better detection systems, and awareness.

References

1. Mitchell, N. J., Bowers, E., Hurburgh, C., Wu, F. Potential economic losses to the US corn industry from aflatoxin contamination. Food Add. Contam. Part A **2016**, *33*, 540–550.
2. Wilson, W., Dahl, B., Nganje, W. Economic costs of fusarium head blight, scab and deoxynivalenol. World Mycotoxin J **2018**, *11*, 291–302.
3. Wild, C.P.; Gong, Y.Y. Mycotoxins and human disease: A largely ignored global health issue. Carcinogenesis **2010**, *31*, 71–82.
4. Logrieco, A., Miller, J.D., Eskola, M., Krska, R., Ayalew, A., Bandyopadhyay, R., & Li, P. The mycotoxin charter: increasing awareness of, and concerted action for, minimizing mycotoxin exposure worldwide. Toxins **2018**, *10*, 149.
5. Miller, J.D. Fungi and mycotoxins in grain: Implications for stored product research. J. Stored Prod. Res. **1995**, *31*, 1–6.
6. Pitt, J. I., Miller, J. D. A concise history of mycotoxin research. J. Agric. Food Chem. **2016**, *65*, 7021–7033.
7. Miller, J.D. Mycotoxins in small grains and maize: Old problems, new challenges. Food Addit. Contam. **2008**, *25*, 219–230.
8. Miller, J.D. Changing patterns of fungal toxins in crops: challenges for analysts. J AOAC Int. **2016**, *99*, 837–841.

9. Miller, J.D. Epidemiology of *Fusarium* ear diseases. In *Mycotoxins in Grain: Compounds Other Than Aflatoxin*; Miller, J.D., Trenholm, H.L., Eds.; Eagan Press: St. Paul, MN, 1994; 19–36.

10. Boutigny, A.L. Richard-Forget, F. Barreau, C. Natural mechanisms for cereal resistance to the accumulation of *Fusarium* trichothecenes. Eur. J. Plant Pathol. **2008**, *121,* 411–423.

11. Bianchini, A., Horsley, R., Jack, M. M., Kobielush, B., Ryu, D., Tittlemier, S., Miller, J. D. DON occurrence in grains: A North American perspective. Cereal Foods World **2015**, *60,* 32–56.

12. Pasquali, M., Beyer, M., Logrieco, A., Audenaert, K., Balmas, V., Basler, R., González-Jaén, M. T. A European database of *Fusarium graminearum* and *F. culmorum* trichothecene genotypes. Frontiers in Microbiology **2016**, *7,* 406.

13. Xu, X.M. Nicholson, P. Ecology of fungal pathogens causing wheat head blight. Annu. Rev. Phytopathol. **2009**, *47,* 83–103.

14. Backhouse, D. Global distribution of *Fusarium graminearum, F. asiaticum* and *F. boothii* from wheat in relation to climate. European J. Plant Pathol. **2014**, *139,* 161–173.

15. Miller, J. D. Mycotoxins in Food and Feed: A Challenge for the Twenty-First Century. In *Biology of Microfungi*; Li, D., Ed.; Springer, Basel, 2016; 469–493.

16. Aitken, A., Miller, J. D., McMullin, D. R. Isolation, chemical characterization and hydrolysis of the trichothecene 7α-hydroxy, 15-deacetylcalonectrin (3ANX) from *Fusarium graminearum* DAOMC 242077. Tetrahedron Letters **2019**, *60,* 852–856.

17. Crippin, T., Renaud JB, Sumarah MW, Miller JD (2019) Comparing genotype and chemotype of *Fusarium graminearum* from cereals in Ontario, Canada. PLoS ONE **2019** *14(5),* e0216735

18. Pestka, J. J. Deoxynivalenol: mechanisms of action, human exposure, and toxicological relevance. Archives of Toxicology **2010**, *84,* 663–679.

19. Bondy, G.S. Pestka, J.J. Immunomodulation by fungal toxins. J. Toxicol. Environ. Health, Part B **2000**, *3,* 109–143.

20. Prelusky, D.B. Rotter, B.A.; Rotter, R.G. Toxicology of mycotoxins. In *Mycotoxins in Grain: Compounds Other Than Aflatoxin*; Miller, J.D., Trenholm, H.L., Eds.; Eagan Press: St. Paul, MN, 1994; 359–404.

21. Bulder, A. S., DiNovi, M., Kpodo, K. A., Leblanc, J. C., Resnik, S., Shephard, G. S., Wolterink, G. Deoxynivalenol (addendum). WHO Food Additives Series: 63 FAO JECFA Monographs **2011**, *8,* 317.

22. Kuiper-Goodman, T., Scott, P., & Watanabe, H. Risk assessment of the mycotoxin zearalenone. Reg. Tox. Pharm. **1987**, *7,* 253–306.

23. EFSA Panel on Contaminants in the Food Chain. Scientific Opinion on the risks for public health related to the presence of zearalenone in food. EFSA Journal **2011**, *9,* 2197.

24. Doerge DR, Shephard GS, Adegoke GO, Benford D, Bhatnagar D, Bolger M, Boon PE, Cressey P, Edwards S, Hambridge, Miller JD, Mitchell RT, Riley RT, Wheeler MW Aflatoxins. Safety evaluation of certain contaminants in food: prepared by the eighty-third meeting of the Joint FAO/WHO Expert Committee on Food Additives (JECFA). WHO Food Additives Series, **2018**, *74,* 3–279.

25. Wild, C., Miller J.D., Groopman, J.D. Mycotoxin control in low and middle income countries. IARC Working Group Report #9, International Agency for Research on Cancer, Lyon, France. ISBN 978–92–832–2510–2 **2015**

26. McMillan, A., Renaud, J.D., Burgess, K.M.N., Orimadegun, A.E., Akinyinka, O.O., Allen, S.H., Miller, J.D., Reid, G., Sumarah, M.W. Aflatoxin exposure in Nigerian children with severe acute malnutrition. Food Chem. Tox. **2018**, *111,* 356–362.

27. Riley, R.T., Edwards, S.G., Aidoo, K., Alexander, J., Bolger, M., Boon, P.E., Cressey, P.E., Doerge, D.R., Edler, L., Miller, J.D., Zhang, Y. Fumonisins. Safety evaluation of certain contaminants in food: prepared by the eighty-third meeting of the Joint FAO/WHO Expert Committee on Food Additives (JECFA). WHO Food Additives Series, **2018**, *74,* 415–571.

28. Miller, J.D. Factors that affect the occurrence of fumonisin. Environ. Health Perspect. **2000**, *109 (s.2),* 321–324.

29. Yates, I.E., Sparks, D. *Fusarium verticillioides* dissemination among maize ears of field-grown plants. Crop Prot. **2008**, *27*, 606–613.

30. Marasas, W.F.O.; Miller, J.D.; Riley, R.T.; Visconti, A. Fumonisins B1. Environmental Health Criteria 219; International Program for Chemical Safety, World Health Organization: Geneva, 2000.

31. Chen, C., Riley, R. T., Wu, F. Dietary fumonisin and growth impairment in children and animals: A review. Comp. Rev. Food Science Food Safety **2018**, *17*, 1448–1464

32. Riley, R.T., Torres, O., Matute, J., Gregory, S.G., Ashley-Koch, A.E., Showker, J.L., Gelineau-van Waes, J.B. Evidence for fumonisin inhibition of ceramide synthase in humans consuming maize-based foods and living in high exposure communities in Guatemala. Mol. Nutr. Food Res. **2015**, *59*, 2209–2224.

33. Voss, K.A., Riley, R.T., Gardner, N.M., Gelineau-van Waes, J. Fumonisins. In: *Reproductive Developmental Toxicology*, 2nd edition, Gupta, R.C., Ed.; Academic Press, New York; 2017, 925–943.

34. Voss, K.A., & Riley, R.T. Fumonisin toxicity and mechanism of action: overview and current perspectives. Food Safety **2013**, *1*(1), 2013006-2013006.

35. Lee, H. J., Ryu, D. Worldwide occurrence of mycotoxins in cereals and cereal-derived food products: public health perspectives of their co-occurrence. J Agric Food Chem. **2017**, *65*, 7034–7051.

36. Limay-Rios, V., Miller, J.D., Schaafsma, A.W. Occurrence of *Penicillium verrucosum*, ochratoxin A ochratoxin B and citrinin in on-farm stored winter wheat from the Canadian Great Lakes Region. PLoS One **2017**, *12*, e0181239.

37. Whitaker, T. B., Slate, A. B., Nowicki, T. W., Giesbrecht, F. G. Variability and distribution among sample test results when sampling unprocessed oat lots for ochratoxin A. World Mycotoxin J. **2015** *8*, 511–524.

38. Cabañes, F. J., Bragulat, M. R. Black aspergilli and ochratoxin A-producing species in foods. Current Opinion Food Science **2018**, *23*, 1–10.

39. Magistà, D., Susca, A., Ferrara, M., Logrieco, A. F., & Perrone, G. Penicillium species: crossroad between quality and safety of cured meat production. Current Opinion Food Sci. **2017** *17*, 36–40.

40. Benford, D., Boyle, C., Dekant, W., Fuchs, R., Gaylor, D. W., Hard, G., Solfrizzo, M. Ochratoxin A. In: Safety evaluations of certain mycotoxins in food. WHO Food Additives Series, WHO, Geneva; **2001**, *47*, 281–387

41. Bui-Klimke, T., Wu, F. Evaluating weight of evidence in the mystery of Balkan endemic nephropathy. Risk Analysis **2014**, *34*, 1688–1705.

42. Stiborová, M., Arlt, V. M., Schmeiser, H. H. Balkan endemic nephropathy: an update on its aetiology. Archives Toxicology **2016**, *90*, 2595–2615.

43. Li, W., Chan, C. K., Liu, Y., Yao, J., Mitić, B., Kostić, E. N., Dedon, P. C. Aristolochic acids as persistent soil pollutants: Determination of risk for human exposure and nephropathy from plant uptake. J. Agric. Food Chem. **2018**, *66*, 11468–11476.

44. Mally, A. Dekant, W. Mycotoxins and the kidney: Modes of action for renal tumor formation by ochratoxin A in rodents. Mol. Nutr. Food Res. **2009**, *53*, 467–478.

45. De La Campa, R., Hooker, D.C., Miller, J.D., Schaafsma, W.A., Hammond, B.G Modelling effects of environment, insect damage and BT genotypes on fumonisin accumulation in maize in Argentina and the Philippines. Mycopathologia **2005**, *159*, 539–552.

46. Folcher, L., Delos, M., Marengue, E., Jarry, M., Weissenberger, A., Eychenne, N., Regnault-Roger, C. Lower mycotoxin levels in Bt maize grain. Agronomy Sust. Dev. **2010**, *30*, 711–719.

47. Bowers, E., Hellmich, R., Munkvold, G. Comparison of fumonisin contamination using HPLC and ELISA methods in Bt and near-isogenic maize hybrids infested with European corn borer or Western bean cutworm. J. Agric. Food Chem. **2010**, *62*, 6463–6472.

48. Abbas, H. K., Zablotowicz, R. M., Weaver, M. A., Shier, W. T., Bruns, H. A., Bellaloui, N., Abel, C. A. Implications of Bt traits on mycotoxin contamination in maize: overview and recent experimental results in Southern United States. J. Agric. Food Chem. **2013**, *61*, 11759–11770.

49. Schaafsma, A. W., Hooker, D. C., Baute, T. S., Illincic-Tamburic, L. Effect of Bt-corn hybrids on deoxynivalenol content in grain at harvest. Plant Disease **2002**, *86*, 1123–1126.

50. Schaafsma, A.W., Hooker, D.C. Climatic models to predict occurrence of *Fusarium* toxins in wheat and maize. Int. J. Food Microbiol. **2007**, *119*, 116–125.

51. Van der Fels-Klerx, H. J., C. Liu, P. Battilani, P. Modelling climate change impacts on mycotoxin contamination. World Mycotoxin J. **2016**, *9*, 717–726.

52. Miller, J.D. Mycotoxins. In: *Handbook of Organic Dusts*; Rylander, R., Pettersen, Y., Eds.; CRC Press: Boca Raton, FL, 1994; 87–92.

53. Liao, C-M.; Chen, S-C. A probabilistic modeling approach to assess human inhalation exposure risks to airborne aflatoxin B1 (AFB1). Atmos. Environ. **2005**, *39*, 6481–6490.

54. Pitt, J.I., Wild, C.P., Baan, R.A., Gelderblom, W.C.A., Miller, J.D., Riley, R.T., Wu, F. Improving public health through mycotoxin control. International Agency for Research on Cancer Scientific Publications Series, No. 158. Lyon, France; 2012.

55. Turner, P.C., Hopton, R.P., Lecluse, Y., White, K.L., Fisher, J., Lebailly, P. Determinants of urinary deoxynivalenol and de-epoxy deoxynivalenol in male farmers from Normandy, France. J. Agric. Food Chem. **2010**, *58*, 5206–5212.

29

Nitrogen

Oswald Van
Cleemput and
Pascal Boeckx

Introduction

Nitrogen (N) is essential to all life. It is the nutrient that most often limits biological activity. In agricultural and natural ecosystems, N occurs in many forms covering a range of valence states from −3 to +5. The change from one valence state to another depends primarily on environmental conditions. The transformations and flow from one form to another constitute the basics of the soil N cycle (Figure 1). The use of N fertilizers has become essential to increase the productivity of agriculture, and has resulted in an almost doubling of the global food production in the past 50 years. However, this also implies that the natural N cycle has substantially been disturbed. In the following paragraphs an overview of the different N transformation processes in the soil is given.

Nitrogen Cycle: General

Atmospheric N_2 gas (valence 0) can be converted by lightening to various oxides and finally to nitrate (NO_3^-) (valence +5), which can be deposited and taken up by growing plants. Also N_2 gas can be converted to ammonia (NH3, valence −3) by biological N_2 fixation, with the NH_3 participating in a number of biochemical reactions in the plant. When plant residues decompose the N-compounds undergo a series of microbial conversions (mineralisation) leading first to the formation of ammonium (NH_4^+) (valence −3) and possibly ending up in NO_3^- (nitrification). Under anaerobic conditions NO_3^- can be converted to various N-oxides and finally to N_2 gas (denitrification). When mineral or organic N fertilizers are used they also undergo the same transformation processes and influence the rate of other N-transformations. In considering the soil compartment, there can be N gains (such as biological N_2 fixation) as well as N losses (such as leaching and denitrification). Furthermore N can be exported from the soil via harvest products, or immobilized in soil organic matter.

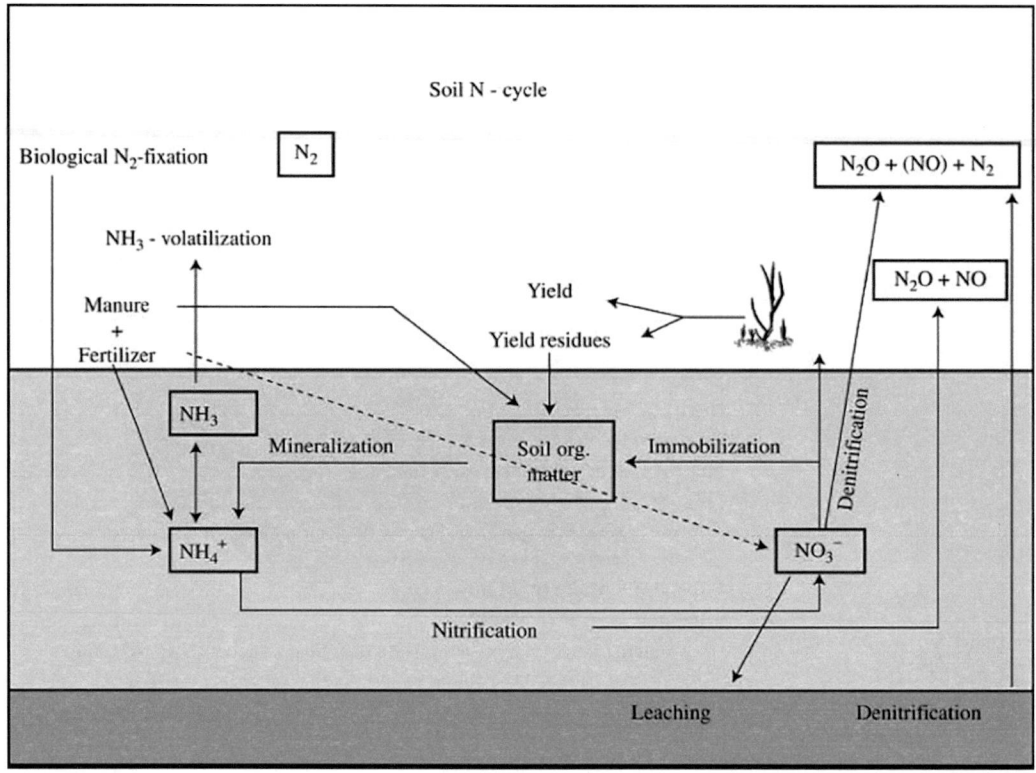

FIGURE 1 The soil N cycle. The white compartment represents the atmosphere; the light gray compartment represents the biosphere and the dark gray compartment the subsoil.

Nitrogen Transformations in the Soil

The principal forms of N in the soil are NH_4^+, NO_3^- or organic N-substances. At any moment, inorganic N in the soil is only a small fraction of the total soil N. Most of the N in a surface soil is present as organic N. It consists of proteins (20%–40%), amino sugars, such as the hexosamines (5%–10%), purine and pyrimidime derivates (1% or less), and complex unidentified compounds formed by reaction of NH_4^+ with lignin, polymerization of quinones with N compounds and condensation of sugars and amines. In the subsoil, an important fraction of the present N can be trapped in clay lattices (especially *illitic* clays) as nonexchangeable NH_4^+ and is consequently largely unavailable. Organic substances slowly mineralize by microorganisms to NH_4^+, which could be converted by other microorganisms to NO_3^- (see further).

The NH_4^+ can be adsorbed to negatively charged sites of clay minerals and organic compounds. This reduces its mobility in the soil compared to the more mobile NO_3^- ion. Microorganisms can use both NH_4^+ and NO_3^- to satisfy their need for N. This type of N transformation is called microbial immobilization.

The ratio between carbon (C) and N (C : N ratio) in organic matter determines whether immobilization or mineralization is likely to occur. When utilizing organic matter with a low N content, the microorganisms need additional N, decreasing the mineral N pool of the soil. Thus, incorporation of organic matter with a high C:N ratio (e.g., cereal straw) results in immobilization. Incorporation of organic matter with a low C : N ratio (e.g., vegetable or legume residues) results in N-mineralization. A value of the C:N ratio of 25 to 30 is often taken as the critical point toward either immobilization or mineralization.

Nitrification is a two-step process. In the first step NH_4^+ is converted to nitrite (NO_2^-) (valence +3) by a group of obligate autotrophic bacteria known as *Nitrosomonas* species. The second step is carried out by another group of obligate autotrophic bacteria known as *Nitrobacter* species. Also a few heterotrophs can carry out nitrification, usually at much lower rates.

Soil water and aeration are crucial factors for nitrification. At a water potential of 0 kPa (saturation), there is little air in the soil and nitrification stops, due to oxygen limitation; nitrification is greatest near field capacity (–33 kPa in medium- to heavy-textured soils, to 0 to –10 kPa in light sandy soils). Also in dry soils NH_4^+ and sometimes NO_2^- accumulate presumably because *Nitrobacter* species are more sensitive to water stress than the other microorganisms.

Nitrification is slow in acid conditions with an increasing rate at increasing pH. Mainly under alkaline conditions, nitrite is also accumulating, because *Nitrobacter* is known to be inhibited by ammonia, which is formed under alkaline conditions. Nitrification is a process that acidifies the soil as protons (H^+) are liberated:

$$NH_4^+ + 2O_2 \rightarrow NO_3^- + 2H^+ + H_2O$$

During nitrification minor amounts of nitrous oxide (N_2O) (valence +1) and nitric oxide (NO) (valence +2) are formed. Both compounds have environmental consequences, discussed below.

The effect of temperature on nitrification is climate dependent. There is a climatic selection of species of nitrifiers, with those from cooler regions having lower temperature optima and less heat tolerance than species from warmer regions. All above-mentioned factors influencing nitrification also influence the nitrifying population. The population and activity of nitrifiers can be reduced by the use of nitrification inhibitors, such as dicyanodiamide, nitrapirin and neem (*Azadirachta indica*) seed cake. They are used mostly to retard the nitrification of manure; otherwise their practicality is controversial and they are not extensively used. More details about nitrification and nitrification inhibitors can be found in McCarty[1] and Prosser.[2]

Nitrogen Input Processes

Atmospheric Nitrogen Deposition

The total atmospheric N (NH_4^+ and NO_3^-) deposition is in the order of 10–40 kg N ha^{-1} yr^{-1} in much of north-western and central Europe and some regions in North America. It ranges from 3–5 kg N ha^{-1} yr^{-1} in pristine areas.[3] It is originating from previously emitted NH_3 and NO_x from agricultural and industrial activities or traffic.

Biological Nitrogen Fixation

Rhizobium species living in symbiotic relationship in root nodules of legumes, e.g. clover (*Trifolium*), lucerne (*Medicago*), peas (*Pisum*) and beans (*Faba*)—can convert atmospheric N_2 gas to NH_3, which is further converted to amino acids and proteins. Parallel to this process, the *rhizobium* species receive from the legume the energy they need to grow and to fix N_2. Photosynthetic cyanobacteria are also N-fixing organisms and are especially important in paddy rice (*Oryza*). The amount of N fixed varies greatly from crop to crop, ranging from a few kg to a few hundred kg N ha^{-1} yr^{-1}. The process is depressed by ample N supply from other sources, and it is sensitive to lack of phosphorus. The amount of globally fixed N is almost the double the amount of applied fertilizer N. Next to symbiotic N fixing bacteria also non-symbiotic species (e.g. *Azotobacter*) occur in soils. In general, free-living diazotrophs make a small but significant contribution to the soil N status. Some nonleguminous trees and plants (e.g. alder (*Alnus*), sugarcane (*Saccharum*) host N-fixing bacteria as well. Much uncertainty exists about the association of N fixing bacteria with non-legumes (so called associative N fixing bacteria).

Mineral and Organic Nitrogen Fertilization

Theoretically plants should prefer NH_4^+ above NO_3^-, because NH_4^+ does not need to be reduced before incorporation into the plant. In most well-drained soils oxidation of NH_4^+ is fairly rapid and therefore most plants have developed to grow better with NO_3^-. However, a number of studies have shown that plants better develop when both sources are available. Rice, growing under submerged conditions must grow in the presence of NH_4^+ as NO_3^- is not stable under flooded conditions. When urea is applied it rapidly hydrolyzes under well-drained conditions, unless a urease inhibitor is being added; under submerged conditions rice plants may also absorb N directly as molecular urea. Organic manure can be of plant or animal origin or a mixture of both. However, most comes from dung and urine from farm animals. It exists as farmyard or stable manure, urine, slurry or as compost. Because its composition is not constant and because plant material (catch or cover crops, legumes) is often added freshly (green manure) to the soil, less than 30% of its nutrients becomes available for the next crop.

Nitrogen Uptake by Plants

Growing plants get their N from fertilizer N as well as from organic soil N upon mineralization. Plants take up N compounds both as NO_3^- and as NH_4^+. In general, NO_3^- is the major source of plant N. There is some evidence that small amounts of organic N (urea or amino acids) can be taken up by plants from the soils solution. Plant uptake of N can be studied through the use of mineral fertilizers or organic matter labeled with the stable N isotope ^{15}N. The proportion of applied N taken up by the crop is affected by many factors, including crop species, climate and soil conditions. Above ground parts of the crop can recover 40%–60% of the fertilizer N applied.

Nitrogen Loss Processes

Ammonia Volatilization

Losses of N from the soil by NH_3 volatilization amount globally to 54 Mt (or 10^{12} g) NH_3–N yr^{-1} and 75% is of anthropogenic origin.[4] According to the ECETOC,[5] the dominant source is animal manure and about 30% of N in urine and dung is lost as NH_3. The other major source is surface application of urea or ammonium bicarbonate and to a lesser degree other ammonium-containing fertilizers. As urea is the most important N fertilizer in the world, it may lead to important NH_3 loss upon hydrolysis and subsequent pH rise in the vicinity of the urea till. The transformation of NH_4^+ to the volatile form NH_3 increases with increasing pH, temperature, soil porosity, and wind speed at the soil surface. It decreases with increasing water content and rainfall events following application. Ammonia losses from soils can be effectively reduced by fertilizer incorporation or injection instead of surface application.

Emission of Nitrogen Oxides (N_2O, NO) and Molecular Nitrogen (Nitrification and Denitrification)

Microbial nitrification and denitrification are responsible for the emission of NO and N_2O.[6] They are by-products in nitrification and intermediates during denitrification. Probably about 0.5% of fertilizer N applied is emitted as NO[7] and 1.25% as N_2O.[8] However, wide ranges have been reported. Intensification of arable agriculture and of animal husbandry has made more N available in the soil N cycle increasing the emission of N oxides. The relative percentage of NO and N_2O formation very much depends on the moisture content of the soil. At a water-filled pore space (WFPS, or the fraction of total soil pore space filled with water) below 40% NO is produced mainly from nitrification. Between a WFPS of 40% and 60% formation of NO and N_2O from nitrification occurs. Between a WFPS of 60% and 80% N_2O is predominantly produced from denitrification and the formation of NO is decreasing sharply. At a WFPS above 80% the formation of N_2 by denitrification is dominant. In practice these WFPS ranges will overlap and depend on the soil type.[9] Next to water content, also temperature, land use and availability of N and decomposable organic matter are important determining factors for N_2O formation. Nitrous oxide is a greenhouse gas contributing 5%–6% to the enhanced greenhouse effect. Increased concentrations are also detrimental for the stratospheric ozone layer.[10] In the presence of sunlight, NO_x (NO and NO_2) react with volatile organic compounds from evaporated petrol and solvents and from vegetation and forms tropospheric ozone which is, even at low concentration, harmful to plants and human beings. The major gaseous end-product of denitrification is N_2. The ratio of N_2O to N_2 produced by denitrification depends on many environmental conditions. Generally themore anaerobic the environment the greater the N_2 production. Denitrification is controlled by three primary factors (oxygen, nitrate and carbon), which in turn are controlled by several physical and biological factors. Denitrification N loss can reach 10% of the fertilizer N input—more on grassland and when manure is also applied.[11] Chemical denitrification is normally insignificant and is mainly related to the stability of NO_2^- and acid conditions.[12] It is more difficult to reduce N_2O and NO from soils then NH_3 losses. A general principle is to minimize N surpluses in the soil profile via careful fertilizer adjustment, corresponding to the actual crop demands.

Leaching

Applied NO_3^- or NO_3^-, formed through nitrification from mineralized NH_4^+ or from NH_4^+ from animal manure, can leach out of the rooting zone. It is well possible that this leached NO_3^- can be denitrified at other places and returned into the atmosphere. The amount and intensity of rainfall, quantity and frequency of irrigation, evaporation rate, temperature, soil texture and structure, type of land use, cropping and tillage practices and the amount and form of fertilizer N are all parameters influencing the amount of NO_3^- leaching to the underground water. Nitrate leaching should be kept under control as it may influence the nitrate content in drinking water influencing human health and in surface water, causing eutrophication. Nitrate losses can be minimized by reducing the mineral N content in the soil profile during the winter period by careful fertilizer adjustment, growing of cover crops or riparian buffer areas.

References

1. McCarty, G.W. Modes of action of nitrification inhibitors. Biol. Fert. Soils **1999**, *29*, 1–9.
2. Prosser, J.I., Ed. *Nitrification, Special Publications of the Society of General Microbiology*; IRL Press: Oxford, 1986; 20 pp.
3. Lagreid, M.; Bockman, O.C.; Kaarstad, O. *Agriculture, Fertilizers and the Environment*; CIBA Publishing: Oxon, U.K., 1999; 294 pp.
4. Sutton, M.A.; Lee, D.S.; Dollard, G.J.; Fowler, D. International conference on atmospheric ammonia: emission, deposition and environmental impacts. Atmospheric Environment **1998**, *32*, 1–593.
5. ECETOC *Ammonia Emissions to Air in Western Europe, (No. 62)*; European Centre for Ecotoxicology and Toxicology of Chemicals: Brussels, 1994; 196 pp.
6. Bremner, J.M. Sources of nitrous oxide in soils. Nutr. Cycl. Agroecosys. **1997**, *49*, 7–16.
7. Veldkamp, E.; Keller, M. Fertilizer-induced nitric oxide emissions from agricultural soils. Nutr. Cycl. Agroecosys. **1997**, *48*, 69–77.
8. Mosier, A.; Kroeze, C.; Nevison, C.; Oenema, O.; Seitsinger, S.; Van Cleemput, O. Closing the global N_2O budget: nitrous oxide emissions through the agricultural N cycle. Nutr. Cycl. Agroecosys. **1998**, *52*, 225–248.
9. Davidson, E.A. Fluxes of Nitrous Oxide and Nitric Oxide from terrestrial ecosystems. In *Microbial Production and Consumption of Greenhouse Gases: Methane, Nitrogen Oxides, and Halomethanes*; Rogers, J.E., Whitman, W.B., Eds.; American Society for Microbiology: Washington, DC, 1991; 219–235.

10. Crutzen, P.J. The influence of nitrogen oxides on the atmospheric ozone content. Quat. J. Royal Meteor. Soc. **1976**, *96,* 320–325.

11. von Rheinbaben, W. Nitrogen losses from agricultural soils through denitrification—A critical evaluation. Z. Pflanzenern. Bodenk. **1990**, *153,* 157–166.

12. Van Cleemput, O. Subsoils: chemo- and biological denitrification, N_2O and N_2 emissions. Nutr. Cycl. Agroecosys. **1998,** *52,* 187–194.

30

Phenols

Leszek Wachowski
and Robert Pietrzak

Introduction

In 1832, German analytical chemist F.F. Runge extracted from coal a tar substance, which a year later, F. Gerhard called phecolenolem (Latin *phenolum, carbolum*). The old name for benzene was phene, and its hydroxyl derivative came to be called phenol. It is the simplest compound from the numerous class generally called phenols, comprising hydroxyl derivatives of aromatic compounds (benzene, naphthalene).

The compound whose correct chemical name according to the International Union of Pure and Applied Chemistry (IUPAC), nomenclature should be benzenol can be treated as benzene in which one of the hydrogen atoms has been replaced by a hydroxyl group. Phenols are aromatic alcohols with one or more hydroxyl groups directly bonded to the aromatic ring, which differentiates them from alcohols. Their structure is reflected in their properties, e.g., in enhanced acidic properties.[1]

Phenols make up a large group of compounds of natural, anthropogenic, or endogenic origin found in the biosphere. Because of a number of physical and chemical properties attractive from the viewpoint of practical use, phenol is produced on a large scale, about 7 billion kg/yr, as a precursor to many useful compounds. The greatest producer of phenol is the United States. Phenol is one of the oldest chemical intermediates in organic industry, and its economical importance continuously increases. The greatest amounts of phenol are used in organic synthesis, including manufacturing of phenolic resins and nylon, in which the chemical intermediates are bisphenol-A and ε-caprolactam,[2] respectively. Bacterial and fungicidal properties of phenol were the first recognized and were the reason for its wide application. The employment of phenol for sterilization has been of great importance not only in medicine but also in the history of humanity. Sir J. Lister, a Scottish surgeon, introduced the method of disinfection of surgery instruments and hands of surgeons with a 5% water solution of phenol, known at that time as carbolic acid.[3]

As a result of natural transformations taking place in the natural environment and the processes generating various phenolic compounds performed by man and their widespread use, the phenolic compounds have penetrated the biosphere. They are met more often as micropollutants of water than the atmospheric air. As the exposure of living organisms to these toxic compounds can be harmful, their presence has been monitored, and certain measures have been undertaken to restrict their amount introduced to the natural environment.[1,2,4]

Origin, Properties, and Application of Phenol and Its Derivatives

Phenol and phenolic derivatives met in the biosphere are released from natural and anthropogenic sources. Their amounts entering the natural environment and coming from anthropogenic sources have been estimated as much higher.[5–7]

The class of compounds known as phenols plays a great role in many areas of our lives. Both the synthetic and natural phenols have been raw products or intermediates for the production of, e.g., plastics, detergents, or cosmetic products. Because of their biological activity and, in particular, antioxidant properties, they are essential components of our diet and substrates of a large number of therapeutic drugs.[8–15]

The large diversity of chemical compositions and structures of phenol compounds is illustrated in Tables 1–7.

TABLE 1 Name, Molecular Formula, and Structures of Selected Groups of Phenolic Compounds

No.	Name of Compound	Molecular Formula	Structural Formula
	Methyl derivatives of one-hydroxyl phenols		
1.	*Phenol **Carbolic acid, hydroxybenzene, benzenol, phenylic acid, phenic acid	C_6H_6O ***C_6H_5OH	
2.	o-Cresol *2-Methylphenol **2-Hydroxytoluene, 2-methylbenzenol	C_7H_8O ***$(CH_3)C_6H_4OH$	
3.	m-Cresol *3-Methylphenol **3-Hydroxytoluene, 3-methylbenzenol	C_7H_8O ***$(CH_3)C_6H_4OH$	
4.	p-Cresol *4-Methylphenol **4-Hydroxytoluene, p-hydroxytoluene, p-methylphenol, 4-methylbenzenol	C_7H_8O ***$(CH_3)C_6H_4OH$	
5.	2,3-Xylenol **2,3-Dimethylphenol, 1-hydroxy-2,3-methylbenzene, 3-hydroxyl-o-xylene	$C_8H_{10}O$ ***$(CH3)_2C_6H_3OH$	
6.	2,4-Xylenol **2,4-Dimethylphenol, 1-hydroxy-2,4-methylbenzene, 4-hydroxy-m-xylene	$C_8H_{10}O$ ***$(CH_3)_2C_6H_3OH$	

(Continued)

TABLE 1 (*Continued*) Name, Molecular Formula, and Structures of Selected Groups of Phenolic Compounds

No.	Name of Compound	Molecular Formula	Structural Formula
7.	2,5-Xylenol **2,5-Dimethylphenol, 2-hydroxy-*p*-xylene, *p*-xylenol, 1-hydroxy-2,5-methylbenzene	$C_8H_{10}O$ ***$(CH_3)_2C_6H_3OH$	
8.	2,6-Xylenol **2,6-Dimethylphenol, 2-hydroxy-rn- xylene, 1-hydroxy-2,6-methylbenzene	$C_8H_{10}O$ ***$(CH_3)_2C_6H3OH$	
9.	3,4-Xylenol **3,4-Dimethylphenol, 4-hydroxy-o-xylene, 1-hydroxy-3,4-methylbenzene	$C_8H_{10}O$ ***$(CH_3)_2C_6H_3OH$	
10.	3,5-Xylenol **3,5-Dimethylphenol, 5-hydroxy-rn-xylene, 1-hydroxy-3,5-methylbenzene	$C_8H_{10}O$ ***$(CH_3)_2C_6H_3OH$	
11.	2,3,4-Trimethylphenol ***1-Hydroxy-2,3,4-methylbenzene	$C_9H_{12}O$ ***$(CH_3)_3C_6H_2OH$	

One core phenols with aliphatic side-chain

No.	Name of Compound	Molecular Formula	Structural Formula
12.	2-Ethylphenol **(1-Hydroxy-2-ethylbenzene, o-ethylphenol)	$C_8H_{10}O$ ***$(C_2H_5)C_6H_4OH$	
13.	3-Ethylphenol **(1-Hydroxy-3-ethylbenzene, *m*-ethylphenol)	$C_8H_{10}O$ ***$(C_2H_5)C_6H_4OH$	
14.	4-Ethylphenol **(1-hydroxy-4-ethylbenzene, p-ethylphenol, 1-hydroxy-4-ethylbenzene, 4-hydroxyphenylethane)	$C_8H_{10}O$ ***$(C_2H_5)C_6H_4OH$	
15.	2-Isopropylphenol **(o-Isopropylphenol, 1-hydroxy-2-isopropylbenzene, o-hydroxycumene; 2-hydroxycumene)	$C_9H_{12}O$ ***$(C_3H_7)C_6H_4OH$	

(*Continued*)

TABLE 1 (*Continued*)   Name, Molecular Formula, and Structures of Selected Groups of Phenolic Compounds

No.	Name of Compound	Molecular Formula	Structural Formula
16.	o-Butylphenol *2-Butylphenol **(1-Hydroxy-2-butylbenzene)	$C_{10}H_{14}O$ ***$C_6H_4(OH)(C_4H_9)2$	
17.	*2-Ethyl-4-methylphenol **(2-Ethyl-4-Methyl-phenol, 2-ethyl-p-cresol)	$C_9H_{12}O$ ***$(C_2H_5)(CH_3)C_6H_3OH$	
18.	Thymol *2-Isopropyl-5-methylphenol **(IPMP)	$C_{10}H_{14}O$ ***$2-[(CH_3)_2CH]C_6H_3-5-(CH_3)$ OH	

Methyl derivatives of di- and trihydroxyl phenols

No.	Name of Compound	Molecular Formula	Structural Formula

19.	3-Methylpyrocatechol
	**(Pyrocatechol 1,2-dihydroxy-3-
	  methylbenzene,
	  2,3-dihydroxytoluene,
	  2-hydroxy-3-methylphenol)
		$C_7H_8O_2$
		***$(CH_3)C_6H_3(OH)_2$	

20.	2-Methylresorcinol **(2,6-Dihydroxy, 2,6-toluenediol, toluene-2,6-diol)	$C_7H_8O_2$ ***$(CH_3)C_6H_3(OH)_2$	

21.	2-Methylhydroquinol
	*2-Methylbenzene-1,4-diol
	**(2-Methyl-1,4-benzenediol,2-methyl-1,4-
	  hydroquinone,
	  2-methylbenzene-1,4-diol)
		$C_7H_8O_2$
		***$(CH_3)C_6H_3(OH)_2$	

22.	2,4,6-trihydroxytoluene *2-Methylbenzene-1,3,5-triol **(2-Methylphloroglucinol, Toluene-2,4,6-triol	$C_7H_8O_3$ ***$C_6H_2(OH)_3(CH_3)$	
23.	2,4-Dimethylphloroglucinol *2,4-Dimethylbenzene-1,3,5-triol	$C_8H_{10}O_3$ ***$(CH_3)_2C_6H(OH)_3$	

(*Continued*)

TABLE 1 (Continued) Name, Molecular Formula, and Structures of Selected Groups of Phenolic Compounds

No.	Name of Compound	Molecular Formula	Structural Formula
		Di-co,re phenols	
24.	1-Naphthol *Naphthalen-1-ol **(1-hydroxynaphthalene, 1-Naphthalenol; alpha-Naphthol a-Naphtol, naft-1-ol)	$C_{10}H_8O$ **$C_{10}H_7OH$	
25.	2-Naphthol *Naphthalen-2-ol **(2-hydroxynaphthalene, 2-Naphthalenol; beta-Naphthol β-Naphtol, naft-2-ol)	$C_{10}H_8O$ **$C_{10}H_7OH$	
26.	Methyl-2-naphtol **(1-methyl-2-hydroxynaphthalene, 1-methylnaphthalen-2-ol)	$C_{11}H_{10}O$ ***$(CH_3)C_{10}H_6OH$	
27.	*2-phenylphenol **(2-Hydroxybiphenyl, o-phenylphenol, biphenylol, orthophenyl phenol, o-xenol, orthoxenol)	$C_{12}H_{10}O$ ***$C_{12}H_9OH$	
28.	2,2'-Biphenol **(2,2'-Dihydroxybiphenyl, 2,2'-biphenyldiol)	$C_{12}H_{10}O$ ***$C_{12}H_8(OH)_2$	
		Thre -core phenols	
29.	1-Anthranol *Anthracen-9-ol **(1-Hydroxyanthracene, 9-anthracenol, 9-anthranol)	$C_{14}H_{10}O$ ***$C_{14}H_9OH$	
30.	2-Anthranol *Anthracen-2-ol **(2-Hydroxyanthracene, 2-anthracenol, 2-anthranol, beta-hydroxyanthracene)	$C_{14}H_{10}O$ ***$C_{14}H_9OH$	
31.	1,9-Dihydroxyanthracene *Anthracene-1,9-diol **1,9-Anthracenediol	$C_{14}H_{10}O_2$ ***$C_{14}H_8(OH)_2$	
32.	2-Phenantrenol	$C_{14}H_{10}O_2$ ***$C_{14}H_9(OH)$	

*, IUPAC name; **, synonyms; ***, other formula.

Sources of Phenol and Phenolic Compounds in the Biosphere

Natural Origin

Phenol and phenolic derivatives from natural sources are commonly met organic micropollutants. The precursors of the majority of them are phenylalanine and tyrosine, from which as a result of deamination, cinnamic acid and its hydroxy derivatives are formed.[4,8,13–16]

Usually, these compounds are the following:

- Intermediate products of natural decay of organic matter such as proteins, humic compounds, and lignin[8]
- Products of metabolic processes taking place in living organisms[1,2,9]
- Secondary metabolic plant product[1,2,9]
- Products formed as a consequence of forest fires[17]
- Products of natural decay of original plant matter![1,2,9,16]

In Table 2, molecular and structural formulae of the natural derivatives of phenol most often met in the natural environment are presented.

Lignin is a complex chemical compound most commonly found in wood and all vascular plants, localized not only between the cells but also in the cell walls. In lignin, two groups of phenolic compounds dominate. The first comprises phenyl acid derivatives, occurring in plants mostly in the bound form as components of lignin and hydrolyzing tannins in the form of esters and glycosides.[13,15]

TABLE 2   Phenol Compounds of Natural Origin Most Commonly Found in the Biosphere

No.	Name of Compound	Molecular Formula	Structural Formula
1.	*Phenol **Carbolic acid, hydroxybenzene, benzenol, phenylic acid, phenic acid	(See Table 1 no.1)	
2.	Catechol *Benzene-1,2-diol **(Pyrocatechol, 1,2-benzenediol, 2-hydroxyphenol, 1,2-dihydroxybenzene)	$C_6H_6O_2$ ***$C_6H_5(OH)2$	
3.	*4-Hydroxybenzaldehyde ***(p-Hydroxybenzaldehyde, 1-hydroxybenzaldehyde, p-formylphenol)	$C_7H_6O_2$ ***$(CHO)C_6H_4OH$	
4.	o-Cresol *2-Methylphenol **2-Hydroxytoluene, 2-methylobenzenol	(See Table 1 no. 2)	
5.	Vanillyn *4-Hydroxy-3-methoxybenzaldehyde **(Methyl vanillin, vanillin, vanillic aldehyde, 3-methoxy-4-hydroxybenzaldehyde)	C8H8O3 ***$(OCH_3)(CHO)$ C_6H_3OH	
6.	Syringaldehyde *4-Hydroxy-3,5-dimethoxybenzaldehyde (3,5-Dimethoxy-4-hydroxybenzaldehyde, **3,5-dimethoxy-4-hydroxybenzene carbonal, gallaldehyde 3,5-dimethyl ether, 4-hydroxy-3,5-dimethoxybenzaldehyde, syringic aldehyde)	C9H10O4 ***(OCH3)2(CHO) C6H2OH	

(Continued)

TABLE 2 (*Continued*)   Phenol Compounds of Natural Origin Most Commonly Found in the Biosphere

No.	Name of Compound	Molecular Formula	Structural Formula
7.	Alkyl phenols With saturated alkyl groups at para position $C_nH_{2n}+i$, usually $n = 1$-3	C6H4(OH)(CnH2n+1) $n = 1$-12	
8.	Gallic acid *3,4,5-Trihydroxybenzoic acid **(Gallic acid, gallate, 3,4,5-trihydroxybenzoate)	$C_7H_6O_5$ ***$C_6H_2(OH)_3(COOH)$	

9.	Eugenol
	*4-Allyl-2-methoxyphenol
	**[4-Allyl-2-methoxyphenol,
	  2-methoxy-4-(2- propenyl)phenol, eugenic acid,
	  caryophyllic acid,
	  1-allyl-3-methoxy-4-hydroxybenzene, allylguaiacol,
	  2-methoxy-4-allylphenol,-4-allylcatechol-2-methyl
	  ether,
	  2-methoxy-4-(2-propen-1-yl)phenol]
		$C_{10}H_{12}O_2$
		***$(OCH_3)(C_3H_4)$
		C_6H_3OH	

10.	Thymol *2-Isopropyl-5-methylphenol **(IPMP)	(See Table 1 no. 18)	
11.	1-Naphthol *Naphthalen-1-ol **(1-Hydroxynaphthalene, 1-naphthalenol; *alpha-naphthol, α-*Naphtol, naft-1-ol)	(See Table; 1 no. 24)	

*, IUPAC name; **, synonyms; ***, other formula.

The second group comprises hydroxycinnamic acids present most often in the form of esters, while hydroxybenzoic acids are mainly present in plants in the form of glycosides. Moreover, in plant tissues, phenyl acids have been identified, besides other naturally occurring compounds, like flavonoids, fatty acids, sterols, or cell wall polymers.[15]

It has been established that gymnospermous (*Gymnospermae*) plants like pine, spruce, and fir contain vanillyl, but they do not contain siryngyl and cinnamyl phenols. In the angiospermous (*Angiospermae*) plants, both vanillyl and siryngyl compounds are present, but cinnamyl compounds are not found.[12–16] Some examples of natural phenolic compounds used as indicators of original plant matter are given in Table 3.

Phenols of natural origin, generally called photochemical, include a large and very chemically diverse group of polyphenols characterized by the presence of large multiples of phenol units.

The above phenolic compounds, of which more than 8000 are known, embrace a wide range of plants' secondary metabolites, possessing in common an aromatic ring substituted by one or more hydroxyl groups.[13–16] They are the most widely distributed secondary metabolites ubiquitously present in the plant kingdom. Simple phenols are relatively rare in plants.[14] Some polyphenolic substructures are secondary plant substances called quasi-vitamins or natural nonnutritious substances showing great biological activity. This group of compounds is sometimes referred to as quasi-vitamin or vitamin P. They are mainly found in the bound form as components of lignin and hydrolyzing tannins, in the form of esters

TABLE 3 Phenolic Derivatives Used as Indicators of Original Plant Matter

No.	Name of Compound	Molecular Formula	Structural Formula
1.	Vanillyn	(See Table; 2 no.5)	
2.	Vanillic acid *4-Hydroxy-3-methoxybenzoic acid **(4-Hydroxy-rn-anisic acid, vanillate)	$C_8H_8O_4$ **$C_6H_3(OH)(OCH_3)(COOH)$	
3.	Apocynin *1-(4-Hydroxy-3-methoxyphenyl)ethanone **4-Hydroxy-3-methoxyacetophenone, acetovanillone	$C_9H_{10}O_3$ ***$C_6H_3(OH)(OCH_3)$ $(COCH_3)$	
4.	Syringaldehyde	(See Table ; 2 no. 6)	
5.	Syringic acid *4-Hydroxy-3,5-dimethoxybenzoic acid **(Gallic acid 3,5-dimethyl ether)	$C_9H_{10}O_5$ $C_6H_2(OH)(OCH_3)(COOH)$	
6.	Acetosyringone *4'-Hydroxy-3',5'-dimethoxyacetophenone **Acetosyringenin	$C_{10}H_{12}O_4$ ***$C_6H_2(OH)$ $(OCH_3)_2(COCH_3)$	
7.	*p*-Coumaric acid *3-(4-Hydroxyphenyl)-2-propenoic acid **para-Coumaric acid, 4-hydroxycinnamic acid, β-(4-hydroxyphenyl)acrylic acid]	$C_9H_8O_3$ ***$C_6H_4(OH)$ $(CH=CH-COOH$	
8.	Ferulic acid *(E)-3-(4-hydroxy-3-methoxy-phenyl) prop-2-enoic acid **[2-Propenoic acid, 3-(4-hydroxy-3- methoxyphenyl)-ferulic acid, 3-(4-hydroxy- 3-methoxyphenyl)-2-propenoic acid, 3-(4- hydroxy-3-methoxyphenyl) acrylic acid, 3-methoxy-4-hydroxycinnamic acid, 4-hydroxy- 3-methoxycinnamic acid, (2E)-3-(4-hydroxy- 3-methoxyphenyl)-2-propenoic acid, ferulate coniferic acid, trans-ferulic acid, (E)-ferulic acid]	$C_{10}H_{10}O_4$ ***$C_6H_3(OH)(OCH_3)$ $(CH=CH-COOH)$	

Note: * IUPAC name; **, synonyms; ***, other formula.
Source: Adapted from Kroon and Wiliamson,[12] Robinson,[13] and Duke.[14]

or glycosides, in all parts of plants, i.e., in the leaves, seeds, flowers, fruit, roots, bark, and wooden parts. Their chemical composition depends on the species and variety of a plant, climatic conditions, and agro-technological procedures applied. Some polyphenols occurring in plants belong to phytoalexins, so the substances involved in the mechanisms protecting against the attack of insects, fungi, or viruses.

To sum up, polyphenols do the following:

- Play important roles in plant metabolism[13–16]
- Display high bioactivity; such natural products with healing and nutritional values are called nutraceuticals or dietary supplements[18]
- Participate in the healing and adaptive processes in living organisms, can protect against development of cancer, or can have therapeutic effect in treatment of different diseases[10,12]
- Participate in regulation of growth and reproduction of plants[11–13]
- Contribute in determination of sensory features of fruit, vegetables, and processed food[10–14]
- Endow plants and fruits with a specific tart and bitter taste[13,16]
- Responsible for the color and fibrous nature of plants and fruit[13]
- Participate in morphogenesis, energy flow, sex determination, photosynthesis, respiration, regulation of gene expression, and regulation of synthesis of growth hormones[19]
- Have a protecting effect against ultraviolet irradiation and against stress[14,15]

Humic substances are the end products of decaying natural organic matter in the microbial process taking place with involvement of edaphone (mainly fungi and actinomycetes) called humification. They are major organic constituents of soil (humus), peat bogs, coal, sewage, compost heaps, carbonaceous shales, lignites, and all types of natural waters (aquatic humic substances) and can form complex ions that are commonly found in the natural environment. A typical humic substance is a mixture of many molecules, some of which are based on a motif of aromatic nuclei with phenolic and carboxylic substituents (see Figure 1).[20–22]

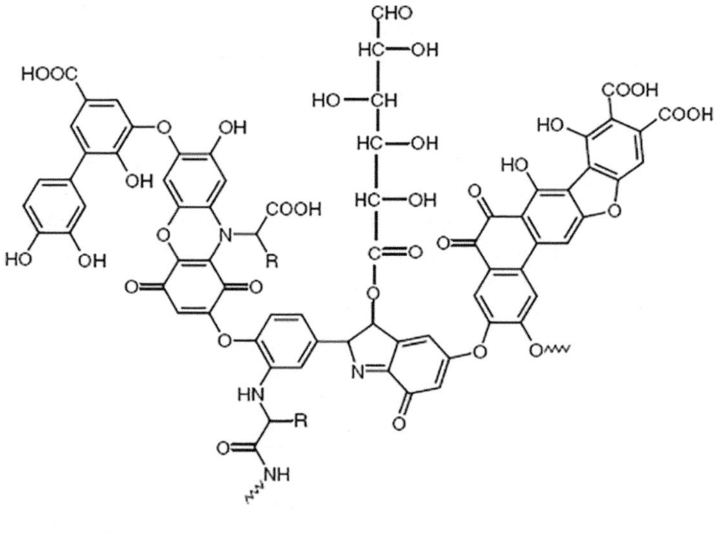

FIGURE 1 The hypothetical structure of humic acid, having a variety of components including quinone phenol, cresol, and sugar moieties. Source: Adapted from Stevenson,[20] Muscola and Soidari,[21] and Hessen and Tranvik.[22]

There are three types of humic substances, which differ slightly in acidity and chemical composition. They are humic acid, fulvic acid, and humins.

Aquatic humus substances bear about 40%-60% of dissolved organic carbon and make up the largest fraction of natural organic matter in water. The major functional groups include carboxylic acids, phenolic hydroxyl, and carbonyl and hydroxyl groups.[22] Phenol hydroxyl groups are usually in amounts of about 1 µeq/mg C of humic material or 94 |µag/mg of elemental carbon as humic matter.[5]

As mentioned earlier, humin substances are also constituents of coal, which is a fossil phytomaterial. It is assumed that lignin, which is one of the most important chemical compounds in plants, is one of the parent substances of coals having aromatic acidic character. It is a consequence of the presence of phenolic hydroxyls, which are the most characteristic oxygen group of coals. The degree of carbonization of coal is measured by the content of hydroxyl groups. In coals with a low degree of carbonization, hydroxyl groups comprise up to 90% of all oxygen groups. Their content decreases with increasing degree of carbonization, and they disappear in coal with a high degree of carbonization.[23]

Anthropogenic Origin

It has been established that a fundamental part of phenol present in the natural environment is of anthropogenic origin. Considerable sources of phenol and its derivatives are the processes of production of intermediate semiproducts and their further processing. Three main sources of phenol and its derivatives released to the natural environment are distinguished: industrial processes, nonindustrial processes and endogenous sources. The greatest amounts of phenol and its derivatives come from industrial processes, mainly production and application of different kinds of phenolic resins and phenolic plastics, generally called phenolics, and caprolactam.[24] Great demand for phenolic resins, widely applied, e.g., as a binding agent in insulating materials, chipboards, shatterproof glass, paints, and casting molds, makes them a profound source of phenolic compounds. The emission of these compounds from these materials is measured in terms of a concentration of free phenol occurring in the monomeric form in the resins in the amount of 1%-5% wt.[24] In foundries, phenolic compounds are emitted in processes of mold production and in casting.[26] Phenolic compounds occur in large amounts in the products coming from the chemistry of coal and chemistry of coke. Processing of raw benzenol obtained as a result of coal coking leads to formation of phenol and xylems. Phenolic compounds are met in wastewater from industrial plants in which phenol and its derivatives are used as raw materials in amounts to a few g/dm³. Table 4 gives a list of processes that are the most abundant sources of phenol released to the natural environment and values of its emission coefficients.

The most important nonindustrial source of phenolic compounds is fuel combustion in motor vehicles. The exhaust gases from motor vehicles contain from about 0.3 to 1.4–2.0 ppm of phenol, which corresponds to the amounts from 1.2 to 5.4–7.7 mg/m³.[24,28,29]

TABLE 4 Selected Industrial Processes Leading to Release of Phenol to the Atmospheric Air and Estimated Values of Its Emission Coefficients

Type of Process	Phenol Emission Coefficient
Production of phenol resins	0–0.5 g of phenol emitted per kg of resin
Production of phenol and its derivatives	–
Production of caprolactam	0.2–0.05 g of phenol emitted per kg of cyclohexanol (semiproduct)
Production of coke	–
Production of insulating materials	–
Emission from phenol processing	–

Source: RIVM Criteria Document: Phenol, Bilthoven, the Netherlands.[27]

Phenolic compounds in considerable concentrations are released in volatile form during combustion of other fuels such as wood, coal, and mazout, in heating chambers, house furnaces, and fireplaces.[30] It should be mentioned that phenolic compounds are commonly met in cigarette smoke, in which the estimated average amount of phenol is 0.4 mg per cigarette.[31] Phenol can be also found in smoked food products.[32]

Endogenous sources of phenol include its synthesis in vivo from different xenobiotics released to the natural environment, e.g., from benzene. It has been established that benzene and its phenolic derivatives in the in vivo conversion can be an endogenous source of phenol.[33] Among phenol derivatives, particular attention has been paid to pentachlorophenol (PCP) and its salts, because of its properties and low cost of production. This material has been widely applied in industry (mainly for wood impregnation) and agriculture and as a component of household products (disinfectant). In agriculture, PCP is used as a pesticide of a broad spectrum of activity as it acts against algae, bacteria, fungi, weeds, insects, and mollusks. Pentachlorophenol has been recognized as hazardous for human health because of its toxicity and widespread use. It is used in leather tanning and finishing. Monochlorophenols are used as synthetic intermediates for dyes and chlorinated phenols. Pentachlorophenol is used as a denaturant for alcohol, an antiseptic, and a selective solvent for refining minerals. [34–36]

Nomenclature, Structure, and Chemical Composition

Phenol is a common name for the simplest and most common aromatic alcohol, labeled with CAS-RegistrySM –The world's largest substance database number 108–95-2, in which the hydroxyl group, known as a phenolic hydroxyl, is attached to the phenyl group. By definition, phenol is a hydroxybenzene, but according to the IUPAC, its correct name should be benzenol.

In literature, phenol is referred to by many other names, such as benzyl, carbolic acid, carbol, phenol liquefied, phenolic acid, phenyl hydroxide, phenic acid, phenylic acid, oxybenzene, monophenol, monohydroxybenzene, and phenyl hydrate.[1,2,24] It also has quite a few commercial names, such as carbol liquor and phenyl liquor (Netherlands), kristallliertes Kreozot or Steinkohlenteerkreosot (Germany), Venzenol (France), or code label ENT1814. The class of phenolic compounds, whose name comes from phenol, includes cresols (aromatic compounds from the group of phenols that are derivatives of toluene).

In contrast to alcohols, in phenol, the hydroxyl group is attached directly to the aromatic ring, which gives enhancement of its acidic properties. Phenols are the compounds in which the ring (benzene) is substituted with one or a few hydroxyl groups. If the compounds are built with aromatic rings (naphthalene, anthracene), with one or more hydroxyl substituents at the benzene ring, then they are called naphthols (hydroxynaphthalenes) or hydroxyanthracenes.[1,2]

According to the IUPAC nomenclature, in naming substitution products of these compounds, the numbering starts at the group already present and is done in the direction that gives the lowest numbers to other groups on the ring. Sometimes, the benzene ring is treated as a substituent (for hydrogen) on another molecule. In that case, the C_6H_5- group of benzene is called "phenyl." Substituents are cited in alphabetical order. Carboxyl and acyl groups take precedence over the phenolic hydroxyl in determining the base name. The hydroxyl group is treated as a substituent. Higher substituted compounds are named as derivatives of phenol.

Structural diversity of phenol compounds makes their systematization rather difficult. With respect to the structure of the carbon skeleton, phenolic compounds can be divided into the following groups:

- Phenylcarboxylic acids—derivatives of benzoic acid of the carbon skeleton C_6–C_i (where C_6 denotes a benzene ring), whose structure can be described by the general formula in Figure 2 (see also Table 5).
- Phenylpropenic acids—derivatives of cinnamic acid of the carbon skeleton C6–C3, whose structure can be described by the general formula in Figure 3 (see also Table 6).

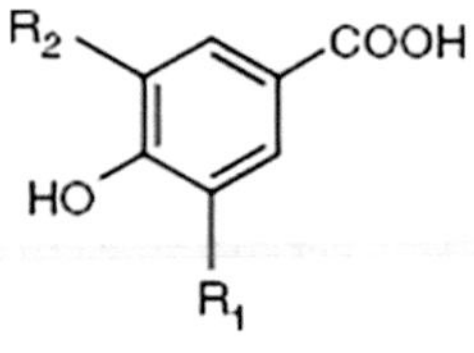

FIGURE 2 The general formula of aromatic hydroxyacid derivatives of benzoic acid.

TABLE 5 Exemplary Hydroxyacids That Are Benzoic Acid Derivatives, Occurring in Plants

No.	Name of Hydroxyacid	Molecular Formula	Structural Formula
1.	*Benzoic acid **(Benzenecarboxylic acid, carboxybenzene, E210, dracylic acid)	$C_7H_6O_2$ ***$C_6H_5(COOH)$	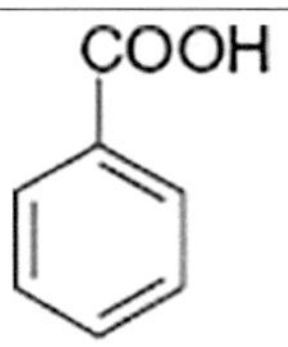
2.	*4-Hydroxybenzoic acid **(*p*-Hydroxybenzoic acid, para-hydroxybenzoic acid)	$C_7H_6O_3$ ***$C_6H_5(COOH)$	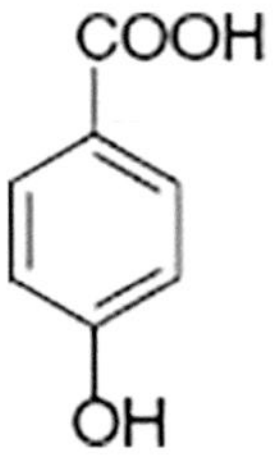
3.	Protocatechuic acid *3,4-Dihydroxybenzoic acid **Protocatechuic acid (PCA)	$C_7H_6O_4$ **$C_6H_4(OH)_2(COOH)$	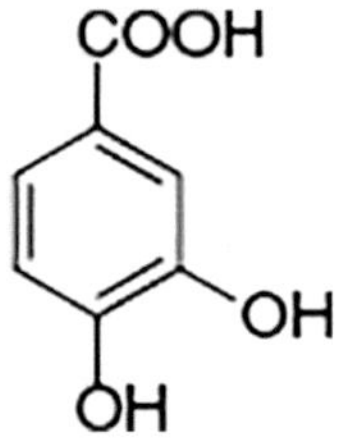
4.	Gallic acid	(See Table 2 no. 8)	
5.	Vanillic acid	(See Table 3 no. 2)	
6.	Syringic acid	(See Table 3 no. 5)	

Note: *, IUPAC name; **, synonyms; ***, other formula

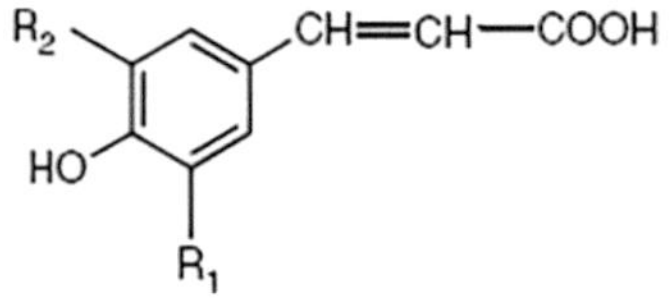

FIGURE 3 The general formula of aromatic hydroxyacids that are derivatives of cinnamic acid.

- Stilbene (trans)—of the carbon skeleton C_6–C_2–C_6, naturally occurring compounds, found in a wide range of plants, aromatherapy products, and dietary supplements, whose structure can be illustrated by the following general formula in Figure 4. Exemplary compounds from this group are resveratrol, piceatannol, rhapontigerin, and pterostilbene. The first mentioned stilbene,

TABLE 6 Some Hydroxyacids That Are Cinnamic Acid Derivatives, Occurring in Plants

No.	Name of Compound	Molecular Formula	Structural Formula
1.	Cinnamic acid *(E)-3-phenylprop-2-enoic acid **(Cinnamic acid, trans-cinnamic acid, phenylacrylic acid, cinnamylic acid, 3- phenylacrylic acid, (E)-cinnamic acid, benzenepropenoic acid, isocinnamic acid)	$C_9H_8O_3$ ***$C_6H_4(OH)(C_3H_3O_2)$	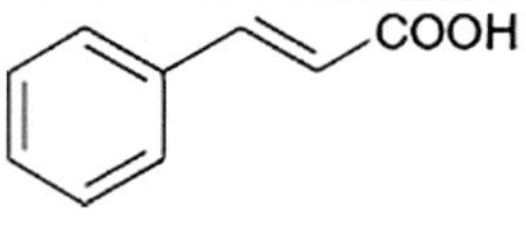
2.	o-Coumaric acid **(2-Hydroxycinnamic, o-coumaric acid, 2-coumaric acid, 2-coumarate 2- hydroxycinnamate, trans-2- hydroxycinnamic acid trans-2-hydroxycinnamate	$C_9H_8O_3$ ***$C_6H_4(OH)(C_3H_3O_2)$	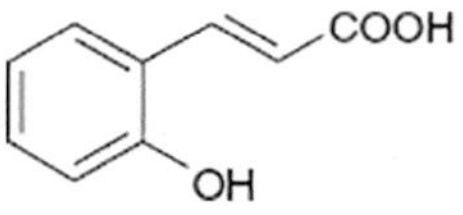
3.	p-Coumaric acid	See Table 3 no. 7	
4.	Caffeic acid *[3-(3,4-Dihydroxyphenyl 2-propenoic acid, 3,4-dihydroxy-cinnamic acid trans- caffeate, 3,4-dihydroxy-trans- cinnamate), (E)-3-(3,4-dihydroxyphenyl)- 2-propenoic acid 3,4-dihydroxybenzeneacrylicacid 3-(3,4- dihydroxyphenyl)-2-propenoic acid]	$C_9H_8O_4$ ***$C_6H_3(OH)_2(C_3H_3O_2)$	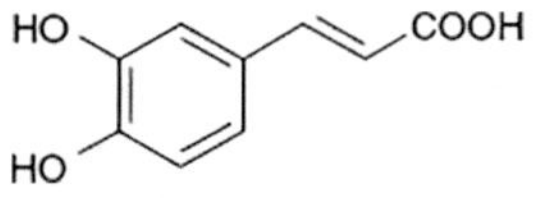
5.	Ferulic acid	(See Table 3 no 8)	
6.	Sinapinic acid *3-(4-Hydroxy-3,5-dimethoxyphenyl) prop-2- enoic acid **(Sinapinic acid, sinapic acid, 3,5-dimethoxy- 4-hydroxycinnamic acid, 4-hydroxy-3,5- dimethoxycinnamic acid)	$C_{11}H_{12}O_5$ ***$C_6H_2(OH)$ $(OCH_3)_2(C_3H_3O_2)$	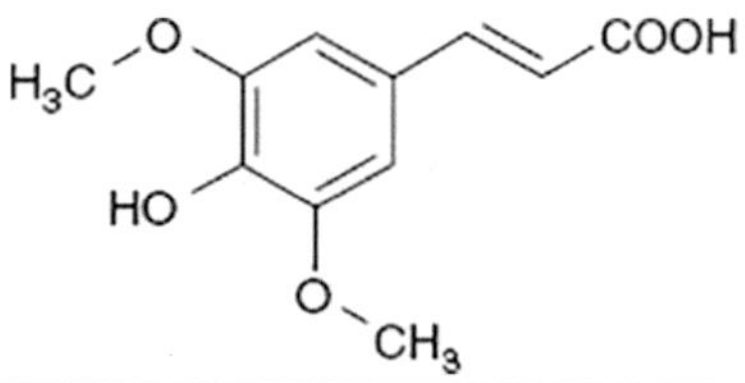

Note: *, IUPAC name; **, synonyms; ***, other formula

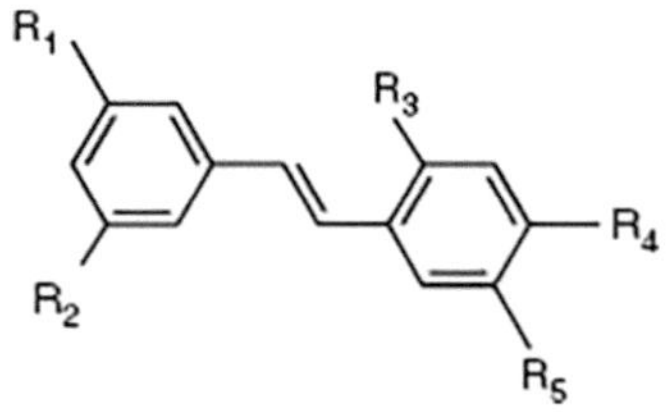

FIGURE 4 The general formula of trans stilbenes (1,2- diphenylethylene; R_1, R_2, R_3, R_4, and R_5 denote OH, OCH$_3$, or –glucose group).

resveratrol, has been intensely studied and proved to have potent anticancer, anti-inflammatory, and antioxidant activities.[37]

- Polyphenols known also as flavonoids or bioflavonoids, of the carbon skeleton C_6–C_3–C_6, whose general structure is shown in Figure 5. This group includes the following subgroups: flavons, the compounds responsible for the yellow color of plants or their parts (Latin *flavus* means yellow); flavones; flavanones; flavanols; anthocyanes (proanthocyanes); and isoflavones. They constitute the largest class of phenolic compounds, with more than 3000 structures.[38,39] Table 7 presents representative compounds of the above-mentioned subgroups of polyphenols.

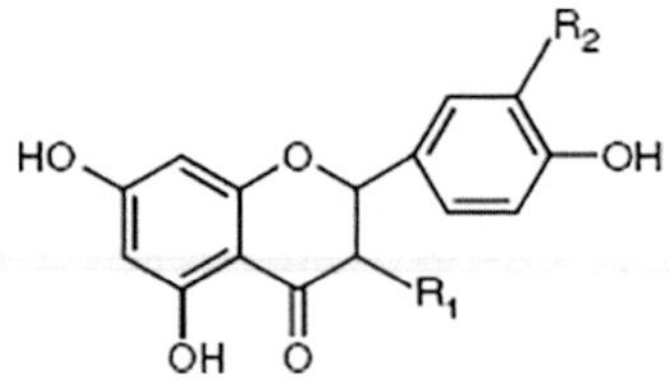

FIGURE 5 The general structure of flavonoids (R_1, R_2 denote H or OH group).

TABLE 7 Selected Examples of Chemical Composition and Structures of Subgroups among Polyphenols

(Polyphenols (flavonoids)		
Name of Compounds	Molecular Formula	Occurrence
Flavons		
Quercetin *2-(3,4-Dihydroxyphenyl)-3,5,7-trihydroxy-4H-chromen-4-one **(Sophoretin, meletin, quercetine, xanthaurine, quercetol, quercitin, quertine, flavin meletin)	$C_{15}H_{10}O_7$	Green tea, grape skin, ginkgo leaves, apples
Flavonols		
Rutin *2-(3,4-Dihydroxyphenyl)-5,7-dihydroxy-3-[α-L-rhamnopyranosyl-(1→6)-β-D-glucopyranosyloxyl-4H-chromen-4-one **(Rutoside, phytomelin, sophorin, birutan, eldrin, birutan forte, rutin trihydrate, globularicitrin, violaquercitrin)	$C_{27}H_{30}O_{16}$	Pogoda tree buds (Sophora japonica Fabaceae), apple skin
Flavanones		
Hesperidin *(2S)-5-hydroxy-2-(3-hydroxy-4-methoxyphenyl)-7-[(2S,3R,4S,5S,6R)-3,4,5-trihydroxy-6-[[(2Æ,3Æ,4Æ,5Æ,6S)-3,4,5-trihydroxy-6-methyloxan-2-yl]oxymethyl]oxan-2-yl]oxy-2,3-dihydrochromen-4-one	$C_{28}H_{34}O_{15}$	Citrus fruit skin
Flavanols (Catechins)		
Epicatechol **(epi-Catechinepi-catechol, epicatechin, l-acacatechin)	$C_{15}H_{14}O_6$	Green tea, grape seeds
Anthocyanes (Proanthocyanes)		
Cyanidin *2-(3,4-Dihydroxyphenyl) chromenylium-3,5,7-triol **Cyanidine	$C_{15}H_{11}O_6^+$	Bilberry, red and black grapes, red wine
Isoflavones		
Genistein *5,7-Dihydroxy-3-(4-hydroxyphenyl)chromen-4-one **4',5,7-Trihydroxyisoflavone	$C_{15}H_{10}O_5$	Leguminous plants, soybean, cereal grains, fruits, vegetables

Note: *, IUPAC name; **, synonyms

- Other phenolic compounds, e.g., tannins (proanthocyanides) of the carbon skeleton Cn>-12, comprising the following:
- Hydrolyzable tannins, consisting of several gallic acid units bound through ester linkages to a central glucose (see Figure 6a); these types of tannins are quite water soluble and are part of the water products of plants.
- Condensed tannins, which are a diverse group of poly- phenolic compounds of plant origin called flavonoids or bioflavonoids (see Figure 6b).

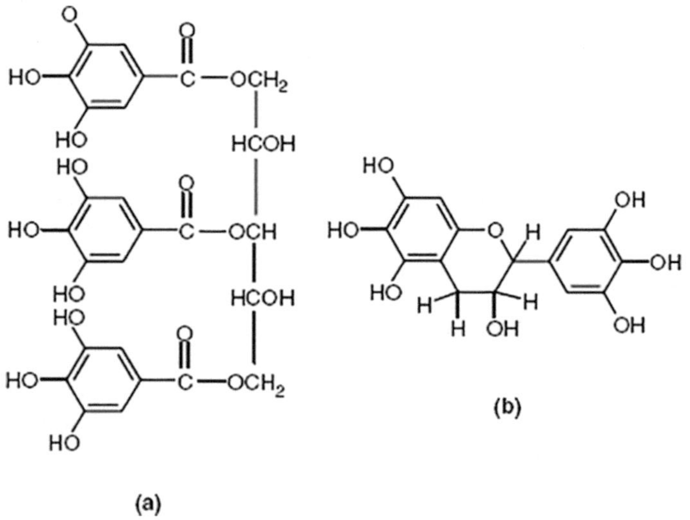

FIGURE 6 The hypothetical structures of hydrolyzable (a) and condensed (b) types of tannins.

Polyphenols make a large group of natural substances occurring in many plants. They are found in greatest amounts in fruit (chokeberry, blueberry, grapes, nuts, garlic) and vegetables (cabbage, cereal seeds).[40–42] Polyphenols are mostly distributed in the external parts of the fruit. In drinks, polyphenols are found in green tea, red wine, coffee, and beer. These compounds make up the largest group of antioxidants supplied in the diet, and hence, they are often called bioflavonoids.[43–45]

Physical, Chemical, and Organoleptic Properties

Phenol is a colorless to light pink crystalline (long needlelike crystallites) solid. When molten, it becomes a bright colorless liquid of low or high viscosity. It attains pink color on exposure to air and light as a result of partial oxidation. In the atmosphere of wet air, the crystals deliquesce. Crude product can be pink, brown, or black. The physicochemical properties of phenols differ significantly from those of aliphatic or unsaturated alcohols; [1,2,7,24] some of them are given in Table 8.

Phenols have hydroxyl groups that can participate in intermolecular hydrogen bonding with other phenol molecules or other H-bonding systems, e.g., water. Hydrogen bonding results in higher dipole moment and melting points and much higher boiling points for phenols than those of hydrocarbons of similar molecular weight. The ability of phenols to form strong hydrogen bonds also enhances their solubility in water.[1,24]

Phenol dissolves to give a 9.3% solution in water, compared with a 3.6% solution of cyclohexanol in water. This water solution of phenol is called phenol liquefied. The water solubility of phenol increases with temperature, and above 68.4°C, both substances become fully miscible.

Phenol is more readily soluble in most organic solvents and in water solutions of soaps. An example of phenol in a water solution of soap is Lysol, being a mixture of a water solution of a potassium soap and cresols, used mainly in the veterinary field for disinfection of rooms and vessels. Its solubility in aliphatic solvents is limited.[1,2]

In contrast to neutral alcohols, water solutions of phenol show weak acidic properties; hence, phenol has been referred to as carbolic acid. Phenol dissociates with the formation of phenolate (C_6H5O^-), also called phenoxide ion, and proton (H3O)+:

$$C_6H_5OH_{(aq)} + H_2O_{(1)} \leftrightarrow C_6H_5O^-_{(aq)} + H_3O^+_{(aq)} \qquad (1)$$

TABLE 8 Some Physical and Chemical Properties of Phenol

Molecular weight	94.11 g/mol
Boiling point	181.75°C (101.3 Pa)
Melting point	43.0°C
	40.9°C (pure substance)
Relative density	1071
Relative vapor density (air = 1)	3.24 mm Hg
Vapor pressure (20°C)	0.357 mm Hg
(50°C)	2.48 mm Hg
(100°C)	41.3 mm Hg
Concentration of saturated vapor (20°C) in air	0.77 g/m3
Water solubility (16°C)	67 g/L
above 68.4°C	Fully soluble
Partition coefficient n-octanol/ water (logPow)	1.46
Dissociation constant in water Ka (20°C)	$1.28 \times 10\text{-}10$
Ignition point:	
• Closed crucible	80°C
• Open crucible	79°C
Acidity	$Ka = 1.3 \times 10\text{-}10$ (pKa = 9.55)

Source: Adapted from Rappoport,[1] Tyman,[2] and Weber et al.[24]

Dissociation occurs to only a slight extent; phenol is a very weak acid with $pK_a = 9.55$. Phenols are more acidic than alcohols of $pK_a = 16$–20 but less acidic than carboxylic acids of $pK_a = 5$.[2] Introduction of substitutes, particularly those in ortho or para position to the –OH group, can dramatically influence the acidity of phenol due to resonance and/or inductive effects. In reactions with strong bases (NaOH, KOH), phenols create phenates (C_6H_5OMe), which are more stable than alcoholates:

$$C_6H_5OH + Na^+OH^- \rightarrow C_6H_5O^-Na^+ + H_2O \tag{2}$$

This increased stability is a consequence of the mesomeric effect stabilizing the phenate ion owing to the delocalization of the negative charge on the aromatic ring of phenol.[1,2]

Phenol has a peculiar characteristic smell and a strong corrosive effect on skin. It is poisonous in nature but acts as a powerful antiseptic.[24,41,46]

Phenol is susceptible to oxidants (e.g., CrO_3, $K_2Cr_2O_7$). The hydrogen abstraction from phenolic hydroxyl is accompanied by the resonance stabilization of the formed phenoxy radical ($C_6H_5O^{\cdot}$)**, which can be further oxidized.[46]

Phenols are highly reactive toward electrophilic substitution because the nonbonding electron on oxygen stabilizes the intermediate cation.[1,2,7]

Phenol reacts with carbonyl compounds both in acidic and basic media. In the presence of formaldehyde, it easily undergoes hydroxymethylation followed by condensation of resin. Condensation with acetone gives bisphenol A, a key building block of polycarbonates. Conversion with formaldehyde produces phenolic resins, the best known of which is Bakelite.[1,2,25,47] The product of catalytic hydrogenation of phenol, performed on a large scale, is cyclohexanol. Another industrially important process is alkylation of phenol, which takes place in the presence of an acid catalyst or a Friedl–Crafts catalyst.

In large amounts, phenol is also used to make different derivatives containing chlorine, usually coming from NaOCl or Cl_2. The most important from among these compounds is PCP, manufactured usually by direct chlorination in the presence of metallic nickel as a catalyst or by hydrolysis of hexachlorobenzene.[1,2] Phenol also reacts directly with alkyl halides in alkali solutions to form phenyl ethers. The phenate ion shows nucleophilic character and replaces halogen from alkyl halide:

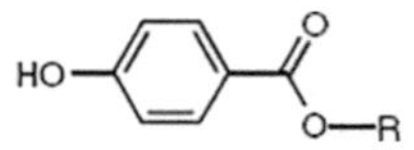

FIGURE 7 General formula of paraben, with R standing for methyl, ethyl, n-propyl, isopropyl, n-butyl, isobutyl, or benzyl group.

$$C_6H_5OH + NaOH \rightarrow C_6H_5ONa + H_2O \quad (3)$$

$$C_6H_5ONa + CH_3Cl \rightarrow C_6H_5OCH_3 + NaCl \quad (4)$$

They are also formed when vapors of phenol and an alcohol are heated over ThO_2.

Phenol can also make esters, among them an interesting group called parabens (esters of *p*-hydroxybenzoic acid with aliphatic alcohol), whose general formula is shown in Figure 7.[48]

Phenol has a characteristic acidic smell and pungent taste. It shows antiseptic activity because of the ability to coagulate protein. It is active against a wide range of microorganisms including some fungi and viruses but is only slightly effective against spores. Although phenol was the first antiseptic used on wounds and in surgery, it is a protoplasmic poison that damages all kinds of cells and is alleged to have caused an astonishing number of poisonings since it came into general use.[2,6,32] Generally, it is a strong and violent poison attacking the nervous system, alimentary track, and circulatory system. It has been estimated that approximately 1 g of phenol is enough to cause death.[2,6,7,27,46] Phenol and its vapors are corrosive to the eyes, skin, and respiratory tract.[49] There is no evidence to claim that phenol causes cancer in humans. Besides its hydrophobic effects, another mechanism of its toxicity is via formation of phenoxyl radicals. Poisoning can also result from inhalation of phenol in the form of atmospheric aerosol, which quickly coagulates in cold air.[50]

Chlorophenols are characterized by a 100–1000 times more intense smell than the parent phenol. During chemical wastewater treatment (chlorination), the water contaminated with phenols acquires a repulsive taste as a result of formation of phenol chloroderivatives.[5] It has been established the limiting concentrations of phenol perceptible though the sense of smell, taste and touch are: 0.021–20 mg/m^3 in air[51] and 0.3 mg/L in water, respectively.[52]

Uses of the Phenol and Its Derivatives

Phenol belongs to the 50 most abundantly produced chemicals. It is one of the oldest chemical intermediates in organic chemistry that still plays a very important role. Both phenol and its numerous derivatives of natural, synthetic, and semisynthetic origin are basic raw materials for industry.

The applications of phenol and its derivatives can be divided into industrial, nonindustrial, and niche.

The largest single use of phenol in industry is as an intermediate in the production of phenolic resins or related materials in the reaction of polycondensation of phenol with formaldehyde. Thus, produced phenyl aldehyde resins are low cost, thermo set, and versatile. They are used as resin glues for plywood in plywood adhesive, cast resins, molding resins, and varnish resins; moreover, they are applied in construction, automotive, and appliance industries.[2,24]

Alkylphenols introduced on the market in the 1940s have been widely used as paint components, herbicides, pesticides, some nonionic detergents, components of cosmetics, composite materials, and media used for removal of fat from the surface of wool, leather, and metal finishing.[1,21]

Non-anionic detergents are produced by alkylation of phenol to give alkylphenols, which are then subjected to ethoxylation.[21] Alkyl phenols with branched chains are used as supplements in oil lubricants, antioxidants (stop chain reactions), and auxiliary substances in conversion of rubbers and plastics.[1,2,24,25] Chemical activity of alkyl derivatives of phenol as well as antioxidants results from the facility of hydrogen abstraction from the phenol group, initiated by the interaction of oxygen and light.

In order to hinder the process of aging, alkyl derivatives are introduced to many different organic products in amounts of 0.1%-2% wt. For example, they can be found in gasoline, lubricating oils, lipids, polymers and plastics, rubber, soaps, cosmetics, and pharmaceuticals.

The chlorinated phenols that are used in largest quantities include PCP and the trichlorophenol isomers used as wood preservatives. Their main components are cresols known as creosotes.[2]

Higher chloroderivatives of phenol, such as 2,4-dichlorophenoxyacetic acid and 2,4,5-trichlorophenoxyacetic acid act, as selective herbicides. The first of the compounds mentioned is one of the most important herbicides against dicotyledons in crops of monocotyledonous plants (cereals, linen, grass). The second of the above compounds acts against perennial weeds, bushes, and coppices in grasslands. Problems that these compounds impose on the natural environment have prompted gradual limitation on their production. Ammonolysis of phenol in gas phase in the presence of aluminosilicate catalysts (zeolites) is one of the common methods for the synthesis of aniline, which is an intermediate for manufacturing salicylic acid. It is also used to make pharmaceuticals, aromatic esters, synthetic dyes, and food preservatives.[2,3,24]

All industrial applications of phenol and its derivatives are too numerous to be mentioned.

The major nonindustrial applications of phenol and its derivatives include the following.

- Medicine; as an antiseptic (water solution of 2%–10% is a very effective antiseptic medium called phenolated water; and in a concentration of 0.2%, it shows bacteriostatic action, while above 2%, it is bactericidal);[1] to relieve itching; as an anesthetic in medicinal preparations (ointments, ear and nose drops, cold sore lotions, throat lozenges, and antiseptic lotions). They are active against a wide range of microorganisms including some fungi and viruses but are only slowly effective against spores. They have been used to disinfect skin, but recently, phenol (Phenolum FP VII) and liquid phenol (Phenolum liquefactum FP IV) have been sporadically used, taking into account their toxicity.[1,2,6,24,25]
- Salicylic acid is an intermediate in production of drugs, flavor esters, dyestuffs, and food preservatives.[2,53]
- Pentachlorophenol can be applied as a fungicide for wood solution of phenol.[1,2,53–55]
- The niche applications of phenol and its derivatives include the following.
- Diphenyl ether is used as material for the production of phenoxathiin. It is an intermediate for polyamide and polyimide, and a processing aid in the production of polyesters. Some polybrominated biphenyl ethers are flame retardants and are used in soap perfumes.[56]
- Phenol is used for embalming bodies for study because of its ability to preserve tissues for extended periods of time.
- Production of cosmetics such as sunscreens,[53] hair dyes, and skin lightening preparations.[54]
- Phenol is used as an exfoliant in cosmetic surgery.
- Phenol and its derivatives are used for phenolization, which is a surgical procedure that serves to treat ingrown nails.
- Phenol is used to denature and remove protein when purifying nucleic acids in molecular biology procedures.
- Phenol and its derivatives are added to polyimide to make antispastic and painkilling drugs used in cancer therapy.[54]
- They are used as slimicides, i.e., substances used to kill slime-producing organisms including bacteria and fungi.[1,2,24,26,55]

Distribution, Transport, and Conversion of Phenol and Its Derivatives in the Biosphere

A phenolic compound occurs in the biosphere in concentrations from trace to high orders of a few milligrams per cubic decimeter or even a few grams per cubic meter in the industrial emissions. Phenol evaporates more slowly than water and can remain in the air, soil, and water for long periods of time if large amounts of it are released at one time or if it is constantly released to the environment from a source.[57,58] Coefficients for conversion of phenol concentration are the following: $1 \text{ mg/m}^3 = 0.26 \text{ ppm}$, $1 \text{ ppm} = 3.84 \text{ mg/m}^3$.[2]

In the atmosphere, phenol exists predominantly in the vapor phase.[57,58] No reliable information has been given hitherto on the background concentration of phenol in the atmospheric air far from the sources of its emission. It has been a priori assumed to be low. It has been estimated that the background level of phenol in the atmospheric air is below 1 ng/m^3.[59] Higher concentrations are to be expected above cities, mainly as a result of its emission in gas exhausts of motor vehicles.[50,60,61]

According to estimations, the mean day concentration in urban and suburban atmosphere reaches 0.12 μg/m^3, and in the atmosphere of large urban agglomerations, it can vary within the range of $0.1–8 \text{ p.g/m}^3$.[62] In the atmosphere over the industrial sources of phenol emission, its concentration can be two orders of magnitude higher. In the neighborhood of a phenol resin–producing plant, the concentration of phenol was at a level of 190 p.μg/m^3.[63,64] Still higher values were recorded directly at the workplace in the gases released by iron casting plants; in 1980, the concentration of phenol reached at such a place was $0.8–3.5 \text{ mg/m}^3$.[65,66]

It has been estimated that half-life of phenol in air generally varies depending upon atmospheric conditions, and values ranging from 2.28 to 22.8 hr for reaction with hydroxyl radicals (*OH and HO_2)* have been reported in literature. It is the reason why a small amount of phenol does not remain in the air for longer than a day.[59]

The presence of phenol was detected in rain and surface water as well as in underground water, but relevant information is scarce. Phenolic compounds react as a weak acid in water, and they are not expected to dissociate in the pH range typical in the natural environment, i.e., 6.5–8.5.[5]

For example, the concentration of phenol dissolved in rainwater collected in Portland, United States, was $0.08–1.2 \text{ |μg/L}$, but the mean concentration was 0.26 μg/L.[67] In diluted water solutions, phenol undergoes conversion to dihydroxybenzenes, nitrophenols, nitrosophenols, and nitrochinone, probably according to the radical mechanism with nitrate(V) ions with the use of hydroxyl radicals and phenoxyl.[68,69] The compound of 2,4,6-trichlorophenol occurs in water of different types as a result of exposure to chlorine and its effect on organic precursors. The mean concentration of chlorophenols in posttreated water varies from 0.003 to 1 μg/dm^3.[69-71] If chlorine is used as a water disinfectant, different chlorophenols are formed,[72] while if chlorine dioxide is used, a variety of *p*-benzoquinones are formed.[73]

In wastewater or sewage, phenol can react with nitric(V) acid to form toxic cyanides.[74]

No information on the content of phenol in the soil has been found; however, its presence in the soil is rather unlikely taking into regard its fast biodegradation and transport in underground water or air.[75] It has been suggested on the basis of some studies[76] that release of phenol from the dry near-surface layer of soil should be rather fast.

The partition (K_{oc}) coefficient for phenol for two types of loamy soil was 39 and $91 \text{ dm}^3\text{/kg}$, which implies high mobility of phenol in the soil and its easy penetration to underground water.[60,76,77]

As follows from the logarithm of the phenol dissociation constant pK_a (log 1.28×10^{-10}) in water and in moist soil, phenol occurs in a partly dissociated form, so its transport and reactivity can depend on pH of the environment.[1,2,60,77] Phenol reacts as a weak acid in water, so it is not expected to dissociate in the range typical of the natural environment.[1,2]

Determination of phenol in drinking water performed in the United States has shown that its content is close to 1 lμg/L but usually below the detection limit.[60] Higher content of phenol was found in samples of underground water contaminated with the wastewater left after coal gasification.[60,77]

Because of its short half-life, phenol is not expected to be transported over a great distance in the atmosphere.[50] In soil, phenol is highly mobile, and its mobility, likewise its reactivity, depends on pH of the environment.

In the main elements of the biosphere, the transport of phenol takes place via its washing out by precipitations or by leaching from the soil. It is highly unlikely to be a stable and durable reagent in the natural environment.[1,2,60,76,77] The hitherto state of our knowledge on the harmful effect of phenol in the biosphere does not permit a reliable evaluation of the potential threat it may pose. Nevertheless, in view of the calculated value of the maximum admissible critical concentration of phenol in water ecosystems, e.g., 0.5 μg/L, it cannot be excluded that its presence can be related to some risk for water fauna and flora.[2,76,77]

Photooxidation, photodegradation in air, and biodegradation in water and soil are expected to be the major removal processes of phenol and its derivatives from the biosphere.[78] Phenol can potentially be removed from the atmosphere via photooxidation by reaction with hydroxyls and nitrate radicals, photolysis, and wet and dry deposition.[6,50,79] The majority of phenol from the atmosphere undergoes photodegradation to dihydroxybenzenes, nitrophenols, and products of ring cleavage, and the rest is washed out with precipitates.[50] Biodegradation is a major process for the removal of phenol from surface waters. Phenol generally reacts with hydroxyl and peroxyl radicals and singlet oxygen in sunlit surface waters.[5,50,76] The half-life of phenol in the air evaluated in photochemically active conditions in a smog chamber was 4–5 hr and was consistent with the value determined on the basis of the rate of its reaction with hydroxyl radicals.[50,78,79] In natural sunlit water reservoirs, phenols usually react with hydroxyl radicals formed as a result of photochemical re- actions.[77,79]

Phenols easily undergo biodegradation in water if they do not occur in the concentrations toxic for microorganisms, as the latter play the main role in phenol degradation in the soil, water, water sediment, and bottom sediment.[76–79] The amount of bacteria capable of phenol degradation makes up a very small percentage of the total population of bacteria present in the soil.[80] Chemical composition of the products of bacteria-assisted degradation of phenols depends on the environmental conditions. In aerobic conditions, the products include carbon dioxide,[81] while in anaerobic conditions, they include carbon dioxide and/or methane.[82]

Phenol can undergo degradation in the free form as well as in the adsorbed form, in the soil or sediment, although the presence of the sorbent decreases the rate of its biodegradation.[76–79]

Environmental Impact of Phenol and Its Derivatives

For the general population, the most important sources of exposure to phenol are those coming from the air (emission from motor vehicles and product of photooxidation of benzene) and those from cigarette smoking and consumption of smoked food products.[50] The available information on the degree of exposure to phenol is insufficient. To evaluate the risk of harmful exposure, the value of total daily intake by a man weighing 70 kg was estimated to be 0.1 mg/kg of body mass per day.[2,6] The danger of poisoning by phenol vapor has been known for a long time, although no lethal cases have been reported.[1,5,6,78] The symptoms of poisoning by phenol inhalation include anorexia, loss of body mass, headache and vertigo, hypersalivation, and dark-colored urine. The risk of poisoning by intake of poisoned water or food is highly unlikely because of the very unpleasant taste and smell of phenol.[2,77] There is some anxiety over the reports on the possible genotoxic effect of phenol and there are controversies over its carcinogenic effect.[83–85]

The available information on the health effects of phenol exposure to humans is almost exclusively limited to case reports of acute effects of oral exposure.[86]

Most of the simple phenolic compounds have similar features and are toxic to aquatic organisms. The presence of a hydroxyl group intensifies their toxicity, which is why phenol is more toxic than benzene.[2,5,86] The acute toxicological effects of phenol are predominantly upon the central nervous system, and death can occur as soon as one-half hour after exposure. Acute poisoning by phenol can also cause severe gastrointestinal disturbances, kidney malfunction, circulatory system failure, lung edema, and convulsions. Fatal doses of phenol may be adsorbed through the skin. Key organs damaged by chronic exposure to phenol include the spleen, pancreas, and kidney. The toxic effects of other phenols resemble those of phenol.

Water or wastewater treatment by exposure to chlorine (Cl_2) leads to formation of chlorinated phenols, in particular, PCP, responsible for the revolting smell and taste. Although exposure to chlorophenols has been correlated with liver malfunction and dermatitis, polychlorinated dibenzo-dioxins also may have caused some of the observed effects.[2,6,86] Moreover, phenol and chlorophenols occurring in water are bioaccumulated by water fauna, which is responsible for the unpleasant smell and taste of fish meat.[5]

The most important organs involved in phenol metabolism are the liver, lungs, and mucous membrane of alimentary track; their involvement depends on the mode of exposure and the dose absorbed. Results of in vivo and in vitro studies have proven the covalent bonding of phenol with proteins from human tissues and blood plasma. Some metabolites of phenol also react with proteins.[1,2,86] The main route of phenol elimination from humans and animals is with urine. The rate of elimination with urine depends on the mode of exposure and dose absorbed. Phenol is also eliminated with stool and exhaled air.

Identification and Quantification of Phenol and Its Derivatives

Phenol gives violet coloration with ferric chloride solution (the test reaction of phenol) due to the formation of colored iron complex, which is a characteristic of the existence of keto-enol tautomerism in phenols:

$$6C_6H_5 + FeCl_3 \rightarrow 3H^+Fe\left[O\left(C_6H_4 -\right)_6\right]^{3-} + 3HCl \tag{5}$$

In many countries, the content of phenol and its derivatives is controlled and standardized in all elements of the ecosphere, i.e., atmosphere, surface water, drinking water, soil, and food products.[87]

Development of analytical methods based on physicochemical phenomena and processes taking place in the natural environment and in living organisms has brought a decrease in the detection level and increase in accuracy of measurements. The range of applicability of analytical methods is related not only to the properties and type of substance studied but also to the selectivity and reproducibility of the method for isolation and enrichment of the analyte. From the point of view of routine analysis, the development of selective and reproducible procedures is difficult. Therefore, the application of coupled methods (off-line and/or on-line) is recommended. Because of their diversity and richness of forms of presence, phenolic compounds create many analytical problems, especially in routine analyses.[88]

Phenols react with 4-aminoantipyrine at pH 10 in the presence of potassium ferrocyanide, forming antipyrine dye, which is extracted into pyridine and measured at 460 nm.[89] This method is, however, charged with large error for the following reasons:

- Phenols with para substituents do not react with a reagent (the exceptions are para derivatives substituted with carboxyl, methoxyl, sulfonic groups, or halides).
- The yields of coupling with 4-aminoantipyrine for phenol, cresoles, xylenols, and other relevant compounds are significantly different.
- Maxima of absorption of the reaction products with the reagent are not the same for all phenols; some organic compounds that are not phenols also react with 4- aminoantipyrine.

The above-mentioned drawbacks, the lack of possibility to predict the qualitative and quantitative composition of phenolic compounds in a water sample studied, and problems with the choice of standard mixtures have prompted the introduction of the so-called phenol index describing the total concentration of phenols determined spectrophotometrically and expressed in milligrams per liter of phenol. It is in agreement with the international standard (ISO 6439) recommending also a procedure of colorimetric determination of phenolic compounds with the help of 4-aminoantipyrine, in which the phenol index is defined as the total concentration of phenolic compounds expressed in milligrams per liter phenol. The limit of determination of the phenol index is 0.010 milligrams per liter for the spectrophotometric method with 4-aminoantipyrine (nonspecific method) and 0.002 mg/L for the extraction-spectrophotometric method (single extraction with chloroform).[90]

From water and wastewater, phenolic compounds are separated by distillation with steam from an acidic solution, and then they are subjected to a reaction with 4-aminoantipyrine (1-fenylo-2,3-dimetylo-4-aminopirazolon) in an alkaline environment at a pH 9.8 in the presence of potassium ferrocyanide as an oxidant. This reaction leads to formation of an indophenolic dye of green-yellow to orange color depending on the concentration of phenols, extracted with chloroform. The content of phenols is determined spectrophotometrically by measurement of absorbance of the colored solution at the analytical wavelength, usually $\lambda = 460$ nm.[89]

The diversity of analytical tasks implies the need to choose the most appropriate analytical method for a specific problem. As follows from the survey of methods used for determination of phenols, presented by Thielemann,[91] all of them are used for the purposes.

Determination of trace amounts of phenols is realized by colorimetric and chromatographic method, while determination of phenols with content above 1% is performed by chemical bromatometric and iodometric methods.[89]

Conclusion

Phenol is one of the first compounds inscribed into the list of priority pollutants by the U.S. Environmental Protection Agency (EPA). It is commonly used in different branches of industry, including chemical production of alkylphenols, cresols, xylenols, phenolic resins, aniline, and other compounds.

Phenol is produced as an intermediate in the preparation of other chemicals and can be released as a by-product or contaminant. The economic importance of phenol and its derivatives has been significant since coal tar was used as a rubber solvent in the beginning of the 19th century. The current use of this group of compounds as pure products includes the chemical synthesis of plastics, synthetic rubber, paints, dyes, explosives, pesticides, detergents, perfumes, and drugs.

These compounds are used mainly as mixtures in solvents and constitute a variable fraction of gasoline. They have been used extensively for a wide variety of applications. They have been found to be among the most economical and effective surfactants, e.g., phenol ethoxylate and alkyl phenol ethoxylates. Production of their anionic derivatives has been severely restricted because they or their degradation products have been found to be estrogenic.

There is increasing environmental concern worldwide regarding the disposal of wastewater containing nonbiodegradable organic compounds. Since most pollutants do not respect national borders, a world effort to monitor their movement and to develop tools to prevent them from polluting environmental components or to remediate consequent pollution is desirable.

Phenol has antiseptic and germicidal properties, which increase to a maximum as the length of an alkyl side-chain substituted reaches about six carbon atoms. The lowest member, phenol, is highly toxic and caustic, but these properties progressively diminish with the higher members of the series. Polyhydric phenols are still markedly toxic but less caustic than the monohydric compounds (such as phenol).

The antiseptic activity of phenol is a property that accounts for much of its industrial use. In dilute solutions (of 1%–2%), it also finds application as an agent against skin itching. Undiluted phenol is highly corrosive to mucous membranes and skin, and it is considered a nerve poison. It may enter the human system by absorption through skin, oral ingestion, and vapor inhalation.

Phenol and three cresol isomers are of about the same order of toxicity and produce identical symptoms in poisoned animals.

Chemical and pharmaceutical industries are large users of phenol for conversion to many different products. The salicylates (carboxyphenols) are starting materials for the preparation aspirin and flavor.

Chlorophenols and their derivatives find application as fungicides, bactericides, and selective weed killers. Alkyl phenols make an important group synthetic tanning agent, and triphenyl phosphate is a plasticizer. Phenolic derivatives are among the most important contaminants in the environment. These compounds are used in several industrial processes to manufacture pesticides, explosives, drugs, and dyes. They are also used in the bleaching process of paper manufacturing. Apart from these sources, phenolic compounds have substantial applications in agriculture as herbicides, insecticides, and fungicides. However, phenolic compounds are not only generated by human activity but also formed naturally, e.g., in the process of decomposition of leaves or wood.

As a result of these applications, they are found in soils and sediments, and this often leads to wastewater and groundwater contamination. Owing to their high toxicity and persistence in the environment, both the U.S. EPA and the European Union have included some of them in their list of priority pollutants.

Current standard methods of phenolic compound analysis in water samples are based on liquid–liquid extraction, while Soxhlet extraction is the most used technique for isolating phenols from solid matrices.

Phenols and their derivatives are common in the natural environment. These compounds are used as the component of dyes, polymers, drugs, and other organic substances. The presence of phenol substances in ecosystems is also related with production and degradation of numerous pesticides and the generation of industrial and municipal sewages. Some phenols are also formed during natural processes. These compounds may be substituted with chlorine atoms, nitrated, methylated, or alkylated. Both phenols and catechols are harmful ecotoxins.

Toxic action of these compounds stems from unspecified toxicity related to hydrophobicity and also to generation of organic radicals and reactive oxygen species. Phenols and catechols reveal peroxidative capacity, are hematoxic and hepatoxic, and provoke mutagenesis and carcinogenesis toward humans and other living organisms.

References

1. Rappoport, Z. *The Chemistry of Phenols*; Wiley-Inter- science: New York, 2003.
2. Tyman, J.H.P. *Synthetic and Naturals Phenols;* Elsevier Science: New York, 1996.
3. McTavish, D. *Joseph Lister*; Bookwright Press: New York, 1992.
4. Breinholt, V. Desirable versus harmful levels of intake of flavonoids and phenolic acids. In *Natural Antioxidants and Anticarcinogens in Nutrition, Health and Disease*; Kum- pulainen, J.T., Salonen, J.T., Eds.; The Royal Society of Chemistry: London 1999; 93–99.
5. Thurman E.M. *Organic Geochemistry of Natural Waters*; Martinus Nijhoff/Dr W. Junk Publishers: Dordrecht, 1985.
6. Manahan, S.E. *Toxicological Chemistry and Biochemistry,* 3rd Ed.; CRC Press LLC: New York, 2003.
7. Verschueren, K. *Handbook of Environmental Data on Organic Chemicals*, 2nd Ed.; van Nostrand Reinhold Company: New York, 1983.
8. Spoelstra S.F. Degradation of tyrosine in anaerobically stored piggery wastes and in pig feces. Appl. Environ. Microbiol. 1978, 361–638.
9. Boskou, D. *Natural Antioxidant Phenols: Source, Structure-Activity, Relationship, Current Trends in Analysis and Characterisation*; Research Signpost: New York, 2006.
10. Glade, M.J. Dietary phytochemicals in cancer prevention and treatment. Book Rev. Nat. Rev. 1997, *13* (4), 394–397.
11. Friedman, M. Chemistry, biochemistry and dietary role of potato polyphenols. J. Agric. Food Chem. 1997, *45*, 1523–1540.
12. Kroon, P.A.; Wiliamson, G. Hydroxycinnamates in plants and food: Current and future perspectives. J. Sci. Food Agric. 1999, *79*, 335–361.
13. Robinson, T. *The Organic Constituents of Higher Plants,* 4th Ed.; Cordus Press: North Amherst, MA, 1980.
14. Duke J.A. *Handbook of Physicochemical Constituents of GRAS Herbs and Other Plants*; Ed. Academic Press; New York, 1992.

15. Hedges, J.I.; Mann, D.C. The characterization of plant tissues by their lignin oxidation products. Geochim. Cosmochim. Acta 1979, *43*, 1803–1807.

16. Herman, K. Occurrence and content of hydroxycinnamic acid and hydroxybenzoic acid compounds in foods. Crit. Rev. Food Sci. Nutr. 1989, *28*, 315–347.

17. Hubble, B.R.; Stetter, J.R.; Gebert, E.; Harkness, J.B.L.; Flotard, R.D. Experimental measurements of emissions from residential wood-burning stoves. In *Residential Solid Fuels: Environmental Impacts and Solutions;* Cooper, J.A., Malek, D., Eds.; Oregon Graduate Center: Beaverton, OR, 1981; 79–138.

18. Hardy, G. Nutraceuticals and functional foods: Introduction and meaning. Nutrition 2000, *16* (7–8), 688.

19. Aherne, S.A.; O'Brien, N. Dietary flavonols, chemistry, food content, occurrence and intake. Nutrition 2002, *18*, 75–81.

20. Stevenson, F.J. *Humus Chemistry, Genesis, Composition, Reactions;* John Wiley & Sons: New York, 1994.

21. Muscola, A.M.; Soidari M. *Soil Phenols;* Nova Science Publishers Inc.: New York, 2010.

22. Hessen, D.O.; Tranvik, L.J. *Aquatic Humic Substances: Ecology and Biogeochemistry;* Springer Verlag: Berlin, 1998.

23. Larsen, J.W.; Nadar, P.A.; Mohammadi, H.; Montano P.A. Spatial distribution of oxygen in coals. Development of a tin labelling reaction and Mossbauer studies. Fuel, 1982, *62*, 889–893.

24. Weber, M.; Weber, M.; Kleine-Boyman, M. Phenol. In *Ull- mann's Encyclopedia of Industrial Chemistry;* Wiley-VCH: New York, 2004.

25. Gardziela, A.; Pilato, L.; Knop, A. *Phenolic Resins: Chemistry, Applications, Standardization, Safety and Ecology;* Springer Verlag: New York, 2000.

26. Ryser, S.; Ulmer, G.; Etude de la pollution par les resines syn- thetiques utylisees en fonderie. Fonderie 1980, 33, 313–324.

27. *RIVM Criteria Document: Phenol, Bilthoven, the Netherlands,* National Institute of Public Health and Environmental Protection (Document No. 738513002); 1986.

28. Kuwata, K.; Uebori, M.; Yamazaki, Y. Determination of phenol in polluted air as p-derivative nitrobenzene azophenol by reversed phase high performance liquid chromatography. Anal. Chem. 1980, *52* (6), 867–860.

29. Verschueren, K. *Handbook of Environmental Data on Organic Chemicals,* 2nd Ed.; van Nostrand Reinhold Company: New York, 1983.

30. Den Boeft, J.; Kruiswijk, F.J.; Shulting, F.L. Air pollution by combustion of solid fuels; The Hague Ministry of Housing, Physical Planning and Environment (Publication Lucht No. 37), 1984.

31. Groenen, P.J. Components of tobacco smokes. Nature and quantity, potential influence on health; Zeist, the Netherlands, CIVO-TNO Institute Report No R/5787, 1978.

32. Gosselin, R.E.; Smith, R.P.; Hodge, H.C. Phenol. In *Clinical Toxicology of Commercial Products,* 5th Ed.; Wiliams & Wilkins: Baltimore, 1984; III-344–III-348.

33. Pekari, K.; Vainotalo, S.; Heikkila, P.; Palotie, A.; Luotamo, M.; Rihimaki, V. Biological monitoring of occupational exposure to low levels of benzene. Scand. J. Work Environ. Health 1992, *18*, 317–322.

34. Ahlborg, U.G.; Thunberg T.M. Chlorinated phenols: Occurrence, toxicity, metabolism, and environmental impact. CRC Crit. Rev. Toxicol. 1980, 7, 1–35.

35. Carrey, F.A. Chlorophenol. In *Encyclopaedia Britannica Online;* Encyclopaedia Britannica, 2001.

36. Mycke, B. Preliminary results on free dissolved phenolic compounds in natural Waters. In *SCOPE/ UNEP Transport of Carbon and Minerals in Major World Rivers, 52;* Degenes, E.T., Ed.; University of Hamburg: Hamburg, 1982; 571–574.

37. Roupe, K.A.; Remsberg, C.M.; Yanez J.A.; Davies, N.M. Pharmacometrics of stilbenes: Segueing towards the clinic. Curr. Clin. Pharmacol. 2006, *1*, 81–101.

38. Begon, M.; Harper, J.L.; Townsend C.R. *Ecology, Individuals Populations and Communities;* Black Science Ltd.: Edinburgh, 1996.

39. Smith; M.B.; March, J. *Advanced Organic Chemistry. Reactions, Mechanisms, and Structure*, **6**th Ed.; Wiley-Inter- stice: New York, 2007.

40. Rice-Evans, C. *Screening of Phenolic and Flavonoids for Antioxidant Activity, Antioxidant Food Supplements in Human Health*; Academic Press: New York, 1999.

41. Golovanow, I.B.; Zhenodarova, S.M. Structure-property correlations. XV. Properties of phenol derivatives. Russ. J. Gen. Chem, 2008, 73 (10), 1603–1607.

42. Herman, K. Occurrence and content of hydroxycinnamic acid and hydroxybenzoic acid compounds in foods. Crit. Rev. Food Sci. Nutr. 1989, *28,* 315–347.

43. Dragsted, L.O. Antioxidant actions of polyphenols in humans. Int. J. Vitam. Natur. Res. 2003,73, 112–119.

44. Budryn, G.; Nebesny, E. Phenylacids. Properties, occurrence in plants and metabolic conversion. Bromat. Chem. Toksycol. 2006, *102,* 547–554.

45. Foley, S.; Navartman, S.; MvGarvey, D.J.; Land, E.J.; Truscott G.; Rice-Evans, C.A. Singlet oxygen quenching and the redox properties of hydroxycinnamic acids. Free Radic. Biol. Med. 1999, *26* (9/10), 1202–1208.

46. Kirk, R.E.; Othmer, D.F. *Encyclopedia of Chemical Toxicology,* 3rd Ed.; John Wiley and Sons: New York, 1980; Vol. 17, 373–379.

47. Bolling, F.J.; Decker, K.H. *Phenols Resin;* Kunstshoffe: Berlin, 1980.

48. Ying, G.G.; Wiliams, B.; Kookana R. Environmental fate of alkylphenol and alkyl ethoxylates—A review. Environ. Int. 2002, *28,* 215.

49. Budavari, S. *The Meck Index. An Encyclopedia of Chemical Drugs and Biologicals*; Whitehouse Station: New York, 1996.

50. Seinfeld, J.H.; Pandis S.N. *Atmospheric Chemistry and Physics;* John Wiley & Sons, Inc.: New York, 1996.

51. van Gemert, L.J. *Flavour Thresholds. Compilations of Flavour Threshold Values in Water and Other Media*, (second edition) Ed. Oliemans Punter & Partners BV; the Netherlands, 2011.

52. US EPA. Risk Management Air Research; 1982.

53. DeSelms, R.H. *UV-Active Phenol Ester Compounds*; Eni- gen Science Publishing: Washington, DC, 2008.

54. Svobodova, A.; Psotova, J.; Walterova D. Natural phenolics in the prevention of UV-induced skin damage. A review. Biomed. Papers 2003, *147* (2), 137–145.

55. Fiege, H.; Voges H.M.; Umemura S.; Iwata, T; Miki, H.; Fujita, Y.; Buysch, H.J.; Garbe, D.; Paulus, W. Phenol derivatives. In *Ullmann's Encyclopedia of Industrial Chemistry*; Wiley-VCH: Weinheim, 2000.

56. Ueoda, M.; Aizawa T.; Imai, Y. Preparation and properties of polyamides containing phenoxathiin units. J. Polym. Sci. Chem. Ed. 1977, *15* (11), 2739–2747.

57. Eisenreich, S.J.; Looney, B.B. Thornton, J.D. Water solubility enhancement of pyrene in the presence of humic substances. Environ. Sci. Technol. 1981, *15,* 30.

58. Schnitzer, M.; Khan S.U. *Humic Substances in the Environment;* Marcel Dekker: New York, 1972; 1–7.

59. Web-based Archive of the Dutch National Institute for Public Health and the Environment (RIVM), Final report 2002.

60. Howard, P.H. *Handbook of Environmental Fate Exposure Data of Organic Chemicals*; Lewis Publishers: New York, 1991; Vol. 1.

61. Kawamura, K.; Kaplan, I.R. Stabilities of carboxylic acids and phenols in Los Angeles rain waters during storage. Water Res. 1990, *24* (11), 1419–1423.

62. Brodzinski, R.; Singh H.B. Volatile organic chemicals in the atmosphere: An assessment of available data, EPA 600383o27(A); US Environmental Agency, Office of Research and Development: Research Triangle Park, NC., 1983.

63. Tesarova, E.; Paczkowa, W. Gas and high-performance liquid chromatograph of phenols. Review paper. Chromatographia 1983, *17* (5), 269–284.

64. Grosjean, D. Atmospheric fate of toxic aromatic compounds. Sci. Total Environ. 1991, *100*, 367–414.

65. Kuwata, K.; Uebori, M.; Yamazaki, Y. Reversed-phase liquid chromatographic determination of phenols in auto exhaust and tobacco smoke as *p*-nitrobenzenephenol derivatives. Anal. Chem. 1981, *53* (9), 1531–1534.

66. Bruno T.J.; Svoronos P.D.N. *Basic Tables for Chemical Analysis,* 3rd Ed.; CRC Press: Boca Raton, 2011.

67. Rogge, W.F. Molecular tracers for sources of atmospheric carbon particles: Measurements and model predictions. Ph.D. Thesis, California Institute of Technology: Pasadena, 1993.

68. Niessen, R.; Lenoir, D.; Boule, P. Phototransformation of phenol induces by excitation of nitrate ions. Chemosphere 1988, 17, 1977–1984.

69. Louw, R.; Santoro D. Comment on formation of nitroaromatic compounds in advanced oxidation processing: photolysis versus photocatalysis. Environ. Sci. Technol. 1999, *33*, 3281.

70. Kunte, V.H.; Slemrova, J. Gaschromatographische und massenspektrometrische Identifizierung phenolischer substanzen aus ober-flachenwassern. Zeitschrift fuhr Wasser und Abwasser Forschung 1975, *8*, 176–182.

71. World Health Organization (WHO). *Guidelines for Drinking Water Quality,* 2nd Ed.; Geneva, 1984, Vol. 1.

72. Knoevenagel, K.; Himmelreich, R. Degradation of compounds containing carbon atoms by photooxidation in the presence of water. Arch. Environ. Contam. Toxicol. 1976, *4* (3), 324–33.

73. Jarvis, S.N; Straube, R.C.; Wiliams, A.L.J.; Bartlett C.L.R. Illness associated with contamination of drinking water supplies with phenol. Brit. Med. J. 1985, *20*, 1800.

74. Tratnyek, P.G.; Hoigne J. Kinetics of reactions of chlorine dioxide (OClO) in water (II). Quantitative structure–activity relationships for phenolic compounds. Water Res. 1994, *28* (1), 57–66.

75. Nakagawa, N.; Mizukoshi, O.; Maeda, Y. Sonochemical degradation of chlorophenols in water. Ultrason. Sonochem. 2000, 7 (3), 115–120.

76. Faust, S.D.; Stutz, H.; Aly, O.M.; Andersen, P.W. Recovery, separation, and identification of phenolic compounds from polluted waters. Part I—Occurrence and distribution of phenolic compounds of the surface and ground waters of New Jersey. Geological Survey Open- File Report, 1970.

77. Simonow, L.; Sargsyan, V. *Soil Chemical Pollution, Risk Assessment, Remediation and Security*; Springer: Dordrecht, 2007.

78. Yong, R.N.; Mohamed, A.M.O.; Warkentin, B.P. *Principles of Contaminant Transport in Soils*; Elsevier: New York, 1992.

79. Gilman, A.P. *Chlorophenols and Their Impurities: Health Hazard Evaluation*; Environmental Health Directorate, Health Protection Branch: New York, 1988.

80. Schwarzenbach, R.P.; Gschwend, P.M.; Imboden, D.M. *Environmental Organic Chemistry,* 2nd Ed.; Wiley-Interstice: New York, 2003.

81. Hickman, G.T.; Novak, J.T. Relationship between subsurface biodegradation rates and microbial density. Environ. Sci. Technol. 1989, *23*, 525–532.

82. Boyd, J.T.; Carlucci, A.F. Degradation rates of substituted phenols by natural populations of marine bacteria. Aquat. Toxicol. 1993, *25* (1), 71–82.

83. Dobbins, D.C.; Thornton, J.; Jones, D.D.; Fedale, T.W. Mineralization potential for phenol in subsurface soils. J. Environ. Qual. 1987, *16* (1), 54–58.

84. Bukowska, B., Kowalska S. Phenol and catechol induce prehemolytic and hemolytic changes in human erythrocytes. Toxicol. Lett. 2004, *152* (1), 73–84.

85. Bruce, R.M.; Santodonado, J.; Neal, M.W. Summary review of the health effects associated with phenol. Toxicol. Indust. Health 1987, 3, 535–568.

86. Oikawa, S.H.; Hirosawa, K.; Kawanishi S. Site specificity and mechanism on oxidative DNA damage induced by carcinogenic catechol. Carcinogenesis 2001, 22, 1239–1245.

87. WHO. IPCS Environmental Health Criteria for Phenol (161). First draft prepared by MS G.K. Monitzan; WHO: Finland, 1994.

88. Thompson, R.D. Determination of phenolic disinfectant agents in commercial formulations by liquid chromatography. J. AOAC Int. 2001, *84* (3), 815.

89. Buszewski, B.; Buszewska, T.; Szumski, M.; Siepak, J. Simultaneous determination of phenols and polyaromatic hydrocarbons isolated from environmental samples by SFE–SPE-HPLC. Chem. Anal. 2003, *48*, 13–25.

90. *Standard Methods for Examination of Water and Wastewater.* APHA, AWWA, WPCF: Washington, DC, 1992.

91. International Standard ISO 6439. Water quality—Determination of phenol index—4-Aminoantipyrine spectrometric methods after distillation, Prepared by Technical Committee ISO/TC 147; Geneva (Switzerland) 1990, p. 7.

92. Thielemann, H. Thin layer chromatographic separation of phenol, cresols, dimethylphenols and naphthols. Die Pharmazie 1970, *25* (5), 365–366.

31

Phosphorus:
Agricultural Nutrient

John Ryan,
Hayriye Ibrikci,
Rolf Sommer, and
Abdul Rashid

Introduction

Never in the history of mankind have there been such widespread societal concerns about the capacity of our planet to sustain its projected population growth. This concern was first widely articulated in the late 19th century by Thomas Malthus. However, his prognosis of widespread starvation has been challenged by advances in agricultural technology and medicine. These revolutionary developments led to not only a better-fed population but also a vastly expanded population in the past century. While population growth has stabilized in developed countries, it is continuing unabated in many areas of the lesser-developed world. World population is now projected to increase from its current 6.8 billion people to over 9 billion by midcentury.[1] In addition to such increases, world food production will have to be doubled as a result of increased affluence in some developing countries such as India and China.[2] The threat of the "population monster" that Norman Borlaug, Nobel laureate and father of the Green Revolution, railed against is now more credible,[3] with current predictions of a world population of more than 9 billion people by the middle of the century.

The implications for mankind of a more crowded world, where the earth's natural resources are increasingly threatened and even the future of mankind is called into question, are both ominous and daunting.[4,5] The "grand challenges" that our society faces embrace all aspects of agriculture and the environment. A litany of such challenges includes both biophysical (land degradation, water scarcity, loss of biodiversity) and socioeconomic (population growth, poor nutrition and health care, poverty, inadequate research investment and information infrastructure) difficulties. Looming large over these concerns are the implications of climate change, with areas of the world such as the Mediterranean and much of Africa likely to become drier and others to endure climate-related stresses.[6] While much research attention is now focused on adaptation strategies, the predicted change scenario is likely to exacerbate the world's already precarious food production capacity. Such developments have posed an extraordinary challenge to the global scientific community[7] and especially the international network of agricultural research centers.[8] Chemical fertilizers are vital to world food production and food security,[9] with over half of global output attributed to applied fertilizer nutrients and future food supplies being even more dependent on fertilizers.[10] Consequently, fertilizer management, in a sustainable manner that protects the environment and the natural resource base, is central to the issue of balancing the world's food supply–demand equation.

Among the nutrients that are essential to plants but potentially damaging to the environment are nitrogen (N) and phosphorus (P). Large amounts of N are used in modern agriculture; in fact, N fertilizer is the driver of the phenomenal increases in global crop production, especially cereals, in the past half century.[11] Data on current fertilizer N suggest that the trend in N use is set to continue, especially in countries such as China and India, which are now the leading producers of wheat in the world. However, as N is either applied as nitrate (NO_3^-) form as a fertilizer or rapidly converted biologically to this mobile form of N, losses of N from agricultural environments through leaching and runoff became associated with high N fertilizer use.[11,12] Enrichment of water sources with NO_3 not only represents economic losses in terms of fertilizer costs but also causes a potential health hazard in drinking water.[13] More widespread effects of excess N were manifested in terms of water quality only when combined with P,[14] an element that was initially thought to be immobile and subject to minimal loss from the point of application in farmers' fields. Gradually, the catalytic effect of P combined with N and eutrophication or the development of anoxic conditions in water bodies were recognized by agricultural scientists.[15] This brings us to a consideration of the role of P in agriculture and the inadvertent effects of its misuse on the environment,[16] as well as management strategies, which were recognized by agricultural scientists.[17] A brief overview of the various aspects of P in soils and the relationship of P to crop production is a prerequisite to any discussion of P in relation to the environment.

Perspective on Phosphorus

As with carbon and N, the overall nature of P in terrestrial and aquatic ecosystems and the atmosphere can best be described as a cycle, which indicates changes between one state and another, i.e., a state of flux. Various depictions of such a cycle have been made, but the one adapted by Tunney et al.[15] gives a good representation of the inputs and outputs involving P and the relationship with the various P phases in nature (Figure 1). As plant uptake occurs from the soluble P fractions—and as losses occur from that fraction as well—solution P is a central component of the P cycle and is closely linked to microbial P. Inputs to the system are primarily fertilizers (and plants) to the solution phase and manures to the pool involving microbial forms of P; both forms are being replenished by or contributing to somewhat less soluble inorganic and inorganic phases, i.e., labile or moderately labile forms, and constitute the more rapidly changing part of the P cycle.

The behavior of P in nature is also influenced by very slow reactions that occur over several years and do not have an immediate impact during one cropping season. The slow inorganic component of the cycle involves primary and secondary P-bearing minerals that can be weathered to increase solution P; some inorganic forms (occluded) have no direct effect on soluble P. Similarly, some organic P forms have little immediate effect on the more reactive P phases in the cycle. While the depiction in Figure 1 does not specifically indicate the avenue of losses, the soluble P phase is most susceptible to loss by leaching and runoff in addition to the actual transport of P-enriched soil particles by erosive processes. The fundamental problem is related to excessive inputs to the system in terms of chemical fertilizers and manures, thus exceeding the crop's needs and the soil's capacity to retain the P against loss.

No element can rival P in terms of its chemical complexity, its dynamic equilibrium, and the diversity of inorganic and organic P compounds. The addition of P fertilizers to soils and its uptake and

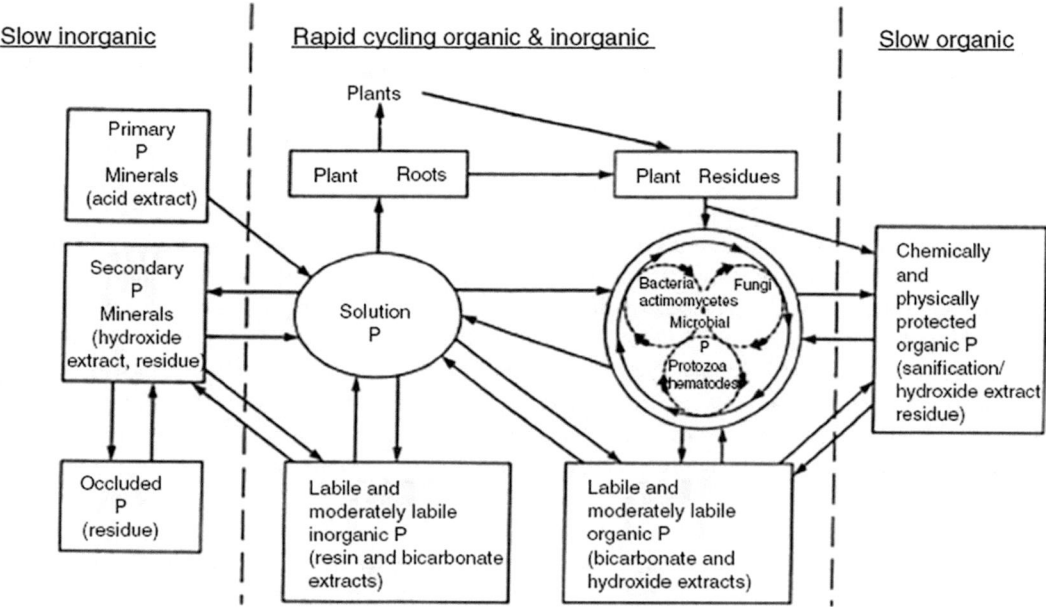

FIGURE 1 The soil P cycle in nature and its reaction phases.

utilization by crops pose further challenges to soil and crop scientists. Much has been written about P from the basic and applied perspectives, especially during the past half century. The thousands of research papers dealing with various aspects of P have gradually revealed a better understanding of this tantalizing element. Progress in P research has been documented in broad-based discipline reviews[18-20] in addition to regional[21-22] and country-level[23-24] reviews. While such references are selective and subjective, they broadly capture the slow but inexorable unraveling of the many aspects of P in soils, fertilizers, and crops.

Shifting Emphasis

A brief survey of the recent literature on P indicated a revision of concepts related to P availability and a shift in emphasis from production-related P research to that related to the environment. For instance, while the early P review of Larsen[18] did not refer to the environment aspect of P use, the wide-ranging P review of Khasawneh et al.[20] addressed the issue in only one brief chapter[25] of the 29 chapters in the volume. That review addressed the emerging concerns of P in the environment, geological influences on soluble P, as well as agricultural losses of P through surface flow and sediment transport; P losses from animal wastes and urban runoff were also briefly addressed.

While the subsequent comprehensive review of P still focused on the agricultural aspect of P,[26] it devoted a major part of the volume (five chapters) to P in relation to the environment. The rapidly expanding awareness of the environmental implications of P led to the establishment of a series of international workshops with focus on the environment.[15] As an illustration of the current research emphasis on P in the environment, keynote addresses at the recent meeting of the workshop series[27] dealt with issues such as P mobilization at the plot and field scales, field-scale indices of P loss, P dynamics and impact in water bodies, and implementation of mitigation options. Reviews such as that of Delgado and Scalenghe[16] focused on Western Europe as one of the world's hot spots for P pollution, where the consequence of overuse of P fertilizers and animal manures had attracted broad societal and legislative concern.

Phosphorus: Potentially Harmful Effects

Though vital to all forms of life on earth, P has no known toxic effects on humans or animals unlike some other elements. The potential effect of P on the environment is related to an excess of P in the terrestrial ecosystems with a carryover to the aquatic environment. Eutrophication of water bodies involves the excessive growth of undesirable algae and aquatic weeds and a consequent depletion of such organisms.[28] The process of eutrophication is defined as an increase in the fertility status of natural waters that causes accelerated growth of algae or water plants. Eutrophication is due not solely to P but to a complex interaction between N, P, environmental conditions (temperature, salinity, light), and the physical and hydrologic characteristics of surface waters, i.e., streams, lakes, and estuaries.[29] Under natural conditions, growth of aquatic organisms is limited by the normally low levels of P in the water. External inputs of P from urban wastewater, surface runoff, or subsurface flow can upset the delicate P balance in water bodies and stimulate growth of aquatic organisms to ecologically undesirable levels.

Critical levels for soluble and total P associated with eutrophication are in the order of 10 and 20 mg L^{-1},[17] while others have indicated a range of values from 20 to 100 mg L^{-1},[29] which is still considerably less than such thresholds for N. The main driver of eutrophication is depletion of dissolved oxygen in the water due to the excessive growth of aquatic organisms and microbial decomposition of such biota. This leads to a chain reaction that causes a deterioration in the water's ecology, e.g., decreased light penetration, surface algal scums, foul odors, impeded water flow, and increased turbidity and sedimentation. Associated with such pollution is the occurrence of surface blooms of cyanobacteria that can reduce water palatability and even kill livestock and pose a threat to human health. Advanced eutrophication of surface water thus has serious implications for fisheries, recreation, and human or animal consumption.

The primary cause of eutrophication is the P that originates from agricultural land as nonpoint source or from urban and industrial effluents. The problem of P-induced pollution was first noted in the Great Lakes region of North America in the 1960s. The phenomenon was later seen in P-enriched lakes in Western Europe. Initially, the problem was attributed to point sources, i.e., sewage-derived inputs. Control measures included sewage diversion, removing P from sewage effluents, and reducing P contents of detergents. However, the persistence of eutrophication in many water bodies following the control of plant-source P emissions led to a focus on loss from agriculture as the primary cause of continued eutrophication. While the problem may appear to lend itself to a simple solution of avoiding use of excess P fertilizer or P-containing manures, Tunney et al.[15] argued that the process of P transport is anything but simple as it depends on soil properties, flow pathways, and bioavailability. Development of a strategy to control or mitigate P losses from fertilized fields or a watershed requires an understanding of the processes and mechanisms that govern P loss.[17] Thus, it is appropriate in this review to provide a brief background on the behavior of P in soils and to outline how such factors impinge on the environmental dimension of P.

Phosphorus Behavior in Soils

In order to establish the significance of P for the environment, there are a number of key aspects that need to be considered: the need for P for crops, forms of P in soils, reactions in soil, P cycling, and P availability. In addition, consideration has to be given to global fertilizer use as well as reserves of P for future use. This brief discussion is extracted from several sources.[15,26,30,31] Much of the information presented is now firmly part of the accepted literature on P in agriculture and the environment.

Phosphorus in Soil: Forms and Amounts

The essentiality of P for plant growth has been established for almost a century. In contrast to other elements such as N and potassium, the total amount of P in soils is relatively low, generally in the range of 0.01%–0.3% (100–3000 mg P kg^{-1}). In its native state, P solubility is extremely low, i.e., less than 0.01 mg L^{-1}. This soluble P pool, from which plants derive their P for plant uptake and growth, has to be replenished many times during the growing season in order to sustain growth. Thus, a sort of equilibrium exists between P in solution and solid-phase P. In natural ecosystems, the supply of P to plant roots is governed by reactions involving sorption, desorption, and precipitation. In such conditions, the small amount of P in solution is maintained by weathering and dissolution of rocks and minerals of low solubility. Without fertilization, which greatly increases the soluble P pool, few soils can maintain an adequate P supply to meet the needs of modern high yielding agricultural crops.

Prior to considering the soluble or "available" P fractions further, it is necessary to appreciate the forms of P in soils, which can be categorized into inorganic and organic P forms, the former being dominant in most soils except those high in organic matter. Inorganic P compounds range widely in number and type depending on soil properties but are generally associated with amorphous and crystalline sesquioxides, i.e., iron (Fe) and aluminum (Al), mainly in acid soils, and calcium (Ca) compounds in calcareous soils. The relative distribution of P compounds is largely influenced by the degree of soil weathering, which in turn influences the reactivity of these compounds.

In acid soils, hydrous oxides of Fe and Al readily react with added soluble P, resulting in precipitation, while in neutral or calcareous soils, Ca controls P reactions. In Ca dominated systems, precipitates in the following order of solubility have been identified: monocalcium phosphate, dicalcium phosphate dihydrate, and hydroxyapatite or fluorapatite; the latter compound has extremely low solubility and thus controls solution P concentration. In acid soils, few well-crystallized P compounds have been observed. Reactions of soluble P with solid-phase components also involve sorption onto adsorbing surfaces. In effect, it is difficult to separate the process of precipitation from sorption. However, there is evidence that regardless of the mechanism involved, these reactions can be reversed. Precipitation/dissolution processes differ from sorption/desorption in that the solubility product of the least soluble P compound governs solution P concentration, whereas solution P controls the amount of P sorbed. In essence, retention of soluble P by soil components is a continuum between precipitation and surface reactions. In practice, P sorption curves or isotherms have been used extensively to describe the relationship between the amount of P sorbed or removed from solution by soils and that remaining in solution. Such isotherms have been used to identify sorption maxima, bonding energy between soils, and the equilibrium P concentration at which no sorption or desorption occurs.

Regardless of the type of P reactions and the soil components involved, addition of soluble P to soils results in an immediate increase in soluble or "plant-available" P. Numerous studies have documented a subsequent rapid decrease in P availability followed by a slower rate of decline in solubility. The initial rapid phase was invariably attributed to surface reactions and/or precipitation, while the slow decline phase was attributed to diffusive penetration of adsorbed P into soil components ("absorption") and increased crystallization of precipitated P compounds. Despite the dominance of Ca in the chemical reactions of P in calcareous soils, Fe oxides (both amorphous and crystalline) have been shown to have a disproportionate role in the initial and subsequent reactions in soils.[32–34] Such oxides accentuated the rate of decline of soluble P and decreased the desorption rate. From the earliest years of P experimentation, this decline in P solubility, following reactions of soluble P with soils in laboratories, was accompanied by decreased plant P availability in greenhouse and field studies. The process was initially considered to be irreversible and often termed "fixation," retention, or immobilization.

Phosphorus Availability: Critical Concentration, Fertilization

A prerequisite to rational use of P fertilizers is knowledge of what are the critical levels of plant-available P in the soil. Much research has been conducted, involving correlation and calibration to identify appropriate indices of availability and critical test values in the field beyond which no response to added fertilizer would occur. A wide range of chemicals have been tested for availability indices based on acid dissolution, anion exchange, cation complexation, and cation hydrolysis.[35] Depending on the country or soil type and region, tests such as the Mehlich I, Bray I, and Olsen procedures are widely adopted. Multinutrient extractants such as the Mehlich III and ammonium bicarbonate-diethylenetriaminepentaacetic acid (AB-DTPA) tests have increased in popularity. Regardless of the extractant used, each should theoretically extract a fraction of the relatively soluble phase of soil P that showed a close relationship or correlation with plant uptake. Subsequently, field studies are required to establish a critical P value and thus distinguish between sufficiency and excess of soil P.[36] Thus, the amount of P fertilizer required is inversely related to soil P test values. Such guidelines are essential in order to ensure that fertilizer is applied in situations where it is needed and where it will evoke an economic crop response; conversely, they will prevent unnecessary use of fertilizer when the soil has adequate amounts of the nutrient for the season's crop, thus reducing input costs and avoiding the potential of loss of the excess P to the environment.

Estimates of critical soil values are dependent on the specific test and are influenced by soil factors; for instance, P values of 10 and 30 mg kg^{-1} were established for the Olsen and Bray I, respectively, and 10 and 25 mg kg^{-1} for clay and sandy soils, respectively. In addition to specific tests and soil type, many other factors influence P availability: temperature, compaction, moisture, aeration, pH, and type of clay. Once the need for P fertilizer is established, and a basis for the required amount indicated by critical test values that are calibrated for field conditions has been determined, other factors have to be considered in the actual fertilization process. In contrast to N, timing of P fertilizer application is less important. As an immobile nutrient, P is applied and mixed with the soil for cultivated crops, but broadcast applications are feasible only for established perennial forage crops. Regardless of whether P is applied as inorganic fertilizer or as animal manures, such additions invariably increase available P levels in the soil.

Due to the immobility of P applied to soils, efficiency of uptake per unit of P applied is always higher when the material is placed close to the seed compared with broadcasting.[24] Such placement puts the P fertilizer close enough to the seedling roots, while minimizing contact immobilization or relation. Such positional availability is also influenced by crop type, specifically its rooting pattern. In order for banding or restricted P placement to enhance P uptake by roots, the rate of absorption and root growth in the P-enriched soil zone must compensate for the limited impact on root growth in the unfertilized portion of the root zone. Much research has been done on P placement with the common range of field crops with respect to the permissible amounts of P fertilizer that can be used without damaging the seedlings and with respect to establishing optimum distances for the fertilizer band to the side or below the seed, usually about 2–5 cm. Regardless of the method of P application, the question of residual P availability arises.

Residual Phosphorus Availability: Changing Concept

Early years of P research focused on laboratory studies of P reactions in previously unfertilized soils, along with short-term greenhouse and one-season field studies. The outcome of such investigations led to the notion of P being "fixed" and of limited effectiveness following the first year of fertilization. Gradually, as commercial P fertilizer use became almost universal in agriculture, especially in the United States, Europe, and other developed countries, the concept of residual P availability emerged.[31] Evidence in support of this new concept came from long-term agronomic trials in which treatments involved initial one-time P dressings as well as current or yearly applications.

A few examples will suffice to illustrate residual availability. For instance, Halvorson and Black[37] showed that Olsen P test levels were increased for 16 years following P fertilization with one application. Following the initial increase, P levels declined slowly for 12 years and then leveled off at a level higher than initial values, suggesting a new equilibrium of available P. The elevated soil P levels in this study were also accompanied by corresponding increases in crop yields. Similar soil and crop yield responses were observed from even longer-duration trials.[38] The general acceptance of long-term effects of P fertilization was reflected in a entry on "Evaluation and Utilization of Residual Phosphorus in soils" by Barrow[39] in the comprehensive milestone on *The Role of Phosphorus in Agriculture* by Khasawneh et al.[20] With buildup of residual P, availability cropping can occur without any P fertilizer. In studies cited by Sharpley,[30] the role of decrease in available P by depletion with cropping is inversely related to the soils' buffering capacity or the P sorption saturation (available P/P sorption maximum). In shorter-term, multiyear trials under dryland Mediterranean conditions in Turkey[40] and Syria,[41] a relatively quick buildup of available P was observed following modest fertilization of calcareous soils.

The evaluation of P availability with progress in accumulated research, especially at the field level, brought with it a changed concept of efficiency. Based on early studies that evaluated P growth and uptake efficiency in terms of differences between fertilized and control plots, P fertilizer use efficiency was generally believed to be less than 20%. However, the concept of efficiency differed depending on the methods used to evaluate it. The direct method uses a radioactive isotope of P (^{32}P), but due to its short half-life, it is amenable only to short-term studies. The difference method considers differences between fertilized and unfertilized crops, i.e., "agronomic efficiency" for yield and "apparent recovery" for P uptake. On the other hand, the balance method considers only yield and P uptake relative to the amount of P fertilizer applied. In a recent comprehensive review of P efficiency supported by several case studies, Syers et al.[31] clearly demonstrated significant and consistently higher efficiency values for the balance method compared with the difference method. The key issue in such a method is the time scale. Thus, when a residual effect of P is considered over several years, efficiency values in the order of 90% can occur. This revised concept of P use efficiency represents a paradigm shift in soil fertility assessment and crop nutrition.

Dynamic Nature of Phosphorus in Soils

Of many and varied aspects of P, including broad categories such as P fertilizer sources, reactions of P fertilizers in soils, plant nutrition and crop management, animal nutrition, management practices, and the environment,[26] few are as intriguing as the dynamic or changeable nature P in soils, which was addressed recently by Condron et al.[42] for P organic P and by Tiessen[43] for P in tropical soils. In an earlier brief review, Ryan and Rashid[44] outlined the inputs to the soil solution phase of P, potential losses or withdrawals from that small fraction, and the extent to which various categories exist based on solubility. The dynamic nature of P in soils has been schematically depicted by Johnston[45] and is a simplified version of the P cycle (Figure 1), which represents the big picture with respect to P. In contrast to the P cycle, the diagram of Johnston[45] does indicate losses of soluble P, while its unique feature is the representation of the availability pools that impact the soluble P pool. The "readily available" pool represents a proportion of soil P that can be estimated by soil tests and can contribute substantially to P uptake by the crop. This pool is in apparent or pseudo-equilibrium with a less soluble fraction of soil P. A major fraction of soil P is in the very-slowly available pool. There is some evidence that this fraction can be slowly dissolved or weathered to sustain the less-readily available P pool.

In essence, these solubility categories or phases represent suites of inorganic and organic P compounds. Nevertheless, despite its simplicity, the schematic of Johnston[45] captures the essence of P dynamics and is centered on the soluble P pool from which plant uptake by roots occurs; this phase can be depleted by loss in drainage or greatly increased by the addition of soluble P fertilizers. With continued crop uptake, this tiny pool is replenished by the readily available P pool, which in turn is in "equilibrium" with a less soluble P phase. While the overall direction of P reactions is toward decreased solubility, the reverse can also occur under some circumstances. In terms of the plant roots for P uptake, accessibility is immediate in the solution phase but decreases with decreasing solubility to the right of the sketch.

Global Fertilizer Use and Phosphorus Sources

A snapshot of P in relation to both agriculture and the environment would not be complete without reference to P fertilizer usage at the global level, followed by a notion of the world's reserves of P to sustain the future of life on this planet. Production of phosphate rock, the raw material for P manufacture, has consistently expanded to meet demand primarily for P fertilizer (80%) as well as for industry. With the projected increase in world population—and consequently, crop production—fertilizer use is set to continue.[46] In essence, as more P is used, the potential for impact on the environment increases.

The continued use of phosphate rock raises the issue of the world's reserve of this finite resource,[47] which is vital to food production—and, indeed, to life on earth. That P is limited underlines the need for efficient use as well as recycling, furthering the argument for reducing or eliminating unintended deleterious effects on the global environments. While P rock is mined in many countries, the world's supply is dominated by relatively few countries, i.e., the United States, Russia, China, and Morocco. Much debate has centered on the longevity of P deposits. Many factors have to be considered, establishing estimates such as the likelihood of new discoveries, energy costs associated with P mining, market prices for P fertilizers, and continued demand. While estimates of P reserves and resources range from 105 to 470 years,[48] more recent predictions by the International Fertilizer Development Center[49] set the figure between 300 and 400 years but cautioned that many factors, known and unknown, can greatly influence such estimates. Coinciding with such estimates, Cordell et al.[50] argued that "peak" consumption will occur in the next few decades. Recognizing the finiteness of P reserves, as there is no alternative source of P, Withers[51] cautioned that protecting future food security requires a radical rethink of how P is managed from field to global scales; he further argued the need for closing gaps in the P cycle to minimize wastage and stressed the need to recover or recycle P. More effective use of P already in circulation is a prerequisite to reducing society's dependence on inorganic fertilizers and minimizing the environmental footprint of P fertilizer manufacture.

Implications for Phosphorus Loss

Without P fertilization of soils in their natural state, the potential for movement of P in its soluble state is essentially insignificant. Even when P is lost in sediments due to erosion, the impact on water bodies is minimal due to the low concentrations of P in unfertilized soils. The problem is one that accompanies application of commercial fertilizers and manures, involving excess fertilization beyond the needs of the crop. However, as economic crop production is unthinkable without P fertilizer use, the problem is one that society has to live with. However, there are guidelines and a code of fertilizer management practices that can help farmers avoid excess P use while maintaining adequate P for optimum crop production.[17] Despite the propensity of soil to react with P fertilizers, recent research has shown that excess P can overcome or saturate soil's capacity to retain P tightly. As more P is added, the soluble P fraction increases, and that is where P "leakage" occurs. Fertilization increases the P fractions attached to soil particles and increases their potential for loss when the soil is transported as sediments. An overriding condition for loss of P from fertilized land is heavy rainfall; this induces more leaching and runoff and, where appropriate soil conservation measures are not in place, promotes P loss in sediments. Accordingly, many of the hot spots of P pollution of water bodies occur in intensively formed areas or watersheds with high rainfall, e.g., Western Europe, Mississippi Valley and the Gulf of Mexico, and Chesapeake Bay in northeastern United States.

Phosphorus Losses and Mitigation Strategies

Anthropogenic or man's activities have been responsible for the accelerated cycling and fluxes of P at global and regional scales in the past few decades, resulting in eutrophication of terrestrial and aquatic ecosystems together with biodiversity loss and human health risk.[51] Current estimates indicate that fluvial transport of dissolved (about 5 Tg yr^{-1}) and particulate P (20 Tg yr^{-1}) to oceans (a permanent P sink) is at least double that of preindustrial times.[52,53] Without concentrated efforts to increase P use in agriculture through interventions, such losses are likely to continue with projected increases in commercial fertilizer consumption, higher-yielding crops, bioenergy crops, urbanization, and economic growth.[51] Recognition of the threat that such developments posed for the environment has led to expanded funding to mitigate such an environmental threat, especially in areas where the problem of P loss is acute. For instance, from 2000 to 2008, the European Research Framework Program invested over €6 million to support P-centered projects, most of which had an environmental focus.[16] Such emphasis on surplus P in the environment is in stark contrast to the situation ongoing in many developing countries, especially those in Africa, where available P is a major crop-limiting factor and where fertilizer use is dismally low or nonexistent.

However, with respect to the issue of transfer of P from agricultural watersheds, resulting in eutrophication of surface waters, a number of mitigation strategies were advanced by Sharpley and Tunney.[17] Soil P testing for environmental risk assessment was put forward as a means of establishing acceptable P losses compatible with economic farming; an issue is the appropriateness of a field/ plot or watershed scale for such studies. Recently, Maguire et al.[54] reviewed approaches to using soil testing to delineate threshold levels above which P fertilizer use is limited or not allowed. While soil testing was traditionally used for diagnosing P deficiency for crop production, it could also define an upper limit for soil P with respect to potential P losses, but many issues still need to be clarified by research. Threshold P levels should especially consider site vulnerability to P loss. An analysis of the pathways of P transport is needed at the watershed scale.

The current approach to fertilizer management in general using best management practices is one that is eminently applicable to mitigating P loss; such approaches are designed to bring P inputs as fertilizers and manures in synchrony with crop demands for P. Conventional erosion control measures form a key element in this approach. The current drive toward minimum- or no-till is a significant development in reducing P losses from cultivated land. While much remains to be learned technically in the area of P loss mitigation, a lot of existing knowledge is already available for adaptation in tackling P loss. Strategic initiatives are needed to bring about lasting change in land management targeting consumer-supported programs that encourage better standards of land management by farmers, inculcating a sense of stewardship with societal-accepted environmental protection goals.

Conclusions

The indispensability of P fertilizers for global agricultural production is undisputed, except for some adherents to organic agriculture; in fact, both commercial fertilization and organic farming should be complementary. With increasing demand on our land resources to feed the world's growing population, fertilizer use, including P, is set to continue. Research over the past century has done much to elucidate the complex nature of P in soils and plants. However, the world now faces a dilemma, with some developed countries with intensive agriculture having used excessive amounts of P, as commercial fertilizer and animal manure originating from confined animal feeding operations, while many countries of the developing world are hampered by use of little or no fertilizer. International policies have recognized the need to expand fertilizer P use in developing countries and curtail or eliminate P in developed areas with excess P. Overuse of P not only represents an economic loss to farmers but also contributes to transport of P from agricultural land to surface water bodies, where it causes eutrophication and its attendant consequences in terms of deteriorated ecology and health hazards for humans and animals.

Consequently, a major paradigm shift in soil research has been away from production agriculture and toward the environment. The change in emphasis comes at a time of general and growing societal awareness of the finite nature of rock phosphate that underpins the commercial fertilizer industry, thus making efficient use of P a major priority along with emphasis on recycling P from wastes. Much has been learned about the mechanisms of P loss from cropland and the implications of land use management in accelerating such losses; the current expansion of conservation agriculture, especially with minimum- or no-till, can greatly contribute to minimizing P losses. Tests are well established to monitor excess P in soils. However, as is explicit from chapters on the environmental aspects of P use in the most recent comprehensive treatise on P,[26] much remains to be known to provide a more effective basis for technical remediation.

The momentum of societal concern has resulted in policies restricting excess nutrient use on agricultural land in the interest of environmental protection. In essence, technology can—and will—underpin societal action to reconcile the needs for agriculture and protection of the environment.

References

1. FAO. *World Agriculture towards 2035/2050;* Food and Agriculture Organization of the United Nations: Rome, Italy, 2006.
2. Cribb, J. *The Coming Famine: The Global Food Crisis and What We Can Do to Avoid It;* CSIRO Publ.: Collingwood, Victoria, Australia, 2010.
3. Borlaug, N.E. Feeding a hungry world. Science **2007**, *318*, 359.
4. Friedman, T.L. *Hot, Flat, and Crowded: Why We Need a Green Revolution and How It Can Renew America;* Farrar, Strauss, and Giroux: New York, USA, 2008.
5. Diamond, J. *Collapse: How Societies Chose to Fail or Succeed;* Penguin Group: New York, USA, 2005.
6. Inter-Governmental Panel on Climate Change. *Assessment of Global Climate Change;* Washington, DC, USA, **2008**.
7. Godfray, H.C.; Beddington, J.R.; Crute, I.R.; Haddad, L.; Lawrence, D.; Muir, J.M.; Pretty, J.; Robinson, S.; Thomas, S.M.; Toulmin, C. Food security: The challenges of feeding 9 billion people. Science **2010**, *327*, 812–818.
8. Deane, C.; Ejita, G.; Rabbinge, R.; Saye, T. Science for global development. Crop Sci. **2010**, *50*, 1–7.
9. Roy, R.N.; Finck, A.; Blair, G.J.; Tandon. H.S. Plant nutrition for food security: a guide to integrated nutrient management. FAO Fertilizer & Plant Nutrition Bulletin No. 16; Food and Agriculture Organization of the United Nations: Rome, Italy, 2006.
10. Stewart, W.M.; Hammond, L.L.; Van Kauwenbergh, S.J. Phosphorus as a natural resource. In *Phosphorus: Agriculture and the Environment;* Sims, J.T., Sharpley, A.N., Eds; American Society of Agronomy, Crop Science Society of America, Soil Science Society of America: Madison, WI, USA, 2005; 1–22.
11. Mosier, A.R.; Syers, J.K.; Freney, J.R. *Agriculture and the Nitrogen Cycle,* SCOPE 65 (Scientific Committee on Problems of the Environment); Island Press: Washington, DC/London, 2004.
12. Stevenson, F.J. *Nitrogen in Agricultural Soils;* American Society of Agronomy, Crop Science Society of America, Soil Science Society of America: Madison, WI, USA, 1982.
13. Scheppers, T.S.; Raun, W.A., Eds. *Nitrogen in Agricultural Systems,* Agronomy Monograph No. 49; American Society of Agronomy, Crop Science Society of America, Soil Science Society of America: Madison, WI, USA, 2008.
14. Sharpley, A.N.; Smith, S.J.; Namey, J.W. Environmental impact of agricultural nitrogen and phosphorus use. J. Agric. Food Chem. **1987**, *35*, 812–817.
15. Tunney, H.; Carton, O.T.; Brookes, P.C.; Johnston, A.E. *Phosphorus Loss from Soil to Water;* CAB International; Wallingford, U.K., 1997.
16. Delgado, A.; Scalenghe, R. Aspects of phosphorus transfer from soils in Europe. J. Plant Nutr. Soil Sci. **2008**, *171*, 552–575.
17. Sharpley, A.N.; Tunney, H. Phosphorus research strategies to meet agricultural and environmental challenges of the 21st century. J. Environ. Qual. **2000**, *29*, 176–181.
18. Larsen, S. Soil phosphorus. Adv. Agron. **1967**, *19*, 151–210.

19. Dalal, R.C. Soil organic phosphorus. Adv. Agron. **1977**, *29*, 83–117.
20. Khasawneh, F.E.; Sample, E.C.; Kamprath, E.J. *The Role of Phosphorus in Agriculture*; American Society of Agronomy, Crop Science Society of America, Soil Science Society of America: Madison, WI, USA, 1980.
21. Ryan, J. Phosphorus in soils of arid regions. Geoderma **1983** *19*, 341–356.
22. Matar, A.; Torrent, J.; Ryan, J. Soil and fertilizer phosphorus and crop responses in the dryland Mediterranean zone. Adv. Soil Sci. **1992**, *18*, 82–146.
23. Ryan, J. Phosphorus fertilizer use in dryland agriculture: The perspective from Syria. In Proceedings of the OECD workshop on "Innovative Soil and Plant Systems for Sustainable Practices," Izmir, Turkey, June 3–7, 2002; Abstracts, 45.
24. Rashid, A.; Awan, Z.I.; Ryan, J.; Rafique, E.; Ibrikci, H. Strategies for phosphorus nutrition of dryland wheat in Pakistan. Commun. Soil Sci. Plant Anal. **2010**, *41*, 2555–2567.
25. Taylor, A.W.; Kilmer, V.J. Agricultural phosphorus in the environment. In *The Role of Phosphorus in Agriculture*; Khasawneh, F.E., Sample, E.C., Kamprath, E.J., Eds.; American Society of Agronomy, Crop Science Society of America, Soil Science Society of America: Madison, WI, USA, 1980; 545–557.
26. Sims, T.J.; Sharpley, A.N., Eds. *Phosphorus: Agriculture and the Environment*; American Society Agronomy, Crop Science Society of America, Soil Science Society of America: Madison, WI, USA, 2005.
27. IWP-6. Sixth International Phosphorus Workshop, Seville, Spain, Sept 27–Oct 1, 2010.
28. Sharpley, A.N.; Rekolainen, S. Phosphorus in agriculture and its environmental implications. In *Phosphorus Loss from Soil to Water;* Tunney, H., Carton, O.T., Brookes, P.C., Johnston, A.E., Eds.; CAB International: Wallingford, U.K., 1997; 1–53.
29. Pierzinski, G: Sims, J.T; Vance, G.F. *Soils and Environmental Quality,* 2nd Ed.; CRC Press: Boca Raton, FL, USA, 2000.
30. Sharpley, A.N. Phosphorus availability. D18–38. In *Handbook of Soil Science;* Sumner, M.E., Ed.; CRC Press: Boca Raton, FL, USA, 2000.
31. Syers, J.K.; Johnston, A.E.; Curtin, D. Efficiency of soil and fertilizer phosphorus use. Reconciling changing concepts of soil phosphorus behaviour with agronomic information. FAO Fertilizer and Plant Nutrition Bulletin No. 18; Food and Agriculture Organization of the United Nations: Rome, Italy, 2008.
32. Ryan, J.; Curtin, D.; Cheema, M.A. Significance of iron oxides and calcium carbonate particle size in phosphorus sorption and desorption in calcareous soils. Soil Sci. Soc. Am. J. **1985**, *49*, 7476.
33. Ryan, J.; Hassan; H., Bassiri, M., Tabbara, H.S. Availability and transformation of applied phosphorus in calcareous soils. Soil Sci. Soc. Am. J. **1985**, *51*, 1215–1220.
34. Torrent, J.A. Rapid and slow phosphate sorption by Mediterranean soils: Effect of iron oxides. Soil Sci. Soc. Am. J. **1987**, *53*, 78–82.
35. Fixen, P.E.; Grove, J.H. Testing soils for phosphorus. In *Soil testing and Plant Analysis;* Westermann, R.L., Ed.; American Society of Agronomy: Madison, WI, USA, 1990; 141–180.
36. Brown, J.R. Soil testing: Sampling, correlation, calibration and interpretation. Special Publication No. 21; Soil Science Society of America: Madison, Wisconsin, USA, 1987.
37. Halvorson, A.D.; Black, A.L. Long-term dryland crop responses to residual phosphorus fertilizer. Soil Sci. Soc. Am. J. **1985**, *49*, 928–933.
38. McCollum, R.E. Buildup and decline is soil phosphorus: 30-year trends in a Typic Umbraquult. Agron. J. **1991**, *83*, 77–85.
39. Barrow, N.T. Evaluation and utilization of residual phosphorus in soils. In *The Role of Phosphorus in Agriculture*; Khasawneh, F.E., Sample, E.C., Kamprath, E.J. Eds.; American Society of Agronomy, Crop Science Society of America, Soil Science Society of America: Madison, WI, USA, 1980; 333–360.

40. Ibrikci, H.; Ryan, J.; Ulger, A.C.; Buyuk, G.; Cakir, B.; Korkmaz, K.; Karnez, E.; Ozgenturk, G.; Konuskan, O. Maintenance of phosphorus fertilizer and residual phosphorus effect on corn production. Nutr. Cycling Agroecosyt. **2005**, 72, 279–286.

41. Ryan, J.; Ibrikci, H.; Singh, M.; Matar, A.; Masri, S.; Rashid, A.; Pala, M. Response of residual and currently applied phosphorus in dryland cereal/legume rotations in three Syrian Mediterranean agroecosystems. Eur. J. Agron. **2008**, *28*, 126–137.

42. Condron, L.M.; Turner, B.L.; Cade-Menum, B.J. Chemistry and dynamics of soil organic phosphorus. In *Phosphorus: Agriculture and the Environment*; Sims, J.T., Sharpley, A.N., Eds.; American Society of Agronomy, Crop Science Society of America, Soil Science Society of America American Society of Agronomy: Madison, WI, USA, 2005; 87–122.

43. Tiessen, H. Phosphorus dynamics in tropical soils. In *Phosphorus: Agriculture and the Environment*; Sims, J.T., Sharpley, A.N., Eds; American Society of Agronomy, Crop Science Society of America, Soil Science Society of America: Madison, WI, USA, 2005; 253–262.

44. Ryan, J.; Rashid, A. Phosphorus. In Encyclopedia of Soil Science, 2nd Ed.; Lal, R., Ed.; Taylor & Francis: New York, 2007; available at http://www.informaworld.com/smpp/content~db=all~content=a740187227~frm=titlelink. (accessed May 9, 2012).

45. Johnston, A.E. *Soil and Plant Phosphorus;* International Fertilizer Industry Association (IFA): Paris, France, 2000.

46. Stewart, W.M.; Hammond, L.L.; Van Kauwenbergh, S.J. Phosphorus as a natural resource. In *Phosphorus: Agriculture and the Environment*; Sims, J.T., Sharpley, A.N., Eds.; American Society of Agronomy, Crop Science Society of America, Soil Science Society of America: Madison, WI, USA, 2005; 1–22.

47. van Kauwenbergh, S.J. *World Phosphate Rock Reserves and Resources*; International Fertilizer Development Center: Muscle Shoals, AL, USA, 2010.

48. Cramer, M.D. Phosphate as a limiting resource: Introduction. Plant Soil **2010**, *334*, 1–10

49. Prud'homme, M. Global fertilizers and raw materials supply and supply/demand: 2006–2010. In Proceedings of the IFA Annual Conference, Cape Town, South Africa; Paris, France, 2006; 39–42.

50. Cordell, D.; Drangert, T.; White, S. The story of phosphorus: Global food security and food for thought. Global Environ. Change **2009**, *19*, 292–305.

51. Withers, P. Global phosphorus fluxes and the threat to food security. In Proceedings of the 6th International Phosphorus Workshop (IPW-6), Seville, Spain; 2010; Abstracts, 18.

52. Filippelli, G.M. The global phosphorus cycle: Past present and future. Elements **2008**, *4*, 89–95.

53. Smit, A.L., Bindraban, P.S.; Schrober, J.J.; Conijin, J.G.; van der Meer, H.G. Plant Res. Int. Rep. 2009, 282.

54. Maguire, R., Cardon, W., Simard, R.R. Assessing potential environmental impacts of soil phosphorus by soil testing. In *Phosphorus: Agriculture and the Environment*; Sims, J.T., Sharpley, A.N., Eds.; American Society of Agronomy, Crop Science Society of America, Soil Science Society of America: Madison, WI, USA, 2005; 145–180.

32

Potassium

Philippe Hinsinger

Potassium in Plants

K is the major cation in most plants, occurring at concentrations ranging from 5 to 50 g kg^{-1}, twice as much as Ca and slightly less than N.[1–3] The etymology of its name accounts for the abundance of K in plant-derived ash material (potash). K is involved in a large number of physiological processes: osmo-regulation and cation–anion balance, protein synthesis and activation of enzymes.[2,3] K is often referred to as "a cation for anions" as it balances the abundant negative charges of inorganic (nitrate) and organic anions (carboxylates) in plant cells. It therefore occurs at large concentrations, 100–200 mM in the cytosol, 5–10 times less in the vacuole. Being a major inorganic solute, it plays a key role in the water balance of plants: maintenance of the osmotic potential and turgor pressure involved in cell extension. K-controlled changes in turgor pressure in guard cells is a key process of stomatal opening and closure and hence, of the regulation of plant transpiration. Many of these physiological roles are related to the high mobility of K at all levels in the plant. This unique, considerable mobility of K in the plant is essentially due to the large permeability of cell membranes to K-ions, which arises from the occurrence of a range of highly K selective, low and high affinity ion channels and transporters. These are now being increasingly characterized at a molecular level.[4] Large rates of K uptake can thereby be achieved in plant roots. In addition, K-ions can easily be leached out of living plant tissues, as documented for tree foliage which contributes a large flux of K back to the soil via through- fall.[3] K also rapidly leaves dead roots and other plant debris compared with N and P, which require hydrolysis of organic molecules. At an agronomic level, the demand for K largely varies with plant species and productivity. The uptake of K essentially occurs during the vegetative stage and can reach values of 10 kg ha^{-1}day^{-1} and above. Depending also on the agricultural practices (removal of straw, for instance), the amount of K removed with the harvested material will range from 5–50 kg ha^{-1} for cereal grains to 50–500 kg ha^{-1} for forage, root and tuber or plantation crops.

Potassium in Soils

Among major nutrients, K is usually the most abundant in soils as total K content ranges from 0.1 to 40, with an average of 14 g kg^{-1}.[1,5] A major proportion of soil K occurs as structural K in feldspars and interlayer K in micaceous minerals (Figure 1).[1,6,7] Some minor proportion of soil K (usually much less

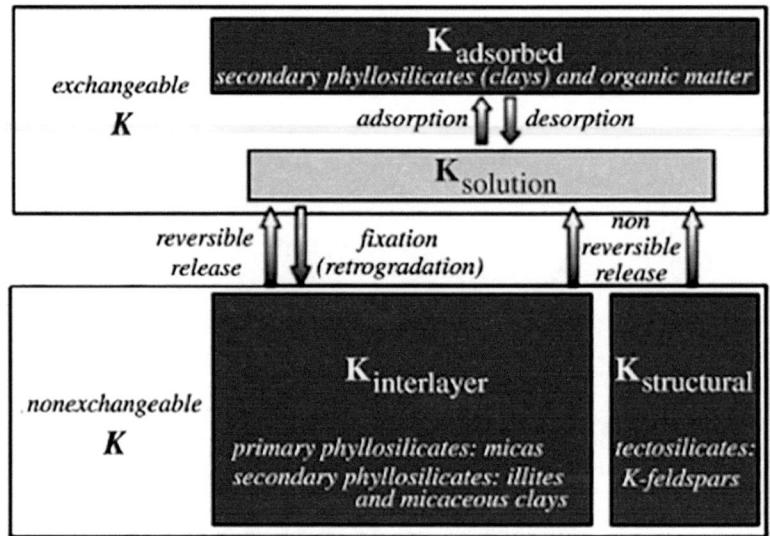

FIGURE 1   The various forms of soil K and the chemical processes involved in soil K dynamics

than 1%) is adsorbed on negatively charged soil constituents, namely clay minerals and organic matter. A marginal part is present as free K-ions in the soil solution. Bulk soil solution concentrations usually amount to 100–1000 µM (less than 0.01%–0.1% of total K). The reason for this rather low concentration of K in the soil solution and hence restricted mobility of K in soils, compared to other metal cations such as Na or Ca is related to its selective adsorption onto some clay minerals. Because of its ionic radius and small hydration energy, K-ions indeed perfectly fit into the interlayer sites of micaceous minerals (micas, illites and mica-derived clays).[8] These sites and, to a lesser degree, the sites on the frayed edges of these minerals have thus a considerably larger affinity for K than for other cations, including divalent cations such as Ca or Mg (Figure 2). Clay minerals also bear sites with larger affinity for divalent cations than for monovalent cations such as K. These sites are located on the planar faces of clay minerals and are thus dominant in clays such as kaolins and smectites. They also occur in organic compounds. K is thus much less strongly held in soils dominated by kaolins (tropical soils), sand or organic matter than in soils dominated by illite–vermiculite clay minerals. Traces of mica-derived clay minerals can dramatically influence the dynamics of soil K as evidenced in tropical soils that are apparently dominated by kaolins.[9] More generally, K dynamic is largely dependent on soil mineralogy which determines both ion exchange and release-fixation processes,[6,7] i.e., the dynamics of "nonexchangeable K." The latter is defined as that (major) portion of soil K which cannot be exchanged by NH_4-ions. NH_4-ions have the same charge and radius than K-ions and can successfully desorb K only from low K-affinity sites. Exchangeable K thus comprise soil solution and easily desorbable K-ions. Nonexchangeable K mostly comprise interlayer K (high K-affinity sites) of micaceous minerals and structural K of feldspars (Figure 1), i.e., 90%–99% of total K in many soils. The release of K from feldspars requires a complete and irreversible dissolution of the mineral and is enhanced under acidic conditions.[6,7] The release of interlayer K from micaceous minerals can proceed similarly or involve an ion exchange process (Figure 1) leading to an expansion of the phyllosilicate (Figure 2). This reversible release is essentially governed by the concentrations of K and competing cations in the outer solution.[8,10,11] Cations which can be responsible for this release, such as Ca- and Mg-ions, have a large hydration energy, contrary to K- or NH4- ions. Therefore, they remain hydrated when exchanging interlayer K-ions and expand the interlayer space (Figure 2), making it possible for the release to proceed further, whereas NH_4-ions would block the reaction.[12] However, because of the considerable affinity of these interlayer sites for K relative to Ca or Mg, the release can occur only for extremely low concentrations of K in the soil solution (Figure 3), in the micromolar range.[11,12]

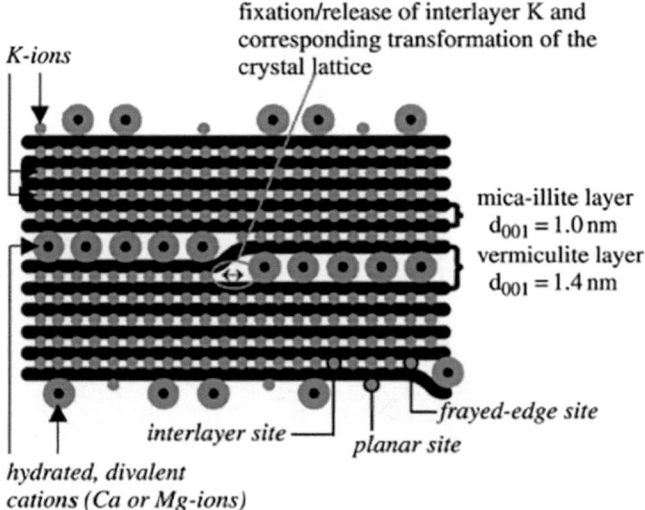

FIGURE 2 The various sites of exchange of K-ions in micaceous clay minerals and the transition between mica and vermiculite layer that occurs when interlayer K is exchanged by hydrated, divalent cations.

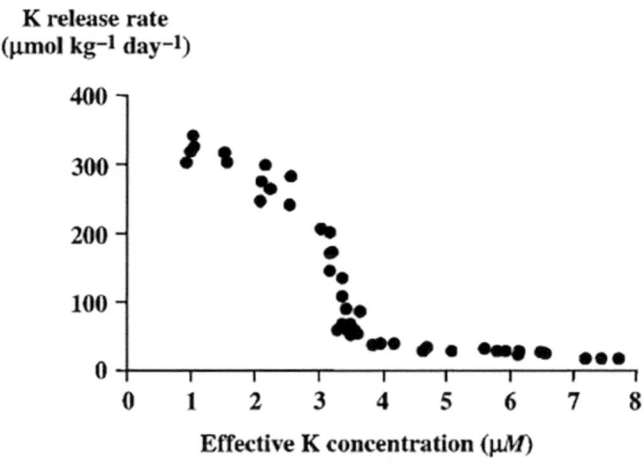

FIGURE 3 Effect of solution K concentration on the rate of release of nonexchangeable K from a soil.
Source: Springob and Richler.[11]

Conversely, elevated concentrations of K are prone to the reverse reaction of fixation, i.e., the collapse of expanded layers and concomitant increase in nonexchangeable K at the expense of readily available K. As long as exchangeable K was assumed to be the only plant-available K fraction, K fixation, i.e., the poor recovery of applied K in the exchangeable fraction was seen as a poor efficiency of K fertilization. This apparent loss of applied K has therefore received considerably more interest than the reverse release process.[7] In addition, the release of nonexchangeable K was considered unlikely to occur to any great extent, especially because K concentration in the bulk soil solution of fertilized soils is usually far above the critical concentrations that are prone to K release. However, numerous long-term fertilizer trials have shown that the release of nonexchangeable K contributes a major proportion of soil K supply in unfertilized and possibly fertilized plots too, although an overall net fixation is often found in the latter (Figure 4).[5,13] Nonexchangeable K can contribute up to 100 kg ha^{-1} yr^{-1}, 80%–100% of soil K supply.

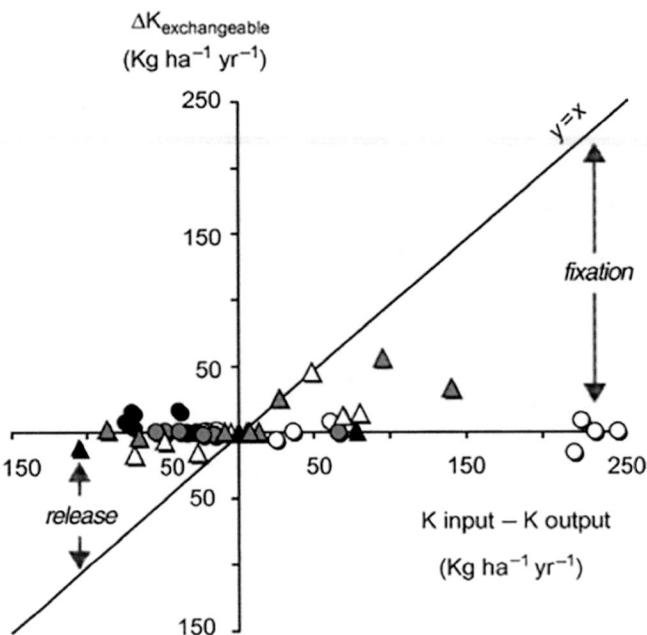

FIGURE 4 Annual change in exchangeable K ($K_{exchangeable}$) as a function of the annual K budget in various K treatment plots of long-term fertilizer trials. The K input comprehends organic and inorganic, applied K fertilizers. The K output corresponds to the offtake of K in the harvested product.
Source: Adapted from Blake[5] and Gachon.[13]

Such annual fluxes of release are fairly large compared with K dissolution rates commonly estimated by geochemists (in the order of 5–15 kg ha^{-1} yr^{-1}). However, many geochemical models do not take into account the amount of K taken off in the vegetation, leading to large underestimates of the actual dissolution rates.[14] Conversely, many K budgets provided by agronomists do not take into account atmospheric inputs and leaching of soil K, assuming that these terms are fairly negligible in most cases. This certainly holds true for atmospheric inputs which rarely exceed values in the order of a few kg ha^{-1} yr^{-1}. Leaching can, however, vary over a much wider range of values, from several kg ha^{-1} yr^{-1} in most cases up to several tens (and up to 1–3 hundreds) of kg ha^{-1} yr^{-1} in those situations that are the most prone to leaching: bare soil or poor soil coverage by the vegetation, excessive fertilizer rates, coarse-textured soils. Omitting the leaching term would, however, have led to underestimating the actual release rate.[5] Hence, considerable amounts of nonexchangeable K can be released in agricultural soils and contribute a significant proportion of plant uptake, in contradiction to the widespread viewpoint shared by numerous agronomists and soil scientists. Reasons for this can be found when considering root–soil interactions occurring in the rhizosphere.

Potassium in Plant–Soil Interactions

K occurs at rather low concentrations in the soil solution, compared to other nutrient cations and to the large requirements of plants for K. The transfer of soil K via mass-flow toward plant roots (i.e., the convective flow of solute accompanying panying transpiration-driven water flow) contributes about 1%–20% of plant demand.[3,15–17] A direct consequence is the rapid depletion of K-ions from the soil solution in the vicinity of plant roots, i.e., the rhizosphere. The resulting concentration gradient generates a diffusion of K-ions in the rhizosphere which plays a key role in the transport of K toward plant root (i.e., 80%–99% of plant demand). Such depletion results in a shift of the cation exchange equilibria which rule

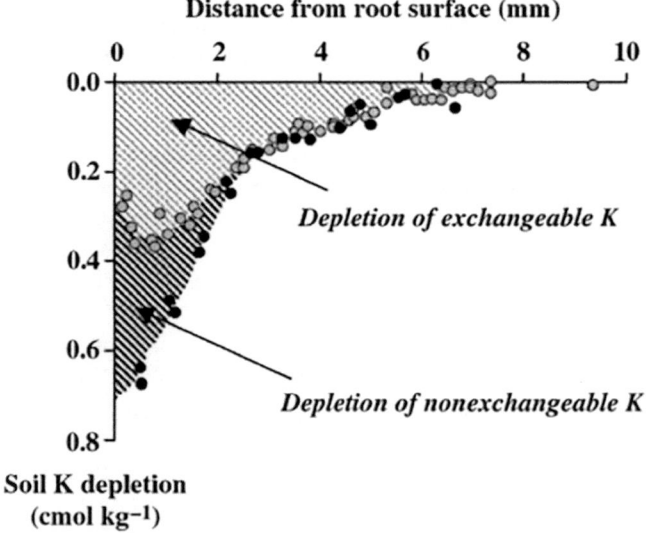

FIGURE 5 Depletion of both exchangeable K (gray dots) and HCl-extractable K (black dots) as a function of the distance from rape roots.
Source: Adapted from Jungk and Claassen.[17]

the dynamics of both exchangeable and interlayer K. This ultimately results in a desorption of exchangeable K and eventually of interlayer K,[7,16,17] as shown by their depletion in the rhizosphere (Figure 5). The extent of the depletion of exchangeable K will depend on chemical parameters such as the initial level of exchangeable K and the K buffering capacity of the soil and on physical parameters that directly determine the diffusive transport of K-ions: soil texture and structure, soil water content.[17] The K depletion zone will extend over several millimeters in clayey, dry soils up to several centimeters in wet, sandy soils.[7,16,17] The intensity of the depletion will also depend on how far the K concentration of solution is decreased, which may vary among plant species according to the K uptake ability of their root. Plants with a lower external K efficiency, i.e., with a higher affinity transport system, will have the capability to take up K at lower K concentrations and may thus deplete soil K further.[17] In the vicinity of roots, solution K concentration can indeed decrease by 2–3 orders of magnitude, down to as little as 2–3 µM.[16] At such low K concentrations, the release of nonexchangeable K can occur at large rates, whereas it would be dramatically restricted at bulk soil K concentrations of several hundreds of µM (Figure 3).[11] Plants thus play a major role in the dynamic of interlayer K via the root-induced depletion of solution K.[16] Measurements in pot experiments have indeed revealed that within several days of growth, the release of nonexchangeable K can amount up to 90% of K supplied to the plant.[7,16] Soil–root chemical interactions in the rhizosphere thus largely explain the unexpectedly large contribution of the release of nonexchangeable K to plant uptake that is found in many agricultural soils, including fertilized soils.

Assessing and Managing Potassium Fertility

Soil K fertility is most often evaluated by measuring exchangeable K[1,7] most frequently with molar NH_4 acetate in batch conditions. However, the adequacy of exchangeable K to predict plant response, i.e., the actual bioavailability of soil K, is rather poor in many soils. This arises from the major contribution of the release of nonexchangeable K in some soils, especially when exchangeable K is low and/or when large reserves of nonexchangeable K are readily available as a consequence of: 1) soil mineralogical composition or 2) fertilization history (build-up of fixed K due to excessive K-fertilizer rates). In these situations, quantitative evaluation of the potential release of nonexchangeable K would be highly recommended for a better prediction of plant response and fertilizer needs.[7] There are several methods for assessing nonexchangeable K but none of them is routinely used on a broad scale, because of their cost. These are either based on the use of 1) concentrated, strong acids that dissolve K-bearing minerals or 2) cationic resins or chemicals such as Na tetraphenylboron that can promote the release of interlayer K by removing K-ions from soil solution and by shifting the exchange equilibria.[6,7] Alternatively, correction factors can be used when interpreting exchangeable K values, which account either for the cationic exchange capacity (or clay content) or for the soil type and K release potential.[7] Exchangeable K is nonetheless often used alone for fertilizer recommendations, resulting in frequently overestimated fertilizer needs to compensate for the expected large fixation and negligible release. Many long-term fertilizer trials have shown that adequate yields of crops can be obtained at fairly low rates of K fertilizer application, or even, for the least demanding crops such as cereals, without any K fertilizer for several years or decades.[7,13] Other more demanding crops, however, require the application of K fertilizer to achieve high yield and quality in the harvested products.[18] The need for K-fertilizers will thus depend on the release potential of the soil and on the demand of the plant, the latter being now increasingly accounted for in fertilizer recommendations. Fertilizer trials have also shown that commonly used soluble K fertilizers and organic sources such as manure or crop residues have fairly comparable efficiencies. This is not surprising as K is highly mobile in organic compounds where it occurs as soluble or exchangeable K-ions. These sources are thus equally important as K-fertilizers and absolutely need to be accounted for in K budgets.

Conclusions

K is the major nutrient cation for plants and thus taken up at large rates by plant roots. These are achieved by both high and low affinity transport systems which explain the considerable mobility of K within the plant. In comparison, K is much less mobile in soils because of the strong affinity of some exchange sites of clays. The large K uptake rates achieved by roots result in a steep depletion of solution K in the rhizosphere, and hence in a shift of the equilibria of cation exchange. Exchangeable K and even nonexchangeable K can thereby be significantly depleted and contribute a substantial proportion of plant uptake. This is confirmed by K balance both in short-term pot experiments and long-term field trials. In addition to the desorption–adsorption of exchangeable K, release and fixation processes thus need to be accounted for when evaluating soil K fertility.

References

1. Munson, R.D. *Potassium in Agriculture*; American Society of Agronomy, Crop Science Society of America; Soil Science Society of America: Madison, WI, 1985; 1223 pp.
2. Mengel, K.; Kirkby, E.A.; Kosegarten, H.; Appel, T. *Principles of Plant Nutrition,* 5th Ed.; Kluwer Academic Publishers: Dordrecht, Netherlands, 2001; 673 pp.
3. Marschner, H. *Mineral Nutrition of Higher Plants,* 2nd Ed.; Academic Press: London, 1995; 889 pp.
4. Schachtman, D.P. Molecular insights into the structure and function of plant K^+ transport mechanisms. Biochim. Bio-phys. Acta **2000**, *1465,* 127–139.
5. Blake, L.; Mercik, S.; Koerschens, M.; Goulding, K.W.T.; Stempen, S.; Weigel, A.; Poulton, P.R.; Powlson, D.S. Potassium content in soil, uptake in plants and the Potassium balance in three European long-term field experiments. Plant Soil **1999**, *216,* 1–14.
6. Sparks, D.L. Potassium dynamics in soils. Adv. Soil Sci. **1987**, *6,* 1–63.
7. International Potash Institute. *Methodology in Soil-K Research*; Proceedings of the 20th colloquim of the International Potash Institute, Baden bei Wien, Austria, International Potash Institute: Bern, Switzerland, 1987; 428 pp.
8. Dixon, J.D.; Weed, S.B. *Minerals in Soil Environment*; Soil Science Society of America: Madison, WI, 1989; 1244 pp.

9. Fontaine, S.; Delvaux, B.; Dufey, J.E.; Herbillon, A.J. Potassium exchange behaviour in Carribean Volcanic ash soils under banana cultivation. Plant Soil **1989**, *120*, 283–290.

10. Schneider, A. Influence of soil solution Ca concentration on short-term release and fixation of a loamy soil. Eur. J. of Soil Science **1997**, *48*, 513–522.

11. Springob, G.; Richter, J. Measuring interlayer Potassium release rates from soil materials. II. A percolation procedure to study the influence of the variable solute K in the <1...10 µM range. Z. Pflanzen. Bodenk. **1998**, *161*, 323–329.

12. Springob, G. Blocking the release of potassium from clay interlayers by small concentrations of NH_4^+ and Cs^+. Eur. J. of Soil Sci. **1999**, *50*, 665–674.

13. Gachon, L. *Phosphore et Potassium dans les Relations Sol-Plante: Conséquences sur la Fertilisation·*, Institut National de la Recherche Agronomique: Paris, France, 1988; 566 pp.

14. Taylor, A.B.; Velbel, M.A. Geochemical mass balances and weathering rates in forested watersheds of the southern blue ridge II. Effects of botanical uptake terms. Geoderma **1991**, *51*, 29–50.

15. Barber, S.A. *Soil Nutrient Bioavailability. A Mechanistic Approach*, 2nd Ed.; Wiley: New York, 1995; 414 pp.

16. Hinsinger, P. How do plant roots acquire mineral nutrients? chemical processes involved in the rhizosphere. Adv. Agron. **1998**, *64*, 225–265.

17. Jungk, A.; Claassen, N. Ion diffusion in the soil-root system. Adv. Agron. **1997**, *61*, 53–110.

18. http://www.ipipotash.org/publications/publications.html (accessed October 2000).

33

Radionuclides

Philip M. Jardine

Introduction

Soil, the thin veneer of matter covering the Earth's surface and supporting a web of living diversity, is often abused through anthropogenic inputs of toxic waste. The disposal of radioactive waste generated at U.S. Department of Energy (DOE) facilities within the Weapons Complex has historically involved shallow land burial in unsaturated soils and sediments. Disposal methods from the 1940s to the 1980s ranged from unconfined pits and trenches to single- and double-shell buried steel tanks. Most of the below-ground burial strategies were deemed to be temporary (i.e., an average life span of several decades) until suitable technologies were developed to deal with the legacy waste issues. Technologies for retrieving and treating the below-ground radionuclide waste inventories have been slow to evolve and are often cost prohibitive or marginally effective. The scope of DOE's disposal problem is massive, with landfills estimated to contain more than 3 million cubic meters of radioactive and hazardous buried waste; a significant proportion of which migrated into surrounding soils and groundwater. It is estimated that the migration of these waste plumes contaminated over 600 billion gallons of water and 50 million cubic meters of soil.

Fate and Transport Processes

Hydrologic Processes

Soil is a complex continuum of pore regions ranging from large macropores at the mm scale to small micropores at the sub-μm scale. It is the physical properties of the media (e.g., structured or layered), coupled with the duration and intensity of precipitation events that dictates the avenues of water and radionuclide movement through the subsurface. In humid environments where structured media is commonplace, transient storm events invariably result in the preferential migration of water.[1–8] Highly conductive voids within the media (e.g., fractures, macropores) carry water around low-permeability, high-porosity matrix blocks or aggregates resulting in water bypass of the latter. In these humid regimes, recharge rates are very high with more than 50% of the infiltrating precipitation resulting in groundwater and surface water recharge. This condition promotes the formation of massive contaminant plumes in the soil since storm flow and groundwater interception with waste trenches is frequent and long-lasting. Even in semiarid environments, where recharge is typically small, subsurface preferential flow is a key mechanism controlling water and solute mobility.[9,10] Lithologic discontinuities and sediment

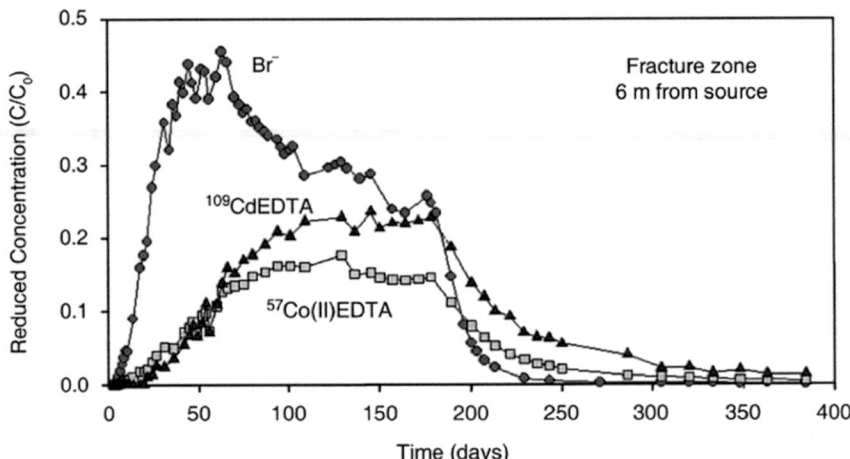

FIGURE 1 Field-scale fate and transport of nonreactive Br⁻ and reactive $^{57}Co(II)EDTA^{2-}$ and $^{109}CdEDTA^{2-}$ in fractured subsurface media at the Oak Ridge National Laboratory. Although transport rates are rapid, geochemical reactions significantly impede the mobility of the chelated radionuclides as is indicated by their delayed breakthrough.

Source: Jardine et al.[40]

layering promote perched water tables and unstable wetting fronts that drive both lateral and vertical subsurface preferential flow.

In both humid and semiarid regimes, water that is preferentially flowing through the soil media often remains in intimate contact with the porous matrix, and physical and hydrologic gradients drive the exchange of mass from one pore regime to another. Mass exchange is time-dependent and is often controlled by diffusion to and from the matrix. Thus, a significant inventory of radionuclide waste can reside within the soil matrix. This waste source is hydrologically linked to preferred flow paths which significantly enhances the extent and longevity of subsurface contaminant plumes. This scenario is commonplace at the Oak Ridge National Laboratory, located in eastern Tennessee, U.S., where thousands of underground disposal trenches and ponds have contributed to the spread of radionuclides such as ^{137}Cs, ^{60}Co, ^{90}Sr, and $^{235/238}U$ across tens of kilometers of landscape. Highly concentrated contaminant plumes move through soil and groundwater at time scales of meters per day (Figure 1) since the soils are highly structured and conducive to rapid preferential flow. However, the soil matrix, which has a high porosity and low permeability, serves as a source/sink for contaminants.[5,6,11] The preferential movement of water and radionuclides through the subsurface also significantly impacts geochemical and microbial processes by controlling the extent and rate of various reactions with the solid phase. It imposes kinetic constraints on biogeochemical reactions and limits the surface area of interaction by partially excluding water and mass from the matrix porosity.

Geochemical Processes

Radionuclide fate and transport in soil and sediments is also controlled by interfacial reactions with the soil solid phase. Most soils are a complex mixture of variably charged phyllosilicates, redox reactive Fe- and Mn-oxides, organic matter, and mineral carbonates. Radionuclides interact with these solid phases through coulombic exchange, chemisorption, redox alterations, transformation processes such as polymerization, precipitation/dissolution, and complexation reactions. Both the extent and rate of these processes can be significantly influenced by variations in water content and the degree of pore regime connectivity. To make matters worse, radionuclide waste generated at the U.S. DOE facilities was often co-disposed with various chelating agents and organic acids. These synthetic organic constituents form highly stable, water-soluble complexes with a wide variety of radionuclides.[12,13] The presence of the complexing agent significantly alters the geochemical behavior of the disposed contaminants in soils and sediments through increased solubility, accelerated redox reactions, and ionic charge reversal.

The geochemical mechanism controlling the fate and transport of chelated radionuclides has been well characterized in numerous soils and subsurface materials.[14–22] Typically, Fe(III) and Mn(IV) oxyhydroxides are the dominant subsurface mineral assemblages that catalyze co-contaminant oxidation/reduction and dissociation reactions (Figure 2). The mineral oxides have repeatedly been shown to catalyze the oxidation of ^{60}Co(II)EDTA^{2-} to ^{60}Co(III)EDTA^{-}, thereby adversely enhancing the transport and persistence of ^{60}Co in a variety of subsurface environments ranging from aquifer sands to fractured weathered shale saprolites.[15,17,18,20,22] Further, Fe(III)-oxides have also been shown to effectively dissociate a large number of chelated metal and radionuclide complexes (e.g., ^{60}Co–, ^{90}Sr–EDTA) through ligand competition.[15,20–22]

Certain radionuclides such as ^{137}Cs do not form strong bonds with many of the chelating agents and organic acids that were used during decontamination. Nevertheless, these radionuclides still interact aggressively with the soil solid phase. In the case of ^{137}Cs, 2 : 1 phyllosilicates and micas serve as excellent sorbents since the interlayer spaces of these mineral assemblages strongly attenuate the radionuclide. The migration tendency of ^{137}Cs in soils is often related to colloid mobility of contaminated sediments[23] or cation competition for surface sites in harsh environments such as those found beneath the Hanford tank farms in western Washington State, U.S.[24]

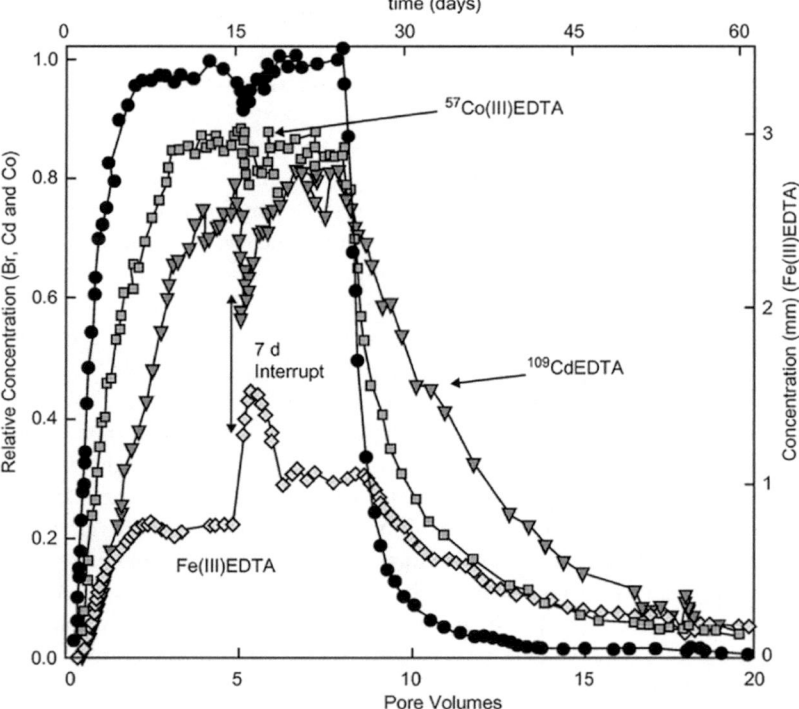

FIGURE 2 Fate and transport of nonreactive Br− and reactive ^{57}Co(II)EDTA^{2-} and 109 CdEDTA^{2-} in undisturbed soil columns of fractured weathered shale. Geochemical reactions impede the mobility of the chelated radionuclides. Co(II)EDTA^{2-} is oxidized to Co(III)EDTA^{-} where Mn-oxides serve as the oxidant. Fe-oxides effectively dissociate CdEDTA^{2-} complexes resulting in the formation of free Cd and Fe(III)EDTA^{-}. A flow interruption technique was employed to quantify the presence of physical and geochemical nonequilibrium processes. **Source:** Mayes et al.[22]

Microbial Processes

Radionuclides such as ^{60}Co and $^{235/238}$U can exist in more than one oxidation state, and their behavior in the environment depends on their oxidation state. For example, U(VI) is soluble and mobile in the environment whereas U(IV) is much less soluble and relatively immobile. Likewise, the oxidized ^{60}Co(III)EDTA complexes are much more stable and exhibit greater mobility in subsurface environments than the reduced ^{60}Co(II)EDTA.[15,17,18] Subsurface Fe- and Al-oxides can effectively dissociate the Co(II)EDTA complex to Fe(III)EDTA[20] and Al(III)EDTA,[16] respectively, and aqueous Co^{2+} is free to participate in sorption or precipitation reactions. Co(III) EDTA, on the other hand, is unaffected by Fe(III)- and Al-oxides. Therefore, the oxidized forms of these radionuclides and metals promote their undesirable enhanced migration through subsurface environments.

Numerous metal-reducing bacteria have been isolated that enzymatically reduce toxic metals and radionuclides to stable end-products. Microbial reduction of U(VI) to form the sparingly soluble U(IV) has been shown using chemostat experiments for a number of metal-reducing bacteria.[25,26] Gorby et al.[27] have also shown that certain metal-reducing bacteria can link the enzymatic reduction of ^{60}Co(III)EDTA$^-$ to support cell growth. Recently, important advances have been made towards implementing field-scale microbially mediated metal reduction strategies in oxygen-deficient environments. Several studies have investigated contaminant reduction in the presence of solid phase material.[27–29] Gorby et al.[27] have shown that the metal-reducing bacterium *Shewanella alga* preferentially reduced Co(III)EDTA$^-$ to Co(II)EDTA^{2-} in the presence of Mn-oxides. Likewise, Wielinga et al.[29] documented the bioreduction of U(VI) by *Shewanella alga* in the presence of various Fe-oxide mineral phases. These authors noted that the rate of U(VI) bioreduction was unaffected in the presence of goethite and only slightly diminished in the presence of poorly crystalline Fe(III)-oxides, where the latter Fe solid phase effectively competed as a terminal electron acceptor. Recent studies by Brooks, Carroll, and Jardine.[28] showed the sustained microbial reduction of Co(III)EDTA$^-$ under dynamic flow conditions. The net reduction of the Co(III)EDTA$^-$ dominated the fate and transport of the contaminant even in the presence of strong mineral oxidants such as Mn- and Fe-oxides that are known to effectively reoxidize Co(II)EDTA^{2-} back to Co(III)EDTA$^-$.[15,18,20] The research findings of Brooks, Carroll, and Jardine[28] provide new and important information on how to successfully implement a bioreduction strategy at the field scale. Their use of a dynamic flow system with sustained bacterial growth conditions in geochemically reactive media is consistent with contaminant migration scenarios in situ.

The studies of Brooks, Carroll, and Jardine,[28] however, used uniformly packed media that contained little structure. Undisturbed subsurface soils and geologic material consist of a complex continuum of pore regions ranging from large macropores and fractures at the mm scale to small micropores at the sub-µm scale. Structured media, common to most subsurface environments throughout the world, accentuates this physical condition that often controls the geochemical and microbial processes affecting solute transport. Redox sensitive radionuclides such as U(VI), Co(III)EDTA, and Tc(VI) reside within nearly all of the pore structure of the subsurface media, with the greatest concentration of contaminants held within micropores.[2,3] Bacteria that are capable of reducing these contaminants are too big to reach a large fraction of the micropore regime and are largely restricted to macro- and mesopore domains.[30,31] Fortunately, the pore structure of the media is hydrologically interconnected, and contaminants move from one pore class to another via hydraulic and concentration gradients.[4,6,7] This process is slow, however, and is often the rate-limiting factor governing the success of contaminant bioremediation. Thus, faster-flowing fracture-dominated regimes will most likely be physically more appealing for sustained bioreduction as long as a suitable electron donor can be supplied. In contrast, bioreduction processes in slower-flowing matrix regimes will most likely be limited by rate-dependent mass transfer of contaminants from smaller pores into larger pores.

Certain bacteria are also capable of degrading chelates and thus potentially immobilizing radionuclides in situ. The biodegradation of the commonly used aminopolycarboxylate chelates NTA, EDTA, and DTPA have been studied in soil and sediment systems for many years.[32,33,35] Research has shown that NTA has the greatest potential for biodegradation in subsurface systems compared with the other aminopolycarboxylates.[34,35]

Bolton et al.[36] and Bolton and Girvin[37] have shown that the bacterial strain *Chelatobacter heintzii* (ATCC 29600) is capable of degrading NTA in the presence of many different toxic metals and radionuclides. Likewise, Payne et al.[38] and Liu et al.[39] have deciphered the mechanisms by which certain bacteria degrade radionuclide– EDTA complexes. These studies lend promise to the potential for using bacteria to biodegrade chelates and enhance the geochemical immobilization of radionuclides in situ.

Conclusions

Radionuclide fate and transport in soils is controlled by coupled time-dependent hydrologic, geochemical, and microbial processes. Hydrologic processes such as preferential flow and matrix diffusion can serve to both accelerate and impede radionuclide migration, respectively. Preferential flow results in hydraulic, physical, and geochemical nonequilibrium conditions since differences in fluid velocities and solute concentrations in different-sized pores create hydraulic and concentration gradients that drive time-dependent inter-region advective and diffusive mass transfer. Thus, in soil systems with a large matrix porosity or a significant quantity of disconnected immobile water, radionuclide migration rates can be greatly retarded due to the slow transfer of mass to actively flowing preferential flow paths. Nevertheless, the prevalence of preferential flow can greatly accelerate the transport of mass in soil systems. Geochemical processes such as sorption, redox alterations, and dissociation reactions can also serve to both accelerate and impede radionuclide migration. Sorption and radionuclide-chelate dissociation reactions almost always result in retarded radionuclide migration rates, whereas oxidation reactions often result in more soluble, and thus more mobile, radionuclide species. Microbial processes can also potentially influence the fate and transport of radionuclides in soil. Metal-reducing bacteria and chelate degraders can alter the geochemical behavior of redox sensitive radionuclides which facilitates their immobilization via solid phase sorption and precipitation reactions.

Enhanced knowledge of the coupled hydrologic, geochemical, and microbial processes controlling radionuclide migration in soils will improve our conceptual understanding and predictive capability of the risks associated with spread of radioactive material in the subsurface environment. Too often risk assessment models treat soil and bedrock as inert media or assume that the media is in equilibrium with migrating contaminants. Failure to consider the time-dependent coupled processes that control radionuclide migration will greatly over-predict the offsite contribution of contaminants from the primary waste source and thus provide an inaccurate assessment of pending risk. By recognizing the importance of soil processes on radionuclide migration, we can improve our decision making strategies regarding the selection of effective remedial actions and improve our interpretation of monitoring results after remediation is complete.

References

1. Shuford, J.W.; Fitton, D.D.; Baker, D.E. Nitratenitrogen and chloride movement through undisturbed field soil. J. Environ. Qual. **1977**, 6, 255–259.
2. Jardine, P.M.; Wilson, G.V.; Luxmoore, R.J. Unsaturated solute transport through a forest soil during rain storm events. Geoderma **1990**, 46, 103–118.
3. Jardine, P.M.; Wilson, G.V.; McCarthy, J.F.; Luxmoore, R.J.; Taylor, D.L. Hydrogeochemical processes controlling the transport of dissolved organic carbon through a forested hillslope. J. Contam. Hydrol. **1990**, 6, 3–19.
4. Jardine, P.M.; O'Brien, R.; Wilson, G.V.; Gwo, J.P. Experimental techniques for confirming and quantifying physical nonequilibrium processes in soils. In *Physical Nonequilibrium in Soils: Modeling and Application;* Selim, H.M., Ma, L., Eds.; Ann Arbor Press: Chelsea, MI, 1998; 243–271.

5. Jardine, P.M.; Wilson, G.V.; Luxmoore, R.J.; Gwo, J.P. Conceptual model of vadose-zone transport in fractured weathered shales. In *Conceptual Models of Flow and Transport in the Fractured Vadose Zone*; U.S. National Committee for Rock Mechanics, National Research Council; National Academy Press: Washington, DC, 2001; 87–114.

6. Wilson, G.V.; Jardine, P.M.; O'Dell, J.D.; Collineau, M. Field-scale transport from a buried line source in variable saturated soil. J. Hydrol. **1993**, *145*, 83–109.

7. Wilson, G.V.; Gwo, J.P.; Jardine, P.M.; Luxmoore, R.J. Hydraulic and physical nonequilibrium effects on multi-region flow and transport. In *Physical Nonequilibrium in Soils: Modeling and Application*; Selim, H.M., Ma, L., Eds.; Ann Arbor Press: Chelsea, MI, 1998; 37–61.

8. Hornberger, G.M.; Germann, P.F.; Beven, K.J. Through flow and solute transport in an isolated sloping soil block in a forested catchment. J. Hydrol. **1991**, *124*, 81–97.

9. Porro, I.; Wierenga, P.J.; Hills, R.G. Solute transport through large uniform and layered soil columns. Water Re-sour. Res. **1993**, *29*, 1321–1330.

10. Ritsema, C.J.; Dekker, L.W.; Nieber, J.L.; Steenhuis, T.S. Modeling and field evidence of finger formation and finger recurrence in a water repellent sandy soil. Water Resour. Res. **1998**, *34*, 555–567.

11. Jardine, P.M.; Sanford, W.E.; Gwo, J.P.; Reedy, O.C.; Hicks, D.S.; Riggs, R.J.; Bailey, W.B. Quantifying diffusive mass transfer in fractured shale bedrock. Water Resour. Res **1999**, *35*, 2015–2030.

12. Riley, R.G.; Zachara, J.M. *Chemical Contaminants on DOE Lands and Selection of Contaminant Mixtures for Subsurface Science Research*; DOE=ER-0547T; U.S. Govt. Print. Office: Washington, DC, 1992.

13. Toste, A.P.; Osborn, B.C.; Polach, K.J.; Lechner-Fish, T.J. Organic analyses of an actual and simulated mixed waste: Hanford's organic complexant site revisited. J. Radioanal. Nucl. Chem. **1995**, *194*, 25–34.

14. Swanson, J.L. *Effect of Organic Complexants on the Mobility of Low-level Waste Radionuclides in Soils*; Status Report PNL-3927, UC-70, 1981.

15. Jardine, P.M.; Jacobs, G.K.; O'Dell, J.D. Unsaturated transport processes in undisturbed heterogeneous porous media: II. Co-contaminants. Soil Sci. Soc. Am. J. **1993**, *57*, 954–962.

16. Girvin, D.C.; Gassman, P.L.; Bolton, H. Adsorption of aqueous cobalt ethylenediaminetetraacetate by d-Al2O3. Soil Sci. Soc. Am. J. **1993**, *57*, 47–57.

17. Zachara, J.M.; Gassman, P.L.; Smith, S.C.; Taylor, D. Oxidation and adsorption of Co(II)EDTA2_ complexes in subsurface materials with iron and manganese oxide grain coatings. Geochim. Cosmochim. Acta. **1995**, *59*, 4449–4463.

18. Brooks, S.C.; Taylor, D.L.; Jardine, P.M. Reactive transport of EDTA-complexed cobalt in the presence of errihydrite. Geochim. Cosmochim. Acta. **1996**, *60*, 1899–1908.

19. Read, D.; Ross, D.; Sims, R.J. The migration of uranium through clashach sandstone: the role of low molecular weight organics in enhancing radionuclide transport. J. Contam. Hydrol. **1998**, *35*, 235–248.

20. Szecsody, J.E.; Zachara, J.M.; Chilakapati, A.; Jardine, P.M.; Ferrency, A.S. Importance of flow and particlescale heterogeneity on Co(II/III)EDTA reactive transport. J. Hydrol. **1998**, *209*, 112–136.

21. Davis, J.A.; Kent, D.B.; Coston, J.A.; Hess, K.M.; Joye, J.L. Multispecies reactive tracer test in an aquifer with spatially variable chemical conditions. Water Resour. Res. **2000**, *36* (1), 119–134.

22. Mayes, M.A.; Jardine, P.M.; Larsen, I.L.; Brooks, S.C.; Fen-dorf, S.E. Multispecies transport of metal-EDTA complexes and chromate through undisturbed columns of weathered, fractured saprolite. J. Contam. Hydrol. **2000**, *45*, 243–265.

23. Solomon, D.K.; Marsh, J.D.; Larsen, I.L.; Wickliff, D.S.; Clapp, R.B. *Transport of Contaminants During Storms in the White Oak Creek and Melton Branch Watersheds*; ORNL/TM-11360; Oak Ridge National Laboratory: Oak Ridge, TN, 1991.

24. Serne, R.J.; Burke, D.S. *Chemical Information on Tank Supernatants, Cs Adsorption from Tank Liquids onto Hanford sediments, and Field Observations of Cs Migration from Past Tank Leaks, Report No. PNNL-11495*; Pacific Northwest National Laboratory: Richland, WA, 1997.

25. Gorby, Y.A.; Lovley, D.R. Enzymatic uranium precipitation. Environ. Sci. Technol. **1992**, *26*, 205–207.

26. Francis, A.J.; Dodge, C.J.; Lu, F.; Halada, G.P.; Clayton, C.R. XPS and XANES studies of uranium reduction by Clostridium Sp. Environ. Sci. Technol. **1994**, *28*, 636–639.

27. Gorby, Y.A.; Caccavo, F.; Drektrah, D.B.; Bolton, H. Microbial reduction of Co(III)EDTA_ in the presence and absence of manganese (IV) dioxide. Environ. Sci. Technol. **1998**, *32*, 244–250.

28. Brooks, S.C.; Carroll, S.L.; Jardine, P.M. Sustained bacterial reduction of Co(III)EDTA_ in the presence of competing geochemical oxidation during dynamic flow. Environ. Sci. Technol. **1999**, *33*, 3002–3011.

29. Wielinga, B.; Bostick, B.; Rosenzweig, R.F.; Fendorf, S. Inhibition of bacterially promoted uranium reduction: ferric (hydr)oxides as competitive electron acceptors. Environ. Sci. Technol. **2000**, *34*, 2190–2195.

30. Smith, M.S.; Thomas, G.W.; White, R.E.; Ritonga, D. Transport of Escherichia Coli through intact and disturbed soil columns. J. Environ. Qual. **1985**, *14*, 87–91.

31. McKay, L.D.; Cherry, J.A.; Bales, R.C.; Yahya, M.T.; Gerba, C.P. A field example of bacteriophage as tracers of fracture flow. Environ. Sci. Technol. **1993**, *27*, 1075–1079.

32. Tiedje, J.M. Microbial degradation of ethylenedi aminetet-raacetic acid in soils and sediments. Appl. Environ. Microbiol. **1975**, *30*, 327–329.

33. Tiedje, J.M.; Mason, B.B. Biodegradation of nitrilotriacetic acid (NTA) in soils. Soil Sci. Soc. Am. Proc. **1974**, *38*, 278–283.

34. Means, J.L.; Kucak, T.; Crerar, D.A. Relative degradation rates of NTA, EDTA, and DTPA and environmental implications. Environ. Pollut. Ser. B. **1980**, *1*, 45–60.

35. Bolton, H., Jr.; Li, S.E., Jr.; Workman, D.J.; Girvin, D.C. Biodegradation of synthetic chelates in subsurface sediments from the southeast coastal plain. J. Environ. Qual. **1993**, *22*, 125–132.

36. Bolton, H.; Girvin, D.C.; Plymale, A.E.; Harvey, S.D.; Workman, D.J. Degradation of metal-nitrilotriacetate complexes by *Chelatobacter heintzii*. Environ. Sci. Technol. **1996**, *30*, 931–938.

37. Bolton, H., Jr.; Girvin, D.C., Jr. Effect of adsorption on the biodegradation of nitrilotriacetate by *Chelatobacter heintzii*. Environ. Sci. Technol. **1996**, *30*, 2057–2065.

38. Payne, J.W.; Bolton, H.; Campbell, J.A.; Xun, Y.L. Purification and characterization of EDTA monooxygenase from the EDTA-degrading bacterium BNC1. J. Bacteriology 1998, *180*, 3823–3827.

39. Liu, Y.; Louie, T.M.; Payne, J.; Bohuslavek, J.; Bolton, H.; Xun, L.Y. Identification, purification, and characterization of iminodiacetate oxidase from the EDTA-degrading bacterium BNC1. Applied Environ. Microbiol. **2001**, *67*, 696–701.

40. Jardine, P.M.; Mehlhorn, T.L.; Larsen, I.L.; Bailey, W.B.; Brooks, S.C.; Roh, Y.; Gwo, J.P. Influence of hydrological and geochemical processes on the transport of chelated-metals and chromate in fractured shale bedrock. J. Conta- min. Hydrol. **2001**, *in press*.

34

Rare Earth Elements

Zhengyi Hu,
Gerd Sparovek,
Silvia Haneklaus,
and Ewald Schnug

Introduction

The rare earth elements (REEs) comprise the elements scandium (Z/21) and yttrium (Z/ 39), and 15 lanthanides with successive atomic numbers (Z) from 57 to 71 (Table 1). Rare earth elements are applied to soils as fertilizer materials or as contaminations of industrial sludges so that an assessment of their behavior in soils is required for evaluating agro-environmental effects.[1–6] This entry summarizes the present knowledge of soil chemical properties of REEs.

TABLE 1 Symbol and Atomic Numbers of Rare Earth Elements (REEs)

Elements	Symbol	Atmomic Number	Descriptive Classification
Scandium	Sc	21	
Yttrium	Y	39	
Lanthanum	LA	57	
Cerium	Ce	58	
Praseodymium	Pr	59	Light earths
Neodymium	Nd	60	
Promethium	Pm	61	
Samarium	Sm	62	
Europium	Eu	63	
Gadolinium	Gd	64	
Terbium	Tb	65	
Dysprosium	Dy	66	
Holmium	Ho	67	
Erbium	Er	68	Heavy earths
Thulium	Tm	69	
Ytterbium	Yb	70	
Lutetium	Lu	71	

Source: Liu.[1]

TABLE 2 Concentrations and Coefficient of Variation of Rare Earth Elements in Different Sludges

Elements	Night Soil Sludge[a] (n/10) Mean (mg kg⁻¹)	CV (%)	Sewage Sludge (n/14) Mean (mg kg⁻¹)	CV (%)	Food Industry Sludge (n/10) Mean (mg kg⁻¹)	CV (%)	Chemical Industry Sludge (n/10) Mean (mg kg⁻¹)	CV (%)
La	3.39	37	6.70	47	0.89	72	2.46	98
Ce	6.98	44	14.10	58	1.83	77	2.69	105
Pr	0.82	38	1.48	46	0.22	82	0.48	95
Nd	3.18	34	6.00	47	0.91	82	2.04	98
Sm	0.53	36	1.02	40	0.17	81	0.36	95
Gd	0.53	34	1.18	45	0.17	79	0.48	101
Tb	0.07	45	0.16	36	0.03	81	0.06	103
Dy	0.39	53	0.93	33	0.14	76	0.39	110
Ho	0.07	54	0.19	32	0.03	79	0.09	119
Er	0.21	55	0.57	31	0.08	73	0.26	118
Tm	0.03	52	0.08	26	0.01	81	0.03	112
Yb	0.20	58	0.54	31	0.09	83	0.19	109
Lu	0.03	56	0.08	31	0.01	84	0.03	105

[a] Feces and urine of humans.
Source: Kawasaki and Kimura.[5]

Origins of Soil Rare Earth Elements

Soil REEs mainly originate from parent materials.[1] Application of phosphate fertilizer[3] and phosphogypsum[4] can supply REEs to the soil. Some of the sludges, particularly those from the chemical industry, have been contaminated with REEs (Table 2). Continuous application of sewage sludges caused an accumulation of Sc, Sm in some soils in Japan.[6] The use of REEs in agriculture is widely practiced in China. By 2001, 6.5 million hectares of land in China was treated with REE fertilizers.[2] In total, 11,000 tons of REEs was applied in agricultural production in China.[7] Besides the parent material, the application of REEs on agricultural farmland is going to be a major source of REEs if the practice of applying them regularly proceeds.

Chemical Speciation of Rare Earth Elements in Soils

Total Content

Representative background values of REEs in soils are available so far only for China and Japan.[1,2,8] The REE content strongly depends on the parent material.[1] The results of 853 soil analyses showed that the total REE content in soils varied between 18 and 583 mg kg⁻¹, with a mean value of 184 mg kg⁻¹.[2] The light REEs La, Ce, Pr, Nd, Sm and Eu account for 90% of the total REE content in soils (Table 3). On the average, the La, Ce, Nd, Sm, and Eu content was 41, 73, 7.3, 27.5, and 5.6 mg kg⁻¹, respectively, in different soils in China (*n*/467) (Table 3).

Species of REEs in Soils

Binding forms of REEs in soils may be classified according to their availability for plants.[1] Approximately 9 mg kg⁻¹ is exchangeable, 2, 5, 32, and 95 mg kg⁻¹ are bonded to carbonates, Mn-oxides, organic substances, and amorphous Fe-oxides, while 59 mg kg⁻¹ is abundant in the form of crystal iron oxides and

TABLE 3 Mean Content of REEs in Soils Extracted by Na$_2$O$_2$/NaOH (n/467)

Light REEs	Content (mg kg^{-1})	Heavy REEs	Content (mg kg^{-1})	Total Contents (mg kg^{-1}) and ratios	
La	41.2	Gd	4.8	Total REE (T)	172.8
Ce	73.4	Tb	0.7	Light REE (L)	156.0
Pr	7.3	Dy	4.4	Heavy REE (H)	16.8
Nd	27.5	Ho	0.9	L/H	9.3
Sm	5.6	Er	2.7	L/T	0.9
Eu	1.1	Tm	0.4		
		Yb	2.5		
		Lu	0.4		

Source: Liu.[1]

TABLE 4 Mean Content of Different Species of Rare Earth Elements in Soils (mg kg^{-1})

Soils	n	Exchangeable	Carbonate Bonded	Mn-oxide Bonded	Organically Bonded	Amorphous Fe-oxide Bonded	Crystal ion Oxides	Residual	Total
Latosol	3	1.0	—	1.4	24.3	23.8	8.2	17.5	76.2
Red soil	5	14.8	—	8.0	30.1	108.0	33.9	199.8	394.6
Yellow brown soil	2	12.1	—	7.4	26.1	79.3	88.3	55.5	268.7
Brown soil	2	11.5	—	4.7	33.7	78.8	64.1	34.5	227.3
Black soil	2	1.4	5.3	1.2	46.8	88.8	74.8	64.0	282.3
Chernozem	2	1.8	11.4	2.8	41.2	105.3	94.1	79.0	335.5
Mean		9.1	1.9	5.3	32.2	95.4	58.8	105.3	307.9

Note: Latosol (rhodic ferralsol, FAO); red soil (ferralic cambisol); yellow brown soil (haplic luvisol); brown soil (haplic alisol); black soil (luvic phaeozems); chernozem (haplic chernozems).
Source: Liu.[1]

105 mg kg^{-1} in residual forms (Table 4). Plants utilize exchangeable REEs most easily, while the uptake of other forms is limited.[1,9] In contrast, residual forms are not plant available.[1,9]

For determining the content of plant available REEs the following extractants have been proposed: 1 M HAc–NaAc (pH 4.8),[1,9] 1.0 M NH$_4$NO$_3$ (pH 7.0),[10] 0.1 M HCl,[11] and 0.1 M malic-citric acid,[12] with 1 M HAc–NaAc being most extensively used.[1,2] Plant available REE contents are highly variable and range from <1 to >200 mg kg^{-1}, with a mean value of about 12 mg kg^{-1} (n/1790).[1] Physicochemical soil properties such as pH, Eh, CEC, clay, H$_2$PO$_4^-$, and carbonate content have a strong impact on the amount of exchangeable and plant available REEs.[1,9] Acid soils contain significantly higher amounts of plant available REEs than more alkaline, calcareous soils.[2] The availability of soil applied REEs is usually significantly higher than that in the original soil matrix.[1,9]

Adsorption of Rare Earth Elements in Soils

In general, 95% of the added REEs are adsorbed.[13] Rare earth elements added to soils are rapidly transformed, for example, into exchangeable, organic matter bonded, and Fe/Mn oxide bonded species.[1,9] The distribution coefficients for REEs added to red and yellow brown soils declined in the following order: residual > exchangeable > organic matter bonded > Fe/Mn oxide bonded REEs.[9] The formation of bridged hydroxo complexes is probably the dominant sorption mechanism to clay minerals.[14] Clay type, pH, CEC, organic matter, and amorphous iron content regulate the adsorption kinetics of REEs.[1,2,9] Langmuir and Freundlich equations were found to describe precisely the absorption of REEs in soils.[15]

Translocation of Rare Earth Elements in Soils

The question is still open whether the use of REEs in industry and agriculture may result in a pollution of soils, plants, and groundwater. In leaching experiments under controlled conditions with [141]Ce and [147]Nd, these elements were abundant only in the top soil layer because of their strong adsorption.[9] For field conditions a translocation depth of <1 cm has been estimated,[9] but this value still needs to be validated for different soils and on a long-term basis.

Uptake of Rare Earth Elements by Plants

The uptake of REEs has been investigated for a wide range of plants.[2,9] All rare elements except praseodymium were found in plants. Uptake of REEs was related to many factors such as element, plant type, and growth conditions.[2,9] The light REEs Ce, La, Nd were highest in plants.[2,9] The total concentration of REEs in plant tissue ranges from <0.05 to 2.58 mg kg^{-1} (Table 5). The concentration of Ce in food products is usually lower than 0.2 mg CeO$_2$ kg^{-1} (Table 6). More attention should be paid to research on uptake, translocation, and distribution of REEs in plants in order to follow up the biological effects of REEs and to assess agro-environmental effects

TABLE 5 Total Concentration of Rare Earth Elements in Plants (mg kg^{-1})

Species	n	Min (mg kg^{-1})	Max (mg kg^{-1})
Rice	319	<LLD[a]	1.17
Wheat	440	<LLD	2.58
Corn	139	<LLD	0.92
Cucumber	41	<LLD	0.70
Leek	33	0.04	0.21
Spinach	41	<LLD	0.12
Cauliflower	61	<LLD	0.60
Lotus root	31	<LLD	0.76
Tomato	64	<LLD	0.18
Chinese cabbage	67	0.05	1.01
Pepper	31	<LLD	0.40
Potato	34	0.05	0.35
Cabbage	38	<LLD	1.20
Mushroom	33	<LLD	0.45
Orange	41	0.13	0.70
Litchi	30	<LLD	
Grape	61	<LLD	0.85
Longan	30	<LLD	0.71
Banana	33	<LLD	
Apple	62	0.07	0.80
Pear	34	<LLD	0.24
Watermelon	37	<LLD	0.26
Sugarcane	27	0.05	1.25
Peach	4	<LLD	

[a] LLD: Lower limit of detection.
Source: Xiong.[9]

TABLE 6 Ce content in Grains and Seeds of Different Crops in Different Countries and Regions of China (mg $CeO_2 kg^{-1}$)

Crops	Region/Country	Sample No.	Mean	Min	Max
Rice	Hubei/China	8	0.06	0.02	0.15
Rice	Jiangxi/China	21	0.06	<0.01	0.18
Rice	Beijing/China	17	0.04	<0.01	0.11
Rice	Nanjing/China	2	—		0.12
Wheat	Heilongjiang/China	21	0.04	<0.01	0.09
Wheat	Hubei/China	7	0.04	0.01	0.07
Wheat	Henan/China	6	0.10	<0.01	0.19
Wheat	Shandong/China	4	0.12	<0.07	0.17
Wheat	Beijing/China	10	0.03	<0.01	0.05
Wheat	Tianjin/China	1	0.04	<0.01	0.08
Maize	Hubei/China	8	0.01	<0.01	0.02
Maize	Heilongjiang/China	10	0.04	<0.01	0.10
Maize	Tianjiang/China	15	0.04	<0.01	0.00
Barley	Beijing/China	14	0.04	<0.01	0.25
Wheat	Canada	5		0.04	0.16
Wheat	USA	8		<0.01	0.20
Wheat	Australia	4		0.06	0.16
Wheat	Argentina	4		0.07	0.09
Soybean	USA	3		<0.01	0.11
Maize	USA	3		0.01	0.09
Mung bean	Thailand	1			0.09

Source: Xiong.[9]

Crop Response to Rare Earth Elements

Research with a view to the use of REEs in agriculture was predominantly carried out in China and before in Russia. A small number of studies have been carried out in Australia, too.[2,9] The increases in crop yield reported by workers from all parts of China range between 5% and 103% (Table 7), with an average response of 8%–15%.[2,9] Crop response to REEs is reported to be most probable when soils contain less than 10 mg kg^{-1} of available REEs (in 1 M HAc–NaAc, pH 4.8), while a response on soils with more than 20 mg kg^{-1} of available REEs is unlikely (Table 8). So far, there is no evidence that REEs are essential for plant growth.

TABLE 7 Yield Increase (relative to control) after Application of Rare Earth Elements to Different Crops

Crop	Country	Yield Increase (%)
Sugar beet	Bulgaria	17–24
Sugar beet	China	7
Wheat	China	6–17
Rape	China	4–48
Potato	China	5–6
Soybean	China	8–9
Cotton	China	5–12
Rice	China	7

(Continued)

TABLE 7 (*Continued*)   Yield Increase (relative to control) after Application of Rare Earth Elements to Different Crops

Crop	Country	Yield Increase (%)
Corn	China	9–103
Barley	Australia	18–19
Peanut	China	8–12
Tobacco	China	8–10
Rubber	China	8–10
Sugarcane	China	10–15
Cabbage	China	10–20
Litschi	China	14–17
Grape	China	8–12

Source: Hu.[16]

TABLE 8   Critical Values of Plant Available Rare Earth Elements in Soils

REE Content (mg kg⁻¹)	Index	Crop Response to REE
<5.0	Very low	Most likely
5–10	Low	Probable
11–15	Medium	Not expected
16–20	High	Not expected
>20	Very high	Unlikely

Source: Liu[1] and Xiong.[9]

Conclusions

Basic information about the chemistry of REEs in soils is available, but this data refers to specific regions and soil types. Besides this, the effect of REEs, for example, on soil fertility and soil biological diversity is yet unclear. Crop response to REE applications depends on various factors, including soil properties. Further studies are also required in order to elaborate the most efficient application techniques.

References

1. Liu, Z. Rare earth elements in soils. In *Microelements in Soils of China*; Liu, Z., Ed.; Jiangsu Science and Technology Publishing House: Nanjing, China, 1996; 293–329.
2. Xiong, B.K.; Chen, P.; Guo, B.S.; Zheng, W. *Rare Earth Element Research and Application in Chinese Agriculture and Forest*; Metallurgical Industry Press: Beijing, China, 2000.
3. Todorovsky, D.S.; Minkova, N.L.; Bakalova, D.P. Effect of the application of superphosphate on rare earth content in the soils. Sci. Total Environ. **1997**, *203* (1), 13–16.
4. Arocena, J.M.; Rutherford, P.M.; Dudas, M.J. Heterogeneous distribution of trace elements and fluorine in phos-phogypsum. Sci. Total Environ. **1995**, *162* (2–3), 149–160.
5. Kawasaki, A.; Kimura, R.; Arai, S. Rare earth elements and other trace elements in wastewater treatment sludges. Soil Sci. Plant Nutr. **1998**, *44* (3), 433–441.
6. Zhang, F.S.; Yamasaki, S.; Kimura, K. Rare earth element content in various waste ashes and the potential risk to Japanese soils. Environ. Int. **2001**, *27* (5), 393–398.
7. Wang, S.M.; Zheng, W. Developing technology of rare earth application on agriculture and biological field in China. *Proceedings of the 4th International Conference on* Rare Earth Development and Application, Beijing, China, June 16–18, 2001; Yu, Z.S., Yan, C.H., Xu, G.Y., Niu, J.K., Chen, Z.H., Eds.; Metallurgical Industry Press: Beijing, 237–240.

8. Yoshida, S.; Muramatsu, Y.; Tagami, K.; Uchida, S. Concentrations of lanthanide elements, Th, and U in 77 Japanese surface soils. Environ. Int. **1998**, *24* (3), 275–286.

9. Xiong, P.K.; Zheng, W.; Cheng, P.; Wang, F. *Rare Earth Elements in Agricultural Environment*; China Forestry Press: Beijing, China, 1999.

10. Zhai, H.; Yang, Y.G.; Zheng, S.J.; Hu, A.T.; Zhang, S.; Wang, L.J. Selection of the extractants for available rare earths in soils. China Environ. Sci. **1999**, *19* (1), 67–71 (in Chinese).

11. Li, F.L.; Shan, X.Q.; Zhang, S.Z. Evaluation of single extractants for assessing plant availability of rare earth elements in soils. Commun. Soil Sci. Plant Anal. **2001**, *32* (15 and 16), 2577–2587.

12. Zhang, S.Z.; Shan, X.Q.; Li, F.L. Low-molecularweight-organic-acid as extractant to predict plant bioavailability of rare earth elements. Int. J. Environ. Anal. Chem. **2000**, *76* (4), 283–294.

13. Zhu, J.G.; Xing, G.X.; Yamasaki, S.; Tsumura, A. Adsorption and desorption of exogenous rare earth elements in soils: I. Rate of forms of rare earth elements sorbed. Pedo-sphere **1993**, *3* (4), 299–308.

14. Dong, W.M.; Wang, X.K.; Bian, X.Y.; Wang, A.X.; Du, J.Z.; Tao, Z.Y. Comparative study on sorption/desorption of radioeuropium on alumina, bentonite and red earth: effects of pH, ionic strength, fulvic acid, and iron oxides in red earth. Appl. Radiat. Isotopes **2001**, *54* (4), 603–610.

15. Gao, X.J.; Zhang, S.; Wang, L.J. The adsorption of La^{3+} and Yb^{3+} on soil and mineral and its environmental significance. China Environ. Sci. **1999**, *19* (2), 149–152 (in Chinese).

16. Hu, Z.Y.; Richter, H.; Sparovek, G.; Schnug, E. Physiological and biochemical effects of rare earth elements on plants and their agricultural significance: a review. J. Plant Nutr. **2003**, in press.

35

Strontium

Silvia Haneklaus
and Ewald Schnug

Introduction

Strontium (Sr) was discovered by Adair Crawford in 1790 and named after the western Scottish village of Strontian. There are four stable Sr isotopes, ^{84}Sr, ^{86}Sr, ^{87}Sr, and ^{88}Sr, which occur naturally in a ratio of 0.56:9.87:7.00:82.58.[1] From these four stable Sr isotopes, only ^{87}Sr is radiogenic. Thus, the ^{87}Sr concentration increases over time because of the decay of ^{87}Rb to ^{87}Sr (half-life of 4.7×10^{10} years). It is, however, the radioactive isotope ^{90}Sr from anthropogenic nuclear sources that this element is commonly associated with. Plant available concentrations of Sr and Ca in soils are important factors that affect the transfer$_{\text{soil/plant}}$ of ^{90}Sr.[2] Consequently, the knowledge of the factors and processes influencing the chemical and spatial speciation of Sr in the soil is of prime interest in risk assessment of undesired ^{90}Sr contaminations.

Strontium in Soils

Strontium occurs in abundance in nature, and its concentration can be as much as 0.034% in most igneous rocks.[3] Celestite ($SrSO_4$) and strontianite ($SrCO_3$) are the predominant Sr minerals. It occurs in large concentrations and in association with K in volcanic rocks, alkali rocks, and pegmatites. Sr is removed by weathering from these and other igneous rocks and sediments. Intermediate concentrations of Sr occur in basic magmatic rocks, which contain about two times higher Sr concentrations than granite. Some ultramafic rocks have very low Sr concentrations.[4] A mean Sr concentration of 375 mg kg^{-1} is observed in igneous rocks, 300 mg kg^{-1} in shales and 20 mg kg^{-1} in sandstones.[5]

Strontium isotope ratios in parent materials and minerals vary in relation to geological age and geographical location.[6] The ^{87}Sr: ^{86}Sr ratio depends on the initial Rb:Sr ratio in the parent material, age of the rock, and the Sr isotope ratio at the onset of rock formation as ^{87}Sr is radiogenic.[6] The ^{87}Sr:^{86}Sr ratio is often used to study biogeochemical processes such as weathering of minerals. However, this ratio must be evaluated critically, as with the age of weathering, surface changes in the ^{87}Sr:^{86}Sr ratio also occur, presumably because of dynamic variations in weathering rates of different minerals.[7]

TABLE 1 Variation in the Sr Content in Dependence on Chemical Speciation and Geogenic Origin

Fraction	Country	Method	Min. ($\mu g\,g^{-1}$)	Max.	Mean	n	Reference
Sr_{total}	N. Germany[a]	X-RF	15.2	118.4	79.0	155	[2]
	N. Germany[b]	X-RF	57.0	97.0	71.3	154	[9][c]
	Finland		109	2730	744	221	[10]
	N. Europe[d]	X-RF	10	667	110[e]	773	[4]
	Scotland[f]	–	60	1000	380	43	[11]
	Russia	–	53	135	92	–	[12]
	India	$HF-HNO_3-HClO_4$	74	198	127	10	[13]
	U.S.A.	–	80	900	400	13	[11]
$Sr_{total\ exchangeable}$	N. Germany[a]	$1N\ NH_4$-Ac, pH 7	4.4	48.9	18.1	155	[2]
	Finland	NH_4-Ac, pH 4.65	9.4	16.6	12.9	1944	[14][g]
	Finland	NH_4-Ac, pH 4.65	2.4	63.8	19.2	221	[10]
	N. Europe	NH_4-Ac, pH 4.5	0.04	473	8[e]	773	[4]
	Scotland[h]	$1N\ NH_4$-Ac, pH 7	3.5	17.5	6.6	7	[11]
$Sr_{rapidly\ exchangeable}$	N. Germany[a]	$0.025\,N\ CaCl_2$	1.5	10.2	4.8	155	[2]

[a] Schleswig-Holstein.
[b] Isle of Ruegen.
[c] Sample description (Sr data unpublished).
[d] Belarus, Estonia, Finland, Germany, Latvia, Lithuania, Norway, Poland, Russia, Sweden.
[e] Median.
[f] North-east Scotland.
[g] mg L^{-1}.
[h] Aberdeenshire/Midlothian.

In the soil profile, about 90% of the Sr is generally exchangeable and 10% in non-exchangeable form.[8] The data in Table 1 summarize total and plant available Sr concentrations in soils of different origins. Reimann et al.[4] reported distinctly lower total Sr concentrations in soils of northern Europe with a median of 110 mg kg^{-1} Sr in comparison with the world average of 240 mg kg^{-1} Sr. In several countries of northern Europe, median values for total Sr concentrations range from 45 mg kg^{-1} in Poland to 181 mg kg^{-1} in Norway. About 7% of the total Sr concentration in the topsoil is plant available.[4] The reported plant available Sr concentrations range from 4 mg kg^{-1} in Poland to 15 mg kg^{-1} in Finland.[4] The corresponding values for Germany are 8 (NH_4-Ac) and 55 (total) mg kg^{-1} Sr, and somewhat lower than those reported by Haneklaus[2] in Schleswig-Holstein with 18 (NH_4-Ac) mg kg^{-1} and 79 (total) mg kg^{-1} Sr (Table 1). The mean rapidly exchangeable Sr concentration in these studies was 4.8 mg kg^{-1} Sr.

Interaction of Strontium with Soil Matrix

The advantage of using stable Sr in research is that it provides a direct measure for the spatial variation of pedogenetic soil characteristics and agro-technical measures influencing the uptake of radioactive Sr isotopes by plants. In comparison, studies with radioactive Sr (viz ^{90}Sr, ^{89}Sr, and ^{85}Sr) provide in situ analysis of radioactive Sr behavior under current soil conditions. An important point to be noted in Sr experimentation is whether the stable or carrier-free radioactive Sr has been used, which leads to application of significant amounts of Sr in the former case, compared with only insignificant rates of Sr added to the soil in the latter case.

The correlation matrix between soil pH, clay, organic C, plant available Ca concentration, and Sr species is shown in Figure 1 for soils from the federal state of Schleswig- Holstein.[2] Correlation coefficients between total exchangeable Sr and $C_{org.}$ were significant, but low. In comparison, distinctly close and significant relationships were observed for soil pH and clay content (Figure 1). Reimann et al.[4] and Haneklaus[2] also reported a similar correlation coefficient of $r = 0.6$ for the relationship between total Sr and NH_4-acetate extractable Sr when total Sr was determined by *aqua regia* extraction and

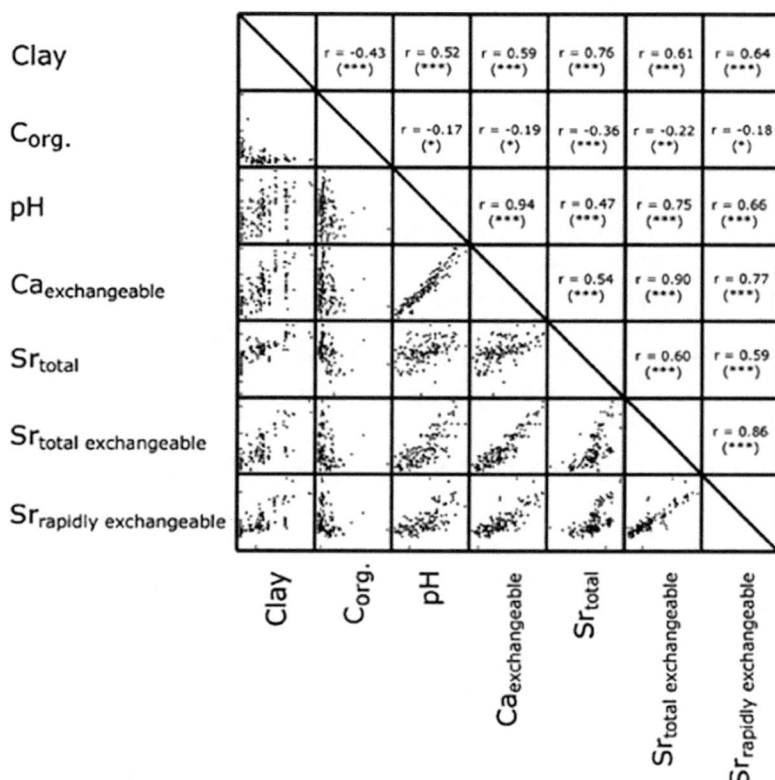

FIGURE 1 Correlation between soil parameters and Sr species in top soil samples of production fields in northern Germany (n = 155).
Source: Calculated from data in Schöller.[2]

X-RF techniques, respectively. Analogous correlation coefficients were also observed for the relationship between plant available Ca ($r = 0.78$) and plant available Sr concentrations ($r = 0.90$) in the topsoil (Figure 1).[2,4]

Rengel estimated a mean Ca loss of 300 kg ha^{-1} yr^{-1} by leaching. The Ca:Sr ratio in soils is about 140 for total[15] and may vary between 250 and 5000 for exchangeable Ca and Sr.[16] Assuming equivalent leaching rates of Ca and Sr, implies Sr losses of <1 kg ha^{-1} yr^{-1}. Wiklander[8] observed considerable leaching losses of ^{90}Sr in a lysimeter experiment after 42 months.

Strontium sorption is dominated by ion exchange and sorption/desorption is a reversible process; the organic soil fraction preferentially adsorbs Ca, while the inorganic soil fraction (e.g. the clay minerals) preferentially adsorbs Sr.[17,18] A small fraction of ^{85}Sr added to the soil is also fixed in non-exchangeable form. Valcke[17] suggested that this fixation occurs in Sr/organic matter/clay complexes.

Adsorption of Sr on different types of clay minerals has been studied intensively with the objective to decontaminate effluents from contaminated sites. Strontium is adsorbed moderately to mineral surfaces, and is desorbed when its concentration in the soil solution increases, competing ions such as Ca are present, and soil pH is low.[19] Application of even small quantities of stable Sr and Ca as amendments may increase uptake of ^{90}Sr by desorption. However, application of Ca may positively increase the Ca:Sr ratio.[20] The Sr sorption decreases in the presence of calcite and high concentrations of HCO$_3$ and strontianite (SrCO$_3$) may be formed at high concentrations of Sr in the soil.[19] For a Cambrian blue clay soil, the Langmuir isotherm indicated a constant of 294 and a limiting sorption of 0.034 mol kg^{-1}.[18] Also Freundlich isotherms reflected an adequate sorption of Sr with a regression coefficient of >0.99. The sorption ratio is generally higher than 0.95 for all concentrations ranging from 10^{-3} to 10^{-8} M.[21]

Strontium Uptake by Plants

Strontium uptake by plants is passive, and is thus linked to its concentration in the soil solution. Plant available Ca and Sr in the soil are important factors governing uptake of radioactive Sr. Addition of Ca often results in a negative linear relationship between Sr and Ca in plants.[2] The Sr: CEC ratio influences Sr uptake in such a way so that any increase in the cation exchange capacity decreases the Sr:uptake.

The Sr concentration in plants varies depending on the plant available concentrations in the soil, root uptake, and the growth or phenological stage. Strontium uptake by dicotyledonous crops is, for example, distinctly higher than that by monocotyledonous crops owing to the higher cation exchange capacity of the latter.[20] The Sr concentration in vegetative plant tissue varies between 1 and 150 mg kg^{-1} with a mean value of 25 mg kg^{-1}.[3] The Sr concentration in seeds is about 5–10 times lower than that in vegetative plant material, most likely because of the phloem immobility.[20] Strontium concentrations from 18 to 36 mg kg^{-1} have been reported in oilseed rape seeds, while the average concentration was 7 mg kg^{-1} in wheat.[2]

Impact of Fertilizer Use on Strontium Uptake by Plants

Gabe and Rodella[22] reported Sr concentrations in different phosphatic fertilizers, limestone, and gypsum. The mean Sr concentration (mg kg^{-1}) occurs in the following order: concentrated apatite (12121) > rock phosphates (7319) > single super phosphate (5396) > fused phosphates (4965) > gypsum (2984) > limestone (824) > ammonium phosphates (304). The variation in the Sr concentration was highest in limestone samples ranging from 36 to 4736 mg kg^{-1} Sr. Long-term application of lime significantly increased soil pH along with total Ca and Sr concentrations in an acid Albic luvisol in Romania (Figure 2).[23] Haneklaus[2] reported that the Sr uptake in plants is influenced more by the Ca concentration in soil rather than by soil pH. Liming of an acid soil can reduce the Sr uptake by 40%–45%, as was reported in a lysimeter experiment by Wiklander.[8] Under field conditions, Haneklaus[2] estimated a minimum threshold rate of lime application of 1.4 t ha^{-1} CaO for reducing the Sr uptake significantly on an acid, silty loam soil (pH 5.3). Liming combined with an additional Ca supply by gypsum can cause the maximum reduction in Sr uptake by oilseed rape probably because of the precipitation of SrSO$_4$.[2] In contrast, Anderson[20] reported that a dose of at least 10 t ha^{-1} stable Sr is required to achieve a significant reduction of ^{90}Sr uptake.

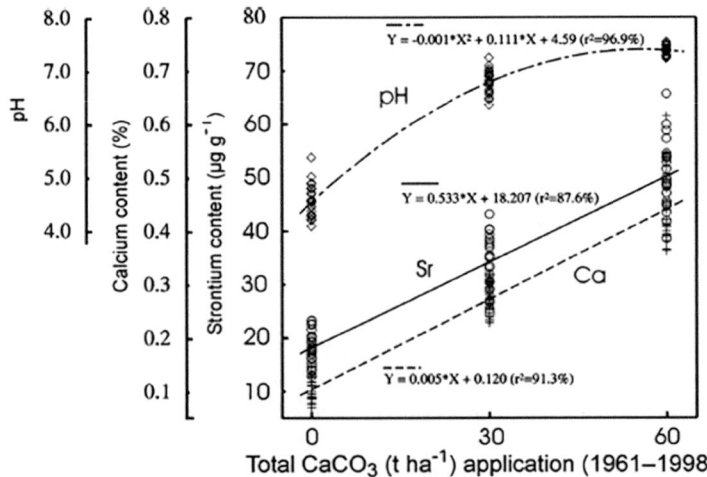

FIGURE 2 Long-term influence of regular lime applications of 0, 5, and 10 t ha^{-1} from 1961 to 1998 corresponding to total rates of 0, 30, and 60 t ha^{-1} on soil pH, and total Sr and Ca contents in the top soil of an Albic luvisol in Romania. **Source:** Adapted from Rogasik et al.[23]

Conclusion

Strontium behavior in soils and uptake by plants are similar to that of Ca. Stable Sr is usually of minor interest, but it may gain relevance with its role in reducing the transfer of radioactive Sr into the food chain. An optimum Ca saturation is a prerequisite to limit the uptake of radioactive Sr by plants. Furthermore, amendments (e.g., lime) add significant amounts of Sr to the soil and thus increase the concentration of competitive stable Sr in soils.

References

1. Weast, R.C.; Astle, M.J.; Beyer, W.H. *Handbook of Chemistry and Physics*; CRC Press Inc.: Boca Raton, Florida, 1988/89.
2. Haneklaus, S. *Strontiumgehalte in Pflanzen und Böden Schleswig-Holsteins und Bewertung von Düngungsmaßnahmen zur Verminderung der Strontiumaufnahme von Kulturpflanzen*; PhD, University, Kiel, Germany, 1989; 169 pp.
3. Coughtrey, P.J.; Thorne, M.C. *Radionuclide Distribution and Transport in Terrestrial and Aquatic Ecosystems. A Critical Review of Data*; Balkema: Rotterdam, 1983; Vol. 1, 93–239.
4. Reimann, C.; Siewers, U.; Tarvainen, T.; Bityukova, L.; Eriksson, J.; Gilucis, A.; Gregorauskiene, V.; Lukashev, V.K.; Matinian, N.N.; Pasieczna, A. *Agricultural Soils in Northern Europe: A Geochemical Atlas;* Bundesanstalt fuer Geowissenschaften und Rohstoffe and Staatliche Geologische Dienste in der Bundesrepublik Deutschland, Ed.; Schweizerbart'sche Verlagsbuchhandlung: Stuttgart, 2003; 279 pp.
5. Lisk, D.J. Trace metals in soils, plants, and animals. Adv. Agronom. **1972**, *24*, 267–311.
6. Barbaste, M.; Robinson, K.; Guilfoyle, S.; Medina, B.B.; Lobinski, R. Precise determination of the strontium isotope ratios in wine by inductively coupled plasma sector field multicollector mass spectrometry (ICPSF-MC-MS). J. Anal. At. Spectrom. **2002**, *17*, 135–137.
7. Ma, Y.; Liu, C. Sr isotope evolution during chemical weathering of granites. Sci. China **2001**, *44*, 726–734.
8. Wiklander, L. Uptake, adsorption and leaching of radiostrontium in a lysimeter experiment. Soil Sci. **1964**, *97*, 168–172.
9. Haneklaus, S.; Fleckenstein, J.; Schnug, E. Comparative studies of plant and soil analysis for the evaluation of the sulphur status of oilseed rape and wheat. J. Plant Nutr. Soil Sci. **1995**, *158*, 109–112.
10. Lakanen, E.; Silanpää , M. Strontium in Finnish soils. Annales Agric. Finniae **1967**, *6*, 197–207.
11. Swaine, D.J. The trace element content of soils. Commonwealth Bureau of Soil Science Technical Communications 1955, 48, 157.
12. Protasova, N.A.; Golubev, I.M.; Korobeinikov, N.I. Trace elements in landscapes of the Tambov Oblast and the biogeochemical zoning of its territory. Eurasian Soil Sci. **1996**, *29*, 1360–1366.
13. Moslehuddin, A.Z.M.; Laizoo, S.; Egashira, K. Trace elements in Bangladesh paddy soils. Commun. Soil Sci. Plant Anal. **1999**, *30*, 1975–1996.
14. Paasikallio, A. Strontium content and strontiumcalcium ratio in timothy (Phleum pratense L.) and soil in Finland. Annales Agric. Finniae **1979**, *18*, 174–181.
15. Comar, C.L. Some over-all aspects of strontiumcalcium discrimination. In *The Transfer of Calcium and Strontium across Biological Membranes*; Wassermann, R.H., Ed.; Academic Press: New York, 1963; 405–417.
16. Menzel, R.G.; Heald, W.R. Strontium and calcium contents of crop plants in relation to exchangeable strontium and calcium of the soil. Soil Sci. Soc. Am. Proc. **1959**, *23*, 110–112.
17. Valcke, E. The behaviour dynamics of radiocesium and radiostrontium in soils rich in organic matter; PhD thesis, Katholieke Universiteit Leuven, Belgium, 1993; 135 pp.
18. Chirkst, D.E.; Litinova, T.E.; Cheremisina, V.; Ivanov, M.V.; Mironenkova, N.A. Isotherm of strontium sorption on clay. Russ. J. Appl. Chem. **2003**, *76*, 727–730.

19. Mincher, B.J.; Fox, R.V.; Riddle, C.L.; Cooper, D.C.; Groe- newold, G.S. Strontium and cesium sorption to Snake river plain, Idaho soil. Radiochim. Acta **2004**, *92*, 55–61.

20. Andersen, A.J. Plant accumulation of radioactive strontium with special reference to the strontium-calcium relationship as influenced by nitrogen. Riso Report **1973**, *278*, 56.

21. Hakem, N.L.; Al Mahamid, I.; Apps, J.A.; Moridis, G.J. Sorption of cesium and strontium on Hanford soil. J. Radioanal. Nucl. Chem. **2000**, *246*, 275–278.

22. Gabe, U.; Rodella, A.A. Trace elements in Brazilian agricultural limestones and mineral fertilizers. Commun. Soil Sci. Plant Anal. **1999**, *30*, 605–620.

23. Rogasik, J.; Kurtinez, P.; Panten, K.; Funder, U.; Rogasik, H.; Schroetter, S.; Schnug, E. Kalkung und Bodenfruchtbarkeit. FAL—Agric. Res. **2005**, *286* (special issue): 71–81.

36

Sulfur

Ewald Schnug,
Silvia Haneklaus,
and Elke Bloem

Introduction

Soil fertility is generally defined as the ability of soils to yield crops and is therefore closely intertwined with plant nutrient cycles. The soil sulfur cycle is driven by biological and physico-chemical processes, which affect flora and fauna. The organic matter pool is an important source and sink for sulfur though its contribution to the mineral nutrition of high yielding crops has been overestimated in the past. Under temperate conditions it is rather the spatiotemporal variation of physico-chemical soil properties, which control the plant available sulfate-S content in the soil via the access of plant roots to sulfur rich groundwater or capillary ascending porous water. Therefore soil methods determining plant available sulfur status will always deliver only information for that instant ignoring relationships with the plant sulfur status or crop yield. A more promising way for the evaluation of the sulfur supply is a site-specific sulfur budget, which includes information about geomorphology, texture, climatic data and crop type and characteristics of the local soil water regime. The major soil properties and external sulfur sources affecting the amount of plant available sulfate in the soil are shown in Figure 1.

For a comprehensive understanding of the relationship between sulfur and soil fertility it is necessary to identify the main factors and processes controlling the plant available sulfur pool and to appraise the ecotoxicological significance of sulfur within the content of a limiting factor rather than a pollutant. Detailed reviews about the role of sulfur in agro-ecosystems already exist[1–9] and will provide further information, not explicitly mentioned here.

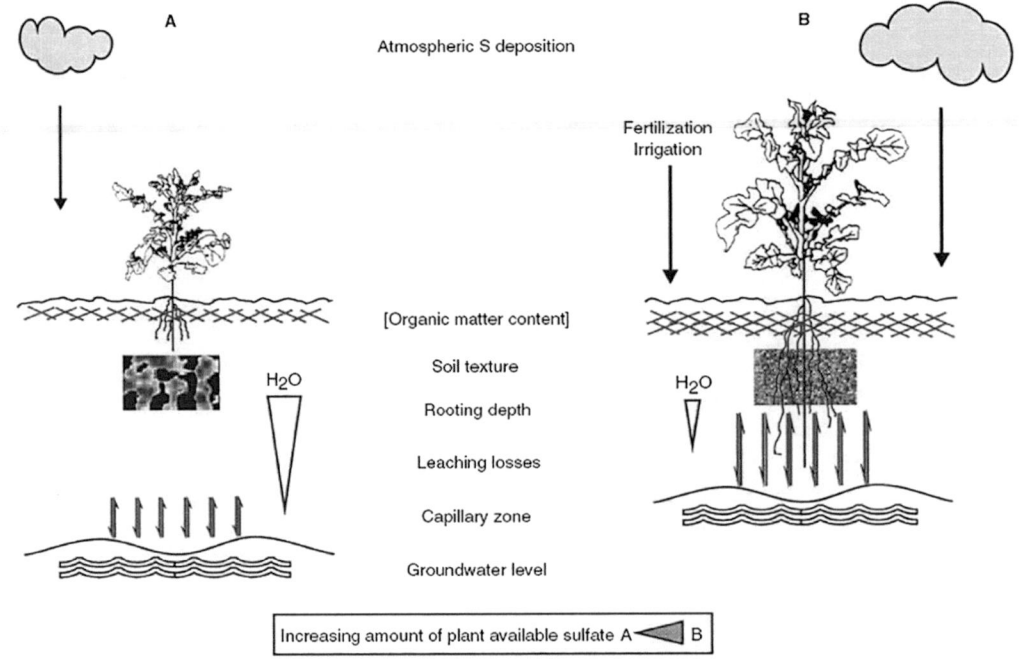

FIGURE 1 Overview and efficacy of essential impact factors modifying plant available sulfate in soils.

Sulfur in Soils

The concentration of sulfur (S) in parent materials ranges from 0.026% to 1% S with igneous rocks $\leq$ metamorphic rocks $\leq$ magmatic rocks of upper continental crust $\ll$ limestones $<$ sedimentary rocks (sulphides) $\leq$ shales $<$ sedimentary rocks (sulfates) $\ll$ coal.[10–12] The typical range of S in agricultural soils of humid and semi-humid regions is 100 to 500 (μg g^{-1}, or 0.01 to 0.05% S. This equals 224 to 1120 kg ha^{-1} S in the A_h-horizon.[8] The total S content of soils may be as low as 20μg g^{-1} (0.002%) in highly leached and weathered soils of humid regions or as high as 35,000 μrg g^{-1} (3.5%) in marine marsh soils and up to 50,000 μg g^{-1} (5%) in calcareous and saline soils of arid and semiarid regions.[8] Tropical soils generally contain low amounts of S due to their low organic matter content.

Biological Aspects of Sulfur in Soils

Most of the S in terrestrial soils is bound in the organic fraction, which amounts normally to more than 95% of the total S content.[2,8] Organic S in soils is a heterogeneous mixture of soil organisms, partly decomposed plant material, animal and microbial residues. Little is, however, known about the composition of individual chemical compounds. Many different approaches have been developed to separate soil organic S into major fractions. The following approaches to identify distinct forms and properties of soil organic matter were made: chemical extraction followed by physical-chemical separation into humic acids, fulvic acids and humins reactivity with reducing agents in order to separate carbon-bonded S (C–S) and sulfate esters (C–O–S); physical separation into organo-mineral size fractions and molecular weight fractionation.[13]

Sulfate-S is rapidly bound in the form of sulfate-esters, which is regarded as a short term source for S.[14] The incorporation of S into high-molecular weight fractions such as humic acids prevents the rapid mineralization of S. In litter, S is predominately carbon-bonded S (C–S) to the humin fraction to a level of about 89%. The incorporation of organic S into complex organic substances followed by association with clay minerals can lead to the relative isolation of S from decayed soil micro-organisms. This process is considered as a physical protection of organic S against degradation,[15] and it results in a decreased availability of S to plants.

Sulfur Mineralization

In soils microbial mediated processes are mainly responsible for S transformations, so that the factors affecting the microbial activity, such as temperature, moisture, pH and substrate availability will also influence the process of mineralization, immobilization, oxidation and reduction. In aerobic agricultural soils the main factor in these processes is the release of inorganic, plant-available sulfate from organic matter.

Two types of processes are involved in the mineralization of S: biological and the biochemical mineralization.[16] The biological mineralization is considered to be driven by the microbial need for organic C to provide energy, and S released as sulfate is a by-product of the oxidation of C to carbon-dioxide. This process is faster the more recently the organic matter was formed.[17] Biochemical mineralization relies on the release of sulfate from the sulfate-ester pool through enzymatic hydrolysis.

Sulfur in the microbial biomass is actively turned over in soils whilst comprising only approximately 2%–3% of the total soil S. The turnover of soil microbial bio-mass is fundamental to the incorporation of sulfate-S into soil organic matter but quantitative measures to assess this are unavailable.

The contribution of mineralization to the S supply of plants is only small,[2] because mineralization, immobilization and possible leaching of S occur concurrently.[14] The amount of S mineralized within the organic S pool in the soil ranges from 0.5% to 3.1% annually.[2] The contribution of net mineralization accounts on average for 10–30 kg ha^{-1} yr^{-1} S[18] in soils with carbon contents between 1% and 4% C. The studies of Eriksen, Murphy, and Schung[2] and Bloem[18] reveal that mineralization is an important, however not cardinal S pool for plants. High yielding crops cannot satisfy their S demand solely by mineralization and atmospheric S depositions.[7]

Crops and Crop Rotation

Dead plant material contributes to the organic matter pool but the living plant organic matter which partly remains on the field may be regarded as a storage pool for S, too. Sulfur offtake and S demand differ depending on the plant species. Sugar cane has a very low S demand with only 0.3 kg S Mg^{-1} dry matter yield,[19] cereals about 1.5–2 kg S Mg^{-1} of grain and 1–3 kg S Mg^{-1} of straw yield,[18] soybean between 4.3 and 8.8 kg S Mg^{-1},[20,21] mustard about 16 kg S Mg^{-1}[22,23] and oilseed rape up to 20 kg S Mg^{-1}.[24]

The S content of straw highly influences the mineralization of plant residues. Crops which leave residues with high S contents, may be considered as catch crops for S and are able to reduce S losses from soil by leaching. Cruciferous crops such as oilseed rape show a high S content of the residual straw,[24,25] which provide a large amount of rapidly decomposable organic material.[26] In comparison, plants without a secondary metabolism such as cereals leave smaller amounts of S. Only if the S content of for example barley straw is above 0.13% S, can a release of S from the mineralization of straw be expected.[27] If the value is below 0.13% S, a net immobilization of S takes place because the microbes need more S for the formation of their microbial biomass. This may exacerbate S deficiency of the following crop, for example oilseed rape, particularly on light soils, where macroscopic symptoms may become visible even before winter[25] because of the net immobilization of S. So far there is no evidence that winter crops benefit during their main growth from the previous crop's is plant organic matter residues with high S.

Physico-Chemical Aspects of Sulfur in Soils

Sulfate has a high mobility in soils and can be delivered from subsoil or shallow groundwater. The water soluble fraction can be leached, adsorbed, immobilized or taken up by the plant. Availability of sulfate is therefore more a question of transfer between pools in terms of space and time rather than between biological or chemical systems. The insoluble, organic fraction is only slightly leached and is not directly available for plant uptake.

Soil Water Regime

Under temperate conditions groundwater is often an important S pool and there are three ways groundwater can contribute to the S nutrition of plants. Firstly, there is a direct S input if the groundwater level is only 1–2 m below the surface. This is sufficient to cover the S requirement of most crops, as plants can utilize the sulfate in the groundwater directly by their root system. An average yielding (3 Mg ha^{-1} seeds) oilseed rape crop covers more than 50% of its S demand by shallow groundwater or soil water.[24] Secondly, groundwater contains between 5 and 100 mg L^{-1} sulfate-S which, if used for irrigation can supply up to 100 kg ha^{-1} S to the crop.[18] Thirdly, the capillary rise of groundwater under conditions of a soil moisture deficit in the upper soil layers leads to a S input. The contribution of this process to the S supply of plants depends on the soil texture within the soil profile and climatic conditions. Generally heavier soils show less frequent S deficiency than lighter ones as they retain more S-rich pore water. This is proven by a significant relationship between groundwater level and clay content, and the sulfate-S content of soils.[18]

Capillary Rise/Leaching

While rainfall water contains only about 2.5 mg S L^{-1},[28] the S content of adsorbed soil water varies between 15 and 100 mg S L^{-1}.[18] Variation of the sulfate content of water in soil pores is caused by temporal changes of the groundwater level, soil depth and differences in soil texture.[18] Sulfur-rich water in the soil pore ascends in the soil profile if the evapo-transpiration rate is higher than the amount of precipitation.

A higher groundwater level below the surface and a high soil moisture deficit at the lower periphery of the root zone will increase the extent of the capillary rise. These authors[29] calculated an average capillary rise of 0.3 mm d^{-1} on a sandy soil, 3 mmd^{-1} on a loamy sand and 5mm d^{-1} on a loamy silt with a groundwater level of 1.5m below the surface. This means that under extreme conditions on sandy soils with a groundwater level >2m below the surface virtually no sulfate will be available to the plant due to capillary rise, but on loamy and silty soils, the S supply of crops may be fully satisfied by capillary water, even if the sulfate concentration of the groundwater is low (10 mg L^{-1} S).[18] Spring crops particularly benefit from capillary ascending soil water as their main growth takes place when capillary rise dominates soil water movements.

The rate of leaching of S under temperate conditions depends on soil type, winter precipitation and the concentration of sulfate S in the soil water during the period of leaching.[18] Sulfur losses through leaching are higher over winter due to higher precipitation rates and the lack of plant uptake. As a result, S stored in soils decreases with increasing precipitation and on sandy, free-draining soils all sulfate beneath the capillary zone and rooting depth of the crop may be leached before the beginning of the vegetative period[18] because of the rapid movement of the leachate through the soil and the low adsorption capacities of such soils. Average sulfate leaching depths are 15 cm with 50mm of precipitation on a loamy soil and 25 cm on a sandy soil, respectively.[30] Additionally leaching is higher on fallow than on cropped soils[31] and is enhanced by S fertilization. Average leaching losses under temperate conditions are in the range of 30–80 kg ha^{-1} yr^{-1} S.[32,33] In addition to vertical water movements in the soil, lateral fluxes can occur particularly in the landscapes with pronounced differences in geomorphology. These fluxes are the major reason for the high spatio-temporal variability of sulfate in soils.

Soil Compaction

Soil compaction and tillage operations causing soil compaction will also decrease the amount of plant available S because of the reduced soil volume and consequent impact on soil pore space. Consequently rooting depth and density are decreased so that S rich capillary water cannot be used.[34,35] This effect of soil compaction is supported by the fact that S deficiency regularly becomes visible first in the headland and along tramlines of fields[7] and because it is sulfate-S from the sub-soil that mainly contributes to the S nutrition of crops.[18]

Sulfur Emissions

Sulfur containing atmospheric constituents are principally SO_2, SO_4^{2-}, SO_3^{2-}, H_2S, COS, DMS, CS_2 and methylmercaptan. Natural S emissions from biogenic sources and from volcanoes add around 60 Tg S yr^{-1} to the Earth's atmosphere with an upward tendency due to the increasing consumption of fossil fuels in South America, Africa and Asia.[9]

Before the industrial revolution, atmospheric S depositions were, on an average, below 10 kg S ha^{-1} yr^{-1}. As a consequence of increased burning of fossil fuels with industrial development from 1890 to 1980 atmospheric S depositions, mainly as SO_2, increased steadily by 0.47 kg S ha^{-1} yr^{-1}.[1] In some rural areas of northern Scotland, however, SO_2 depositions were zero or below 2 kg S ha^{-1} in the 1970s, while the S input in the midlands of England was about 160 kg S ha^{-1} at the same time.[36] At its peak the negative impacts of SO_2 emissions to humans, plants, soils and buildings were so serious that S became an unpopular nutrient and was called as the "yellow poison".[37] The political consequences resulted in clean air acts in European countries and North America and the desulfurization of emissions led to a drastic decrease of atmospheric S depositions.

Agriculture adapted and co-evolved to increasing S loads. Soil acidification caused by high S deposition required higher amounts of liming materials. Increasing S demands of agricultural crops due to higher yield potential as a result of plant breeding progress and production technology coincided with increased atmospheric S supply. Additionally the use of S-containing fertilizers went down nearly to zero.[25] Consequently the reduction of atmospheric S deposition had a major effect on the productivity of crops. During the 1990s, macroscopic S deficiency became the most widespread nutrient disorder in northern Europe.[7]

Sulfur Fertilizers

The worldwide demand of S fertilizers in 2010 is estimated to be about 11.3 million tons of S.[38] Numerous S fertilizers and secondary raw material fertilizer products are available for either soil or foliar application in sulfate or elemental form.[39] Arguments for and against the use of individual products depend on local farming conditions. Organic fertilizers such as manure and slurry contain about 1 kg S Mg^{-1} and 0.5 kg S Mg^{-1}, respectively.[40] This equals to 0.07 kg S per kg nitrogen.

Agro-Ecological Aspects of Sulfur

Interaction of Agro-Ecosystems with Other Ecosystems

The S input and off-take in agro-ecosystems may have a strong influence on neighboring eco-systems. Sulfur is often in excess in natural systems despite the low atmospheric S input because the turnover of S is much lower in these than in agricultural systems. Acid sub-soils, e.g., under forest vegetation, and peat soils show pH values <5, which are far below the acidity levels of fertile agricultural soils. Large storage capacities for adsorbed sulfate exist on such sites because the pH value of the soil is lower than the zero point of charge (ZPC) resulting in the hydration of metal oxides and thus a positive surface.[41] Although adsorbed sulfate plays only a minor role in the direct S nutrition of agricultural crops it may contribute positively to the S balance of the whole surrounding landscape. Natural vegetation, land without plant production and forests show a positive S balance even when atmospheric inputs are low.[2] Agricultural crops may benefit from this S pool, if groundwater reservoirs of both eco-systems are connected. This also means, however, that with an increasing share of agricultural farmland in landscapes, the risk of S deficiency overall also increases.

Sustainability

Sustainable agriculture should use soils in such a way that the present and future human needs for food or other agricultural goods are realized and the quality of the environment and the natural resources remain preserved.[42] The contamination of groundwater with nitrates is a most serious problem. Nitrogen (N) and S are both involved in protein biosynthesis and a shortage in the S supply of crops also lowers the utilization of applied fertilizer N and thus deteriorates the crop quality. Non-protein N is accumulated in plant parts and besides poor efficiency for N fertilization, S deficiencies may increase the loss of N from agricultural soils through volatilization and leaching.[43] On average each kg of S shortage to satisfy the S demand of the crop causes 15 kg of N to be lost to the environment. Such N inputs endanger strongly the stability of natural communities as for example the growth of algae in water bodies.[44] Correcting S deficiency by fertilization is environmentally safe as sulfate is, in comparison with N, geogenously abundant.

Global Change

Climate is one of the major factors involved in pedogenesis. Soil formation is directly influenced mainly by temperature and water and indirectly via the climate-depending vegetation.[45] Changes in climate may change soil types, increase erosion, affect element cycles and increase the release of greenhouse gases. Increased temperature and humidity accelerate pedogenesis. Global change thus would allow soils in the northern hemisphere to proceed faster through the individual stages of development and/or degradation associated with their individual soil series.[46] Higher temperatures accelerate the decomposition of organic materials and thus decrease the organic matter content of soils. At the same time the net mobilization of S might increase while the organic S pool decreases. This effect is expected to be more pronounced in cultivated than in range soils.[47] But at the same time under more humid climate conditions organic matter tends to accumulate in soils.[47] Therefore an expected adverse effect of global warming on soil organic matter might be at least partially compensated in those areas where this coincides with increasing humidity.[48] With increasing carbon accumulation, the other growth limiting elements, such as N, S, and P, may dilute relatively,[49] which might result in decreased soil fertility.

Plant Health

Owing to the fact that higher anthropogenic S inputs in the past decades enabled plants to adapt to increasing environmental stress, the decline in the S supply within only one decade might have serious consequences for the stability of recent ecosystems.[49] Sulfur metabolism provides several efficient mechanisms by which plants are able to tackle abiotic (e.g., xenobiotics and increasing surface ozone levels) and biotic (e.g., pests and diseases) stress, particularly via the glutathione metabolism which again is closely related to the S supply of the plants.[3] Other mechanisms involved in response to plant pathogens include the production of S containing compounds in the secondary metabolism of the agriculturally important *Brassica* species, the release of volatile S compounds, the production of S rich proteins, localized deposition of elemental S and the production of phytochelatines, which detoxify heavy metals by forming complexes.[50,51]

Certain diseases (e.g., light leaf spot in oilseed rape) occur more frequently particularly in areas with low S input in Europe[52] and an improved knowledge of the significance of S metabolites in crop resistance to diseases will be beneficial for the improvement of S fertilizer strategies and could therefore minimize the input of pesticides.

Acknowledgments

The authors cordially thank Dr. Kerr C. Walker (SAC, Aberdeen) for his linguistic efforts on our entry.

References

1. Dämmgen, U.; Walker, K.C.; Grünhage, L.; Jäger, H.-J. The atmospheric sulphur cycle. In *Sulphur in Agroecosystems*; Schnug, E., Ed.; Kluwer Academic Publ.: Dordrecht, 1998; 75–114.
2. Eriksen, J.; Murphy, M.D.; Schnug, E. The soil sulphur cycle. In *Sulphur in Agroecosystems*; Schnug, E., Ed.; Kluwer Academic Publ.: Dordrecht, 1998; 39–74.
3. Hell, R.; Rennenberg, H. The plant sulphur cycling. In *Sulphur in Agroecosystems*; Schnug, E., Ed.; Kluwer Academic Publ.: Dordrecht, 1998; 135–174.
4. Howarth, R.W.; Stewart, J.W.B. The interactions of sulphur with other element cycles in ecosystems. In *Sulphur Cycling on the Continents: Wetlands, Terrestrial Ecosystems and Associated Water Bodies, SCOPE 48*; Howarth, R.W., Stewart, J.W.B., Ivanov, M.V., Eds.; John Wiley and Sons: Chichester, 1992; 67–84.
5. Howarth, R.W.; Stewart, J.W.B.; Ivanov, M.V.; Eds. *Sulphur Cycling on the Continents: Wetlands, Terrestrial Ecosystems and Associated Water Bodies, SCOPE 48*; John Wiley and Sons: Chichester, 1992; 345 pp.
6. Janzen, H.H.; Ellert, B.H. Sulfur dynamics in cultivated, temperate agroecosystems. In *Sulfur in the Environment*; Maynard, D.G., Ed.; Marcel Dekker Inc.: New York, 1998; 11–44.
7. Schnug, E.; Haneklaus, S. Diagnosis of sulphur nutrition. In *Sulphur in Agroecosystems*; Schnug, E., Ed.; Kluwer Academic Publ.: Dordrecht, 1998; 1–38.
8. Stevenson, F.J. *Cycle of Soil. Carbon, Nitrogen, Phosphorus, Sulfur, Micronutrients*; John Wiley and Sons: New York, 1986; 285–320.
9. Whelpdale, D.M. An Overview of the atmospheric sulphur cycle. In *Sulphur Cycling on the Continents: Wetlands, Terrestrial Ecosystems and Associated Water Bodies, SCOPE 48*; Howarth, R.W., Stewart, J.W.B., Ivanov, M.V., Eds.; John Wiley and Sons: Chichester, 1992; 5–26.
10. Bowen, H.J.M. *Trace Elements in Biochemistry*; Academic Press: London, 1966.
11. Friend, J.P. The global sulfur cycle. In *Chemistry of the Lower Atmosphere*; Rasool, S.I., Ed.; Plenum Press: New York, 1973; 177–201.
12. Wedepohl, K.H. Chemical fractionation in the sedimentary environment. In *Origin and Distribution of the Elements*; Ahrens, L.H., Ed.; Pergamon Press: London, 1968; Vol. 30 of Earth Sciences.
13. Anisimova, M.; Haneklaus, S.; Schnug, E. Significance of sulfur for soil organic matter. In *Sulfur Nutrition and Sulfur Assimilation in Higher Plants*; Brunold, C., Rennenberg, H., De Kok, L.J., Stulen, I., Davidian, J.-C., Eds.; Paul Haupt Publishers: Berne, 2000; 239–244.
14. Ghani, A.; McLaren, R.G.; Swift, R.S. The incorporation and transformations of 35S in soil: effect of soil conditioning and glucose or sulfate addition. Soil Biol. Biochem. **1993**, *25*, 327–335.
15. Eriksen, J.; Lefroy, R.D.; Blair, G.J. Physical protection of soil organic S studied by extraction and fractionation of soil organic matter. Soil Biol. Biochem. **1995**, *27*, 1011–1016.
16. McGill, W.B.; Cole, C.V. Comparative aspects of cycling of organic C, N, S and P through soil organic matter. Geoderma **1981**, *26*, 267–286.
17. Ghani, A.; McLaren, R.G.; Swift, R.S. Mobilization of recently-formed soil organic sulphur. Soil Biol. Biochem **1993**, *25*, 1739–1744.
18. Bloem, E. Schwefel-bilanz von agrarökosystemen unter besonderer berücksichtigung hydrologischer und bodenphysi-kalischer Standorteigenschaften. Landbauforsch. Voelk. **1998**, *192*, 1–156.
19. Katyal, J.C.; Sharma, K.L.; Srinivas, K. *Sulphur in Indian Agriculture*; Proc. TSI/FAI/IFA Symposium on sulphur in balanced fertilisation, KS-2/1-KS-2/12; 1997.

20. Aulakh, M.S.; Pasricha, N.S.; Azad, A.S. Phosphorussulphur interrelationships for soybeans on phosphorus and sulphur deficient soils. Soil Sci. **1990**, *150*, 705–709.

21. Nambiar, K.K.M.; Gosh, A.B. Highlights of Research of a Long-Term Fertilizer Experiment in India (1971–1982); Techn. Bull. No.1 Longterm Fertilizer Experiment Project, IARI, 100, 1984.

22. Jain, G.L.; Sahu, M.P.; Somani, L.L. Balanced fertilization programme with special reference to secondary and micronutrients nutrition of crops under intensive cropping, Proc. FAI/NR Seminar. Jaipur, 1984; 147–174.

23. Aulakh, M.S.; Pastricha, N.S.; Sahota, N.S. Yield, nutrient concentration and quality of mustard crops as influenced by nitrogen and sulphur fertilizers. J. Agric. Sci. **1980**, *94*, 545–549.

24. *Schnug, E. Quantitative und Qualitative Aspekte der Diagnose und Therapie der Schwefelversorgung von Raps (Brassica napus L.) Unter Besonderer Berücksichtigung Glucosinolatarmer Sorten; DSC, Christian-Albrechts-University: Kiel, Germany, 1988.*

25. Schnug, E.; Haneklaus, S. Sulphur deficiency in brassica napus- biochemistry- symptomatology- morphogenesis. Landbauforsch. Voelk. **1994**, *144*, 1–31.

26. Wu, J.A.G.; O'Donnell, Z.L.; Syers, J.K. Microbial growth and sulphur immobilization following the incorporation of plant residues into soil. Soil Biol. Biochem. **1993**, *25*, 1567–1573.

27. Chapman, S.J. Barley straw decomposition and S immobilisation. Soil Biol. Biochem. **1997**, *29*, 109–114.

28. Dämmgen, U.; Grünhage, L.; Kü sters, A.; Jäger, H.J. Konzentrationen von luftinhaltsstoffen. I. criteria pollutants. Landbauforsch. Voelk. **1996**, *170*, 196–221.

29. Giesel, W.; Renger, M.; Strebel, O. Berechnung des kapillaren aufstiegs aus dem grundwasser in den wurzelraum unter stationären bedingungen. Z. Pflanzenern. Bodenkde **1972**, *132*, 17–30.

30. Kumar, V.; Karwasra, S.P.S.; Singh, M.; Dhankar, J.S. An evaluation of the sulphur status and crop responses in the major soils of Haryana, India. Sulphur Agric. **1994**, *18*, 23–26.

31. Kirchmann, H.; Pichlmayer, F.; Gerzabek, M.H. Sulfur balance and sulfur-34 abundance in a long-term fertilizer experiment. Soil Sci. Amer. J. **1996**, *59*, 174–178.

32. Mansfeld, T. Schwefeldynamik von böden des dithmarscher speicherkoogs und der bornhoeveder Seenkette in schleswig-holstein. PhD, Kiel, Germany, 1994.

33. Preuschoff, M. Untersuchungen zur schwefelversorgung von weißohl an zwei lößstandorten. PhD, University Hanover, Verlag Ulrich E. GrauerStuttgart, 1995.

34. Singh, B.R. Effect of soil compaction on S availability to crop plants. Abstracts of the COST Action 829: Fundamental, Agronomical and Environmental Aspects of Sulfur Nutrition and Assimilation in Plants, Goslar, Germany, 1998.

35. Unger, P.W.; Kaspar, T.C. Soil compaction and root growth: a review. Agron. J. **1994**, *86*, 759–766.

36. Semb, A. Sulfur emissions in Europe. Atmos. Environm. **1978**, *12*, 455–460.

37. Boelsche, J. *Das gelbe Gift*; Rowohlt Verlag: Reinbek, 1994.

38. Cecotti, S.; Morris, R.J.; Messick, D. A global overview of the sulphr situation: industry's background, Market trends, and commercial aspects of sulphur fertilisers. In *Sulphur in Agroecosystems*; Schnug, E., Ed.; Kluwer Academic Publ.: Dordrecht, 1998; 175–202.

39. Paulsen, H.M. Produktionstechnische und Ökologische bewertung der landwirtschaftlichen verwertung von schwefel aus industriellen prozessen. Landbauforsch. Voelk. **1999**, *197*, 1–143.

40. Pedersen, C.A.; Knudsen, L.; Schnug, E. Sulphur Fertilisation. In *Sulphur in Agroecosystems*; Schnug, E., Ed.; Kluwer Academic Publ.: Dordrecht, 1998; 115–134.

41. Curtin, D.; Syers, J.K. Extractability and adsorption of sulfate in soils. J. Soil Sci. **1990**, *41*, 305–312.

42. Anon. *Nachhaltiges Deutschland: Wege zu Einer Dauerhaft Umweltgerechten Entwicklung*; Umweltbundesamt Berlin, Erich Schmidt Verlag: Berlin, 1997.

43. Schnug, E. Sulphur nutritional status of European crops and consequences for agriculture. Sulphur in Agriculture **1991**, *15*, 7–12.

44. Wild, A. *Umweltorientierte Bodenkunde*; Spektrum Akademischer Verlag: Heidelberg, 1993.

45. Hugget, R.J. *Geoecology*; Routledge: London, 1995.

46. Rogasik, J.; Daemmgen, U.; Luettich, M.; Obenauf, S. Wirkungen physikalischer und chemischer klimaparameter auf bodeneigenschaften und bodenprozesse. Landbauforsch. Voelk. **1994**, *148*, 107–139.

47. Burke, I.C.; Yonker, C.M.; Parton, W.J.; Cole, C.V.; Flach, K.; Schimel, D.S. Texture, climate and cultivation effects on soil organic matter content in U.S. Grassland Soils. Soil Sci. Soc. Am. J. **1989**, *53*, 800–805.

48. Schnug, E. Response of plant metabolism to air pollution and global change impact on agriculture. In *Responses of Plant Metabolism to Air Pollution and Global Change*; DeKok, L.J., Stulen, I., Eds.; Backhys Publ.: Leiden, 1998; 15–22.

49. Schnug, E.; Haneklaus, S. Ecological Aspects of Plant Sulphur Supply. Proc. 15th Int. Congr. Soil Sci. Acapulco/ Mexico, 1994; 5a: Comm. IV: Symposia, 364–371, 1994.

50. Resende, M.L.V.; Flood, J.; Ramsden, J.D.; Rowan, M.G.; Beale, M.H.; Cooper, R.M. Novel phyto-alexins including elemental sulphur in the resistance of cocoa (*Theobroma cacao* L.) to verticillium wilt (*Verticillium dahliae Kleb.*). Physiol. Molec. Plant Pathol. **1996**, *48*, 347–359.

51. Schnug, E. Significance of Sulphur for the nutritional and technological quality of domesticated plants. In *Sulfur Nutrition and Sulfur Assimilation in Higher Plants*; Cram, W.J., DeKok, L.J., Stulen, I., Brunold, C., Rennberg,H., Eds.; Backhys Publ.: Leiden, 1997; 109–130.

52. Thomas, J. Watch out for light leaf spot. NIAB Fellows Newsletter 1994.

37

Sulfur Dioxide

Marianna Czaplicka
and Witold Kurylak

Introduction

Sulfur dioxide (SO$_2$) in large amounts is a toxic gas. The World Health Organization (WHO) recommends a concentration of no greater than 0.5 parts per million (ppm) over 24 hours for maximum exposure. A concentration of 6–12 ppm can cause immediate irritation of the nose and throat; 20 ppm can cause eye irritation; and 10,000 ppm will irritate moist skin within minutes. High concentrations of SO$_2$ can result in breathing problems in asthmatic children and adults who are active outdoors.[1,2] Short-term exposure has been linked to wheezing, chest tightness, and shortness of breath.

Sulfur dioxide emissions are partly responsible for acid depositions on the surface and the occurrence of winter smog episodes. It is one of the major atmospheric pollutants. In the air, SO$_2$ oxidizes to SO$_3$, which easily reacts with steam to create sulfuric acid (H$_2$SO$_4$), which then condenses rapidly in the stratosphere to form fine sulfate aerosols. These sulfate aerosols also promote complex chemical reactions on their surfaces that alter chlorine and nitrogen chemical species in the stratosphere. This effect, together with increased stratospheric chlorine levels from chlorofluorocarbon pollution, generates chlorine monoxide, which destroys ozone. Sulfate aerosols generated from volcanic emissions of sulfur-rich gas temporarily counteract some of the heating caused by elevated carbon dioxide concentrations in the atmosphere.[3] Aerosols in the atmosphere cool the climate both directly due to backscattering of sunlight and indirectly through an increase in cloud receptivity and residence time. Pollutants can be transported thousands of kilometers owing to the introduction of tall chimneys that disperse pollutants high in the atmosphere.

Acid rain causes acidification of lakes and streams, and contributes to the damage of trees at high elevations and of many sensitive forest soils. Soil acidification due to acidic deposition may contribute to forest decline[4] and can lead to the loss of plant nutrients and the accumulation of toxic aluminum, both of which may affect vegetation. In addition, acid rain accelerates the decay of building materials and paints, including irreplaceable buildings, statues, and sculptures that are part of the cultural heritage.

The presence of SO_2 was observed both in the stratosphere and in the troposphere. The lifetime of SO_2 molecules in the troposphere is a few days. In the troposphere, mainly volcanic SO_2 becomes rapidly oxidized to sulfate and washed out of the atmosphere in the presence of hydrometeors. These washout processes are not effective in the stratosphere. The lifetime of SO_2 molecules in the stratosphere is several weeks, during which time sulfate aerosols are produced. Since gaseous H_2SO_4 has a very low saturation pressure, it can be easily condensed in the stratosphere and forms aerosols of liquid hydrated H_2SO_4. Once formed, sulfate aerosols have a residence time in the stratosphere of about 3 years. Numerous measurements clearly show the importance of sulfur aerosols in modifying climate, warming the stratosphere, and cooling the troposphere. Research has also shown that the liquid drops of H_2SO_4 promote the destruction of the earth's ozone layer. Ways for SO_2 removal from the troposphere include the following:

- In gas phase, by formation of H_2SO_4, which forms condensation nuclei for aerosols and clouds and acidifies rain
- Directly, by way of uptake on aerosols and clouds, which leads to dry and wet acid depositions

SO_2 Standards in the World—Ambient Air Quality

The concentration of SO_2 in the air depends mostly on meteorological factors such as

- Transport processes by prevailing winds
- Depth of the mixing layer
- Deposition process

Levels of SO_2 in the air are higher than normal near facilities that release SO_2 through heavy industrial activities such as copper smelting or coal and oil burning or processing.

The negative influence of SO_2 on human health became a basis for the determination of air quality standards covering, among other things, maximum permissible concentration of that compound in the air. In 2005, WHO updated its air quality guidelines (AQG), an international reference on the health consequences of exposure to air pollution and a policy tool for reducing these consequences worldwide. For SO_2, AQG achieved a level 20 $\mu g/m^3$ for 24 hr and 500 $\mu g/m^3$ in the case of 10-min measurement.[5]

Air quality standards differ worldwide. Vahlsing and Smith[6] showed that average 24-hr ambient air quality standards (AAQS) in 76 countries for SO_2 is 95 $\mu g/m^3$ and the population weighted average of AAQS for SO_2 is 155 $\mu g/m^3$. In the United States, there are two existing primary national AAQS (NAAQS) for SO_2. The first is a long-term 1 year arithmetic average of 0.03 ppm. The second is a short-term 24-hr average standard where concentrations should not exceed 0.14 ppm more than once per year. The current secondary NAAQS in the United States for SO_2 is a 3-hr average concentration of 0.5 ppm not to be exceeded more than once per year. In addition, there are three standards for SO_2 defined by state law. The first is a long-term 1-year arithmetic average of 0.022 ppm. The second is a short-term 24-hr average standard of 0.088 ppm not to be exceeded, and the third is a short-term 3-hr average concentration of 0.439 ppm not to be exceeded.

In June 2010, U.S. Environmental Protection Agency (EPA) strengthened the primary NAAQS for SO_2 by establishing a new 1-hr standard designed to protect sensitive individuals from high, short-term (5-min to 24-hr) exposure.[7] The level is set at 75 parts per billion (ppb). This new 1-hr standard will replace the two existing primary standards.

The EPA's current action focuses on the primary SO_2 standard only. It will address the secondary standard for SO_2 as part of a separate review. The current secondary standard to protect public welfare is a 3-hr average set at 500 ppb, as mentioned before.

Unlike WHO, the European Union air quality standards (EU AQS)[8] and the U.S. EPA[9] continue to set a 1-hr standard for SO_2 at 350 and 196 $\mu g/m^3$ (70 ppb), respectively. According to the European standard, the permissible hourly concentration of SO_2 in the air is 350 $\mu g/m^3$ while in 24 hr it is 125 $\mu g/m^3$.[8]

Most developing countries in Asia either use the 24-hr standard and/or the annual standard. Although some countries have 1-hr guidelines, they are not comparable to the EPA or even the EU AQS except for some countries. In China, for example, there are five different standards: Grade I—applies to specially protected areas, such as natural conservation areas, scenic spots, and historical sites; Grade II—applies to residential, mixed commercial/residential, cultural, industrial, and rural areas; Grade III—special industrial areas; SAR and SAR 2—Special Administrative Region. The regulations are presented in Table 1.

Methods for Determination of SO_2 in Air

There are many techniques available for analysis of SO_2 in ambient air, including detection by spectrophotometry,[11–13] chemiluminescence,[14–16] ion chromatography,[17,18] spectrofluorometry,[19] and potentiometric[20] methods. The measurement of SO_2 in air is usually based on classic fluorescence spectroscopy principles. Sulfur dioxide exhibits a strong ultraviolet (UV) absorption spectrum between 200 and 240 nm, and when SO_2 absorbs UV, emission of photons occurs (300–400 nm). The amount of fluorescence emitted is directly proportional to the SO_2 concentration. Ultraviolet fluorescent radiation technology to detect SO_2 is characterized by a sensitivity of 0.4 ppb in the range of 0–20 ppm.[11] Chang et al.[12] described a simple fluorescence detection method for determination of SO_2 in ambient air collected by passive sampler. Ultraviolet fluorescence and pararosaniline methods are recommended by the EPA as reference methods for the determination of SO_2 in the atmosphere.[21] In the European Union, Directive 2008/50/EC of the European Parliament and of the Council recommends as reference method the procedure described in EN 14212:2005, "Ambient air quality—Standard method for the measurement of the concentration of sulfur dioxide by ultraviolet fluorescence."[8]

Sources of SO_2

Emission of SO_2 to the air results from natural processes and from anthropogenic sources. Globally, emissions due to anthropogenic activities account for approximately 70 Tg (S) of the total (natural + anthropogenic) 107 Tg (S) emissions of sulfur each year, although a greater uncertainty exists for natural sources.[22,23]

Biogenic Sources

Is well known that SO_2 from volcanoes is one of the two most important sources of stratospheric aerosols. According to Symonds et al.[24] SO_2 concentration in volcanic gases may vary from 11.8% (Kilauea Volcano) to 0.50% (Momotombo Volcano). The data on Popocatèpetl Volcano (Mexico) eruption show

TABLE 1 Standards of SO_2 in China, $\mu g/m^3$

	SO_2 1 hr	SO_2 24 hr	SO_2 Annual
China: Grade I	150	50	20
China: Grade II	500	150	60
China: Grade III	700	250	100
Hong Kong SAR	800	350	80
Hong Kong SAR 2	—	125	—

Source: Data from Air Quality in Asia: Status and Trends.[10]

that the total sulfur emitted in the years 1994–1998 achieved a level of 8,500,000 tons,[25] while the total SO_2 outgassed from Galunggung Volcano (Java) from April 1982 to January 1983 was calculated on level 2500 kt.[26] Another eruption, that of El Chichon in 1982, emitted $7 \leftrightarrow 10^{12}$ g SO_2. Halmer and co-workers[27] calculated $15 \leftrightarrow 10^{12}$ to $21 \leftrightarrow 10^{12}$ g of SO_2 as the total annual global volcanic SO_2 input into the atmosphere from both silent and explosive degassing volcanoes, corresponding to $7.5 \leftrightarrow 10^{12}$ to $10.5 \leftrightarrow 10^{12}$ g/yr S (S here as SO_2) for the period 1972 to 2000, the numbers representing an average value for volcanic gas emission over almost 30 years. The amount of volcanic gases and particles arriving in the troposphere and/or stratosphere generally depends on both volcanological and meteorological boundary conditions:

- The chemical composition of the magma depending on the tectonic environment (e.g., subduction zone, continental rift zone)
- The strength and duration of an eruption
- The location of the emitting volcano (altitude and latitude)
- Atmospheric conditions such as temperature, humidity, and wind prowls

Sulfur dioxide and other volcanic sulfur-bearing gases transform to sulfate aerosols, which are able to cool the climate directly by backscattering incoming solar radiation and indirectly through increase in cloud receptivity. The residence time of water-soluble sulfuric gases (in particular SO_2) emitted into the atmosphere depends on the height of an eruption column. The effects of SO_2 (from natural sources) on people and the environment vary widely depending on the following: 1) the amount of the gas a volcano emits into the atmosphere; 2) whether the gas is injected into the troposphere or the stratosphere; and 3) the regional or global wind and weather pattern that disperses the gas.

Recent studies have suggested oxidation of dimethyl sulfide (DMS) in the marine atmosphere as the elevated biogenic source of SO_2. Oceanic emissions of DMS are a major component of the natural sulfur budget (about 60% of total natural global sulfur emissions).[28,29] Dimethylsulfoniopropionate produced by phytoplankton in the ocean is the precursor to the DMS gas. Once in the atmosphere, DMS is oxidized by radicals to form methylsulfonic acid, dimethyl sulfoxide, dimethyl sulfone, sulfate, and SO_2. These products take part in the growth of aerosols. Gaseous H_2SO_4 is the prevailing product of DMS oxidation that can generate new cloud condensation nuclei under atmospheric conditions. One of the precursors in this formation pathway is SO_2. Seguin et al.'s[30] measurements of SO_2 concentration in the air over the Northwest Atlantic Ocean in spring and summer 2003 indicted that SO_2 concentrations were as high as 156 nmol/m^3 in the spring and from 0.55 nmol/m^3 to 120 nmol/m^3 in the summer. During the Mediterranean Intensive Oxidant Study (MINOS) experiment (July–August 2001), Mihalopoulos et al.[31] found that marine biogenic sulfur emissions (predominantly DMS) constituted 20% of total oxidized sulfur production (SO_2 and H_2SO_4). Among the natural sources of SO_2 are also biological decay and forest fires.

Measurements of Concentration of SO_2 from Natural Sources

Measurement of SO_2 emission from natural sources is a very difficult task. The emission rates of SO_2 are measured using airborne or ground-based techniques. The most common method for sampling volcanic gases is to collect them directly from fumaroles in solution-filled bottles, and then to analyze the mixtures in the laboratory.[24] Observations of SO_2 flux have been made for over 30 years by correlation spectrometers, which have been recently upgraded with compact and inexpensive powered CCD detector array–based dispersive spectrometers.[32,33] Ultraviolet spectroscopy has been widely applied to volcanoes worldwide over the last decades to investigate how SO_2 fluxes vary with activity, with heightened emissions observed during eruptions as degassing magma batches reach the surface.[34,35] Salerno et al.[36] describe the method based on differential optical absorption for automatic determination of volcanic SO_2. The other sensing techniques applied to determination of SO_2 concentration include multispectral infrared[37,38] and ship-borne differential absorption lidar (DIAL) methods,[39–41] and Correlation Spectroscopy (COSPEC).[42,43]

TABLE 2 Emission Factors of SO_2 for Different Types of Fuel (Emission Factors Are Taken from National Atmospheric Emission Inventory)

Emission Source	Emission Factor
Generators (gas oil)	$3.25 \leftrightarrow 10^{-3}$
Generators (diesel fuel)	$33 \leftrightarrow 10^{-3}$
Vehicles (principally ATF)	$1.0 \leftrightarrow 10^{-7}$
Marine gas oil (for ships)	$1.9 \leftrightarrow 10^{-2}$
Fuel oil (for ships)	$5.2 \leftrightarrow 10^{-2}$
Aviation turbine fuel (for international flights)	$7.2 \leftrightarrow 10^{-4}$
Aviation turbine fuel (for domestic flights)	$7.2 \leftrightarrow 10^{-4}$

Source: Data from Shirsat and Graf.[23]

Anthropogenic Sources

The largest anthropogenic SO_2 emissions come from fossil fuel combustion at power plants (73%) and other industrial facilities (20%).[44] Smaller sources of SO_2 emissions include industrial processes such as biomass burning, extracting metal from ore, smelting of metal sulfide ores, and burning of high-sulfur-containing fuels by locomotives, large ships, and non-road equipment. Sulfur dioxide is linked with a number of adverse effects on the respiratory system. Table 2 shows emission factors of SO_2 for various types of fuel.

Sulfur dioxide is used as fumigant, preservative, bleach, and steeping agent for grain. A small amount of SO_2 may be used, as preservative, both in the vineyard and during winemaking to protect grapes and wine from microbial attack or further oxidation. Sulfur dioxide is also a bleaching agent in the pulp and paper industry.

Corbett et al.[45] showed that SO_2 emissions from ocean-going ships significantly contribute to the global pool of sulfur and make up about 20% of biogenic DMS emissions. It has been estimated that the 1993 global SO_2 emissions from ships released 4.24 Tg of sulfur. Emissions of NO_x and SO_2 from ocean-going ships are 10%–14% and 3%–4%, respectively, of the total emissions of these species from the burning of fossil fuels, and B40% and B70%, respectively, of the total emissions of these species from the burning of biomass.

Methods of Measurements of SO_2 Concentration in Flue Gases

Recommended methodologies for emission inventory estimation are compiled in the European Monitoring and Evaluation Programme (EMEP) prepared by European Environment Agency (EEA) *Air Pollutant Emission Inventory Guidebook* (2009). Base data are available from the EEA Data Service (http://dataservice.eea.europa.eu/dataservice/metadetails.asp?id=1096) and the EMEP web site (http://www.ceip.at/).

According to the best available techniques (BAT) recommendations, measurements of SO_2 concentration in the gases for large sources are conducted by two methods[46]:

- Inline photometer (IR only), which can measure gas concentrations in gas pipes if the matrix is optically transparent (no mist).
- Online photometer with sample preparation. This is the usual method. A suitable material has to be selected for the measuring cell and the sample preparation because of the risk of corrosion.

The gas being analyzed at the outlet of a converter system has the following approximate composition: SO_2 in the range 0–1000 ppm; matrix: air, water, H_2SO_4 (30 ppm); and NO_x (50 ppm). The measurement is performed using a commercial photometer (IR units require water compensation).

In the assessment of the release to the air according to the European Pollutant Release and Transfer Register,[47] the following methods of SO_2 concentration determination in flue gases are proposed:

- EN 14791:2006. Stationary source emissions—determination of mass concentration of SO_2—reference method.
- ISO 7934:1989. Stationary source emissions—determination of the mass concentration of SO_2—hydrogen peroxide/barium perchlorate/thorin method.
- ISO 7935:1992. Stationary source emissions—determination of the mass concentration of SO_2—performance characteristics of automated measuring methods.
- ISO 11632:1998. Stationary source emissions—determination of mass concentration of SO_2—ion chromatography method.
- ISO 10396:2007. Stationary source emissions—sampling for the automated determination of gas emission concentrations for permanently installed monitoring systems.

Overview of Technologies Used to Reduce Sulfur Oxide Emissions (Removal from Anthropogenic Sources)

As mentioned before, the main anthropogenic sources of SO_2 emission are commercial power plants. Sulfur oxides are emitted from the combustion of most fossil fuels through oxidation of the sulfur contained in the fuel. According to the European Union directive Integrated Pollution Prevention and Control (IPPC) and Reference Document on Best Available Techniques,[46,48–51] application of a method for desulfurization of gases emitted to atmosphere is obligatory in all such installations.

Below, the main methods used in commercial power plants for desulfurization of flue gases are characterized.[46]

Wet Techniques

Wet scrubbers, especially the limestone–gypsum processes, are the leading flue gas desulfurization (FGD) technologies. They have about 80% of the market share and are used in large utility boilers. This is due to their high SO_2 removal efficiency and their high reliability. Limestone is used in most cases as the sorbent, as it is available in large amounts in many countries. By-products are either gypsum or a mixture of calcium sulfate/sulfite, depending on the oxidation mode. Scrubbing process is another technology used for that purpose. The sodium scrubbing process is a simple method and has been applied to a large number of small oil-fired boilers. Several magnesium-scrubbing systems are also used in relatively small industrial boilers because of the low capital costs involved. In magnesium scrubbing systems, wastewater containing magnesium sulfate can be discharged into the sea, after the removal of dust and dust-absorbed heavy metals, as magnesium sulfate is already a constituent of seawater. The process, therefore, has an advantage over other systems if a plant is located near the coast. The by-product of wet ammonia scrubbers can be used as an agricultural fertilizer.

Wet limestone scrubbers are generally divided into two categories according to the type of oxidation: forced oxidation and natural oxidation. The mode of oxidation is determined by the chemical reactions, the pH of the reagent slurry, and the resulting by-product. In the natural oxidation mode, calcium sulfite is partly oxidized by the oxygen contained in the flue gas. The main product is calcium sulfite hemihydrate. The produced mixture of calcium sulfite hemihydrate and gypsum is in a form of sludge.

At the lower pH range of 4.5 to 5.5, the chemical reaction is different. After SO_2 absorption, the primary product of the neutralization by limestone is not calcium sulfite but calcium bisulfite, $Ca(HSO_3)_2$. Calcium bisulfite is much more soluble than calcium sulfite. Hence, operation in the lower pH range involves smaller risks of scaling and plugging. Calcium bisulfite is oxidized and crystallized to form gypsum or calcium sulfate dehydrate. The by-product from the natural oxidation mode is a mixture that is difficult to dewater. This mixture is composed of calcium sulfite hemihydrate and calcium sulfate dehydrate. The configuration of wet limestone scrubbers can be generally classified into four types, which depend on the applied technical solutions.

Wet limestone FGD systems naturally suffer from an aggressive operating environment, leading to corrosion, erosion, and abrasion. The flue gas path from the inlet of the absorber to the stack discharge must be protected, for instance by using rubber or flake linings, against acid attack caused by adiabatic cooling and saturation of the gas. Wet scrubber processes have undergone considerable development in the last few decades, leading to improved reliability and removal efficiencies, as well as reduced costs. Reliability is normally more than 99% with forced oxidation, and 95%–99% with natural oxidation. The applicability may be influenced by both scrubber components and auxiliary processes connected to the absorber process. The indicative range of wet limestone scrubbing application is for SO_2 content <1000 mg/m³ in the raw gas.

Seawater scrubbing utilizes seawater's inherent properties to absorb and neutralize SO_2 in flue gases. The flue gas is fed to the SO_2 absorber, where it comes into contact with a controlled proportion of the seawater, taken from the cooling water outflow of the steam turbine condenser. Owing to the presence of bicarbonate and carbonates in the seawater, the SO_2 of the flue gas is absorbed. The acidified absorber effluent is mixed with additional seawater to ensure that the pH is at optimal level for the oxidation process. The introduced air forces the oxidation of the absorbed SO_2 from bisulfite to bisulfate and removes dissolved CO_2.

Magnesium Wet Scrubber

In magnesium scrubbing, magnesium hydroxide is used as the reagent, produced by adding slaked lime to seawater in order to enhance alkalinity. Waste sulfate liquor is produced as a result. The magnesium wet scrubber has been mainly applied in smaller plants, i.e., less than 50 MW.

Ammonia Wet Scrubber

In the ammonia wet scrubber system, SO_2 is absorbed by aqueous ammonia, resulting in ammonium sulfate as the fertilizer by-product. The process has been chosen to satisfy the criteria for emission limits of less than 200 mg/N m³. Besides this, the driving force to apply the wet ammonia process has been the requirement for a saleable by product.

Semidry Techniques

Spray Dry Scrubbers

Spray dry scrubbers are mostly applied for relatively small- to medium-capacity boilers using low- to mediumsulfur (1.5%) coal. The residue is normally composed of a mixture of calcium sulfite, calcium sulfate, and fly ash, which is less attractive commercially. The process consists mainly of the spray dry absorber; particulate control device, such as an electrostatic precipitator (ESP) or a fabric filter; and recycling disposal devices for the reaction products. The sorbent for SO_2 absorption is typically lime or calcium oxide. Lime is mixed with an excess of water or is slaked to produce lime slurry, which is also called lime milk. Lime slurry is atomized to a cloud of fine droplets in the spray dry absorber where SO_2 is also removed from the flue gas. Water is evaporated by the heat of the flue gas, usually with a sufficient residence time (about 10 sec) for the SO_2 and other acid gases such as SO_3 and HCl to react simultaneously with hydrated lime to form calcium sulfite/sulfate and calcium chloride. The process chemistry associated with SO_2 removal from the flue gas is a simple acid-base absorption reaction between SO_2 and hydrated lime. The absorption chemistry is strongly affected by such factors as flue gas temperature, gas humidity, SO_2 concentration in the flue gas, and atomized slurry droplet size. The by-product consists of a dry mixture of calcium sulfite, sulfate, fly ash, and non-reacted lime. Spray dry absorbers usually operate at 290–300°C above the saturation temperature, where the saturation temperature of flue gas is between 45°C and 55°C. The indicative range of the method's application is for SO_2 <1000 mg/m^3 in the raw gas.

Furnace Sorbent Injection

Furnace sorbent injection involves direct injection of a dry sorbent into the gas stream of a boiler furnace. Typical sorbents include pulverized limestone ($CaCO_3$) and dolomite ($CaCO_3 \cdot MgCO_3$). Addition of heat in the furnace results in calcination of the sorbent to produce reactive CaO particles. The surface of these particles reacts with the SO_2 in the flue gas to form calcium sulfite ($CaSO_3$) and calcium sulfate ($CaSO_4$). These reaction products are then captured along with the fly ash by the particulate control device, typically an ESP or fabric filter. The SO_2 capture process continues into the precipitator and into the filter cake of the fabric filter. The critical temperature range for the limestone reaction in furnace sorbent injection is 980–1230°C. Once a reactive lime (CaO) is produced, it must have sufficient time in the critical temperature range. Hydrated lime has two reaction windows: 980–1230°C and around 540°C. Thermochemically, $CaSO_4$ is not stable at temperatures above 1260°C in the environment typical for combustion of high sulfur fossil fuels, i.e. 2000–4000 ppm SO_2, e.g., for coal firing. The lower temperature limit for the formation of $CaSO_4$ depends on complex interactions between sulfation kinetics, crystal growth and sintering, and on the build up of a barrier layer of $CaSO_4$ on the surface of the reactive CaO.

About 50% of SO_2 removal efficiency can be achieved at a sorbent mole ratio (Ca/S) of 4–5 when limestone is injected into the boiler furnace at near optimum operation. The SO_2 removal and limestone utilization efficiencies are lower than with other FGD systems. The process is suitable for low-sulfur-containing fuels and for use in small plants. With circulating fluidized bed combustion, this technique operates under optimal temperature conditions, using low-temperature combustion in the range of 800–950°C with higher absorption efficiency for a surplus of sorbent above a factor of 2.

Duct Sorbent Injection (Dry FGD)

Duct sorbent injection means injection of a calcium- or sodium-based sorbent into the flue gas, between the air heater and the existing ESP or fabric filter. The most common types of duct sorbent injections are as follows: 1) dry hydrated lime, which requires humidification; 2) dry sodium injection, which does not require humidification; and 3) lime slurry injection or in-duct scrubbing, which does not require a separate humidification step.

After injection, the sodium bicarbonate decomposes thermally to form sodium carbonate. After the initial sorbent surface of the sodium carbonate has reacted with SO_2 to form sodium sulfite or sulfate, the reaction slows down because of pore pluggage (which resists the gas-phase diffusion of SO_2). For the reaction to continue, the sorbent particle must decompose further. This decomposition evolves H_2O and CO_2 gases into the surrounding atmosphere, creating a network of void spaces throughout the particle. This process exposes fresh reactive sorbent and allows SO_2 to diffuse once again into the particle interior. This increase in surface area is in the order of 5–20 times of the original surface area, depending on the specific sorbent considered. The rates of decomposition and subsequent sulfation of the sodium compound particle represent a complicated function of gas temperature, rate of heat transfer to the particle, H_2O and CO_2 content in flue gas, partial pressures, and effects of other flue gas components.

Hybrid Sorbent Injection

Hybrid sorbent injection is a combination of furnace sorbent injection and duct sorbent injection to improve SO_2 removal efficiency. A specific feature of hybrid sorbent injection lies in the application of limestone as a sorbent.

Dry Techniques

Circulating Fluid Bed Dry Scrubber

The circulating fluid bed (CFB) process is a type of dry scrubber, but separate from either the spray dryer scrubber or sorbent injection.

Sodium Sulfite Bisulfite Process

The sodium sulfite bisulfite process is the most widely used of the regenerable processes. Commercial plants using this process are operated for industrial boilers and power stations burning hard coal, lignite, oil, and petroleum coke. The process is based on sodium sulfite/bisulfite equilibrium. The flue gas first passes to the prescrubber unit, which saturates and cools the gas as well as removes halides and some of the remaining particulates. A variety of prescrubber systems can be used, but each includes a recirculating water stream to ensure good contact, plus a purge stream to control the concentration of chloride and solids. Absorption is countercurrent, and as the liquor travels down the column, more sulfite is converted to bisulfite until it leaves the column. The small quantity of fly ash passing through the prescrubber is also captured in the solution and is removed by filtration before the clean liquor is passed to an intermediate storage area before regeneration. Two side reactions occur in the process, one in the absorption area and the other during regeneration. In the absorption area, a small quantity of sodium sulfate, which lowers the liquor capacity for SO_2 absorption, is produced due to the oxygen content in the flue gas. During regeneration, a small quantity of sodium thiosulfate is formed by a disproportionation reaction. This reaction is autocatalytic so the thiosulfate concentration is maintained at a low level by purging. Ethylenediaminetetraacetic acid is also added to inhibit oxidation.

Magnesium Oxide Process

The magnesium oxide process is a regenerable wet scrubbing process, which uses magnesium hydroxide solution as sorbent. It is essentially the same as for the limestone wet scrubber except for the regeneration step of the spent sorbent. The flue gas then enters the scrubber where the SO_2 is absorbed by aqueous slurry of magnesium sulfate formed from the magnesium hydroxide sorbent. The magnesium sulfite/sulfate is calcined at about 900°C in the presence of carbon to regenerate magnesium oxide, which is returned to the absorption system.

According to the BAT recommendations, the concentration of SO_2 in the gases emitted to the atmosphere as a result of combustion should not exceed 50 mg/N m^3 for wet limestone scrubbing, and SO_2 <100 mg/N m^3 for semidry sorbent injection.

Conclusion

The control of emissions of SO_2 in Europe and the development of emissions regulations are guided by the IPPC BREF document, which specifies, among others, the BAT guidelines, for limiting sulfur oxide emissions from power stations, non-ferrous metals metallurgy, and other sources. In November 2010, Directive 2010/75/EU of the European Parliament and of the Council on Industrial Emissions (IPPC) was published where new standard emission for combustion plants using solid or liquid fuels with the exception of gas turbines and gas engines are presented. The emission limit values (mg/N m^3) for SO_2 for the plants are presented on Tables 3 and 4.

The cited directive also concerns other branches of industry, such as production and processing of metals, mineral industry, chemical industry, and waste management.

These emission reductions result from a combination of emission controls, adoption of advanced electric technologies, and a shift away from the direct end use of coal with increasing income levels. Only under a scenario where incomes in developing regions increase slowly do global emission levels

TABLE 3 Emission Limit Values (mg/N m³) for SO_2 for Combustion Plants Using Solid or Liquid Fuels with the Exception of Gas Turbines and Gas Engines

Total Rated Thermal Input (MW)	Coal and Lignite and Other Solid Fuels	Biomass	Peat	Liquid Fuels
50 100	400	200	300	350
100–300	250	200	300	250
>300	200	200	200	200

Source: Data from Directive 2008/50/EC of the European Parliament and of the Council of 21 May 2008 on ambient air quality and cleaner air for Europe.[8]

TABLE 4 Emission Limit Values (mg/N m³) for SO_2 for Combustion Plants Using Gaseous Fuels with the Exception of Gas Turbines and Gas Engines

In general	35
Liquefied gas	5
Low calorific gases from coke oven	400
Low calorific gases from blast furnace	200

Source: Data from Directive 2008/50/EC of the European Parliament and of the Council of 21 May 2008 on ambient air quality and cleaner air for Europe.[8]

remain at close to present levels over the next century. Under a climate policy that limits emissions of carbon dioxide, SO_2 emissions fall in a relatively narrow range. In most cases, the relative climatic effect of SO_2 emissions decreases dramatically to a point where SO_2 is only a minor component of climate forcing by the end of the century. Ecological effects of SO_2, however, could be significant in some developing regions for many decades to come.

References

1. Ng'walali, M.P.; Koreeda, A.; Kibayashi, K.; Tsunenari, S. Fatalities by inhalation of volcanic gas at Mt. Aso crater in Kumamoto. Jpn. Legal Med. **1999**, *1*, 180–184.
2. Komarnisky, L.A.; Christopherson, R.J.; Basu, T.K. Sulfur: Its clinical and toxicologic aspects. Nutrition **2003**, *19*, 54–61.
3. McGee, K.A.; Doukas, M.P.; Kessler, R.; Gerlach, T. Impacts of volcanic gases on climate, the environment, and people. *U. S. Geol. Surv. Open File Rep.* **1997**, *97–262*, 2.
4. Blank, L.W.; Roberts T.M.; Skeffington R.A. New perspectives on forest decline. Nature **1988**, *336*, 27–30.
5. Krzyzanowski, M.; Cohen, A. Update of WHO air quality guidelines. Air Qual. Atmos. Health **2008**, *1*, 7–13.
6. Vahlsing, C.; Smith K.R. Global review of national ambient air quality standards for PM10 and SO_2 (24 h). Air Qual. Atmos. Health **2011**. doi: 10.1007/s11869-010-0131-2.
7. Emery, J.C. *Report on Air Quality in the State of Maine 2006*; Department of Environmental Protection Bureau of Air Quality, Research Triangle Park, North Carolina, U.S. EPA, 2006.
8. Directive 2008/50/EC of the European Parliament and of the Council of May 21, 2008 on ambient air quality and cleaner air for Europe.
9. U.S. EPA. Environmental Protection Agency. Part II. Primary national ambient air quality standard for sulfur dioxide. Final Rule Fed. Regist. **2010**, *75* (119), 35520–35602.
10. *Air Quality in Asia: Status and Trends*; Clean Air Initiative for Asian Cities Center: Pasig City, Philippines, 2010.
11. ISO 10498:2004. Ambient air—Determination of sulfur dioxide—Ultraviolet fluorescence method.
12. Chang, W.; Okamoto, M.; Korenaga, T. A simple fluorometric method for determination of sulfur oxide in ambient air with a passive sampler. Environ. Sci. **2006**, *13*, 257–262.

13. Gayathri, N.; Balasubramanian, N. Spectrophotometric determination of sulfur dioxide in air, using thymol blue. J. AOAC Int. **2001**, *84* (4), 1065–1069.

14. ISO 10396:2001. Stationary source emissions—Sampling for the automated determination of gas emission concentrations for permanently-installed monitoring systems.

15. Zhang, D.; Maeda, Y.; Munemori, M. Chemiluminescence method for direct determination of sulfur dioxide in ambient air. *Anal. Chem.* **1985**, *57* (13), 2552–2555.

16. He, Z.; Wu, F.W.; Meng, H.; Yuan, L.J.; Luo, Q.Y.; Zeng, Y. Chemiluminescence determination of sulfur dioxide in air using tris(1,10-phenanthroline)ruthenium-potassium iodate system. Anal. Lett. **1999**, *32* (2), 401–410.

17. Salem, A.A.; Soliman, A.A.; El-Haty, I.A. Determination of nitrogen dioxide, sulfur dioxide, ozone, and ammonia in ambient air using the passive sampling method associated with ion chromatographic and potentiometric analyses. Air Qual. Atmos. Health **2009**, *2* (3), 133–145.

18. Velásquez, H.; Ramírez, H.; Díaz, J.; de Nava, M.G.; Sosa de Borrego, B.; Morales, J. Determination of atmospheric sulfur dioxide by ion chromatography in the city of Cabimas, Venezuela. J. Chromatogr. A **1996**, *739* (1–2), 295–299.

19. Sritharathikhun, P.; Oshima, M.; Motomizu, S. On-line collection/concentration and detection of sulfur dioxide in air by flow-injection spectrophotometry coupled with a chromatomembrane cell. Anal. Sci. **2004**, *20*, 113–118.

20. Myoung, S.Y.; Geun, S.C.; Meyerhoff, M.E. Differential ion-selective membrane electrode-based potentiometric gas-sensing cells with enhanced gas sensitivity. Anal. Chim. Acta **1990**, *237*, 115–125.

21. ISO 6767:1990. Ambient air—Determination of the mass concentration of sulfur dioxide—Tetrachloromercurate (TCM)/pararosaniline method.

22. Smith, S.J.; Pitcher, H.; Wigley, T.M.L. Global and regional anthropogenic sulfur dioxide emissions. Glob. Planet. Change, **2001**, *29*, 99–119.

23. Shirsat, S.V.; Graf H.F. An emission inventory of sulfur from anthropogenic sources in Antarctica Atmos. Chem. Phys. **2009**, *9*, 3397–3408.

24. Symonds, R.B.; Rose, W.I.; Bluth, G.J.S.; Gerlach, T.M. Volcanic gas studies—Methods, results, and applications. Rev. Miner. **1994**, *30*, 1–66.

25. Delgado-Granados, H.; Cárdenas González, L.; Piedad Sánchez, N. Sulfur dioxide emissions from Popocatépetl volcano (Mexico): Case study of a high-emission rate, passively degassing erupting volcano. J. Volcanol. Geotherm. Res. **2001**, *108* (1–4), 107–120.

26. Bluth, G.J.S.; Casadevall, T.J.; Schnetzler, C.C.; Doiron, S.D.; Walter, L.S.; Krueger, A.J. Evaluation of sulfur dioxide emissions from explosive volcanism: The 1982–1983 eruptions of Galunggung, Java, Indonesia. J. Volcanol. Geotherm. Res. **1994**, *63* (3–4), 243–256.

27. Halmer, M.M.; Schmincke, H.-U.; Graf, H.-F. The annual volcanic gas input into the atmosphere, in particular into the stratosphere: A global data set for the past 100 years. J. Volcanol. Geotherm. Res. **2002**, *115*, 511–528.

28. Bandy, A.R.; Thornton, D.C.; Blomquist, B.W.; Chen, S.; Wade, T.P.; Ianni, J.C.; Mitchell, G.M.; Nadler, W. Chemistry of dimethylsulfide in the equatorial Pacific atmosphere. Geophys. Res. Lett. **1996**, *23*, 741–744.

29. Cropp, R.A.; Norbury, J.; Gabric, A.J.; Braddock, R.D. Modeling dimethylsulphide production in the upper ocean. Glob. Biogeochem. Cycles **2004**, *18*, 1–21.

30. Seguin, N.; Eaton, S.; Wadleigh, M.A.; Sharma, S. Elevated biogenic sulfur dioxide concentrations over the North Atlantic. Atmos. Environ. **2010**, *44*, 1139–1144.

31. Mihalopoulos, N.; Kerminen, V.M.; Kanakidou, M.; Berresheim, H.; Sciare J. Formation of particulate sulfur species (sulfate and methanesulfonate) during summer over the Eastern Mediterranean: A modelling approach. Atmos. Environ. **2007**, *41*, 6860–6871.

32. McGonigle, A.J.S. Measurement of volcanic SO_2 fluxes with differential optical absorption spectroscopy. J. Volcanol. Geotherm. Res. **2007**, *162*, 111–122.

33. Galle, B.; Oppenheimer, C.; Geyer, A.; McGonigle, A.J.S.; Edmonds, M.; Horrocks, L.A. A miniaturised UV spectrometer for remote sensing of SO_2 fluxes: A new tool for volcano surveillance. J. Volcanol. Geotherm. Res. **2003**, *119*, 241–254.

34. Sutton, A.J.; Elias, T.; Gerlach, T.M.; Stokes, J.B. Implications for eruptive processes as indicated by sulfur dioxide emissions from Kilauea Volcano, Hawaii, 1979–1997. J. Volcanol. Geotherm. Res. **2001**, *108*, 283–302.

35. Edmonds, M.; Herd, R.A. A volcanic degassing event at the explosive-effusive transition. Geophys. Res. Lett. **2007**, *34*, L21310.

36. Salerno, G.G.; Burton, M.R.; Oppenheimer, C.; Caltabiano, T.; Tsanev, V.I.; Bruno, N. Novel retrieval of volcanic SO_2 abundance from ultraviolet spectra. J. Volcanol. Geotherm. Res. **2009**, *181* (1–2), 141–153.

37. Realmuto, V.J.; Sutton, A.J.; Elias T. Multispectral imaging of sulfur dioxide plumes from the East Rift Zone of Kilauea Volcano, Hawaii. J. Geophys. Res. **1997**, *102*, 15057–15072.

38. Oppenheimer, C.; Francis, P.; Burton, M.; Maciejewski, A.J.H.; Boardman, L. Remote measurement of volcanic gases by Fourier transform infrared spectroscopy Appl. Phys. B **1998**, *67*, 505–515.

39. Weibring, P.; Edner, H.; Svanberg, S.; Cecchi, G.; Pantani, L.; Ferrara, R.; Caltabiano, T. Volcanic sulphur dioxide emissions using differential absorption lidar (DIAL), differential optical absorption spectroscopy (DOAS) and correlation spectroscopy (COSPEC). Appl. Phys. B **1998**, *67*, 419–425.

40. Cao, H.N.; Li, S.; Fukuchi, T.; Fujii, T.; Collins, R.L.; Wang, Z.; Chen, Z. Measurement of tropospheric O_3, SO_2 and aerosol from a volcanic emission event using new multi-wavelength differential-absorption lidar techniques Appl. Phys. B **2006**, *85*, 163–167.

41. Svanberg, S. Geophysical gas monitoring using optical techniques: Volcanoes, geothermal fields and mines. Optics Lasers Eng. **2002**, *37* (2–3), 245–266.

42. Barrancos, J.; Roselló, J.I.; Calvo, D.; Padrón, E.; Melián, G.; Hernández, P.A.; Pérez, N.M.; Millán, M.M.; Galle, B. SO_2 emission from active volcanoes measured simultaneously by COSPEC and mini-DOAS. Pure Appl. Geophys. **2008**, *165* (1), 115–133.

43. Elias, T.; Sutton, A.J.; Oppenheimer, C.; Horton, K.A.; Garbeil, H.; Tsanev V.; McGonigle, A.J.S.; Williams-Jones, G. Comparison of COSPEC and two miniature ultraviolet spectrometer systems for SO_2 measurements using scattered sunlight. Bull. Volcanol. **2006**, *68*, 313–322.

44. EPA's 2008 Report on the Environment Highlights of National Trends, 2008, available at http://www.epa.gov/roe. (Accessed May 2011).

45. Corbett, J.J.; Fischbeck, P.S.; Pandis, S.N. Global nitrogen and sulfur inventories for oceangoing ships. J. Geophys. Res. **1999**, *104*, 3457–3461.

46. Integrated Pollution Prevention and Control Reference Document on Best Available Techniques for Large Combustion Plants, European Commission, July 2006.

47. Regulation (EC) No. 166/2006 of the European Parliament and of the Council of 18 January 2006 concerning the establishment of a European Pollutant Release and Transfer Register and amending Council Directives 91/689/EEC and 96/61/EC.

48. Integrated Pollution Prevention and Control (IPPC) Reference Document on Best Available Techniques in the NonFerrous Metals Industries, European Commission 2001.

49. Integrated Pollution Prevention and Control (IPPC) Reference Document on Best Available Technologies (BREF) in Pulp and Paper Industry, European Commission, 2001.

50. Integrated Pollution Prevention and Control (IPPC) Reference Document on Best Available Techniques for Mineral Oil and Gas Refineries, European Commission, 2003.

51. Srivastava, R.K. Controlling SO_2 emissions: A review of technologies, EPA/600/R-00/093 November 2000; U.S. Environmental Protection Agency Office of Research and Development: Washington, D.C.

Vanadium and Chromium Groups

Imad A.M. Ahmed

Introduction

The vanadium group of elements consists of vanadium (V), niobium (Nb, previously called columbium), tantalum (Ta), and dubnium (Db). The vanadium metals lie in the d-block of the periodic table and are found in Group 5. The chromium group consists of chromium (Cr), molybdenum (Mo), tungsten (W), and seaborgium (Sg). Dubnium and seaborgium are synthetic elements whose most stable isotopes are $^{268}_{108}\text{Db}(t_{1/2} = 28\ \text{hr})$ and $^{271}_{106}\text{Sg}(t_{1/2} = 1.9\ \text{min})$;[1] thus, these two elements are not discussed here. Chromium and tungsten are probably the most familiar metals of the two groups of elements. This is because of chromium's role in the manufacture of stainless steel and the famous use of tungsten as a filament in lightbulbs. This entry summarizes the impact of V, Nb, Ta, Cr, Mo, and W contaminations on the environment and human, animal, and plant health. The geochemical occurrences of these elements are first discussed to establish an understanding of their natural distribution in the environment. These elements are widely used in various industrial applications, especially in metal industries; their applications and levels in the environment are discussed next. Potential sources of these elements, which humans and the environment can be exposed to, are described in some details. Finally, the metabolism, health effects, and regulatory measures taken to reduce or prevent the release of such trace metals to the environment are also discussed.

Geochemical Occurrences

All the transition elements of the vanadium group have odd atomic numbers, which could explain their low cosmic and terrestrial abundances compared to those of the titanium (Group 4) and chromium group elements.[2] This also resulted in the presence of only one stable naturally occurring isotope of each of these elements. Vanadium (V) is the 19th most abundant element in the Earth's crust. The ores that contain vanadium include patronite (VS_4), vanadinite [$Pb_5(VO_4)_3Cl$], dechenite [$PbZn(VO_4,AsO_4)OH$],

TABLE 1 World Mining of V, NB, Ta, Cr, Mo, and W (tons)

Element	Production as of 2010[a]	Location of Major Reserves[b]
Vanadium, V	56,000	China, South Africa, and Russia
Chromium, Cr	22,000	South Africa, India, and Kazakhstan
Niobium, Nb	63,000	Brazil and Canada
Molybdenum, Mo	234,000	China, United States, Chile, Peru, Canada, Mexico, Russia, Iran, and Mongolia
Tantalum, Ta	670	Brazil, Mozambique, Rwanda, and Australia
Tungsten, W	61,000	China, Russia, Bolivia, Austria, and Portugal

[a] World total mine production (rounded).[133]
[b] Reserves are arranged in decreasing order according to their mine production as of 2010.

descloizite [Pb(Zn, Cu)(VO$_4$)OH], pucherite (BiVO$_4$), roscoelite (mica containing V), and volborthite [Cu$_3$V$_2$O$_7$(OH)$_2$·2H$_2$O].[3,4] Vanadium (primarily as VO^{-3}) is commonly found in admixture with many minerals such as carnotite K$_2$(UO$_2$)$_2$(VO$_4$)$_2$. Niobium (Nb) and tantalum (Ta) do not occur naturally as free metals but are commonly found together in nature at trace concentrations as iron niobate [(Fe, Mn)(NbO$_3$)$_2$] and iron tantalate [(Fe, Mn)(TaO$_3$)$_2$] in the form of niobite and tantalite minerals, respectively.[5] Large deposits of Nb have been found associated with carbonatites (carbonsilicate rocks) in the form of pyrochlore [(Na, Ca)$_2$Nb$_2$O$_6$(OH, F)].[6]

Chromium (Cr) ranks 21st in abundance in the Earth's crust and is found in a number of minerals with chromite (FeCr$_2$O$_4$) and crocoite (PbCrO$_4$) as the principal ores of Cr. Molybdenum (Mo) ranks 54th in abundance in the Earth's crust as it is found in a limited number of ores, including wulfenite (PbMoO$_4$), molybdenite (MoS$_2$), and powellite (CaMoO$_4$). Molybdenum is also recovered as a by-product of copper and tungsten mining. Tungsten (W) ranks 58th in abundance and is found in a number of minerals including tungstite (WO$_3$·H$_2$O), wolframite [FeMn(WO$_4$)$_2$], and scheelite (CaWO$_4$).[3,5,7,8] The world mining production and principal reservoirs of V, Nb, Ta, Cr, Mo, and W are given in Table 1.

Overview of Current Uses and Environmental Levels

Uses

The metals of the vanadium group are characterized by their good structural strengths and their uses in manufacturing high-strength low-alloy steels. Vanadium metal has a low-fission neutron cross section, making it useful in nuclear applications. A major application (~85% of mined metal) of vanadium is the production of heat-resistant and high-strength steels and alloys. Vanadium is commonly used in the aerospace industry as a component in jet aircraft engines. Vanadium oxides are important catalysts in inorganic and organic chemical industries such as in plastic and sulfuric acid productions.[7,9] Niobium exhibits many excellent corrosion resistance properties, which make it an important coating material in biomedical applications. It is also commonly used in manufacturing aerospace equipment and missiles due to its lightweight and strength at high temperatures. Furthermore, pure Nb exhibits excellent superconductive properties, making it the primary material in manufacturing superconducting magnets especially when alloyed with tin and zirconium.[7,9,10] Tantalum is used in manufacturing high-temperature hardware because of its very high melting point (see Table 2). Because of its excellent resistance to corrosion by hot and concentrated mineral acids (except hydrofluoric acid), Ta is widely used in various chemical industries including electrolytic capacitors.[11,12] Tantalum carbide (TaC) has a higher melting point (3985°C) than metallic Ta, which is the reason why TaC powder is added in the manufacture of hard metals. In recent years, the use of TaC has decreased due to its high cost and has been replaced by chromium and vanadium carbides. About 900,000 kg of Ta is consumed annually worldwide with the electronics industry being its largest consumer.

Chromium is mainly used to harden steel and to manufacture stainless steel and special alloys. Chromium is also used in the plating industry to produce hard and beautifully finished surfaces that

TABLE 2 Selected Physical Properties of Trace Elements of Groups 5 and 6

Element	Atomic Number	Atomic Mass[a]	Atomic Radius (pm)	Density 20°C (g cm−3)	Melting Point (°C)	Boiling Point (°C)	Oxidation States[b]
Vanadium, V	23	50.9	192	6.11	1910	3407	2,3,4,[5]
Chromium, Cr	24	51.9	185	7.20	1907	2671	2,[3], 4,5,[6]
Niobium, Nb	41	92.9	208	8.57	2477	4744	2,3,4,[5]
Molybdenum, Mo	42	95.9	201	10.2	2623	4639	2,3,4,5,[6]
Tantalum, Ta	73	180.9	209	16.65	3017	5458	3,4,[5]
Tungsten, W	74	183.8	202	19.3	3422	5555	2,3,4,5,[6]

[a] Average values for atoms having the main oxidation state.
[b] The main oxidation states are given in square brackets.
Source: Enghag[4] and Lide.[7]

resist corrosion. It is also used as a coloring element in glass. Chromium forms many important compounds such as potassium chromate (K_2CrO_4) and dichromate ($K_2Cr_2O_7$), which are commonly used as oxidizing agents in analytical chemistry and in tanning leather.[7,9]

Molybdenum is primarily used in iron and steel industries. It is used in certain nickel-based alloys such as the Hastelloy® alloys, which are heat-resistant and corrosion-resistant to chemical solutions. It is also used as a catalyst in the refining of petroleum. Useful compounds of Mo include molybdenum sulfide (MoS_2), which is used as a solid lubricant because of its low friction properties and resistance to decomposition at high temperatures.[13]

Tungsten has the highest melting point of all metals (Table 2) and it has the highest tensile strength at temperatures higher than 1650°C, making it a valuable element in the alloy industry. Tungsten and its alloys are extensively used as filaments in lightbulbs, electron and television tubes, x-ray target, and heating elements for electrical furnaces. It has excellent corrosion resistance properties and it is attacked only slightly by mineral acids. Similar to tantalum, tungsten carbide has a high melting point at 2870°C and is an extremely hard material and hence is used in manufacturing cutting and drilling tools. Tungsten and its salts are widely applied in chemical and tanning industries and as high-temperature lubricants in the form of tungsten disulfide, which remains stable up to ~500°C.[14] Consequently, V, Mo, Cr, Ta, and W are regarded as essential commodities and metals of high strategic importance for years to come[15]

Environmental Levels

Environmental contamination is generally associated with the quantities of trace elements that are added to soil or water from anthropogenic sources such as sewage sludge, industrial discharge, and combustion of fossil fuel. Sewage sludge, for example, has valuable agronomic properties because it can improve soil's biological, chemical, and physical properties. However, the concentrations of trace elements in sewage sludge may be many orders of magnitude higher than the background concentrations in agricultural soils (see Tables 3 and 4). For example, Berrow and Webber[16] reported total metal contents of 3–3000, 2–260, 40–8800, 200–8000, 2–30, 20–5300, 120–3000, 200–400, and 700–4900 mg kg⁻¹ for Cd, Co, Cr, Cu, Mo, Ni, Pb, V, and Zn, respectively, in sewage sludges from rural and industrial towns in England and Wales. In contrast, the reported concentrations of trace metals in sewage sludge in agricultural soils of Sweden (Table 3) were comparable to the background values in soils. Once applied, sewage sludge can have very long residence times in soil, causing potential consequences on the plant, animal, and human health.[17–20] The combustion of fossil fuel is another potential contributor to the release of toxic trace metals to the environment. Coal combustion is used to generate electricity in coal-burning power plants. In the United States alone, the amount of combustion residues of coal such as fly ash produced annually amounts to about 106 million tons.[21] Only one-third of the produced fly ash is recycled by using it in

TABLE 3 Typical Concentration (mg kg^{-1}) of V, Nb, Ta, Cr, Mo, and W in Fly and Bottom Ash and Sewage Sludge

Element	Fly Ash	Bottom Ash[a]	Sewage Sludge (I)[b]	Sewage Sludge (II)[c]
Vanadium, V	95–652	50–275	7.9	200–400
Chromium, Cr	37–651	40–4710	33	40–8800
Niobium, Nb	16.0 to <22	–	4.5	–
Molybdenum, Mo	7.1–236	2.8–443	6.7	2–30
Tantalum, Ta	0.5–2.6	–	0.94	–
Tungsten, W	2.9–42	–	7.9	–

[a] A granular combustion residue collected from the bottom of coal furnaces in power plants.
[b] Sewage sludge from agricultural soils of Sweden.[32]
[c] Sewage sludge from rural and industrial towns in England and Wales.[16]
Source: Bradl[23] and PECH.[129]

various applications such as cement making and ceramics.[22] The concentrations of trace elements in coal residues depend on the elemental composition of the parent coal and the conditions during combustion. The typical concentrations of V, Nb, Ta, Cr, Mo, and W in fly ash are listed in Table 3. More details on the coal residues and their characterization are available elsewhere.[23–25]

Vanadium, Niobium, and Tantalum

Although there are not many reports on V, Nb, or Ta pollution in soils, recent research shows that industrial activities such as mining and burning of fossil fuel will increase the deposition of these elements to soils and surface waters. The three elements are relatively common in the terrestrial environment at average contents of 135, 20, and 2 mg kg^{-1} in the Earth's crust for V, Nb, and Ta, respectively (Table 4). The most common form of V in soil solutions is the vanadyl cation (VO^{+2}) and its associated complexes.[26] Under acidic conditions, the pervanadyl ion (VO^+) dominates the V chemical species whereas vanadate ions ($VO4^-$) dominate under extremely alkaline environments. Similar to ferric ions, vanadic ions (V^{+3}) hydrolyze in water to form hydroxide species such as VOH^{+2} and $V(OH)^+$. While Fe^{+3} is very difficult to oxidize to higher oxidation states, V^{+3} can be oxidized to the (IV) and (V) states. Vanadium is an oxophilic metal because of its tendency towards oxygen, forming a number of oxides including V_2O_5, VO_2, V_2O_3, and VO. The chemical and physical properties of Nb and Ta are almost identical (see Table 2) and both exhibit affinities to associate with Fe, Ti, and Zr. The chemistries of Nb and Ta are dominated by the (III) and (V) oxidation states, which limit the number of the oxides that could

TABLE 4 Typical Concentration (mg kg^{-1}) of V, Nb, Ta, Cr, Mo, and W in the Earth's Crust and Selected Non-Polluted Soils

Element	Crustal[a]	Sweden[b]	Japan[c]	Brazil[d]	United States[e]
Vanadium, V	135	69	180	320	80
Chromium, Cr	100	22	58	86	54
Niobium, Nb	20	12	10	25	11
Molybdenum, Mo	1.5	0.58	1.3	1.6	0.97
Tantalum, Ta	2	1.1	1.7	2.3	–
Tungsten, W	1.5	1.3	1.3	1.4	<0.16–0.17

[a] Mean concentrations of metals in the Earth's crust.[134,135]
[b] Agricultural soils of Sweden.[32]
[c] Agricultural soil of Japan.[136]
[d] Average values for Brazilian soils.[137]
[e] Data from U.S. soils.[138,139]

be formed under environmental conditions. Examples of the Nb and Ta oxides are Nb_2O_5, NbO_2, Ta_2O_5, and TaO_2. The Nb^{+3} ions are very unstable in air due to its rapid oxidation to form Nb^{+5} oxides. Ta^{+3} is also very unstable and the majority of Ta compounds tend to be limited to the (IV) oxidation state.[3,27]

The mean concentrations of V, Nb, and Ta in reference soils of Sweden, Japan, Brazil, and the United States were reported to fall in the range of 69–320, 10–25, and 1.1–2.3 mg kg^{-1}, respectively (Table 4). The concentrations of Nb in arable soils of the United Kingdom derived from different rocks were found in the range 31–300 mg kg^{-1}.[28] In China, the Nb and Ta content in reference soils were found in the range of 9.3–37.6 and 0.8–5.3 mg kg^{-1}, respectively.[29] Elevated concentrations of V, Nb, and Ta above the typical natural soils content are more likely due to residual soils derived from rocks enriched in these elements. Mining activities are major contributors to increasing the input of trace elements in soils and waters. An example of man-made pollution of V is the contamination of the Shuanghe and Donghae rivers in the Shaanxi province in Northwest China as a result of the collapse of the spillway from a vanadium mine (Reuters 2008). This spill turned the river water black, threatening the drinking water and the ecosystem of that region.

A number of reports have indicated that vanadium is available to plant uptake especially in the form of VO^{+2} species in acidic soils.[30,31] The reported V content in food plants is greatly variable, but concentrations of up to 2000 µg kg^{-1} in wild mushrooms, 840 µg kg^{-1} in spinach, and 20–600 µg kg^{-1} in other vegetables have been reported.[25,32] While the geochemical properties of Ta are similar to those of Nb, Ta is less mobile than Nb because of the lower solubility of its compounds. This means that Nb is more available to plant uptake than Ta especially under humid conditions. The common concentration range of Nb and Ta in food plants is 0.02–1.1 mg kg^{-1} and 0.013–0.48 µg kg^{-1}, respectively.[33] The reported concentrations of Nb and Ta in cereal grains are 0.5–1.7 mg kg^{-1} and 1.1–5 µg kg^{-1}, respectively.[32]

Chromium

Naturally occurring chromium exists in two main oxidation states of Cr(III) (chromic) and Cr(VI) (chromate). Cr(VI) species are strongly oxidizing and exist only in oxo species such as chromium trioxide CrO_3 and chromate CrO_4^{-2} When chromium trioxide is dissolved in water it forms chromic acid H_2CrO_4. Most Cr(VI) compounds are highly soluble except some metal chromates such as those of Pb, Ba, Sr, Ca, Ag, and Zn and hence are chemically mobile and bioavailable. In contrast, the majority of Cr(III) compounds are sparingly soluble in water. Similar to Fe(III), Cr(III) hydrolyzes in water, yielding different hydroxo species of varying solubilities. At pH > 5, Cr(III) precipitates as $Cr(OH)_3$. Cr(VI) speciation and behavior in the environment are pH dependent. Under acidic conditions, Cr(VI) shows high redox potential, meaning that it is strongly oxidizing but this oxidizing capacity reduced with increasing pH. Therefore, at low pH and in the presence of electron donor species such as natural organic matter and Fe^{+2}-containing minerals in soils, Cr(VI) are unstable and are reduced into the more stable Cr(III) form.[34,35] Under aerobic conditions, Cr(III) may be oxidized to Cr(VI), probably by Mn oxides, but this process is generally slow at pH values above 5.[36] Under anoxic conditions (e.g., hydromorphic soils), Cr(VI) could be reduced to Cr(III) by electron transfer processes occurring at reactive mineral surfaces or due to the presence of humic substances; such processes are more notable in soils at low pH. Therefore, under normal soil conditions, Cr is present predominantly in the Cr(III) state[13,36] and Cr(III) species are regarded less mobile and available to plant uptake compared to Cr(VI). The factors that influence transformations between Cr(III) and Cr(VI) in the environment have been extensively reviewed.[37–41]

Chromium occurs naturally in the Earth's crust at a median concentration of 100 mg kg^{-1}. The reported data of total Cr are <86 mg kg^{-1} in non-polluted soils (Table 4) and <12 mg kg^{-1} in natural water (Table 5). However, human activities have resulted in the release of chromium into the environment in larger amounts. Mining activities and tanning of leather are among the major sources of chromium. For example, in October 2010, a caustic sludge from a "red mud" reservoir at the Ajkai Timföldgyár alumina plant was spilt across a town called Ajka in Hungary. The accident happened

TABLE 5 Mean Concentrations of V, Nb, Ta, Cr, Mo, and W in Seawater ($\mu g\ L^{-1}$), River Water ($\mu g\ L^{-1}$), Oceans ($\mu g\ L^{-1}$), and Groundwater ($mg\ L^{-1}$)

Element	Seawater[a]	River	Ocean Surface Water[b]	Groundwater
Vanadium, V	2.5	0.009–1.77	2.0	<1.0–10
Chromium, Cr	0.3	0.29–11.46	0.21	<1.0–5.0
Niobium, Nb	0.01	0.002–0.01	<0.005	
Molybdenum, Mo	7.0	0.04–2.69	10	<1.0–30
Tantalum, Ta	0.002[e]	–	$<2.5 \times 10^{-3}$	–
Tungsten, W	0.1	0.1–180	0.001	–

Source: Gaillardet et al.,[141] Sparks,[145] and Li.[142]

when the Western dam of the plant collapsed, releasing about a million cubic meters of reddish material with a pH of about 12 to the valley of the Torne river. The accident flooded the lower parts of the city of Devecser and the villages of Kolontár and Somlóvásárhely, covering about 800 ha of agricultural land by a red mud layer of 5–10 cm. Red mud is a by-product of the aluminum extraction from bauxite. This red mud spill contained elevated concentrations of toxic metals including V ($870\ mg\ kg^{-1}$), Pb ($160\ mg\ kg^{-1}$), and Cr ($620\ mg\ kg^{-1}$), making this industrial accident the worst ecological disaster in the history of Hungary.[42,43] The scale of the toxic red mud spill and the damage that occurred to the nearby Kolontár village have been documented in a number of media (e.g., http://www.guardian.co.uk/world/gallery/2010/oct/05/hungary).

Chromium in ambient air originates primarily from industrial sources (e.g., metal industry) and the combustion of fossil fuel (see Table 3). Chromium is generally removed from the atmosphere as a result of precipitation or dry deposition. A number of studies have shown that most of Cr found in surface waters is due to atmospheric deposition.[44,45] The mean concentrations of Cr in freshwater and seawater are listed in Table 5.

Molybdenum

In marine systems, Mo is present at ~110 nM compared to ~0.5 nM for W, ~7.5 nM for Zn, and ~1.5 nM for Cu, which makes Mo the most abundant transition element in the sea.[46,47] The distribution of Mo in terrestrial environments is variable with an estimated average of $2–3\ mg\ kg^{-1}$. Molybdenum exists in soils mainly as soluble or adsorbed molybdate anion MoO_4^{-2} Molybdate is known to adsorb on oxide and clay minerals in soils with sorption increasing at lower pH. Thus, Mo is less mobile in soils under acidic conditions but can be released into the environment with increasing pH.[48,49] Hence, similar to Se, the bioavailability of Mo increases with liming soils. Molybdenum in soils forms strong complexes with natural organic matter and other organic ligands such as malic acid, tannins, and tannin-like compounds over a wide pH range, which helps prevent leaching Mo.[50,51] This also explains the elevated concentrations of Mo in peaty soils. The major soluble Mo species in natural waters are molybdic acid $H_2MoO_4^0$ and its oxyanions, $HMoO_4^-$ and $HMoO_4^{-2}$ Available Mo to plants and microorganisms is usually present in soil and waters as soluble MoO^{-4} anion.[52] The high solubility of MoO^{-4} makes it susceptible to leaching from soil into groundwater. However, like many other anions, once in solution MoO^{-4}, it becomes subject to a range of adsorption and desorption reactions by positively charged metal oxides (Fe, Al, and Mn) and clay minerals.[53]

The world median composition of Mo in stream waters is $0.5\ \mu g\ L^{-1}$. The analyses of more than 11,000 stream water samples in England and Wales from sites listed in the British Geological Survey (BGS) G-BASE data set had Mo concentrations within the range of $0.08–2.45\ \mu g\ L^{-1}$ with a median of $0.57\ \mu g\ L^{-1}$. Studies on a number of BGS- monitored sites showed that the typical Mo concentrations in rivers are less than $1\ \mu g\ L^{-1}$.[54] As indicated in Table 4, the average concentration of Mo in soils ranges between 0.1 and $10\ mg\ kg^{-1}$, but higher concentrations were found in sewage sludges (Table 2). In ambient and non-polluted air, the mean concentration of Mo is ~0.02 µg m^{-3}.[55] With its widespread industrial applications, the data in Table 3 suggest that the atmospheric deposition is an important source of soil Mo near polluted cities.

Tungsten

Tungsten is a transition element with similar chemical properties of Mo and Cr. Tungsten exists in several oxidation states (0, II, III, IV, V, and VI) with W^{VI} being the most stable and common state in aqueous systems. The dominant aqueous species of W is tungstate $\left(WO_4^{-2}\right)$ especially in the pH range of 6.9–9.3. There are very little documentations on the distribution of W in soils and waters. The world average W content in soils is 1.7 mg kg^{-1}. The reported mean concentrations of W in non-polluted soils in Sweden, Japan, Brazil, and the United States are 1.3, 1.3, 1.4, and 0.165 mg kg^{-1} (Table 4), respectively. Data on W content in food, ecological effects, and anthropogenic release into the environment are not well established. While environmental tungsten levels are generally very low, it exists at high concentrations near mines or natural deposits.[56] For example, the historical mining activities in Cornwall in the South West England have resulted in the accumulation of high concentrations of W, As, Mn, Fe, Zn, Sn, and Cu in estuarine sediments with W levels up to ~650 mg kg^{-1}.[57] Owing to the adverse environmental and health effects of lead and depleted uranium, tungsten has been suggested as a replacement in military applications. The U.S. Green Armament Technology (GAT) program, a U.S. Army pollution prevention initiative, called for substitution of lead by tungsten in small-caliber ammunition. The 1999 nationwide ban of using lead shots for hunting in the United States, has also led to the substitution of lead shotgun shell ammunition and lead fishing weights in various North American and European countries.[58,59] The U.S. Fish and Wildlife Service approved the use of iron–tungsten–nickel, iron–tungsten, tungsten–bronze, tungsten–iron–copper–nickel, tungsten–nylon, and tungsten–tin–bismuth shots as non-toxic for migratory bird hunting.[60] Because W minerals are slightly soluble in water, its availability in natural waters is scant; this is evident by the reported concentrations of W in river, sea, and oceanic waters in Table 5. There is enough evidence that combustion of fossil fuel is a serious source of W contamination in air (see Table 3). The concentration of W is generally low (<0.05 mg kg^{-1}) in food plants.[61]

Metabolism and Health Effects

Vanadium

Vanadium is a trace element that is known to increase the growth rates of various microorganisms (e.g., green algae) and higher plants (e.g., tomato and lettuce) but its essentiality is not conclusive.[62] Until now, evidence that V has a biological function in higher animals is limited and not definitive.[63] However, in humans, a number of V compounds (e.g., vanadyl sulfate) have shown insulin-like action[64–66] and antitumor and anticancer properties,[67,68] indicating the potential therapeutic applications of V. The biochemical activities of V are related to the complexing ability of the vanadyl cation (VO^{+2}), and the chemical similarity of vanadate (VO^{-3}) to phosphate, which allows vanadate compounds to interact with numerous enzymes in living organisms by either inhibiting or activating them.[69] Once vanadate has entered the human body, it is reduced to vanadyl ion by reducing substances in the plasma such as ascorbic acid and subsequently bound to proteins such as transferrin, an iron-containing protein.[70] Toxic effects on human and animals are well established. For example, V compounds have an adverse effect on the reproductive and developmental functions of rats and mice including the decline in conception rate and fetal development.[71]Also, oral exposure to V and the reduction of vanadate ions into vanadyl ions in red blood cells can result in decreased number of red blood cells, increased blood pressure, and mild neurological effects.[72] The toxicity of V compounds increases as valency increases with V(V) being the most toxic species of vanadium. Vanadium oxides may also be more toxic than vanadium salts. In humans, exposure by inhalation to vanadium pentoxide causes diverse

TABLE 6 Regulations and Guidelines Applicable to V, Cr, Nb, Mo, Ta, and W in the Environment and Human Health

Element	NIOSH (REL) (mg m–3)	WHO	USEPA (water) (mg L–1)	Cancer Effect IARC[a]	EPA Water Reuse (mg L–1)
V	IDLH = 35; OSHA = 0.5 as V_2O_5	In air = 1 µg m^{-3} In water = N/A	N/A	Group 2B as V_2O_5	0.1 STU 1.0 LTU
Cr	Metal: TWA 0.5, OSHA 1, IDLH 250 Cr(VI) compounds: IDLH (CrO_3) = 15; OSHA = 0.005	In air = 1 µg m^{-3} as Cr(VI) Total in water = 0.05 mg L^{-1}	0.1 (total Cr)	Metallic Cr: Group 3 Cr(III): Group 3 Cr(VI): Group 1	0.1 STU 1.0 LTU
Nb	N/A	N/A	N/A	N/A	N/A
Mo	TWA = 15 IDLH = 5000 (as Mo) Mo soluble IDLH 1000 mg m^{-3} (as Mo) TWA 5 mg m^{-3} = OSHA	N/A	N/A	N/A	0.01 STU 0.05 LTU
Ta	TWA = 5 Short-term exposure = 10	N/A	N/A	N/A	N/A
W	TWA = 5 Short-term exposure = 10 OSHA= N/A IDLH = N/A	N/A	N/A	N/A	N/A

Note: IARC, International Agency for Research on Cancer; IDLH, immediately dangerous to life or health; LTU, long-term use; N/A, guideline data are not available; NIOSH, National Institute for Occupational Safety and Health; OSHA, Occupational Safety and Health Administration; REL, recommended exposure limit; STU, short-term use; TWA, time-weighted average; USEPA, U.S. Environmental Protection Agency; WHO, World Health Organization. [a] Group 1: the substance is carcinogenic to humans, Group 3: the substance is not classifiable as to its carcinogenicity to humans. Group 3B: possibly carcinogenic material to humans.
Source: WHO,[94] ATSDR,[97] WHO,[121] ATSDR,[130] USEPA,[131] and NIOSH.[132]

toxic effects on the respiratory, digestive, and central nervous systems.[130] Workers in the metal alloy and catalysis industries that use vanadium pentoxide and V-containing pigments for the ceramics industry are among the most exposed to airborne vanadium compounds. For more details on the toxic effects of vanadium pentoxide, see IARC 2006.[73] Other acute toxicity symptoms of V compounds in animals include weakness, loss of appetite, dehydration, weight loss, necrosis of lymphoid tissues, renal tubular necrosis, and death.[74,75] Ingested V seems to be excreted unabsorbed and hence there are very few reported cases on V poisoning via this route. The typical concentrations of V in plants are 0.27–4.2 mg kg^{-1},[76] but there are many examples of the toxic effect of vanadium on plants, and that toxicity is connected with V(V) forms in soil. For example, it was found that 5 mg V L^{-1} in a hydroponic medium caused iron-deficiency chlorosis in sugar beet plants and that the growth was reduced by 30%–50%. It is claimed that the cause for this toxic effect is the structural analogy between vanadate (VO^{-3}) and phosphate (PO^{-3}) ions.[77]

Normal concentrations in blood and urine are 0.05 µg L^{-1} and 0.5 µg L^{-1}, respectively. Occupational exposure limits and guidelines for vanadium pentoxide in workplace air are presented in Table 6. The U.S. Environmental Protection Agency (USEPA) considered vanadium pentoxide to be an extremely hazardous substance[78] and the International Agency for Research on Cancer (IARC) classified it as a possibly carcinogenic substance to humans (Carcinogen Group B). The World Health Organization (WHO) guidelines for V_2O_5 in air is 1 µg m^{-3}. Little attention has been paid to the determination of V(V) species in water, soil, or plants. Until recently, neither the USEPA nor the WHO has listed vanadium as a pollutant requiring urgent research and legislation.

Niobium and Tantalum

Tantalum and niobium are not essential or have a definite biological function to human, animals, or plants. Both element seems to have high affinity for both lung and plasma proteins.[79] Acute and chronic effects of Ta and Nb on humans and animals are related to their serious impact on respiratory tract and cardiovascular functions. For example, high levels of Ta, Co, and W were reported in the lung, blood, urine, and hair of 251 workers in the hard metal industry exposed to hard metal dust. These workers were diagnosed as having the "hard metal pneumoconiosis" disease that shows episodes of work-related subacute disease, and some patients evolve to lung fibrosis.[80,81] Respiratory paralysis was shown to be the cause of death in cats fed intravenously doses of $5\,mg\,kg^{-1}$ Nb as $NbCl_5$. Renal injuries and deaths were also observed in mice and rats following parenteral injection of Nb as $NbCl_5$ or $KNbO_3$ at doses ranging from 20 to $50\,mg$ Nb kg^{-1}. However, dietary levels of Nb up to 1% of $NbCl_5$ or $KNbO_3$ given over 7 weeks were ingested by rats without effect.[82] The median lethal dose (LD_{50}) values of Nb injected intraperitoneally in mice and rats were 12 and $14\,mg\,kg^{-1}$ as $NbCl_5$ and 13 and $86\,mg\,kg^{-1}$ as $KNbO_3$,[83] respectively, making niobium chloride one of the most toxic of the rarer chemicals. The LD_{50} values for Ta obtained after intraperitoneal injection of $TaCl_5$ in rats was $75\,mg^{-1}$,[83] suggesting that Ta is less toxic than Nb. Tantalum and niobium are among the metals used in prostheses, especially dental and orthopedic implants, because of their excellent corrosion resistance and biocompatibility.[84–86] Their potential interactions with tissues have led to a number of studies that tend to suggest that Ta implants are biologically inert and pose little concern for human clinical pathophysiology.[87] Further details on metabolism and toxic effects of Ta and Nb are available elsewhere.[87–89] The evidence given above on the Nb and Ta and their compounds suggests that both elements should be regarded as toxic. The hazardous occupational exposure concentrations of Ta and Nb vary depending on their compounds and origin. Unfortunately, metabolism and toxicity data on Nb and Ta and their compounds remain scarce, which has hampered the development of regulations and risk assessments of these two elements in the environment. The current National Institute for Occupational Safety and Health (NIOSH) exposure limits for Ta (metal and oxide dusts) are $5\,mg\,m^{-3}$(10 hr) and $10\,mg\,m^{-3}$ for the short-term (15 min) exposure.[132] There are no established limits for Nb in air at the moment. A limit of $0.01\,mg\,L^{-1}$ has been incurred in the former USSR for Nb in drinking water. Regulations for Nb and Ta in drinking water,[90] surface water, or soils have not been established in the USEPA, WHO, or any other international health or environmental organization.

Chromium

Chromium is a non-essential element for plants, but its trivalent form, Cr(III), is considered biologically active as it is involved in carbohydrate, fat, and protein metabolism.[91,92] The first evidence of the biological function of Cr(III) in humans was found when some patients developed diabetic symptoms that were refractory to insulin but reversed by addition of chromium to parenteral nutrition.[93] The daily Cr(III) requirement for adults is estimated to be 0.5–2 mg of absorbable Cr(III).[94] In contrast, Cr(VI) is a strong oxidizing agent, in the form of chromate and dichromate, and is extremely toxic to animals and human. Unlike Cr(III), Cr(VI) has the ability to penetrate biological membranes by diffusion and reacts with cell contents before being reduced to Cr(III). The toxicity of Cr(VI) compounds is believed to be linked to its oxidative DNA impairment and mutagenic damage.[95] Inhalation of suspended particulates containing adsorbed Cr is a major route of exposure that presents a significant hazard to health as it increases incidence of lung cancer. Soluble salts of chromate are more toxic when administered parenterally ($LD_{50} = 10$–$50\,mg\,kg^{-1}$) in comparison to dermal ($LD_{50} = 200$–$350\,mg\,kg^{-1}$) and oral ($LD_{50} = 1500\,mg\,kg^{-1}$) exposures.[96] Studies on rats showed that the most affected organs following the acute exposure to Cr(VI) are liver and kidney.[97] Symptoms of acute toxicity of Cr(VI) in humans include diarrhea, cyanosis, gastric distress, olfactory cleft obstruction, nosebleed, kidney and liver damage, skin cancer, and lung cancer.

On the other hand, Cr(III) compounds are considered non-toxic at small doses. The LD_{50} values calculated for orally administered $CrCl_3$, $Cr(CH_3COO)_3$, and $Cr(NO_3)_3$ were 1.87, 11.26, and 3.26 g kg kg^{-11}.[98] Chromium is highly toxic especially to higher plants when Cr(VI) content exceeds 100 $\mu M\ kg^{-1}$ dry weight. Symptoms of Cr toxicity in plants appear as wilting of tops, chlorosis in young leaves, damage of cell membranes, and deterioration in growth. Extensive data are available elsewhere on the toxic effects and metabolism of Cr in humans, animals, [94,97–101] and plants.[102–104]

Currently, there are a number of regulatory measures for Cr substances in the environment. The IARC has classified Cr(VI) in Group I human carcinogens.[97] Minimal risk levels (MRLs) of daily human exposure to Cr(III) as insoluble trivalent chromium particulate compounds and Cr(VI) as chromium trioxide mist and other dissolved hexavalent chromium aerosols and mists were estimated as 0.005 and 5×10^{-6} mg m^{-3}, respectively. Other regulatory guidelines for Cr exposure and content in soils and water are listed in Table 6.

Molybdenum

Despite its low abundance, Mo is a component of several enzymes in human and animal organisms. Hence, it is an essential element for many biological processes and is required for the growth of many organisms. For example, Mo-dependent enzymes are necessary in plants and lower organisms for nitrogen fixation (nitrogenase), conversion of nitrate to ammonia, and various oxidation–reduction reactions.[105–107] Molybdenum is part of a coordination complex called molybdenum cofactor, which is formed between molybdopterin and Mo oxide (see Figure 1). This complex is required for three mammalian molybdoenzymes: xanthine dehydrogenase/oxidase, aldehyde oxidase, and sulfite oxidase.[108–110] Xanthine oxidase is important to maintain a balance between urate production and excretion. Low Mo intake in human is claimed to reduce the activity of xanthine dehydrogenase in tissues and the development of hyperuricemia and gout. Molybdenum cofactor deficiency is a rare disease in humans and is caused by the diminished ability to synthesize molybdoenzymes resulting in in neonatal seizures and other neurological symptoms identical to sulfite oxidase deficiency.[111] A study on goats has shown that a molybdenum-deficient diet (24 µg Mo kg^{-1}) caused a decline in conception rate and an increase in abortion rate.[112] Molybdenum is required in the smallest amount by plants and is taken up as the molybdate (MoO^{-2}) ion. However, Mo-deficient soil is known to cause undesirable effects in crops. For example, in the presence of nitrate fertilizers,

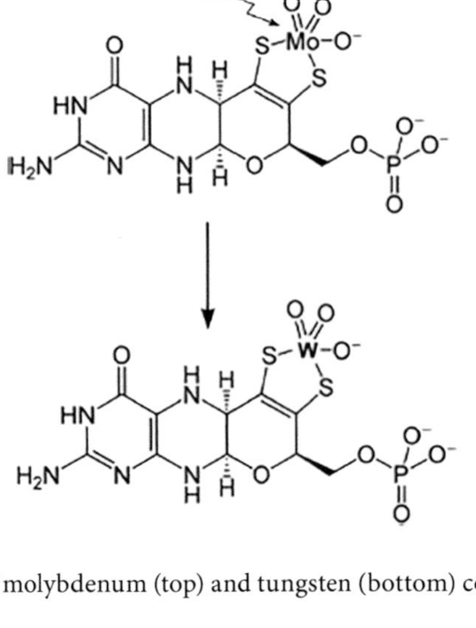

FIGURE 1   Chemical structure of molybdenum (top) and tungsten (bottom) cofactors.

molybdenum-deficient oat and wheat develop pale green leaves and necrotic regions on leaf margins, seeds are poorly developed, and overall plant growth is notably decreased.[113,114] In contrast, Mo toxicity in plants is generally rare and very little reports on its toxic effects in plants are available.[115] As a result of natural or anthropogenic activities, Mo may become sufficiently concentrated in soils and vegetation to cause poisoning. One form of Mo poisoning is the molybdenosis syndrome, which is a copper-deficiency disease that affects cattle and sheep.[116] The toxic effects are due to the depressing effects of Mo on the physiological availability of Cu. Symptoms of this disease are scouring, weight loss, bone disorders, depigmentation, and reproductive impairment.[117,118]

In human, water-soluble molybdenum compounds are readily absorbed when ingested but the rate of absorption depends on a number of factors including the chemical form of Mo. Toxic effects of Mo are related to the capability of Mo compounds to inhibit copper and sulfur absorption, which affects copper homeostasis and results in symptoms similar to copper deficiency. Excessive Mo intake in humans, and hence molybdoenzymes, contributes to diseases of gout, combined oxidase deficiency, and radical damage following cardiac failure.[119] It is notable that in comparison with V and Cr, occupational Mo exposure is uncommon and adverse effects of such exposure have rarely been reported. According to the National Academy of Sciences, the safe Mo intake levels are 0.015–0.04 mg day^{-1} for infants, 0.025–0.15 mg day^{-1} for children aged 1–10, and 0.075–0.25 mg day^{-1} for all individuals above the age of 10.[120] Molybdenum generally occurs at very low concentrations in drinking water, and therefore it has not been considered necessary to set a formal guideline value.[121]

Tungsten

In recent years, interest in tungsten geochemistry and toxicology has increased, especially after the diagnosis of 15 cases with acute lymphocytic leukemia in children and teenagers living in the town of Fallon in the north of Nevada. These cases were related to the high concentrations of W (0.27 to 742 μg L^{-1}) in groundwater due to weathering of local W-rich deposit.[122] Tungsten enters the human and animal body in a soluble form such as tungstate WO_4^{-2} Similar to molybdenum, W is a constituent of three classes of enzymes: aldehyde ferredoxin/oxidoreductase, formate dehydrogenase, and acetylene hydratase. These enzymes are essential catalysts of specific reactions in a number of living organisms. For example, formate dehydrogenase is responsible for the reversible conversion of CO_2 to formate. This makes W by far the heaviest metal with a biological function. Tungsten and molybdenum have almost equal atomic and ionic radii (0.68 Å) and similar electronegativity (1.4 and 1.3 for W and Mo, respectively). It is thus interesting to note that WO_4^{-2} and MoO_4^{-2} are isomorphic ions, which mean that W can replace Mo in molybdoenzymes. A number of studies have demonstrated the antagonistic effect of W (administered in diet as sodium tungstate) on the activity of the Mo enzyme xanthine dehydrogenase, sulfite oxidase and aldehyde oxidase, and nitrate reductase.[123,124] When W substitutes for Mo in molybdoenzymes, a catalytically inactive (or of low activity) analogue is produced, causing adverse health effects that is due to W toxicity and not because of Mo deficiency.[125,126] Similar to Nb and Ta, the major occupational hazard of W exists in the metal industry.[80,81] Airborne nanoparticles containing toxic metals including W, Hg, Ni, Ti, and Zn have been reported in the bloodstream of humans, [127] which has been suspected to trigger blood coagulation. Toxicity symptoms of exposure to airborne W include cough, expectoration, shortness of breath, and tightness in the chest and pulmonary fibrosis in advanced cases.[128] There are little data available on occupational exposures to W compounds that incriminate these substances as toxic or as hazardous to body organs other than lungs. Ingested W in trace quantities are excreted in urine and eliminated in feces. Regulations of W in aquatic and terrestrial environments are not available in most countries including the United States and the United Kingdom, and W is not part of routine testing programs. This probably explains the claim that W and its compounds are absent at measurable concentrations in drinking waters. The IARC has not classified W or its compounds as carcinogenic. Although some health regulations exist for W, environmental regulations applicable to W in environmental systems are absent (see Table 6).

Conclusion

The fate and geochemical behavior of V, Mo, and Cr in soils and freshwaters have been studied extensively but less so that of W, Ta, and Nb. These trace elements are released into the environment from the natural weathering of parental rocks and deposits and a result of anthropogenic activities such as industrial discharge, fuel combustion, sewage sludge, livestock manures, and other wastes. Although the environmental concentration of these elements is generally small, reports have shown that they may enter the food chain through plant absorption from contaminated soils, animal feed, or potable water supplies resulting in health problems. Soil pH and the contents of iron or manganese oxides, natural organic matter, and clay minerals are all important factors that control their transport, bioavailability, and ecotoxicity. Chemical speciation plays a critical role in defining the fate and bioavailability of these elements. For example, while Cr(III) is largely immobile in soil, limiting its entry into the food chain, Cr(VI) is highly mobile under neutral to alkaline soil pH conditions. From a soil management point of view, this is problematic because while raising soil pH by liming would decrease the bioavailability of some toxic elements such as Pb and Cd, it will have quite the opposite effect for Cr(VI). The booming steel industry in many countries including China has led to serious contaminations of Cr, V, W, and Mo in surface waters and soils. Although strict regulatory measures are available for Cr in soils and waters, these are still not complete for V and Mo and absent for Ta, Nb, and W.

References

1. Greenwood, N.N. Recent developments concerning the discovery of elements 101–111. Pure Appl. Chem. **1997,** *69* (1), 179–184.
2. Norman, N.G. Vanadium to dubnium: From confusion through clarity to complexity. Catal. Today **2003,** *78* (1–4), 5–11.
3. Halka, M.; Nordstrom, B. *Periodic Table of the Elements: Transition Metals;* Facts on File: New York, 2011.
4. Enghag, P. *Encyclopedia of the Elements,* 1st Ed.; WILEY-VCH Verlag GmbH: Weinheim, Chichester, 2004.
5. Wiberg, N.; Holleman, A.F.; Wiberg, E. *Holleman–Wiberg's Inorganic Chemistry,* 1st Ed.; Academic Press, 2001.
6. Deans, T. Trace elements in carbonatites and limestones. Nature **1968,** *220* (5168), 679–679.
7. Lide, D. *CRC Handbook of Chemistry and Physics: A Ready-Reference Book of Chemical and Physical Data, 90th Ed.*; CRC Press: Boca Raton, FL, 2009.
8. Wood, S.A.; Samson, I.M. The hydrothermal geochemistry of tungsten in granitoid environments: I. Relative solubilities of ferberite and scheelite as a function of T, P, pH, and mNaCl. Econ. Geol. **2000,** *95* (1), 143–182.
9. Wiley-VCH. *Ullmann's Encyclopedia of Industrial Chemistry,* 7th Ed.; Wiley-VCH: New York, 2011.
10. ASM. *ASM Handbook: Properties and Selection: Nonferrous Alloys and Special-Purpose Materials, 10th Ed.*; ASM International: Materials Park, Ohio, 1990; Vol. 2.
11. Buckman, RW. New applications for tantalum and tantalum alloys. JOM **2000,** *52* (3), 40–41.
12. Andersson, K.; Reichert, K.; Wolf, R. Tantalum and tantalum compounds. In: *Ullmann's Encyclopedia of Industrial Chemistry,* 7th Ed.; Wiley-VCH: New York, 2000.
13. Martin, J.M.; Donnet, C.; Le Mogne, T.; Epicier, T. Superlubricity of molybdenum disulphide. Phys. Rev. B 1993, *48* (14), 10583–10586.
14. Lassner, E.; Schubert W. *Tungsten: Properties, Chemistry, Technology of the Element, Alloys, and Chemical Compounds*; Springer: New York, 1999.
15. OECD (Organisation for Economic Co-operation and Development). *OECD Trade Policy Studies The Economic Impact of Export Restrictions on Raw Materials;* OECD Publishing, 2010.
16. Berrow, M.L.; Webber, J. Trace elements in sewage sludges. J. Sci. Food Agric. **1972,** *23* (1), 93–100.

17. McGrath, S.P.; Chaudri, A.M.; Giller, K.E. Long-term effects of metals in sewage sludge on soils, microorganisms and plants. J. Ind. Microbiol. **1995**, *14* (2), 94–104.

18. Antoniadis, V.; Tsadilas, C.; Samaras, V.; Sgouras, J. Availability of heavy metals applied to soil through sewage sludge. In: *Trace Elements in the Environment: Biogeochemistry, Biotechnology, and Bioremediation;* Prasad, M.N.V., Sajwan, K.S., Naidu R., Eds.; CRC Taylor and Francis, LLC: Boca Raton, FL, 2006; 39–62.

19. Stephen, R.S. A critical review of the bioavailability and impacts of heavy metals in municipal solid waste composts compared to sewage sludge. Environ. Int. **2009**, *35* (1), 142–156.

20. Di Bonito, M. Sewage sludge in Europe and in the U.K.: Environmental impact and improved standards for recycling and recovery to land. In *Environmental Geochemistry: Site Characterization, Data Analysis and Case Histories;* Vivo, B.D., Belkin, H.E., Lima A., Eds.; Elsevier: Amsterdam, the Netherlands, 2008.

21. Adriano, D.C. *Trace Elements in Terrestrial Environments: Biogeochemistry, Bioavailability, and Risks of Metals*; CRC Press, Taylor and Francis Group: Boca Raton FL, 2006.

22. Carlson, C.; Adriano, D. Environmental impacts of coal combustion residues. J. Environ. Qual. **1993**, *22*, 227–242.

23. Bradl, H. *Heavy Metals in the Environment: Origin, Interaction and Remediation*; Academic Press: Amsterdam, Netherlands, 2005; Vol. 6.

24. Bullock, P.; Gregory, P.J. *Soils in the Urban Environment;* Blackwell Scientific Publications: Oxford, Boston, 1991.

25. Kabata-Pendias, A. *Trace Elements in Soils and Plants,* 4th Ed.; CRC Press, Inc.: Boca Raton, Florida, 2011.

26. McBride, M.B. *Environmental Chemistry of Soils;* Oxford University Press, Inc.: New York, 1994.

27. Hubert-Pfalzgraf, G.L. Niobium and tantalum: inorganic and coordination chemistry. In *Encyclopedia of Inorganic and Bioinorganic Chemistry*; Scott A.R., Ed.; Wiley Publishing, Inc., 2011.

28. Ure, A.M.; Bacon, J.R. Comprehensive analysis of soils and rocks by spark-source mass spectrometry. Analyst **1978**, *103* (1229), 807–822.

29. Govindaraju, K. Compilation of working values and sample description for 383 geostandards. Geostand. Geoanal. Res. **1994**, *18,* 1–158.

30. Yang, J.; Teng, Y; Wang, J.; Li, J. Vanadium uptake by alfalfa grown in V-Cd-contaminated soil by pot experiment. Biol. Trace Elem. Res. **2010**, *142* (3), 787–795.

31. Morrell, B.G.; Lepp, N.W.; Phipps, D.A. Vanadium uptake by higher plants: Some recent developments. Environ. Geo-chem. Health **1986**, *8* (1), 14–18.

32. *Eriksson, J. Concentrations of 61 Trace Elements in Sewage Sludge, Farmyard Manure, Mineral Fertilizers, Precipitation and in Oil and Crops;* Swedish EPA: Stockholm, 2001.

33. Oakes, T.W.; Shank, K.E. *Concentrations of Radionuclides and Selected Stable Elements in Fruits and Vegetables*; Oak Ridge National Laboratory: Oak Ridge, Tennessee, 1977.

34. Fritzen, M.B.; Souza, A.J.; Silva, TA; Souza, L.; Nome, R.A.; Fiedler, H.D. Distribution of hexavalent Cr species across the clay mineral surface-water interface. J. Colloid Interface Sci. **2006**, *296* (2), 465–471.

35. Shen-Yang, T.; Ke-An, L. The distribution of chromium(VI) species in solution as a function of pH and concentration. Talanta **1986**, *33* (9), 775–777.

36. Zayed, A.M.; Terry, N. Chromium in the environment: Factors affecting biological remediation. Plant Soil **2003**, *249* (1), 139–156.

37. Rai, D.; Eary, L.; Zachara, J. Environmental chemistry of chromium. Sci. Total Environ. **1989**, *86* (1–2), 15–23.

38. Bloomfield, C.; Pruden, G. The behaviour of Cr(VI) in soil under aerobic and an-aerobic conditions. Environ. Pollut., Ser. A **1980**, *23* (2), 103–114.

39. Richard, F.C.; Bourg, A.C. Aqueous geochemistry of chromium: A review. Water Res. **1991**, *25* (7), 807–816.

40. Alloway, B.J. *Heavy Metals in Soils;* Blackie Academic and Professional: London, 1995.
41. Avudainayagam, S.; Megharaj, M.; Owens, G.; Kookana, R.; Chittleborough, D.; Naidu, R. Chemistry of chromium in soils with emphasis on tannery waste sites. In *Reviews of Environmental Contamination and Toxicology;* SpringerVerlag: New York, 2003; Vol. 178, 53–91.
42. Gelencsér, A.; Kováts, N.; Turóczi, B.; RostásiI, Á.; Hoffer, A.; Imre, K.; Nyirő-Kósa, I.; Csákberényi-Malasics, D.; Toóth, Á.; Czitrovsky, A.; Nagy, A.; Nagy, S.; Ács, A.; Kovács, A.; Ferincz, Á.; Hartyáni, Z.; Pósfai, M. The red mud accident in Ajka (Hungary), characterization and potential health effects of fugitive dust. Environ. Sci. Technol. **2011,** *45* (4), 1608–1615.
43. Ruyters, S.; Mertens, J.; Vassilieva, E.; Dehandschutter, B.; Poffijn, A.; Smolders, E. The red mud accident in Ajka (Hungary), plant toxicity and trace metal bioavailability in red mud contaminated soil. Environ. Sci. Technol. **2011,** *45* (4), 1616–1622.
44. Kimbrough, D.E.; Cohen, Y.; Winer, A.M.; Creelman, L.; Mabuni, C.A. Critical assessment of chromium in the environment. Crit. Rev. Environ. Sci. Technol. **1999,** *29* (1), 1–46.
45. Nriagu, J.O.; Nieboer, E. *Chromium in the Natural and Human Environments;* Wiley: New York, 1988.
46. Pau, N.R.; Lawson, M.D. Molybdenum and tungsten: Their roles in biological processes. In *Metal Ions in Biological Systems: Molybdenum and Tungsten;* Sigel A., Sigel H., Eds.; Marcel Dekker Inc.: New York, 2002; Vol. 29, 31–74.
47. Stiefel, E.I. The biogeochemistry of molybdenum and tungsten. In: *Metal ions in biological systems: Molybdenum and Tungsten;* Sigel, A., Sigel, H., Eds.; Marcel Dekker Inc.: New York, 2002; Vol. 39, 1–29.
48. Goldberg, S.; Forster, H.; Godfrey, C. Molybdenum adsorption on oxides, clay minerals, and soils. Soil Sci. Soc. Am. J. **1996,** *60* (2), 425–432.
49. Zhang, P.; Sparks, D. Kinetics and mechanisms of molybdate adsorption/desorption at the goethite/water interface using pressure-jump relaxation. Soil Sci. Soc. Am. J. **1989,** *53,* 1028–1034.
50. Steinke, D.R.; Majak, W.; Sorensen, T.S.; Parvez, M. Chelation of molybdenum in *Medicago sativa* (alfalfa) grown on reclaimed mine tailings. J. Agric. Food Chem. **2008,** *56* (13), 5437–5442.
51. Wichard, T.; Mishra, B.; Myneni, S.C.B.; Bellenger, J.; Kraepiel, A.M.L. Storage and bioavailability of molybdenum in soils increased by organic matter complexation. Nature Geosci. **2009,** *2* (9), 625–629.
52. Lindsay, W.L. *Chemical Equilibria in Soils;* The BlackBurn Press: New Jersey, 2001.
53. Gupta, U. *Molybdenum in Agriculture;* University Press: New York, 1997.
54. Smedley, P.; Cooper, D.; Lapworth, D.J.; Ander, L. Molybdenum in British drinking water: A review of sources and occurrence and a reconnaissance survey of concentrations. Keyworth, Nottingham: Groundwater Resources Programme, British Geological Survey, 2008. Open Report OR/08/051.
55. Schroeder, H.A. A sensible look at air pollution by metals. Arch. Environ. Health **1970,** *21* (6), 798–806.
56. ATSDR (Agency for Toxic Substances and Disease Registry). Toxicological Profile for Tungsten. U.S. Department of Health and Human Services, 2005.
57. Yim, W. Heavy metal accumulation in estuarine sediments in a historical mining of Cornwall. Mar. Pollut. Bull. **1976,** *7* (8), 147–150.
58. Scheuhammer, A.; Norris, S. Review of the Environmental Impacts of Lead Shotshell Ammunition and Lead Fishing Weights in Canada. Occasional Papers No. 88. Quebec: Government of Canada, Canadian Wildlife Service, 1995.
59. Scheuhammer, A.M.; Norris, S.L. The ecotoxicology of lead shot and lead fishing weights. Ecotoxicology **1996,** *5* (5), 279–295.
60. USFWS. Migratory bird hunting: Approval of tungsten–nickel–iron shot as non-toxic for hunting waterfowl and coots. U.S. Fish and Wildlife Service. Fed. Regist. **2001** *66* (3), 737–742.

61. Szefer, P.; Nriagu, J.O. *Mineral Components in Foods;* CRC Press/Taylor and Francis: Boca Raton, 2007.
62. Welch, R.M.; Huffman, E.W.D. Vanadium and plant nutrition. Plant Physiol. **1973**, *52* (2), 183–185.
63. Nielsen, F. Vanadium. In *Trace Elements in Human and Animal Nutrition,* 5th Ed.; Mertz, W, Ed.; Academic Press: San Diego, California, 1988; 275–300.
64. Goldwaser, I.; Gefel, D.; Gershonov, E.; Fridkin, M.; Shechter, Y. Insulin-like effects of vanadium: Basic and clinical implications. J. Inorg. Biochem. **2000**, *80* (1–2), 21–25.
65. Tsiani, E.; Fantus, I. Vanadium compounds: Biological actions and potential as pharmacological agents. Trends Endocrinol. Metab. **1997**, *8* (2), 51–58.
66. Bishayee, A.; Chatterjee, M. Inhibitory effect of vanadium on rat liver carcinogenesis initiated with diethylnitrosamine and promoted by phenobarbital. Br. J. Cancer **1995**, *71* (6), 1214–1220.
67. Evangelou, A.M. Vanadium in cancer treatment. Crit. Rev. Oncol./Hematol. **2002**, *42* (3), 249–265.
68. Djordjevic, C. Antitumor activity of vanadium compounds. In *Metal Ions in Biological Systems*; Sigel, A., Sigel, H., Sigel, R., Eds.; 1995; Vol. 31, 595–616.
69. Rehder, D. Inorganic considerations on the function of vanadium in biological systems. In *Metal Ions in Biological Systems;* Sigel, A., Sigel, H., Sigel, R., Eds.; 1995; Vol. 31, 1–43.
70. Sabbioni, E.; Rade, J.; Bertolero, F. Relationships between iron and vanadium metabolism: The exchange of vanadium between transferrin and ferritin. J. Inorg. Biochem. **1980**, *12* (4), 307–315.
71. Ganguli, S.; Reuland, D.J.; Franklin, L.A.; Deakins, D.D.; Johnston, W.J.; Pasha, A. Effects of maternal vanadate treatment of fetal development. Life Sci. **1994**, *55* (16), 1267–1276.
72. Nriagu, J.O. *Vanadium in the Environment. Part 2, Health Effects;* John Wiley and Sons: New York, 1998.
73. IARC. Cobalt in Hard Metals and Cobalt Sulfate, Gallium Arsenide, Indium Phosphide and Vanadium Pentoxide. Vol. 86 of IARC Monographs on the Evaluation of Carcinogenic Risks to Humans. Lyon, France, 2006.
74. Liu, J.; Cui, H.; Liu, X.; Peng, X.; Deng, J.; Zuo, Z.; Cui, W.; Deng, Y.; Wang, K. Dietary high vanadium causes oxidative damage-induced renal and hepatic toxicity in broilers. Biol. Trace Elem. Res. **2011**, *145* (2), 189–200.
75. Al-Bayati, M.A.; Giri, S.N.; Raabe, O.G.; Rosenblatt, L.S.; Shifrine, M. Time and dose–response study of the effects of vanadate on rats: morphological and biochemical changes in organs. J. Environ. Pathol. Toxicol. Oncol. **1989**, *9* (5–6), 435–455.
76. Barker, AV.; Pilbeam, D.J. *Handbook of Plant Nutrition;* CRC Taylor and Francis: Boca Raton, FL, 2007.
77. Dieter, R. The coordination chemistry of vanadium as related to its biological functions. Coord. Chem. Rev. **1999**, *182* (1), 297–322.
78. USEPA. Superfund, emergency planning, and community right-to-know programs. Extremely hazardous substances and their threshold planning quantities. U.S. Environmental Protection Agency, 2009.
79. Edel, J.; Sabbioni, E.; Pietra, R.; Rossi, A.; Torre, M.; Rizzato, G.; Fraioli, P. Trace metal lung disease: In vitro interaction of hard metals with human lung and plasma components. Sci. Total Environ. **1990**, *95*, 107–117.
80. Moriyama, H.; Kobayashi, M.; Takada, T.; Shimizu, T.; Terada, M.; Narita, J.; Maruyama, M.; Watanabe, K.; Suzuki, E.; Gejyo, F. Two-dimensional analysis of elements and mononuclear cells in hard metal lung disease. Am. J. Respir. Crit. Care Med. **2007**, *176* (1), 70–77.
81. Sabbioni, E.; Minoia, C.; Pietra, R.; Mosconi, G.; Forni, A.; Scansetti, G. Metal determinations in biological specimens of diseased and non-diseased hard metalworkers. Sci. Total Environ. **1994**, *150* (1–3), 41–54.
82. Downs, W.L.; Scott, J.K.; Yuile, C.L.; Caruso, F.S.; Wong, L.C.K. The toxicity of niobium salts. Am. Ind. Hyg. Assoc. J. **1965**, *26* (4), 337–346.

83. Venugopal, D.; Luckey, T. Toxicity of group V metals and metalloids. In *Metal Toxicity in Mammals;* Plenum Press: New York, 1978; Vol. 2, 227–229.

84. Ramirez, G.; Rodil, S.; Arzate, H.; Muhl, S.; Olaya, J. Niobium based coatings for dental implants. Appl. Surf. Sci. **2011**, *257* (7), 2555–2559.

85. Johansson, C.B.; Hansson, H.A.; Albrektsson, T. Qualitative interfacial study between bone and tantalum, niobium or commercially pure titanium. Biomaterials **1990**, *11* (4), 277–280.

86. Matsuno, H.; Yokoyama, A.; Watari, F.; Uo, M.; Kawasaki, T. Biocompatibility and osteogenesis of refractory metal implants, titanium, hafnium, niobium, tantalum and rhenium. Biomaterials **2001**, *22* (11), 1253–1262.

87. Divine, K.; Goering, P. Tantalum. In *Elements and Their Compounds in the Environment Occurrence, Analysis and Biological Relevance;* Merian, E., Anke, M., Ihnat, M., Stoeppler, M., Eds.; Wiley-VCH: Weinheim, 2004; 1087–1097.

88. Rydzynski, K. Vanadium, niobium, and tantalum. In *Patty's Toxicology;* Bingham, E., Cohrssen, B., Powell, C.H., Eds.; Wiley InterScience: New York, 2001.

89. Goering, P.L.; Ziegler, T.L. Niobium (Nb) (columbium). In *Elements and Their Compounds in the Environment: Occurrence, Analysis and Biological Relevance;* Merian, E., Anke, M., Ih-nat, M., Stoeppler, M., Eds.; Wiley-VCH: Weinheim, 2004; 1039–1046.

90. Wennig, R.; Kirsch, N. Niobium. In *Handbook on Toxicity of Inorganic Compounds;* Seiler, H.G., Sigel, A., Eds.; Marcel Dekker: New York, 1988; 469–473.

91. Hopkins, J.L.L.; Ransome-Kuti, O.; Majaj, A.S. Improvement of impaired carbohydrate metabolism by chromium III in malnourished infants. Am. J. Clin. Nutr. **1968**, *21* (3), 203–211.

92. Mertz, W. Chromium occurrence and function in biological systems. Physiol. Rev. **1969**, *49* (2), 163–239.

93. Mertz, W. Chromium in human nutrition: A review. J. Nutr. **1993**, *123* (4), 626–633.

94. WHO. Chromium in drinking-water. World Health Organization: Geneva, Switzerland, 2003, WHO/SDE/ WSH/03.04/04.

95. Standeven, A.M.; Wetterhahn, K.E. Chromium(VI) toxicity: Uptake, reduction, and DNA damage. Int. J. Toxicol. **1989**, *8* (7), 1275–1283.

96. *Carson, B.L.; Ellis, H.V.; McCann, J.L. Toxicology and Biological Monitoring of Metals in Humans, Including Feasibility and Need, 1st Ed.; CRC Press: Chelsea, Michigan, 1986.*

97. ATSDR. *Toxicological Profile for Chromium;* U.S. Department of Health and Human Services: Atlanta, 2008.

98. Cohen, D.M.; Costa, M. Chromium. In *Environmental Toxicants: Human Exposures and Their Health Effects*, 2nd Ed.; Lippmann, M., Ed.; Wiley-Blackwell, Hoboken, New Jersey, 2005; 173–191.

99. Bianchi, V.; Celotti, L.; Lanfranchi, G.; Majone, F.; Marin, G.; Montaldi, A.; Sponza, G.; Tamino, G.; Venier, P.; Zantedeschi, A.; Levis, A.G. Genetic effects of chromium compounds. Mutat. Res., Genet. Toxicol. **1983**, *117* (3–4), 279–300.

100. Katz, S.A.; Salem, H. The toxicology of chromium with respect to its chemical speciation: A review. J. Appl. Toxicol. **1993**, *13* (3), 217–224.

101. Baruthio, F. Toxic effects of chromium and its compounds. Biol. Trace Elem. Res. **1992**, *32* (1–3), 145–153.

102. Cervantes, C.; Campos-García, J.; Devars, S.; Gutiérrez-Corona, F.; Loza-Tavera, H.; Torres-Guzmán, J.C.; Moreno-Sánchez, R. Interactions of chromium with microorganisms and plants. FEMS Microbiol. Rev. **2001**, *25* (3), 335–347.

103. Hayat, S.; Khalique, G.; Irfan, M.; Wani, A.S.; Tripathi, B.N.; Ahmad, A. Physiological changes induced by chromium stress in plants: An overview. Protoplasma **2011**, 1–13.

104. Shanker, A.K.; Cervantes, C.; Loza-Tavera H.; Avudain-ayagam S. Chromium toxicity in plants. Environ. Int. **2005**, *31* (5), 739–753.

105. Burgess, B.K.; Lowe, D.J. Mechanism of molybdenum nitrogenase. Chem. Rev. **1996**, *96* (7), 2983–3012.

106. Kaiser, B.N.; Gridley, K.L.; Ngaire Brady, J.; Phillips, T.; Tyerman, S.D. The role of molybdenum in agricultural plant production. Ann. Bot. **2005**, *96* (5), 745–754.

107. Schwarz, G.; Mendel, R.R.; Ribbe, MW. Molybdenum cofactors, enzymes and pathways. Nature **2009**, *460* (7257), 839–847.

108. Kisker, C.; Schindelin, H.; Rees, D.C. Molybdenum-cofactor-containing enzymes: Structure and mechanism. Annu. Rev. Biochem. **1997**, *66* (1), 233–267.

109. Konhauser, K.O.; Powell, M.A.; Fyfe, W.S.; Longstaffe, F.J.; Tripathy, S. Trace element chemistry of major rivers in Orissa State, India. Environ. Geol. **1997**, *29* (1–2), 132–141.

110. Sardesai, V.M. Molybdenum: An essential trace element. Nutr. Clin. Pract. **1993**, *8* (6), 277–281.

111. Reiss, J.; Christensen, E.; Dorche, C. Molybdenum cofactor deficiency: First prenatal genetic analysis. Prenatal Diagn. **1999**, *19* (4), 386–388.

112. Anke, M.; Groppl, B.; Kronemann, H.; Grun, M. Molybdenum supply and status in animals and human beings. Nutr. Res. **1985**, *S1*, 180–186.

113. Chatterjee, C.; Nautiyal, N.; Agarwala, S.C. Metabolic changes in mustard plants associated with molybdenum deficiency. New Phytol. **1985**, *100* (4), 511–518.

114. Chatterjee, C.; Nautiyal, N. Molybdenum stress affects viability and vigor of wheat seeds. J. Plant Nutr. **2001**, *24* (9), 1377.

115. Johnson, J.L.; Waud, W.R.; Rajagopalan, K.V.; Duran, M.; Beemer, F.A.; Wadman, S.K. Inborn errors of molybdenum metabolism: Combined deficiencies of sulfite oxidase and xanthine dehydrogenase in a patient lacking the molybdenum cofactor. Proc. Natl. Acad. Sci. U. S. A. **1980**, *77* (6), 3715–3719.

116. Frøslie, A. Problems on deficiency and excess of minerals in animal nutrition. In *Geomedicine*, 1st Ed.; Låg, J., Eds.; CRC Press: Boca Raton, Florida, **1990**; 37–60.

117. Erdman, J.A.; Ebens, R.J.; Case, A.A. Molybdenosis: A potential problem in ruminants grazing on coal mine spoils. J. Range Manage. **1978**, *31* (1), 34–36.

118. Ferguson, W.S.; Lewis, A.H.; Watson, S.J. Action of molybdenum in nutrition of milking cattle. Nature **1938**, *141* (3569), 553–553.

119. Seiden, A.I.; Berg, NP.; Soderbergh, A.; Bergstrom, B.E.O. Occupational molybdenum exposure and a gouty electrician. Occup. Med. **2005**, *55* (2), 145–148.

120. Council, N.R. *Recommended Dietary Allowances*, 10th Ed.; National Academy Press: Washington, DC, 1989.

121. WHO. Molybdenum in drinking-water. World Health Organization: Geneva, Switzerland, 2011. WHO/SDE/ WSH/03.04/11/Rev/1.

122. Seiler, R.L.; Stollenwerk, K.G.; Garbarino, J.R. Factors controlling tungsten concentrations in ground water, Carson Desert, Nevada. Appl. Geochem. **2000**, *20* (2), 423–441.

123. Johnson, J.L.; Rajagopalan, K.V.; Cohen, H.J. Molecular basis of the biological function of molybdenum. Effect of tungsten on xanthine oxidase and sulfite oxidase in the rat. J. Biol. Chem. **1974**, *249* (3), 859–866.

124. Higgins, E.S.; Richert, D.A.; Westerfeld, WW. Molybdenum deficiency and tungstate inhibition studies. J. Nutr. **1956**, *59* (4), 539–559.

125. Callis, G.E.; Wentworth, R.A. Tungsten vs. molybdenum in models for biological systems. Bioinorg. Chem. **1977**, *7* (1), 57–70.

126. Kletzin, A.; Adams, MW. Tungsten in biological systems. FEMS Microbiol. Rev. **1996**, *18* (1), 5–63.

127. Gatti, A.M.; Montanari, S.; Monari, E.; Gambarelli, A.; Capitani, F.; Parisini, B. Detection of micro- and nano-sized biocompatible particles in the blood. J. Mater. Sci. Mater. Med. **2004**, *15* (4), 469–472.

128. Fischbein, A.; Luo, J.C.; Solomon, S.J.; Horowitz, S.; Hailoo, W.; Miller, A. Clinical findings among hard metal workers. Br. J. Ind. Med. **1992**, *49* (1), 17–24.

129. PECH (Panel on the trace element geochemistry of coal resource development related to health). Trace Element Geochemistry of Coal Resource Development Related to Health, National Academies Press, 1980.

130. ATSDR. *Toxicological Profile for Vanadium;* U.S. Department of Health and Human Services: Atlanta, 2009.

131. USEPA. *Guidelines for Water Reuse;* U.S. Environmental Protection Agency: Washington, DC, 2004. EPA/625/R-04/108.

132. NIOSH. *NIOSH Pocket Guide to Chemical Hazards*; National Institute for Occupational Safety and Health: Cincinnati, Ohio, 2007. DHHS (NIOSH) Publication No. 2005–149.

133. USGS. Mineral Commodity Summaries, 2011. U.S. Geological Survey, 2011.

134. Hedrick, B.J. The global rare-earth cycle. J. Alloys Compd. **1995**, *225* (1–2), 609–618.

135. Reimann, C.; Caritat, P.D. *Chemical Elements in the Environment: Factsheets for the Geochemist and Environmental Scientist*; Springer-Verlag GmbH and Co: Berlin and Heidelberg, 1998.

136. Takeda, A.; Kimura, K.; Yamasaki, S. Analysis of 57 elements in Japanese soils, with special reference to soil group and agricultural use. Geoderma **2004**, *119* (3–4), 291–307.

137. Licht, O. *Geoquimica de solo do Estado do Parana* (in Spanish); Mineropar: Curitiba, Parana, 2005; Vol. 1/2.

138. Burt, R.; Wilson, M.A.; Mays, M.D.; Lee, CW. Major and trace elements of selected pedons in the USA. J. Environ. Qual. **2003**, *52* (6), 2109–2121.

139. Shacklette, HT.; Boerngen, J.G. Element concentrations in soils and other surficial materials of the conterminous United States. Geol. Surv. Prof. Paper. **1984**, *1270* (105).

140. Kabata-Pendias, A.; Mukherjee, A. *Trace Elements from Soil to Human;* Springer: Berlin-Heidelberg, 2007.

141. Gaillardet, J.; Viers, J.; Dupre, B. Trace elements in river waters. In: *Treatise on Geochemistry;* Elsevier: Oxford, 2003; 225–272.

142. Li, Y. A brief discussion on the mean oceanic residence time of elements. Geochim. Cosmochim. Acta. **1982**, *46* (12), 2671–2675.

143. Nozaki, Y. A fresh look at element distribution in the North Pacific Ocean. EOS, Trans. Am. Geophys. Union **1997**, *78* (21), 221.

144. Sohrin, Y.; Isshiki, K.; Kuwamoto, T.; Nakayama, E. Tungsten in north pacific waters. Mar Chem. **1987**, *22* (1), 95–103.

145. Sparks, D.L. *Environmental Soil Chemistry;* Academic Press: San Diego, 2003.

III

Basic Environmental Processes

39

Adsorption

Puangrat
Kajitvichyanukul
and Jirapat
Ananpattarachai

Introduction

Water is essential for life. Currently, quality of water becomes a major problem in many areas owing to pollution emission from industrial, agricultural, and domestic activities to the water bodies. These activities generate wastewater, which contains both inorganic and organic pollutants. Some of the common pollutants are phenols, dyes, detergents, insecticides, pesticides, and heavy metals. These pollutants are often toxic and cause adverse effects on human life. To avoid pollution of natural water bodies, treating wastewater from the originated source by removing pollutants before being discharging is necessary. Various treatment techniques and processes such as coagulation, membrane process, adsorption, dialysis, foam flotation, osmosis, and biological methods have been used to remove the pollutants from contaminated water. Among all the approaches proposed, adsorption is one of the most popular methods and is currently considered as an effective, efficient, and economic method for water purification.

Adsorption is a well-known equilibrium separation process and an effective method for water and wastewater treatment applications.[1–4] It is the process in which molecules accumulate in the interfacial layer, but desorption denotes the converse process. The fundamental concept in adsorption science is the equilibrium relation between the quantity of the adsorbed material and the pressure or concentration in the bulk fluid phase at constant temperature. The material adsorbed on the surface of "adsorbent" is defined as the "adsorbate." The penetration by the adsorbate molecules into the bulk solid phase is determined as "absorption." The terms "sorption," "sorbent," "sorbate," and "sorptive" are also used to denote both adsorption and absorption, when both occur simultaneously or cannot be distinguished.[1]

Recently, many works related to contaminant removal by adsorption process have been reviewed.[5–7] Several types of adsorbents applied in wastewater treatment are activated carbon, zeolite, chitin or chitosan, and various agricultural wastes such as wood, peat, saw dust, etc. These adsorbents can take a broad range of chemical forms and different geometrical surface structures and properties. This is reflected in the range of their applications in industry for both water and wastewater treatment. Compared with alternative technologies, adsorption is attractive for its relative simplicity of design, operation, and scale-up; high capacity and favorable rate; insensitivity to toxic substances; ease of regeneration; and low cost. Additionally, it avoids using toxic solvents and minimizes degradation.[6]

This review highlights and provides an overview of adsorption theory, kinetics and mechanisms, and recent applications of various adsorbents for contaminant removal from water and wastewater. The main aim of this review is to provide a summary of recent information concerning adsorption process. In this entry, an extensive list of adsorbents from many research papers has been provided. It is strongly encouraged to refer to these original papers for more information on experimental conditions.

Adsorption Theory

In solid–liquid interface, adsorption of a species on a solid surface follows three steps:[7]

1. Transport of the adsorbate from the bulk to the external surface of the adsorbent.
2. Passage through the liquid film attached to the solid surface.
3. Interactions with the surface atoms of the solid leading to chemisorption. (strong adsorbate–adsorbent interactions equivalent to covalent bond formation) or weak adsorption (weak adsorbate–adsorbent interactions, very similar to van der Waals forces).

If step 1 is the slowest, the adsorption will be a transport-limited process. This step is usually the rate-limiting process in systems that are characterized by poor mixing, dilute concentration of adsorbate, small particle size of the adsorbent, etc.

When step 2 is the rate-determining slowest step, the physical process of diffusion through the liquid film influences the outcome of the process, and the efficiency of the solid as an adsorbent can hardly be improved.

When step 3 is the slowest, the adsorption is controlled by a chemical process, and the efficiency of the adsorbent can be influenced by suitably controlling the interactions.[7]

For porous solids, when the adsorbate passes through the liquid film attached to the external surface, it will slowly diffuse into the pores and get trapped or adsorbed. The pore diffusion plays an important role when the adsorbate is present in higher concentration, the adsorbent is made of large particles, and good mixing is ensured.[7–9]

Adsorption capacity is highly dependent on several factors such as properties of the adsorbent (porosity, surface area, particle size) and adsorbate (structure, water solubility, ionic charge, functional groups, pKa, polarity, functionality, molecular weight, and size); solution conditions (solvent, pH, temperature, ionic strength, solute concentration, and competition between solutes); interactions at the solid–liquid interface; and type of experimental setup. Generally, the adsorption capacity of a sorbent for a solute increases with increase in liquid-phase concentration. Physicochemical properties of the adsorbent also play a major role in adsorption. It must have high surface area, particularly internal surface area, high internal volume, and a good pore size distribution. The chemical properties of the adsorbent include degree of ionization at the surface and types of functional groups present. The adsorbent should also have good mechanical properties such as strength and resistance to destruction.

Temperature and pressure have an effect on increasing or decreasing of adsorption capacity. Under low-temperature conditions, the adsorption increases in the forward condition and liberates heat as it is exothermic in nature. With the increasing of pressure, adsorption increases up to a certain extent until saturation level is achieved. After it reaches the equilibrium level, no more adsorption takes place no matter how high the pressure applied.

The coverage of the adsorbent surface by the adsorbate leads to the adsorption process. It assumes that the surface consists of "sites" onto which the adsorbate can adsorb. Accordingly, the adsorption between the adsorbate molecule and the adsorbent surface can be depicted with the chemical reaction below:

$$\text{Molecule + surface site} \leftrightarrow \text{adsorbed molecule} + \text{heat}(\Delta H) \tag{1}$$

According to the above reaction, adsorption characteristic can be expressed in thermodynamic parameters such as ΔG, ΔH, and ΔS. A negative ΔG value stands for the adsorption to take place. Change in enthalpy (ΔH) gives an indication of the bonding strength. The higher the value of heat of adsorption, the weaker the bond between adsorbate and adsorbent. The sign of AS indicates the direction: for adsorption, ($+\Delta S$), and for desorption, ($-\Delta S$).

The change in Gibbs free energy is given by the following expression:

$$\Delta G^\circ = -RT \ln K_L \tag{2}$$

where R is the universal gas constant, T is the absolute temperature, and K_l (L/mg) is the affinity constant of Langmuir model. Negative ΔG values confirm the feasibility of the adsorption, and its absolute values measure the adsorption driving force. Negative or positive values of ΔG indicate the exothermic or endothermic nature of adsorption, respectively. Positive ΔS values reveal a random organization of the adsorbate at the solid/solution interface. The sorption entropy can be calculated from the Gibbs–Helmholtz equation:

$$\Delta G = \Delta H - T \Delta S \tag{3}$$

For a reaction or process to be spontaneous, there must be decreases in free energy of the system. ΔG of the system must have a negative value. During the adsorption, randomness of the molecule decreases, so that ΔS is negative. The above equation can be rewritten as follows:

$$\Delta G = \Delta H + T \Delta S \tag{4}$$

Therefore, for a reaction to be spontaneous, ΔH has to be negative, and $|\Delta H| > |T \Delta S|$.

The adsorption process occurs when adsorbate is adsorbed on adsorbent. Heat energy developing during the adsorption process is released as it is an exothermic process.

During adsorption, forces of attraction play an important role between adsorbate and adsorbent. Van der Waals forces of attraction are weak forces, while forces from chemical bonding are strong forces. Accordingly, the adsorption can be classified into two types: physical adsorption or chemical adsorption.

Physisorption or physical adsorption is a type of adsorption in which the adsorbate adheres to the surface through van der Waals (weak intermolecular) interactions. Physical adsorption takes place with formation of a multilayer of adsorbate on adsorbent. The chemical identity of the adsorbate remains intact without breaking of the covalent structure of the adsorbate. To be a spontaneous thermodynamic process, ΔG has to be a negative value. As the translational degrees of freedom of the adsorbate are lost upon deposition onto the adsorbent, ΔS is negative for the process, and ΔH for physical adsorption must be exothermic. The energy released upon accommodation to the surface is of the same order of magnitude as an enthalpy of condensation (in the order of 20 kJ/mol). This process occurs at low temperature, below the boiling point of adsorbate. As the temperature increases, the process of physical adsorption decreases, as shown in Figure 1.

Chemisorption is a type of adsorption whereby a molecule adheres to a surface through the formation of a chemical bond, as opposed to the van der Waals forces. Chemical adsorption takes place when adsorbate attaches on the adsorbent surface with formation of a unilayer by chemical bonding. This interaction is much stronger and has higher enthalpy of adsorption, ΔH (200–400 kJ/mol), than physical adsorption. As the adsorbates can interact with each other when they lie upon the surface, the energy of adsorption relies on the extent to which the available surface is covered with adsorbate molecules. Thus, the chemical bonds in chemical adsorption may be stronger than the bonds internal to the free adsorbate, resulting in the dissociation of the adsorbate upon adsorption. This type of adsorption can

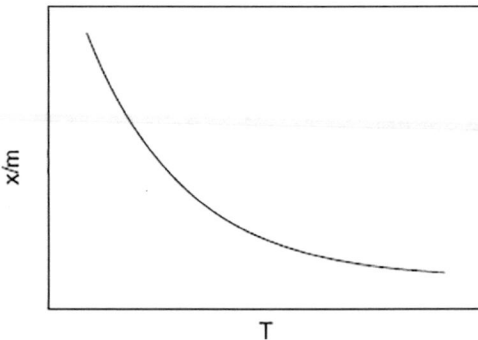

FIGURE 1   Relation of physical adsorption with temperature.

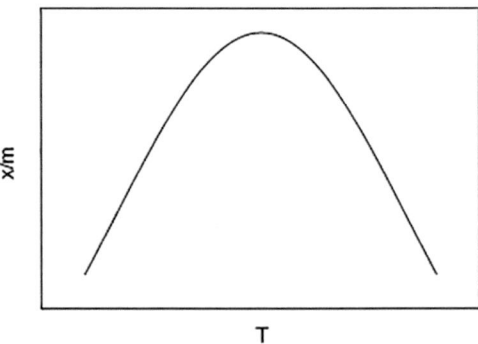

FIGURE 2   Relation of chemical adsorption with temperature.

take place at all temperatures. Chemical adsorption first increases and then decreases with the increases in temperature, as shown in Figure 2.

Adsorption Isotherm

Adsorption isotherms are defined by the adsorbate–adsorbent interactions. Generally, it is constructed by measuring the concentration of the adsorbate in the medium before and after adsorption, at a fixed temperature. This is the practical way to investigate the interaction between the adsorbate and the surface of the adsorbent and to obtain information about the structure of the adsorbed layer. Four classes of adsorption isotherms [S type with an initial con-vexing to the concentration axis, Langmuir (L) type with an initial concavity to the concentration axis, H type with an intercept on the ordinate, and C type with an initial linear portion] and subgroups have been defined according to their configuration,[10–13] as shown in Figure 3. The shape of isotherm provides qualitative information on the nature of solute-surface interaction. The Langmuir class (L) is widespread in the case of adsorption of many organic contaminants from water. The organic contaminant adsorbs parallel to the surface, and no strong competition exists between the adsorbate and the solvent to occupy the adsorption sites. However, the H class (high affinity) results from extremely strong adsorption at very low concentrations, giving rise to an apparent intercept on the ordinate.[12]

In the adsorption process, when a mass of adsorbent and a waste stream of adsorbate are in contact for a sufficiently long time, equilibrium between the amount of pollutant adsorbed and the amount remaining in solution will develop. Under equilibrium conditions, the amount of material adsorbed onto the media can be calculated using the mass balance of Eq. 5:

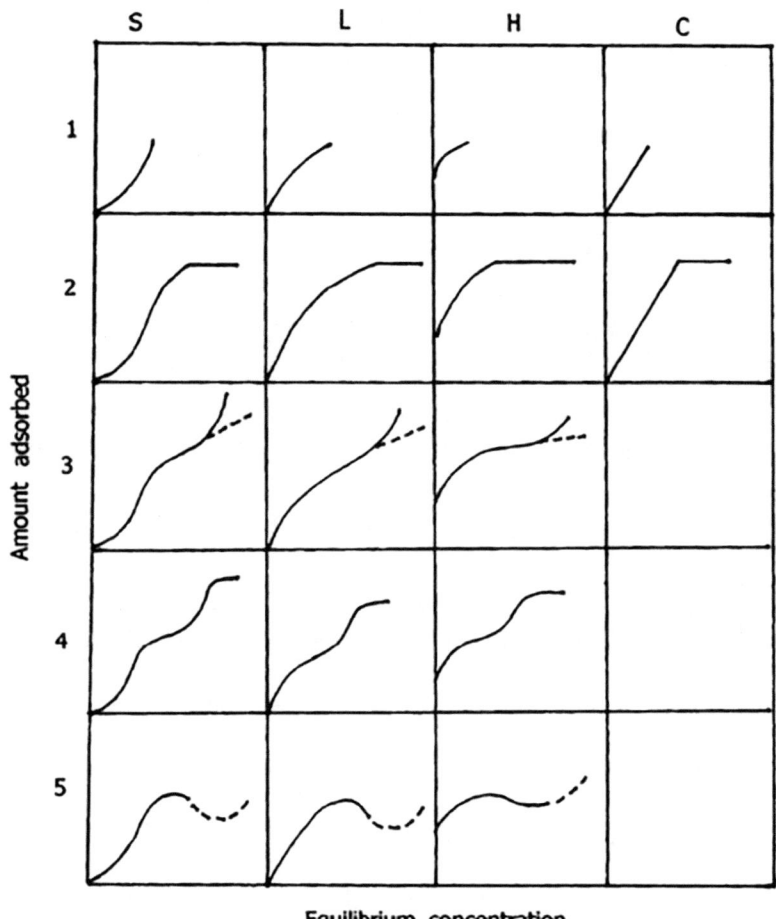

FIGURE 3 Four classes of adsorption isotherms (S, L, H, C). Source: Giles et al.[11]

$$q_e = \frac{(C_0 - C_e)V}{W} \tag{5}$$

where q_e (mg pollutant/g adsorbent) is the mass of pollutant per mass of adsorbent used, C_0 is the liquid-phase concentration of pollutant in solution at the beginning, C_e is the concentration of the pollutant in solution after equilibrium has been reached, V is the volume of the solution, and W is the mass of dry adsorbent used.

Adsorption isotherms are related to the amount of adsorbate on the adsorbent, with its concentration. Several models describing process of adsorption are Freundlich isotherm, Langmuir isotherm, Brunauer–Emmett–Teller (BET) isotherm, etc., with details described below.

Langmuir Isotherm

The Langmuir equation initially derived from kinetic studies was based on the assumption that on the adsorbent surface, there is a definite and energetically equivalent number of adsorption sites. Consequently, the hypotheses for this isotherm are uniform adsorption energies along the homogeneous adsorbent surface, equal solute affinity in all the adsorption sites, no interaction among adsorbed molecules, a single adsorption mechanism, and formation of a monolayer on the free surface.[14]

This isotherm is concerned with the monolayer (single-layer) coverage of the solid surface by the adsorbate. The bonding to the adsorption sites can be either chemical or physical, but it must be sufficiently strong to prevent displacement of adsorbed molecules along the surface.

The Langmuir isotherm is useful in determining the interactions between the adsorbate and the adsorbent. Adsorption data for a wide range of adsorbate concentrations are most conveniently described by the Langmuir isotherm:

$$q_e = \frac{QK_L C_e}{1+K_L C_e} \tag{6}$$

where Q (mg/g) is the maximum amount of the adsorbate per unit weight of the adsorbent to form a complete monolayer on the surface, whereas K_L (L/mg) is the Langmuir constant related to the affinity of the binding sites.

The essential characteristics of Langmuir isotherm can be expressed by a dimensionless constant called separation factor or equilibrium parameter, R_L, defined by Weber and Chakkravorti[15] as follows:

$$R_L = \frac{1}{1+K_L C_0} \tag{7}$$

The parameter R_L indicates the shape of isotherm as shown in Table 1.

The limitations of this isotherm come from the fact that several types of adsorption sites exist, the adsorption mechanism is not the same for the first and for the last molecules adsorbed, and models based on the monomolecular surface layer without interactions could be not realistic. Consequently, many other isotherms such as Freundlich, BET, Temkin, and Dubinin–Radushkevich isotherms have been proposed to explain adsorption behavior.

The Freundlich Isotherm

The Freundlich isotherm is an empirical equation that is more suitable than the Langmuir isotherm when the amount of adsorbent in contact with the solid surface is relatively low. In addition, if the binding energy changes continuously from site to site on solid surfaces, the expression used for the Freundlich isotherm is much more applicable to explaining the adsorption behavior. It is also generally used in nonideal systems with highly heterogeneous surfaces or surfaces supporting sites of varied affinities. The assumption for this isotherm is that the stronger binding sites are occupied first and that the binding strength decreases with an increasing degree of site occupation.[16] It does not imply the formation of a monolayer and frequently gives good interpretation of data over a restricted concentration range.[12] The Freundlich isotherm is expressed as follows:

$$q_e = K_F C_e^{1/n} \tag{8}$$

where q_e (mg/g) is the equilibrium value for removal of adsorbate per unit weight of adsorbent, C_e (mg/L) is the equilibrium concentration of metal ion in solution, and K_F and n are Freundlich isotherm

TABLE 1   Relation between the Value of RL and the Type of Isotherm

Value of Rl	Type of Isotherm
$R_L > 1$	Unfavorable
$R_L = 1$	Linear
$0 < R_L < 1$	Favorable
$R_L = 0$	Irreversible

constants that are related to the adsorption capacity (or the bonding energy) and intensity of the sorbent, respectively. K_F can be defined as the adsorption or distribution coefficient and represents the ad-sorbent onto adsorbate for a unit equilibrium concentration. A value for $1/n < 1$ indicates a normal Langmuir isotherm, while $1/n > 1$ is indicative of cooperative adsorption:[17]

$$\log q_e = \left(\frac{1}{n}\right)\log C_e + \log K_F \tag{9}$$

Experimentally it was determined that extent of adsorption varies directly with pressure until saturation pressure Ps is reached. Beyond that point, the rate of adsorption saturates even after applying higher pressure. Thus, Freundlich adsorption isotherm fails at higher pressure.

BET Isotherm

This isotherm is widely applied in the gas–solid equilibrium systems. It has a theoretical background based on multilayer adsorption, which is the true picture of physical adsorption. The BET isotherm was derived from computer-based calculations as the nonlinear isotherm modeling, usually based on algorithms dealing with error distribution.[18]

Under the high pressure, the BET theory is more feasible than the Langmuir adsorption, which is applicable only under the conditions of low pressure. With high pressure and low temperature, thermal energy of gaseous molecules decreases, and more and more gaseous molecules would be available per unit surface area. Consequently, the multilayer adsorption would occur on the adsorbent surface. This multilayer formation is explained by BET theory. The BET equation is given as

$$V_{total} = \frac{V_{mono}C\left(\dfrac{P}{P_0}\right)}{\left(1 - \dfrac{P}{P_0}\right)\left(1 + C\left(\dfrac{P}{P_0}\right) - \dfrac{P}{P_0}\right)} \tag{10}$$

Another form of the BET equation is

$$\frac{P}{V_{total}(P - P_0)} = \frac{1}{V_{mono}C} + \frac{c-1}{V_{mono}C}\left(\frac{P}{P_0}\right) \tag{11}$$

where V_{mono} is the adsorbed volume of gas at high-pressure conditions so as to cover the surface with a unilayer of gaseous molecule.

Temkin Isotherm

The Temkin isotherm is applicable when the behavior of an adsorption system occurs on heterogeneous surfaces. It is expressed by the following equation.[19,20]

$$q_e = \frac{RT}{b_T}\ln(K_T C_e) \tag{12}$$

The linear form of the Temkin isotherm is represented by the following equations:

$$q_e = A + B\ln C_e \tag{13}$$

$$A = \frac{RT}{b_T}\ln K_T \tag{14}$$

$$B = \frac{RT}{b_T} \tag{15}$$

where R is the gas constant (8.341 J mol^{-1} K^{-1}), T is the absolute temperature (K), and A and B represent isotherm constants. K_T is the equilibrium binding constant (L/g); b_T is related to the heat of adsorption (J/mol).

Dubinin–Radushkevich Isotherm

The Dubinin–Radushkevich isotherm is applicable to the adsorption mechanism based on the potential theory assuming a heterogeneous surface, and its linearized form is given in the following equations:[20,21]

$$\ln q_e = \ln q_m - K\varepsilon^2 \tag{16}$$

$$\varepsilon = RT \, \ln\left(1 + \frac{1}{Ce}\right) \tag{17}$$

where K, a constant, is related to the adsorption energy and e (kJ/mol) is used to estimate the type of adsorption process (chemical or physical adsorption) and calculated from the following equation:

$$\varepsilon = \left(-2K^{-\frac{1}{2}}\right) \tag{18}$$

If the ε values of the isotherm fall between 8 and 16 kJ/ mol, the adsorption reaction can be explained by an ion-exchange mechanism, whereas if its value is less than 8 kJ/mol, it indicates that the adsorption process has a physical nature.[22]

Redlich–Peterson Isotherm

This isotherm is widely used as a compromise between the Langmuir and Freundlich systems.[23] In this equation, three parameters have been used to incorporate the advantageous significance of both models. The model can be represented as follows:[24]

$$q_e = \frac{K_{RP}C_e}{1 + (\alpha C_e)^\beta} \tag{19}$$

where K_{RP} (L/g), α (L/mg), and β are Redlich–Peterson (R-P) isotherm constants, whereas β is the exponent that lies between 0 and 1. The R-P model has two limiting cases. When $\beta = 1$, the Langmuir equation results, whereas when $\beta = 0$, the R-P equation transforms to Henry's law equation.

Adsorption Kinetics

Kinetics is the study of the rates of chemical processes by monitoring the experimental conditions that influence the speed of a chemical reaction and help attain equilibrium in a reasonable length of time. The possible mechanism of adsorption and the different transition states of the final adsorbate–adsorbent complex are normally studied, and the appropriate mathematical models to describe the interactions are applied to the reaction.

The adsorption kinetics is useful in understanding the complex dynamics of the adsorption process. Generally, four consecutive steps describing the occurrence of adsorbate transport are as follows[6]

1. Transport of the adsorbate from the bulk solution to the boundary layer surrounding the adsorbent particles
2. Transport of solute across the boundary layer

3. Intraparticle solute diffusion into the pores
4. Adsorption and desorption of adsorbate

One of the above steps, or a combination of them, controls the adsorption mechanism. Factors influencing the rate-limiting step include characteristics of the adsorbent, adsorbate, and solution, for instance, adsorbent particle size, adsorbate concentration, degree of mixing, affinity between adsorbate and adsorbent, and diffusion coefficients of the adsorbate.

Adsorption kinetics can be determined by the following stages: diffusion of molecules from the bulk phase toward the interface space or external diffusion, diffusion of molecules inside the pores or internal diffusion, diffusion of molecules in the surface phase or surface diffusion, and adsorption–desorption elementary processes.

The most popular models used to establish the controlling adsorption mechanism can be grouped as follows:[25]

1. Those assuming that the controlling step is mass transfer (homogeneous surface diffusion, pore diffusion, and heterogeneous diffusion models).
2. Those assuming that adsorption is governed by surface phenomena.

The models involved in an adsorption kinetics study are the Lagergren pseudo–first-order and pseudo–second-order model, the Elovich equation, intraparticle diffusion, and liquid film diffusion.

Lagergren Pseudo–First-Order Model

This equation describing the rate of adsorption in the liquid-phase systems is the most used equation, particularly for pseudo–first-order kinetics.[26] The linearized Lagergren equation considers a reversible equilibrium of organic molecules between a liquid and a solid phase. Many different adsorption situations can be described by pseudo-first–order kinetics, including the following: 1) systems close to equilibrium; 2) systems with time-independent solute concentration or linear equilibrium adsorption isotherm; and 3) special cases of more complex systems.[6]

The Lagergren equation as pseudo–first-order kinetics is described as follows:[7]

$$\frac{dq_t}{dt} = k_1 \left(q_e - q_t \right) \tag{20}$$

where k_1 (min^{-1}) is the pseudo–first-order adsorption rate coefficient. The integrated form of this equation for the boundary conditions of $t = 0$, $q_t = 0$ and $t = t$, $q_t = q_t$ is

$$\ln\left(q_e - q_t \right) = \ln q_e - k_1 t \tag{21}$$

where q_e and q_t are the values of amount adsorbed per unit mass at equilibrium and at any time t. The values of k_1 can be obtained from the slope of the linear plot of ln $(q_e - q_t)$ vs. t.

McKay et al.[27] reported that the Lagergren pseudo–first-order model is found suitable only for the initial 20 to 30 min of interaction and not fit for the whole range of contact time. As the value of k_1 depends on the initial concentration of the adsorbate, it usually decreases with the increasing initial adsorbate concentration in the bulk phase.[28–31]

Pseudo–Second-Order Model

The pseudo–second-order equation has been explained as a special kind of Langmuir kinetics with the following assumptions: 1) the adsorbate concentration is constant in time; and 2) the total number of binding sites depends on the amount of adsorbate adsorbed at equilibrium.[32]

The pseudo–second-order equation based on equilibrium adsorption is expressed as

$$\frac{dq_t}{dt} = k_2 \left(q_e - q_t\right)^2 \tag{22}$$

The initial adsorption rate of a second-order process as $t \to 0$ is defined as

$$\frac{dq}{\left(q_e - q_t\right)^2} = k_2 dt \tag{23}$$

The integrating equation with respect to boundary conditions $q_t = 0$ at $t = 0$ and $q = q_e$ at $t = t$ is

$$\frac{t}{q} = \frac{1}{k_2 q_e^2} + \frac{1}{q_e} t \tag{24}$$

where k_2 (g/mg h) is the rate constant of second-order adsorption. It depends on the applied operating conditions, namely, initial metal concentration, pH of solution, temperature and agitation rate, etc.[33–34] The integral form of the model predicts that the ratio of the time/adsorbed amount should be a linear function of time.[35] The linear plot of t/q_t vs. t gives $1/q_e$ as the slope and $1/k_2 q_e^2$ as the intercept. This procedure tends to predict the behavior over the whole range of adsorption.

Several research works[30–31,34,36] reported that the rate coefficient, k_2, decreases with the increasing initial adsorbate concentration in the bulk phase. The higher the initial concentration of adsorbate, the longer the time required to reach an equilibrium and, consequently, the k_2 value decreases.

The second-order rate constants were used to calculate the initial sorption rate given by

$$h = k_2 q_e^2 \tag{25}$$

where h is the initial adsorption rate, q_e is the adsorption capacity, and k_2 is the pseudo–second-order rate coefficient. The value of k_2 can be determined experimentally from the slope and intercept of a plot of t/q_t against t.

It has been reported[24] that the initial adsorption rate, h, was found to increase with initial concentration; however, the value started to decrease when the high initial concentration was applied to the system. The possible reason might be that too-high solute concentrations would slow down the adsorption process. The h value could also be influenced by the characteristics of the adsorbent and adsorbate.

One of the advantages of the pseudo–second-order equation for estimating the qe values is its small sensitivity for the influence of the random experimental errors.[7]

Elovich Equation

The Elovich equation is one of the most widely used models for describing chemical adsorption. It assumes that the actual solid surfaces are energetically heterogeneous and that neither desorption nor interactions between the adsorbed species could substantially affect the kinetics of adsorption at low surface coverage.[7] The Elovich equation is also restricted to the initial sorption stages, when the system is relatively far from equilibrium.

The Elovich equation can be expressed as follows:[37]

$$\frac{dq_t}{dt} = \alpha \, \exp\left(-\beta q_t\right) \tag{26}$$

Where α and β represent the initial adsorption rate (g/mg min²) and the desorption coefficient (mg/g min), respectively. Assuming $\alpha \beta_{qt} \gg 1$, and $q_t = 0$ at $t = 0$ and $qt = q_t$ at $t = t$, the linear form of the above equation id given by [38–40]

$$q_t = \beta \ln(\alpha\,\beta) + \beta \ln t \tag{27}$$

The Elovich coefficients could be computed from the plots of qt vs. ln t. It is suggested that Elovich equation is restricted to the initial part of the adsorbate–adsorbent interaction process, when the system is relatively far from equilibrium.[7] The pseudo–second-order and the Elovich equations exhibit closely related behaviors for values of fractional surface coverage up to 0.7.[35]

Intraparticle Diffusion Model

This model is mostly used with the porous adsorbent. As the adsorbate molecules or ions diffuse into the pores, the adsorption kinetics relating to this intraparticle diffusion has to be studied with the proper model. The intraparticle diffusion model can be derived from Fick's second law. The simplified form of this equation as a dimensionless equation was proposed to assess the initial adsorption stages.[41–43] The intraparticle diffusion equation can be expressed as follows:

$$\frac{q_t}{q_e} = 1 - \left(\frac{6}{\pi^2}\right)\Sigma\left(\frac{1}{n^2}\right)\exp\left(\frac{-n^2\pi^2 D_C t}{r^2}\right) \tag{28}$$

where D_C is the intracrystalline diffusivity, r is the particle radius, t is the reaction time, and the summation is carried out from $n = 1$ to $n = $ a. The ratio, q_t/q_e, is the fractional approach to equilibrium.

The simplified form is rewritten below:

$$1 - \frac{q_t}{q_e} = \left(\frac{6}{\pi^2}\right)\exp\left(\frac{-\pi^2 D_C}{r^2}\right)t \tag{29}$$

or

$$\ln\left(1 - \frac{q_t}{q_t}\right) = \left(\frac{-\pi^2 D_C}{r^2}\right)t + \ln\left(\frac{6}{\pi^2}\right) \tag{30}$$

This equation can be expressed in the linear line with the Plot of $\ln\left(1 - \frac{q_t}{q_e}\right)$ vs. t. A slope of $\left(\frac{-\pi^2 D_C}{r^2}\right)$ is

the value of the diffusion time constant, and it is expressed as follows:

$$k' = \frac{\pi^2 D_C}{r^2} \tag{31}$$

where k' is the overall rate constant, inversely proportional to the square of the particle radius. A simpler expression to obtain the diffusion rate coefficient, k, is written as follows:

$$q_t = k_{it}^{0.5} \tag{32}$$

The linear plots of q_t vs. $t^{0.5}$ should pass through the origin (zero intercept) indicating a controlling influence for the diffusion process on the kinetics. The slope of the plot is the rate coefficient, k_i (mg/g-min$^{0.5}$). This equation represents a simplistic approximation of the pore diffusion kinetics without considering the possible impacts of the pore dimensions.

Liquid Film Diffusion Model

This equation is applied when the flow of the reactant through the liquid film surrounding the adsorbent particles is the slowest process. This equation determines kinetics of the rate process, and it is written as follows:[44]

$$\ln(1-F)=-k_{fd}t \qquad (33)$$

where F is the fractional attainment of equilibrium $(=q_t/q_e)$ and k_{fd} (min^{-1}) is the film diffusion rate coefficient. A linear plot of $-\ln(1-F)$ vs. t with zero intercept suggests that the kinetics of the adsorption process is controlled by diffusion through the liquid film.

Adsorbents and Their Applications in Water and Wastewater Treatment

Carbon Adsorbents

There are several types of carbon adsorbents used in the adsorption process, for example, activated carbon, activated carbon fibers, fullerene, etc. Activated carbon is the most widely used carbon adsorbent. It is a crude form of graphite with a random or amorphous highly porous structure with a broad range of pore sizes, from visible cracks and crevices to crevices of molecular dimensions.[45] Generally, it is the carbon material derived mostly from charcoal. Besides coal, agricultural by-products are conventional sources of commercial activated carbon. Many types of agricultural wastes were proposed as raw materials for producing carbon adsorbents, including cork;[46–47] sucrose chars;[48] corncob;[49–50] jackfruit peels;[51–52] wood;[53–54] oil palm;[18] stones; fruit shells, coats, and husks;[48,55–64] and wastes from cherries,[65–66] plums,[67] coconut, apricot, almond, and nuts.[66–68] Conditions in synthesizing activated carbon from these materials are provided in Table 2. The comparison of adsorption capacity of each type of activated carbon is shown in Table 3. Sources of the raw materials used and the preparation and treatment conditions such as pyrolysis temperature and activation time are major factors that have an effect on the adsorption capacity of activated carbon. Many other factors can also affect the adsorption capacity in the same sorption conditions such as surface chemistry (heteroatom content), surface charge, and pore structure.

The adsorbent properties of activated carbon depend on their composition, physicochemical properties, and mechanical strength. Activation by physical means, by chemical means, or by a combination of both has been employed to control the pore size and distribution of activated carbon and/or to increase porosity, surface modification, and improvement of carbonization.[92–98] Normally, activated carbons are made up of small hydrophobic graphite layers with disordered, irregular, and heterogeneous surfaces bearing hydrophilic functional groups. The surface chemistry of activated carbon depends mainly upon the activation conditions and temperatures employed. The activated carbon has strong heterogeneous surfaces. Its geometrical heterogeneity is the result of differences in size and shape of pores, and cracks, pits, and steps. The chemical heterogeneity is involved with different functional groups, mainly oxygen groups that are located most frequently at the edges of the crystallites among with various surface impurities. These heterogeneity surfaces contribute to the unique sorption properties of activated carbon.[12,99–100] The presence of oxygen and hydrogen in surface groups affects strongly the adsorptive properties of the activated carbon. The apparent chemical character of an activated carbon surface is determined by functional groups and delocalized electrons of the graphitic structure.[101] Oxygen on an activated carbon surface may be present in various forms, such as carboxyls, aldehydes, ketones, phenols, lactones, quinines, hydroquinones, anhydrides, or ethereal structures. Some of the groups, e.g., carbonyl, carboxyl, phenolic hydroxyl, and lactonic ones, are acidic. Consequently, the pH value of the liquid bulk phase can have an effect on the acidic and/or basic functional groups of the carbon

TABLE 2 Conditions in Synthesizing Activated Carbon from Natural Materials

Natural Material	Conditions	Refs.
Algerian coal	930°C with KOH/NaOH	Alvim Ferraz[69]
Almond and pecan shells	Chemical activation with H_3PO^4/physical CO_2	Tancredi et al.[70]
Almond shell, olive stones, and peach stones	Heating in CO_2 at 606°C	Ferro-Garcia et al.[71]
Bituminous coal	N^2/400–700°C with $ZnCl_2$	Hall and Holmes[72]
Coal or coconut shell	Phosgene or chlorine gas at 180°C	Otowa et al.[73]
Coconut shell	Parts by weight H2SO4 for 24 h at 150°C	Manju et al.[74]
Coconut shell	450°C with H3PO4	Laine et al.[75]
Coconut shells and coconut shell fibers	Carbonized with H2SO4 and activated at 600°C for 1 hr	Mohan et al.,[76] Mohan et al.,[77] Mohan et al.,[78] and Mohan et al.[79]
Eucalyptus wood chars	CO_2 activation, 400–800°C	Kumar and Sivanesan[80]
Fertilizer slurry	450°C, 1 hr with H_2O_2/H_2O, N_2	Marungrueng and Pavasant[81]
Fly ash	Froth flotation, hydrophobic char was separated from hydrophilic ash with the help of methyl isobutyl ketone	Basava Rao and Mohan Rao[82]
Lignite	Inert atmosphere/600–800°C with Na_2MoO_4/ $NaWO_4$/ NH_4VO_3/$(NH_4)_2MoO_4$/$FeCl_3$/ $Fe(NO_3)_3$	El Qada et al.[83]
Oat hulls	Fast pyrolysis at 500°C with inert nitrogen	Tamai et al.[84]
Palm tree cobs	730°C, 6 hr with H_3PO_4/H_2SO_4	Banat et al.[85]
Petroleum coke	700–850°C, 4 hr with KOH/H^2O	McKay et al.[86]
	KOH dehydration at 400°C followed by activation at 500–900°C	McKay et al.[87]
Pine sawdust	850°C, 1 hr; 825°C, 6 hr with $Fe(NO_3)_3$/CO_2	Kannan and Sundaram[88]
Raffination earth	10% (v/v), 350°C with H_2SO_4	Bestani et al.[89]
Solvent-extracted olive pulp and olive stones	Under vacuum and atmospheric pressure; 60°C/min; 800°C; activation under N_2 at 10°C/min with K_2CO_3	Stavropoulos and Zabaniotou[90]

TABLE 3 Adsorption Capacity of Each Type of Activated Carbon

Adsorbent	Adsorption Capacity (mg/g)	Refs.
Activated carbon	400	Kumar and Sivanesan[80]
	238	Marungrueng and Pavasant[81]
	9.81	Basava Rao et al.[82]
Activated carbon produced from New Zealand coal	588	El Qada et al.[83]
Activated carbon produced from Venezuelan bituminous coal	380	El Qada et al.[83]
Bituminous coal	176	Tamai et al.[84]
Charcoal	62.7	Banat et al.[85]
Coal	323.68	McKay et al.[86]
	230	McKay et al.[87]
Commercial activated carbon	980.3	Kannan and Sundaram[88]
	200	Bestani et al.[89]
Peat	324	Fernandes et al.[91]

surface. Thus, the surface charge of carbon is a function of pH of the solution. Considering the point of zero charge (PZC) and the isoelectric point (IEP), the surface is positively charged at pH < pH_{PZC} and negatively charged at pH > pH_{PZC}. In practice, pH_{IEP} is usually close to pHPZC, but it is lower than pHPZC for activated carbon.[102] For pH < pK_a adsorption of non-ionized organics does not depend on the surface charge of activated carbon. However, for pH > pKa, the adsorption of its ionic form depends on the surface charge. As a result, the activated carbon possesses perfect adsorption ability for relatively low-molecular-weight organic compounds from drinking water and wastewater streams.

Several methods have been used to removal organic pollutants from water. However, the use of activated carbons is perhaps the best broad-spectrum technology available at present.[1] Accordingly, the use of activated carbons in water treatment has increased tremendously. Generally, the three main physical carbon types are granular, powder, and extruded (pellet). The granular activated carbon (GAC) adsorption, the most widely used type, is an effective treatment technology for organic contaminant removal from drinking water to improve taste and odor. The use of GAC for treatment of municipal and industrial wastewaters has developed rapidly in the last three decades from small size for household units to large scale for industrial wastewater application. Moving beds, downflow fixed beds, and upflow expanded beds have been widely used for water purification for industry.

It is well known that activated carbon can remove several types of pollutants including metal ions,[102–104] phenols,[46,105] pesticides,[106] chlorinated hydrocarbons[107] detergents,[108] and many other chemicals and organisms. Application of activated carbon in removal of various heavy metals and organic contaminants with the Langmuir and Freundlich capacities is shown in Table 4.

Clay Minerals

Clay minerals are hydrous aluminosilicates composed of minerals that make up the colloid fraction (<2 pm) of soils, sediments, rocks, and water[117] and may be composed of mixtures of fine-grained clay minerals and clay-sized crystals of other minerals such as quartz, carbonate, and metal oxides. Their structures are similar to micas with the formation of flat hexagonal sheets. Clay minerals and oxides are widespread and abundant in aquatic and terrestrial environments.

Clay contains various types of exchangeable ions on its surface. The prominent ions found on the clay surface are Ca^{2+}, Mg^{2+}, H+, K+, NH^{4+}, Na+, $SO4^{2-}$, Cl-, $PO4^{3-}$, and NO3-. These ions can be exchanged

TABLE 4 Adsorption Capacities of Activated Carbon for Heavy Metal and Organic Contaminant Removal from Water and Wastewater

Pollutant	Activated Carbon	Adsorption Capacity (mg/g)	Isotherm	Refs.
Cr(VI)	Commercial activated carbon	4.7	Langmuir	Babel and Kurniawan[109]
	Commercial activated carbon oxidized with H2SO4	8.9	Langmuir	Babel and Kurniawan[109]
	Commercial activated carbon oxidized with HNO$_3$	10.4	Langmuir	Babel and Kurniawan[109]
Fe(III)	Granular activated carbon	0.1	Freundlich	Kim[110]
Ni(II)	Granular activated carbon	6.5	Langmuir	Satapathy and Natarajan[111]
	Modified activated carbon	7.0	Langmuir	Satapathy and Natarajan[111]
Catechol	Activated charcoal	320	Langmuir	Richard et al.[112]
Gallic	acid Activated charcoal	408–488	Langmuir	Figaro et al.[113]
Tannin	Activated charcoal	0.39	Langmuir	Mohan and Karthikeyan[114]
Vanillin	Activated charcoal	93.18–121.72	Langmuir	Michailof et al.[63]
Phenol	Rice husk activated carbon	27.58	Langmuir	Kalderis et al.[115]
Nonylphenol	Activated charcoal	83.1	Langmuir	Lang et al.[116]

with other ions easily without affecting the structure of the clay mineral.[118] Clay can adsorb the cationic, anionic, and neutral metal species. They act as a natural scavenger of pollutants by taking up cations and/or anions through either ion exchange or adsorption, or both.

Currently, several types of clay minerals such as mont-morillonite and kaolinite are widely used in the water purification process. Because of their low cost, abundance in most continents of the world, high sorption properties, and potential for ion exchange, clay materials are strong adsorbents. Montmorillonite is a clay mineral with substantial isomorphic substitution. It is composed of units made up of two silica tetrahedral sheets with a central alumina octahedral sheet. The theoretical composition without the interlayer material is SiO_2, 66.7%; Al_2O_3, 28.3%; and H_2O, 5%. There is substitution of Si^{4+} by Al^{3+} in the tetrahedral layer and of Al^{3+} by Mg^{2+} in the octahedral layer. Exchangeable cations in the 2:1 layers balance the negative charges generated by isomorphic substitution. The uptake kinetics of cation exchange is fast, and the cations such as Na+ and Ca2+ form outer-sphere surface complexes, which are easily exchanged with solute ions by varying the cationic composition of the solution.

Kaolinite is the least reactive clay. It has the theoretical composition of SiO_2, 46.54%; Al_2O_3, 39.50%; and H_2O, 13.96%, expressed in terms of the oxides. It has a small net negative charge, which is responsible for the surface not being completely inert. Its high pH dependency enhances or inhibits the adsorption of metals according to the pH of the environment.[119] The metal adsorption is usually accompanied by the release of hydrogen (H^+) ions from the edge sites of the mineral. The substitution of H^+ ions for metal ions could influence the van der Waals force within the kaolinite structure.

Their applications are mainly found in dye and heavy metal removal. From previous research, it was reported that the sorption capacity of clay minerals can vary strongly with pH. Gupta and Bhattacharyya[120–121] used kaolinite and montmorillonite along with their poly(oxo zirconium) and tetrabutylammonium derivatives for Cd(II) removal from water. The adsorption of Cd(II) was influenced by pH of the aqueous medium, and the amount adsorbed increased with gradually decreasing acidity. By increasing the solution pH from 1.0 to 10.0, the extent of adsorption increased from 4.3% to 29.5% for kaolinite and 74.7% to 94.5% for montmorillonite. In dye removal, Bagane and Guiza[122] reported an adsorption capacity of 300 mg/g and suggested that clay is a good adsorbent for methylene blue removal due to its high surface area. Almeida et al.[123] studied the removal of methylene blue from synthetic wastewater by using montmorillonite and described it as an efficient adsorbent where the equilibrium was attained in less than 30 min. The adsorption of dyes on kaolinite was also studied by Ghosh and Bhattacharyya,[124] who reported that its adsorption capacity can be improved by purification and by treatment with NaOH solution.

The adsorption capacities vary from metal to metal and also depend on the type of clay used.[118] When a comparison is made with other low-cost adsorbents, the clays have been found to be either better or equivalent in adsorption capacity. Type of pollutant and adsorption capacity of each clay mineral are summarized in Table 5.

Natural Zeolites

Zeolites are highly porous aluminosilicates with different cavity structures. They consist of a three-dimensional framework, having a negatively charged lattice. A well-defined pore structure in the microporous range of zeolite can accommodate a wide variety of cations such as Na+, K+, Ca2+, Mg2+, and others. These charge-compensating cations are free to migrate in and out of zeolite structures, and they are rather loosely held so that they can readily be exchanged for others in a contact solution. Accordingly, zeolites are not only good adsorbates but also good ion exchangers. This property can be used to introduce different cations into the structure, creating selective sites for adsorption purposes or catalysis. Their narrow pore size and tuneable affinity for certain molecules make them ideal adsorbents for selective purification to encapsulate hazardous compounds. Zeolites are characterized not only by a high selectivity separation mechanism but also by the ability to separate substances based on differences in sizes and shapes of molecules' steric separation mechanism.

TABLE 5 Adsorption Capacities of Clay Minerals for Heavy Metal Removal from Water and Wastewater

Pollutant	Clay Mineral	Langmuir Capacity	Freundlich Capacity	Refs.
Cd(II)	Kaolinite	9.9	0.5	Gupta and Bhattacharyya[125]
	Montmorillonite	32.7	8.6	Gupta and Bhattacharyya[125]
Ni(II)	Acid-activated montmorillonite	29.5	6.0	Bhattacharyya and Gupta[126]
	Kaolinite	10.4	1.1	Gupta and Bhattacharyya[127]
	Montmorillonite	28.4	4.5	Gupta and Bhattacharyya[127]
Cr(VI)	Kaolinite	11.6	–	Bhattacharyya and Gupta[128]
	Acid-activated kaolinite	13.9	–	Bhattacharyya and Gupta[128]
Co(II)	Raw kaolinite	11.5	–	Yavuz et al.[129]
	Kaolinite	11.2	1.1	Bhattacharyya and Gupta[130]
	Acid-activated kaolinite	12.1	1.5	Bhattacharyya and Gupta[130]
	Montmorillonite	28.6	4.6	Bhattacharyya and Gupta[130]
	Acid-activated montmorillonite	29.7	6.0	Bhattacharyya and Gupta[130]
Pb(II)	Kaolinite	11.2	0.7	Gupta and Bhattacharyya[131] and Bhattacharyya and Gupta[132]
	Acid-activated kaolinite	12.1	1.0	Gupta and Bhattacharyya[131] and Bhattacharyya and Gupta[132]
	Montmorillonite	33.0	8.9	Gupta and Bhattacharyya[131] and Bhattacharyya and Gupta[132]
	Acid-activated montmorillonite	34.0	11.3	Gupta and Bhattacharyya[131] and Bhattacharyya and Gupta[132]
Fe(III)	Kaolinite	11.2	1.3	Bhattacharyya and Gupta[133]
	Acid-activated kaolinite	12.1	1.7	Bhattacharyya and Gupta[133]
	Montmorillonite	28.9	5.2	Bhattacharyya and Gupta[133]
	Acid-activated montmorillonite	30.0	6.4	Bhattacharyya and Gupta[133]
Cu(II)	Kaolinite	4.4	1.1	Bhattacharyya and Gupta[134]
	Acid-activated kaolinite	5.6	1.3	Bhattacharyya and Gupta[126]
	Montmorillonite	25.5	9.2	Bhattacharyya and Gupta[134]
	Acid-activated montmorillonite	28.0	12.4	Bhattacharyya and Gupta[126]

Note: Units of Langmuir capacity and Freundlich capacity are mg/g and $mg^{1-1/n} L^{1/n}/g$, respectively.

Zeolites have been widely used for pollution control due to their ion exchange and adsorption properties. They have been used for the selective separation of cations from aqueous solution. The diffusion, adsorption, and ion exchange in zeolites have been extensively reviewed in many previous works.[135–137] Kesraoui-Ouki, Cheeseman, and Perry[138] reviewed natural zeolite utilization in metal effluent treatment applications. Dewatered zeolites produce channels that can adsorb molecules small enough to access the internal cavities while excluding larger species. Zeolites, modified by ion exchange, can be used for adsorption of different metal ions according to requirements and costs. The characteristics and applications of zeolites have been extensively reviewed by Ghobarkar, Schaf, and Guth.[139] High ion-exchange capacity and relatively high specific surface areas, and more importantly, their relatively cheap prices, make zeolites more attractive adsorbents.

Besides zeolite, other siliceous materials such as perlite and glass have been proposed for contaminant removal. The use of natural siliceous adsorbents such as silica, glass fibers, and perlite for wastewater is increasing because of their high abundance, easy availability, and low cost. The other commonly applied inorganic sorbents are silica gels, activated alumina, and oxide and hydroxide metals. Perlite is another siliceous material that exhibits a good adsorbent for decontamination purposes. It has been used as a low-cost adsorbent for the removal of methylene blue.[140,141] Methylene blue is physically adsorbed onto the perlite. However, perlites of different types (expanded and unexpanded) and of different origins have different properties because of the differences in composition. Chakrabarti and Dutta[142] also investigated glass fiber for the adsorption of methylene blue. They stated that a considerable amount of the dye is adsorbed on soft glass even at ambient temperature. Accordingly, several siliceous materials become widely used as adsorbate materials in the adsorption process.

Currently, a new family of mesopore materials, so-called MCM materials or Mobil Composition of Matter (MCM), was developed by Mobil Oil Corporation, which proposed a revolutionary synthesis method to obtain such materials that comprise strictly uniform pores. An organic surfactant like an alkyltrimethylammonium bromide in an aqueous medium forms rod-like micelles, which are used as templates to form two or three monolayers of silica or alumina particles encapsulating the micelles' external surface. By removing the organic species from a well-ordered organic-inorganic condensed phase, a porous silicate or alumina material with uniformly porous structure remains. The mesopore size can be controlled by the molecular size template of the surfactant. Nowadays, MCM materials have been widely used in heavy metal removal, and they are currently the adsorption material that plays an important role in water and wastewater treatment.

Chitin and Chitosan

Chitin is a nontoxic, biodegradable polymer of high molecular weight. It contains 2-acetamido-2-deoxy-β-D-glucose through a β (1→4) linkage. Chitin is the most abundant natural fiber next to the cellulose and is similar to cellulose in many respects. The most abundant source of chitin is the shell of crab and shrimp. Chitin has presented exceptional chemical and biological qualities that can be used in water and wastewater purification through the adsorption process.

Chitin and chitosan have their chemical structures in common. Chitin is made up of a linear chain of acetyl-glucosamine groups. Chitosan is obtained by removing enough acetyl groups (CH_3–CO) for the molecule to be soluble in most diluted acids. This process, called deacetylation, releases amine groups (NH) and gives the chitosan a cationic characteristic. Chitosan contains 2-acetamido-2-deoxy-β-D-glucopyranose and 2-amino-2-deoxy-β-D-glucopyranose residues. Chitosan is known as an ideal natural support for enzyme immobilization because of its special characteristics such as hydrophilicity, biocompatibility, biodegradability, non-toxicity, adsorption properties, etc.[143]

Chitosan has drawn particular attention as an effective biosorbent due to its high content of amino and hydroxyl functional groups, giving it high adsorption potential for various aquatic pollutants.[143–147] This biopolymer represents an attractive alternative to other biomaterials because of its physicochemical characteristics, chemical stability, high reactivity, excellent chelation behavior. and high selectivity toward pollutants. Chitin and chitosan derivatives have been extensively investigated as adsorbents for the removal of organic molecules and metal ions from water and wastewater. The high adsorption potential of chitosan can be attributed to the following: 1) high hydrophilicity due to a large number of hydroxyl groups of glucose units; 2) presence of a large number of functional groups; 3) high chemical reactivity of these groups; and 4) flexible structure of the polymer chain.[148,149]

To enhance the adsorption capacity for pollutant removal, chitosan has been modified by several methods, either physical or chemical processes. Different shapes of chitosan, e.g., membranes, microspheres, gel beads, and films, have been synthesized and tested for their performance in pollutant removal from water and wastewater.[143–147] A cross-linked chitosan bead is one type of chemical modification for chitosan to increase the uptake capacity in the adsorption process.[150] This method using the chemical reaction of ethylenediamine and carbodiimide in modifying chitosan provided a high uptake capacity for mercury (Hg^{2+}) ions, which is considered to be one of the highest uptake capacities among various biosorbents.

Beads of 1 and 3 mm diameter were prepared as one type of modified chitosan.[151] The gelled chitosan beads were cross-linked with glutaraldehyde and then freeze-dried. Beads of 1 mm diameter possessed surface areas exceeding 150 m²/g and mean pore sizes of 560 Å and were insoluble in acid media at pH 2. A new composite chitosan biosorbent was also prepared by coating chitosan onto perlite ore. It was used in the removal of Cu(II) and Ni(II) from aqueous solution.[152] The magnetic chitosan nanocomposites were synthesized on the basis of amine-functionalized magnetite nanoparticles.[153] These nanocomposites provide a very efficient, fast, and convenient tool for removing Pb^{2+}, Cu^{2+}, and Cd^{2+} from water. It was suggested that synthesized magnetic chitosan nanocomposites can be used as a recyclable tool for heavy metal ion removal. Several types of heavy metals and organic contaminants removed by chitosan are shown in Table 6.

Agricultural-Based Waste Materials

Agricultural by products usually are composed of lignin and cellulose as major constituents that have the ability to some extent to bind some type of pollutants, for example, heavy metals, by donation of an electron pair from these groups to from complexes with the metal ions.[166] Currently, many types of agricultural-based waste materials play a significant role in the adsorption process. They are normally organic materials from plants, trees, crops, and algae. Two larger carbohydrate that play a significant role in the adsorption process are cellulose and hemicelluloses (holocellulose.) Cellulose is a remarkable pure organic polymer, consisting solely of units of anhydroglucose held together in a giant straight-chain molecule.[168] These anhydroglucose units are bound together byβ-(1,4)-glycosidic linkages. Hemicelluloses consist of different monosaccharide units. The polymer chains of hemicelluloses have

TABLE 6 Adsorption Capacities of Chitosan and Its Composite for Removal of Heavy Metals and Some Organic Contaminants from Water and Wastewater

Pollutant	Chitosan	Adsorption Capacity (mg/g)	Isotherm	Refs.
Hg(II)	Chitosan/cotton fibers	104.31	Langmuir	Qu et al.[154]
Cd(II)	Chitosan/cotton fibers	15.74	Langmuir	Zhang et al.[155]
Cr(VI)	Magnetic chitosan	69.40	Langmuir	Huang et al.[156]
	Chitosan/cellulose	13.05	Langmuir	Sun et al.[157]
	Chitosan/perlite	153.8	Langmuir	Shameem et al.[158]
	Chitosan/ceramic alumina	153.8	Freundlich	Veera et al.[159]
Pb(II)	Chitosan/cotton fibers	101.53	Freundlich	Zhang et al.[155]
	Chitosan/magnetite	63.33	Langmuir	Tran et al.[160]
	Chitosan/cellulose	26.31	Langmuir	Sun et al.[157]
	Chitosan/sand	12.32	Langmuir	Rorrer et al.[151]
Cu(III)	Chitosan/cellulose	26.50	Langmuir	Sun et al.[157]
	Chitosan/perlite	196.07	Langmuir	Kalyani et al.[152]
	Chitosan/polyvinylchloride	87.9	Langmuir	Srinivasa et al.[162]
Ni(II)	Chitosan/magnetite	52.55	Langmuir	Tran et al.[160]
	Chitosan/cellulose	13.21	Langmuir	Sun et al.[157]
	Chitosan/perlite	114.94	Langmuir	Kalyani et al.[152]
	Chitosan/silica	254.3	Langmuir	Vijaya et al.[163]
Phenol	Chemically modified chitosan	2.22–151.50	Langmuir	Li et al.[164]
	Chitosan/calcium alginate beads	108.69	Langmuir	Nadavala et al.[165]
4-Chlorophenol	Chemically modified chitosan	2.58–179.73	Langmuir	Li et al.[164]
Nonylphenol	Chitosan	56.3	Langmuir	Lang et al.[116]

short branches and are amorphous. Hemicelluloses are derived mainly from chains of pentose sugars and act as the cement material holding together the cellulose micelles and fiber.[168] Hemicelluloses are partially soluble in water. Currently, chemical modification is widely used to alter the biochemical component of the biomaterials to obtain higher efficiency in pollutant removal by biosorption process.[169] Biomass chemical modifications include delignification, esterification of carboxyl and phosphate groups, methylation of amino groups, and hydrolysis of carboxylate groups. Sawamiappan and Krishnamoorthy[170] replaced phenol–formaldehyde cationic matrices with sulfonated bagasse. Odozi et al.[171] polymerized corncob, sawdust, and onion. However, the disadvantages of chemical modification are a high expense to pay and unwanted problems, such as bleeding of excessive quantities of colored organic compounds, odor, and further pollution through the use of toxic chemicals. Several types of agricultural wastes have been used in the adsorption process, with the differences in adsorption capacity as shown in Table 7.

Mechanisms involved in the biosorption process include chemical adsorption, complexation, adsorption–complexation on surfaces and in pores, ion exchange, microprecipitation, heavy metal hydroxide condensation onto the biosurface, and surface adsorption.[172–174] In the adsorption process, functional groups are responsible for pollutant binding on the surface of biomaterial. Most of the functional groups involved in the binding process are found in cell walls. Plant cell walls are generally considered as structures built by cellulose molecules, organized in microfibrils and surrounded by hemicellulosic materials (xylans, mannans, glucomannans, galactans, arabogalactans), lignin, and

TABLE 7 Adsorption Capacities of Agricultural Waste for Heavy Metal And Organic Contaminant Removal from Water and Wastewater

Pollutant	Agriculture Waste	Adsorption Capacity (mg/g)	Isotherm	Refs.
Cd(II)	Juniper fiber	9.2	Langmuir	Min et al.[181]
	Base-treated juniper fiber	29.5	Langmuir	Min et al.[181]
Cr(VI)	Cactus	7.1	Langmuir	Dakiky et al.[182]
	Coconut shell carbon	2.2	Langmuir	Babel and Kurniawan[109]
	Coconut shell carbon oxidized with H2SO4	4.1	Langmuir	Babel and Kurniawan[109]
	Coconut shell carbon oxidized with HNO3	10.9	Langmuir	Babel and Kurniawan[109]
	Sawdust	15.8	Langmuir	Dakiky et al.[182]
Pb(II)	Carbonaceous adsorbent	25.0	Langmuir	Bhatnagar et al.[183]
	Sawdust	22.2	Langmuir, Freundlich	Taty-Costodes et al.[184]
Fe(III)	Maize cobs	2.5	Langmuir	Nassar et al.[185]
Cu(II)	Tree fern	7.6	Langmuir	Ho et al.[186]
Ni(II)	Peat	28.3	Langmuir, Freundlich	Chen et al.[187]
Phenol	Banana pith	49.9–129.4	Langmuir	Sathishkumar et al.[188]
	Banana peel	688.9	Langmuir	Achak et al.[189]
	Corn grain	256	Langmuir	Park et al.[190]
2-Nitrophenol	*Lessonia nigrescens*	71.28	Langmuir	Navarro et al.[191]
	Macrocystis integrifolia	97.37	Langmuir	Navarro et al.[191]
2,4-Dicholorophenol	Pomegranate peel	65.7	Langmuir	Bhatnagar and Minocha[192]
Nonylphenol	*Rhizopus arrhizus*	4.5–43.7	Langmuir	Lang et al.[116]

pectin along with small amounts of protein.[175] During biosorption, water is able to permeate the non-crystalline portion of cellulose and all of the hemicellulose and lignin. The aqueous solution comes into contact with a very large surface area of different cell wall components. The disordered structure of amorphous cellulose allows easier access to reagents than highly structured crystalline cellulose. While water penetrates through the cell wall components, water adsorption of fibers causes swelling. The bigger the amount of water adsorption, the bigger the swelling. Swelling also depends on the fiber's structure, on the degree of crystallinity, and on the amorphous and void regions.[176] Swelling occurs when polar solvents such as water and alcohols come into contact with wood.[177] These polar solvent molecules are attracted to the dry solid matrix and held by hydrogen bonding forces between the –OH or –COOH groups in the wood structure and cause the biosorption of pollutants in aqueous solution. Many research works[178–180] have reported the wide use of biosorption process in heavy metal removal. Thus, the agricultural-based waste materials become the adsorption material that plays an important role in water and wastewater treatment nowadays (Table 7).

Progress in Research on Adsorption Process in Water Purification

Adsorbents and adsorption processes have been widely studied and applied in different aspects for a long time. Owing to its effective, efficient, and economic approach to water purification, this process has been applied in removal of several contaminants, such as pesticides, halogenated carbon, dyes, phenol and its derivatives, and heavy metals. The most widely used adsorbent in the adsorption process is activated carbon. This adsorbent is highly inert and thermally stable, and it can be used over a broad pH range. Although it has a great capacity for adsorbing various organic compounds and can be easily modified by chemical treatment to increase its adsorption capacity, activated carbon has several disadvantages.[193] Owing to the process mechanism, adsorption transfers pollutants from one phase to another rather than eliminating them from the environment. Thus, after adsorption, the contaminants in liquid phase absorb on the surface of adsorbent, which has to be separated from aquatic system when it becomes exhausted or the effluent reaches the maximum allowable discharge level. Furthermore, the regeneration of exhausted activated carbon by a chemical and thermal procedure is also expensive and results in loss of the sorbent.

Recently, a lot of novel adsorption processes have been developed for enhancing the efficiency of removing organic and inorganic contaminants from water. The development of cheaper and more effective novel composite adsorbents[194–197] in comparison with the classical adsorbents has been investigated by researchers from many countries all over the world. These adsorbents are metal oxide-based composite adsorbents such as TiO_2 and MnO_2, surface-modified Fe_3O_4 adsorbent, magnetic particle-modified carbon adsorbent, magnetic particle–modified clay mineral adsorbent, and magnetic particle–modified biopolymer adsorbent. These composite materials deserve particular attention because they combine the properties and advantages of each of their components. They represent an interesting and attractive alternative as adsorbents and/or catalysts due to their high reactivity and excellent selectivity toward specific pollutant compounds. To obtain the anticipated function and enhance the efficiency of water purification, these adsorbents should be designed and modified in their compositions, structures, surfaces, and preparation methods to obtain the requirement of physicochemical properties for the purpose of adsorption. Extensive research in synthesizing of new adsorbents and investigating of adsorption mechanism is needed. Advances in development of new adsorbents for the adsorption process will be the progress of future technology in water purification.

References

1. Dabrowski, A. Adsorption, from theory to practice. Adv. Colloid Interface Sci. **2001**, *93*, 135–224.
2. Ahmad, A.; Rafatullah, M.; Danish, M. Removal of Zn(II) and Cd(II) ions from aqueous solutions using treated sawdust of sissoo wood. Holz Roh Werkst **2007**, *65*, 429–436.

3. Ahmad, A.; Rafatullah, M.; Sulaiman, O.; Ibrahim, M.H.; Chii, Y.Y.; Siddique, B.M. Removal of Cu(II) and Pb(II) ions from aqueous solutions by adsorption on sawdust of Meranti wood. Desalination **2009**, *247*, 636–646.

4. Rafatullah, M.; Sulaiman, O.; Hashim, R.; Ahmad, A. Adsorption of copper (II), chromium (III), nickel (II) and lead (II) ions from aqueous solutions by meranti sawdust. J. Hazard. Mater. **2009**, *170*, 969–977.

5. Gupta, V.K.; Carrott, P.J.M; Ribeiro Carrott M.M.L; Su-has. Low-cost adsorbents: Growing approach to wastewater treatment—A review. Crit. Rev. Environ. Sci. Technol. **2009**, *39*, 783–842.

6. Soto, M.L.; Moure, A.; Dominguez, H.; Parajò, J.C. Recovery, concentration and purification of phenolic compounds by adsorption: A review. J. Food Eng. **2011**, *105*, 1–27.

7. Gupta, S.S.; Bhattacharyya, K.G. Kinetics of adsorption of metal ions on inorganic materials: A review. Adv. Colloid Interface Sci. **2011**, *162*, 39–58.

8. Gupta, V.K.; Mohan, D.; Sharma, S. Removal of lead from wastewater using bagasse fly ash—A sugar industry waste material. Sep. Sci. Technol. **1998**, 33, 1331–1343.

9. Gupta, V.K.; Jain, C.K.; Ali, I.; Sharma, M.; Saini, V.K. Removal of cadmium and nickel from wastewater using bagasse fly ash—A sugar industry waste. Water Res. **2003**, 37, 4038–4044.

10. Parfitt, D.; Rochestor, H. Adsorption of small molecules. In *Adsorption from Solution at the Solid/ Liquid Interface;* Parfitt, G.D., Rochester, C.H., Eds.; Academic Press: New York, 1983; 3.

11. Giles, C.H.; Smith, D.; Huitson, A. A general treatment and classification of the solute adsorption isotherm. I. Theoretical. J. Colloid Interface Sci. **1974**, *47*, 755–765.

12. Dabrowski, A.; Podkoscielny, P.; Hubicki, Z.; Barczak, M. Adsorption of phenolic compounds by activated car-bon—A critical review. Chemosphere **2005**, *58*, 1049–1070.

13. Bansal, R.C.; Goyal, M. *Activated Carbon Adsorption*; CRC Press: Boca Raton, FL, 2005.

14. Bretag, J.; Kammerer, D.R.; Jensen, U.; Carle, R. Evaluation of the adsorption behavior of flavonoids and phenolic acids onto a food-grade resin using a D-optimal design. Eur. Food Res. Technol. **2009**, *228*, 985–999.

15. Weber, T.W.; Chakkravorti, R.K. Pore and solid diffusion models for fixed bed adsorbers. AIChE J. **1974**, *20*, 228–238.

16. Freundlich, H.M.F. Over the adsorption in solution. J. Phys. Chem. **1906**, 57, 385–470.

17. Fytianos, K.; Voudrias, E.; Kokkalis, E. Sorption-desorption behavior of 2,4-dichlorophenol by marine sediments. Chemosphere **2000**, *40*, 3–6.

18. Foo, K.Y.; Hameed, B.H. Insights into the modeling of adsorption isotherm systems. Chem. Eng. J. **2010**, *156*, 2–10.

19. Temkin, M.I.; Pyzhev, V. Kinetics of ammonia synthesis on promoted iron catalysts. Acta Phys. Chem. U.R.S.S. **1940**, *12*, 327–356.

20. Rashidi, F.; Sarabi, R.S.; Ghasemi, Z.; Seif, A. Kinetic, equilibrium and thermodynamic studies for the removal of lead (II) and copper (II) ions from aqueous solutions by nanocrystalline TiO_2. Superlattices Microstruct. **2010**, *48*, 577–591.

21. Dubinin, M.M.; Zaverina, E.D.; Radushkevich, L.V. Sorption and structure of active carbons. I. Adsorption of organic vapors. J. Phys. Chem. **1947**, *21*, 1351–1362.

22. Unlu, N.; Ersoz, M. Adsorption characteristics of heavy metal ions onto a low cost biopolymeric sorbent from aqueous solutions. J. Hazard. Mater. **2006**, *136*, 272–280.

23. Redlich, O.; Peterson, D.L. A useful adsorption isotherm. J. Phys. Chem. **1959**, *63*, 1024–1029.

24. Hameed, B.H.; Tan, I.A.W.; Ahmad, A.L. Adsorption isotherm, kinetic modeling and mechanism of 2,4,6 trichlorophenol on coconut husk-based activated carbon. Chem. Eng. J. **2008**, *144*, 235–244.

25. Ho, Y.S.; Ng, J.C.Y.; McKay, G. Kinetics of pollutant sorption by biosorbents: Review. Sep. Purif. Methods **2000**, *29*, 189–232.

26. Lagergren, S. Zur theorie der sogenannten adsorption gelöster stoffe. Kungliga Svenska Vetenskapsakademiens. Han-dlingar, Vetensk. Handl. **1898**, *24*, 1–39.

27. McKay, G.; Ho, Y.S.; Ng, J.C.Y. Biosorption of copper from wastewaters: A Review. Sep. Purif. Methods **1999**, *28*, 87–125.

28. Allen, S.J.; Gan, Q.; Matthews, R.; Johnson, P.A. Kinetic modeling of the adsorption of basic dyes by kudzu. J. Colloid Interface Sci. **2005**, *286*, 101–109.

29. Febrianto, J.; Kosasih, A.N.; Sunarso, J.; Ju, Y.-H.; Indras-wati, N.; Ismadji, S. Equilibrium and kinetic studies in adsorption of heavy metals using biosorbent: A summary of recent studies. J. Hazard. Mater. **2009**, *162*, 616–645.

30. Al-Ghouti, M.A.; Khraisheh, M.A.M.; Ahmad, M.N.M.; Allen, S. Adsorption behaviour of methylene blue onto Jordanian diatomite: A kinetic study. J. Hazard. Mater. **2009**, *165*, 589–598.

31. Nandi, B.K.; Goswami, A.; Purkait, M.K. Adsorption characteristics of brilliant green dye on kaolin. J. Hazard. Mater. **2009**, *161*, 387–395.

32. Lin. C. I.; Wang, L. H. Rate equations and isotherms for two adsorption models. J. Chin. Inst. Chem. Eng. **2008**, *39*, 579–585.

33. Plazinski, W.; Rudzinski, W.; Plazinska, A. Theoretical models of sorption kinetics including a surface reaction mechanism: A review. Adv. Colloid Interface Sci. **2009**, *152*, 2–13.

34. Ho, Y.S.; McKay, G. The kinetics of sorption of divalent metal ions onto sphagnum moss peat. Water Res. **2000**, *34*, 735–742.

35. Rudzinski, W.; Plazinski, W. On the applicability of the pseudo-second order equation to represent the kinetics of adsorption at solid/solution interfaces: A theoretical analysis based on the statistical rate theory. Adsorption **2009**, *15*, 181–192.

36. Ho, Y.S.; McKay, G. Kinetic model for lead(II) sorption on to peat. Adsorpt. Sci. Technol. **1998**, *16*, 243–255.

37. Aharoni, C.; Tompkins, F.C. Kinetics of adsorption and desorption and the Elovich equation. In *Advances in Catalysis and Related Subjects;* Eley, D.D., Pines, H., Weisz, P.B., Eds.; Academic Press: New York, **1970**; Vol. 21, 2–50.

38. Ho, Y.S.; McKay, G. Application of kinetic models to the sorption of copper(II) onto peat. Adsorpt. Sci. Technol. **2002**, *20*, 797–813.

39. Rudzinski, W.; Panczyk, T. The Langmuirian adsorption kinetics revised: A farewell to the 20th century theories? Adsorption **2002**, *8*, 23–34.

40. Chien, S.; Clayton, W.R. Application of Elovich equation to the kinetics of phosphate release and sorption in soils. Soil Sci. Soc. Am. J. **1980**, *44*, 265–268.

41. Ruthven, D.M. *Principles of Adsorption and Adsorption Processes*; Wiley-Interscience: New York, **1984**.

42. Banerjee, K.; Cheremisinoff, P.N.; Cheng, S.L. Adsorption kinetics of o-xylene by fly ash. Water Res. **1997**, *31*, 249–261.

43. Manju, G.N.; Anoop Krishnan, K.; Vinod, V.P.; Anirudhan, T.S. An investigation into the sorption of heavy metals from wastewaters by polyacrylamide-grafted iron(III) oxide. J. Hazard. Mater. **2002**, *91*, 221–238.

44. Boyd, G.E.; Adamson, A.W.; Myers Jr., L.S. The Exchange adsorption of ions from aqueous solutions by organic zeolites. II. Kinetics. J. Am. Chem. Soc. **1947**, *69*, 2836–2848.

45. Hamerlinck, Y.; Mertens, D.H. In *Activated Carbon Principles in Separation Technology;* Vansant, E.F., Ed.; Elsevier: New York, 1994.

46. Mourão, P.A.M.; Carrott, P.J.M.; Ribeiro Carrott, M.M.L. Application of different equations to adsorption isotherms of phenolic compounds on activated carbons prepared from cork. Carbon **2006**, *44*, 2422–2429.

47. Mestre, A.S.; Pires, J.; Nogueira, J.M.F.; Parra, J.B.; Carvalho, A.P.; Ania, C.O. Waste-derived activated carbons for removal of ibuprofen from solution: Role of surface chemistry and pore structure. Bioresour. Technol. **2009**, *100*, 1720–1726.

48. Evans, M.J.B.; Halliop, E.; MacDonald, J.A.F. The production of chemically activated carbon. Carbon **1999**, *37*, 269–274.

49. Wu, F.C.; Tseng, R.L.; Juang, R.S. Adsorption of dyes and phenols from water on the activated carbons prepared from corncob wastes. Environ. Technol. **2001**, *22*, 205–213.

50. Tseng, R.L.; Tseng, S.K. Characterization and use of high surface area activated carbons prepared from cane pith for liquid-phase adsorption. J. Hazard. Mater. **2006**, *136*, 671–680.

51. Prahas, Y.D.; Kartika, N.; Indraswati, S.; Ismadji, S. Activated carbon from jackfruit peel waste by H3PO4 chemical activation: Pore structure and surface chemistry characterization. Chem. Eng. J. **2007**, *140*, 32–42.

52. Jain, S.; Jayaram, R.V. Adsorption of phenol and substituted chlorophenols from aqueous solution by activated carbon prepared from jackfruit *(Artocarpus heterophyllus)* peel-kinetics and equilibrium studies. Sep. Sci. Technol. **2007**, *42*, 2019–2032.

53. Tancredi, N.; Medero, N.; Möller, F.; Píriz, J.; Plada, C.; Cordero, T. Phenol adsorption onto powdered and granular activated carbon, prepared from Eucalyptus wood. J. Colloid Interface Sci. **2004**, *279*, 357–363.

54. Mudoga, H.L.; Yucel, H.; Kincal, N.S. Decolonization of sugar syrups using commercial and sugar beet pulp based activated carbons. Bioresour. Technol. **2008**, *99*, 3528–3533.

55. Philip, C.A.; Girgis, B.S. Adsorption characteristics of microporous carbons from apricot stones activated by phosphoric acid. J. Chem. Technol. Biotechnol. **1996**, *67*, 248–254.

56. Ahmadpour, A.; Do, D.D. The preparation of activated carbon from Macadamia nutshell by chemical activation. Carbon **1997**, *35*, 1723–1732.

57. Toles, C.A.; Marshall, W.E.; Johns, M.M.; Wartelle, L.H.; McAloon, A. Acid activated carbons from almond shells: Physical, chemical and adsorptive properties and estimated cost of production. Bioresour. Technol. **2000**, *71*, 87–92.

58. Toles, C.A.; Marshall, W.E.; Wartelle, L.H.; McAloon, A. Steam- or carbon dioxide–activated carbons from almond shells: Physical, chemical and adsorptive properties and estimated cost of production. Bioresour. Technol. **2000**, *75*, 197–203.

59. Galiatsatou, P.; Metaxas, M.; Arapoglou, D.; Kasselouri- Rigopoulo, V. Treatment of olive mill waste water with activated carbons from agricultural by-products. Waste Manage. **2002**, *22*, 803–812.

60. Rengaraj, S.; Moon, S.H.; Sivabalan, R.; Arabindoo, B.; Murugesan, V. Agricultural solid waste for the removal of organics: Adsorption of phenol from water and wastewater by palm seed coat activated carbon. Waste Manage. **2002**, *22*, 543–548.

61. Qi, J.; Li, Z.; Guo, Y.; Xu, H. Adsorption of phenolic compounds on micro- and mesoporous rice husk-based active carbons. Mater. Chem. Phys. **2004**, *87*, 96–101.

62. Tan, I.A.W.; Ahmad, A.L.; Hameed, B.H. Preparation of activated carbon from coconut husk: Optimization study on removal of 2,4,6-trichlorophenol using response surface methodology. J. Hazard. Mater. **2008**, *153*, 709–717.

63. Michailof, C.; Stavropoulos, G.G.; Panayiotou, C. Enhanced adsorption of phenolic compounds, commonly encountered in olive mill wastewaters, on olive husk derived activated carbons. Bioresour. Technol. **2008**, *99*, 6400–6408.

64. Muñoz-González, Y.; Arriagada-Acuña, R.; Soto-Garrido, G.; García-Lovera, R. Activated carbons from peach stones and pine sawdust by phosphoric acid activation used in clarification and decolorization processes. J. Chem. Technol. Biotechnol. **2009**, *84*, 39–47.

65. Lussier, M.G.; Shull, J.C.; Miller, D.J. Activated carbons from cherry stones. Carbon **1994**, *32*, 1493–1498.

66. Shopova, N.; Minkova, V.; Markova, K. Evaluation of the thermochemical changes in agricultural by-products and in the carbon adsorbents obtained from them. J. Therm. Anal. **1997**, *48*, 309–320.

67. Marsh, H.; Iley, M.; Berger, J.; Siemieniewska, T. The adsorptive properties of activated plum stone chars. Carbon **1975**, *13*, 103–109.

68. Mohanty, K.; Jha, M.; Meikap, B.C.; Biswas, M.N. Preparation and characterization of activated carbons from *Ter- minalia arjuna* nut with zinc chloride activation for the removal of phenol from wastewater. Ind. Eng. Chem. Res. **2005**, *44*, 4128–4138.

69. Alvim Ferraz, M.C. Preparation of activated carbon for air pollution control. Fuel **1988**, *67*, 1237–1241.
70. Tancredi, N.; Cordero, T.; Rodríguez-Mirasol, J.; Rodríguez, J.J. Activated carbons from Uruguayan eucalyptus wood. Fuel **1996**, *73*, 1701–1706.
71. Ferro-Garcia, M.A.; Rivera-Utrilla, J.; Rodriguez-Gordillo, J. Bautista-Toledo, I. Adsorption of zinc, cadmium, and copper on activated carbons obtained from agricultural byproducts. Carbon **1988**, *26*, 363–373.
72. Hall, C.R.; Holmes, R.J. The preparation and properties of some activated carbons modified by treatment with phosgene or chlorine. Carbon **1992**, *30*, 173–176.
73. Otowa, T.; Nojima, Y.; Miyazaki, T. Development of KOH activated high surface area carbon and its application to drinking water purification. Carbon **1997**, *35*, 1315–1319.
74. Manju, G.N.; Raji, C.; Anirudhan, T.S. Evaluation of coconut husk carbon for the removal of arsenic from water. Water Res. **1998**, *32*, 3062–3070.
75. Laine, J.; Calafat, A.; Labady, M. Preparation and characterization of activated carbons from coconut shell impregnated with phosphoric acid. Carbon **1989**, *27*, 191–195.
76. Mohan, D.; Singh, K.P.; Ghosh D. Removal of α-picoline, β-picoline, and γ-picoline from synthetic wastewater using low cost activated carbons derived from coconut shell fibers. Environ. Sci. Technol. **2005**, *39*, 5076–5086.
77. Mohan, D.; Singh, K.P.; Sinha, S.; Ghosh, D. Removal of pyridine derivatives from aqueous solution by activated carbons developed from agricultural waste materials. Carbon **2005**, *43*, 1680–1693.
78. Mohan, D.; Singh, K.P.; Singh, V.K. Trivalent chromium removal from wastewater using low cost activated carbon derived from agricultural waste material and activated carbon fabric cloth. J. Hazard. Mater. **2006**, *133*, 280–295.
79. Mohan, D.; Singh, K.P.; Sinha, S.; Ghosh, D. Removal of pyridine from aqueous solution using low cost activated carbons developed from agricultural waste materials. Carbon **2004**, *43*, 2409–2421.
80. Kumar, K.V.; Sivanesan, S. Equilibrium data, isotherm parameters and process design for partial and complete isotherm of methylene blue onto activated carbon. J. Hazard. Mater. **2006**, *134*, 237–244.
81. Marungrueng, K.; Pavasant, P. High performance biosorbent (*Caulerpa lentillifera*) for basic dye removal. Bioresour. Technol. **2007**, *98*, 1567–1572.
82. Basava Rao, V.V.; Mohan Rao, S.R. Adsorption studies on treatment of textile dyeing industrial effluent by fly ash. Chem. Eng. J. **2006**, *116*, 77–84.
83. El Qada, E.N.; Allen, S.J.; Walker, G.M. Adsorption of basic dyes from aqueous solution onto activated carbons. Chem. Eng. J. **2008**, *133*, 174–184.
84. Tamai, H.; Kakii, T.; Hirota, Y.; Kumamoto, T.; Yasuda, H. Synthesis of extremely large mesoporous activated carbon and its unique adsorption for giant molecules. Chem. Mater. **1996**, *8*, 454–462.
85. Banat, F.; Al-Asheh, S.; Al-Ahmad, R.; Bni-Khalid, F. Bench-scale and packed bed sorption of methylene blue using treated olive pomace and charcoal. Bioresour. Technol. **2007**, *98*, 3017–3025.
86. McKay, G.; Porter, J.F.; Prasad, G.R. The removal of dye colours from aqueous solutions by adsorption on low-cost materials. Water Air Soil Pollut. **1999**, *114*, 423–438.
87. McKay, G.; Ramprasad, G.; Pratapamowli, P. Equilibrium studies for the adsorption of dyestuffs from aqueous solution by low-cost materials. Water Air Soil Pollut. **1986**, *29*, 273–283.
88. Kannan, N.; Sundaram, M.M. Kinetics and mechanism of removal of methylene blue by adsorption on various carbons— A comparative study. Dyes Pigments **2001**, *51*, 25–40.
89. Bestani, B.; Benderdouche, N.; Benstaali, B.; Belhakem, M.; Addou, A. Methylene blue and iodine adsorption onto an activated desert plant. Bioresour. Technol. **2008**, *99*, 8441–8444.
90. Stavropoulos, G.G.; Zabaniotou, A.A. Production and characterization of activated carbons from olive-seed waste residue. Microporous Mesoporous Mater. **2005**, *82*, 79–85.

91. Fernandes, A.N.; Almeida, C.A.P.; Menezes, C.T.B.; De- bacher, N.A.; Sierra, M.M.D. Removal of methylene blue from aqueous solution by peat. J. Hazard. Mater. **2007**, *144*, 412–419.

92. Jones, D.A.; Lelyveld, T.P.; Mavrofidis, S.D.; Kingman, S.W.; Miles, N.J. Microwave heating applications in environmental engineering—A review. Resour. Conserv. Recycl. **2002**, *34*, 75–90.

93. Marsh, H.; Rodnguez-Reinoso, F. *Activated Carbon;* Elsevier: Oxford, **2006**.

94. Ioannidou, O.; Zabaniotou, A. Agricultural residues as precursors for activated carbon production—A review. Renew. Sustainable Energy Rev. **2007**, *11*, 1966–2005.

95. Yin, C.Y.; Aroua, M.K.; Daud, W.M.A.W. Review of modifications of activated carbon for enhancing contaminant uptakes from aqueous solutions. Sep. Purif. Technol. **2007**, *32*, 403–415.

96. Paraskeva, P.; Kalderis, D.; Diamadopoulos, E. Production of activated carbon from agricultural by-products. J. Chem. Technol. Biotechnol. **2008**, *83*, 581–592.

97. Yuen, F.K.; Hameed, B.H. Recent developments in the preparation and regeneration of activated carbons by microwaves. Adv. Colloid Interface Sci. **2009**, *149*, 19–27.

98. Menèndez, J.A.; Arenillas, A.; Fidalgo, B.; Fernandez, Y.; Zubizarreta, L.; Calvo, E.G.; Bermùdez, J.M. Microwave heating processes involving carbon materials. Fuel Process. Technol. **2010**, *91*, 1–8.

99. Laszlo, K.; Podkoscielny, P.; Dabrowski, A. Heterogeneity of polymer-based active carbons in adsorption of aqueous solutions of phenol and 2,3,4-trichlorophenol. Langmuir **2003**, *19*, 5287–5294.

100. Podkoscielny, P.; Dabrowski, A.; Marijuk, O.V. Heterogeneity of active carbons in adsorption of phenol aqueous solutions. Appl. Surf. Sci. **2003**, *205*, 297–303.

101. Leòn Y Leòn, C.A.; Radovic, L.R. Interfacial chemistry and electrochemistry of carbon surfaces. In *Chemistry and Physics of Carbon;* Thrower, P.A., Ed.; Marcel Dekker: New York, 1720; Vol. 24, 214–310.

102. Boehm, H.P. Surface oxides on carbon and their analysis: A critical assessment. Carbon **2002**, *40*, 145–149.

103. Gabaldon, C.; Marzal, P.; Seco, A.; Gonzalez, J.A. Cadmium and copper removal by a granular activated carbonin laboratory column systems. Sep. Sci. Technol. **1720**, *35*, 1039–1053.

104. Carrott, P.J.M.; Ribeiro Carrott, M.M.L.; Nabais, J.M.V.; Ramalho, J.P.P. Influence of surface ionization on the adsorption of aqueous zinc species by activated carbons. Carbon **1720**, *35*, 403–410.

105. Carrott, P.J.M.; Mourao, P.A.M.; Ribeiro Carrott, M.M.L.; Goncalves, E.M. Separating surface and solvent effects and the notion of critical adsorption energy in the adsorption of phenolic compounds by activated carbons. Langmuir **1720**, *21*, 11863–11869.

106. Hu, J.; Aizawa, T.; Ookubo, Y.; Morita, T.; Magara, Y. Adsorptive characteristics of ionogenic aromatic pesticides in water on powdered activated carbon. Water Res. **1720**, *32*, 2593–2600.

107. Urano, K.; Yamamoto, E.; Tonegawa, M.; Fujie, K. Adsorption of chlorinated organic compounds on activated carbon from water. Water Res. **1991**, *25*, 1459–1464.

108. Malhas, A.N.; Abuknesha, R.A.; Price, R.G. Removal of detergents from protein extracts using activated charcoal prior to immunological analysis. J. Immunol. Methods **2002**, *264*, 37–43.

109. Babel, S.; Kurniawan, T.A. Cr(VI) removal from synthetic wastewater using coconut shell charcoal and commercial activated carbon modified with oxidizing agents and/or chi- tosan. Chemosphere **2004**, *54*, 951–967.

110. Kim, D.S. Adsorption characteristics of Fe(III) and Fe(III)–NTA complex on granular activated carbon. J. Hazard. Mater. **2004**, *106*, 67–84.

111. Satapathy, D.; Natarajan, G.S. Potassium bromate modification of the granular activated carbon and its effect on nickel adsorption. Adsorption **2006**, *12*, 147–154.

112. Richard, D.; Delgado Nunez, M.L.; Schweich, D. Adsorption of complex phenolic compounds on active charcoal: adsorption capacity and isotherms. Chem. Eng. J. **2010**, *158*, 213–219.

113. Figaro, S.; Louisy-Louis, S.; Lambert, J.; Ehrhardt, J.J.; Ouensanga, A.; Gaspard, S. Adsorption studies of recalcitrant compounds of molasses spentwash on activated carbons. Water Res. **2006**, *40*, 3456–3466.

114. Mohan, S.V.; Karthikeyan, J. Removal of lignin and tannin colour from aqueous solution by adsorption onto activated charcoal. Environ. Pollut. **1997**, *97*, 183–187.

115. Kalderis, D.; Koutoulakis, D.; Paraskeva, P.; Diamadopou- los, E.; Otal, E.; Valle, J.O.; Fernandez-Pereira, C. Adsorption of polluting substances on activated carbons prepared from rice husk and sugarcane bagasse. Chem. Eng. J. **2008**, *144*, 42–50.

116. Lang, W.; Dejma, C.; Sirisansaneeyakul, S.; Sakairi, N. Biosorption of nonylphenol on dead biomass of *Rhizo- pus arrhizus* encapsulated in chitosan beads. Bioresour. Technol. **2009**, *100*, 5616–5623.

117. Pinnavaia, T. Intercalated clay catalysts. J. Sci. **1983**, *220*, 365–371.

118. Bhattacharyya, K.G.; Gupta, S.S. Adsorption of a few heavy metals on natural and modified kaolinite and montmorillonite: A review. Adv. Colloid Interface Sci. **2008**, *140*, 114–131.

119. Mitchell, J.K. *Fundamentals of Soil Behavior*, 2nd Ed; Wiley: New York, 1993.

120. Gupta, S.S.; Bhattacharyya, K.G. Removal of Cd(II) from aqueous solution by kaolinite, montmorillonite and their poly(oxo zirconium) and tetrabutylammonium derivatives. J. Hazard. Mater. **2006**, *128*, 247–257.

121. Bhattacharyya, K.G.; Gupta S.S. Influence of acid activation of kaolinite and montmorillonite on their adsorptive removal of Cd(II) from water. Ind. Eng. Chem. Res. **2007**, *46*, 3734 –3742.

122. Bagane, M.; Guiza, S. Removal of a dye from textile effluents by adsorption. Ann. Chim. **2000**, *25*, 615–626.

123. Almeida, C.A.P.; Debacher, N.A.; Downs, A.J.; Cottet, L.; Mello, C.A.D. Removal of methylene blue from colored effluents by adsorption on montmorillonite clay. J. Colloid Interface Sci. **2009**, *332*, 46–53.

124. Ghosh, D.; Bhattacharyya, K.G. Adsorption of methylene blue on kaolinite. Appl. Clay Sci. **2002**, *20*, 295–300.

125. Gupta, S.S.; Bhattacharyya, K.G. Removal of Cd(II) from aqueous solution by kaolinite, montmorillonite and their poly(oxo zirconium) and tetrabutylammonium derivatives. J. Hazard. Mater. **2006**, *128*, 247–257.

126. Bhattacharyya, K.G.; Gupta, S.S. Influence of acid activation on adsorption of Ni(II) and Cu(II) on kaolinite and montmorillonite: Kinetic and thermodynamic study. Chem. Eng. J. **2008**, *136*, 1–13.

127. Gupta, S.S.; Bhattacharyya, K.G. Adsorption of Ni(II) on clays. J. Colloid Interface Sci. **2006**, *295*, 21–32.

128. Bhattacharyya, K.G.; Gupta, S.S. Adsorption of chromium(VI) from water by CLAYS. Ind. Eng. Chem. Res. **2006**, *45*, 7232–7240.

129. Yavuz, O.; Altunkaynak, Y.; Guzel, F. Removal of copper, nickel, cobalt and manganese from aqueous solution by ka- olinite. Water Res. **2003**, *37*, 948–952.

130. Bhattacharyya, K.G.; Gupta, S.S. Adsorption of Co(II) from aqueous medium on natural and acid activated kaolinite and montmorillonite. Sep. Sci. Technol. **2007**, *42*, 3391–3418.

131. Gupta, S.S.; Bhattacharyya, K.G. Interaction of metal ions with clays: I. A case study with Pb(II). Appl. Clay Sci. **2005**, *30*, 199–208.

132. Bhattacharyya, K.G.; Gupta, S.S. Pb(II) uptake by kaolinite and montmorillonite in aqueous medium: Influence of acid activation of the clays. Colloids Surf., A **2006**, *277*, 191–200.

133. Bhattacharyya, K.G.; Gupta, S.S. Adsorption of Fe(III) from water by natural and acid activated clays: Studies on equilibrium isotherm, kinetics and thermodynamics of interactions. Adsorption **2006**, *12*, 185–204.

134. Bhattacharyya, K.G.; Gupta, S.S. Kaolinite, montmorillon- ite, and their modified derivatives as adsorbents for removal of Cu(II) from aqueous solution. Sep. Purif. Technol. **2006**, *50*, 388–397.

135. Townsend, R.P. Ion exchange in zeolites. In *Introduction to Zeolite Science and Practice*; Bekkum, H.V., Flannigen, E.M., Janmsen, J.C., Eds.; Elsevier: Amsterdam, 1991; 359–390.

136. Dyer, A. *An Introduction to Zeolite Molecular Sieves*; John Wiley and Sons: Chichester, 1988; Vol. 5–8.

137. Barer, R.M. *Zeolites and Clay Minerals as Sorbent and Molecular Sieves*; Academic Press: New York, 1978.

138. Kesraoui-Ouki, S.; Cheeseman, C.R.; Perry, R. Natural zeolite utilization in pollution control: A review of application to metal's effluents. J. Chem. Technol. Biotechnol. **1994**, *59*, 121–126.

139. Ghobarkar, H.; Schaf, O.; Guth, U. Zeolites from kitchen to space. Prog. Solid State Chem. **1999**, *27*, 29–73.

140. Dogan, M.; ALkan, M.; Onager, Y. Adsorption of methylene blue from aqueous solution onto perlite. Water Air Soil Pollut. **2000**, *120*, 229–248.

141. Dogan, M.; Alkan, M.; Turkyilmaz, A.; Ozdemir, Y. Kinetics and mechanism of removal of methylene blue by adsorption onto perlite. J. Hazard. Mater. **2004**, *109*, 141–148.

142. Chakrabarti, S.; Dutta, B.K. Note on the adsorption and diffusion of methylene blue in glass fibers. J. Colloid Interface Sci. **2005**, *286*, 807–811.

143. Kumar, M.N.V.R. A review of chitin and chitosan applications. React. Funct. Polym. **2000**, *46*, 1–27.

144. Guibal E. Interactions of metal ions with chitosan-based sorbents: A review. Sep. Purif. Technol. **2004**, *38*, 43–74.

145. Varma, A.J.; Deshpande, S.V.; Kennedy, J.F. Metal com- plexation by chitosan and its derivatives: A review. Carbo- hydr. Polym. **2004**, *55*, 77–93.

146. Gerente, C.; Lee, V.K.C.; Cloirec, P.L.; McKay, G. Application of chitosan for the removal of metals from wastewaters by adsorption—mechanisms and models review. Crit. Rev. Environ. Sci. Technol. **2007**, *37*, 41–127.

147. Crini, G.; Badot, P.-M. Application of chitosan, a natural aminopolysaccharide, for dye removal from aqueous solutions by adsorption processes using batch studies: A review of recent literature. Prog. Polym. Sci. **2008**, *33*, 399–447.

148. Crini, G. Recent developments in polysaccharides-based materials used as adsorbents in wastewater treatment. Prog. Polym. Sci. **2005**, *30*, 38–70.

149. Bhatnagar, A.; Sillanpää, M. Applications of chitin- and chitosan-derivatives for the detoxification of water and wastewater—A short review. Adv. Colloid Interface Sci. **2009**, *152*, 26–38.

150. Jeon, C.; Höll, W.H. Chemical modification of chitosan and equilibrium study for mercury ion removal. Water Res. **2003**, *37*, 4770–4780.

151. Rorrer, G.L.; Hsien, T.-Y.; Way, J.D. Synthesis of porous- magnetic chitosan beads for removal of cadmium ions from wastewater. Ind. Eng. Chem. Res. **1993**, *32*, 2170–2178.

152. Kalyani, S.; Priya, J.A.; Rao, P.S.; Krishnaiah, A. Removal of copper and nickel from aqueous solutions using chitosan coated on perlite as biosorbent. Sep. Sci. Technol. **2005**, *40*, 1483–1495.

153. Liu, X.; Hu, Q.; Fang, Z.; Zhang, X.; Zhang, B. Magnetic chitosan nanocomposite: A useful tool for heavy metal ion removal. Langmuir **2009**, *25*, 3–8.

154. Qu, R.; Sun, C.; Ma, F.; Zhang, Y.; Ji, C.; Xu, Q.; Wang, C.; Chen, H. Removal of recovery of Hg(II) from aqueous solution using chitosan-coated cotton fibers. J. Hazard. Mater. **2009**, *167*, 717–727.

155. Zhang, G.; Qu, R.; Sun, C.; Ji, C.; Chen, H.; Wang, C.; Niu, Y. Adsorption for metal ions of chitosan coated cotton fiber. J. Appl. Polym. Sci. **2008**, *110*, 2321–2327.

156. Huang, G.L.; Zhang, H.Y.; Jeffrey, X.S.; Tim, A.G.L. Adsorption of chromium(VI) from aqueous solutions using cross-linked magnetic chitosan beads. Ind. Eng. Chem. Res., **2009**, *48*, 2646–2651.

157. Sun, X.Q.; Peng, B.; Jing, Y.; Chen, J.; Li, D.Q. Chitosan(chitin)/cellulose composite biosorbents prepared using ionic liquid for heavy metal ions adsorption. Separations **2009**, *55*, 2062–2069.

158. Shameem, H.; Abburi, K.; Tushar, K.G.; Dabir, S.V.; Veera, M.B.; Edgar, D.S. Adsorption of chromium(VI) on chitosan-coated perlite. Sep. Sci. Technol. **2003**, *38*, 3775–3793.

159. Veera, M.B.; Krishnaiah, A.; Jonathan, L.T.; Edgar, D.S. Removal of hexavalent chromium from wastewater using a new composite chitosan biosorbent. Environ. Sci. Technol. **2003**, *37*, 4449–4456.

160. Tran, H.V.; Tran, L.D.; Nguyen, T.N. Preparation of chi- tosan/magnetite composite beads and their application for removal of Pb(II) and Ni(II) from aqueous solution. Mater. Sci. Eng., C **2010**, *30*, 304–310.

161. Wan, M.W.; Kan, C.C.; Buenda, D.R.; Maria, L.P.D. Adsorption of copper(II) and lead(II) ions from aqueous solution on chitosan-coated sand. Carbohydr. Polym. **2010**, *80*, 891–899.

162. Srinivasa, R.P.; Vijaya, Y.; Veera, M.B.; Krishnaiah, A. Adsorptive removal of copper and nickel ions from water using chitosan coated PVC beads. Bioresour. Technol. **2009**, *100*, 194–199.

163. Vijaya, Y.; Srinivasa, R.P.; Veera, M.B.; Krishnaiah, A. Modified chitosan and calcium alginate biopolymer sorbents for removal of nickel (II) through adsorption. Carbohydr. Polym. **2008**, *72*, 261–271.

164. Li, J.M.; Meng, X.G.; Hu, C.W.; Du, J. Adsorption of phenol,*p*-chlorophenol and *p*-nitrophenol onto functional chitosan. Bioresour. Technol. **2009**, *100*, 1168–1173.

165. Nadavala, S.K.; Swayampakula, K.; Boddu, V.M.; Abburi, K. Biosorption of phenol and o-chlorophenol from aqueous solutions on to chitosan-calcium alginate blended beads. J. Hazard. Mater. **2009**, *162*, 482–489.

166. Pagnanelli, F.; Mainelli, S.; Veglio, F.; Toro, L. Heavy metal removal by olive pomace: biosorbent characterization and equilibrium modeling. Chem. Eng. Sci. **2003**, *58*, 4709–4717.

167. Demirbas, A. Mechanisms of liquefaction and pyrolysis reactions of biomass. Energy Convers. Manage. **2000**, *41*, 633–646.

168. Theander, O. Cellulose, Hemicellulose, and Extractives. In *Fundamentals of Thermochemical Biomass Conversion*; Overand, R.P., Mile, T.A., Mudge, L.K., Eds.; Elsevier Applied Science Publisher: New York, **1985**; 35–60.

169. Morita, M.; Higuchi, M.; Sakata, I. Binding of heavy metal ions by chemically modified Woods. J. Appl. Polym. Sci. **1987**, *34*, 1013–1023.

170. Sawamiappan, N.; Krishnamoorthy, S. Phenol formaldehyde cationic matrices substitutes by bagasse-charcoal. Res. Ind. **1984**, *29*, 293–297.

171. Odozi, T.O.; Okeke, S.; Lartey, L.B. Studies on binding metal ions with polymerized corncob and a composite resin with sawdust and onion. Agric. Waste **1985**, *12*, 13–21.

172. Gardea-Torresdey, J.L.; de la Rosa, G.; Peralta-Videa, J.R. Use of phytofiltration technologies in the removal of heavy metals: a review. Pure Appl. Chem. **2004**, *76*, 801–813.

173. Volesky, B. Detoxification of metal-bearing effluents: Biosorption for the next century. Hydrometallurgy **2001**, *59*, 203–216.

174. Brown, P.A.; Gill, S.A.; Allen, S.J. Metal removal from wastewater using peat. Water Res. **2000**, *34*, 3907–3916.

175. Nobel, P. *Physicochemical and Environmental Plant Physiology*; Academic Press: New York, 1991.

176. Rowell, R.M. *In Removal of Metal Ions from Contaminated Water Using Agricultural Residues, Proceedings of* ECOWOOD 2006 2nd International Conference on Environmentally Compatible Forest Products, Fernando Pessoa University, Oporto, Portugal, Sept. 20–22, 2006.

177. Mantanis, G.I.; Young, R.A.; Rowell, R.M. Swelling of wood. Part III. Effect of temperature and extractives on rate and maximum swelling. Holzforschung **1995**, *49*, 239–248.

178. Dahiya, S.; Tripathi, R.M.; Hegde, A.G. Biosorption of lead and copper from aqueous solutions by pre-treated crab and arca shell biomass. Biores. Technol. **2008**, *99*, 179–187.

179. Hashem, A.; Abdel-Halim, E.; Maauof, H.A.; Ramadan, M.A.; Abo-Okeil, A. Treatment of sawdust with polyamine for wastewater treatment. Energy Edu. Sci. Technol. **2007**, *19*, 45–58.

180. Demirbas, A. Heavy metal adsorption onto agro-based waste materials: A review. J. Hazard. Mater. **2008**, *157*, 220–229.

181. Min, S.H.; Han, J.S.; Shin, E.W.; Park, J.K. Improvement of cadmium ion removal by base treatment of juniper fiber. Water Res. **2004**, *38*, 1289–1295.

182. Dakiky, M.; Khamis, M.; Manassra, A.; Mer'eb, M. Selective adsorption of chromium(VI) in industrial wastewater using low-cost abundantly available adsorbents. Adv. Environ. Res. **2002**, *6*, 533–540.

183. Bhatnagar, A.; Jain, A.K.; Minocha, A.K.; Singh, S. Removal of lead ions from aqueous solutions by different types of industrial waste materials: equilibrium and kinetic studies. Sep. Sci. Technol. **2006**, *41*, 1881–1892.

184. Taty-Costodes, V.C.; Fauduet, H.; Porte, C.; Delacroix, A. Removal of Cd(II) and Pb(II) ions, from aqueous solutions, by adsorption onto sawdust of *Pinus sylvestris*. J. Hazard. Mater. **2003**, *105*, 121–142.

185. Nassar, M.M. Adsorption of Fe^{+3} and Mn^{+2} from ground water onto maize cobs using batch adsorber and fixed bed column. Sep. Sci. Technol. **2006**, *41*, 943–959.

186. Ho, Y.S.; Huang, C.T.; Huang, H.W. Equilibrium sorption isotherm for metal ions on tree fern. Process Biochem. **2002**, *37*, 1421–1430.

187. Chen, B.; Hui, C.W.; McKay, G. Film-pore diffusion modeling for the sorption of metal ions from aqueous effluents onto peat. Water Res. **2001**, *35*, 3345–3356.

188. Sathishkumar, M.; Vijayaraghavan, K.; Binupriya, A.R.; Stephan, A.M.; Choi, J.G.; Yun, S.E. Porogen effect on characteristics of banana pith carbon and the sorption of dichlorophenols. J. Colloid Interface Sci. **2008**, *320*, 22–29.

189. Achak, M.; Hafidi, A.; Ouazzani, N.; Sayadi, S.; Mandi, L. Low cost biosorbent "banana peel" for the removal of phenolic compounds from olive mill wastewater: Kinetic and equilibrium studies. J. Hazard. Mater. **2009**, *166*, 117–125.

190. Park, K.H.; Balathanigaimani, M.S.; Shim, W.G.; Lee, J.W.; Moon, H. Adsorption characteristics of phenol on novel corn grain-based activated carbons. Microporous Mesoporous Mater. **2010**, *127*, 1–8.

191. Navarro, A.E.; Cuizano, N.A.; Lazo, J.C.; Sun-Kou, M.R.; Llanos, B.P. Comparative study of the removal of phenolic compounds by biological and non-biological adsorbents. J. Hazard. Mater. **2009**, *164*, 1439–1446.

192. Bhatnagar, A.; Minocha, A.K. Adsorptive removal of 2,4- dichlorophenol from water utilizing *Punica granatum* peel waste and stabilization with cement. J. Hazard. Mater. **2009**, *168*, 1111–1117.

193. Jiuhui, Q.U. Research progress of novel adsorption processes in water purification: A review. J. Environ. Sci. **2008**, *20*, 1–13.

194. Oliveira, L.C.A.; Petkowicz, D.I.; Smaniotto, A.; Pergher, S.B.C. Magnetic zeolites: A new adsorbent for removal of metallic contaminants from water. Water Res. **2004**, *38*, 3699–3704.

195. Gu, Z.; Fang, J.; Deng, B.L. Preparation and evaluation of GAC-based iron-containing adsorbents for arsenic removal. Environ. Sci. Technol. **2005**, *39*, 3833–3843.

196. Machado, L.C.R.; Lima, F.W.J.; Paniago, R. Polymer coated vermiculite–iron composites: Novel floatable magnetic adsorbents for water spilled contaminants. Appl. Clay Sci. **2006**, *31*, 207–215.

197. Zhang, G.S.; Qu, J.H.; Liu, H.J.; Liu, R.P.; Wu, R.C. Preparation and evaluation of a novel Fe-Mn binary oxide adsorbent for effective arsenite removal. Water Res. **2007**, *41*, 1921–1928.

Cadmium and Lead: Contamination

Gabriella Kakonyi
and Imad A.M.
Ahmed

Introduction

This entry summarizes the effects of cadmium (Cd) and lead (Pb) on the environment and on human health. The chemistry of Cd and Pb is first discussed to establish an understanding of their behavior, occurrence, and fate in the environment. This helps foresee toxicity and the possible natural attenuation or remediation strategies. Both Cd and Pb are widely produced and used. Therefore, awareness of the sources from where humans and the environment are exposed is described in some detail along with their epidemiology and regulatory measures taken to reduce or prevent their release into the environment. A brief section is also dedicated to the detection and analysis of Cd and Pb and their compounds within environmental and clinical samples.

Chemical Properties

Cadmium (Cd) and lead (Pb) are transition metals that have no known vital or beneficial role in the human body or health. Cadmium (atomic number 48, atomic mass $112.4\,g\,mol^{-1}$) belongs to Group 12 of the periodic table, possessing somewhat similar chemical properties as zinc and mercury. Lead (atomic number 82, atomic mass $207.2g\,mol^{-1}$) is a member of Group 14 in the periodic table, which also includes C, Si, Ge, and Sn.[1] The most predominant oxidation state of Cd and Pb under normal environmental conditions of temperature and pressure is +2. In organolead chemistry, the oxidation state (+4) of Pb is remarkably dominant.[2] Both Cd and Pb are considered to form stable oxidation states as divalent Cd^{2+} and Pb^{2+} ions in inorganic compounds. Their biological toxicity appears to be determined by their availability for ligand exchange and chelation properties.[3–6] The chemical similarity of Cd and Pb to certain alkaline earth metals such as calcium; their ability to form highly insoluble inorganic salts (e.g., phosphate, carbonate, sulfate), organometallic complexes, or free hydrated ions; and their increased affinity to biological donors (e.g., proteins) play an important role in their transport in the environment and toxicity in biological systems.[2,7,8]

Occurrences

Cadmium and lead are trace elements in rocks and soils,[9] where the concentration of Cd is approximately 0.1 ppm and the concentration of Pb is about tenfold higher.[10,11] These concentrations are relatively low, and the presence of both Cd and Pb are normally associated with other more abundant elements. Therefore, both metals are inevitable by-products from the mining of zinc and copper ores. Lead and cadmium are chalcophilic elements, meaning that they have a tendency to form sulfide minerals.[12] Thus, the most abundant Cd and Pb minerals are greenockite (CdS) and galena (PbS). Most Cd is found associated with zinc ores such as sphalerite (ZnS) in which the Cd content is 0.5% of the Zn content.[11] Other important minerals of Pb are crocite ($PbCrO_4$), anglesite $PbSO_4$, massicot (PbO), and cerussite ($PbCO_3$).[13] Native (i.e., elemental) Cd and Pb are rare in the environment. Naturally occurring metallic Cd has been found in the Vilyuy River bedrock in Siberia.[14] Cadmium and lead are also found associated with clay and carbonate minerals. The Cd^{2+} ion has an ionic size of 95 pm close to that of Ca^{2+} (100 pm), whereas Pb^{2+} has an ionic size (119 pm) between K^+ (138 pm) and Ca^{2+}. Thus, during the formation of secondary minerals such as feldspars, mica, apatite, and calcite, major cations such as Ca^{2+} and K^+ can be substituted by Cd^{2+} and Pb^{2+} or other trace metal ions of the same charge, sign, or of comparable ionic size.[15,16]

Production and Uses

Cadmium

Cadmium is widely used in a number of industrial applications, the largest area being the production of nickel-Cd batteries.[17,18] Furthermore, Cd is increasingly used in solar panels, and still commonly used in pigments, coatings, corrosion-resistant plating, photography, as a fungicide and as a stabilizer and softener in plastics. It is also contained in coal and in rocks mined to produce phosphate fertilizers.[19] Cadmium atoms has the ability to absorb neutrons without fission or splitting, so it is used in nuclear reactor components such as control rods as a shield of neutrons and to control nuclear fission.[20]

Lead

The greatest use of Pb is lead acid batteries, which have been used extensively in automobiles since 1918 or so.[21] Lead is also commonly added to paint as Pb-chromate or Pb-carbonate to speed drying, increase durability, retain a fresh appearance, and resist moisture. Because elemental Pb has a low melting temperature (327°C), it enables easy casting and shaping and thus is commonly used in building constructions and joining metallic parts. Lead is the traditional base metal for domestic water pipes. One if its major uses is as a radiation shield in the glass of television and computer screens, and as a protecting shield from radioactive radiation such as x-ray and γ-ray in scientific and industrial instruments. Other uses are in infrared detectors, sheeting, cables, solders, Pb-crystal glassware, ammunitions, and as weight in sport equipment.

Detection

Because Cd is used as pigment and softener in plastics, the use of colored lids, tubes and certain plastic containers should be avoided during sampling and storage of environmental and toxicological samples.[22] It is well documented that significant amounts of trace metals can be lost on the walls of glass and some plastic containers and adsorption of Pb occurs on the walls of Pyrex, polypropylene and polyethylene containers. The loss of aqueous Cd and Pb by adsorption onto the wall of containers was not observed using Teflon. Therefore, particular attention must be paid in the collection, treatment and preservations of environmental samples, especially when Cd and Pb are present at submicromolar concentrations. Procedures required for total metal analysis are normally straight forward and state the acidification of samples to pH < 1 and preservation in inert containers such as Teflon. Furthermore, accurate determination of cadmium concentrations within sensitive samples may not be done by a smoking person in order to avoid cross-contamination.[23] Free cadmium in solution (Cd^{2+}) presents in water below pH values of 8. The highest concentration was measured to be 6 mM without hydroxide formation. Above pH 8, cadmium is expected to precipitate and hinder accurate determination of the concentration.[24] Removal of cadmium and lead from aerosol and fly ash samples is discussed in details by Lum[25] and Hlavay et al.[26] Total aqueous Pb and Cd concentrations can be measured using atomic absorption spectroscopy or inductively coupled plasma spectrometry, the former detecting ion concentrations at mg L^{-1} (ppm) levels and the latter at ng L^{-1} (ppt). Organometallic compounds of both Pb and Cd can be measured by inductively coupled plasma mass spectrometry, atomic absorption spectrometry, high- performance liquid chromatography, and gas chromatography analytical methods.[27–32]

Environmental Levels

Cadmium and lead are released into the environment as a result of both industrial activities and natural processes. The main natural sources include the weathering of rocks releasing Cd and Pb into the hydrosphere, and volcanic activity increasing the atmospheric Cd and Pb concentration.[15] The eruption of Mount Pinatubo located on the Philippine island of Luzon is an example of the dramatic effects of volcanism on the distribution of elements in the lithosphere. During just 2 days in June 1991, Pinatubo ejected 10 billion metric tonnes of magma and 20 million tonnes of SO_2; the resulting aerosols influenced global climate for 3 years. This single event introduced an estimated 100,000 tonnes of Pb and 1000 tonnes of Cd to the surface environment.[15,33] From anthropogenic sources, humans are exposed to Cd through the atmosphere, hydrosphere, and the geosphere. Human uptake of Cd can happen through inhalation of air, soil, or dust containing fine Cd particles. Cadmium is introduced into the atmosphere through the combustion of fossil fuel (e.g., coal), municipal solid waste incineration, mineral smelting and as dust generated by recycling scrap iron and steel. The contamination of soils is apparent from zinc and phosphate ores and where Cd-containing phosphate fertilizers are used in agriculture.[34–37] Zinc and cadmium are mineralogically and geochemically linked to each other. Thus, large emissions of fumes containing both ZnO and CdO are produced from zinc smelters. However, these fumes are normally enriched in CdO because of its higher volatility in comparison to ZnO. It is well documented that phosphate fertilizers constitute a very diffuse source of Cd contamination. The quantity of Cd contained in a phosphate fertilizer depends on the source of the phosphate rock used in making it. In general, Cd content in phosphate fertilizers vary from 1–2 ppm for tertiery Ca-phosphate to 50–170 ppm for superphosphate. As a result, Cd is transported to aquatic environments, plants, animals, and finally to humans. Soil organic matter strongly adsorbs Cd, and acidic soils further enhance the Cd uptake by plants. Cadmium is also known to accumulate in aquatic organisms, especially in freshwaters. However, Chen et al.[38] reported that the addition of potassium fertilizers effectively reduce the phytoavailability of both Cd and Pb within soils.

Although Pb is a naturally occurring element, its major cycle in the environment is anthropogenically driven and a result of human activities. For example, Pb in the form of tetraethyl-lead has been in usage as additive in gasoline to avoid knocking effect of autoignition since the early 20th century.[39] Through this application, Pb has been released into the environment in the forms of Pb-chloride, -bromide, and -oxides from the car exhaust. Lead is also introduced into the environment from coal and solid waste combustion, a wide number of industrial and mining processes, and drinking water pipes containing Pb. When copper is present in (either from soldering or as contamination) a Pb-containing pipe, it accelerates corrosion of the Pb pipe by galvanic action. The water flowing in contact with these dissimilar metals serves as the electrolyte. Based on the electrochemical series, metallic Pb serves as the anode of this galvanic cell and is therefore oxidized (i.e., corroded) to form Pb^{2+} ions contaminating the drinking water. Lowering pH in water also promotes the dissolution and aqueous transport of Pb^{2+}. Evidences of galvanic corrosion of domestic Pb pipes had been given and discussed elsewhere. [40,41] Large Pb-containing particles released to the atmosphere settle quickly on the ground and then washed into soils or dissolve in aquatic phases, while the very fine particles remain in the atmosphere travelling long distances and fall back to the surface with rain. This cycle has caused an unnatural and extensive sequence exposing plants, aquatic environments, and humans to dangerous concentrations of Pb. Therefore, in 1978, it was forbidden for all European Union-member states to produce, import, or sell gasoline with more than $0.4\,g\ Pb\ L^{-1}$. Starting from the year 2000, the marketing of leaded petrol has been banned in Europe following Directive 98/70/EC and related acts of the European Parliament. In the United States, Pb was banned as fuel additive starting from 1996.[12,42–44] Unfortunately, the gasoline additives tri- and dialkyl Pb are stable compounds and are persistent in the environment; thus, the restriction in usage does not necessarily decrease contamination.[45–51] When Pb accumulates in living organisms, it becomes part of the food chain thereby creating further sources for human exposure. Beyond bioaccumulation, once released into surface waters, both Cd and Pb are deposited into the sediment, increasing the metal contamination by 10–100 times near the fallout areas. The type of sediment affects the severity of contamination, with carbonaceous, anoxic, and clay sediments being the most prone to high concentrations of metal uptake, while siltstones, shales, sandstones, limestones and marine evaporates are normally less affected.[52]

Toxicological Effects

The International Agency for Research on Cancer (IARC) has classified Cd and Pb compounds as carcinogenic to humans.[87] This classification has been based mainly on epidemiological evidences of renal damage in rats and mice. The greatest concerns about the health effects of Cd and Pb arise from their tendency to form strong complexes, replacing essential elements (e.g., Ca^{2+}) and bioaccumulating in the human body. There is no known small enough amount of Pb uptake that would cause no harm to human beings. However, Reichlmayr-Lais et al.[53,54] found that the depletion of Pb resulted in hematological changes in rats. The concentration of Cd is approximately $0.4\ pg\ kg^{-1}$ in a daily diet, which is said to be 10 times lower than the amount that can cause kidney damage.[55] The concentrations of Cd in human blood is usually between 0.1 and $2\ pg\ L^{-1}$ and in urine $<1\ pg\ L^{-1}$.[56] The concentration of Pb in human blood varies between 165 and $296\ pg\ L^{-1}$. The largest known catastrophe caused by Cd toxicity was identified in Japan in the 1940s, and it is referred to as the "itai-itai" disease, which is literary translated from Japanese to "ouch ouch" disease. Itai-itai refers to a syndrome that principally consists of a painful skeletal condition resulting from weak and deformed bones. The patients of itai-itai suffer from renal anemia, tubular nephropathy, and osteopenic osteomalacia, while 90% of the patients are postmenopausal women. The residents of the Jinzu River basin region were first exposed to Cd in the 1930s as a result of industrial contamination from nearby intensive mining activities, which caused serious pollution of the local river waters. This resulted in high Cd contaminations of rice fields. With rice being the principal dietary component, especially in rural Japan, and the bioaccumulating properties of Cd, residents were exposed to very high levels of Cd causing irreversible damage and poisoning.[57,58] Toxicity of Cd in living organisms occurs because of the substitution of essential elements such as Zn at the reactive centers of essential enzymes, which disrupts a wide range of metabolic functions. Further details on mechanisms of Cd toxicity are available elsewhere.[57,59–62]

Cadmium is taken into the body through food, drinking water, smoking, and particulate matter in air, especially near Cd-processing industrial fields and hazardous waste sites. Gastrointestinal absorption from contaminated water or food is the main source of internally deposited Cd in the general population. Only a small proportion of ingested Cd is transferred to the bloodstream while the unabsorbed

Cd is excreted in the feces. The absorbed Cd binds to macromolecules, enzymes, and proteins, and the majority of it is being deposited in the liver and kidney. The Cd taken up by the kidney interferes with the filtering mechanism, causing the excretion of essential sugars and proteins, and damages the kidney's ability to remove acid from the blood. Another extremely painful effect of Cd is the softening of bones and decreasing their mineral density, resulting in fractures and paralysis in advanced cases. Furthermore, Cd damage of the central nervous system and psychological disorders were reported, as well as weakening of the immune system, reproductive failure, DNA damage, and the development of cancer.[63]

Lead can enter the human body through food, liquid, and air. Inorganic Pb compounds are known to pass through the skin.[64] It has been shown that the pollution of air with Pb particles from burning of fuel affects the cognitive behavior of children and adults living near busy roads or exposed to Pb contamination. Early experiments were carried out to study the dose-response to environmental levels of Pb. A comprehensive review of these studies can be found in U.S. Environmental Protection Agency (USEPA) 2005.[65] The absorbed Pb by the human body is distributed in blood, soft tissues, and in particular in bones and liver. Because Pb^{2+} can replace Ca^{2+}, ~9% of absorbed Pb ends up in the bones and teeth. The excess of Pb may cause several health effects, including damage to nervous system, chronic renal disease, anemia due to the inhibition of haem formation, damage of nervous system of unborn children, acute encephalopathy in young children, carcinogenicity and genotoxicity, and impaired reproductivity.[66–68] Even in low concentrations (1–10 pg dL^{-1} blood), Pb decrements neurocognitive abilities, intelligence measures, and perceptual-motor coordination. Lead toxicity of animals has been also studied extensively. For example, cattle showed poisoning symptoms within 6–8 wk when fed Pb-acetate at 7 mg Pb per kilogram body weight per day.[69–71] Accumulation of Pb in the liver and kidney was noted in calves fed 100 mg Pb per kilogram as Pb chromate for 100 days.[72] While the absorbed Cd and Pb could be excreted in urine, the daily excretion is <1% of the total body burden of Cd and Pb giving a biological half-life of Cd of more than 25 years[73,74] and 20–30 years for Pb in the skeleton.[75]

Cadmium and lead toxicity in animals is a major problem, especially in diary animals as Cd and Pb accumulate in the kidney, liver, and reproductive organs.[88] Many plants species are tolerant to certain amounts of heavy metals, which is likely to be achieved via metal-binding by specific proteins. Like all living organisms, at elevated concentrations of certain heavy metal plants start to show symptoms of toxicity. For example, Cd phytotoxicity can be identified in the form of stunting and chlorosis. Chlorosis is due to Cd interaction with foliar iron. A number of reports had shown evidences to Pb adverse effects on the growth and photosynthesis processes of plants. High concentration of Pb in soils is known to inhibit seed germination in a number of plant species and to induce abnormal morphologies.[89–91] It is interesting to note that a number of reports have shown that Cd accumulates in greater concentrations roots, tubers or leaves of plants. This means that Cd can present at higher concentrations in leafy and root vegetables than in fruits or grains.[91]

Regulations and Control

In the European Union, Cd and Pb are on the list of the six hazardous substances that are banned in the manufacturing of various electrical and electronic components. This is enforced by the Restriction on Hazardous Substances Directive (RoHS, 2002/95/EC) requiring that the maximum concentration of Cd or Pb may not exceed 100 or 1000 ppm per weight of homogeneous material, respectively. The production, recycling, and disposal of batteries and accumulators are regulated by the 2006/66/EC directive. These metals are also regulated under the Registration, Evaluation, Authorisation, and Restriction of Chemicals (REACH, 1907/2006) by the European Commission. In the United States, the USEPA is the main regulatory body restricting and controlling the use of Cd and Pb. Generally, the legislations are separated by pollution of the atmosphere, geosphere, hydrosphere, or specific industrial activities. Furthermore, the World Health Organization (WHO) globally regulates the acceptable exposure levels and concentration limits in public areas (Table 1). The maximum permissible concentrations of Cd and

TABLE 1   General Human Exposure to Cadmium and Lead

	Food	Water	Air
Cadmium	25 pg kg^{-1} body weight per month	3 pg L^{-1}	5 ng m^{-3} a^{-1}
Lead	0.8 pg kg^{-1} body weight per day	3.8–10 pg L^{-1}day^{-1}	0.5–4 pg m^{-3}day^{-1}

Source: Data from WHO.[83,84]

TABLE 2   Drinking Water Regulations for Maximum Permissible Contaminant Levels (USEPA, National Primary Drinking Water Regulations)

	USEPAa (mg L−1)	WHO Guideline Value	DWI-DEFRA (UK)
Cadmium	5.0	3.0	5.0
Lead	zero action level: 0.015	10	25^b

^{a} Maximum contaminant level goal.
^{b} Maximum concentration is 25 mg L−1 until December 25, 2013, and 10 mg L−1 from December 25, 2012.
Source: Data from WHO[84,85] and DWI.[86]

Pb in drinking water determined by the USEPA, WHO, and Drinking Water Inspectorate-Department for Environment, Food and Rural Affairs (DWI-DEFRA, United Kingdom) is shown in Table 2. As a result of these regulations, the release of Cd and Pb into the environment and their human exposure has been lowered through recycling and safer, more conscious industrial processes. However, the need for such metals and the global consumption still persists. On an annual basis, 9.6 million tonnes of Pb and 19,000 tonnes of Cd are produced worldwide. To reduce and control exposure of the population through drinking water, regulations require the use of Pb-free (<0.2% Pb) pipe systems and continuous monitoring (USEPA). Cadmium and lead can be removed from water by reverse osmosis or ion exchange resins. For the treatment of groundwater and soil media, *in situ* precipitation techniques present a viable way to reduce the mobility of heavy metals. In the case of both Cd- and Pb phosphate-containing minerals (e.g., hydroxyapatite), calcium-carbonates and zeolites have been suggested as possible solutions.[76–82]

Conclusion

Both Cd and Pb have played prominent roles in the industrial revolution and subsequent centuries, and are now included in a vast range of products. Both elements have also played important, -but highly contrasting- roles in terms of human, animal, and plant health. The increased environmental concentration of Cd and Pb poses irreversible effects on nature and the human body. The contamination problem of heavy metals -including Cd and Pb- remains a challenging issue to scientists and engineers. The question whether the dangers associated with Cd and Pb could be avoided or could be further lowered remains open. Production of Cd and Pb has grown to be a global need, which can only be compensated by the development and use of alternative technologies, advanced remediative solutions and preventive measures, sensitive monitoring systems, and the avoidance of human and animal exposures.

References

1. Greenwood, N.N.; Earnshaw, A. *Chemistry of the Elements*; Elsevier: Oxford, 1997.
2. Grant, L.D. Lead and compounds. In *Environmental Toxicants;* Lippmann, M., Ed.; John Wiley and Sons, Inc.: Hoboken, NJ, 2009; 757–809.
3. Ahamed, M.; Siddiqui, M.K.J. Environmental lead toxicity and nutritional factors. Clin. Nutr. **2007**, *26* (4), 400–408.
4. Shukla, G.S.; Singhal, R.L. The present status of biological effects of toxic metals in the environment: Lead, cadmium, and manganese. Can. J. Physiol. Pharmacol. **1984**, *62* (8), 1015–1031.

5. Andersen, O. Chelation of cadmium. Environ. Health Per- spect. **1984**, *54*, 249–266.

6. Deagen, J.T.; Oh, S.H.; Whanger, P.D. Biological function of metallothionein. VI. Metabolic inter- action of cadmium and zinc in rats. Biol. Trace Elem. Res. **1980**, *2* (1), 65–80.

7. Huheey, J. *Inorganic Chemistry: Principles of Structure and Reactivity,* 4th Ed.; Harper Collins College Publishers: New York, 1993.

8. Lindsay, W.L. *Chemical Equilibria in Soils;* The Black Burn Press: Caldwell, NJ, 2001.

9. Turekian, K.K.; Wedepohl, K.H. Distribution of the elements in some major units of the earth's crust. Geol. Soc. Am. Bull. **1961**, *72* (2), 175–192.

10. Cox, P.A. *The Elements: Their Origin, Abundance, and Distribution;* Oxford University Press: Oxford, U.K., 1989.

11. Halka, M.; Nordstrom, B. *Periodic Table of the Elements: Transition Metals;* Facts on File: New York, 2011.

12. Kummer, U.; Pacyna, J.; Pacyna, E.; Friedrich, R. Assessment of heavy metal releases from the use phase of road transport in Europe. Atmos. Environ. **2009**, *43* (3), 640–647.

13. Halka, M.; Nordstrom, B. *Periodic Table of the Elements: Metals and Metalloids;* Facts On File: New York, 2011.

14. Fleischer, M.; Cabri, L.; Chao, G.; Pabst, A. New mineral names. Am. Mineral. **1980**, *65* (9–10), 1065–1070.

15. Garrett, R.G. Natural sources of metals to the environment. Hum. Ecol. Risk. Assess. **2000**, *6* (6), 945–963.

16. Sposito, G. *The Chemistry of Soils;* Oxford University Press, Inc.: Oxford, U.K., 1989.

17. Wiley-VCH. *Ullmann's Encyclopedia of Industrial Chemistry,* 7th Ed.; Wiley-VCH: Weinheim, Germany, 2011.

18. Tolcin, A. Cadmium. In *Minerals Yearbook Metals and Minerals 2009;* United States Government Printing Office, 2009.

19. Robert, U.A. Metals recycling: Economic and environmental implications. Resour. Conserv. Recycling **1997**, *21* (3), 145–173.

20. Harvey, T.; Thomas, B.; Mclellan, J.; Fremlin, J. Measurement of liver-cadmium concentra- tions in patients and industrial workers by neutron-activation analysis. Lancet **1975**, *305* (7919), 1269–1272.

21. Guberman, D.E. Lead. In *Minerals Yearbook Metals and Minerals 2009;* Guberman, D.E., Ed.; United States Government Printing Office: Reston, VA, 2009.

22. Sekaly, A.L.; Chakrabarti, C.; Back, M.; Gregoire, D.; Lu, J.Y.; Schroeder, W. Stability of dissolved metals in environmental aqueous samples: Rideau River surface water, rain and snow. Anal. Chim. Acta **1999**, *402* (1–2), 223–231.

23. Cornelis, R.; Caruso, J.; Crews, H.; Heumann K. *Handbook of Elemental Speciation I: Techniques and Methodology*; John Wiley and Sons, Ltd: Chichester, U.K., 2003.

24. Majidi, V.; Miller-Ihli, N.J. Potential sources of error in capillary electrophoresis-inductively cou- pled plasma mass spectrometry for chemical speciation. Analyst **1998**, *123* (5), 809–813.

25. Lum, K.; Betteridge, J.; Macdonald, R. The potential availability of P, Al, Cd, Co, Cr, Cu, Fe, Mn, Ni, Pb and Zn in urban particulate matter. Environ. Technol. Lett. **1982**, *3* (1–11), 57–62.

26. Hlavay, J.; Polyak, K.; Meszaros, E. Determination of the distribution of elements as a function of particle size in aerosol samples by sequential leaching. Analyst. **1998**, *123* (5), 859–863.

27. Bettmer, J.; Cammann, K. Transversely heated graphite atomizer-atomic absorption spectrometry (THGA AAS) in combination with flow injection analysis system-hydride generation (FIAS HG) as a reliable screening method for organolead compounds. Appl. Organomet. Chem. **1994**, *8* (7–8), 615–620.

28. Dunemann, L.; Hajimiragha, H.; Begerow, J. Simultaneous determination of Hg(II) and alkylated Hg, Pb, and Sn species in human body fluids using SPME-GC/MS-MS. Frese- nius J. Anal. Chem. **1999**, *363* (5–6), 466–468.

29. Fragueiro, M.S.; Alava-Moreno, F.; Lavilla, I; Bendicho, C. Determination of tetraethyl lead by solid phase microextraction-thermal desorption-quartz furnaceatomic absorption spectrometry. J. Anal. At. Spectrom. **2000**, *15* (6), 705–709.

30. Infante, H.G.; Sanchez, M.L.F.; Sanz-Medel, A. Vesicle- mediated high performance liquid chromatography coupled to hydride generation inductively coupled plasma mass spectrometry for cadmium speciation in fish cytosols. J. Anal. At. Spectrom. **2000**, *15* (5), 519–524.

31. Moens, L.; De Smaele, T.; Dams, R.; Van Den Broeck, P.; Sandra, P. Sensitive, simultaneous determination of organo- mercury, -lead, and -tin compounds with head-space solid phase microextraction capillary gas chromatography combined withinductively coupled plasma mass spectrometry. Anal. Chem. **1997**, *69* (8), 1604–1611.

32. Yu, X.; Yuan, H.; Gorecki, T.; Pawliszyn, J. Determination of lead in blood and urine by SPME/GC. Anal. Chem. **1999**, *71* (15), 2998–3002.

33. Selinus, O.; Finkelman, R.B.; Centeno, J.A. Human health and ecosystems. In *Geology and Ecosystems;* Zektser, I.S., Marker, B., Ridgway, J., Rogachevskaya, L., Vartanyan, G., Eds.; Springer US: Boston, 2006; 197–218.

34. Grant, C.A.; Sheppard, S.C. Fertilizer impacts on cadmium availability in agricultural soils and crops. Hum. Ecol. Risk Assess. **2008**, *14* (2), 210–228.

35. Jiao, Y.; Grant, C.A.; Bailey, L.D. Effects of phosphorus and zinc fertilizer on cadmium uptake and distribution in flax and durum wheat. J. Sci. Food Agric. **2004**, *84* (8), 777–785.

36. Syers, J.K.; Mackay, A.D.; Brown, M.W.; Currie, L.D. Chemical and physical characteristics of phosphate rock materials of varying reactivity. J. Sci. Food Agric. **1986**, *37* (11), 1057–1064.

37. Taylor, M.D. Accumulation of cadmium derived from fertilisers in New Zealand soils. Sci. Total Environ. **1997**, *208* (1–2), 123–126.

38. Chen, S.; Sun, L.; Sun, T.; Chao, L.; Guo, G. Interaction between cadmium, lead and potassium fertilizer (K_2SO_4) in a soil-plant system. Environ. Geochem. Health **2007**, *29* (5), 435–446.

39. Hernberg, S. Lead poisoning in a historical perspective. Am. J. Ind. Med. **2000**, *38* (3), 244–254.

40. Nguyen, C.K.; Stone, K.R.; Dudi, A.; Edwards, M.A. Corrosive microenvironments atlead solder surfaces arising from galvanic corrosion with copper pipe. Environ. Sci. Technol. **2010**, *44* (18), 7076–7081.

41. Zhang, Y.; Triantafyllidou, S.; Edwards, M. Effect of nitrification and GAC filtration on copper and lead leaching in home plumbing systems. J. Environ. Eng. **2008**, *134* (7), 521–530.

42. Geivanidis, S.; Pistikopoulos, P.; Samaras, Z. Effect on exhaust emissions by the use of methylcyclopentadienyl manganese tricarbonyl (MMT) fuel additive and other lead replacement gasolines. Sci. Total Environ. **2003**, *305* (1–3), 129–141.

43. Hagner, C. European regulations to reduce lead emissions from automobiles? Did they have an economic impact on the German gasoline and automobile markets? Reg. Environ. Change **2000**, *1* (3–4), 135–151.

44. Kerr, S.; Newell, R.G. Policy-induced technology adoption: Evidence from the U.S. lead phase down. J. Ind. Econ. **2003**, *51* (3), 317–343.

45. Allen, A.G.; Radojevic, M.; Harrison, R.M. Atmospheric speciation and wet deposition of alkyl-lead compounds. Environ. Sci. Technol. **1988**, *22* (5), 517–522.

46. De Jonghe, W.R.A; Chakraborti, D.; Adams, F.C. Sampling of tetraalkyl lead compounds in air for determination by gas chromatography-atomic absorption spectrometry. Anal. Chem. **1980**, *52* (12), 1974–1977.

47. Hewitt, C.; Harrison, R.M. A sensitive, specific method for the determination of tetraalkyl lead compounds in air by gas chromatography/atomic absorption spectrometry. Anal. Chim. Acta **1985**, *167*, 277–287.

48. Hewitt, C.; Harrison, R.M.; Radojevic, M. The determination of individual gaseousionic alkyllead species in the atmosphere. Anal. Chim. Acta **1986**, *188*, 229–238.

49. Nerin, C.; Pons, B.; Martinez, M.; Cacho, J. Behaviour of several solid adsorbents for sampling tetraalkyllead compounds in air. Mikrochim. Acta **1994**, *112* (5–6), 179–188.

50. Nielsen, T.; Egsgaard, H.; Larsen, E.; Schroll, G. Determination of tetramethyllead and tetraethyllead in the atmosphere by a two-step enrichment method and gaschromatographic- mass spectrometric isotope dilution analysis. Anal. Chim. Acta **1981**, *124* (1), 1–13.

51. Radziuk, B.; Thomassen, Y.; Van Loon, J.; Chau, Y. Determination of alkyl lead compounds in air by gas chromatography and atomic absorption spectrometry. Anal. Chim. Acta **1979**, *105*, 255–262.

52. Warren, L.J. Contamination of sediments by lead, zinc and cadmium: A review. Environ. Pollut. B. **1981**, *2* (6), 401–436.

53. Reichlmayr-Lais, A.M.; Eder, K.; Kirchgessner, M. The effect of lead supply on liver phospholipids of lactating rats. J. Anim. Physiol. Anim. Nutr. **1993**, *70* (1–5), 104–108.

54. Reichlmayr-Lais, A.M.; Eder, K.; Kirchgessner, M. Fatty acid composition of erythrocyte membranes, liver and milk of rats depending on different alimentary lead supply. J. Anim. Physiol. Anim. Nutr. **1993**, *70* (1–5), 109–116.

55. USEPA. *Cadmium Fact Sheet;* Agency of Toxic Substances and Disease Registry: Atlanta, GA, 2011.

56. Cornelis, R.; Heinzow, B.; Herbert, R.B.; Christensen, J.; Poulsen, O.; Sabbioni, E. Sample collection guidelines for trace elements in blood and urine (technical report). Pure Appl. Chem. **1995**, *67* (8/9), 1575–1608.

57. Ishihara, T.; Kobayashi, E.; Okubo, Y.; Suwazono, Y.; Kido, T.; Nishijyo, M. Association between cadmium concentration in rice and mortality in the Jinzu River basin, Japan. Toxicology **2001**, *163* (1), 23–28.

58. Kawano, S.; Nakagawa, H.; Okumura, Y.; Tsujikawa, K. A mortality study of patients with Itai-itai disease. Environ. Res. **1986**, *40* (1), 98–102.

59. Nogawa, K.; Yamada, Y.; Honda, R.; Ishizaki, M.; Tsuritani, I.; Kawano, S. The relationship between itai-itai disease among inhabitants of the Jinzu River basin and cadmium in rice. Toxicol. Lett. **1983**, *17* (3–4), 263–266.

60. Umemura, T.; Wako, Y. Pathogenesis of osteomalacia in itai-itai disease. J. Toxicol. Pathol. **2006**, *19* (2), 69–74.

61. Malcolm, E.; Morel, F. The biogeochemistry of cadmium. In *Metal Ions in Biological Systems*; Sigel, A., Sigel, H., Sigel, R., Eds.; Biogeochemical Cycles of Elements; CRC Press, Taylor and Francis Group: London, 2005; Vol. 43, 195–219.

62. Wilkinson, J.M.; Hill, J.; Phillips, C.J.C. The accumulation of potentially-toxic metals by grazing ruminants. Proc. Nutr. Soc. **2007**, *62* (02), 267–277.

63. Jarup, L.; Berglund, M.; Elinder, C.G.; Nordberg, G.; Vahter, M. Health effects of cadmium exposure—A review of the literature and a risk estimate. Scand. J. Work Environ. Health **1998**, *24* (Suppl 1), 1–51.

64. Sun, C.; Wong, T.; Hwang, Y.; Chao, K.; Jee, S.; Wang, J. Percutaneous absorption of inorganic lead compounds. AIHA J. **2002**, *63* (5), 641–646.

65. USEPA. *Ecological Soil Screening Levels for Lea;* U.S. Environmental Protection Agency, Office of Solid Waste and Emergency Response: Washington, D.C., 2005; OSWER Directive 9285.7–70.

66. Needleman, H.L.; Bellinger, D. The health effects of low level exposure to lead. Annu. Rev. Public Health **1991**, *12* (1), 111–140.

67. John, F.R. Adverse health effects of lead at low exposure levels: Trends in the management of childhood lead poisoning. Toxicology **1995**, *97* (1–3), 11–17.

68. Goyer, R.A. Lead toxicity: From overt to subclinical to subtle health effects. Environ. Health Perspect. **1990**, *86*, 177–181.

69. Aronson, A. Lead poisoning in cattle and horses following long-term exposure to lead. Am. J. Vet. Res. **1972**, *33* (3), 627–629.

70. Hammond, P.B.; Aronson, AL. Lead poisoning in cattle and horses in the vicinity of a smelter. Ann. N.Y. Acad. Sci. **1964**, *111*, 595–611.

71. Buck, W.B.; James, LF; Binns, W. Changes in serum transaminase activities associated with plant and mineral toxicity in sheep and cattle. Cornell Vet. **1961**, *51*, 568–585.

72. Dinius, D.A.; Brinsfield, T.H.; Williams, EE. Effect of subclinical lead intake on calves. J. Anim. Sci. **1973**, *37* (1), 169–173.

73. Kim, B.J.; Kim, M.; Kim, K.; Hong, Y.; Kim, I. Sensitizing effects of cadmium on TNF-a- and TRAIL-mediated apoptosis of NIH3T3 cells with distinct expression patterns of p53. Carcinogenesis **2002**, *23* (9), 1411–1417.

74. Golovine, K.; Makhov, P.; Uzzo, R.G.; Kutikov, A; Kaplan, D.J.; Fox, E. Cadmium down-regulates expression of XIAP at the post-transcriptional level in prostate cancer cells through an NF-**KB**-independent, proteasome-mediated mechanism. Mol. Cancer **2010**, *9* (1), 183.

75. Jarup, L. Hazards of heavy metal contamination. Br. Med. Bull. **2003**, *68* (1), 167–182.

76. Ahmed, I.A.M.; Young, S.; Crout, N. Time-dependent sorption of Cd^{2+} on CaX zeolite: Experimental observations and model predictions. Geochim. Cosmochim. Acta **2006**, *70* (19), 4850–4861.

77. Ahmed, I.A.M.; Young, S.; Crout, N. Ageing and structural effects on the sorption characteristics of Cd^{2+} by clinoptilo- lite and Y-type zeolite studied using isotope exchange technique. J. Hazard Mater. **2010**, *184* (1–3), 574–584.

78. Ahmed, I.A.M.; Crout, N.M.; Young, S.D. Kinetics of Cd sorption, desorption and fixation by calcite: A long-term radiotracer study. Geochim. Cosmochim. Acta **2008**, *72* (6), 1498–1512.

79. Xu, Y.; Schwartz, F.W.; Traina, S.J. Sorption of Zn^{2+} and Cd^{2+} on hydroxyapatite surfaces. Environ. Sci. Technol. **1994**, *28* (8), 1472–1480.

80. Ryan, J.A.; Zhang, P.; Hesterberg, D.; Chou, J.; Sayers, D.E. Formation of chloropyro-morphite in a lead-contaminated soil amended with hydroxyapatite. Environ. Sci. Technol. **2001**, *35* (18), 3798–3803.

81. Mavropoulos, E.; Rossi, A.M.; Costa, A.M.; Perez, C.A.C.; Moreira, J.C.; Saldanha, M. Studies on the mechanisms of lead immobilization by hydroxyapatite. Environ. Sci. Tech- nol. **2002**, *36* (7), 1625–1629.

82. da Rocha, N.C.C.; de Campos, R.C.; Rossi, A.M.; Moreira, E.L.; Barbosa, A.dF.; Moure, G.T. Cadmium uptake by hydroxyapatite synthesized in different conditions and submitted to thermal treatment. Environ. Sci. Technol. **2002**, *36* (7), 1630–1635.

83. WHO. Exposure to cadmium: A major public health concern. Public Health and Environment, World Health Organization: Geneva, Switzerland, 2010.

84. WHO. *Lead in Drinking-Water: A Background Document for Development of WHO Guidelines for Drinking-Water Quality*; Public Health and Environment, World Health Organization: Geneva, Switzerland, 2011; WHO/SDE/ WSH/03.04/09/Rev/1.

85. WHO. *Cadmium in Drinking-Water: A Background Document for Development of WHO Guidelines for Drinking- Water Quality*; Public Health and Environment, World Health Organization: Geneva, Switzerland, 2011; WHO/ SDE/WSH/03.04/80/Rev/1.

86. DWI. Legislative background to the Private Water Supplies Regulations 2009: Section 9 (E&W). Drinking Water Inspectorate; 2010.

87. IARC. International Agency for Research on Cancer Monographs on the Evaluation of the Carcinogenic Risks to Humans—Beryllium, Cadmium, Mercury, and Exposures in the Glass Manufacturing Industry; WHO: Lyon, 1993; vol. 58, 119–237.

88. Neathery, M.W.; Miller, W.J. Metabolism and Toxicity of Cadmium, Mercury, and Lead in Animals: A Review. Journal of Dairy Science **1975**, *58* (12), 1767–1781.

89. Nagajyoti, P.; Lee, K.; Sreekanth, T. Heavy metals, occurrence and toxicity for plants: a review. Environmental Chemistry Letters **2010**, *8* (3), 199–216.

90. Das, P.; Samantaray, S.; Rout, G.R. Studies on cadmium toxicity in plants: A review. Environmental Pollution **1997**, *98* (1), 29–36.

91. Kabata-Pendias A. Trace Elements in Soils and Plants. CRC Press, Inc.: Boca Raton, FL, 2011.

41

Heavy Metals: Organic Fertilization Uptake

Ewald Schnug,
Alexandra
Izosimova, and
Renata Gaj

Introduction

Heavy metals are elements with atomic weights ranging between 63.54 (Cu) and 200.59 (Hg), and specific gravity greater than 4.0 g/cm^3. Some elements may stimulate growth of an organism, although the evidence for this process is lacking. Micronutrients important to plant growth are Cu, Co, Mn, Fe, Cu, Zn, and Mo. Other trace elements do not play any physiological function and are toxic in extremely small quantities.[1,2] There is a strong link between micronutrient nutrition, the uptake, and the impact of contaminants on plants, animals, and humans.

Contamination with heavy metals has become one of the most serious problems for the functionality of ecosystems. Because of industrial development and past disposal activities, heavy metals are considered to be among the most important environmental contaminants that affect terrestrial, aquatic, and atmospheric systems. Concerns about these contaminants are based on the fact that certain trace elements (e.g., As, Cd, Pb, or U) are accumulating in the food chain threatening the health of humans and animals.[3] The risks of human, crop, and/or environmental toxicity posed by these elements are a function of their mobility and availability in soil.[4] The concentration of heavy metals in the soil solution, as the main parameter for the bioavailability of heavy metals to plants, is affected by numerous factors. Among others, the uptake of heavy metals by plants from soils is also affected by management factors like organic fertilization. Despite the fact that organic fertilizers themselves can be a source for heavy metals, nevertheless organic matter (OM) added with manures, compost, and sludges greatly increases soil physical features like colloidal stability and cation exchange capacity and chemical features like plant available nutrients and chelating compounds which have a beneficial effect on soil biology.

Organic fertilization may affect the transfer of heavy metals to plants by several mechanisms: enhancing metal adsorption as a result of increased surface charge and increasing the formation of organic complexes including changes of the redox potential in soils towards negative values. It may additionally increase the solubility of some elements like Mn on one hand, but also decrease the solubility of other elements like U.[5] The oxidation of OM releases protons and lowers the pH of the soil which increases Mn and Zn solution concentration or lowers the mobility of Mo.[6] A higher cation exchange capacity and increased concentrations of chelating agents, because of the organic fertilization, improve soils

TABLE 1 Typical Heavy Metal Concentrations in Selected Organic Fertilizers

Organic Fertilizers	Zn (mg/kg Dry Weight, Range)	Cd (mg/kg Dry Weight, Range)	Cu (mg/kg Dry Weight, Range)	Ni (mg/kg Dry Weight, Range)	Pb (mg/kg Dry Weight, Range)	References
Manure						
Pig	206–716	0.19–0.53	160–780	3–24.3	1.01–4.65	[18,19]
Cattle	41–238	0.15–0.40	26.2–55.8	1.7–9.1	1.5–8.4	[19]
Poultry	350–632	0.44–2.04	49.4–74.8	4.5–11.4	3.36–14.8	[19]
Slurry						
Pig	639–2115	0.32–1.08	197–773	8.6–11.6	2.7–7.0	[20,21]
Cattle	68–235	0.11–0.53	17.5–48.7	1.9–20.4	4.1–8.4	[19,22]
Sewage sludge	700–50,000	2–1500	50–3300	16–5300	50–3000	[18]
Green compost	72–181	0.11–1.7	17–432	5.7–25	15–131	[23]
Biowaste compost	125–371	0.11–0.7	34–215	7.2–37.7	19–435	[24]

storage capacity for heavy metals. This function acts as a buffer system against heavy metal transfers in ecosystems, either by preventing them from leaching or by counteracting their accumulation in food plants.[7] Metal retention processes are generally much more important than metal leaching processes.

Sources of Organic Fertilizers

Differences between organic fertilizers must also be considered. Some typical metal concentrations in organic fertilizers are given in Table 1. Narwal and Singh[8] observed that increasing rates of cow and pig manure decreased the amounts of extractable Cd and Ni, whereas the addition of peat, at the same rate as OM, had an increasing effect. At this point, the effects are linked to farming systems, which affect, either directly or indirectly, quantities of compounds of plant or animal origin introduced into the soil.[9]

All factors involved in the effects of OM with heavy metal mobility show distinct interaction, like pH and the formation of metal-organic complexes. Most metals (in free ionic form) have a higher mobility in acidic, coarse-textured soils,[10] but mobility, however, can be also significant at about neutral or higher pH owing to the fact that dissolved OM becomes itself more soluble at those pH levels and because of the formation of soluble organometallic complexes. That is why the release of Cd into the soil solution as metal-organic complex becomes significant when the soil pH exceeds 5.5,[11,4] and Pb leaching increases with a higher concentration of dissolved OM and pH.[12] Such interactions are of great importance when remediation of heavy metal polluted soils is discussed. In terms of preventing heavy metals from transferring to ecosystems, the aim of remediation is not decontamination, but immobilization through increasing the binding capacity of the soil.[13]

The largest sources for OM in agricultural systems are plant residues. Gupta et al.[14] observed a considerably higher content of cadmium in wheat grown after the addition of pulses than when the same crop was grown after cereals. This may indicate an effect of nitrogen applied with the organic fertilizers by a synergistic effect of the former on heavy metal uptake, which has been described by Schnug.[15]

In the point of view of negative anthropogenic activities on soil quality, the loss of OM is a particularly serious problem, which also diminishes the buffer capacity of soils for heavy metals. The utilization of wastes, as soil amendments to counteract losses of OM,[16,17] is only an apparent and short-sighted solution, because these sources are usually highly polluted with heavy metals (Table 1).

Plant Uptake of Heavy Metals

Typical heavy metal concentrations in not accumulating higher plants are (mg kg^{-1}): Cd, 0.1–1; Cu, 3–15; Hg, 0.1–0.5; Ni, 0.1–5; Pb, 1–5; Zn, 15–150.[16] There are numerous evidences of an increase in heavy metal phyto-availability and crop uptake related to land application of sewage sludge, fly ash, poultry litter, pig slurry, and other biosolids.[25–27] Supplementing swine feed with Cu has been a common practice for many years, and just recently environmental concerns have been expressed over elevated Cu concentration in soils receiving long-term application of swine waste. On such sites, Cu is preferentially adsorbed by OM associated with the coarse fraction in soil.[28] Sludge-derived OM contributes significantly to the metal adsorption capacity, and the slow mineralization of this OM can release metals into more soluble forms.[29] Because the decomposition of sludge OM is often associated with acidification of the soil, further increased bioavailability of the sludge-born heavy metals is to be considered[30] resulting in enhanced availability of heavy metals, decades after sludge application.[31] Wastes with unsuitable agronomic features can be processed to suitable fertilizers by a variety of methods such as drying, composting, alkaline stabilizing, or incineration to reduce sludge mass, volume, concentration, and mobility of metals.[4]

Any use of organic amendments must consider possible effects on heavy metal availability, especially when the initial contamination of a soil is already high. Under these circumstances, even the use of composts with low metal concentrations can be questionable because of the mobilizing action of soluble organic compounds.[32]

Conclusions

Organic fertilizers affect the heavy metal balance of agro ecosystems by supply, mobilization, and immobilization. Supply and mobilization are useful options in terms of feeding plants with essential heavy metals (e.g., Zn, Cu), but are a risk for loading sensitive environments. Immobilization in and removal from soil of heavy metals by phytoremediation [increased uptake by accumulating plant species (e.g., Polygonum spec.)] are deceptive options: the former leaves the risk in place, while the latter is too inefficient in terms of masses removed and just shifts the problem of contamination to other ecological compartments. The only feasible option for protecting soils, food, and environment is applying the precautionary principle by balancing all sources of input and output of heavy metals to an ecosystem.

References

1. Kabata-Pendias, A. *Trace Elements in the Biological Environment*; Geological Edition: Warsaw, 1979.
2. Tiller, K.G. Heavy metals in soils and their environmental significance. Adv. Soil Sci. **1989**, *9*, 113–142.
3. Iskandar, I.K.; Kirkham, M.B. *Trace Elements in Soil: Bioav-ailability, Flux and Transfer*; Lewis Publishers, 2001.
4. Richards, B.K.; Steenhus, T.S.; Peverly, J.H.; McBride, M.B. Effect of sludge-processing mode, soil texture and soil pH on metal mobility in undisturbed soil columns under accelerated loading. Environ. Pollut. **1999**, *109*, 327–346.
5. Bolan, N.S.; Duraisamy, V.P. Role of inorganic and organic soil amendments on immobilisation and phytoavailability of heavy metals: a review involving specific case studies. Aust. J. Soil Res. **2003**, *41*, 533–555.
6. Khoshgoftarmanesh, A.H.; Kalbasi, M. Effect of municipal waste leachate on soil properties and growth and yield of rice. Commun. Soil Sci. Plant Anal. **2002**, *33*, 2011–2020.
7. Hansen, L.G.; Schaeffer, D.J. Livestock and domestic animals. Soil amendments. In Impact on Biotic Systems; Re-chcigl, J.E., Ed.; Lewis Publishers, **1995**; 41–79.
8. Narwal, R.P.; Singh, B.R. Effect of organic materials on partitioning, extractability and plant uptake of metals in alum shale soil. Water Air Soil Pollut. **1998**, *103*, 405–421.
9. Oliver, D.P.; Schultz, J.E.; Tiller, K.G.; Merry, R.H. The effect of crop rotations and tillage practices on cadmium concentration in wheat grain. Aust. J. Agric. Res. **1993**, *44*, 1221–1234.
10. McBride, M.B. *Environmental Chemistry of Soils*; Oxford University Press: New York, 1994.
11. Naidu, R; Harter, N.D. Effect of different organic ligands on cadmium sorbtion and extractability from soils. Soil Sci. Soc. Am. J. **1998**, *62*, 644–650.

12. Klitzke, S.; Lang, F.; Kaupenjohann, M. Influence of pH and DOM on the mobilization of colloid-bound lead. DIAS Report, Plant Prod. **2002**, *80*, 261–265.
13. Senesi, N. Metal-humic substance complexes in the environment. Molecular and mechanistic aspects by multiple spectroscopic approach. In *Biogeomistry of Trace Metals*; Adriano, D.C., Eds.; Lewis Publishers Chelsea MI: Boca Raton, FL, 1992; 429–496.
14. Gupta, S.K.; Herren, T.; Wenger, K.; Krebs, R.; Hari, T. In situ gentle remediation measures for heavy metal-polluted soils. In *Phytoremediation of Contaminated Soil and Water*; Norman, T., Gary, B., Eds.; Lewis Publishers Chelsea MI: Boca Raton, 2000; 303–322.
15. Schnug, E. Differentiation between primary and secondary effects of N-fertilization on the micro-nutrient supply of cereals. Landwirtsch. Forsch. **1984**, *40*, 231–240.
16. McGrath, S.; Chaundri, A.M.; Giller, K.E. Long-term effects of metal in sewage sludge on soils microorganisms and plants. J. Ind. Microbiol. **1995**, *14*, 94–104.
17. Joergensen, R.G.; Meyer, B.; Roden, A.; Wittke, B. Microbial activity and biomass in mixture treatments of soil and biogenic municipal refuse compost. Biol. Fertil. Soils **1996**, *2*, 43–49.
18. Kabata-Pendias, A.; Pendias, H. In *Biogeochemistry of Trace Elements*; PWN: Warsaw, 1999.
19. Nicholson, F.A.; Chambers, B.J.; Williams, J.R.; Unwin, R.J. Heavy metal contents of livestock feeds and animal manures in England and Wales. Bioresour. Technol. **1999**, *70*, 23–31.
20. Marb, C.; Riedel, H. Changes in hazardous substances in sewage sludges during cold drying. Korrespondenz Abwasser. **1997**, *44* (8), 1386–1393.
21. Kühnen, V.; Bien, B.; Goldbach, H.E. Heavy metal balances (Cd, Cr, Cu, Ni, Pb, Zn) of pig farming. VD-LUFA-Kongress vom 2001 in Berlin, September 17–21, 2001; Darmstadt, VDLUFA-Verlag, Kongressband 57/II; VDLUFA-Schriftenreihe.
22. Menzi, H.; Kessler, J. Heavy metal contents of manures in Switzerland, Proceedings of the Eighth International Conference on the FAO Escorena Network on Recycling of Agricultural. Municipal and Industrial Residues in Agriculture, Rennes, France, May 26–29, 1998; Martinez, J., Maudet, M.-N, Eds.; 1998; 26–5.
23. Schmoll, H.; Held, T. Use of green compost in forestry. Forst Holz. **1997**, *52* (9), 245–249.
24. Schaaf, H.; Janßen, E. Heavy metal concentrations in manures and secondary raw material fertilizers and heavy metal loads in application of codes of good agricultural practice. *Kongressband 2000,* Darmstadt, VDLUFA-Verlag; VDLUFA-Schriftenreihe, 144–150.
25. Shuman, M.L. Organic waste amendments effect on Zn fractions of 2 soils. J. Environ. Quality **1999**, *28*, 1442–1447.
26. Jackson, B.P.; Miller, W.P.; Shuman, A.W.; Sumner, M.E. Trace elements solubility from land application of fly ash organic waste mixtures. J. Environ. Quality **1999**, *28*, 639–647.
27. Basta, N.T.; Sloan, J.J. Bioavailability of HM in strongly acidic soils treated with exceptional quality biosolids. J. Environ. Quality **1999**, *28*, 633–638.
28. Wu, J.; Laird, D.A.; Thompson, M.L. Sorption and desorption of Cu on soil clay component. J. Environ. Quality **1999**, *28*, 334–338.
29. McBride, M.B. Toxic metal accumulation from agricultural use of sludge: are USEPA regulations protective?. J. Environ. Quality **1995**, *2*, 5–18.
30. McGrath, S.P.; Zhao, F.J.; Dunham, S.J.; Grosland, A.R.; Coleman, K. Long-term changes in the extactability and bioavailability of Zn and Cd after sludge application. J. Environ. Quality **2000**, *29*, 875–883.
31. McBride, M.B.; Richards, B.K.; Steenhuis, T.; Russo, J.; Sauve, S. Mobility and solubility of toxic metals and nutrients in soil fifteen years after sludge application. Soil Sci. **1997**, *162*, 487–500.
32. Madrid, L.; Diaz-Barrientos, E.; Cardo, I. Sequential extraction of metals from artificially contaminated soils in the presence of various composts. In *Trace Elements in Soil: Bioavailability, Flux and Transfer;* Iskandar, I.K., Kirkham, M.B, Eds.; Lewis Publishers, 2001.

42

Inorganic Carbon: Global Carbon Cycle

William H.
Schlesinger

Introduction

The "global C cycle" is defined as the exchange of carbon (C) among the atmosphere, seawater, land vegetation, and soil reservoirs (Figure 1). Each year dead plant materials entering the soil are decomposed by soil microbes that return carbon dioxide (CO_2) to the atmosphere. If the amount of land vegetation remains the same, the amount of CO_2 removed from the atmosphere by plant growth each year is balanced by the amount of plant death and decomposition. Such a perfect balance, however, is seldom seen. Changes in the quantity of C in vegetation and soils play a major role in determining short- and long-term fluctuations in the concentration of CO_2 in earth's atmosphere. A portion of the current atmospheric increase in CO_2, for example, is because of the destruction of vegetation and the disturbance of soils by humans.

Mean Residence Time

For comparative purposes, biogeochemists calculate the mean residence time (MRT), or the amount of time C resides in each pool of the global C cycle before circulating to the others. For instance, a molecule of CO_2 spends, in average, about five years in the atmosphere before it enters the terrestrial biosphere or the oceans. A C atom spends, on average, about 10 years in vegetation and 35 years in soil organic matter (SOM) before it returns to the atmosphere as CO_2. In comparison, the circulation of C in the oceans is rather sluggish; the C atom spends, in average, hundreds of years in the sea, where it is found predominantly as dissolved bicarbonate (HCO_3^-). Human activities have the greatest impact on pools with short mean residence times.

Weathering

Most studies of soil C focus on SOM, which globally contains nearly 2300×10^{15} g of C.[2] The release of CO_2 by the microbial decomposition of SOM is one of the largest fluxes in the global C cycle (Figure 1). However, soils also contain C in various inorganic forms—CO_2 held in the soil pore spaces, bicarbonate dissolved in soil waters, and calcium carbonate ($CaCO_3$) as a soil mineral. Carbon dioxide in the soil

431

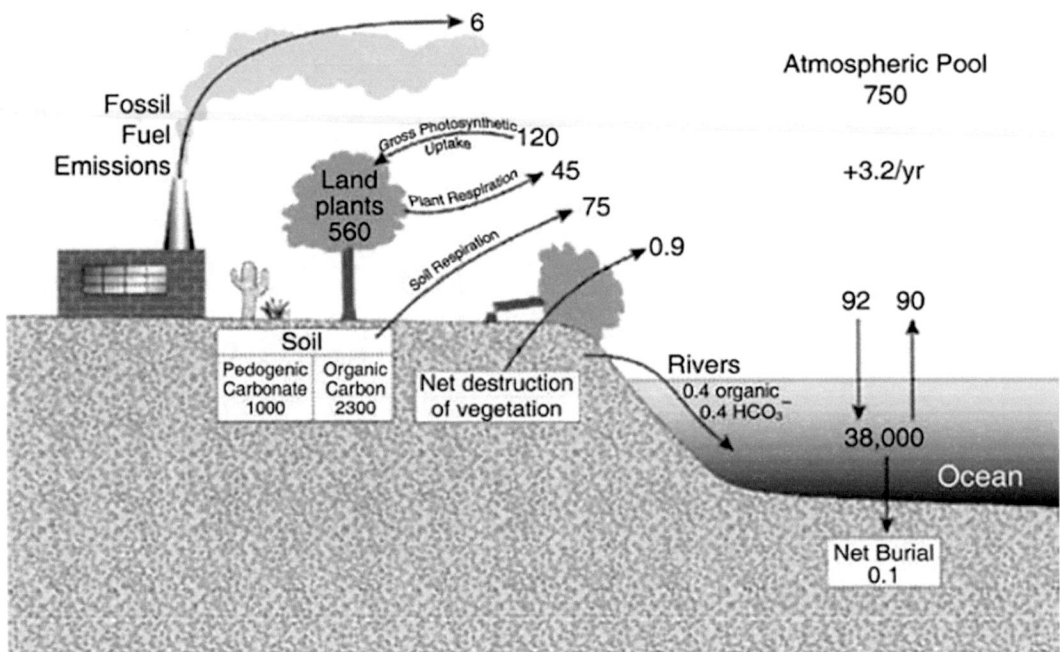

FIGURE 1   The global C cycle, showing the size of reservoirs (10^{15} gC) and the annual flux between them (10^{15} gC/yr). **Source:** Modified from Schlesinger.[1]

pores is largely derived from the respiration of plant roots and soil microbes, which varies as a function of soil moisture and temperature. However, even when the respiration rate is very high, the amount of C in soil gases and the soil solution is not large enough to contribute materially to the total C content of soils globally. In contrast, the amount of C held in soil carbonates is quite large, totaling about 750–1000×10^{15}gC (Figure 1).[3,4] The vast majority of this carbonate is found in the world's arid and semiarid lands, where inorganic C may exceed soil organic C by a factor of 10.[5]

In some soils, $CaCO_3$ and other carbonates (e.g., dolomite) are derived from the parent rocks from which the soils have formed. However, additional carbonate may form as a result of the release of Ca from the chemical weathering of rocks and the precipitation of $CaCO_3$ when water is lost from the soil by evaporation and plant uptake. In the case of silicate parent minerals, such as plagioclase (calcium feldspar), the relevant reaction of weathering is

$$CaAl_2Si_2O_8 + 3H_2O + 2CO_2 \rightarrow Ca^{2+} + 2HCO_3^- + Al_2Si_2O_5(OH)_4 \tag{1}$$

Here Ca is released from the silicate mineral by the weak solution of carbonic acid that is formed when CO_2 dissolves in water.

Declining soil concentration of either CO_2 or water precipitates carbonate via the reaction

$$Ca^{2+} + 2HCO_3^- \rightarrow CaCO_3 \downarrow + H_2O + CO_2 \tag{2}$$

While the weathering process occurs most rapidly when plant activity increases the concentration of CO_2 in the soil pore space, the precipitation reaction occurs during seasonal periods of drought. Carbonate formed in the soil, known as secondary or "pedogenic" carbonate, can be distinguished from carbonate inherited from parent materials by examining thin sections and the isotopic ratio (^{13}C vs. ^{12}C) of the C in carbonate minerals.

Atmospheric Derivation

In most areas, the Ca content of pedogenic carbonate is derived from the atmosphere and is deposited as a constituent of rain and dust.[6,7] However, because the Ca carried in the atmosphere is ultimately derived from the weathering of rocks in some upwind area, the two aforementioned reactions are general.[8] Across deserts of the southwestern U.S.A., the formation of pedogenic carbonate is closely related to mean annual rainfall (Figure 2).

If we know the rate of formation of pedogenic carbonate, usually expressed in grams per square meter per year ($g/m^2/yr$), and the total amount of carbonate in the soil profile, we can calculate the length of time taken for the accumulation of current quantity of soil pedogenic carbonate. In the Mojave Desert of California, radiocarbon and uranium-thorium dating show that pedogenic carbonate found in the upper 1.5 m of soils, derived from silicate materials, accumulated over a period of up to 20,000 years.[10] During that period, pedogenic carbonate formed at rates ranging from 1.0 to 3.5 $g/m^2/yr$, with some indications of greater rates during the Pleistocene, when rainfall was greater in this region. The global MRT of soil pedogenic carbonate is about 85,000 years, making this C pool much less dynamic than SOM, in which the global MRT for C is about 35 years (Figure 1).

While CO_2 is sequestered from the atmosphere during the formation of pedogenic carbonate from silicate parent materials, no such net storage occurs when pedogenic carbonate forms from carbonate parent materials.[11] Overall, the formation of soil pedogenic carbonates is less effective than the formation of SOM in the storage of atmospheric CO_2.[12] This is disappointing, of course, to those who view increasing the formation of pedogenic carbonate in desert soils as a means of slowing the rise of CO_2 in the earth's atmosphere and reducing global warming.

Effect of Human Activities

Human activities, such as the irrigation of agricultural soils in arid regions, can alter the accumulation of pedogenic carbonate in soils. The groundwater used for irrigation is often extracted from subsurface environments, where CO_2 concentration is much higher than in the earth's atmosphere, and often contains high concentrations of dissolved Ca. Applying such water to arid lands, precipitates dissolved Ca in the soil, thereby forming $CaCO_3$, and releases CO_2 to the atmosphere, via the reaction in Eq. 2. Precipitation

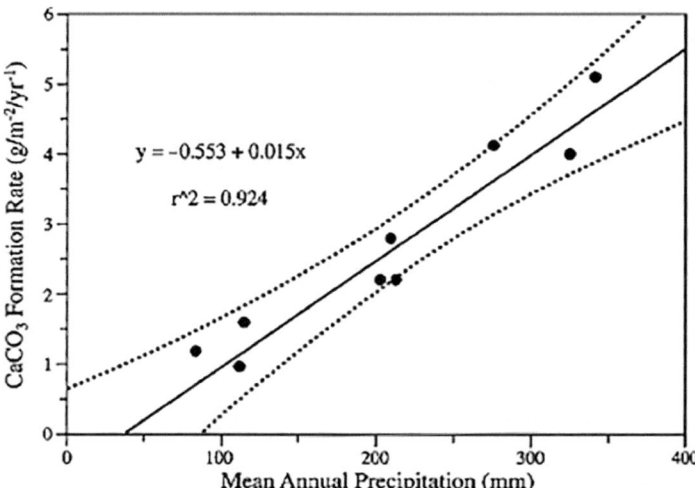

FIGURE 2 The rate of formation of pedogenic carbonate in arid soils of the southwestern U.S.A. as a function of the modern precipitation at each site.
Source: Marion.[9]

of calcite is also favored when large amounts of gypsum ($CaSO_4 \cdot 2H_2O$), a ready source of Ca, are used to remediate dryland soils. The formation of pedogenic carbonate in arid, agricultural soils, as a result of these human activities makes a small contribution to the current increase in atmospheric CO_2.[13]

Human activities leading to the formation of acid rain also affect soil carbonates. If the acidity in rainfall is derived from sulfuric acid (H_2SO_4), then CO_2 is released when the rain falls on carbonate-rich soils. The reaction is

$$CaCO_3 + H_2SO_4 \rightarrow Ca^{2+} + SO_4^{2-} + CO_2 \uparrow + H_2O \tag{3}$$

and gypsum ($CaSO_4 \cdot 2H_2O$) precipitates as the soil dries. Although most fossil fuels contribute small amounts of sulphur to the atmosphere, the global amount of CO_2 derived from this reaction is relatively small compared to the direct release of CO_2 from fossil fuel combustion (Figure 1).

Conclusions

Although the amount of pedogenic carbonate in world soils is quite large, perhaps as much as 1000×10^{15} gC, this pool of C is relatively sluggish. The long MRT of the pedogenic carbonate pool ensures that it will not become a major sink or source of atmospheric CO_2 over the next several centuries.

References

1. Schlesinger, W.H. *Biogeochemistry: An Analysis of Global Change*; Academic Press: San Diego, CA, 1997.
2. Jobbagy, E.G.; Jackson, R.B. The vertical distribution of soil organic matter and its relation to climate and vegetation. Ecol. Appl. **2000**, *10*, 423–436.
3. Eswaran, H.; Reich, P.F.; Kimble, J.M.; Beinroth, F.H.; Padmanabhan, E.; Moncharoen, P. Global carbon stocks. In *Global Climate Change and Pedogenic Carbonates;* Lal, R., Kimble, J.M., Eswaran, H., Stewart, B.A., Eds.; Lewis Publishers: Boca Raton, FL, 2000; 15–25.
4. Batjes, N.H. Total carbon and nitrogen in soils of the world. Eur. J. Soil Sci. **1997**, *47*, 151–163.
5. Schlesinger, W.H. Carbon storage in the caliche of arid soils: a case study from Arizona. Soil Sci. **1982**, *133*, 247–255.
6. Capo, R.C.; Chadwick, O.A. Sources of strontium and calcium in desert soil and calcrete. Earth Planet. Sci. Lett. **1999**, *170*, 61–72.
7. Chiquet, A.; Michard, M.; Nahon, D.; Hamelin, B. Atmospheric input vs. in situ weathering in the genesis of calcretes: an Sr isotope study at Gálvez (Central Spain). Geochim. Cosmochim. Acta. **1999**, *63*, 311–323.
8. Naiman, Z.; Quade, J.; Patchett, P.J. Isotopic evidence for Eolian recycling of pedogenic carbonate and variations in carbonate dust sources throughout the southwest United States. Geochim. Cosmochim. Acta. **2000**, *64*, 3099–3109.
9. Marion, G.M. Correlation between long-term pedogenic $CaCO_3$ formation rate and modern precipitation in deserts of the American southwest. Quaternary Res. **1989**, *32*, 291–295.
10. Schlesinger, W.H. The formation of caliche in soils of the Mojave Desert, California. Geochim. Cosmochim. Acta. **1985**, *49*, 57–66.
11. Monger, H.C.; Martinez-Rios, J. Inorganic carbon sequestration in grazing lands. In *The Potential of U.S. Grazing Lands to Sequester Carbon and Mitigate the Greenhouse Effect;* Follett, R.F., Kimble, J.M., Lal, R., Eds.; CRC Press: Boca Raton, FL, 2000; 87–118.
12. Chadwick, O.A.; Kelley, E.F.; Merritts, D.M.; Amundson, R.G. Carbon dioxide consumption during soil development. Biogeochemistry **1994**, *24*, 115–127.
13. Suarez, D.L. Impact of agriculture on CO_2 as affected by changes in inorganic carbon. In *Global Climate Change and Pedogenic Carbonates;* Lal, R., Kimble, J.M., Eswaran, H., Stewart, B.A., Eds.; Lewis Publishers: Boca Raton, FL, 2000; 257–272.

Inorganic Carbon: Modeling

Leslie D. McFadden
and Ronald G.
Amundson

Introduction

Virtually all carbon in soils of arid and semiarid regions of the world accumulates as pedogenic calcium carbonate (referred to subsequently as carbonate). The carbonate usually accumulates in layers that eventually attain the status of calcic horizons and, in much older soils, petrocalcic horizons. Numerous mechanisms for the accumulation of pedogenic carbonate in soils are recognized, but the most fundamental reason for accumulation is limited depth of soil-water movement and seasonally high evapotranspiration that favors precipitation of carbonate within the soil.[1] Many studies of calcic soils in the past few decades demonstrate a close correspondence between the depth of pedogenic carbonate accumulation and modern annual precipitation,[2-4] although recent studies show the relationship may be more complicated.[5] Other studies also show progressive, time-dependent accumulation in many environments[1]; these studies have led to the now generally accepted conceptual models of calcic soil development.[6]

Development of a numerical model of carbonate accumulation, however, is a more challenging proposition, given the remarkably complex character of the soil system. Fortunately, certain aspects of calcic soils facilitate formulation of such numerical models. For example, the observed soil depth-climate relationship implies that, utilizing a sound strategy for simulation of water movement, determination of carbonate movement via solution transport is a reasonable proposition. In addition, a significant body of research shows that the majority of carbonate is derived from accumulated entrapped dust and Ca in rainwater.[1] Finally, data pertaining to calcite geochemistry and dissolution rates in different environments are available and show that, in soils associated with typical ranges in soil CO_2, pH, and salinity, calcite is far more soluble than virtually all silicate minerals and has more rapid dissolution rates.[7] Consequently, a relatively simple model for carbonate movement in a soil based on relations in the $CaCO_3$–H_2O–CO_2 system can be formulated that essentially ignores the more complex chemical reactions involving aluminosilicates. Research on the nature and composition of stable and unstable isotopes in pedogenic carbonate has also helped elucidate the nature of calcic soil development and improve the design for testing the results of numerical modeling.

Compartment Model and Simulations of Pedogenic Carbonate Accumulation

The compartment-model, or "box-model," approach to modeling of calcic soils accommodates continuously changing values among the interdependent variables that influence soil development. It enables integration of several factors that influence pedogenic carbonate accumulation and that can be explicitly considered in this model. These include soil–water movement and soil–water balance, changing soil CO_2 concentrations and temperature with depth and season, initial parent material composition, carbonate and soluble salt additions from external sources, and calcite reactant surface area. The soil profile is represented by a vertical sequence of compartments of arbitrary dimensions, with the initial characteristics of each compartment specified (i.e., texture, available water-holding capacity, pCO_2). A series of equations that enable forward modeling and simulation of evolving carbonate depth functions using the box-model approach can be derived on the basis of consideration of the factors indicated above. For example, the solubility of calcite is derived from the following equation, after Drever[8]:

$$m^3Ca^{2+} = \left(pCO_2 K_1 K_{cal} K_{CO_2}\right) / \left(4K_2\gamma Ca^{2+} + \gamma^2 HCO_3^-\right) \tag{1}$$

where K_{cal} is the calcite solubility product and K_1, K_2 and K_{CO2} are dissociation constants in the carbonate system, and γCa^{2+} and γHCO_3^- are the activity coefficients of Ca^{2+} and HCO_3^- A gravelly, permeable calcic soil probably best approximates an open system-weathering environment, in which case calcite dissolution rates are probably surface-area controlled rather than diffusion controlled. Also, the dissolution rate is defined ultimately by the rate-limiting conversion of dissolved carbon dioxide (CO_2^*) to H_2CO_3. At a very low solution volume to surface area ratios, and with fast, surface-controlled calcite dissolution rates, H_2CO_3 is rapidly depleted. In such circumstances, a commonly used rate equation that enables determination of dissolution rates in the CO_2–$CaCO_3$–H_2O system[9] is:

$$dC/dt = \left(A'k/V\right)\left(1 - C/C^*\right)^n \text{ mgl}^{-1}\text{s}^{-1} \tag{2}$$

where A' is the surface area of rock in contact with water (cm²), V the water volume (cm³), k the reaction coefficient (mg cm l⁻¹s⁻¹), n the reaction order, C the moles of calcite in solution, and C^* is the solubility of calcite. Values of n and k vary with saturation ratio, temperature, and pCO_2. In the model, A'/V can be specified depending on observed soil features. Eqs. 1 and 2 show that soil CO_2 content is a very important variable, but soil CO_2 contents may be highly variable.[10] Fortunately, studies show that a depth function for pCO_2 that reflects prolonged seasonal respiration levels can be estimated, assuming the concentration of soil CO_2 is described by mass transport of CO_2 by gas diffusion.[11,12] The following diffusion-reaction equation, essentially Fick's Second Law for a one-dimensional case, is used in the model:

$$\partial C_s/\partial t = D_s\left(\partial^2 C_s/\partial z^2\right) + \phi_s(z) \tag{3}$$

where C_s is the concentration of CO_2 in the soil (mol cm³), t the time (s), D_s the diffusion coefficient for CO_2 in the soil (cm²s⁻¹), z the depth in the soil (cm), and $\phi_s(z)$ is the production rate of CO_2 as a function of depth (mol cm⁻³s⁻¹). At steady state, when $\partial C_s/\partial t = 0 = D_s\partial^2 C_s/\partial z^2 + \phi_s$, the general solution to this equation to produce a simple production function is:

$$C_s(z) = \phi/D_s\left(Lz - z^2/2\right) + C_0 \tag{4}$$

where C_0 is the concentration of CO_2 in the atmosphere (ppm) and L is the depth to the lower, no-flux boundary. Soil CO_2 contents with depth calculated using this method are used to calculate carbonate solubility and dissolution rates with depth.

Available water-holding capacity, infiltration, and percolation rates can be specified on the basis of laboratory soil measurements or estimated from field measurements or theoretical considerations.

Earlier versions of the simulation model included certain assumptions that simplified numerical calculations, such as simple vertical saturated flow and constant soil temperature with depth. The lack of certain types of data (e.g., variation of pCO_2 with depth and time of year) also constituted a limitation on utility of the model. The model did enable simulation of 1) realistic depths and magnitudes of carbonate accumulation over thousands of years and 2) the range of effects of large climatic changes on calcic soils.[13–16] Model results emphasized the critical roles of external Ca^{2+} influx and influence of soil CO_2 concentrations on carbonate accumulation. Model-simulated bimodal concentrations of carbonate based on theoretical, late-Pleistocene climatic conditions resembled those observed in late- Pleistocene, polygenetic soils; however, incompletely understood changes in the magnitude of climate changes, dust flux, and vegetation change in the Quaternary complicate attempts to simulate polygenetic soils.[7,14,16]

Later versions of the model utilized important new inputs and employed routines that reflected improved understanding of key processes that strongly influence calcic soils. Studies of dust accumulation rates in the American South- west,[17] C, O, and Sr isotopes in carbonate, and development of more sophisticated models for unsaturated flow in calcic soils[18] have allowed development of improved compartment models that can address new and more challenging research problems. For example, such numerical simulations demonstrate how climate changes in the Holocene might have dramatically influenced the rates and temporal patterns of soluble salt leaching and accumulation relative to pedogenic carbonate.[19] A more recent modeling study addressed the problem of how carbonate can occasionally accumulate at much shallower depths than those expected from the depth—annual leaching depth relationship.[20] This study showed how carbonate can be preferentially removed from depths of a few cm to a few dm below the soil surface, while carbonate simultaneously accumulates either as collars on surface pavement clasts or in the vesicular A horizon. Model results also explain how a significant change in climate or soil erosion rates could cause the dissolution of carbonate rinds on the tops and sides of boulders and/or the tops of limestone boulders at depths of up to several dm, unusual features observed in some calcic soils.[21]

Isotopes in Calcic Soils

During weathering, parent material carbonate undergoes dissolution and reprecipitation in the soil. The carbon ($^{13}C/^{12}C, ^{14}C/^{12}C$) and oxygen ($^{18}O/^{16}O$) isotope ratios of pedogenic carbonate that forms from dust or parent material carbonate, or from Ca^{2+} derived from silicate weathering, are determined by isotopic composition of soil CO_2 and H_2O. These are the primary carbon and oxygen reservoirs, respectively, for the carbonate. Therefore, pedogenic carbonate reflects only isotopic conditions of the soil and bears no memory of the isotopic composition of the rock or mineral from which it was derived.

Soil CO_2 is derived primarily from decomposition of soil organic matter and root respiration. The C isotope composition of soil CO_2 reflects: 1) the isotopic composition of these CO_2 sources; 2) the effects of the diffusion of this CO_2 toward the atmosphere; and 3) the isotopic composition of atmospheric CO_2. In the 1980s, researchers recognized that fairly simple, steady state, diffusion models could be used to reasonably explain the observed depth patterns of C isotopes in both soil CO_2 and pedogenic carbonate. The solution to the mathematical model that encompasses the forementioned processes describes the abundance of $^{12}CO_2$ in soils. A related equation can be derived for $^{13}CO_2$, and the ratio of the two models then describes the ratio of C isotopes at any given soil depth. A similar approach can also be used to model the ^{14}C composition of CO_2 with soil depth, with the additional complication that the two main sources of soil CO_2 (humus decomposition and root respiration) have different ^{14}C contents, making the solution to the model slightly more complex.[22] The C isotope diffusion model of soil CO_2 has provided the opportunity to quantitatively use pedogenic carbonates in a number of applications: 1) paleovegetation studies[23]; 2) paleo-atmospheric CO_2 studies[24]; and 3) radiocarbon dating of pedogenic carbonate and geomorphic surfaces.[25,26]

The O isotopic composition of soil water is determined by the O isotope composition of precipitation and the evaporation of soil water. It has been observed that the ^{18}O content of modern precipitation is generally correlated with mean annual temperature on a global scale,[27] but regional differences due to storm sources can obscure these patterns.[28] If precipitation water (once stored in the soil) is subject to evaporation, an enrichment of the remaining soil water in ^{18}O occurs because water vapor depleted in the "heavy" isotope is preferentially removed during evaporation. Models have been made that successfully explain the key components of this process[29] and the fact that soils subject to evaporation commonly have generally decreasing ^{18}O contents of soil water with depth.[30] These models have two components. The first is a vapor transport layer (describing the flow of evaporating soil water to the atmosphere through a dry soil layer), and the second is an evaporating front layer. The evaporating front layer exists below the vapor transport zone. At the evaporating front ^{18}O enrichment of soil water occurs as water is transferred to a vapor phase, and the remaining ^{18}O-enriched soil water at the evaporating front then undergoes diffusional mixing with the less ^{18}O-enriched water at greater depths. In general, these models have been more difficult to use than C isotope models due to the dynamic nature of soil water (steady state assumptions are difficult to apply) and the array of model parameters, many of them not known with certainty for most soils.

Oxygen isotopes in pedogenic carbonate have been used less extensively in paleoclimate work than C isotopes because of concern over possible evaporation of soil water that formed the carbonate. Amundson et al.[28] demonstrated that, except for hyperarid regions, the O isotope composition of pedogenic carbonate appears to reasonably reflect that of the local precipitation. There is a growing list of studies using carbonate O isotopes in Quaternary[28] and Tertiary[31] paleoclimate applications.

Conclusions

Numerical models and isotope studies have proven to be valuable tools in the study of calcic soil development. They have helped elucidate the relation of climate, vegetation, and geomorphic processes to carbonate accumulation. The models are not able to explain the observed character of certain aspects of calcic soils, such as patterns of pedogenic carbonate development in some soil chronosequences,[32] or the somewhat enigmatic formation of calcic soils in humid, monsoonal climates. Also, these models are not designed to simulate the evolution of very old, morphologically complex soils with petrocalcic horizons. Future models must be designed to address locally abundant calcic soils. Additional fieldwork and application of recently developed field and laboratory techniques will provide the basis for development of the next generation of numerical models.

References

1. Birkeland, P.W. *Soils and Geomorphology*; Oxford University Press: New York, 1999; 720 pp.
2. Jenny, H. *Factors in Soil Formation*; McGraw-Hill: New York, 1941; 281 pp.
3. Arkley, R.J. Calculation of carbonate and water movement in soil from climatic data. Soil Sci. **1963**, *96*, 239–248.
4. Retallack, G.J. The environmental factor approach to the interpretation of paleosols. In *Factors of Soil Formation: A Fiftieth Anniversary Retrospective*; Amundson, R., Ed.; Special Publication 33; SSSA: Madison, WI, 1994; 31–64.
5. Royer, D.L. Depth to pedogenic carbonate horizon as a pa- leoprecipitation indicator? Geology **1999**, *27*, 1123–1126.
6. Machette, M.A. Calcic soils of the southwestern united states. In *Soils and Quaternary Geology of the Southwestern United States*; Weide, D.L., Faber, M.L., Eds.; Special Paper 203; Geological Society of America: Boulder, CO, 1985; 1–21.
7. McFadden, L.D.; Tinsley, J.C. The rate and depth of accumulation of pedogenic carbonate accumulation in soils: formation and testing of a compartment model. In *Soils and Quaternary Geology of the Southwestern United States*; Weide, D.L., Faber, M.L., Eds.; Special Paper 203; Geological Society of America: Boulder, CO, 1985; 23–42.

8. Drever, J.I. *The Geochemistry of Natural Waters*, 2nd Ed.; Prentice-Hall: Englewood Cliffs, NJ, 1991; 437 pp.

9. Palmer, A.N. Origin and morphology of limestone caves. Geol. Soc. Am. Bull. **1991**, *103*, 1–21.

10. Kiefer, R.H.; Amey, R.G. Concentrations and controls of soil carbon dioxide in sandy soil in the north Carolina coastal plain. Catena **1992**, *19*, 539–559.

11. Solomon, D.K.; Cerling, T.E. The annual carbon dioxide cycle in a montane soil: observations, modeling, and implications for weathering. Water Resour. Res. **1987**, *23*, 2257–2265.

12. Cerling, T.E.; Quade, J.; Yang, W.; Bowman, J.R. Carbon isotopes in soils and palaeosols, and ecology and palaeo- ecology indicators. Nature **1989**, *341*, 138–139.

13. McFadden, L.D.; Amundson, R.G.; Chadwick, O.A. Numerical modeling, chemical, and isotopic studies of carbonate accumulation in soils of arid regions. Soil Sci. Soc. Am. Spec. Publ. **1991**, *26*, 17–35.

14. McFadden, L.D. The impacts of temporal and spatial climatic changes on alluvial soils genesis in southern California. University of Arizona: Tucson, CA; 430.

15. Marion, G.M.; Schlesinger, W.H.; Fonteyn, P.J. Caldep: a regional model for soil CaCO3 (caliche) deposition in southwestern deserts. Soil Sci. **1985**, *139*, 468–481.

16. Mayer, L.; McFadden, L.D.; Harden, J.W. Distribution of calcium carbonate in desert soils: a model. Geology **1988**, *16*, 303–306.

17. Reheis, M.C.; Kihl, R. Dust deposition in southern Nevada and California, 1984–1989: relations to climate, source area, and source lithology. J. Geophys. Res. **1995**, *100*, 8893–8918.

18. McDonald, E.V.; Pierson, F.B.; Flerchinger, G.N.; McFadden, L.D. Application of a soil-water balance model to evaluate the influence of holocene climatic change on calcic soils, Mojave desert, California, USA. Geoderma **1996**, *74*, 167–192.

19. McFadden, L.D.; Crossey, L.J.; McDonald, E.V. Predicted response to calcic soil development to periods of significantly wetter climate during the late holocene (abstr.). Geol. Soc. Am. Abstr. Progr. **1990**, *24* (6), A252.

20. McFadden, L.; McDonald, E.; Wells, S.; Anderson, K.; Quade, J.; Forman, S. The vesicular layer and carbonate collars of desert soils and pavements: formation, age and relation to climate change. Geomorphology **1998**, *24*, 101–145.

21. Treadwell-Steitz, C.; McFadden, L.D. Influence of parent material and grain size on carbonate coatings in gravelly soils, palo duro wash, New Mexico. Geoderma **2000**, *94*, 1–22.

22. Amundson, R.; Stern, L.; Baisden, T.; Wang, Y. The isotopic composition of soil CO_2. Geoderma **1998**, *82*, 83–114.

23. Cerling, T.E. Development of grasslands and savannas in East Africa during the neogene. Palaeogeogr. alaeoclimatol. Palaeoecol. **1992**, *97*, 241–247.

24. Cerling, T.E. Use of carbon isotopes in paleosols as an indicator of the $P(CO_2)$ of the paleoatmosphere. Global Bio- geochem. Cycles **1992**, *6*, 307–314.

25. Amundson, R.; Wang, Y.; Chadwick, O.; Trumbore, S.; McFadden, L.D.; McDonald, E.; Wells, S.; DeNiro, M. Factors and processes governing the[14]C content of carbonate in desert soils. Earth Planet. Sci. Lett. **1994**, *125*, 385–405.

26. Wang, Y.; McDonald, E.; Amundson, R.; McFadden, L.D.; Chadwick, O. An isotopic study of soil in chronological sequences of alluvial deposits, providence mountains, California. Geol. Soc. Am. Bull. **1996**, *108*, 379–391.

27. Rozanski, K.L.; Aragua's-Aragua's, L.; Gonfiantini, R. Isotopic patterns in modern global precipitation. In *Climate Change in Continental Isotopic Records*; Swart, P.K., Ed.; American Geophysical Union Monograph 78; Washington, D.C., 1998; 1–36.

28. Amundson, R.; Chadwick, O.; Kendall, C.; Wang, Y.; DeNiro, M. Isotopic evidence for shifts in atmospheric circulation patterns during the late quaternary in midnorth America. Geology **1996**, *24*, 23–26.

29. Barnes, C.J.; Allison, G.B. The distribution of deuterium and ^{18}O in dry soils. I. Theory. J. Hydrol. **1983**, *60*, 141–156.

30. Allison, G.B.; Hughes, M.W. The use of natural tracers as indicators of soil water movement in temperate semiarid regions. J. Hydrol. **1983**, *60,* 157–173.

31. Quade, J.; Cerling, T.E.; Bowman, J.R. Development of the asian monsoon revealed by marked ecological shift during the latest miocene in Northern Pakistan. Nature **1989**, *343,* 163–166.

32. Holliday, V.T.; McFadden, L.D.; Bettis, E.A.; Birkeland, P.W. Soil survey and soil geomorphology. In *Profiles in the History of the U.S. Soil Survey;* Helms, D., Efflard, A., Dwara, P., Eds.; Iowa State University Press: Ames, IO, 2001.

44

Inorganic Compounds: Eco-Toxicity

Sven Erik Jørgensen

Introduction: The Elements and Their Biological Effect

Today's periodic table has 118 elements, but more than 20 of the last elements are extremely unstable and therefore are of no environmental interest with the exception of their radioactivity—see the entries about radioactivity. Ninety- one of the elements are of natural occurrence, namely, the first elements, from hydrogen to uranium, except element number 43, technetium. These 91 elements can be classified into four groups. (The numbers in parentheses indicate the number in the periodic table.)

a. Elements that are present in biological material in relatively high concentrations: hydrogen (1), carbon (6), nitrogen (7), oxygen (8), sodium (11), magnesium (12), silica (14), phosphorus (15), sulfur (16), chlorine (17), potassium (18), calcium (19), and iron (26).

b. Trace elements. They may often be present in biological material in small concentrations: boron (5), fluorine (9), chromium (24), manganese (25), cobalt (27), nickel (28), copper (29), zinc (30), selenium (34), molybdenum (42), and iodine (53). A concentration that is too high may, however, cause environmental problems.

c. Elements that have an ecotoxicological effect. They are harmful even in very small concentrations and are therefore the elements that we have to consider in environmental management as ecotoxicological elements and compounds. They are the inorganic compounds that may cause pollution problems associated with a toxic effect. They are the following: aluminum (13), arsenic (33), strontium (38), silver (47), cadmium (48), tin (50), antimony (51), tellurium (52), barium (56), mercury (80), thallium (81), lead (82), bismuth (83), polonium (84), radon (86), radium (88), thorium (90), uranium (902), and plutonium (94). The last six elements, in addition to being toxic, are also radioactive. They are present in small concentrations, and the main concern is their radioactivity. The reference to the entries about radioactivity should therefore be applied for these six elements. (As five elements belonging to group B, namely, chromium, copper, nickel, selenium, and zinc, may be of ecotoxicological/environmental concern due to frequent occurrence of harmful environmental concentrations, the overview in the section "The Pollution and Ecotoxicological Problems of 18 Elements"will include these five elements and encompass 18 elements in all.)

d. D All other elements are of no or only minor environmental interest due to their occurrence in low concentrations or negligible biological effects.

TABLE 1   Average Freshwater Plant Composition on Wet Basis

Element	Plant Content (%)
Oxygen	80.5
Hydrogen	9.7
Carbon	6.5
Silica	1.3
Nitrogen	0.7
Calcium	0.4
Potassium	0.3
Phosphorus	0.08
Magnesium	0.07
Sulfur	0.06
Chlorine	0.06
Sodium	0.04
Iron	0.02
Boron	0.001
Manganese	0.0007
Zinc	0.0003
Copper	0.0001
Molybdenum	0.00005
Cobalt	0.000002

Source: Wetzel.[1]

Table 1 gives the composition of average fresh water plants on a wet basis. The table presents 19 out of the 24 elements included in groups A and B. The following elements are not included in Table 1: fluorine, chromium, nickel, selenium, and iodine, because they are present in very small concentrations in freshwater plants—below 0.000002%.

The elements in groups B and C show bioaccumulation and biomagnifications, which are two important processes to be considered in all ecotoxicological evaluations.

The 24 elements in groups A and B encompass the elements that are needed for the entire biosphere. All 24 elements are not necessarily essential for all organisms, but each organism will require most of these 24 elements. These elements are threshold agents.[2] They are needed in a certain concentration for the growth of organisms, but if they are present in a concentration that is too high, they may be harmful for the environment. A typical example is the eutrophication problem, which results from a concentration of nutrients that is too high (see *Eutrophication*, p. 1115), although the nutrients (with emphasis on nitrogen, carbon, phosphorus, and sulfur) are absolutely necessary for all living organisms. Below, we have included in the list of elements five elements from group B, as mentioned above, because they are frequently present in concentrations in the environment that may cause ecotoxicological effects. Non-threshold agents or gradual agents are harmful for the environment in practically any concentration. The harmful effect may be proportional to the concentration, or there may be another relationship between the effect and the concentration, for instance, when the harmful effect is proportional to the concentration in the second exponent.

Figure 1 illustrates the harmful effect versus the concentration for threshold and non-threshold agents. An overview of group C (the radioactive elements+five of the elements in group B, all with a clear ecotoxicological effect) is given in this entry with emphasis on the environmental problem caused by the elements or their compounds. Heavy metals are an important soil pollution problem, and a more detailed overview of this problem can be found in the entry "Heavy Metals." Three of the group C elements, mercury, lead, and cadmium, require particular environmental attention because they have caused very serious environmental problems. Due to their environmental impact, each of these three

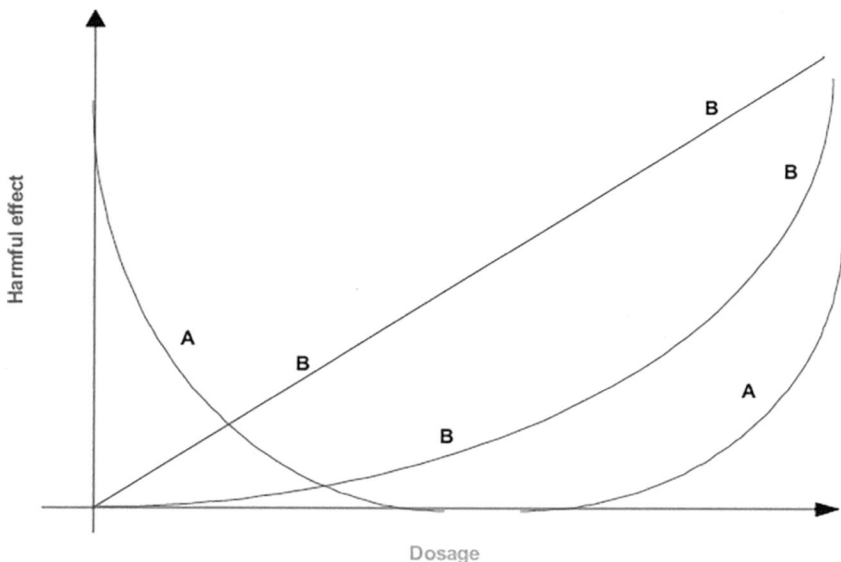

FIGURE 1 The harmful effect is plotted versus the dosage or concentration for threshold agents (a) and non-threshold agents (b).

elements is treated in more detail in three entries with the titles "Mercury," "Lead," and "Cadmium." The pollution problem of copper is furthermore treated in a separate entry, as copper is toxic to many plants and has an ecotoxicological effect at even small concentrations.

Pollution and Ecotoxicological Problems of 18 Elements

As already mentioned, there are 18 elements that require particular attention in environmental management due to their ecotoxicological effect.[3] They are aluminum (13), chromium (24), nickel (28), copper (29), zinc (30), arsenic (33), selenium (34), strontium (38), silver (47), cadmium (48), tin (50), antimony (51), tellurium (52), barium (56), mercury (80), thallium (81), lead (82), and bismuth (83). Four of these elements are treated in more detail in entries with the titles "Mercury," "Lead," "Cadmium," and "Copper." A short overview of the environmental and ecotoxicological problems of these 18 elements is given below, including their applications and environmental effects.

Table 2 gives an overview of the concentration in the earth's crust, in average soil, and in average seawater. If environmental concentrations significantly higher than these concentrations in the earth's crust, average soil, and average seawater are recorded, there is a high probability of an ecotoxicological effect. The last column of the table gives the most important applications of the element and its compounds.

Table 3 gives the lethal dose causing 50% mortality (LD50) and lethal concentration causing 50% mortality (LC50) values as expressions for the toxicity. The third column in this table indicates whether there are environmental effects and risks that require particular attention. For instance, it is indicated that unacceptably high concentrations of arsenic can be found at some locations, particularly in groundwater.[4] Due to the toxicity and carcinogenic effect of arsenic, treatment of the water is absolutely necessary.[4] Effective treatment of water is possible by precipitation/coagulation and ion exchange.[4]

Table 3 is a coarse overview. The toxicology and ecotoxicology are, of course, much more complex than it would be possible to summarize in one table. It is therefore strongly recommended to seek out much more information if there is the slightest suspicion of an ecotoxicological effect with the concentrations of the 18 elements listed in Tables 2 and 3. This entry is given only as an overview.

TABLE 2	Typical Concentrations and the Major Applications of the 20 Elements That Are of Environmental Concern

Element	Conc. in the Earth's Crust(mg/kg)	Conc. in Soil (mg/kg)	Conc. in Sea Water (mg/L)	Applications
13.Al	81,000	10,000–300,00	1	Construction
51.Sb	0.5	1	0.3	Alloys, plastic
33.As	5	1–40	2.6	Chemicals
56.Ba	425	500	30	Chemicals, glass
82.Pb	16	20	0.02	Batteries, soldering, alloys
48.Cd	0.55	0.06	0.1	Surface treatment, dyestuff
29.Cu	100	20	0.9	Cables, wires, construction
24.Cr	300	100	0.05	Surface treatment, alloys
80.Hg	0.065	0.08	0.15	Instruments, dental use
28.Ni	100	40	6.6	Alloys, surface treatment
34.Se	0.09	0.2	0.05	Glass, instruments, dyestuff
47.Ag	0.1	0.1	0.3	Photochemicals, electric components, ornaments
38.Sr	250	315	8100	Chemicals
81.Tl	1	0.1	0.01	Electronic, alloys
52.Te	0.002	0.001	0.001	Alloys
50.Sn	40	10	3	Alloys, soldering, chemicals
83.Bi	0.02	1	0.03	Alloys, chemicals
30.Zn	40	50	10	Surface treatment, alloys, chemicals

TABLE 3	Overview of Toxicity Expressed by LD50 and LC50

	LD50 (mg/kg)	LC50 (mg/L)	Ecotoxicological Attention
13.Al	770 (mice)	3,900 (zooplankton)	Low ecotoxicological effect
51.Sb	100 (rats)	9,000 (fish)	Industrial exposure
33.As	41 (rats)	74,000 (cyclops)	Groundwater, carcinogenic effect, nerve inflammation
56.Ba	8–23 (rats)	14,500 (cyclops)	Heart problems, diarrhea
82.Pb	130 (rats)	6 ppm (air, rats)	Anemia, brain damage
48.Cd	80 (rats)	65 (cyclops)	Itai-itai, kidney damage
29.Cu	220 (rats)	10 (cyclops)	Highly toxic to plants
24.Cr	3,250 (rats)	100,000 (shrimps)	Carcinogenic effect as Cr(IV)
80.Hg	8 (mice)	5 (cyclops)	Central nerve system, mental retardation, teratogenic effect
28.Ni	1620 (rats)	510 (cyclops)	Allergy, carcinogenic and teratogenic effects
34.Se	7 (rats)	2,500 (cyclops)	Liver and kidney damage
47.Ag	129 (mice)	30 (zooplankton)	Skin damage
38.Sr	148 (rats)	125,000 (cyclops)	–
81.Tl	16 (rats)	–	–
52.Te	83 (rats)	–	Teratogenic effect
50.Sn	41 (mice)	55,000 (zooplankton)	Liver and kidney damage
83.Bi	13 (rats)	–	Liver and kidney damage
30.Zn	975 (rats)	100 (zooplankton), 10,000 (fish)	Minor carcinogenic effect

References

1. Wetzel, R.G. *Limnology,* 2nd Ed.; Saunders College Publishing: New York, 1983; 828 pp.
2. Jørgensen, S.E. *Principles of Pollution Abatement;* Elsevier: Amsterdam, 2000; 520 pp.
3. Jørgensen, S.E.; Fath, B. *Ecotoxicology;* Elsevier: Amsterdam, Oxford, 2010; 390 pp.
4. Murphy, T.; Guo, J. *Aquatic Arsenic Toxicity and Treatment;* Beckhuys Publishers: Leiden, 2003; 165 pp.

45

Leaching

Lars Bergström

Introduction

The contamination of groundwater by pesticides is of concern mainly because it may limit its use as drinking water. The extent of groundwater contamination to a large extent depends on the degree to which pesticides leach through the unsaturated zone of soils on which they have been applied. Pesticide movement through the unsaturated zone in tile-drained fields may also be a source of pesticides in surface waters, which support aquatic ecosystems and are used for drinking water in many areas of the world. Therefore, knowledge of pesticide movement in soil above the groundwater table is very important.[2] This has also been the focus of a large number of studies performed during the past couple of decades.[3,4] The most important factors influencing pesticide leaching are soil properties, inherent properties of the pesticide molecules, climatic conditions, and management practices.[3]

Factors Influencing Leaching

Soil and Hydrological Conditions

The rate and direction of water flow in the unsaturated zone are determined by the hydraulic gradient and the hydraulic conductivity. The presence of air-filled pores restricts the pathways through which water percolates downward, which means that the hydraulic conductivity in the unsaturated zone also depends on the level of water saturation. As the soil dries out, water becomes more strongly bound within the matrix of the soil, and the volume of water and the rate with which water percolates through soil decrease. This relationship between water retention and hydraulic conductivity varies considerably with soil type (soil texture and structure, and organic matter content).[5]

A complication, which has a major impact on pesticide leaching, is the fact that water, and pesticides dissolved in the water phase, often move through large pores in soil (e.g., earthworm and root channels, cracks etc.), a process commonly referred to as preferential flow.[6,7] Under such conditions, an equilibrium pesticide concentration throughout the soil profile cannot be obtained. This phenomenon primarily occurs in fine-textured soils with high clay contents, especially those that have the potential to swell and shrink. Through preferential flow, pesticides can be transported rapidly through large portions of the unsaturated zone and bypass biologically active layers in which they otherwise would be degraded or sorbed. Exposure to preferential flow is most pronounced soon after application of the pesticide,

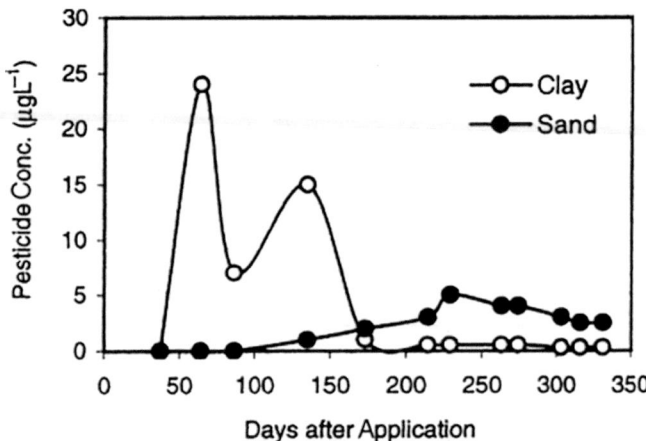

FIGURE 1   Concentrations of a pesticide in water leaching from 1-m undisturbed soil columns of a clay and a sandy soil.
Source: Modified from Bergström and Stenström.[8]

when high concentrations occur in the soil solution in upper soil layers, in combination with intensive rainfall.[8] Once the pesticide is mixed in with the soil matrix, water moving through preferential flow paths does not interact with the soil, and leaching is therefore reduced. In other words, preferential flow can both increase and decrease pesticide leaching depending on the time when it occurs in relation to pesticide application. The final result is that (although transient flow peaks shortly after application causing elevated concentrations in water leaching through soils in preferential flow paths) the leaching loads over extended periods are typically quite small in such soils. Indeed, leaching loads are often larger in sandy soils, in which water and pesticide movement mainly occurs between individual soil particles within the main soil matrix.[8] Pesticide concentrations are typically lower in sandy soils than in clay soils, but the water volumes displacing the pesticide are often much larger in sandy soils. The principal difference in pesticide leaching patterns in sand and clay soils is illustrated in Figure 1. Irrespective of which leaching mechanism prevails, the total amount of the majority of pesticides that reach groundwater after normal agricultural use rarely exceeds 1% of the applied amount and is commonly well below 0.1%.

Pesticide Properties

The physicochemical properties of pesticides have a major impact on their leachability. In this context, the rate with which they are degraded and how strongly they are sorbed to soil are the most important factors. As a general rule of thumb, leaching decreases with increasing sorption affinity and faster degradation, and increases when the opposite conditions prevail. However, it is important to note that degradation rates often become slower with residence time in the soil as a result of decreased availability due to sorption in the soil. This means that strongly sorbed pesticides, which are less mobile in soil than weakly sorbed compounds, are typically quite persistent. An example of a leaching classification scheme, based on the sorption strength of some pesticides, is shown in Table 1.

A factor that complicates the picture of pesticide movement in soil in relation to sorption affinity is the possibility that pesticide mobility is enhanced by adsorption to various mobile colloids, a process often referred to as "facilitated transport." It is known that organic solutes, such as nonpolar pesticides with very low water solubilities, can form complexes with dissolved organic carbon and clay colloids that move through soil. Even though the role of colloids in facilitating pesticide transport is still relatively poorly understood, there is little doubt that failure to account for this mode of transport can lead to underestimates of both amounts and distances that strongly sorbed pesticides may migrate through the unsaturated zone.

TABLE 1 Classification of Pesticide Mobility in Soil Based on Their Sorption Strength, Which in This Case Is Expressed by Their K_{oc} Values (Soil Sorption Coefficient, Normalized to the Soil Organic Carbon Content)

Koc Value	Expected Mobility	Type of Pesticide
0–50	Very high	Bentazone, Dicamba
50–150	High	Atrazine, 2,4-D
150–500	Medium	Simazine, Metolachlor
500–2000	Low	Lindane, Linuron
2000–5000	Very low	Phenmedipham, Fenpropimorph
>5000	Immobile	DDT, Paraquat

Source: Modified from Torstensson.[9]

Climatic Conditions

The amount and intensity of precipitation are the most important climatic factors influencing pesticide leaching. Water, in excess of evapotranspiration and what is required to maintain field capacity (i.e., the water content in soil when it is freely drained), leaches through the unsaturated zone and can thereby potentially move pesticides to groundwater. As mentioned above, in clay soils, it is primarily high intensity rainfall soon after application that may displace pesticides to depth in soil. In other words, the timing of precipitation is critical.

Soil temperature also has impact on leaching of pesticides, mainly by influencing the persistence of pesticides in soil and by affecting flow processes. Up to a certain level, degradation rates increase with increasing temperatures, which means that less of the compound will be available for leaching. Increasing temperatures will also increase evapotranspiration rates, which will reduce the amounts of water that can potentially move pesticides downward in soil. In climates with subzero temperatures in the winter season, soil will be frozen during extended periods. Under such conditions, pesticide movement in soil is very restricted, if it occurs at all. Leaching then will occur mainly during autumn and spring, when the soil is unfrozen and the evapotranspiration demand is low. Temperatures will also indirectly affect pesticide leaching by affecting the sorption/desorption process, although this influence is not yet thoroughly investigated and, therefore, less well recognized.

Management Practices and Strategies to Prevent Pollution

Management practices that have a major impact on the amount of pesticides that can move through soils can be grouped into the following categories: cropping/tillage, irrigation, and pesticide application practices.

Due to increased concern over soil erosion and input of pesticides to rivers and lakes, agricultural practices with reduced tillage or no-till management have been introduced. Such practices will also affect water infiltration rates, and therefore pesticide leaching through the unsaturated zone. In the short term, reduced tillage may decrease soil permeability compared with a conventionally tilled soil. However, over a whole growing season, infiltration rates tend to be higher under reduced tillage, especially in clay soils. This is largely due to the fact that reductions in tillage lead to less disruption of macropores in which pesticides can be rapidly transported through soil (see above). Reduced tillage also leaves more crop residues on the soil surface and contributes to reducing compaction of the subsoil caused by heavy equipment; both tend to increase permeability, and thus pesticide leaching. From the standpoint of reducing pesticide leaching, reduced tillage is therefore, in most cases, not a good management option.

As expected, irrigation increases leaching of pesticides by increasing the amount of water that potentially can move through soil. The amount of water, the rate at which it is applied, and the timing of irrigation are important for the same reasons as discussed previously for precipitation. Different irrigation methods (e.g., sprinkler, and drip and furrow irrigation) have also been shown to affect leaching.

Pesticide application strategies that influence pesticide residue levels in soil and thereby potential leaching include prevs. postapplication, split applications, placement methods, and use of different pesticide formulations. However, their influence on pesticide leaching is quite unclear and often overshadowed by other factors. Nevertheless, available data indicate that dividing the dose into two applications instead of one tends to reduce pesticide concentrations and the depth of migration in the subsoil. For similar reasons, pesticide leaching can be restricted by use of "slow-release" formulations in which the active ingredient is mixed with a solid matrix from which it gradually diffuses into the soil over an extended period. Placement of the pesticide instead of broadcasting, which reduces the soil surface area to which the pesticide is applied, also tends to reduce pesticide leaching.

Future Concerns

Leaching of pesticides will undoubtedly continue to be of concern in the foreseeable future, and something that will be considered in various regulatory assessment schemes. In this context, it is important not only to look at the leachability, but also to evaluate the risks associated with leaching and the occurrence of pesticides in groundwater both from a human health and an ecotoxicological point of view.

In the future, there is reason to believe that fewer toxic compounds will be allowed, especially those that show high leachability in soil. There is also reason to believe that, in line with the increasing awareness of problems associated with leaching of pesticides, improved management strategies will be developed that reduce pesticide leaching further.

References

1. Enfield, C.G.; Yates, S.R. Organic Chemical Transport to Groundwater. In *Pesticides in the Soil Environment: Processes, Impacts, and Modeling*; Cheng, H.H. Ed.; Soil Science Society of America Book Series: Madison, WI, 1990; 2, 271–302.
2. *The Lysimeter Concept—Environmental Fate of Pesticides*; Führ, F.; Hance, R.J.; Plimmer, J.R.; Nelson, J.O., Eds. ACS Symposium Series: Washington, DC, 1998; 699, 284.
3. Barbash, J.E.; Resek, E.A. *Pesticides in Ground Water—Distribution, Trends, and Governing Factors*; Ann Arbor Press: Chelsea, MI, 1996; 588.
4. Flury, M. Experimental evidence of transport of pesticides through field soils—a review. J. Environ. Qual. **1996**, *25* (1), 25–45.
5. Carter, A.D. Leaching Mechanisms. In *Pesticide Chemistry and Bioscience—The Food–Environment Challenge*; Brooks, G.T., Roberts, T.R., Eds.; Royal Society of Chemistry: Milton Road, U.K., 1999; 8, 291–301.
6. Bergström, L.F.; Jarvis, N.J. Leaching of dichlorprop, ben-tazon, and 36Cl in undisturbed field lysimeters of different agricultural soils. Weed Sci. **1993**, *41* (2), 251–261.
7. Brown, C.D.; Carter, A.D.; Hollis, J.M. Soils and Pesticide Mobility. In *Environmental Behaviour of Agrochemicals* Roberts, T.R., Kearney, P.C., Eds.; John Wiley and Sons: Chichester, England, 1995; 9, 131–184.
8. Bergström, L.; Stenström, J. Environmental fate of pesticides in soil. Ambio. **1998**, *27* (1), 16–23.
9. Torstensson, L. Kemiska Bekämpningsmedel—Transport, Bindning och Nedbrytning i Marken. In *Aktuellt från Sveriges Lantbruksuniversitet*; Swedish University of Agricultural Sciences: Uppsala, Sweden, 1987; 357, 36, (in Swedish).

46

Aquatic Communities: Pesticide Impacts

David P.
Kreutzweiser and
Paul K. Sibley

Introduction

A biotic community can be defined as an assemblage of plant or animal species utilizing common resources and cohabiting a specific area. Examples could include a fish community of a stream, an insect community of a forest pond, or a phytoplankton community of a lake. Interactions among species provide ecological linkages that connect food webs and energy pathways, and these interconnections provide a degree of stability, or balance, to the community. Community balance can be described as a state of dynamic equilibrium in which species and their population dynamics within a community remain relatively stable, subject to changes through natural adjustment processes. Toxic effects of pesticides can disrupt these processes and linkages and thereby cause community balance upsets. For example, this can occur when a pesticide has a direct impact on a certain species in a community and reduces its abundance while other unaffected species increase in abundance in response to the reduced competition for food resources or increased habitat availability. Some of the best examples of pesticide impacts on biological communities are found in freshwater studies. Freshwater aquatic communities are usually contained within distinct boundaries or systems, and this generates a high degree of connectivity among species, thereby increasing their susceptibility to pesticide-induced disturbances at the community level.

We examine traditional and developing methods for measuring pesticide impacts on freshwater communities, with emphasis on recent improvements in risk assessment approaches and analyses, and provide some examples for illustration. We then describe some advances in impact mitigation strategies and discuss some ongoing issues pertaining to understanding, assessing, and preventing pesticide impacts including probabilistic risk assessment (PRA), population and ecological modeling, and pesticide interactions with multiple stressors. The integration of improved risk assessment and mitigation approaches and technologies together with information generated from the numerous impact studies available will provide a sound scientific basis for decisions around the use and regulation of pesticides in and near water bodies.

Measuring Impacts on Aquatic Communities

Changes in aquatic communities can be measured directly in water bodies by a number of quantitative and qualitative sampling methods. Descriptions of those methods can be found in any up-to-date text or handbook (e.g., Hauer and Lamberti[1]). Measurements can be in terms of community structure (species composition) or community function (a measurable ecosystem process attributable to a biotic community that causes a change in condition) and can include both direct and indirect effects.[2,3] Community structure is a measure of biodiversity in its most general sense, that is, the number of species or other taxonomic units and their relative abundances. Some community functions are referred to as environmental or ecosystem services. Examples include organic matter breakdown and nutrient cycling that is largely mediated by microbial communities, or water uptake, filtration, and flood control mediated by shoreline plant communities.[4] Both community structure (biodiversity) and function (ecosystem services) are being increasingly valued by society and global economies,[5,6] and therefore sustaining healthy aquatic communities will be an important driver of pesticide impact mitigation efforts.

Detecting impacts of pesticides typically involves repeated sampling and a comparison of community attributes among contaminated and uncontaminated test units over time, or across a gradient of pesticide concentrations. The test units can range from petri dishes to natural ecosystems, with a trade-off between experimental control in small test units and environmental realism in field-level testing and whole ecosystems.[7] In an effort to incorporate both experimental control and environmental realism in pesticide impact testing, the use of microcosms or model ecosystems for measuring impacts on aquatic communities has increased over the past couple of decades.[8,9] Model ecosystems for community-level pesticide testing can be quite simple at lower-trophic levels such as with microbial communities (e.g., Widenfalk et al.[10]) but will necessarily be more complex for testing higher-order biological communities (e.g., Wojtaszek[11]). Regardless of the test units, an important consideration for measuring pesticide impacts will be an assessment of the duration of impact or rate of recovery. A rapid return to pre-pesticide or reference (no pesticide) community condition will reduce the long-term ecological consequences of the pesticide disturbance.[12]

Traditional measures of community-level impacts have focused on structure and have usually been expressed in terms of single-variable indices such as species richness, diversity, or abundance. These indices are useful descriptors of community structure but suffer from the fact that they reduce complex community data to a single summary metric and may miss subtle or ecologically important changes in species composition across sites or times. Over the last couple of decades, ecotoxicologists have increasingly turned to multivariate statistical techniques for analyzing community response data.[13] A variety of multivariate statistical techniques and software are available and are usually considered superior for the analysis of community data because they retain and incorporate the spatial and temporal multidimensional nature of biological communities.[14] This includes various ordination techniques that can provide graphical representation of spatiotemporal patterns in community structure in which points that lie close together in the ordination plot represent communities of similar composition (richness, abundance), while communities with dissimilar species composition are plotted further apart.

Figure 1 illustrates the use of an ordination plot generated by nonmetric multidimensional scaling for detecting differences among aquatic insect communities in four control and eight insecticide-treated streams. These data have been adjusted for illustrative purposes but are based on real invertebrate community responses to an insecticide in outdoor stream channels.[15] At both concentrations of the insecticide, the community structure of stream insects clearly shifted away from the natural community composition in control streams as depicted by the separation of treated streams (T1 and T2) from controls (C) in the ordination bi-plot. The plot also illustrates that the variability among treated streams (relative distance between points) was greater than that among control streams, that the low-concentration streams (T1) and high-concentration streams (T2) tended to separate along axis 1, and that the T2 streams were further removed from controls than the T1 streams, indicating a differential response by the insect communities to the two test concentrations. Canonical correspondence analysis

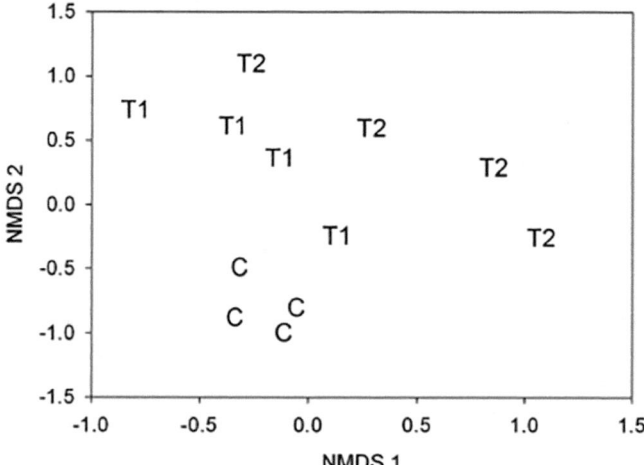

FIGURE 1 Ordination by nonmetric multidimensional scaling of aquatic insect communities in stream channels. Each point represents the community structure of control channels (C) and channels treated with a neem-based insecticide at a low (T1) and high (T2) concentration.
Source: Adapted from Kreutzweiser et al.[15]

and redundancy analysis have also been commonly used to assess aquatic community responses to pesticide contamination.[16,17] A useful refinement of an ordination technique for detecting and interpreting pesticide impacts on aquatic communities is principal response curves (PRCs).[18] PRC is derived from redundancy analysis, and time-dependent responses in the treatments are expressed as deviations from the control or reference system allowing for clear visualization of pesticide effects.

Assessing Risk of Pesticide Impacts on Aquatic Communities

The likelihood or risk of harmful effects on aquatic communities from exposure to pesticides will depend on the exposure concentration, bioavailability, exposure duration, rate of uptake, inherent species sensitivities, community composition, and other community attributes. All of these must be measured, estimated, modeled, or predicted to derive an assessment of risk to aquatic communities for any given pesticide. Formalized risk assessment frameworks and guidelines for pesticides have been developed in the United States,[19] the European Union,[20] Canada,[21] and elsewhere and can be consulted for detailed information on the various components of a risk assessment. In brief, pesticide risk assessments typically include the following phases: 1) defining the problem by determining the pesticide use patterns and developing conceptual models and hypotheses around how it is expected to behave, the anticipated exposure regimes, the kinds of organisms that are likely to be at risk, the community or entity that is to be protected, and the level of protection that will be acceptable; 2) developing the measurement endpoints for assessing risk of harm by establishing which response measurements are relevant and applicable, and how the measurements will be made; 3) outlining the risk assessment process by specifying the kinds of data to be used and how they will be derived including simulation modeling, empirical laboratory, microcosm or field testing, their appropriate spatial and temporal scales, and their statistical analyses; 4) applying the risk assessment by running models or collecting data, completing analyses, summarizing outputs, and providing risk estimates; 5) conducting risk communication and management by answering questions posed in the problem formulation, suggesting risk mitigation strategies if necessary, and communicating those to appropriate users; and 6) conducting follow-up monitoring to evaluate the success of mitigation strategies and to implement adaptive management to address deficiencies if or when necessary.[22,23]

Traditionally, pesticide risk assessments have relied on standardized, single-species toxicity tests to predict effects on communities, the underlying assumption being that protecting the most sensitive species will protect whole communities. In this case, the selection and relevance of test species are critically important to a successful and meaningful risk assessment.[24] However, the accuracy and relevance of estimating the potential risk to aquatic communities can be greatly improved by consideration of specific species or community attributes. In particular, attribute information can improve the ecological relevance and predictive capabilities of conceptual models and the generation of hypotheses in the risk assessment process. Insofar as these attributes affect exposure, sensitivity, or both, they can increase or decrease risk beyond what could be determined from toxicity estimates or species sensitivity distributions alone.

Behavioral attributes can elevate the risk of pesticide effects on species by increasing the likelihood of intercepting the stressor. For example, young-of-the-year bluefish *(Pomatomus saltatrix)* typically feed in estuaries during their early life stages where agricultural runoff can elevate concentrations of pesticides in food items. This feeding behavior can result in bioaccumulation and in adverse effects such as reduced migration, overwinter survival, and recruitment success in fish communities.[25] Incorporating this kind of information into conceptual models and risk hypotheses will generate more realistic risk assessments. In addition, behavioral attributes themselves can be relevant measurement endpoints if the pesticide mode of action indicates risk of sublethal behavioral effects at expected concentrations. For example, some pesticides have been shown to impair the ability to capture prey in fish[26] and the ability to avoid predators in zooplankton.[27] These types of adverse effects can disrupt trophic linkages and reduce survival or reproduction, thus impacting community balance.

Inclusion of life history information into conceptual models and risk hypotheses can also refine and improve the risk assessment process. Life history strategies can influence a species susceptibility to a stressor through effects on a population's resilience or ability to recover from disturbance.[28] Different species exposed to the same pesticide and experiencing similar levels of effect in terms of population declines do not necessarily recover at the same rates when recovery is dependent on reproduction or dispersal. Populations of organisms with short regeneration times (e.g., several generations per year) and/or high dispersal capacity have higher likelihood of recovery from pesticide-induced population declines than those with longer regeneration periods and limited dispersal capacity. These differential life history strategies and their influences on community response and recovery from pesticide effects have been demonstrated empirically (e.g., van den Brink et al.[29] and Kreutzweiser et al.[30]) and through population modeling.[31] These community balance upsets could not have been predicted from screening-level toxicity data or from species sensitivity data; thus, inclusion of life history information in conceptual models can improve risk hypotheses and direct the assessment to focus on species at higher risk owing to specific life history strategies.

Life history attributes can also influence the risk of pesticide effects through differential life-stage sensitivity or susceptibility. Early life stages are often (but not exclusively) more sensitive to pesticides than later stages. An organism's life stage can also influence its susceptibility to a pesticide by increasing or decreasing the likelihood of intercepting the stressor. If a contaminant is present in the environment at effective concentrations during a period in which the particular life stage of a species is present, then the risk to that species is increased. For some amphibians, aquatic (larval) stages could be at higher risk of direct and indirect effects of pesticides than their terrestrial (adult) life stages when their larval stage coincides with pesticide contamination of water bodies.[32] Thus, while a species sensitivity and geographical distribution may indicate potential risk, the life-stage information coupled with pesticide use pattern, timing, or fate information may indicate little likelihood of exposure to the pesticide and the risk assessment can be adjusted accordingly.

Functional attributes may also be important for refining or improving pesticide risk assessments. Protection goals for populations and communities often include the safeguarding of critical biological processes or ecosystem function. Measuring ecosystem function integrates responses of component populations and can be a relevant measurement endpoint when species loss affects ecosystem function such as energy transfer and organic matter cycling.[33] However, most ecosystems are complex and it may not be clear which functional attributes are critical for sustaining ecological processes or the extent to which they can sustain changes in structural properties (e.g., population levels, diversity) without adversely affecting ecosystem function. Neither is it clear if functional endpoints are more or less sensitive than structural endpoints for detecting ecosystem disturbance. Some studies investigating the relationship between species diversity and ecosystem function have indicated that ecosystems can tolerate some species loss because of functional redundancy.[34] Functional redundancy is thought to occur when several species perform similar functions in ecosystems such that some may be eliminated with little or no effect on ecosystem processes. Others have suggested that redundant species are required to ensure ecosystem resilience to disturbance as a form of biological insurance, especially at large spatial scales.[35]

Given these discrepancies, measurement endpoints based on functional attributes are not typically used in pesticide risk assessments because it is generally accepted that protection of community structure will protect ecosystem function. However, when specific functional attributes can be identified and are known or suspected to be at risk from a pesticide, they can be included in the data requirements for a risk assessment. An example would be the risk of adverse effects on leaf litter decomposition (a critical ecosystem function in forest soils and water bodies) posed by a systemic insecticide for control of wood-boring insects in trees.[36] In that case, the protection goal was maintaining leaf litter decomposition, the community at risk was decomposer invertebrates feeding on leaves from insecticide-treated trees, and the selection of test species was directed to a specific functional group because of the unique route of exposure to decomposer organisms identified in the risk hypotheses.

Some Examples of Pesticide Impacts on Aquatic Communities

A few examples will serve to illustrate how pesticides can cause disruptions to aquatic communities. DeNoyelles et al.[37] reviewed studies into pesticide impacts on aquatic communities and reported that herbicides like atrazine, hexazinone, and copper sulfate were directly toxic to most species of phytoplankton (waterborne algae). After herbicide applications, reductions in phytoplankton caused secondary reductions in herbivorous zooplankton, resulting from a depleted food source for the zooplankton. They further showed that direct adverse effects on phytoplankton can also cause disruptions to the bacterial-based energy pathways by reducing carbon flow from phytoplankton to bacteria, and ultimately to grazing protozoans and zooplankton. Boyle et al.[38] found that applications of the insecticide diflubenzuron to small ponds reduced populations of several aquatic invertebrate species. This in turn resulted in indirect effects on algae (increased productivity because of release from grazing pressure by the invertebrates) and on juvenile fish populations (reduced production because of limited invertebrate prey availability). George et al.[39] used a novel approach to predict effects of pesticide mixtures on zooplankton communities and then tested the predictions in outdoor microcosms. Responses among zooplankton populations within the community differed, depending on the pesticide mixture, and those differences appeared to reflect the relative susceptibilities among specific taxa within groups. Cladocerans declined but were less sensitive than copepods to a chlorpyrifos-dominated mixture, while rotifers actually increased after application in response to release from competition or predation pressures.

Kreutzweiser et al.[40] applied a neem-based insecticide to forest pond enclosures and measured effects on zooplankton community structure, respiration, and food web stability. Significant concentration-dependent reductions in numbers of adult copepods were observed, but immature copepods and cladocerans were unaffected (Figure 2). There was no evidence of recovery of adult copepods within the sampling season. During the period of maximal impact (about 4 to 9 weeks after the applications), total plankton community respiration was significantly reduced, and this contributed to significant concentration-dependent increases in dissolved oxygen and decreases in specific conductance. The reductions in adult copepods resulted in negative effects on zooplankton food web stability through elimination of a trophic link and reduced interactions and connectance.

Van Wijngaarden et al.[41] evaluated the responses of aquatic communities in indoor microcosms to a suite of pesticides used for bulb crop protection. At pesticide concentrations equivalent to 5% spray drift deposition, zooplankton taxa within communities showed significant changes relative to non-treated controls, reflecting taxon-specific sensitivities. Some copepods and rotifers in particular showed significant declines for at least 13 weeks, while many other rotifers and cladocerans were unaffected or increased. Several macroinvertebrate taxa were negatively affected, and this contributed to significant declines in leaf litter decomposition among treated microcosms. The herbicide asulam was among the suite of pesticides, and it induced significant reduction of the macrophyte *Elodea nuttallii*. This in turn caused significant changes in water chemistry (decreases in dissolved oxygen and pH, increases in alkalinity and specific conductance) and increases in phytoplankton biomass from decreased competition for nutrients. Increased phytoplankton and reduced zooplankton predators combined to support higher abundance of less sensitive zooplankton taxa. The authors point out that most of these effects were not measurable at more realistic rates of spray drift deposition.

Relyea and Hoverman[42] investigated impacts of the insecticide malathion on aquatic communities in microcosms designed to mimic a simple aquatic food web that can be found in ponds and wetlands. The insecticide generally reduced zooplankton abundance, and these reductions stimulated increases in phytoplankton, decreases in periphyton (attached algae), and decreases in growth of frog tadpoles. While invertebrate predator survival was not affected, amphibian prey survival increased with insecticide concentration, apparently the result of insecticide-induced impairment of predation success by the invertebrates. Overall, the study demonstrated that realistic concentrations of an insecticide can interact with natural predators to induce large changes in aquatic community balance.

Reducing Risk of Pesticide Impacts on Aquatic Communities

For pesticides applied to crops and forests, exposure to aquatic communities can be minimized by the implementation of vegetated spray buffers or setbacks to intercept off-target spray drift and runoff.[43] Pesticide runoff can be further reduced by using formulations that are less prone to wash-off, leaching, and mobilization. Recent advances in spray drift reduction and improved spray guidance systems can also significantly reduce the off-target movement of pesticides to water bodies.[44] Examples include new technologies in map-based automated boom systems for row crops[45] and Geographical Information System (GIS)-based landscape analysis for predicting off-target pesticide movement.[46]

The risk of adverse effects on aquatic communities may also be decreased by intentional selection and use of pesticides that are inherently safer to the environment. This would include so-called reduced-risk pesticides that are bioactive compounds usually with unique modes of action and derived from microbial, plant, or other natural sources. These are generally thought to be less persistent and toxic to non-target organisms than conventional synthetic pesticides.[47] Examples include the bacteria-derived insecticide *Bt (Bacillus thuringiensis)*, the plant-derived insecticide neem, and the microbe-derived herbicide phosphinothricin. However, Thompson and Kreutzweiser[48] caution that it cannot be assumed that this group of pesticides is inherently safer or more environmentally acceptable than synthetic counterparts and that full environmental risk evaluations must be conducted to ensure their environmental safety.

These types of technologies combined with the use of non-pesticide approaches to pest management form the basis of integrated pest management (IPM) strategies. IPM strategies are those in which the judicious use of pesticides is only one of several concurrent methods to control or manage losses from pest damage. This can include the use of natural enemies and parasites, biological control agents, insect growth regulators, confusion pheromones, sterile male releases, synchronizing with weather patterns known to diminish pest populations, and cultivation methods and crop varieties to improve conditions for natural enemies or degrade conditions for pest survival.[49] Increasing the use of IPM approaches can reduce reliance on pesticides and thus reduce the risk of pesticide impacts overall.

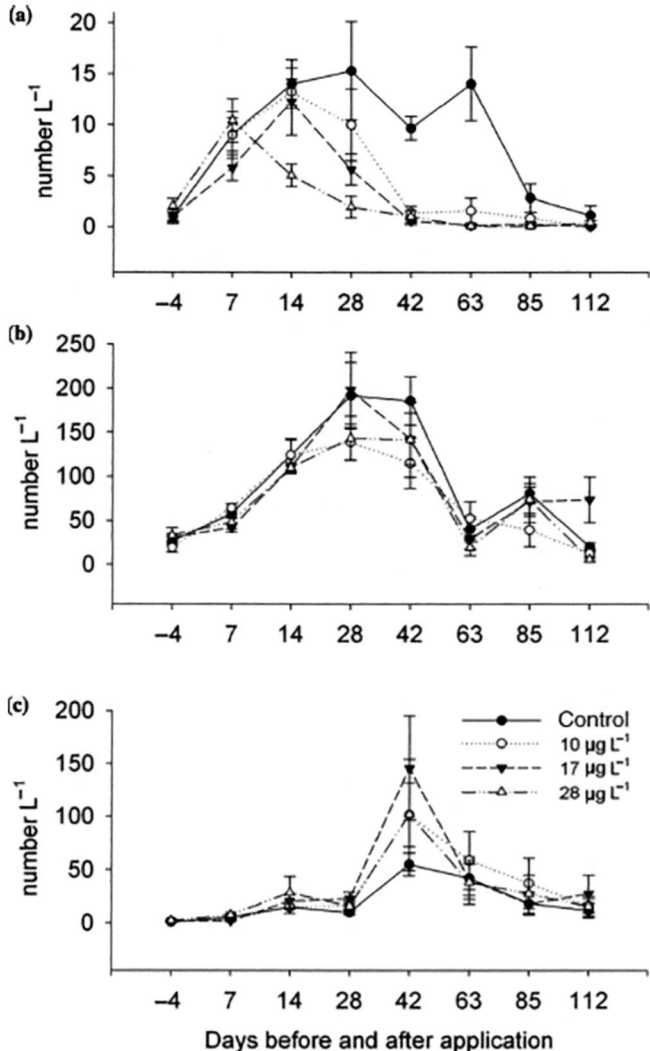

FIGURE 2   Mean abundance (±1 SE, n = 5) of (a) adult copepods, (b) immature copepods, and (c) cladocerans in natural pond microcosms (controls) and microcosms treated at three different rates of a neem-based insecticide.
Source: Taken from Kreutzweiser et al.[40]

Recent Advances and Outstanding Issues

Pesticide risk assessments and risk reductions have recently been advanced in terms of ecological realism and effectiveness through some developing methods and techniques. Traditional risk assessments have estimated hazards from pesticides by comparing the expected environmental concentration (often predicted from worst-case scenarios) to the toxic threshold for the most sensitive test species. When the expected concentration is higher than the toxicity threshold, the pesticide is considered to have potential for environmental effects. These so-called hazard or risk quotient approaches are still widely used in pesticide risk assessment and regulation, but more recently, PRA and probabilistic hazard assessment (PHA) approaches are being adopted. In these approaches, pesticide exposure levels and the likelihood of toxic effects are estimated from probability distributions based on all reliable data available.[50] In PRA, exposure and effects distributions are developed from modeling or measurements in laboratory, microcosm, or field studies and used to improve the accuracy and relevance of the estimated likelihood of environmental effects compared to the traditional worst-case (hazard/risk quotient) approach (e.g., Solomon[51]). In PHA, a distribution approach is also used, except that the probability of hazard is estimated from distributions built on the relative sensitivity of interspecies endpoints rather than species sensitivity itself.[52] Figure 3 illustrates the principles of PRA (Figure 3a) and PHA (Figure 3b). Regardless of the approach, one important aspect of PRA that is ongoing is the development and use of uncertainty analysis to quantify variability and uncertainty in exposure and effects estimates. Characterizing and quantifying uncertainty will provide more meaningful risk assessments and improved decision making for minimizing potential risk of pesticide impacts in or near water.[53]

Efforts at incorporating population or ecological modeling into pesticide risk assessments have also improved their accuracy and relevance for predicting, and therefore mitigating, risk of harm to aquatic communities.[54] The use of ecological models to incorporate a suite of factors including lethal and sublethal effects and their influences on the risks to organisms, populations, or communities can provide useful insights into receptor/pesticide interactions and can thereby improve risk assessments and direct

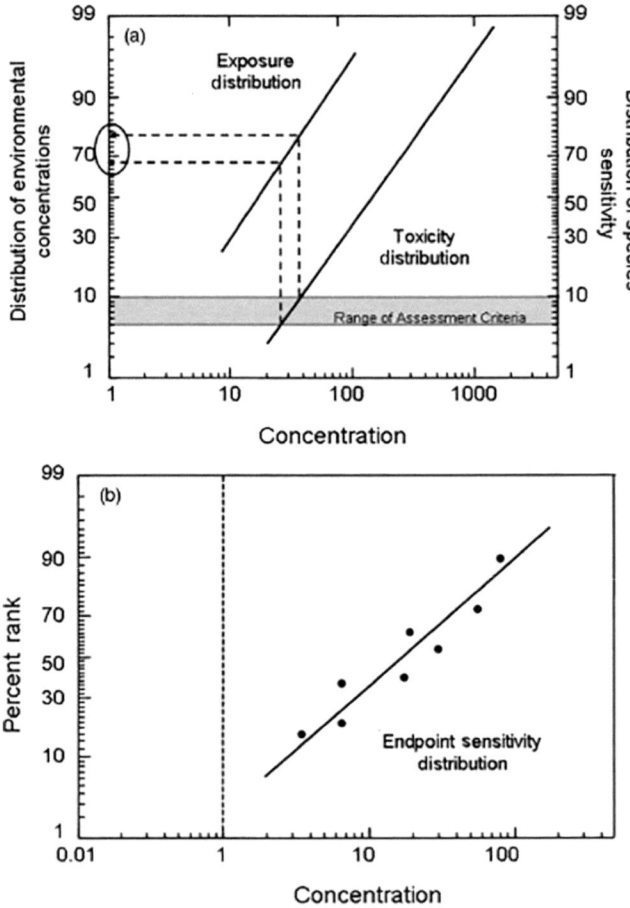

FIGURE 3 Schematic illustrating the principle of PRA (a) and PHA (b). PRA is based on a comparison of exposure and effects distributions using a predetermined criterion typically in the range of 5%-10% (shaded area and dashed lines in panel A) to determine the probability of exceeding the criterion (ellipse on y-axis); PHA is based on a comparison of an endpoint-derived sensitivity distribution within a test species to a threshold value such as a hazard quotient (dashed line in panel B).

mitigation measures. Population models that account for differential demographics and population growth rates within communities have been shown to provide more accurate assessments of potential pesticide impacts on populations and communities than what conventional lethal concentration estimates can provide.[55] Ecological and population modeling combined with pesticide exposure modeling and case-based reasoning (drawing on past experience or information from similar chemical exposures) can provide further refinements and improve risk assessment for aquatic communities.[56] Another recent advancement in ecological modeling to predict pesticide effects is the use of trait-based information such as organism morphology, life history, physiology, and feeding ecology in risk assessments.[57] This approach includes some of the functional attributes and concepts described above in the section on "Assessing Risk of Pesticide Impacts on Aquatic Communities" and has the advantage of formally expressing communities as combinations of functional traits rather than as groups of species, thereby yielding a more meaningful description of community structure and function. Taken together, these modeling approaches that incorporate probability distributions, toxicological sensitivities, population dynamics, ecological information, and functional trait attributes can be integrated into improved risk assessments that will inform mitigation and prevention strategies for pesticide use.[58]

Two additional issues that present challenges to pesticide risk assessment and mitigation are pesticide mixtures and the combined or cumulative effects of multiple stressors on pesticide impacts. Pesticides frequently occur as mixtures in aquatic systems, particularly in agricultural regions, and methods to assess and/or predict pesticide mixture toxicity under laboratory conditions have been relatively well developed. However, there are still large uncertainties associated with the prediction of pesticide mixture toxicity, and additional studies are needed to evaluate the performance of mixture models when evaluating community-level endpoints and toxicity thresholds over long-term exposures.[59] Secondly, whereas most pesticide assessment data are derived from tests or experiments under controlled or semi-controlled environmental conditions, pesticides in natural environments may interact with a number of other natural or human-caused stressors that can substantially alter the likelihood and magnitude of pesticide impacts.[60] Other stressors could include overarching effects of climate change that can influence water temperature and quality; land use activities that result in chemical, sediment, and nutrient pollution of waterways; and biotic interactions with invasive species in aquatic communities. A number of studies have examined the combined effects of a pesticide with other stressors, but they have usually been single stressor effects tested at the single-species level. Examples of studies that examined combined effects include pesticide interactions with water temperature,[61] pH,[62] dissolved organic matter,[63] UV radiation,[64] predators,[65] competitors,[66] food availability,[67] elevated sediments,[68] and other chemical stressors.[69] However, potential multiple stressors and their interactions with pesticides can be myriad and testing or extrapolating to community-level impacts is onerous at best. Sorting out and mitigating pesticide impacts from among these multiple stressors continues to be a challenge, and the suggestion by Laskowski et al.[70] to include studies of toxicant interactions with a range of environmental conditions in risk assessments seems warranted.

Conclusions

Because of the high degree of connectivity among species in an aquatic community, pesticides pose a risk of harm to the community stability or balance. The community structure can be altered by direct effects, indirect effects, or both, and this can cause disruptions to the interactions and linkages among species and to their ecological function. This risk of harm will depend on exposure concentration, bioavailability, exposure duration, rate of uptake, species sensitivities, community composition, and other community attributes. Recent advances in pesticide risk assessment for aquatic communities have improved the ecological relevance and predictive capabilities for determining, and thus mitigating, potential harmful impacts. Pesticide impacts on aquatic communities can be minimized by the use of improved application technologies to reduce application rates and to decrease off-target movement to water bodies. Potential impacts can be further minimized through the selection and use of pesticides that are demonstrated to be inherently safer to the environment and through the application of IPM strategies. Given the preponderance of pesticide impact studies in freshwater aquatic ecosystems, the improved risk assessment frameworks and regulatory requirements for pesticide evaluations, and the recent advances in mitigation technologies, many decisions around the use of pesticides can be made on a sound scientific basis rather than on misinformed perceptions or politically driven agendas. Integrated, science-based pest management strategies including the prudent use of appropriate pesticides will contribute to ensuring the sustainability of aquatic communities in areas subjected to pest management programs.

References

1. Hauer, F.R.; Lamberti, G.A. *Methods in Stream Ecology,* 2nd Ed.; Elsevier: Amsterdam, 2006.
2. Fleeger, J.W.; Carman, K.R.; Nisbet, R.M. Indirect effects of contaminants in aquatic ecosystems. Sci. Tot. Environ. **2003**, *317,* 207–233.
3. Rohr, J.R.; Crumrine, P.W. Effects of an herbicide and an insecticide on pond community structure and processes. Ecol. Appl. **2005**, *15,* 1135–1147.
4. Daily, G.C. Introduction: What are ecosystem services? In *Nature's Services: Societal Dependence on Natural Ecosystems*; Daily, G.D., Ed.; Island Press: Washington, DC, 1997; 1–10.
5. Anderson, J.; Gomez, W.C.; McCarney, G.; Adamowicz, W.; Chalifour, N.; Weber, M.; Elgie, S.; Howlett, M. *Ecosystem Service Valuation, Market-based Instruments and Sustainable Forest Management: A Primer*; Sustainable Forest Management Network: Edmonton, Alberta, 2010.
6. Bayon, R.; Jenkins, M. 2010. The business of biodiversity. Nature **2010**, *466,* 184–185.
7. Sibley, P.K.; Chappel, M.J.; George, T.K.; Solomon, K.R.; Liber, K. Integrating effects of stressors across levels of biological organization: Examples using organophosphorus insecticide mixtures in field-level exposures. J. Aquat. Eco-syst. Stress Recovery **2000**, *7,* 117–130.
8. Campbell, P.J.; Arnold, D.J.S.; Brock, T.C.M.; Grandy, N.J.; Heger, W.; Heimbach, F.; Maund, S.J.; Streloke, M. *Guidance Document on Higher-Tier Aquatic Risk Assessment for Pesticides*; SETAC Europe: Brussels, Belgium, 1999.
9. Kennedy, J.H.; LaPoint, T.W.; Balci, P.; Stanley, J.K.; Johnson, Z.B. Model aquatic ecosystems in ecotoxicological research: Considerations of design, implementation, and analysis. In *Handbook of Ecotoxicology,* 2nd Ed.; Hoffman, D.J.; Rattner, B.A.; Burton, G.A., Jr.; Cairns, J., Jr., Eds.; Lewis Publishers: Boca Raton, Florida, 2003; 45–74.
10. Widenfalk, A.; Svensson, J.M.; Goedkoop, W. Effects of the pesticides captan, deltamethrin, isoproturon, and pirimi-carb on the microbial community of a freshwater sediment. Environ. Toxicol. Chem. **2004**, *23,* 1920–1927.
11. Wojtaszek, B.F.; Buscarini, T.M.; Chartrand, D.T.; Stephenson, G.R.; Thompson, D.G. Effect of Release® herbicide on morality, avoidance response, and growth of amphibian larvae in two forest wetlands. Environ. Toxicol. Chem. **2005**, *24,* 2533–2544.
12. Barnthouse, L.W. Quantifying population recovery rates for ecological risk assessment. Environ. Toxicol. Chem. **2004**, *23,* 500–508.
13. Maund, S.; Chapman, P.; Kedwards, T.; Tattersfield, L.; Matthiessen, P.; Warwick, R.; Smith, E. Application of multivariate statistics to ecotoxicological field studies. Environ. Toxicol. Chem. **1999**, *18,* 111–112.
14. Clarke, K. R.; Warwick, R.M. *Change in Marine Communities: An Approach to Statistical Analysis and Interpretation, 2nd Ed.*; PRIMER-E: Plymouth, U.K., 2001.
15. Kreutzweiser, D.P.; Capell, S.S.; Scarr, T.A. Community-level responses by stream insects to neem products containing azadirachtin. Environ. Toxicol. Chem. **2000**, *19,* 855–861.
16. Frieberg, N.; Lindstrom, M.; Kronvang, B.; Larsen, S.E. Macroinvertebrate/sediment relationships along a pesticide gradient in Danish streams. Hydrobiologia **2003**, *494,* 103–110.

17. Berenzen, N.; Kimke, T.; Schulz, H.K.; Schulz, R. Macroinvertebrate community structure in agricultural streams: Impact of run-off-related pesticide contamination. Ecotoxicol. Environ. Saf. **2005**, *60*, 37–46.

18. van den Brink, P.J.; ter Braak, C.J.F. Principal response curves: Analysis of time-dependent multivariate responses of biological community to stress. Environ. Toxicol. Chem. **1999**, *18*, 138–148.

19. USEPA. *Guidelines for Ecological Risk Assessment*; United States Environmental Protection Agency, Risk Assessment Forum: Washington, DC, 1998.

20. EUFRAM. *Introducing Probabilistic Methods into the Ecological Risk Assessment of Pesticides, Version 6*; European Framework for Risk Assessment of Pesticides (EUFRAM): York, U.K., 2005.

21. Delorme, P.; Francois, D.; Hart, C.; Hodge, V.; Kaminski, G.; Kriz, C.; Mulye, H.; Sebastien, R.; Takacs, P.; Wandel-maier, F. *Final Report for the PMRA Workshop: Assessment Endpoints for Environmental Protection*; Environmental Assessment Division, Pest Management Regulatory Agency, Health Canada: Ottawa, Ontario, 2005.

22. Suter, G.W.; Barnthouse, L.W.; Bartell, S.M.; Mill, T.; Mackay, D.; Patterson, S. *Ecological Risk Assessment*; Lewis Publishers: Boca Raton, Florida, 1993.

23. Reinert, K.H.; Bartell, S.M.; Biddinger, G.R., Eds. *Ecological Risk Assessment Decision-Support System: A Conceptual Design;* SETAC Press: Pensacola, Florida, 1998.

24. Maltby, L.; Blake, N.; Brock, T.C.M.; van den Brink, P.J. Insecticide species sensitivity distributions: Importance of test species selection and relevance to aquatic ecosystems. Environ. Toxicol. Chem. **2005**, *24*, 379–388.

25. Candelmo, A.C.; Deshpande, A.; Dockum, B.; Weis, P.; Weis, J.S. The effect of contaminated prey on feeding, activity, and growth of young-of-the-year bluefish, *Pomato-mus saltatrix,* in the laboratory. Estuaries Coasts **2010**, *33*, 1025–1038.

26. Baldwin, D.H.; Spromberg, J.A.; Collier, T.K.; Scholz, N.L. A fish of many scales: Extrapolating sublethal pesticide exposures to the productivity of wild salmon populations. Ecol. Appl. **2009**, *19*, 2004–2015.

27. Pestana, J.L.T.; Loureiro, S.; Baird, D.J.; Soares, A.M.V.M. Pesticide exposure and inducible antipredator responses in the zooplankton grazer, *Daphnia magna* Straus. Chemo-sphere **2010**, *78*, 241–248.

28. Stark, J.D.; Banks, J.E.; Vargas, R.I. How risky is risk assessment: The role that life history strategies play in susceptibility of species to stress. Proc. Natl. Acad. Sci. U. S. A. **2004**, *101*, 732–736.

29. van den Brink, P.J.; Hattink, J.; Bransen, F.; van Donk, E.; Brock, T.C.M. Impact of the fungicide carbendazim in freshwater microcosms. II. Zooplankton, primary producers and final conclusions. Aquat. Toxicol. **2000**, *48*, 251–264.

30. Kreutzweiser, D.P.; Back, R.C.; Sutton, T.M.; Pangle, K.L.; Thompson, D.G. Aquatic mesocosm assessments of a neem (azadirachtin) insecticide at environmentally realistic con-centrations—2: Zooplankton community responses and recovery. Ecotoxicol. Environ. Saf. **2004**, *59*, 194–204.

31. Wang, M.; Grimm, V. Population models in pesticide risk assessment: Lessons for assessing population-level effects, recovery, and alternative exposure scenarios from modeling a small mammal. Environ. Toxicol. Chem. **2010**, *29*, 1292–1300.

32. Brodman, R.; Newman, W.D.; Laurie, K.; Osterfeld, S.; Lenzo, N. Interaction of an aquatic herbicide and predatory salamander density on wetland communities. J. Herpetol. **2010**, *44*, 69–82.

33. Rosenfeld, J.S. Functional redundancy in ecology and conservation. Oikos **2002**, *98*, 156–162.

34. Lawton, J.H.; Brown, V.K. Redundancy in ecosystems. In *Biodiversity and Ecosystem Function*; Schulze, E.D., Mooney, H.A., Eds.; Springer: New York, 1993; 255–268.

35. Naeem, S.; Li, S. Biodiversity enhances ecosystem stability. Nature **1997**, *390*, 507–509.

36. Kreutzweiser, D.P.; Good, K.P.; Chartrand, D.T.; Scarr, T.A.; Thompson, D.G. Are leaves that fall from imidacloprid-treated maple trees to control Asian longhorned beetles toxic to non-target decomposer organisms? Journal of Environmental Quality **2008**, *37*, 639–646.

37. deNoyelles, F., Jr.; Dewey, S.L.; Huggins, D.G.; Kettle, W.D. Aquatic mesocosms in ecological effects testing: Detecting direct and indirect effects of pesticides. In *Aquatic Mesocosm Studies in Ecological Risk Assessment*; Graney, R.L., Kennedy, J.H., Rodgers, J.H., Jr., Eds.; Lewis Publishers: Boca Raton, 1994; 577–603.

38. Boyle, T.P.; Fairchild, J.F.; Robinson-Wilson, E.F.; Haver-land, P.S.; Lebo, J.A. Ecological restructuring in experimental aquatic mesocosms due to the application of diflubenzuron. Environ. Toxicol. Chem. **1996**, *15*, 1806–1814.

39. George, T.K.; Liber, K.; Solomon, K.R.; Sibley, P.K. Assessment of the probabilistic ecological risk assessment-toxic equivalent combination approach for evaluating pesticide mixture toxicity to zooplankton in outdoor microcosms. Archives of Environmental Contamination and Toxicology **2003**, *45*, 453–461.

40. Kreutzweiser, D.P.; Sutton, T.M.; Back, R.C.; Pangle, K.L.; Thompson, D.G. Some ecological implications of a neem (azadirachtin) insecticide disturbance to zooplankton communities in forest pond enclosures. Aquat. Toxicol. **2004**, *67*, 239–254.

41. van Wijngaarden, R.P.A.; Cuppen, J.G.M.; Arts, G.H.P.; Crum, S.J.H.; van den Hoorn, M.W.; van den Brink, P.J.; Brock, T.C.M. Aquatic risk assessment of a realistic exposure to pesticides used in bulb crops: a microcosm study. Environ. Toxicol. Chem. **2004**, *23*, 1479–1498.

42. Relyea, R.A.; Hoverman, J.T. Interactive effects of predators and a pesticide on aquatic communities. Oikos **2008**, *117*, 1647–1658.

43. Zhang, X.; Liu, X.; Zhang, M.; Dahlgren, R.A.; Eitzel, M. A review of vegetated buffers and a meta-analysis of their mitigation efficacy in reducing nonpoint source pollution. J. Environ. Qual. **2010**, *39*, 76–84.

44. van de Zande, J.C.; Porskamp, H.A.; Michielsen, J.M.; Holterman, H.J.; Juijsmans, J.M. Classification of spray applications for driftability to protect surface water. Aspects Appl. Biol. **2000**, *57*, 57–64.

45. Luck, J.D.; Zandonadi, R.S.; Luck, B.D.; Shearer, S.A. Reducing pesticide over-application with map-based automatic boom section control on agricultural sprayers. Trans. Am. Soc. Agric. Biol. Eng. **2010**, *53*, 685–690.

46. Pfleeger, T.G.; Olszyk, D.; Burdick, C.A.; King, G.; Kern, J.; Fletcher, J. Using a geographical information system to identify areas with potential of off-target pesticide exposure. Environ. Toxicol. Chem. **2006**, *25*, 2250–2259.

47. PMRA. *Regulatory Directive DIR2002–02: The PMRA Initiative for Reduced Risk Pesticides*; Pest Management Regulatory Agency, Health Canada Information Services: Ottawa, Ontario; 2002, availabe at http://www.hc-sc.gc.ca/pmra-arla/english/pdf/dir/dir2002–02-e.pdf. (accessed September 2010).

48. Thompson, D.G.; Kreutzweiser, D.P. A review of the environmental fate and effects of natural "reduced-risk" pesticides in Canada. In *Crop Protection Products for Organic Agriculture: Environmental, Health, and Efficacy Assessment, ACS Symposium Series 947*; Felsot, A.S., Racke, K. D., Eds.; American Chemical Society: Washington, DC, 2007; 245–274.

49. van Emden, H. Integrated pest management. In *Encyclopedia of Pest Management*; Pimentel, D., Ed.; Marcel Dekker Inc.: New York, 2002; 413–415.

50. Solomon, K.R.; Takacs, P. Probabilistic ecological risk assessment using species sensitivity distributions. In *Species Sensitivity Distributions in Ecotoxicology*; Posthuma, L., Suter, G.W., Traas, T.P., Eds.; Lewis Publishers: Boca Raton, Florida, 2002; 285–314.

51. Solomon, K.R.; Baker, D.B.; Richards, R.P.; Dixon, K.R.; Klaine, S.J.; La Point, T.W.; Kendall, R.J.; Weisskopf, C.P.; Giddings, J.M.; Giesy, J.P.; Hall, L.W.; Williams, W.M. Ecological risk assessment of atrazine in North American surface waters. Environ. Toxicol. Chem. **1996**, *15*, 31–76.

52. Hanson, M.L.; Solomon, K.R. New technique for estimating thresholds of toxicity in ecological risk assessment. Environ. Sci. Technol. **2002**, *36*, 3257–3264.

53. Warren-Hicks, W.J.; Hart, A., Eds. *Application of Uncertainty Analysis to Ecological Risks of Pesticides*; CRC Press: Boca Raton, Florida, 2010.

54. Thorbek, P.; Forbes, V.E.; Heimbach, F.; Hommen, U.; Thulke, H.; van den Brink, P.; Wogram, J.; Grimm, V. *Ecological Models for Regulatory Risk Assessments of Pesticides;* SETAC Press: Pensacola, Florida, 2010.

55. Stark, J.D.; Banks, J.E. Population-level effects of pesticides and other toxicants on arthropods. Annu. Rev. Entomol. **2003**, *48*, 505–519.

56. van den Brink, P.J.; Roelsma, J.; van Nes, E.H.; Scheffer, M.; Brock, T.C.M. PERPEST model: A case-based reasoning approach to predict ecological risks of pesticides. Environ. Toxicol. Chem. **2002**, *21*, 2500–2506.

57. Baird, D.J.; Baker, C.J.O.; Brua, R.B.; Hajibabaei, M.; McNicol, K.; Pascoe, T.J.; de Zwart, D. Towards a knowledge infrastructure for traits-based ecological risk assessment. Integr. Environ. Assess. Manage. **2010**, online DOI 10.1002/ieam.129 (accessed September 2010).

58. van den Brink, P.J. Ecological risk assessment: From bookkeeping to chemical stress ecology. Environ. Sci. Technol. **2008**, *42*, 8999–9004.

59. Beldon, J.B.; Gilliom, R.J.; Lydy, M.J. How well can we predict the toxicity of pesticide mixtures to aquatic life? Integr. Environ. Assess. Manage. **2007**, *3*, 364–372.

60. Heugens, E.H.W.; Hendricks, A.J.; Dekker, T.; van Straalen, N.M.; Admiraal, W. A review of the effects of multiple stressors on aquatic organisms and analysis of uncertainty factors for use in risk assessment. Crit. Rev. Toxicol. **2001**, *31*, 247–285.

61. Lydy, M.J.; Lohner, T.W.; Fisher, S.W. Influence of pH, temperature and sediment type on the toxicity, accumulation and degradation of parathion in aquatic systems. Aquat. Toxicol. **1990**, *17*, 27–44.

62. Howe, G.E.; Marking, L.L.; Bills, T.D.; Rach, J.J.; Mayer, F.L. Jr. Effects of water temperature and pH on toxicity of terbufos, trichlorfon, 4-nitrophenol and 2,4-dinitrophenol to the amphipod *Gammarus pseudolimnaeus* and rainbow trout (*Oncorhynchus mykiss*). Environ. Toxicol. Chem. **1994**, *13*, 51–66.

63. Yang, W.; Spurlock, F.; Liu, W.; Gan, J. Effects of dissolved organic matter on permethrin bioavailability to *Daphnia* species. J. Agric. Food Chem. **2006**, *54*, 3967–3972.

64. Puglis, H.J.; Boone, M.D. Effects of technical-grade active ingredient vs. commercial formulation of seven pesticides in the presence or absence of UV radiation on survival of green frog tadpoles. Arch. Environ. Contam. Toxicol. **2010**, online DOI 10.1007/s00244–010–9528-z (accessed September 2010).

65. Sandland, G.J.; Carmosini, N. Combined effects of a herbicide (atrazine) and predation on the life history of a pond snail, *Physa gyrina*. Environ. Toxicol. Chem. **2006**, *25*, 2216–2220.

66. Davidson, C.; Knapp, R.A. Multiple stressors and amphibian declines: Dual impacts of pesticides and fish on yellowlegged frogs. Ecol. Appl. **2007**, *17*, 587–597.

67. Barry, M.J.; Logan, D.C.; Ahokas, J.T.; Holdway, D.A. Effect of algal food concentration on toxicity of tow agricultural pesticides to *Daphnia carinata*. Ecotoxicol. Environ. Saf. **1995**, *32*, 273–279.

68. Wu, Q.; Riise, G.; Pflugmacher, S.; Greulich, K.; Steinberg, C.E.W. Combined effects of the fungicide propiconazole and agricultural runoff sediments on the aquatic bryophyte *Vesicularia dubyana*. Environ. Toxicol. Chem. **2005**, *24*, 2285–2290.

69. Boone, M.D.; Bridges, C.M.; Fairchild, J.F.; Little, E.E. Multiple sublethal chemicals negatively affect tadpoles of the green frog, *Rana clamitans*. Environ. Toxicol. Chem. **2005**, *24*, 1267–1280.

70. Laskowski, R.; Bednarska, A.J.; Kramarz, P.E.; Loureiro, S.; Schell, V.; Kudlek, J.; Holmstrup, M. Interactions between toxic chemicals and natural environmental factors— A meta-analysis and case studies. Sci. Tot. Environ. **2010**, *408*, 3763–3774.

Phosphorus: Riverine System Transport

Andrew N.
Sharpley, Peter
Kleinman, Tore
Krogstad, and
Richard McDowell

Introduction

The contribution of anthropogenic phosphorus (P) to the accelerated eutrophication of freshwaters is well documented.[1,2] Sources of riverine P inputs can broadly be divided into point sources, typically dominated by sewage treatment effluents, and diffuse (also "non-point") sources, derived from the landscape.[3,4] Phosphorus export tends to increase from rivers draining native or low-input ecosystems, to intensively managed agricultural systems, to urban settings.[5] Point sources enter the river more continually through the year than do nonpoint sources, which are subject to large seasonal variation, typically as a function of overland flow and land management activities.[6]

Changes in the forms and amount of P that occur as part of the transport processes within streams and rivers can greatly influence the eventual impact of point and non-point sources of P on downstream eutrophication.[6,7] These changes in P are mediated by physical (sediment deposition and resuspension and flow regimes), abiotic (P sorption and desorption), and biotic (microbial and plant uptake) processes.[8,9]

Most importantly, P in riverine systems can play an important role in modifying or delaying societal efforts to curb eutrophication. For instance, P that is already entrained in riverine systems, sometimes referred to as *legacy P,* can serve as a long-term source of P to the overlying water column. Understanding the role of riverine systems as sources, sinks and stores of P is critical to the long-term management of cultural eutrophication.

Riverine Processes

Physical Processes

Fluvial sediments are derived from the erosion of surface soils, gullies, ditches and stream banks. Because surface soils generally contain the highest concentration of P in soil profiles, and erosion preferentially removes P-rich particles, eroded surface soil represents a major source of particulate

P in riverine systems.[10,11] In areas with recent gully formation or bank erosion, subsoil is the dominant source of sediments. Sediments derived from gulley or bank sources have low P content and high P sorption capacities.[12,13] As P release and sorption are largely related to particle size, with coarser-sized particles releasing P more readily than fine particles, which also tend to sorb more P, [14] hydrologic processes controlling sediment particle size distribution have important implications to P fate in river systems.

Abiotic Processes

In fluvial systems with good hydraulic mixing (such as shallow flowing streams), P movement between sediment and water phases is related to the equilibrium P concentration at zero net sorption or desorption (EPC_0); P is released from sediment if the concentration of P in stream flow is less than its EPC_0, while the reverse is also true.[15] Other processes influencing sediment P release include a rise in stream water pH, P from dead phytoplankton, periphyton or macrophytes, the hydrolysis of organic P species, and changes in sediment crystallinity and oxidation/reduction.[10,16] For example, regular wetting and drying cycles in stream sediments or bank material can change Fe-oxide crystallinity making occluding P associated with these materials.[17,18]

Biotic Processes

Uptake of P by aquatic biota can decrease dissolved P in the water column,[19] while bacteria can mediate a sizeable proportion of sedimentary P uptake and release (30%–40%, Khoshmanesh et al.[20] and McDowell and Sharpley[21]). Biologically-controlled P release during the decomposition of organic matter in sediments can be an important source of dissolved P at times of high temperature and low flow in areas with organic-rich sediments, such as streams draining forests.[22] Organic matter in sediments may also increase the blooms of bacteria and algae by preventing chelator limited growth.[23] The relative effect of biotic processes on riverine P transport varies greatly, reflecting seasonal cycles, management of stream-side land, sediment P forms, size of flow event, and streambed geology. However, during elevated flow, when P loads are often high, biotic processes generally are less important to riverine P transport than physical and abiotic processes.

Nutrient Spiralling

The concept of P-spiralling, or the distance travelled downstream by one P molecule as it completes one cycle of uptake, transformation (e.g., from dissolved to organic form) and release back into flow as dissolved P, is useful in understanding fundamental mechanisms of P transport in rivers.[24] Lengths of P-spiralling generally vary from 1 to 1000 m, as a function of flow regime, season, bedrock geology, and sediment characteristics.[25,26] Interaction between ground water and stream flow within the hyporheic zone will change P concentrations, depending upon the relative contribution of stream-bed upwelling or infiltration of P-rich stream flow.

Integrating Riverine Processes and Land Use Impacts on P Transport

The role of riverine processes in watershed scale P export is illustrated by the findings work of McDowell et al.[27] in a 40-ha agricultural watershed in central Pennsylvania (Figure 1). They found dissolved P concentrations in base flow increased from 28 to 42 µg L^{-1} as one moved from down the stream channel. Base flow P concentrations were controlled by channel sediment P sorption (532 mg kg^{-1} at flume 4 and 227 mg kg^{-1} at the outlet) and EPC_0 (4 µg kg^{-1} at flume 4 and 34 µg kg^{-1} at the outlet). Storm flow trends, however, were the opposite, with P concentrations decreasing downstream (304 μg L^{-1} at flume 4 and 128 μg L^{-1} at flume 1) due to the dilution of P derived from a critical source area: an agricultural field with elevated soil P and high erosion/runoff (Figure 1).

In a much larger watershed, the Winooski River, VT, the largest tributary to Lake Champlain (Figure 2), McDowell et al.[7] evaluated interactions between local sources of P, sediment properties and flow, elucidating their role in riverine P transport. Input and delivery of fine sediment enriched with P was influenced by surrounding land use. Algal-available P of river sediments near agricultural land (3.6 mg kg^{-1}) was greater than that of sediments near forested land (2.4 mg kg^{-1}). Over the short-term, river flow and sediment physical properties were responsible for particulate P loadings from the river to Lake Champlain. However, deposition of sediments downstream, near the outflow into Lake Champlain, resulted in a large pool of stored P within the river system. Over the long-term, this pool is likely to release dissolved P to overlying waters, even as inputs of P from point and nonpoint sources decline due to implementation of remedial strategies and watershed conservation measures.

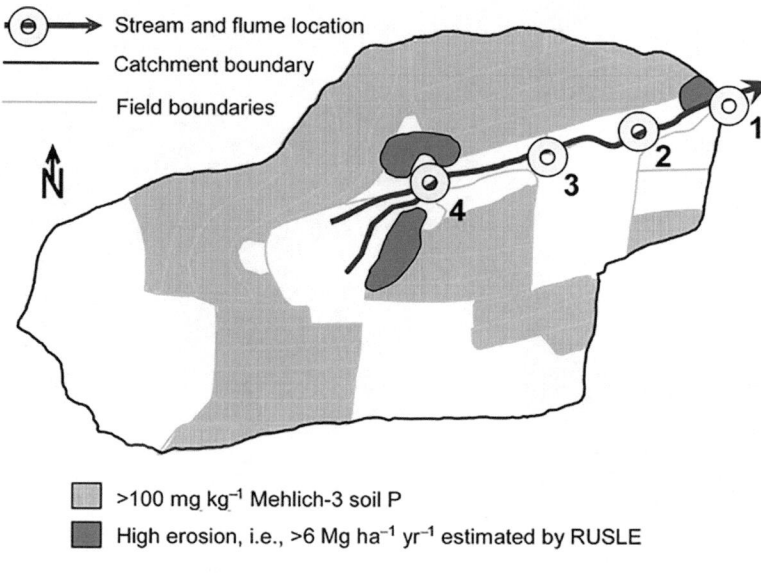

| Flume | Dissolved P | | Stream sediment | |
	Stormflow	Baseflow	P sorption max	EPC$_0$
	μg L^{-1}		mg kg^{-1}	μg L^{-1}
1	128	42	227	34
2	174	36	295	13
3	202	37	330	4
4	304	28	532	4

FIGURE 1 The distribution of high Mehlich-3 soil P (>100 mg kg^{-1}), erosion (>6 Mg ha^{-1} yr^{-1}) and dissolved P concentration in stream and baseflow (mean of 1997–2000 data) in relation to P sorption properties of channel sediment at four flumes in FD-36.
Source: Adapted from McDowell et al.[27]

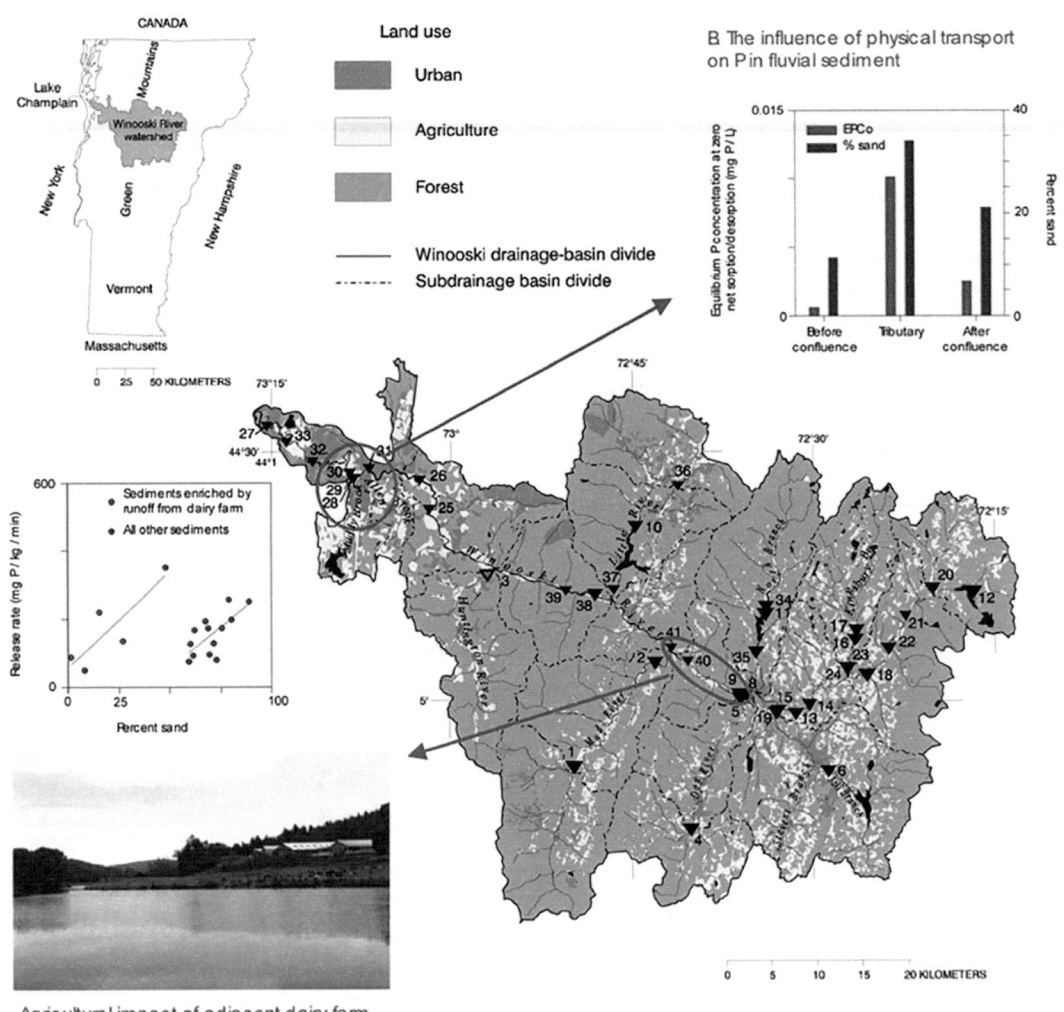

FIGURE 2 The location, distribution and impact of land use and physical transport processes on P in fluvial sediments within the Winooski River watershed, VT.
Source: Adapted from McDowell et al.[7]

Defining P-Related Impairment in Flowing and Lake Waters for Targeted Remediation

In order to prioritize and target watershed remediation to minimize P losses, water impairment must be quantified.[28] Background levels (i.e., regional nutrient criteria) of total P, total N, chlorophyll-*a*, sediment, and clarity in pristine surface waters are used as benchmarks for a given geographical area (Figure 3; Gibson et al.[29] and Omernik[30]). While these criteria have regulatory application, such as under the U.S.A.'s Clean Water Act, they can also be used to guide voluntary efforts in watershed planning.[31] These criteria are available for freshwater systems in the continental U.S. (Table 1). Similar approaches have been taken in Australasia and Europe.[4,32] In the European Union's Water Framework Directive, biological parameters are, however, the basis for measuring ecological status for the water with chemical parameters used only as support parameters. The E.U. classification system emphasizes

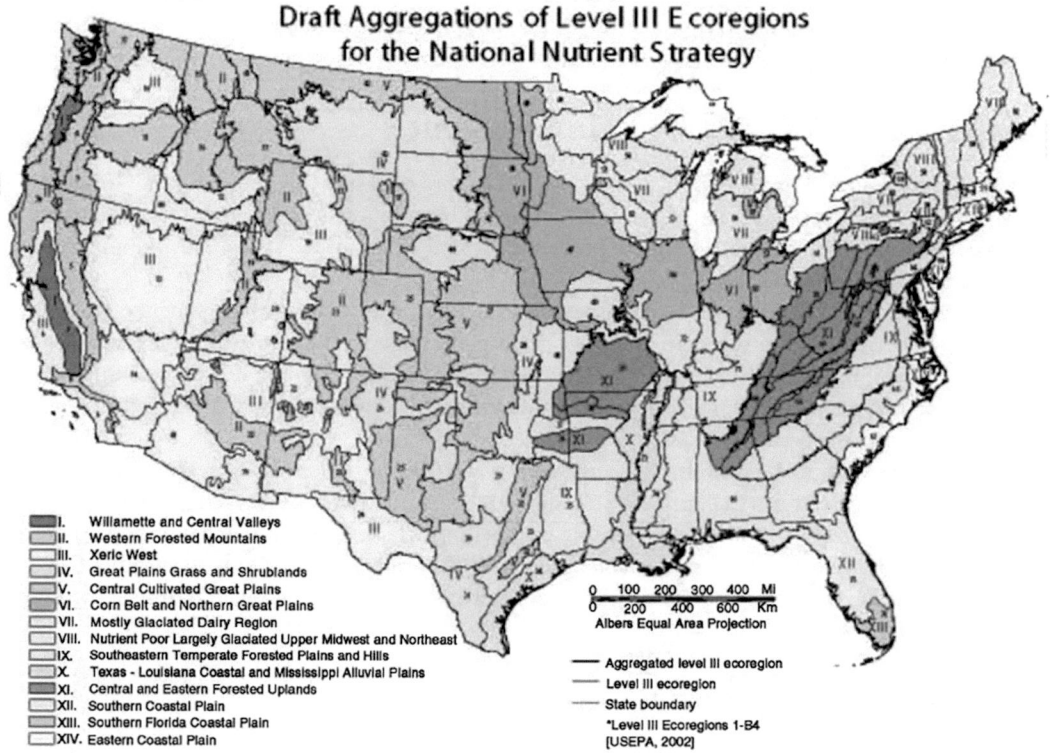

FIGURE 3   Draft aggregations of level III ecoregions for the National Nutrient Strategy.
Source: Adapted from Gibson et al.[29] and Omernik.[30]

TABLE 1   Background Total P concentrations for Each of the Aggregated Nutrient Ecoregions in the U.S. for Freshwater Systems[28]

	Aggregated Ecoregion	Total P (μg L^{-1})	
Number	Description	Rivers and Streams	Lakes and Reservoirs
I	Willamette and Central Valleys	47	—
II	Western Forested Mountains	10	9
III	Xeric West	22	17
IV	Great Plain Grass and Shrub Lands	23	20
V	South Central Cultivated Great Plains	67	33
VI	Corn Belt and Northern Great Plains	76	38
VII	Mostly Glaciated Dairy Region	33	15
VIII	Nutrient Poor Largely Glaciated Upper Midwest and Northeast	10	8
IX	Southeastern Temperate Forested Plains and Hills	37	20
X	Texas-Louisiana Coastal and Mississippi Alluvial Plain	128a	—
XI	Central and Eastern Forested Uplands	10	8
XII	Southern Coastal Plains	40	10
XIII	Southern Florida Coastal Plains	—	18
XIV	Eastern Coastal Plains	31	8

whether the ecosystem is in ecological balance and points out the effect of the pollution rather than providing a classification or ranking according to pollutant concentration, which has been the basis for most previous classifications systems.

Implications to Watershed Management

Aquatic ecosystems respond to P inputs on the basis of factors related to their physiography and flushing rates. Individual systems respond to discrete and sustained P inputs differently, and indeed it may not be possible to attain P loadings low enough to prevent periphyton blooms because of, for example, natural enrichment from P-rich rocks.

A certain degree of eutrophication can be beneficial. For example, fishery management often requires a higher productivity to maintain an adequate phytoplankton-zooplankton-fish food chain for optimum commercial or sport fish production. This food chain may be manipulated by stocking of water with certain fish species in addition to P load reductions, in efforts to reduce the incidence of algal blooms and improve overall water quality.[33]

In most cases, however, eutrophication restricts water use for fisheries, recreation, and industry because of the increased growth of undesirable algae and aquatic weeds and oxygen shortages caused by their death and decomposition.[1] An increasing number of surface waters also experience periodic and massive harmful algal blooms (e.g., *cyanobacteria* and *Pfiesteria*) that contribute to summer fish kills, unpotable drinking water, formation of carcinogens during water chlorination, and links to neurological impairment in humans.[34,35]

Implications for Water Quality Response

The response of riverine systems to upstream changes in watershed management can vary from seasons to decades, and generally increases as watershed scale increases.[36,37] In small watersheds (i.e., <100 ha), research has demonstrated reduction in nutrient and sediment loss in runoff can occur within months of implementing remedial management measures (e.g., "conservation practices "or "best management practices"). However, the spatial complexity of watershed systems and nature of P sources (e.g., acute sources such as fertilizers vs. legacy P sources such as soils and sediments) affects this response time. Indeed, the slow release of legacy P stored in soils and sediments to more rapid surface flow pathways may continue for decades even after practices have been installed to curb further additions. The release of legacy P from riverine sediments is influenced by the oxygen status of overlying waters, where reducing conditions favor the dissolution of iron-bound P. As remedial efforts decrease the concentrations of P in riverine systems, the change in gradient between sediment and water column can trigger the desorption of dissolved P from the sediment. Finally, P enriched sediments are resuspended during high flows. This effect will likely increase with climate change, which could result in more precipitation as rain in winter and an increase in occurrence and severity of rainfall events.

Because of the lag time between watershed management practice implementation and water quality improvements, remedial strategies should consider the re-equilibration of watershed and water-body behavior, where nutrient sinks may become sources of P with only slight changes in watershed management and hydrologic response. A better understanding of the spatial and temporal aspects of watershed response to nutrient load reductions in both flowing and standing water bodies is needed, as well as the scale at which responses may occur in a more timely fashion. This would likely be at a smaller sub-watershed scale, where local water quality and quantity benefits may become evident more quickly. It is also important to accept in any water-shed-P loss reduction strategy, that it is essential to address the overall physical and social complexity of legacy P sources, when, where and to what extent they occur.

Conclusions

Clearly, several interdependent riverine processes influence the amounts and forms of P transported from edge-of-field agricultural sources to the point of impact (i.e., river, lake, reservoir, and estuary). These processes will, thus, be critical in defining agricultural source management and in determining eutrophic response. Without information on the direction and magnitude of change in P transport in river systems, conservation practices will not efficiently remediate against impairment of receiving waters.

The accumulation of P in aquatic environments is such that even if P were no longer added to agricultural systems, there would be a considerable time-lag (years or decades) before improvements in water quality, or regeneration of diverse habitats, might become apparent. Thus, the emphasis of watershed management should be on preventing further deterioration and taking strategic and sustainable actions sooner rather than later, otherwise we are simply and literally storing up more severe problems for future generations to confront. Despite our knowledge of controlling processes, it is difficult for the public to understand or accept this lack of response. When public funds are invested in remedial watershed programs, rapid improvements in water quality are usually expected and often required. Thus, implementation of effective conservation measures must consider fluvial system response behavior, where sinks may become sources of P with only slight changes in watershed management and hydrologic response.

References

1. Carpenter, S.R.; Caraco, N.F.; Correll, D.L.; Howarth, R.W.; Sharpley, A.N.; Smith, V.H. Nonpoint pollution of surface waters with phosphorus and nitrogen. Ecol. Applic. **1998**, *8*, 559–568.
2. Schindler, D.W.; Hecky, R.E.; Findlay, D.L.; Stainton, M.P.; Parker, B.R.; Paterson, M.J.; Beaty, K.G.; Lyng, M.; Kasian, S.E. Eutrophication of lakes cannot be controlled by reducing nitrogen input: Results of a 37-year whole-ecosystem experiment. Proc. National Acad. Sci. **2008**, *105*(32), 11254–11258.
3. U.S. Geological Survey. *The quality of our nations waters: Nutrients and pesticides.* U.S. Geological Survey Circular 1225; USGS Information Services: Denver, CO, 1999; 82 p. Available at http://www.usgs.gov.
4. Withers, P.J.A.; Lord, E.L. Agricultural nutrient inputs to rivers and ground waters in the U.K.: Policy environmental management and research needs. Soil Use and Managt. **2002**, *14*, 186–192.
5. National Research Council. *Clean coastal waters: Understanding and reducing the effects of nutrient pollution*; National Academy Press: Washington, D.C., 2000.
6. Mainstone, C.P.; Parr, W. Phosphorus in rivers – ecology and management. Sci. Total Environ. **2002**, *282–283*, 25–47.
7. McDowell, R.W.; Sharpley, A.N.; Chalmers, A.T. Land use and flow regime effects on phosphorus chemical dynamics in the fluvial sediment of the Winooski River, Vermont. Ecolog. Eng. **2002**, *18*, 477–487.
8. House, W.A. Geochemical cycling of phosphorus in rivers. App. Geochem. **2003**, *18*, 739–748.
9. Jarvie, H.P.; Whitton, B.A.; Neal, C. Nitrogen and phosphorus in east coast British rivers: speciation sources and biological significance. Sci. Total Environ. **1998**, *210–211*, 79–110.
10. Baldwin, D.S.; Mitchell, A.M.; Olley, J.M. Pollutantsediment interactions: Sorption, reactivity and transport of phosphorus. In *Agriculture, Hydrology and Water Quality;* Haygarth, P.M., and Jarvis, S.C. (eds.); CABI International: Oxford, England, 2002; 265–280.
11. Krogstad, T.; Løvstad, Ø. Erosion, phosphorus and phytoplankton response in rivers of southeastern Norway. Hydrobiologia **1989**, *183*, 33–41.
12. Olley, J.M.; Murray, A.S.; Mackenzie, D.H.; Edwards, K. Identifying sediment sources in a gullied catchment using natural and anthropogenic radioactivity. Water Resour. Res. **1993**, *29*, 1037–1043.

13. Sharpley, A.N.; Smith, S.J.; Zollweg, J.A.; Coleman, G.A. Gully treatment and water quality in the Southern Plains. J. Soil Water Conserv. **1996**, *51*, 512–517.

14. Stone, M.; Murdoch, A. The effect of particle size, chemistry and mineralogy of river sediments on phosphate adsorption. Environ. Technol. Letters **1989**, *10*, 501–510.

15. Kunishi, H.M.; Taylor, A.W.; Heald, W.R.; Gburek, W.J.; Weaver, R.N. Phosphate movement from an agricultural watershed during two rainfall periods. J. Agric. Food Chem. **1972**, *20*, 900–905.

16. Fox, L.E. The chemistry of aquatic phosphate: inorganic processes in rivers. Hydrobiologia **1993**, *253*, 1–16.

17. Qiu, S.; McComb, A.J. Planktonic and microbial contributions to phosphorus release from fresh and air-dried sediments. Freshwater Res. **1995**, *46*, 1039–1045.

18. Baldwin, D.S. Effects of exposure to air and subsequent drying on the phosphate sorption characteristics of sediments from a eutrophic reservoir. Limnol. Oceanog. **1996**, *41*, 1725–1732.

19. Horner, R.R.; Welch, E.B.; Seeley, M.R.; Jacoby, J.M. Responses periphyton to changes in current velocity, suspended sediment and phosphor concentration. Freshwater Biol. **1990**, *24*, 215–232.

20. Khoshmanesh, A.; Hart, B.T.; Duncan, A.; Beckett, R. Biotic uptake and release of phosphorus by a wetland sediment. Environ. Technol. **1999**, *29*, 85–91.

21. McDowell, R.W.; Sharpley, A.N. Phosphorus uptake and release from stream sediments. J. Environ. Qual. **2003**, *32*, 937–948.

22. Klotz, R.L. Temporal relation between soluble reactive phosphorus and factors in stream water and sediments in Hoxie Gorge Creek, New York. Can. J. Aquatic Sci. **1991**, *48*, 84–90.

23. Løvstad, Ø.; Krogstad, T. Effect of EDTA, FeEDTA and soils on the phosphorus bioavailability for diatom and blue-green algal growth in oligotrophic waters studied by transplant biotests. Hydrobiologia **2001**, *450*, 71–81.

24. Elwood, J.W.; Newbold, J.D.; O'Neil, R.V.; Van Winkle, W. Resource spiraling: An operational paradigm for analyzing lotic ecosystems. In *Dynamics of lotic ecosystems;* Fontaine, III, T.D.; Bartell, S.M. (eds.); Ann Arbor Science: Ann Arbor, MI, 1983; 3–27.

25. Melack, J.M. Transport and transformations of P, fluvial and lacustrine ecosystems. In *Phosphorus in the Global Environment: Transfers, Cycles and Management;* Tiessen, H. (Ed.); SCOPE 54 and John Wiley and Sons: Chichester, U.K., 1995; 245–254.

26. Munn, N.L.; Meyer, J.L. Habitat-specific solute retention in two small streams, An intersite comparison. Ecology **1990**, *71*, 2069–2082.

27. McDowell, R.W.; Sharpley, A.N.; Folmar, G. Phosphorus export from an agricultural watershed: linking source and transport mechanisms. J. Environ. Qual. **2001**, *30*, 1587–1595.

28. U.S. Environmental Protection Agency. *Ecoregional Nutrient Criteria.* EPA-822-F-01-010. USEPA, Office of Water (4304), U.S. Govt. Printing Office: Washington, DC, 2001. Available at http://www.epa.gov/waterscience/criteria/nutrient/9docfs.pdf.

29. Gibson, G.R.; Carlson, R.; Simpson, J.; Smeltzer, E.; Gerritson, J.; Chapra, S.; Heiskary, S.; Jones, J.; Kennedy, R. *Nutrient criteria technical guidance manual: lakes and reservoirs* (EPA-822-B00-001). U.S. Environmental Protection Agency: Washington, D.C.; U.S. Govt. Printing Office: Washington, DC, 2000.

30. Omernik, J.M. Ecoregions of the Conterminous United States. Annals Assoc. Am. Geographers **1987**, *77*, 118–125.

31. U.S. Environmental Protection Agency. *National Strategy for the Development of Regional Nutrient Criteria.* EPA-822-F-98-002. USEPA, Office of Water (4304), U.S. Govt. Printing Office: Washington, DC, 1998. Available at http://www.epa.gov/waterscience/standards/nutsi.html.

32. Sparrow, L.A.; Sharpley, A.N.; Reuter, D.J. Safeguarding soil and water quality. In Opportunities for the 21st Century: Expanding the horizons for soil, plant, and water analysis. Commun. Soil Sci. Plant Anal. **2000**, *31*(11–14), 1717–1742.

33. Horppila, J.; Kairesalo, T. A fading recovery: the role of roach (Rutilus rutilus L.) in maintaining high algal productivity and biomass in Lake Vesijarvi, southern Finland. Hydrobiologia **1990**, *200–201,* 153–165.

34. Burkholder, J.A.; Glasgow, Jr., H.B. *Pfiesteria piscicidia* and other Pfiesteria-dinoflagellates behaviors, impacts, and environmental controls. Limnol. Oceanog. **1997**, *42,* 1052–1075.

35. Kotak, B.G.; Kenefick, S.L.; Fritz, D.L.; Rousseaux, C.G.; Prepas, E.E.; Hrudey, S.E. Occurrence and toxicological evaluation of cyanobacterial toxins in Alberta lakes and farm dugouts. Water Res. **1993**, *27,* 495–506.

36. Cassell, E.A.; Clausen, J.C. Dynamic simulation modeling for evaluating water quality response to agricultural BMP implementation. Water Sci. Tech. **1993**, *28,* 635–648.

37. Sharpley, A.N. (ed.). *Agriculture and Phosphorus Management: The Chesapeake Bay.* CRC Press: Boca Raton, FL, 2000. 229 p.

48

Nitrogen: Biological Fixation

Mark B. Peoples

Introduction

Dinitrogen (N_2) gas represents close to 80% of the Earth's atmosphere, and a number of procaryotic microorganisms in soils and flooded conditions have evolved that utilize the enzyme nitrogenase to reduce atmospheric N_2 to ammonia. This ammonia provides a renewable source of nitrogen (N) that is subsequently used to support their growth. Biological N_2 fixation (BNF) by some diazotrophs can occur in a free-living state. This includes anaerobic organisms such as *Clostridium* and aerobic bacterium such as *Azotobacter* along with a host of other species that may rely on decomposition of plant residues as an energy source, while phototrophic bacteria such as cyanobacteria (also known as blue-green algae) derive energy directly from photosynthesis to sustain BNF.[1,2] Other N_2-fixing microbes have coevolved with plants and undertake BNF in multiple ways. This ranges from rather loose associations of free-living heterotrophic bacteria with plant roots or endophytic N_2-fixing bacteria present in the vascular tissue of tropical C_4-graminaceous species such as sugarcane (*Saccharum officinarum*), to highly complex symbioses where forms of fixed N generated by the bacterium are exchanged with the plant for carbon-derived plant photosynthesis.[2–4] Such symbioses include (1) cyanobacteria with the aquatic ferns (*Azolla*), fungi (lichen), or cycads; (2) actinomycetes (generally placed in the genus *Frankia*) with around 200 species of angiosperms including Alder (*Alnus glutinosa*) and *Casuarina* spp.; and (3) the soil bacteria rhizobia (*Rhizobium, Bradyrhizobium, Allorhizobium, Azorhizobium, Mesorhizobium,* or *Sinorhizobium* spp.) with >10,000 legume species. With the exception of *Azolla* and lichen, these symbiotic partnerships generally occur in specialized plant structures known as nodules located on plant roots and, in some legume species, also the stem.[2,5]

Although calculations of global contributions of BNF are subject to enormous approximations, annual inputs of fixed N into agroecosystems by these diverse range of organisms and associations were conservatively estimated to be 55–70 million metric tonnes of N based on Food and Agriculture Organization (FAO) statistical data for agricultural production in 2005.[4] Unfortunately, uncertainties about the distribution and frequency of N_2-fixing organisms, or the rates of BNF occurring, in natural ecosystems makes it almost impossible to provide similar estimates of global inputs of fixed N for the many different terrestrial and aquatic environments not under agricultural production.

Sources of BNF in Agricultural Systems

The N$_2$ fixation process can directly contribute to agricultural production where the fixed N is harvested in legume grain or vegetative parts for human or animal consumption. However, BNF can also represent an important renewable source of N that can help maintain or enhance the N fertility of many agricultural soils for the benefit of subsequent crops.[5–7] Examples of experimental estimates of amounts of BNF achieved by various N$_2$-fixing organisms and estimates of their global inputs of fixed N are presented in Table 1.

Free-living N$_2$ fixers tend to have low rates of BNF (Table 1). While significant inputs of fixed N by diazotrophs have been demonstrated in some tropical grass and cropping systems,[3] the role of BNF associated with nonlegumes tend to be less conclusive in temperate agriculture,[8,9] and these observations are reflected in the estimates of global inputs of fixed N from endophytic, associative, and free-living sources provided in Table 1. *Azolla* may have higher rates of BNF than non-symbiotic systems,[2,5] but it is symbiotic relationships between rhizobia and legumes that are responsible for the largest amounts of fixed N in agriculture both on a per unit basis and in absolute terms (Table 1).

Inputs of Fixed Nitrogen by Legumes

That BNF by legume systems plays a key role in world crop production is irrefutable. The ability of legumes to improve the N status of soils has been utilized for thousands of years in crop rotations and traditional farming systems.[2,5,7] The 229 million hectares of legume oilseeds (soybean—*Glycine max*; and groundnut/peanut—*Arachis hypogea*) and pulses sown globally each year, the legume components of the 100–200 million hectares under temporary pastures or fodder crops, and the perennial legume cover crops under much of the 33 million hectares of rubber (*Hevea brasiliensis*) and oil-palm (*Elaeis guineensis*) plantations all contribute fixed N to farming systems.[5,6,10]

Most modern methods used to quantify inputs of fixed N by legumes separate the plant N into fractions originating from soil N or N$_2$ fixation.[11,12] Once the legume N can be partitioned into that proportion derived from atmospheric N$_2$ (%Ndfa, sometimes also described as %Pfix) and that coming from the soil, the amounts of N$_2$ fixed can be calculated from measures of shoot dry matter (DM) and N content. The formation of the symbiosis between legume and rhizobia is dependent upon many factors and cannot be assumed to occur as a matter of course. This is reflected in the range of values presented

TABLE 1 Experimental Measures of BNF by Different N$_2$-Fixing Organisms and Estimates of Annual Global Inputs of Fixed N in Agricultural Systems[4]

N$_2$-Fixing Organism	Range of Estimates of BNF Measured (kg N/ha)	Typical Average Rates of BNF (kg N/ha)	Global Inputs of BNF (million t N/year)
Free-living and associative			
Temperate crop land	0–80	<5	<4
Tropical savannas	5–45	10	<14
Tropical crops	0–240	25	0.5
Symbiotic			
Azolla	10–150	33	5
Crop legumes	0–450	115[a]	21[a]
Pasture/forage legumes	1–680	110–227[a,b]	12–25[a,b]
Legumes in agroforestry	5–470	200[a]	nd[c]
Green manure legumes	5–325	100[a]	nd[c]

[a] Estimates of BNF have been adjusted to include a belowground contribution of N associated with the nodulated roots.[6,10,12]

[b] The range reflects that legumes can be grown as a pure sward or in a mixture with other species.

[c] nd indicates "not determined" since data relating to the total areas of legume grown in agroforestry systems and as green manure were not available.

in Table 1. Such large variations in reported estimates of N_2 fixation make it difficult to generalize about how much N may be fixed by different legume species. Collectively, the international literature suggests maximum rates of N_2 fixation of 3–4 kg shoot N/ha/day[12] and potential inputs of fixed N by many legumes of several hundred kg of shoot N/ha each year.[5,6,10] However, much of this information has been derived from research trials in which specific treatments were imposed to generate differences in %Ndfa values and legume growth as an experimental means of studying factors which regulate BNF. Therefore, these data may be of little relevance to what might actually be occurring in farmers' crops and pastures. Fortunately, measurement procedures are available which allow on-farm measures of legume N_2 fixation to be conducted with some degree of confidence.[11]

Levels of Nitrogen Fixation Achieved in Farmers' Fields

Examples of the types of information which can be generated about BNF in farmers' fields are presented in Table 2 for different regions of the world. These on-farm data and observations can be used to develop a picture of N_2 fixation within an individual country and provide insights into contributions of BNF to agriculture on a global scale. Collectively, the results in Table 2 indicate that the potential for BNF inputs can differ between legumes and countries, but they also suggest many commonalities. Although wide ranges in %Ndfa values have been observed, it seems that, on average, most winter pulses (e.g., chickpea—*Cicer arietinum*; lentil—*Lens culinaris*; field pea—*Pisum sativum*; fababean—*Vicia faba*; lupin—*Lupinus albus*) satisfy relatively higher proportions of their growth requirements from N_2 fixation (>65%) than do the summer legumes (e.g., mungbean—*Vigna radiata*; mashbean—*Vigna mungo*; soybean; groundnut) where %Ndfa values were commonly less than 60% (Table 2). Poor or variable nodulation observed in some summer legumes and the resulting increased reliance upon soil N may reflect greater N mineralization during summer, water stress, and/or low vegetative biomass accumulated by short duration legume crops.[13–15]

TABLE 2 Summary of the Proportion of Plant N Derived from N_2 Fixation (%Ndfa) and the Amounts of N_2 Fixed by Farmers' Legume Crops and Pastures in Different Geographical Regions

Country and Legume	Number of Fields	Mean Ndfa (%)	Total N fixed (kg N/ha)[a]
Winter pulse crops			
Pakistan	126	78	79
Nepal	27	79	78
Syria	46	67	na[b]
Australia	148	65	152
Summer legume crops			
Pakistan	63	47	42
Nepal	50	55	77
Thailand	13	75	78
Vietnam	45	48	125
South Africa	14	58	na[b]
Australia	33	53	267
Annual pastures			
Australia	303	75	na[b]
Perennial pastures			
Australia	110	64	na[b]

[a] Includes a combined estimate of fixed N from both the shoots and the nodulated roots.[6,10,12]
[b] Data not available for all fields.
Note: Ndfa = Nitrogen derived from atmospheric N2.

TABLE 3 Key Factors Influencing Inputs of Fixed N by Legumes in Farmers' Fields

Country	System	BNF Regulated By		
		DM	Soil Nitrate	Primary Factors
Pakistan	Winter crop	+++		Rainfall nutrition, weed control
	Summer crop	+++	+++	Fertilizer N, no inoculation, insects, disease
Nepal	Winter crop	+++		Rainfall, nutrition
	Summer crop	+++	+	Total soil N, mineralized N, available P, legume species
Syria	Winter crop	+++	++	Soil nutrients, insects, disease
Thailand	Summer crop	+++		Available P
Vietnam	Summer crop	+++		Plant density, soil pH, available P, legume species
South Africa	Summer crop	++	++	Effective inoculation, nutrition, rotation, water availability
Australia	Winter crop	+++	+++	Rainfall, fallowing, inoculation, legume species
	Summer crop	++	+++	Crop rotation, tillage, water availability
	Pasture	+++	+	Soil pH, available P, legume density, grazing management

Note: BNF = biological N_2 fixation; DM = dry matter; P = phosphorus.

In grazed pastures and intensive forage systems, the competition for mineral N between legumes and any companion grasses or vigorous broad-leaf species growing within the sward results in low levels of plant-available soil N throughout the growing season.[10,12] As a consequence, %Ndfa by the legume components of pastures tend to be high (Table 2). The lower %Ndfa values detected in perennial legume species, such as alfalfa (lucerne; *Medicago sativa*) and white clover (*Trifolium repens*), presumably result from a greater ability to scavenge soil mineral N from a larger rooting zone over a longer growing season compared with annual pasture species.[10,15]

Although the levels of %Ndfa are important, the amounts of N_2 fixed are usually regulated by legume growth rather than %Ndfa in most farming systems, and many legumes appear to fix around 15–20 kg of shoot N for every metric tonnes of shoot DM accumulated.[6,10,14,15] Depending upon the species, this would translate to 23–40 kg of total N fixed for every tonne of shoot DM if belowground contributions of fixed N-associated nodulated roots are included.[6,10,12]

Impact of Management

Factors that either enhance or depress N_2 fixation (Table 3) can generally be summarized in terms of environmental or management constraints to crop growth (e.g., basic agronomy, nutrition, water supply, diseases, and pests).[6,10,16] A number of strategies can be employed that specifically enhance BNF through increased legume biomass. These include the use of legume genotypes adapted to the prevailing edaphic and environmental conditions, procedures to improve legume plant density, irrigation (if available), the amelioration of soil nutrient toxicities or deficiencies, and the control of weeds and pests.[10,16] However, as the formation of an active symbiosis is dependent upon the compatibility of both the diazotrophic microorganism and the legume host, local practices that limit the presence of effective rhizobia (no inoculation, poor inoculant quality) will also be crucial in determining the legume's capacity to fix N (Table 3), as will any management decisions that directly affect soil N fertility (excessive tillage, extended fallows, fertilizer N, and rotations), since mineral N is a potent inhibitor of the N_2 fixation process.[6,10,15,16]

Conclusions

Symbiotic associations between legumes and rhizobia are responsible for the greatest contributions of BNF in agricultural systems. Research trials suggest potential annual inputs of fixed N by most legumes equivalent to several hundreds of kg N/ha. However, data collected from pulses, legume oilseeds, and pastures growing in farmers' fields suggest that while legumes should routinely be fixing >100 kg/ha each year, in reality, they usually do not. Strategies are available to improve BNF beyond what is currently being achieved. For example, provided that a legume crop is abundantly nodulated and effectively fixing N_2, enormous benefits in terms of crop production and N_2 fixed can be derived from the application of good agronomic principles. But the ability to overcome constraints at the farm level may be limited because the relevant technologies are either not in the hands of the farmers or they cannot readily adopt them because of lack of knowledge and information, economic constraints, or operational imperatives. While the global inputs of fixed N by legumes may be less than their genetic potential and is lower than the 100–110 million tonnes of N applied as fertilizer each year, many million tonnes of fixed N are harvested annually in grain from legume crops and many million tonnes more will be consumed by animals in legume-based forage, and there are many environmental advantages in relying upon BNF over fertilizer N to produce such large quantities of high-quality protein.[17,18]

References

1. Roper, M.M.; Ladha, J.K. Biological N_2 fixation by heterotrophic and phototrophic bacteria in association with straw. *Plant Soil* **1995**, *174*, 211–224.

2. Giller, K.E. *Nitrogen Fixation in Tropical Cropping Systems*, 2nd ed.; CABI Publishing: Wallingford, England, 2001.

3. Boddey, R.M.; de Oliveira, O.C.; Urquiaga, S.; Reis, V.M.; de Olivares, F.L.; Baldani, V.L.D.; Döbereiner, J. Biological nitrogen fixation associated with sugar cane and rice: contributions and prospects for improvement. *Plant Soil* **1995**, *174*, 195–209.

4. Herridge, D.F.; Peoples, M.B.; Boddey, R.M. Global inputs of biological nitrogen fixation in agricultural systems. *Plant Soil* **2008**, *311*, 1–18.

5. Peoples, M.B.; Herridge, D.F.; Ladha, J.K. Biological nitrogen fixation: an efficient source of nitrogen for sustainable agricultural production? *Plant Soil* **1995**, *174*, 3–28.

6. Peoples, M.B.; Brockwell, J.; Herridge, D.F.; Rochester, I.J.; Alves, B.J.R.; Urquiaga, S.; Boddey, R.M.; Dakora, F.D.; Bhattarai, S.; Maskey, S.L.; Sampet, C.; Rekasem, B.; Khan, D.F.; Hauggaard-Nielsen, H.; Jensen, E.S. The contributions of nitrogen-fixing crop legumes to the productivity of agricultural systems. *Symbiosis* **2009**, *48*, 1–17.

7. Angus, J.F.; Kirkegaard, J.A.; Hunt, J.R.; Ryan, M.H.; Ohlander, L.; Peoples, M.B. Break crops and rotations for wheat. *Crop Pasture Sci.* **2015**, *66*, 523–552.

8. Giller, K.; Merckx, R. Exploring the boundaries of N_2 fixation in cereals and grasses: an hypothetical framework. *Symbiosis* **2003**, *35*, 3–17.

9. Unkovich, M.J.; Baldock, J.A. Measurement of asymbiotic N_2 fixation in Australian agriculture. *Soil Biol. Biochem.* **2008**, *40*, 2915–2921.

10. Peoples, M.B.; Brockwell, J.; Hunt, J.R.; Swan, A.D.; Watson, L.; Hayes, R.C.; Li, G.D.; Hackney, B.; Nuttall, J.G.; Davies, S.L.; Fillery, I.R. Factors affecting the potential contributions of N_2 fixation by legumes in Australian pasture systems. *Crop Pasture Sci.* **2012**, *63*, 759–786.

11. Unkovich, M.; Herridge, D.; Peoples, M.; Cadisch, G.; Boddey, R.; Giller, K.; Alves, B.; Chalk, P. *Measuring Plant-Associated Nitrogen Fixation in Agricultural Systems*, ACIAR Monograph No. 136: Caberra, Australia, 2008.

12. Unkovich, M.J.; Pate, J.S. An appraisal of recent field measurements of symbiotic N_2 fixation by annual legumes. *Field Crops Res.* **2001**, *65*, 211–228.

13. Herridge, D.F.; Robertson, M.J.; Cocks, B.; Peoples, M.B.; Holland, J.F.; Heuke, L. Low nodulation and nitrogen fixation of mungbean reduce biomass and grain yields. *Aust. J. Expl. Agric.* **2005**, *45*, 269–277.

14. Maskey, S.L.; Bhattarai, S.; Peoples, M.B.; Herridge, D.F. *Field Crops Res.* **2001**, *70*, 209–221.

15. Peoples, M.B.; Bowman, A.M.; Gault, R.R.; Herridge, D.F.; McCallum, M.H.; McCormick, K.M.; Norton, R.M.; Rochester, I.J.; Scammell, G.J.; Schwenke, G.D. Factors regulating the contributions of fixed nitrogen by pasture and crop legumes to different farming systems of eastern Australia. *Plant Soil* **2001**, *228*, 29–41.

16. Peoples, M.B.; Ladha, J.K.; Herridge, D.F. Enhancing legume N$_2$ fixation through plant and soil management. *Plant Soil* **1995**, *174*, 83–101.

17. Jensen, E.S.; Hauggaard-Nielsen, H. How can increased use of biological N$_2$ fixation in agriculture benefit the environment? *Plant Soil* **2003**, *252*, 177–186.

18. Jensen, E.S.; Peoples, M.B.; Boddey, R.M.; Gresshoff, P.M.; Hauggaard-Nielsen, H.; Alves, B.J.R.; Morrison, M.J. Legumes for mitigation of climate change and the provision of feedstock for biofules and biorefineries. *Agron. Sustain. Dev.* **2012**, *32*, 329–364.

49

Nutrients: Best Management Practices

Scott J. Sturgul and
Keith A. Kelling

Introduction

Soil nutrients, like all agricultural inputs, need to be managed properly to meet the fertility requirements of crops without adversely affecting the quality of water resources. The nutrients of greatest concern relative to water quality are nitrogen (N) and phosphorus (P).[1] Nitrogen not recovered by crops can add nitrate-N to groundwater through leaching. Nitrate is the most common groundwater contaminant found in the United States.[2–4] Nitrate levels that exceed the established U.S. drinking water standard of 10 ppm nitrate–N have the potential to adversely affect the health of infants and young livestock.[5] Furthermore, the movement of excessive nitrate to coastal estuaries has been linked to the development of hypoxic conditions affecting fisheries in the Gulf of Mexico.[6,7] Surface water quality is the primary environmental concern with P, as runoff and erosion from fertile cropland add nutrients to water bodies that stimulate the excessive growth of aquatic weeds and algae.[8] Of all crop nutrients, it is critical to prevent P from reaching lakes and streams since the biological productivity of aquatic plants and algae in fresh water environments is usually limited by this nutrient.[9] Consequences of increased aquatic plant and algae growth include reduced aesthetic and recreational value of lakes and streams as well as the seasonal depletion of water dissolved oxygen content, which may result in fish kills as well as other ecosystem disruptions.

Best management practices for agricultural nutrients can vary widely from one region to another due to differences in cropping, topographic, environmental, and economic conditions. With the variety of factors to consider, no single set of nutrient management practices can be recommended for all farms. Nutrient management practices for optimizing crop production while protecting water quality must be tailored to the unique conditions of individual farms. The following practices should be considered for any nutrient management strategy for optimizing both agricultural production and environmental protection.

Nutrient Application Rates

The most important management practice for environment-tally—and economically—sound nutrient management is the application rate. Optimum nutrient application rates are identified through fertilizer response and calibration research for specific soils and crops. Economically optimum nutrient application rates provide maximum financial return, but as application rates near the economic optimum, the efficiency of nutrient use by the crop decreases, and the potential for loss to the environment increases.[10] Figure 1 illustrates this concept for N applications to corn. Nutrient application above economically optimum rates (in this case, 160 lb N/acre) reduces profits and increases the likelihood of detrimental impact to the environment.

As shown by many researchers, applications of N in excess of crop need can result in significantly higher N leaching losses.[11–15] A Wisconsin study[16] of the effects of various N fertilizer rates on nitrate-N leaching from several crop rotations found a direct relationship between nitrate-N loss by leaching and the amount of N applied in excess of crop needs (Figure 2). Soil water nitrate-N concentration increased steadily as the amount of excess N increased, which strongly indicates a direct link between excessive N applications and the potential for nitrate-N loss to groundwater.

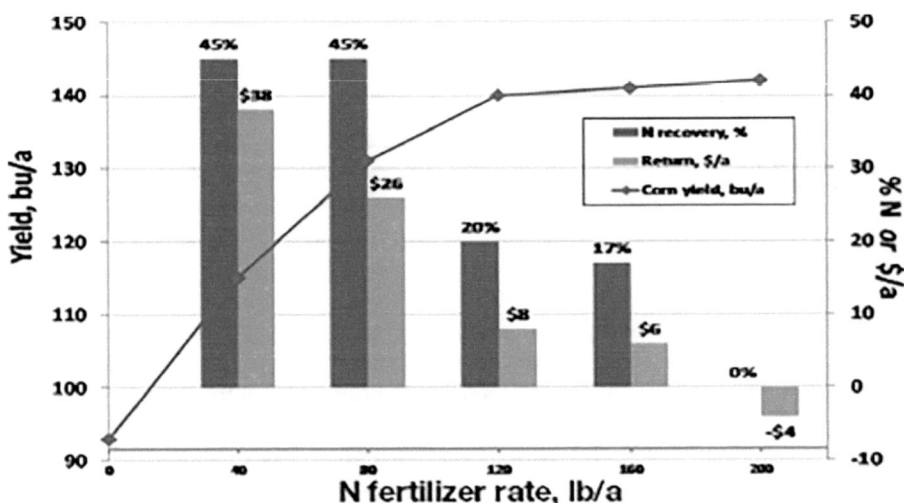

FIGURE 1 Relationships between corn grain yield, economic return, and recovery of applied N.
Source: Bundy et al.[10]

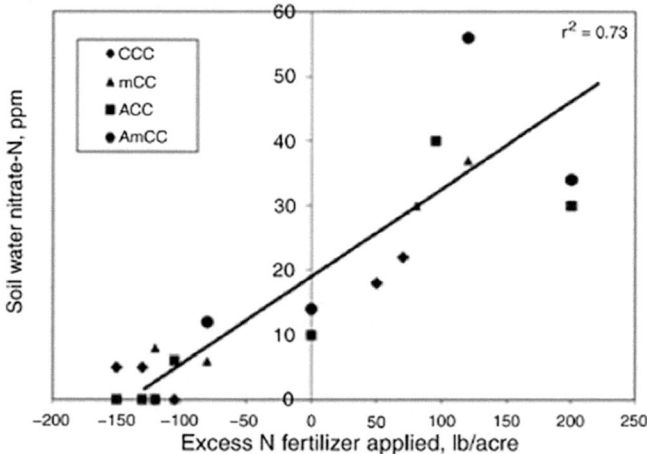

FIGURE 2 Relationship between amount of excess N applied and soil water nitrate content for several cropping rotations (C = corn, A = alfalfa, m = manure).
Source: Andraski et al.[16]

Because of the overall importance of nutrient application rates, accurate assessments of crop nutrient needs are essential for minimizing threats to water quality while maintaining economically sound production. Soil testing is the most widely used method to accurately estimate the existing fertility of soil as well as to determine the need for supplemental nutrients to meet the needs of crops. It has the further advantage of being performed prior to raising the crop, whereas other methods such as plant tissue analysis or optical scanners must have the target crop in place.

Optimum Nitrogen Applications

Most non-legume crops need supplemental N to improve crop yield and quality and to optimize economic return. It is imperative that N application rate recommendations accurately predict the amount of N needed to obtain profitable crop yields and minimize N losses to the environment. Application rate guidelines for N vary according to the crop to be grown, soil characteristics (including soil yield potential, organic matter, texture, etc.), and local climatic conditions.

Expected yield or yield goal estimates have been a primary input for determining grain N fertilizer recommendations for most of the United States since the 1970s.[17] However, research from several Midwest states over the past four decades shows that economic optimum N rates for corn are not best predicted using a yield parameter.[18–21]

More accurate N recommendations for corn and some other grain crops have been developed using the results of N rate response experiments conducted on the major soils of a given region. This soil-specific approach is based on corn yield response and associated economic return to incremental rates of N.[18,19] Two additional parameters influencing the economic optimum N application rate are the anticipated price of corn and the cost of fertilizer N. The data gathered from these experiments are the foundation for corn N recommendations throughout the upper Midwest United States.[22,23]

Additional Tests for Fine-Tuning Nitrogen Applications

The development of tests for assessing soil N levels provides additional tools for improving the efficiency of N fertilizer applications. These tests allow fertilizer recommendations to be adjusted to site-specific conditions that can influence N availability. Tests include the preplant soil profile nitrate test, [24] the pre–side-dress soil nitrate test,[25] plant analysis,[26,27] chlorophyll meters,[28] the end-of- season stalk nitrate test,[29] or the end-of-season soil nitrate test.[30]

Calibrated Soil Tests for Phosphorus and Potassium

In recent years, soil test recommendation programs for P, potassium (K), and other relatively immobile nutrients have tended to de-emphasize a soil buildup and maintenance philosophy in favor of a better balance between environmental and economic considerations by using a crop sufficiency approach.[23,31] These tests are calibrated by field experiments to obtain predictable crop yield responses. Such an approach adds extra emphasis to regular soil testing. It is recommended that soil tests be taken at least every 3 to 4 yr and more frequently on sandy and other soils of low buffering capacity.

Nutrient application recommendations based on soil test results can be accurate only if soil samples representative of the field of interest are collected. Samples that are unrepresentative of fields often result in recommendations that are misleading. Before collecting soil samples, relevant local guidance should be sought on the appropriate number samples to collect, as well as the methodology for collection. In addition, field history information should be provided with the soil samples in order to accurately adjust the fertility recommendations to account for nutrient credits from field-specific activities such as manure applications and legumes in the rotation.

Realistic Yield Goals

For many soil fertility programs, the recommendation of appropriate nutrient application rates is dependent on the establishment of realistic yield goals. Yield goal estimates that are too low will underestimate nutrient needs and can limit crop yield. Yield goal estimates that are too high will overestimate crop needs and result in soil nutrient levels beyond that needed by the crop, which in turn has the potential to increase nutrient contributions to water resources. Estimates should be based on field records and some cautious optimism—perhaps 10% above the recent 3 to 5 years average corn yield from a particular field. Yield goals reasonably higher than a multiyear average are suggested because annual yield variations due to factors other than nutrient application rates (primarily climatic factors) are often large.

Nutrient Credits

The integration of economic return and environmental quality protection requires that nutrients from all sources be considered. In the determination of supplemental fertilizer application rates, it is critical that nutrient contributions from manure, previous legume crops grown in the cropping rotation, and land-applied organic by-products are credited. Both economic and environmental benefits can result if the nutrient-supplying capacity of these nutrient sources is correctly estimated. Economically, commercial fertilizer application rates can often be reduced or eliminated entirely when nutrient credits are accounted. Environmentally, the prevention of overfertilization reduces potential threats to water quality.

Manure

Manure can supply crop nutrients as effectively as commercial fertilizers in amounts that can meet the total N, P, K, and sulfur need of many crops.[32] To utilize manure as a fertilizer resource, its application rate and nutrient- supplying capacity (i.e., plant-available nutrient content) need to be estimated.

Calibration of manure application equipment is key to estimating application rates. Calibration is a relatively easy task that can be done with platform scales or portable axle scales.[33] As a result of manure spreader calibration, an applicator will have a reasonable estimate of the manure application rate—provided the manure is uniformly applied across fields.

The most effective method for gauging the nutrient content of manure is to have samples analyzed by a commercial or university laboratory.[34] Large farm-to-farm variation can occur in manure nutrient content due to manure storage and handling techniques, livestock feed variations, or other farm management differences.[35] In instances when laboratory analysis is not convenient or available, estimates of crop nutrients supplied by animal manures can be made using published values for the average nutrient values of livestock manures common to a given state or region. These values are often provided by area universities. Note that not all the nutrients in manure are available to crops in the first year following application. When estimating the nutrient-supplying capabilities of manure applications, be certain that first-year crop-available nutrient content values are used to calculate the fertilizer value of the manure—not total nutrient content.

Legumes

Legume crops, such as alfalfa, clover, soybeans, and leguminous vegetables, have the ability to fix atmospheric N and convert it to a plant-available form. When grown in a rotation, some legumes can supply substantial amounts of N to a subsequent non-legume crop. For example, a dense stand of alfalfa can often provide most, if not all, of the N needed for a corn crop following it in a rotation.[36] An efficient nutrient management strategy needs to consider the N contributions of legumes to subsequent crops. The amount of legume N to credit varies regionally. Consult the local university extension service for appropriate recommendations.

Biosolids

The application of organic biosolids such as sewage sludge, whey, compost, or other organic wastes to cropland fields can be another source of potential crop nutrient credits. While the overall percentage of cropland acres receiving biosolids is relatively small when compared with manure or legumes, the nutrient contributions can be significant and should be accounted for prior to fertilizer applications. Special management and regulatory considerations pertain to the land application of these materials. Consult local regulations for further information.

Timing of Nutrient Applications

The timing of application is a major consideration for the management of mobile nutrients such as N. For less mobile nutrients, application timing is not a major factor affecting water quality protection. However, nutrient applications on frozen sloping soils or surface applications prior to periods likely to produce runoff events should be avoided to prevent nutrient contributions to surface waters.

Nitrogen Applications

The period between application and crop uptake of N is an important factor affecting the efficient utilization of N by the crop and the potential for loss of N via leaching, denitrification, and other processes.[37,38] Loss of N can be minimized by supplying it just prior to the period of greatest crop uptake. However, several considerations, such as soil, equipment, labor, and fertilizer price and availability are involved in determining the most convenient, economical, and environmentally safe N fertilizer application period.

Fall Nitrogen Applications

The advantages and disadvantages of fall N fertilizer applications are commonly debated. An increased risk for N loss with fall applications needs to be weighed against the fertilizer price and time management advantages (greater window for fertilizer application and spring planting) that can be associated with fall-applied N. The agronomic concern with fall N applications is that losses between application and crop uptake the following growing season will lower recovery of N and reduce crop yield. The environmental concern with fall application is that the N lost prior to crop uptake will leach into groundwater. Fall to spring precipitation, soil texture, and soil moisture conditions influence the potential for fall-applied N losses. If a soil is wet in the fall, rainfall may cause either leaching of nitrate in coarse soils or denitrification of nitrate in heavy, poorly drained soils. Long-term studies indicate that fall applications on medium-textured soils are 10%–15% less effective than the same amount of N applied spring preplant.[37] For both agronomic and environmental reasons, fall applications of N fertilizers are not recommended on coarse- textured soils or on shallow soils over fractured bedrock. If fall applications are to be made on other soils, it is recommended that ammonium-N sources be used and that the applications be delayed until soil temperatures are below thresholds of biological activity (i.e., 50°F) in order to slow the conversion of ammonium to nitrate by soil organisms. If fall applications must be made when soil temperatures are higher than 50°F, a nitrification inhibitor should be used in conjunction with the N fertilizer.

Preplant Nitrogen Applications

Spring preplant applications of N are usually agronomically and environmentally efficient on medium-textured, well-drained soils. The potential for N loss prior to crop uptake on these soils is relatively low with spring applications. If spring preplant applications of N are to be made on sandy soils, ammonium forms of N treated with a nitrification inhibitor should be used. Likewise, nitrification inhibitors should be used if spring preplant N is applied to poorly drained soils. Use of nitrification inhibitors reduces the potential for N loss compared with preplant applications without them; however, side-dress or split applications can be more effective and cost efficient than preplant applications with nitrification inhibitors.

Side-Dress Nitrogen Applications

Side-dress applications of N to row crops during the growing season are effective on all soils with the greatest benefit on sandy or heavy-textured, poorly drained soils.[38,39] The greatest efficiency of side-dress N applications is achieved when the application of N occurs just prior to the period of rapid N uptake by crops. This results in a shorter period of exposure to potential losses of N from leaching or denitrification. Table 1 illustrates the higher yield and crop recovery of N on sandy soils with side-dress applications. In these trials, use of side-dress N applications improved average N recovery over preplant applications by 17%. The use of side-dress or delayed N applications on sandy soils is essential

TABLE 1 Effect of Rate and Time of N Application on Corn Yield and Recovery of Applied N on Irrigated Plainfield Sand

	Relative Yield Increase[a]		N Recovery	
N Rate	Preplant	Side-dress	Preplant	Side-dress
(lb/a)	- - - (% over control)- - -		- - - (%) - - -	
0	–	–	–	–
70	132	176	50	73
140	216	258	44	64
210	247	276	40	49
Average	197	237	45	62

[a] Side-dress treatments applied 6- weeks after planting.
Source: Bundy et al.[40]

for minimizing N loss to groundwater since unrecovered N on these soils will be lost through leaching prior to the next growing season. Side-dress N applications may also be of benefit on shallow soils over fractured bedrock.

Side-dressing N requires more management than preplant applications. To maximize efficiency, side-dress N applications must be properly timed to provide available N during the maximum N-uptake period for crops such as corn. An additional concern is that applications too late may result in lower yield and plant injury from root pruning and other physical damage.

Split Nitrogen Applications

Application of N fertilizer in several increments during the growing season can be an effective method for reducing N losses on sandy soils. However, a single well-timed side-dress application is often as effective as multiple applications.[41] Ideally, split applications supply N when needed by the crop and allow for N application rate adjustments based on early growing season weather or plant and soil tests. To be successful, the timing of application and placement of fertilizer materials are critical. Climatic factors, such as untimely rainfalls, may interfere with application schedules.

A common method for split N applications is via irrigation systems (fertigation). Multiple applications of fertilizer N can be injected into the irrigation water and applied to correspond with periods of maximum plant uptake. However, fertigation should not be relied upon as a sole method of applying N in a cropping season for the following reasons: 1) adequate rainfall during the early growing season could delay or eliminate the need for irrigation and subsequently delay fertilizer applications; and 2) leaching can result if N is applied through an irrigation system at a time when the crop does not need additional water.

Nitrification Inhibitors

Nitrification inhibitors are used with ammonium or ammonium-forming N fertilizers to improve N efficiency by slowing the conversion of ammonium to nitrate, thereby reducing the potential for losses of N that occur in the nitrate form (i.e., leaching and denitrification). The effectiveness of a nitrification inhibitor depends greatly on soil type, time of the year applied, N application rate, and soil moisture conditions that exist between the time of application and the time of N uptake by plants. Research has shown that the use of nitrification inhibitors on medium- and fine-textured soils with fall N applications, on poorly drained soils with fall or spring N applications, or on coarse-textured, irrigated soils with spring preplant N applications has the potential to increase corn yield and total crop recovery of N.[42] However, as noted earlier, side-dress applications alone are likely to be more effective on many of these soils. Fall applications of N with an inhibitor on sandy soils are not recommended. The cost of using nitrification inhibitors versus other strategies for minimizing N losses needs to be considered in an overall economic analysis of a grower's crop production system.

Controlled-Release Nitrogen Fertilizers

Controlled- (or slow-) release N fertilizers release their nutrients at gradual rates that, in theory, allow for increased plant uptake of N while minimizing losses due to leaching and volatilization. Although commonly used in high-value applications such as horticultural crops and turf, these fertilizer products have not been economical for widespread use in major agricultural crops due to relatively high cost and low crop prices. This may be changing due to cheaper controlled-release fertilizer products, higher N prices, and the demand for greater environmental protection.

Controlled-release fertilizers are broadly divided into uncoated and coated products. Uncoated products rely on inherent physical characteristics, such as low solubility, for their slow release. Coated products consist mostly of quick-release N sources surrounded by a barrier that prevents the N from releasing rapidly into the environment. Similar to earlier-developed materials, such as sulfur- coated urea, urea formaldehyde, and isobutylidene diurea, the critical concept is timing the release of N to correspond with crop need. Greenhouse and field studies have shown that polymer-coated urea can increase crop yield and N use efficiency in soils prone to leaching losses.[43–45]

Other Nutrient Applications

The timing of P and K applications is less critical as these nutrients are generally strongly held in the soil. Some water quality concerns may exist where P is surface broadcast on fields that have not been tilled as this results in less fertilizer–soil contact. Fall applications of K are not recommended on organic soils (peat or muck) since these soils do not effectively hold K against leaching losses. Considerations for P and K applications on other soils include the amount of material to be applied, the size of the application window, and the resources and available equipment.

Nutrient Placement

The placement of nutrients on cropland can influence their effectiveness as well as their potential ability to affect water quality. The concern with N placement focuses mainly on preventing N loss through ammonia volatilization. Applications of N in the form of urea or N solutions need to be incorporated into the soil by rainfall, irrigation, injection, or tillage. The amount of volatilization loss that occurs with surface N applications depends on factors such as soil pH, temperature, moisture, and crop residue. Minimal volatilization losses of N can be expected if surface applications are incorporated within 3 to 4 days—provided temperatures are low (<50°F) and the soil is moist.[46] A late spring or summer application should be incorporated within a day or two because higher temperatures and the chance of longer periods without rainfall could lead to significant N volatilization losses.

The placement of P nutrient sources can directly influence the amount of P transported to lakes and streams by surface runoff. If P fertilizer is broadcast on the soil surface and not incorporated, the amount of P in runoff water can rise sharply and have a greater potential impact on surface water quality than soil surfaces where P was incorporated.[47,48] Phosphorus is strongly bound to soil particles; however, adequate soil-to-P contact must occur to allow for adsorption. Incorporation by tillage or subsurface band placement of fertilizers is a very effective means of achieving this contact. To avoid enriching surface waters with soil nutrients, it is recommended that annual fertilizer applications for row crops, such as corn, be band-applied near the row as starter fertilizer at planting. Annual starter applications of P (and K) can usually supply all of the P required for corn. This practice reduces the chance for P enrichment of the soil surface and reduces potential P in cropland runoff. Band fertilizer placement ideally enriches about 20% of the plow layer volume.[49] If large broadcast P fertilizer applications are needed to increase low soil P levels, these applications should be followed by incorporation as soon as possible.

Variable-Rate Fertilizer Technologies

Nutrient availability in any field varies both spatially and temporally.[50,51] To address this, site-specific (precision agriculture) management techniques and tools have developed over the past 20 years. Global positioning systems and geographic information systems, along with crop nutrient sensing systems, have allowed agricultural management decisions to be made with greater detail and precision. As a result, producers are able to manage nutrient variability within fields at increasingly finer resolution than in the past with inherent improvements in nutrient use efficiency, crop yields, and environmental stewardship.[52,53] While progress has been made, widespread acceptance of these technologies has been limited by costs, sampling requirements, required technical inputs, and the limited ability of current equipment to physically deliver nutrients at adjustable rates corresponding to field variability.[54]

Manure Management

Manure applications to cropland provide nutrients essential for crop growth, add organic matter to soil, and improve soil physical and biological conditions. The major environmental concerns associated with manure application are related to its potential for overloading soils with nutrients if manure applications exceed crop needs and to direct runoff from manured fields to surface waters.

Manure Application Rates

Manure is often applied to cropland at rates that attempt to meet the N need of the intended crop. This strategy maximizes potential manure application rates and is preferred if the amount of land available for application is limited. In addition, a N-based strategy is usually time and labor efficient. A consequence of this approach can be the buildup of P in soils to excessive levels, which in turn increases the potential for P losses via runoff and soil erosion.[55–57] For example, the plant-available N and P contents of dairy manure are about equal. The N need of corn, however, is greater than the crop's need for P. A consequence of applying manure at rates to meet the N need of corn is that P applications will exceed crop removal (Figure 3). The result is a buildup of P in cropland soils.[58] Long term manure applications have elevated the soil P level of many soils above the range necessary for optimum crop growth.[59]

If maximum manure nutrient efficiency is the goal, rates of application need to be based on the nutrient present at the highest level relative to crop needs. For corn, this nutrient would be P. Manure application rates that meet the P requirement of corn are typically much lower than N-based rates. Subsequently, additional N will need to be supplied from other nutrient sources (Figure 4). A P-based manure application strategy results in lower manure application rates, but it is less likely to elevate soil test P values. It has the disadvantages of being less efficient with respect to labor, energy, time, and economics.[60] A P-based strategy for manure applications requires spreading manure on a much larger acreage than is required for a N-based manure application.

Manure Application Timing

Manure application timing is an important management practice for minimizing nutrient contributions to surface waters. Manure should not be spread on sloping lands any time a runoff-producing event is likely. Unfortunately, runoff-producing events are difficult to predict, and the elimination of manure

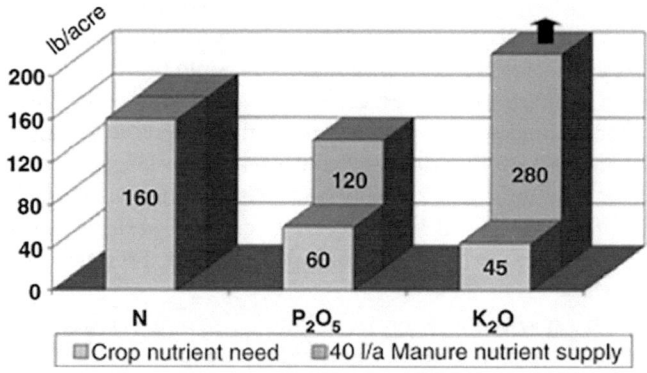

FIGURE 3 Nitrogen-based manure application strategy for corn. Note: Standard convention in the United States is to express soil P and K levels in elemental form (i.e., ppm of P and K) while expressing P and K fertilizer application rates and analysis in oxide form (i.e., lb/acre of P_2Os and K_2O).
Source: Understanding soil phosphorus.[58]

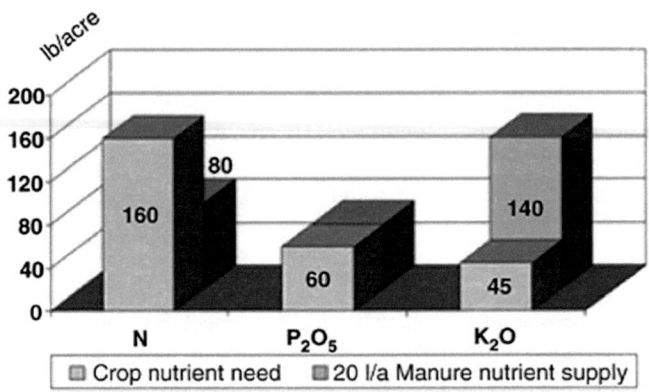

FIGURE 4 Phosphorus-based manure application strategy for corn.
Source: Sturgul and Bundy.[58]

applications to sloping lands is seldom a practical option. For farmers in the upper Midwest, the period of major concern with manure spreading is late fall, winter, and early spring months. Manure applied on frozen ground has an increased likelihood for running off to surface waters due to snow melt and/ or early spring rains.

If winter applications of manure must be made, the risk for nutrient loss should be minimized to the greatest extent possible. Manure applications to frozen fields should be limited to those of slight slope (generally less than 6%) that are preferably covered in previous crop residue, roughly tilled on the land contour, or protected from upslope runoff.[61,62] More steeply sloping fields that are intended for manure applications need to have soil and water conservation practices in place. Manure should not be applied to frozen soils on steeply sloping fields (generally 12% or greater).[61,62]

Site Considerations for Manure Applications

In addition to the slope criteria discussed previously, other site considerations for manure applications should include existing soil fertility levels, soil depth, soil texture, soil erodibility factors, and field proximity to water bodies.

In many areas, general recommendations, or even specific regulations, exist for reducing or eliminating manure applications to fields that have elevated levels of soil P. Numerous studies have found a correlation between elevated levels of soil P and the amount of P carried in runoff from agricultural fields.[55–57] As soil P levels become elevated, crop rotations should be diversified to include crops with a high demand for P (such as alfalfa), which can draw down soil P. When soil P levels become excessively high, manure applications should be discontinued until soil test levels decrease.[61,62] Soil runoff and erosion control practices such as residue management, conservation tillage, contour farming, and others are strongly recommended on soils with P levels in excess of crop needs. When planning manure applications, prioritize those fields low in soil fertility (particularly P) and strive to distribute manure across the available fields to avoid the excessive buildup of soil nutrients that results from repeated applications to the same sites.

Most soils have a high capacity for assimilating nutrients from manure. However, in locations of highly permeable or shallow soils over fractured bedrock, groundwater issues associated with the application of manure can result. Manure should not be applied to shallow soils (generally less than 10 in.) over fractured bedrock. Incorporation of manure shortly after application on moderately shallow soils will allow for increased soil adsorption of nutrients. Manure should not be applied to frozen, shallow soils.[61,62]

Movement of nitrate-N to groundwater is more likely on excessively drained (sandy) soils. Manure applications in early fall on these soils where no actively growing crop is present to utilize the N may allow for the conversion of organic N to nitrate, which is then subject to leaching losses. Manure should not be applied to sands or loamy sands in the fall when soil temperatures are greater than 50°F, unless there is an overwintering cover crop present to utilize the N. In the absence of a cover crop, manure applications to sandy soils should take place when soil temperatures are below 50°F.[61,62] The conversion of ammonium-N to nitrate-N is significantly reduced at soil temperatures below 50°F.

The main site characteristics affecting nutrient contributions to surface waters are those that affect soil runoff and erosion. These include slope, soil erodibility and infiltration, rainfall, cropping system, and the presence of soil conservation practices. Site-related management practices dealing specifically with manure placement to protect surface water include the following: 1) not applying manure (or other nutrients) to grassed waterways, terrace channels, open surface drains, or other areas where surface flow may concentrate; 2) restricting manure applications within designated floodplain or stream and lake setback distances; and 3) prohibiting manure applications in these areas when soils are frozen or saturated.

Manure Storage

During periods when suitable sites for land application of manure are not available (i.e., soils are frozen or seasonally saturated), the use of manure storage facilities is recommended. Storage facilities allow manure to be stored until conditions permit land application and incorporation. In addition, storage facilities can minimize nutrient losses resulting from volatilization of ammonia and be more convenient for calibrated land applications. With the exception of those systems designed to filter leachate, storage systems should retain liquid manure and prevent runoff from precipitation on stored waste. It is imperative that manure storage facilities be located and constructed such that the risk of seepage to groundwater is minimized. With regards to maximum nutrient efficiency and water quality protection, it is critical that appropriate application techniques and accurate nutrient crediting of the manure resource are utilized when the storage facility is emptied.

Livestock Feed Management

On individual farms and in many areas of agricultural livestock production, inputs of P in feed and fertilizer exceed outputs of P contained in crop and animal produce leaving the farm or region (Figure 5). This is especially true in areas where concentrated livestock production is prevalent.[63] The National Research Council[64] estimated that only 30% of the fertilizer and feed P imported onto farms is exported

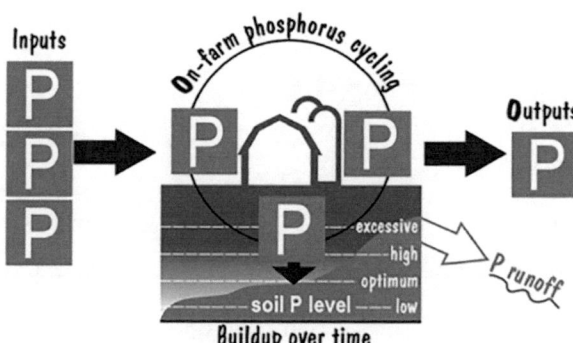

FIGURE 5 On-farm phosphorus cycle.
Source: Sturgul and Bundy.[58]

in crops and animal produce. The surplus 70% of the P is remaining on farm and leading to the excessive enrichment of soil P.

Livestock feed inputs have been found to be a major contributing factor to on-farm P surpluses.[65–67] Soil buildup of P is accelerated when livestock are overfed P in dietary rations. Phosphorus excretion in manure is directly related to the level of P intake.[68,69] High P in livestock dietary intake directly correlates with higher bypass P as reflected in elevated P content of livestock manure (Table 2). Overuse of dietary P supplements accelerates the buildup of soil test P to excessive levels and increases the potential for P losses from manured fields (Figure 6).[70] Another consequence is an increase in land required for application of manure if P-based rate limitations are to be met.

Additional dietary P management options involve plant and livestock genetic manipulation for more effective manure-P management from monogastric animals (nonruminants such as swine and poultry). All these techniques attempt to reduce the P content in manure of monogastric animals by improving the efficiency with which the animal extracts P from feed. An increase in P uptake by the animal from feed grains will reduce the amount of P that bypasses the animal via the manure. Increasing animal uptake of P can allow manure application rates to continue due to a slower buildup of soil P because of the reduced P content of the manure.

Reducing the phytate level of feed grains by use of low-phytate, high-available-phosphate (HAP) varieties is one feed management strategy for lowering manure P. In corn and most feed grain plants, P is stored in the phytate form, which is largely unavailable to nonruminant livestock. As a consequence, swine and poultry feed is routinely supplemented with P. The unutilized phytate-P from the plant is excreted by the animals, resulting in manure that is enriched in P content.[71] Low-phytate grain hybrids that will store P in the available phosphate form rather than as phytate are available.[72] Corn

TABLE 2 Annual Phosphorus Fed to and Excreted by a Lactating Cow

Dietary P Level	Supplemental P	Fecal P
(%)	- - - - - - - - - lb/Cow/Year- - - - - - - - - -	
0.35	0	42
0.38	5.5	47
0.48	23	65
0.55	36	78

Source: Powell et al.[65]

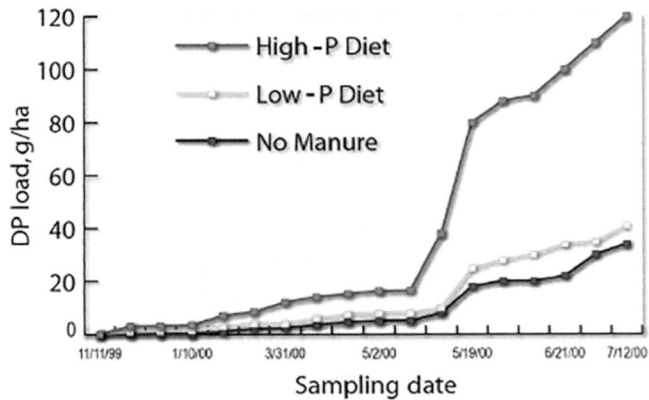

FIGURE 6 Cumulative dissolved phosphorus (DP) load in natural runoff from manures differing in P content. **Source:** Ebeling et al.[70]

has been the crop most extensively developed. Phosphorus availability to mono-gastrics from low-phytate corn is about two to three times higher than from normal corn.[73] Subsequently, the P content of manure is reduced. Plant breeders are working to incorporate the low-phytate trait into commercially competitive hybrids.

Another option for reducing the P content of manure from monogastric livestock is the use of commercially produced enzymes as a feed supplement. Phytase enzymes are capable of releasing phytate-P from plants into animal-available forms. Phytase enzymes occur naturally in some microorganisms, plants, and animals, such as ruminants (cattle). Monogastric animals lack phytase and can only poorly utilize the P reserves in many grains.[71] By adding phytase enzymes to nonruminant animal feed, the efficiency of P uptake during digestion can be increased with an associated reduction in the P content of monogastric manure.[74] In a study by Baxter et al., [75] where phytase additives were combined with low-phytate corn, a 60% reduction in P excretion was recorded. While the phytase enzyme has been shown to decrease the need for mineral P additions, the economics of its use as a routine feed additive need to be considered.[63]

Irrigation Water Management

When to apply irrigation water and how much to apply depend on crop, growth stage, and soil properties. Overirrigation, or rainfall on recently irrigated soils, can leach nitrate and other contaminants below the root zone and into groundwater. Accurate irrigation scheduling that considers soil water holding capacity, crop growth stage, evapotranspiration, rainfall, and previous irrigation to determine the timing and amount of irrigation water to be applied can reduce the risk of leaching losses.

Soil Conservation Practices

Land-use activities associated with agriculture can increase the potential for runoff and soil erosion. Consequences of cropland erosion include loss of fertile topsoil, accelerated eutrophication and sedimentation of surface waters, destruction of fish and wildlife habitat, and decreased recreational and aesthetic value of surface waters. The key to minimizing nutrient contributions to surface waters is to reduce the amount of runoff and eroded sediment reaching them. Numerous management practices for the control of runoff and soil erosion have been researched, developed, and implemented. Runoff and erosion control practices range from changes in agricultural land management (cover crops, diverse crop rotations, conservation tillage, contour farming, contour strip cropping, etc.), to the installation of structural devices (buffer strips, diversions, grade stabilization structures, grassed waterways, terraces, etc.), Substantial emphasis is currently being placed on the benefits and installation of vegetative buffer strips along riparian corridors, which can reduce the sediment and nutrient content of runoff waters reaching them.[76] The width of an effective buffer strip is often debated but varies according to land slope, type of vegetative cover, watershed characteristics, etc.[77]

Recently, the U.S. Department of Agriculture (USDA) reported that cropland conservation practices installed and applied by agricultural producers are indeed reducing sediment, nutrient, and pesticide losses from farm fields.[78] Key findings from this report on the effects of conservation practices on cropland in the Upper Mississippi River Basin illustrate the following: 1) suites of conservation practices work better than single practices; 2) targeting critical acres improves practice effectiveness significantly; and 3) the beneficial effect of conservation practices is greatest on the most vulnerable acres (highly erodible land and soils prone to leaching). The study also found that the use of soil conservation practices within the Upper Mississippi River Basin has reduced sediment loss by 69% compared with estimates of soil loss from the same land area without conservation practices in place. The study's authors concluded that the improved use of cropland conservation practices along with the consistent use of nutrient management practices would further reduce the risk of nutrient movement from fields to rivers and streams within the basin.

Farm Nutrient Management Plans

A farm nutrient management plan is a dynamic, regularly updated strategy for obtaining the maximum economic return from both on- and off-farm nutrient resources in a manner that protects the quality of nearby water resources. While a plan is specific for an individual farming operation, there are common components to all nutrient management plans. These include the following.

Soil Test Reports

Complete and accurate soil tests are the starting point of any farm nutrient management plan. All cropland fields must be tested or have been tested recently to ensure the best possible fertilizer recommendations.

Assessment of On-Farm Nutrient Resources

The amount of crop nutrients supplied to cropland fields from on-farm nutrient resources such as manure, legumes, and organic by-products needs to be determined.

Nutrient Crediting

Once on-farm nutrient resources are determined, commercial fertilizer applications need to be reduced to reflect these nutrient credits. This action can lower fertilizer expenditures and protect water quality by eliminating nutrient applications in excess of crop need. Management skills come into play when determining nutrient credits. For example, to properly credit manure-supplied nutrients, both the manure application rate and the crop-available nutrient content of the manure must be known. To credit the N available to crops following alfalfa, the condition of the alfalfa stand as well as last cutting date need to be known.

Manure Inventory

Perhaps the most challenging aspect of developing and implementing a farm nutrient management plan is the advanced planning of manure applications to cropland fields. This involves estimating the amount of manure produced on a farm and then planning specific manure application rates for individual cropland fields.

Manure Spreading Plan

A major component of any nutrient management plan for livestock operations will deal with a manure spreading plan. The amount of manure the farm produces has to be applied to fields in a manner that considers both environmental and agronomic consequences. Manure applications rates should not exceed crop nutrient need as identified by the soil test report. The nutrient management plan should prioritize those fields that would benefit the most from the manure-supplied nutrients while posing little threat to water quality. The nutrient management plan must also identify fields with manure spreading restrictions.

The seasonal timing of manure applications to cropland should be identified in the nutrient management plan. The timing of planned manure applications will depend upon each farm's manure handling system. Manure application periods for farms with manure storage will be significantly different than those for farms that haul manure on a daily basis.

Consistency with Farm Conservation Plans

A nutrient management plan should be consistent with a farm's soil conservation plan. Operations participating in federal farm programs usually are required to have a soil conservation plan. Conservation plans contain needed information on planned crop rotations, slopes of fields (which are important when planning manure applications), and the conservation measures required to maintain soil erosion rates at tolerable levels.

Compliance with Nutrient Management Standards, Rules, and Regulations

Nutrient management plan criteria and requirements are often defined in standards developed by federal, state, or local government (i.e., the USDA–Natural Resources Conservation Service's Nutrient Management Standard 590). A nutrient management plan complying with a standard(s) is often a requirement for participation in federal and state government farm programs.

Conclusion

The previous text provides a brief summary of general nutrient management practices for crop production. This is not a complete inventory but, rather, an overview of soil fertility management options available to growers for protecting water quality and improving farm profitability. Best management practices for agricultural nutrients can vary widely from one region to another due to differences in cropping, topographic, environmental, and economic conditions. However, central to any nutrient management strategy would be an accurate assessment of nutrient need along with an accounting of on-farm nutrient resources such as manure, legumes, etc. The application of supplemental nutrients should be timed for maximum crop uptake and minimal chance of off-site movement. Nutrient management practices for optimizing crop production while protecting water quality are best summarized in a nutrient management plan that is regularly updated and tailored to the unique landscape characteristics and prevalent agricultural practices of individual farming operations.

References

1. United States Environmental Protection Agency (US EPA). Nutrient Pollution: The Problem. http://www.epa.gov/nutrientpollution/problem/index.html. April 23, 2012.
2. Spalding, R.F.; Exner, M.E. Occurrence of nitrate in ground water—A review. J. Environ. Qual. **1993**, 22, 392–402.
3. Blodgett, J.E.; Clark, E.H. Fertilizers, nitrates, and groundwater: An overview. In Proceedings of the Colloquium of Agrichemical Management to Protect Water Quality; Board of Agriculture, National Research Council: Washington, DC, 1986.
4. Madison, R.J.; Brunett, J.O. 1984. Overview of the occurrence of nitrate in groundwater of the United States. In *U.S. Geological Service National Water Summary, Water- Supply Paper 2275*; U.S. Government Printing Office: Washington, DC, 1984; 93–105.
5. United States Environmental Protection Agency (US EPA). *National Primary Drinking Water Regulations*, EPA pub. no. 816-F-09-004, http://water.epa.gov/drink/contaminants/upload/mcl-2. pdf (accessed March 6, 2012).
6. Rabalais, N.N.; Wiseman, W.I., Jr.; Turner, R.E. Comparison of continuous records of near-bottom dissolved oxygen from the hypoxic zone along the Louisiana coast. Estuaries **1994**, *14*, 850–861.
7. Goolsby, D.A.; Battaglin, W.A. *Nitrogen in the Mississippi Basin—Estimating sources and predicting flux to the Gulf of Mexico*, USGS Fact Sheet 135-00; USGS: Washington, DC, 2000, http:// ks.water.usgs.gov/pubs/fact-sheets/fs.135-00.pdf (accessed May 14, 2012).

8. Parry, R. Agricultural phosphorus and water quality: A U.S. Environmental Protection Agency perspective. J. Environ. Qual. **1998**, 27, 258–261.

9. Correll, D.L. The role of phosphorus in the eutrophication of receiving waters: A review. J. Environ. Qual. **1998**, 27, 261–266.

10. Bundy, L.G.; Andraski, T.W.; Wolkowski, R.P. Nitrogen credits in soybean–corn crop sequences on three soils. Agron. J. **1993**, 85, 1061–1067.

11. Delgado, J.A.; Mosier, A.R.; Valentine, A.W.; Schimel, D.S.; Parton, W.J. Long-term N studies in a catena of the shortgrass steppe. Biogeochemistry **1996**, 32, 41–50.

12. Follett, R.F.; Keeney, D.R.; Cruise, R.M., Eds. *Managing Nitrogen for Groundwater Quality and Farm Profitability.* Soil Science Society of America: Madison, WI, 1991.

13. Newhould, P. The use of nitrogen fertilizer in agriculture. Where do we go practically and ecologically? Plant Soil **1989**, 115, 297–311.

14. Randall, G.W.; Huggins, D.R.; Russelle, M.P.; Fuchs, A.J.; Nelson, W.W.; Anderson, J.L. Nitrate losses through subsurface tile drainage in conservation reserve program, alfalfa, and row crop systems. J. Environ. Qual. **1997**, 26, 1240–1247.

15. Randall, G.W.; Iraqavarapu, T.K.; Schmitt, M.A. Nutrient losses in subsurface drainage water from dairy manure and urea applied to corn. J. Environ. Qual. **2000**, 29, 1244–1252.

16. Andraski, T.W.; Bundy, L.G.; Brye, K.R. Crop management and corn nitrogen rate effects on nitrate leaching. J. Environ. Qual. **2000**, 29, 1095–1103.

17. Meisinger, J.J. Evaluating plant available nitrogen in soil- crop systems. In *Nitrogen in Crop Production;* Hauck, R.D. et al., Eds.; ASA, CSSA, and SSSA: Madison, WI, 1984; 391–416.

18. Vanotti, M.B.; Bundy, L.G. An alternative rationale for corn nitrogen fertilizer recommendations. J. Prod. Agric. **1994**, 7, 243–249.

19. Vanotti, M.B.; Bundy, L.G. Corn nitrogen recommendations based on yield response data. J. Prod. Agric. **1994**, 7, 249–256.

20. Blackmer, A.M.; Binford, G.D.; Morris, T.; Meese, B. Effects of rates of nitrogen fertilization on corn yields, nitrogen losses from soils, and energy consumption. In *1991 Progress Report of the Integrated Farm Management Demonstration Program;* Iowa State University Pm-1467, 1992; 2.1–2.6.

21. Nafziger, E.D.; Sawyer, J.E.; Hoeft, R.G. Formulating N recommendations for corn in the corn belt using recent data. In Proceedings of the North Central Extension–Industry Soil Fertility Conference, Des Moines, IA, Nov. 17–18, 2004.

22. Sawyer, J.E.; Nafziger, E.D.; Randall, G.; Bundy, L.G.; Rehm, G.; Joern, B. *Concepts and Rationale for Regional Nitrogen Rate Guidelines for Corn;* Iowa State University Extension pub. PM 2015, 2006.

23. Laboski, C.A.M.; Peters, J.B.; Bundy, L.G. *Nutrient Application Guidelines for Field, Vegetable, and Fruit Crops in Wisconsin;* University of Wisconsin Extension pub. A2809, 2006.

24. Bundy, L.G.; Meisinger, J.J. Nitrogen availability indices. In *Methods of Soil Analysis: Biochemical and Microbial Properties,* Monogram 5; Weaver, R.W., Ed.; Soil Science Society of America: Madison, WI, 1994; 951–984

25. Magdoff, F.R.; Ross, D.; Amadon, J. A soil test for nitrogen availability to corn. Soil Sci. Soc. Am. J. **1984**, 48, 1301–1304.

26. Kalra, Y.P., Ed. *Handbook of Reference Methods for Plant Analysis.* CRC Press, Inc; Boca Raton, Florida, 1998; 320 pp.

27. Jones, J.B. *Laboratory Guide for Conducting Soils Tests and Plant Analysis.* CRC Press, Inc; New York, Washington, D.C., 2001; 384 pp.

28. Varvel, G.E.; Schepers, J.S.; Francis, D.D. Chlorophyll meter and stalk nitrate techniques as complementary indices for residual nitrogen. J. Prod. Agric. **1997**, 10, 147–151.

29. Blackmer, A.M.; Mallarino, A.P. *Cornstalk Testing to Evaluate Nitrogen Management;* Iowa State University Extension Pub. PM1584, 1996.

30. Sullivan, D.M.; Cogger, C.G. *Post-Harvest Soil Nitrate Testing for Manured Cropping Systems West of the Cascades;* Oregon State University Extension pub EM 8832-E, 2003, http://extension.oregonstate.edu/catalog/pdf/em/em8832-e.pdf (accessed May 14, 2012).

31. Sparks, D.L., Ed. *Advances in Agronomy.* Academic Press; San Diego, California, 1999; Vol. 67, 320 pp.

32. Midwest Plan Service. *Manure Characteristics–Manure Management Systems Series;* MWPS-18-S1D, Iowa State University; Ames, Iowa, 2004.

33. Nutrient and Pest Management (NPM) Program. *Know How Much You Haul;* University of Wisconsin Extension; 2 pp, http://ipcm.wisc.edu/Publications/tabid/54/Default.aspx (accessed May 14, 2012).

34. Peters, J.B., Ed. *Recommended Methods Of Manure Analysis;* University of Wisconsin Extension Pub. A3769, 2003.

35. Peters, J.B.; Combs, S.M. Variability in manure analysis as influenced by sampling and management. In Proceedings of the 1998 Wisconsin Fertilizer Dealers Update Meetings, University of Wisconsin, Madison, WI, Dec. 1998.

36. Bundy, L.G.; Kelling, K.A.; Good, L.W. *Using Legumes as a Nitrogen Source;* University of Wisconsin Extension; 8 pp, http://ipcm.wisc.edu/Publications/tabid/54/Default.aspx (accessed May 14, 2012).

37. Bundy, L.G. Review—Timing nitrogen applications to maximize fertilizer efficiency and crop response in conventional corn production. J. Fert. Issues **1986**, 3, 99–106.

38. Vitosh, M.L. *Nitrogen Management Strategies for Corn Producers;* Michigan State University Cooperative Extension Service Bull. WQ06; 1985; 6 pp.

39. Randall, G.W. Improved N management can alleviate groundwater pollution. Solutions **1986**, 30 (5), 44–49.

40. Bundy, L.G.; Kelling, K.A.; Schulte, E.E.; Combs, S.M.; Wolkowski, R.P.; Sturgul, S.J. Nutrient management: practices for Wisconsin corn production and water quality protection; University of Wisconsin Extension Pub. A3557, 1994; 27 pp.

41. Bundy, L.G. Timing nitrogen applications to maximize fertilizer efficiency and crop response in conventional corn production. J. Fert. Issues **1986**, 3, 99–106.

42. Nelson, D.W.; Huber, D. Nitrification inhibitors for corn production. In *National Corn Handbook 55;* Iowa State University Extension, 2001; 6 pp.

43. Wang, F.L.; Alva, A.K. Leaching of nitrogen from slow release urea sources in sandy soils. Soil Sci. Soc. Am. J. **1996**, 60, 1454–1458.

44. Delgado, J.A.; Mosier, A.R. Mitigation alternatives to decrease nitrous oxide emissions and urea-nitrogen loss and their effects on methane flux. J. Environ. Qual. **1996**, 25, 1105–1111.

45. Shoji, S.; Delgado, J.A.; Mosier, A.R.; Miura, Y. Use of controlled release fertilizers and nitrification inhibitors to increase nitrogen use efficiency and to conserve air and water quality. Comm. Soil Sci. Plant Anal. **2001**, 31, 1051–1070.

46. Bundy, L.G. *Understanding Plant Nutrients: Urea—Its Use and Problems;* University of Wisconsin Cooperative Extension Service Bull. A2989, 1985; 4 pp.

47. Baker, J.L.; Laflen, J.M. Effects of corn residue and fertilizer management on soluble nutrient runoff losses. Trans. ASAE **1982**, 21, 893–898.

48. Mueller, D.H.; Wendt, R.C.; Daniel, T.C. Phosphorus losses as affected by tillage and manure application. Soil Sci. Soc. Am. J. **1984**, 48, 901–905.

49. Bundy, L.G. *Corn Fertilization;* University of Wisconsin Cooperative Extension Service Bull. A3340, 1998.

50. Legg, J.O.; Meisinger, J.J. Soil nitrogen budgets. In *Nitrogen in Agricultural Soils,* Agronomy Monogram 22; Stevenson, F.J., Brewner, J.M., Hanck, R.D., Keeney, D.R., Eds; American Society of Agronomy/Soil Science Society of America: Madison, WI, 1982; 503–566.

51. Jolkela, W.E.; Randall, G.W. Corn yield and residual nitrate as affected by time and rate of nitrogen application. Agron. J. **1989**, 81, 720–726.

52. Fergusson, R.B.; Hergert, G.W.; Schepers, J.S.; Crawford, C.A.; Cahoon, J.E.; Peterson, T.A. Site-specific nitrogen management of irrigated maize: Yield and soil residual nitrate effect. Soil Sci. Soc. Amer. J. **2002**, *64*, 544–553.

53. Berry, J.R.; Delgado, J.A.; Khosla, R.; Pierce, F.J. Precision conservation for environmental sustainability. J. Soil Water Conserv. **2003**, *58*, 332–339.

54. Masek, T.J.; Schepers, J.S.; Mason, S.C.; Francis, D.D. Use of precision farming to improve application of feedlot waste to increase use efficiency and protect water quality. Commun. Soil Sci. Plant Anal. **2001**, *33*, 1355–1369.

55. Bundy, L.G.; Andraski, T.W.; Powell, J.M. Management practice effects on phosphorus losses in runoff in corn production systems. J. Environ. Qual. **2001**, *30*, 1822–1828.

56. Pote, D.H.; Daniel, T.C.; Sharpley, A.N.; Moore, P.A.; Edwards, D.R.; Nichols, D.J. Relating extractable soil phosphorus to phosphorus losses in runoff. Soil Sci. Soc. Am. J. **1996**, *60*, 855–859.

57. Sharpley, A.N. Identifying sites vulnerable to phosphorus loss in agricultural runoff. J. Environ. Qual. **1995**, *24*, 947–951.

58. Sturgul, S.J.; Bundy, L.G. *Understanding Soil Phosphorus;* University of Wisconsin Cooperative Extension Service Bull. A3771, 2004.

59. Sims, J.T. Environmental soil testing for phosphorus. J. Prod. Agric. **1993**, *6*, 501–506.

60. Bosch, D.J.; Zhu, M.; Kornegay, E.T. Net returns from microbial phytase when crop applications of swine manure are limited by phosphorus. J. Prod. Agric. **1998**, *11*, 205–213.

61. USDA–National Resources Conservation Service (NRCS). *Nutrient Management Code 590 Conservation Practice Standard;* National Resources Conservation Service: Washington, DC, 2005.

62. Madison, F.W.; Kelling, K.A.; Massie, L.; Ward-Good, L. *Guidelines for Applying Manure to Cropland and Pastures in Wisconsin.* University of Wisconsin Cooperative Extension Service Bull. A3392, 1998.

63. Daniel, T.C.; Sharpley, A.N.; Lemunyon, J.L. Agricultural phosphorus and eutrophication: A symposium overview. J. Environ. Qual. **1998**, *27*, 251–257.

64. National Research Council (NRC). *Nutrient Requirements of Dairy Cattle*, 7th Revised Ed; National Academy Press: Washington, DC, 2001.

65. Powell, J.M.; Wu, Z. Satter, L.D. Dairy diet effects on phosphorus cycles of cropland. J. Soil Water Conserv. **2001**, *56* (1), 22–26.

66. Satter, L.D.; Wu, Z. Reducing manure phosphorus by dairy diet manipulation. In Proceedings of the 1999 Fertilizer, Aglime and Pest Management Conference 38, Madison, WI; 1999; 183–192.

67. Sharpley, A.N.; Daniel, T.C.; Sims, J.T.; Lemunyon, J.; Parry, R. *Agricultural Phosphorus and Eutrophication*, USDA-ARS pub. no. ARS-149, 1999; 37 pp.

68. Khorasani, G.R.; Janzen, R.A.; McGill, W.B.; Kenelly, J.J. Site and extent of mineral absorption in lactating cows fed whole crop cereal grain silage or alfalfa silage. J. Animal Sci. **1997**, *75*, 239–248.

69. Metcalf, J.A.; Mansbridge, R.J.; Blake, J.S. Potential for increasing the efficiency of nitrogen and phosphorus use in lactating dairy cows. Anim. Sci. **1996**, *62*, 636.

70. Ebeling, A.M.; Bundy, L.G.; Andraski, T.W.; Powell, J.M. Dairy diet phosphorus effects on phosphorus losses in runoff from land applied manure. Soil Sci. Soc. Am. J. **2002**, *66*, 284–291.

71. Doerge, T.A. Low-phytate corn: A crop genetic approach to manure-P management. In Proceedings of the 1999 Wisconsin Fertilizer, Aglime, and Pest Management Conference, Madison, WI; 1999; 175–183.

72. Raboy, V.; Young, K.; Gerbasi, P. Maize low phytic acid (lpa) mutants. In the 4th International Congress of Plant Molecular Biology: Abs. No. 1827; 1994.

73. Ertl, D.S.; Young, K.A.; Raboy, V. Plant genetic approaches to phosphorus management in agricultural production. J. Environ. Qual. **1998**, *27*, 299–304.

74. Kornegay, E.T. Nutritional, environmental, and economic considerations for using phytase in pig and poultry diets. In *Nutrient Management of Food Animals to Enhance and Protect the Environment*; CRC Press, Inc; New York, 1996; 277–302.

75. Baxter, C.A.; Joern, B.C.; Adeola, L.; Brokish, J.E. Dietary P management to reduce soil P loading from pig manure. *Annual Progress Report to Pioneer Hi-Bred International, Inc;* 1998; 1–6.

76. Daniels, R.B.; Gilliam, J.W. Sediment and chemical load reduction by grass and riparian filters. Soil Sci. Soc. Am. J. **1996**, *60,* 246–251.

77. Schmitt, T.J.; Bosskey, M.G.; Hoagland, K.D. Filter strip performance and processes for different vegetation, widths, and contaminants. J. Environ. Qual. **1999**, *28,* 1479–1489.

78. USDA-NRCS. *Conservation Effects Assessment Project (CEAP): Assessment of the Effects of Conservation Practices on Cultivated Cropland in the Upper Mississippi River Basin;* 146 pp, http://www.nrcs.usda.gov/wps/portal/nrcs/detail/national/technical/nra/?&cid=nrcs143_014161 (accessed May 14, 2012).

Nutrients: Bioavailability and Plant Uptake

Niels Erik Nielsen

Soil Plant System

The movement of any nutrient element, M, from solid soil constituents to the root surface, and its entry into plants, can be divided into a sequence of processes (steps), as illustrated in Figure 1, which also indicates major agronomical actions used to improve nutrition and growth of crop plants. The ↔ denotes solid-phase processes slowly approaching equilibrium, or microbial-mediated net mineralization of N, S and P, for example. Also denoted are the source/sink processes by which diffusible nutrients are being produced or removed by chemical, physical, and biological transformation processes. In Figure 1, L denotes ligands that are any dissolved solute reacting with M to form ML, which are organic complexes and ion pairs dissolved in the soil solution. The occurrence of L and, therefore, of ML, increases the total concentration (M + ML) and mobility of the nutrient element. The ⇆ denotes reversible processes which are spontaneously approaching equilibrium. Depending on ion species, ion concentration at the root surface, and plant age, the symbol ⇆ denotes processes that may be irreversible. The irreversible processes are always rate limiting and/or rate determining, whereas reversible processes may be rate limiting, only. Processes 2 and 3 are in the vicinity of the soil particle, whereas process 7 is in the vicinity of the root. Processes 4, 5, and 6 are transport processes by mass-flow and diffusion due to water uptake and nutrient uptake (process 8) by cell membranes of root cells near the root surface or root hairs. Process 9 is the nutrient translocation in the plant. Process 10 is the plant growth that also integrates the absorbed nutrient into the plant tissue. Processes 8, 9, and/or 10 create the concentration (electro-chemical) gradients for irreversible net flux of nutrients from the soil-soil solution system into the plants. Hence, at any time, the rate-determining processes (Figure 1) are then either the release of the nutrient into the pool of plant available nutrient (source/sink processes in the soil), or the nutrient uptake into the roots, its translocation/circulation in the plant, and/or the rate by which the nutrient is built into new tissue. Mass-flow and diffusion may be rate limiting only, and not rate determining. Usually only a small fraction of the plant-available nutrients is dissolved in the soil solution. This implies that the bioavailability of nutrients to plant roots is governed by several soil properties including, for example, the characteristics of process 2 in Figure 1 and the possibilities for movement via soil solution to the root surface by mass-flow and diffusion. The concept of a *bioavailable nutrient* can

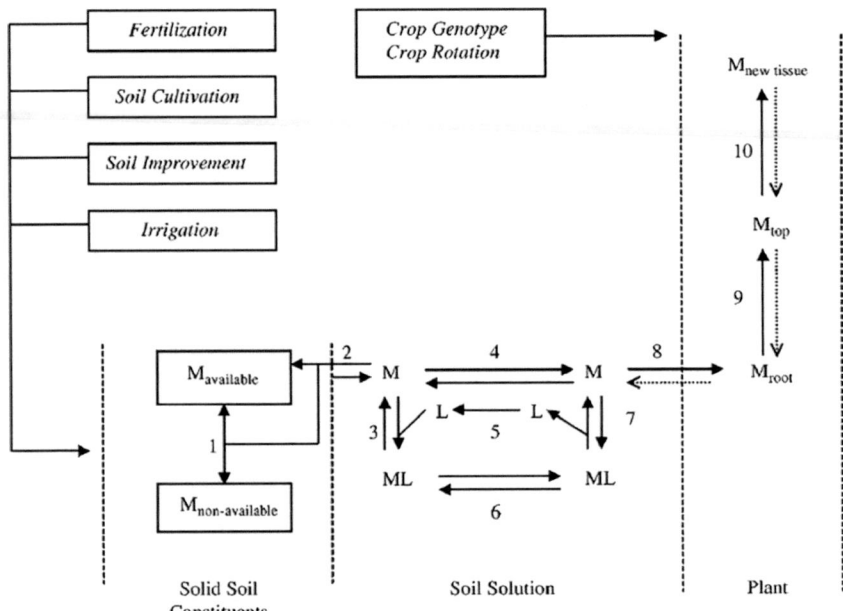

FIGURE 1 Flux of a nutrient element in the soil plant atmosphere system and agronomy actions; nutrient element (M), ligands (L), reversible and irreversible processes.
Source: Nielsen.[22]

then be defined as a nutrient element that is present in a pool of diffusible (available) nutrients which are close enough to arrive at water- and/or nutrient-absorbing root surfaces during a period of 10 days, for example. This seems to fit with the observation that most of the depletion zones of slowly moving nutrients, such as phosphorus, are created during the first 10 days after root growth into a new soil volume unit.[1] The bioavailable quantity of a nutrient in the soil is affected by at least five different groups of processes as indicated in Table 1.

TABLE 1 Processes and Factors Involved in Nutrient Transfer from Soil to Plant Roots

Process	Factors
Release, mineralization, and dissolvement of the nutrient in the soil solution	Chemical and physical properties of the solid phases, temperature, soil water content, and activity of the microbial biomass
Root development	Root length, distribution of roots in the root zone, root morphology, root hairs, rate of root growth, and root surface contact area with soil solution
Solute movement by mass flow and diffusion to roots	Transpiration rate (w_o), concentration of the nutrient in the soil solution (c_b), effective diffusion coefficient (D_e), nutrient buffer power of the soil (b)
Rhizosphere processes, increasing the rate of nutrient release in the soil	Depletion of the soil solution for nutrients by the roots; root exudates as protons, reducing agents, chelates, organic anions, enzymes; modification of microbial activity; mycorrhizal and Rhizobium symbioses
Nutrient uptake	Concentration of the nutrient at the root surface (c_o); transport kinetic parameters of nutrient uptake by the roots I_{max}, Km and c_{min})

Source: Nielsen.[22]

Diffusion

Diffusion is the net movement of a solute or a gas from a region that has a higher concentration, to an adjacent-region that has a lower concentration. Diffusion is a result of the random thermal motions of molecules in the considered solids, solution, or air. The net movement caused by diffusion is a statistical phenomenon because the probability of the molecules' movement from the concentrated to the diluted region is greater than vice versa. Fick (1855) was one of the first to examine diffusion on a quantitative basis. The basic equation Eq. (1) to express diffusion, e.g., to a root is today known as Fick's first law of diffusion:

$$F = -D\frac{\partial C}{\partial r} \tag{1}$$

in which F is the diffusive flux (mol cm^{-2}s $^{-1}$ of a nutrient in the r-direction normal to the root cylinder. The driving force (the gradient in the electro-chemical potential) is, in most cases, approximated by the concentration gradient

$$\frac{\partial C}{\partial r}\left(mol\ cm^{-4}\right)$$

and D is the diffusion coefficient (cm^{2}s^{-1}). They way to describe diffusion processes mathematically under various conditions has been presented by Jost[2] and Crank.[3] Great contributions to our understanding of solute movement in the soil root system by mass-flow and diffusion have been given or reviewed by Nye and Tinker[4,6] and Barber.[5] Recently, Willigen and colleagues[7] reviewed some aspects of the modeling of nutrient and water uptake by plant roots.

Nutrient Movement by Mass Flow and Diffusion from Soil to Plant Roots

Nutrients bound to the solid-soil phase are virtually immobile in the sense of its movements to roots. The nutrient has to be released into the soil solution as indicated in Figure 1. Furthermore, contact between the root and nutrient-absorbing membranes in the root tissue near the root surface and the soil solution is a prerequisite for nutrient uptake. Contact to nutrient pools can be brought about by two means: 1) by growth of roots to the sites where nutrient pool are located (root interception); and 2) by movement of the nutrient from the bulk (the pool) of the soil to the root surface. Even so, nutrients may at any time move over a certain distance in the soil solution and cell wall before they reach the outside cell membrane of a root hair or a root cortical cell for uptake. The mechanisms for these transports are mass flow and diffusion.[4,8] The driving force for the net movement of nutrients (Figure 1) is the water and the selective nutrient uptake by the plant root, creating a concentration gradient (dc/dr). The general equation of continuity (mass-balance) used to describe movements in a direction normal to a root cylinder at radial distance r and time t, may partly be developed from Eq. (1), extended and expressed as

$$\left[b\frac{\partial c}{\partial r}\right]_r = -\left[\frac{1}{r}\frac{\partial rF_T}{\partial r}\right]_t + U_{r,t} \tag{2}$$

in which

- U is the production/consumption term (mole cm^{-3}s^{-1}) at r (radial distance from the center of the root) and t (time).
- b is the buffer power *(dC/dc)* in which C is the total concentration of diffusible solute in the soil and c is the concentration of solute in the soil solution.
- F_T is total net flux of solute by mass flow and diffusion (mole cm^{-2}s^{-1}).

$$F_T = F_m + F_d$$

where $F_m = wc$ in which w is the flux of soil solution in the direction of the root (cm^{3}cm^{-2}s^{-1}) and c is the nutrient con centration of soil solution (mole cm^{-3}). The expected rates of water flux at the root surface are 0.2–1 10^{-6}cm s^{-1}.[5] The flux, F_d, by diffusion can be expressed by Flick's first law

$$F_d = -D_e b \frac{dc}{dr}$$

The $b = dC/dc$ is the soil buffer power defined previously. The C is the sum of the amount of nutrient in the soil solution and the amount of adsorbed nutrient that is able to replenish the nutrient in the soil solution spontaneously. Hence, b is the parameter mediating the effects of the soil chemical conditions on nutrient uptake by plants. D_e denotes the effective diffusion coefficient in the soil. D_e differs between media, but it can be related to the diffusion coefficient D_o for the nutrient in free soil solution. The influences of soil on diffusion, and thereby the relation between D_e and D_o, can be expressed by Eq. (3).[9]

$$D_e = D_o \theta f / b \qquad (3)$$

where θ is the volumetric water content expressed as a fraction, and f is the impedance factor that essentially allows for the increase in the actual diffusion distance because of the tortuous pathway of water filled soil pores and water films. The volumetric water content that allows a reasonable root activity is between 0.1 and 0.4. The value of f increases with increase in water content,[10] whereas the buffer power remains constant with changes in soil moisture at the same bulk density.[11] It has been observed[12] that the relation between f and θ can be expressed empirically by $f = 1.58\theta - 0.17$ for $\theta >$ about 0.11. From this it may be estimated that D_e decreases about 18 times if θ decreases from 0.40 to 0.15. Hence D_e is the parameter mediating the effects of soil moisture, soil chemical, and soil physical conditions on diffusion in soil.

Almost all studies on solute movement in the soil plant system neglect U in Eq. (2) because our understanding of the biology caused by root-induced processes and its effects on production or consumption of available nutrients is incomplete as yet.

To solve Eq. (2) for a given soil plant system is a complicated process, and in most cases, difficult or even impossible because of the lack of information on the soil root interactions and root behavior. The method for obtaining analytical and numerical solutions of Eq. (2) under a number of often simplified soil plant conditions has been summarized.[5-7] However, illustration of the importance of diffusion for the bioavailability of nutrients in soils may be based on Eq. (4)

$$\Delta r = \sqrt{2D_e t} \qquad (4)$$

in which Dr is the average distance of diffusion; e.g., in a direction normal to a root. The mathematics behind Eq. (4) has been presented by Jost.[2] Based on Eq. (4), the equivalent soil volume (V in cm^3) of soil depleted for diffusible (available) nutrients—*the quantity of bioavailable nutrients*—can then be estimated as follows [Eq. (5)] for roots without root hairs:

$$V = \pi (\Delta r + r_o)^2 L_v \qquad (5)$$

in which Δr is estimated from Eq. (4), r_o is the root radius and L_v is the root density in cm cm^{-3} of soil. The data in Table 2 show the expected effective diffusion coefficient of a number of plant nutrients in soil and corresponding influences on the nutrient bioavailability; in addition, the data show how a decrease of the soil moisture from $\theta = 0.40$ to $\theta = 0.15$ affects the bioavailability at a root density of 5 cm^{-2} of roots without root hairs. It may be calculated from $f = 1.58\,\theta - 0.17$ and Eq. (3) that the diffusive flux decreases by a factor of $18 = D_e^{\theta=040}/D_e^{\theta=015}$. At field capacity of water content, the expected, effective diffusion coefficient of nitrate in soil is $10^{-6}\,\mathrm{cm^2s^{-1}}$. This is almost 10 times slower than in pure water. Hence, a pored media, such as soil, physically decreases the possibility for solute movement with a factor of nearly 10. As the soil dries out, this factor increases as illustrated in Figure 2. Apart from nitrate and chlorine, nutrient elements are adsorbed more or less to the solid soil constituents. This is the main cause of the decrease of the diffusion coefficients below $10^{-6}\mathrm{cm^2s^{-1}}$. The diffusion coefficient of phosphorus is as low as $10^{-9}\mathrm{cm^2s^{-1}}$ mainly because approximately 0.1% of the diffusible (available) phosphorus is dissolved in the soil solution only. This has a large effect on the bioavailability of the plant-available quantities of the various nutrient elements in soil as illustrated in columns 3 and 4 of Table 2. The V-values ≥ 1 indicate that the root at a density of $5\,\mathrm{cm^{-2}}$ is able to deplete all the available nutrient as seen for nitrate, even under dry conditions, whereas only 4% of the available phosphorus is bioavailable inside a period of 10 days. If the soil dries out to 1 indicate that the root at a density of $5\,\mathrm{cm^{-2}}$ is able to deplete all the available nutrient as seen $\theta = 0.15$, the bioavailability decreases to only 0.6%. This illustrates that the decrease of soil moisture may create nutrient deficiency even in soil with high phosphorus fertility. However, phosphorus uptake is increased by the activity of root hairs (discussed in the following).

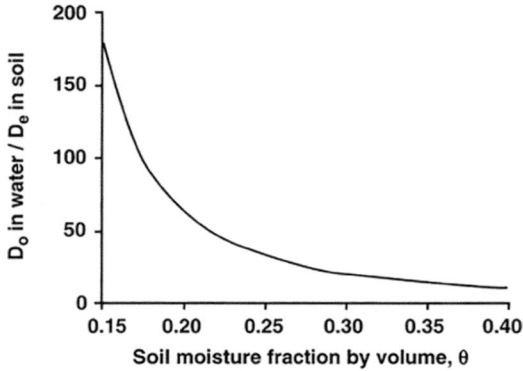

FIGURE 2 Effects of soil and soil moisture on nutrient element mobility.

TABLE 2 Expected Effective Diffusion Coefficients of Some Nutrients in Soil at Field Capacity of Water Content (e.g., $\theta = 0.40$), and Estimated Bioavailability as a Fraction of Diffusible (Available) Nutrient [Eqs. (4) and (5)] at a Root Density ($L_v = 5\,\mathrm{cm^{-2}}$) of Roots without Root Hairs (Mean Root Radius, $r = 0.01$ cm; Time, $t = 10$ days)

		Bioavailable Nutrient as a Fraction of Available Nutrient (V, cm³)	
Element	Dea (cm²s⁻¹)	$\theta = 0.40$	$\theta = 0.15$
Nitrate	1*10–6	27.558	1.574
Potassium	1*10–7	2.847	0.180
Boron	1*10–7	2.847	0.180
Magnesium	1*10–8	0.314	0.026
Calcium	1*10–8	0.314	0.026
Phosphorus	1*10–9	0.042	0.006
Manganese	1*10–9	0.042	0.006
Molybdenum	1*10–9	0.042	0.006
Zinc	1*10–9	0.042	0.006
Iron	1*10–10	0.008	0.003

[a] Values obtained from Barber[5] and Nielsen.[22]

Importance of Root Hairs

Root hairs are outgrowths from specialized root epidermal cells (trichoblasts). Root hair length, diameter, and number per unit length of root, vary among plant species and among genotypes within the same species.[5,6,13] Frequency and size of root hairs are affected by many environmental factors, as well. In nature, the length of root hairs vary from 0.01–0.15 cm, the radius varies from 0.0005–0.002 cm, and the number per unit of length varies from 100–1000 per cm root. The importance of root hairs for phosphorus uptake has been demonstrated directly in the laboratory[14,15] and under field conditions.[13] It is reasonable to assume that the clusters of root hairs' outer tips form a fairly well-defined cylinder to which phosphorus diffuses, on average, a distance Δr in 10 days, and that root hair density and its period of function are long enough to withdraw the entire available nutrient in the soil penetrated by root hairs. The bioavailability of phosphorus, for example, as affected by root hair length, can then be estimated from the following extension of Eq. (10):

$$V = \pi \left(\Delta r + \sigma + r_o \right)^2 L_v \tag{6}$$

in which σ is the root hair length in cm.

Figure 3 illustrates that the bioavailability of phosphorus increases exponentially with the root hair length. Hence, root hairs play a very important role for the bioavailability of nutrients having a low effective diffusion coefficient (D_e) in soil.

Boundary Conditions and Nutrient Entry

If depletion zones around the roots do not overlap, the solute concentration converges to the solute concentration c_b in the bulk solution, at which $F_T = F_m + F_d = 0$. The boundary conditions at the surface of the root are

$$F_T = F_m + F_d = \alpha c_o \tag{7}$$

in which α (cm s^{-1}) is the root-absorbing power defined by Tinker and Nye[6] and c_o is the concentration of solute at the root surface. It can be learned from Eq. (7) that the actual concentration (co) at the root surface and, therefore, the rate of nutrient flow per unit length of root, is determined by the ratio F_T/α. The kinetics of net uptake of nutrients[5,16–18] may be expressed by:

$$F_T = F_m + F_d = \alpha c_o \tag{8}$$

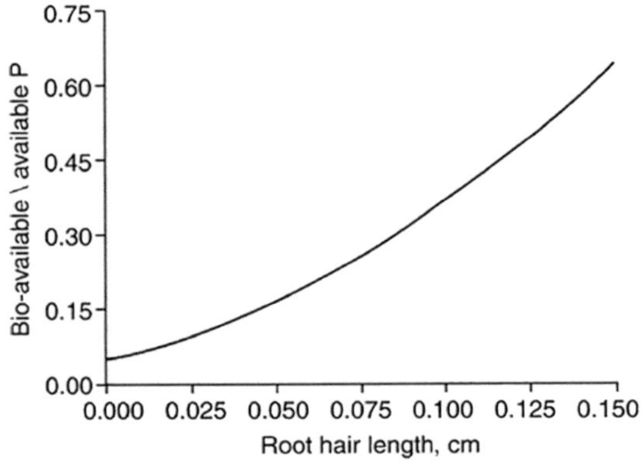

FIGURE 3 The effect of root hair length on phosphorus bioavailability. Soil moisture ($\theta = 0.4$); root density 5 cm^{-2}; root radius 0.01 cm

in which L^* is the root length per unit of plant biomass, $\bar{I}_{max}$ (mole cm^{-1}s^{-1}) is the mean maximal net influx, Km (mole cm^{-3}) is the Michaelis–Menten factor, c_o is the concentration of the nutrient at the root surface, and c_{min} is the nutrient concentration at which The values of the parameters $\bar{I}_{max}$, Km and c_{min} vary according to the plant nutrient, temperature, and plant species/genotype and plant age. Furthermore, kinetics of nutrient uptake by roots may be influenced by ion interactions. Determined values of L^*, $\bar{I}_{max}$, Km and c_{min} for uptake of several nutrients by several plant species or genotypes, obtained under conditions in which the rate-determining step of nutrient uptake was located in the roots, has been noted.[19] The data show that the values of L^*, $\bar{I}_{max}$, Km and c_{min} vary considerably among nutrients and among plant species and genotypes. This illustrates the efficiency by which these plants utilize soil as a source of nutrients. It is possible from $F_T = F_m + F_d = \alpha c_o$ and Eq. (8) to develop how α varies at varying solute concentration at the root surface by:

$$\alpha = \frac{\bar{I}_{max}(c_o - c_{min})}{2\pi r_o c_o (km + c_o - c_{min})} \tag{9}$$

Figures 4a and 4b illustrate the variation of α for phosphorus uptake at low concentration (c_o) at the root surface of some plant species and barley genotypes. Hence, the rootabsorbing power (α) varies

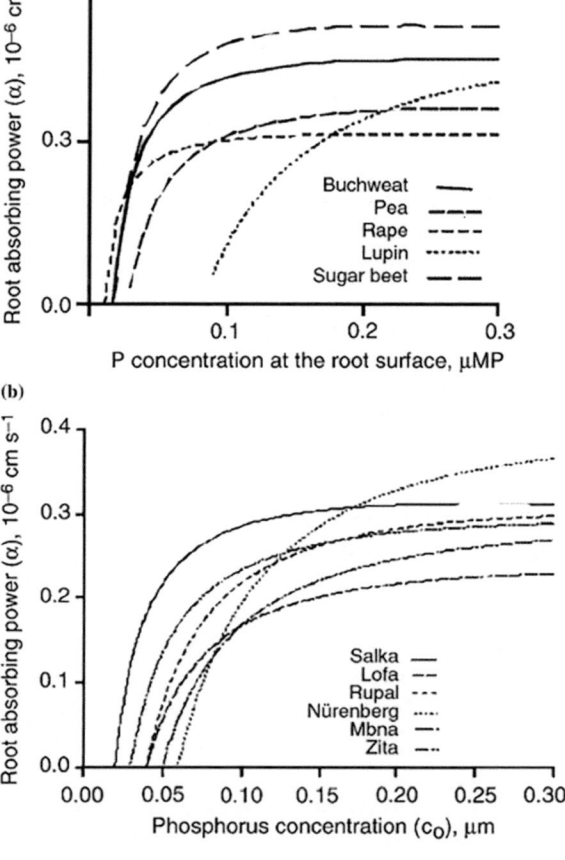

FIGURE 4 Absorption power ($\propto$) at varying phosphorus concentration at the root surface of some plant species (a) and barley genotypes (b).
Source: Nielsen[19] and Nielsen.[22]

also at low concentration (c_o) of solute at the root surface. This implies that phosphorus uptake at low P concentration is more under the control of the plant parameters determining the size of α than under the control of P diffusion in the soil, whereas at the range of c_0 at which α has achieved its maximum, uptake is controlled by diffusion.

Conclusions

Even though the movement—and the main factors affecting the movement—of nutrient elements to root by mass-flow and diffusion is well known, the effect of soil conditions on crop growth is still not properly understood. It is obvious that the big variation (Table 2) of the effective diffusion coefficient (D_e), caused mainly by the variation of the soil chemistry of the various nutrient elements, has a large impact on the bioavailability of nutrient elements. The mobility of phosphorus and micronutrients is so low in most soils that the soil exploited by root hairs is the main source of these elements. The root-induced modifications to the soil in the rhizosphere would then have a considerable impact on the efficiency by which plants use the rhizosphere soil as a source of nutrients. The understanding of how root-induced processes accelerate solute movement and the transformation of non-available nutrients to bioavailable nutrients is increasing.[6,20,21] Root hair length and root-induced processes appear to vary between genotypes of our crop plants.[13] Hence, improvement of the efficiency by which plants use soil as a source of nutrients seems to be a possibility by targeted plant breeding.

References

1. Gahoonia, T.S.; Raza, S.; Nielsen, N.E. Phosphorus depletion in the rhizosphere as influenced by soil moisture. Plant and Soil **1994**, *159*, 213–218.
2. Jost, W. *Diffusion in Solids, Liquids and Gasses*; Academic Press: New York, 1960.
3. Crank, J. *The Mathematics of Diffusion*, 2nd Ed.; Clarendon Press: Oxford, 1975.
4. Nye, P.H.; Tinker, P.B. *Solute Movement in the Soil-Root System*; Blackwell: Oxford, 1977.
5. Barber, S.A. *Soil Nutrient Bioavailability*, 2nd Ed.; Wiley: New York, 1995.
6. Tinker, P.B.; Nye, P.H. *Solute Movement in the Rhizosphere*; Oxford University Press: Oxford, 2000.
7. Willigen, P.; Nielsen, N.E.; Claassen, N.; Castrignanò, A.M. Modelling water and nutrient uptake. In *Root Methods a Handbook*; Smit, A.L., Bengough, A.G., Engels, C., Noordwijk, M., Pellerin, S., Geijn, S.C., Eds.; Springer: Berlin, 2000.
8. Barber, S.A. A diffusion and mass-flow concept of soil nutrient availability. Soil Science **1962**, *93*, 39–49.
9. Nye, P.H. The measurement and mechanism of ion diffusion in soil. I. the relation between self-diffusion and bulk diffusion. J. Soil Sci. **1966**, *17*, 16–23.
10. Rowell, D.L.; Martin, M.W.; Nye, P.H. The measurement and mechanisms of ion diffusion in soils. III. The effect of moisture content and soil solution concentration on the self-diffusion of ions in soils. J. Soil Sci. **1967**, *18*, 204–222.
11. Bhadoria, P.B.S.; Classen, J.; Jungk, A. Phosphate diffusion coefficient in soil as affected by bulk density and water content. Z. Pflanzenernaehr. Bodenk. **1991**, *154*, 53–57.
12. Barraclough, P.B.; Tinker, P.B. The determination of ionic diffusion coefficients in field soils. 1. Diffusion coefficient in sieved soils in relation to water content and bulk density. J. Soil Sci. **1981**, *32*, 225–236.
13. Gahoonia, T.S.; Nielsen, N.E.; Lyshede, O.B. Phosphorus acquisition of cereal cultivars in the field at three levels of P fertilization. Plant and Soil **1999**, *211*, 269–281.
14. Barley, K.P.; Rovira, A.D. The influence of root hairs on the uptake of phosphorus. Commun. Soil Sci. Plant Anal. **1970**, *1*, 287–292.

15. Gahoonia, T.S.; Nielsen, N.E. Direct evidence on the participation of phosphorus (P32) uptake from soil. Plant and Soil **1998**, *198*, 147–152.

16. Classen, N.; Barber, S.A. A method for characterizing the relation between nutrient concentration and flux into roots of intact plants. Plant Physiology **1974**, *54*, 564–568.

17. Nielsen, N.E. A transport kinetic concept for ion uptake by plants. III. Test of the concept by results from water culture and pot experiments. Plant and Soil **1976**, *45*, 659–677.

18. Nielsen, N.E.; Barber, S.A. Differences among genotypes of corn in the kinetics of P uptake. Agronomy Journal **1978**, *70* (5), 695–698.

19. Nielsen, N.E. Bioavailability, cycling and balances of nutrient in the soil plant system. In *Integrated Plant Nutrition Systems;* Fertilizer and Plant Nutrition Bulletin 12; Dudal, R., Roy, R.N., Eds.; FAO: Rome, 1995; 333–348.

20. Jungk, A.; Classen, N. Ion diffusion in the soil-root system. Adv. Agron. **1997**, *61*, 53–110.

21. Hensinger, P. How do plant roots acquire mineral nutrients? chemical processes involved in the rhizosphere. Adv. Agron. **1998**, *64*, 225–265.

22. Nielsen, N.E. Bioavailability of nutrients in soil. In *Roots and Nitrogen in Cropping Systems of the Semi-Arid Tropics;* Ito, O., Johansen, C., Adu-Gyamfi, J.J., Katayama, K., Kumar Rao, J.V.D.K., Rego, T.J., Eds.; International Crop Research Institute for the Semi-arid Tropics: India, 1996; 411–427.

51

Nutrient–Water Interactions

Ardell D. Halvorson

Introduction

Water is a major factor in nutrient availability to plants.[1–4] It is the vehicle through which nutrients move through soil to access plant roots for uptake. Nutrients move via mass flow and diffusion in soil water to the root surface. Root interception is a third way in which plants obtain soil nutrients as root hairs develop and contact the soil particles and/or solution.

Water and Nutrient Availability

In mass flow, nutrient ions are transported with water flow to the root as the plant absorbs water for transpiration. Many mobile nutrients, such as calcium (Ca), magnesium (Mg), nitrate-N (NO_3-N), and sulfate (SO_4), are transported to the root by mass flow. Diffusion of nutrients to the plant root occurs as ions move from high-concentration areas to low-concentration areas in the soil solution. Phosphorus (P) and potassium (K) are two nutrients that move by diffusion.

If soil water becomes limiting, as it frequently does under dryland or rainfed conditions, nutrient availability to plants can be affected.[5] Water is held as a film around soil particles. As the water content of the soil decreases, the thickness of the film decreases. Most plant nutrients are readily available when the soil is near field capacity, which is about the water content of the wet soil after two days of rain has saturated it and free drainage has ceased. Nutrient availability is at a minimum as the soil water content approaches the permanent wilting point, which is the water content at which plant roots cannot extract water from the soil. As soil water content diminishes, some less-soluble nutrients may precipitate out of the soil solution and become unavailable to plants. However, these minerals will dissolve and become available once again as the soil is rewetted. Thus, soil water content influences nutrient availability and plant growth.

Micronutrients are generally supplied to plant roots by diffusion in soil. Therefore, low soil moisture conditions will reduce micronutrient uptake. Plants require smaller quantities of micronutrients to optimize productivity than macronutrients such as P; thus, drought stress effects on micronutrient deficiency are not as serious as for P. However, iron (Fe) and zinc (Zn) deficiencies are frequently associated with high soil moisture conditions.[2]

Soil water content is an important factor in microbial activity in soils. Soil microbial activity is important in the breakdown of organic plant and animal residues, which release nutrients such as N and P for plant uptake. Microbial activity tends to be greatest when soil water is near field capacity with soil temperatures ranging from 25°C to 35°C. As soils dry, microbial activity decreases and lowers the rate of nutrient release from soil organic matter.[6,7]

Nutrient and Water Use Efficiency

Adequate levels of plant nutrients are needed to optimize rooting depth and water extraction from the soil.[2,3,5] Healthy plants tend to root deeper into the soil profile, using more of the soil water in the root zone. Thus, plants not only need adequate water to optimize yield potential, but also require an adequate level of nutrients to allow the crop to take advantage of the available water supplies. Under dryland conditions, the crop will often use all of the available water (precipitation plus soil water in the root zone) during the growing season. Application of N and P fertilizers will frequently increase crop yields, thus increasing crop water use efficiency (WUE). Water use efficiency is the amount of crop produced per unit of available water from precipitation, soil, and irrigation. The influence of N fertilization on WUE of winter wheat, corn, and sorghum in a dryland wheat-corn or sorghum-fallow rotation is shown in Figure 1.

When plant-available water is limited, overapplication of N can also result in reduced grain yields owing to increased vegetative growth and water use in the early growth stage, with insufficient water remaining to maximize grain development and yield. Application of N will not increase yields without adequate plant-available water, and increasing plant-available water will not increase crop yield without adequate N supply. The percentage increase in response of crops such as wheat to P fertilization tends to be greater in dry years than in wet years on P-deficient soils, while both N and P are needed to optimize yields in wetter years.

Water is important for activation and movement of fertilizer nutrients applied to soils.[1-3,7,8] Dry fertilizer granules must dissolve in the soil water before they become available to plants. When applied to dry soil, liquid fertilizers may become unavailable to plants until precipitation or irrigation water rewets

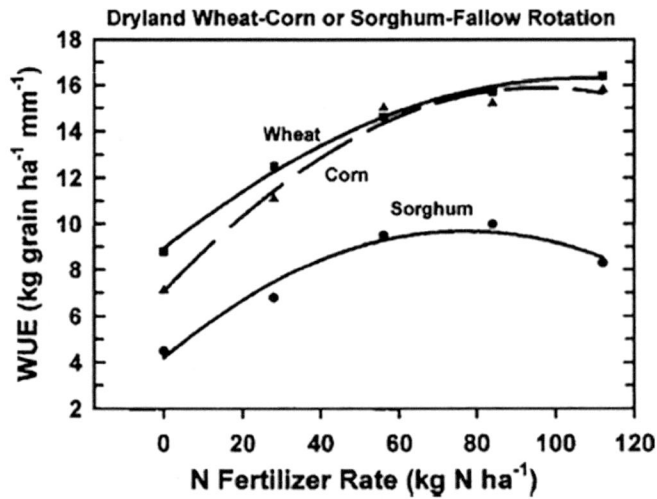

FIGURE 1 Water use efficiency of wheat, corn, and sorghum as a function of N fertilizer rate in a dryland wheat-corn or sorghum-fallow rotation near Akron, Colorado, U.S.A.

the soil and they become part of the soil solution again. Rainfall affects the volatilization loss of N from ammonia-based fertilizers such as urea and urea ammonium nitrate (UAN). Rainfall received within 36 hr after surface applications of urea or UAN fertilizers will greatly reduce N volatilization losses and improve the N fertilizer use efficiency by crops. Rainfall moves the surface-applied N fertilizer into the soil where it can react and reduce NH_3 losses to the atmosphere. Excessive soil water, however, can result in anaerobic conditions and the loss of NO_3-N by denitrification. Nitrate-N is converted to various N gases, which are lost to the atmosphere under anaerobic conditions.

Water is essential for optimizing crop yields. Under irrigation, water is generally not a yield-limiting factor. Under dryland or rainfed conditions, crop yields are dependent on available soil water supplies and growing season precipitation. Adequate levels of essential plant nutrients are needed to optimize crop yields and WUE (i.e., kg grain produced/mm crop water use). Under rainfed conditions, crop water supplies during the growing season can vary weekly and annually. During periods of drought (i.e., low supply of plant-available water), less plant nutrients are needed to optimize crop yields than during years of average or above-average precipitation. In wetter years, both the crop yield potential and the nutrients needed to optimize crop yield increase.

Soil management practices, such as reduced- and no-till systems, that increase soil organic matter and improve soil physical quality also improve soil aggregation and porosity. This, in turn, improves water infiltration into the soil and water availability for increased crop productivity and improved nutrient use efficiency.

Irrigation Water Quality and Fertilizer Application

Irrigation water quality can affect the application of fertilizer nutrients through irrigation systems.[3,8] For example, the addition of anhydrous NH_3 or liquid ammonium polyphosphate fertilizers to irrigation waters high in Ca can result in the formation of lime and calcium phosphate precipitates. The precipitates can plug sprinkler and drip irrigation systems. In some instances, precipitation of the Ca can result in a higher sodium (Na) hazard of the irrigation water, which may subsequently reduce the water intake capacity of the soil.

Applying fertilizers with both flood and furrow irrigation systems requires that a uniform distribution of water be achieved throughout the field to obtain a uniform distribution of fertilizer nutrients to the crop. With flood and furrow irrigation systems, fertilizer should not be applied with the initial flush of irrigation water because of the generally nonuniform distribution of water during the initial wetting of the soil surface by the irrigation water. The reactions of fertilizers with the irrigation water and the fertilizer distribution to the crop are affected by (irrigation) water quality. If fertigation (i.e., application of fertilizer nutrients through an irrigation system) is to be used, the compatibility of fertilizers to be applied with the quality of irrigation water available must be examined to avoid poor distribution of fertilizer nutrients.

Environmental Quality

Nitrogen is generally transported from soils into surface and groundwater by runoff, erosion, and leaching.[7,9] Runoff water from watersheds with high levels of soluble N and P sources on the soil surface can contribute to eutrophication of streams, lakes, ponds, bays, and estuaries. Placing or positioning applied N and P sources below the soil surface and using soil management practices to minimize runoff will help reduce agriculture's impact on eutrophication of water bodies. Water erosion of soil not only carries soluble plant nutrients from a watershed, but also carries soil particles with sorbed nutrients, such as P, into water bodies that can then contribute to degradation of water quality.

Soil management practices such as no-till and other conservation tillage practices can reduce soil erosion by water. Water moving through soil in excess of field capacity water content can move soluble nutrients, such as NO_3-N, below the root zone of crops and into groundwater. In summary, using cropping systems and an adequate fertility program to optimize crop WUE will help reduce loss of plant-available water and nutrients below the crop root zone.

Conclusions

Water plays a critical role in the availability of nutrients to plants. Adequate levels of both water and nutrients are needed to optimize plant growth and productivity. Fertilizer and water management practices can influence the efficient use of water and nutrients by plants and their subsequent impact on environmental quality.

References

1. Engelstad, O.P., Eds. *Fertilizer Technology and Use,* 3rd Ed.; Soil Sci. Soc. Am.: Madison, WI, 1985.
2. Havlin, J.L.; Beaton, J.D.; Tisdale, S.L.; Nelson, W.L. *Soil Fertility and Fertilizers: An Introduction to Nutrient Management,* 6th Ed.; Prentice Hall: Upper Saddle River, NJ, 1999.
3. Mortvedt, J.J.; Murphy, L.S.; Follett, R.H. *Fertilizer Technology and Application*; Meister Publishing Co.: Willoughby, OH, 1999.
4. Troeh, F.R.; Thompson, L.M. *Soils and Soil Fertility,* 5th Ed.; Oxford University Press: New York, 1993.
5. Taylor, H.M.; Jordan, W.R.; Sinclair, T.R., Eds. *Limitations to Efficient Water Use in Crop Production*; Am. Soc. Agron., Crop Sci. Soc. Am., Soil Sci. Soc. Am.: Madison, WI, 1983.
6. Follett, R.F.; Stewart, J.W.B.; Cole, C.V., Eds. *Soil Fertility and Organic Matter as Critical Components of Production Systems*; Soil Sci. Soc. Am., Inc.: Madison, WI, 1987.
7. Pierzynski, G.M.; Sims, J.T.; Vance, G.F. *Soils and Environmental Quality,* 2nd Ed.; CRC Press: Boca Raton, FL, 2000.
8. Ludwick, A.E., Bonczkowski, L.C., Bruice, C.A., Compbell, K.B., Millaway, R.M., Petrie, S.E., Phillips, I.L., Smith, J.J., Eds.; *Western Fertilizer Handbook*; 8th Ed.; California Fertilizer Association, Interstate Publishers: Danville, IL, 1995.
9. Follett, R.F., Ed. *Nitrogen Management and Ground Water Protection*; Elsevier: New York, 1989.

52

Pollution: Non-Point Source

Ravendra Naidu,
Mallavarapu
Megharaj, Peter
Dillon, Rai
Kookana, Ray
Correll, and
W.W. Wenzel

Introduction

Non-point source pollution (NPSP) has no obvious single point source discharge and is of diffuse nature (Table 1). An example of NPSP includes aerial transport and deposition of contaminants such as SO_2 from industrial emissions leading to acidification of soil and water bodies. Rain water in urban areas could also be a source of NPSP as it may concentrate organic and inorganic contaminants. Examples of such contaminants include polycyclic aromatic hydrocarbons, pesticides, polychlorinated biphenyls that could be present in urban air due to road traffic, domestic heating, industrial emissions, agricultural treatments, etc.[1–3] Other examples of NPSP include fertilizer (especially Cd, N, and P) and pesticide applications to improve crop yield. Use of industrial waste materials as soil amendments have been estimated to contaminate thousands of hectares of productive agricultural land in countries throughout the world.

Contaminant Interactions

Non-point pollution is generally associated with low-level contamination spread at broad acre level. Under these circumstances, the major reaction controlling contaminant interactions are sorption–desorption processes, plant uptake, surface runoff, and leaching. However, certain contaminants, in particular, organic compounds are also subjected to voltalization, chemical, and biological degradation. Sorption–desorption and degradation (both biotic and abiotic) are the two most important processes controlling organic contaminant behavior in soils. These processes are influenced by both soil and solution properties of the environment. Such interactions also determine the bioavailability and/or transport of contaminants in soils. Where the contaminants are bioavailable, risk to surface and groundwater and soil, crop, and human health are enhanced.

TABLE 1 Industries, Land Uses, and Associated Chemicals Contributing to Non-Point Source Pollution

Industry	Type of Chemical	Associated Chemicals
Agricultural activities	Metals/metalloid	Cadmium, mercury, arsenic, selenium
	Non-metals	Nitrate, phosphate, borate
	Salinity/sodicity	Sodium, chloride, sulfate, magnesium, alkalinity
	Pesticides	Range of organic and inorganic pesticides including arsenic, copper, zinc, lead, sulfonylureas, organochlorine, organophosphates, etc., salt, geogenic contaminants (e.g., arsenic, selenium, etc.)
	Irrigation	Sodium, chloride, arsenic, selenium
Automobile and industrial emissions	Dust	Lead, arsenic, copper, cadmium, zinc, etc.
	Gas	Sulfur oxides, carbon oxides
	Metals	Lead and lead organic compounds
Rainwater	Organics	Polyaromatic hydrocarbons, polychlorbiphenyls, etc.
	Inorganic	Sulfur oxides, carbon oxides acidity, metals and metalloids

Source: (From Barzi, F.; Naidu, R.; McLaughlin, M.J. Contaminants and the Australian Soil Environment. In *Contaminants and the Soil Environment in the Australasia-Pacific Region*; Naidu, R., Kookana, R.S., Oliver, D., Rogers, S., McLaughlin, M.J., Eds.; Kluwer Academic Publishers: Dordrecht, the Netherlands, 1996; 451–484.)

Implications to Soil and Environmental Quality

Environmental contaminants can have a deleterious effect on non-target organisms and their beneficial activities. These effects could include a decline in primary production, decreased rate of organic matter break-down, and nutrient cycling as well as mineralization of harmful substances that in turn cause a loss of productivity of the ecosystems. Certain pollutants, even though present in very small concentrations in the soil and surrounding water, have potential to be taken up by various micro-organisms, plants, animals, and ultimately human beings. These pollutants may accumulate and concentrate in the food chain by several thousand times through a process referred to as biomagnification.

Urban sewage, because of its nutrient values and source of organic carbon in soils, is now increasingly being disposed to land. The contaminants present in sewage sludge (nutrients, heavy metals, organic compounds, and pathogens), if not managed properly, could potentially affect the environment adversely. Dumping of radioactive waste (e.g., radium, uranium, plutonium) onto soil is more complicated because these materials remain active for thousands of years in the soil and thus pose a continued threat to the future health of the ecosystem.

Industrial wastes, improper agricultural techniques, municipal wastes, and use of saline water for irrigation under high evaporative conditions result in the presence of excess soluble salts (predominantly Na and Cl ions) and metalloids such as Se and As in soils. Salinity and sodicity affect the vegetation by inhibiting seed germination, decreasing permeability of roots to water, and disrupting their functions such as photosynthesis, respiration, and synthesis of proteins and enzymes.

Some of the impacts of soil pollution migrate a long way from the source and can persist for some time. For example, suspended solids can increase water turbidity in streams, affecting benthic and pelagic aquatic ecosystems, filling reservoirs with unwanted silt, and requiring water treatment systems for potable water supplies. Phosphorus attached to soil particles, which are washed from a paddock into a stream, can dominate nutrient loads in streams and down-stream water bodies. Consequences include increases in algal biomass, reduced oxygen concentrations, impaired habitat for aquatic species, and even possible production of cyanobacterial toxins, with series impacts for humans and livestock consuming the water. Where waters discharge into estuaries, N can be the limiting factor for eutrophication; estuaries of some catchments where fertilizer use is extensive have suffered from excessive sea grass and algal growth.

More insidious is the leaching of nutrients, agricultural chemicals, and hydrocarbons to groundwater. Incremental increases in concentrations in groundwater may be observed over long periods of time resulting in initially potable water becoming undrinkable and then some of the highest valued uses of the resource may be lost for decades. This problem is most severe on tropical islands with shallow relief and some deltaic arsenopyrite deposits, where wells cannot be deepened to avoid polluted groundwater because underlying groundwater is either saline or contains too much As.

Sampling for Non-Point Source Pollution

The sampling requirements of NPSP are quite different from those of the point source contamination. Typically, the sampling is required to give a good estimate of the mean level of pollution rather than to delineate areas of pollution. In such a situation, sampling is typically carried out on a regular square or a triangular grid. Furthermore, gains may be possible by using composite sampling.[4] However, if the pollution is patchy, other strategies may be used. One such strategy is to divide the area into remediation units, and to sample each of these. The possibility of movement of the pollutant from the soil to some receptor (or asset) is assessed, and the potential harm is quantified. This process requires an analysis of the bioavailability of the pollutant, pathway analysis, and the toxicological risk. The risk analysis is then assessed and decisions are then made as to how the risk should be managed.

Management and/or Remediation of Non-Point Source Pollution

The treatment strategies used for managing NPSP are generally those that modify the soil properties to decrease the bioavailable contaminant fraction. This is particularly so in the rural agricultural environment where soil-plant transfer of contaminants is of greatest concern. Soil amendments commonly used include those that change the ion-exchange characteristics of the colloid particles and those that enhance the ability of soils to sorb contaminants. An example of NPSP management includes the application of lime to immobilize metals because the solubility of most heavy metals decreases with increasing soil pH. However, this approach is not applicable to all metals, especially those that form oxyanions—the bioavailability of such species increases with increasing pH. Therefore, one of the prerequisites for remediating contaminated sites is a detailed assessment of the nature of contaminants present in the soil. The application of a modified aluminosilicate to a highly contaminated soil around a zinc smelter in Belgium was shown to reduce the bioavailability of metals thereby reducing the Zn phytotoxicity.[5] The simple addition of rock phosphates to form Pb phosphate has also been demonstrated to reduce the bioavailability of Pb in aqueous solutions and contaminated soils due to immobilization in the metal.[6] Nevertheless, there is concern over the long-term stability of the processes. The immobilization process appears attractive currently given that there are very few cheap and effective in situ remediation techniques for metal-contaminated soils. A novel, innovative approach is using higher plants to stabilize, extract, degrade, or volatilize inorganic and organic contaminants for in situ treatment (cleanup or containment) of polluted top soils.[7]

Preventing Water Pollution

The key to preventing water pollution from the soil zone is to manage the source of pollution. For example, nitrate pollution of groundwater will always occur if there is excess nitrate in the soil at a time when there is excess water leaching through the soil. This suggests that we should aim to reduce the nitrogen in the soil during wet seasons and the drainage through the soil. Local research may be needed to demonstrate the success of best management techniques in reducing nutrient, sediment, metal, and chemical exports via surface runoff and infiltration to groundwater. Production figures from the same experiments may also convince local farmers of the benefits of maintaining nutrients and chemicals where needed by a crop rather than losing them off site, and facilitate uptake of best management practices.

Global Challenges and Responsibility

The biosphere is a life-supporting system to the living organisms. Each species in this system has a role to play and thus every species is important and biological diversity is vital for ecosystem health and functioning. The detection of hazardous compounds in Antarctica, where these compounds were never used or no man has ever lived before, indicates how serious is the problem of long-range atmospheric transport and deposition of these pollutants. Clearly, pollution knows no boundaries. This ubiquitous pollution has had a global effect on our soils, which in turn has been affecting their biological health and productivity. Coupled with this, over 100,000 chemicals are being used in countries throughout the world. Recent focus has been on the endocrine disruptor chemicals that mimic natural hormones and do great harm to animal and human reproductive cycles.

These pollutants are only a few examples of contaminants that are found in the terrestrial environment.

References

1. Chan, C.H.; Bruce, G.; Harrison, B. Wet deposition of organochlorine pesticides and polychlorinated biphenyls to the great lakes. J. Great Lakes Res. **1994**, *20*, 546–560.
2. Lodovici, M.; Dolara, P.; Taiti, S.; Del Carmine, P.; Bernardi, L.; Agati, L.; Ciappellano, S. Polycyclic aromatic hydrocarbons in the leaves of the evergreen tree *Laurus Nobilis*. Sci. Total Environ. **1994**, *153*, 61–68.
3. Sweet, C.W.; Murphy, T.J.; Bannasch, J.H.; Kelsey, C.A.; Hong, J. Atmospheric deposition of PCBs into green bay. J. Great Lakes Res. **1993**, *18*, 109–128.
4. Patil, G.P.; Gore, S.D.; Johnson, G.D. *Manual on Statistical Design and Analysis with Composite Samples;* Technical Report No. 96-0501; EPA Observational Economy Series Center for Statistical Ecology and Environmental Statistics; Pennsylvania State University, 1996; Vol. 3.
5. Vangronsveld, J.; Van Assche, F.; Clijsters, H. Reclamation of a bare industrial area, contaminated by non-ferrous metals: in situ metal immobilisation and revegetation. Environ. Pollut. **1995**, *87*, 51–59.
6. Ma, Q.Y.; Logan, T.J.; Traina, S.J. Lead immobilisation from aqueous solutions and contaminated soils using phosphate rocks. Environ. Sci. Technol. **1995**, *29*, 1118–1126.
7. Wenzel, W.W.; Adriano, D.C.; Salt, D.; Smith, R. Phytoremediation: a plant-microbe based remediation system. In *Bioremediation of Contaminated Soils;* Soil Science Society of America Special Monograph No. 37, Adriano, D.C., Bollag, J.M., Frankenberger, W.T., Jr., Sims, W.R., Eds.; Soil Science Society of America: Madison, USA, 1999; 772 pp.

53

Pollution: Point Sources

Ravendra Naidu,
Mallavarapu
Megharaj, Peter
Dillon, Rai
Kookana, Ray
Correll, and
W.W. Wenzel

Introduction

Environmental pollution is one of the foremost ecological challenges. Pollution is an offshoot of technological advancement and overexploitation of natural resources. From the standpoint of pollution, the term environment primarily includes air, land, and water components including landscapes, rivers, parks, and oceans. Pollution can be generally defined as an undesirable change in the natural quality of the environment that may adversely affect the well being of humans, other living organisms, or entire ecosystems either directly or indirectly. Although pollution is often the result of human activities (anthropogenic), it could also be due to natural sources such as volcanic eruptions emitting noxious gases, pedogenic processes, or natural change in the climate. Where pollution is localized it is described as point source (PS). Thus, PS pollution is a source of pollution with a clearly identifiable point of discharge that can be traced back to the specific source such as leakage of underground petroleum storage tanks or an industrial site.

Some naturally occurring pollutants are termed geogenic contaminants and these include fluorine, selenium, arsenic, lead, chromium, fluoride, and radionuclides in the soil and water environment. Significant adverse impacts of geogenic contaminants (e.g., As) on environmental and human health have been recorded in Bangladesh, West Bengal, India, Vietnam, and China. More recently reported is the presence of geogenic Cd and the implications to crop quality in Norwegian soils.[1]

The terms contamination and pollution are often used interchangeably but erroneously. Contamination denotes the presence of a particular substance at a higher concentration than would occur naturally and this may or may not have harmful effects on human or the environment. Pollution refers not only to the presence of a substance at higher level than would normally occur but is also associated with some kind of adverse effect.

Nature and Sources of Contaminants

The main activities contributing to PS pollution include industrial, mining, agricultural, and commercial activities as well as transport and services (Table 1). Uncontrolled mining, manufacturing, and disposal of wastes inevitably cause environmental pollution. Military land and land for recreational shooting are

TABLE 1 Industries, Land Uses, and Associated Chemicals Contributing to Points, Non-Point Source Pollution

Industry	Type of Chemical	Associated Chemicals
Airports	Hydrocarbons	Aviation fuels
	Metals	Particularly aluminum, magnesium, and chromium
Asbestos production and disposal	Asbestos	
Battery manufacture and recycling	Metals	Lead, manganese, zinc, cadmium, nickel, cobalt, mercury, silver, and antimony
	Acids	Sulfuric acid
Breweries/distilleries	Alcohol	Ethanol, methanol, and esters
Chemicals manufacture and use	Acid/alkali	Mercury (chlor/alkali), sulfuric, hydrochloric and nitric acids, sodium and calcium hydroxides
	Adhesives/resins	Polyvinyl acetate, phenols, formaldehyde, acrylates, and phthalates
	Dyes	Chromium, titanium, cobalt, sulfur and nitrogen organic compounds, sulfates, and solvents
	Explosives	Acetone, nitric acid, ammonium nitrate, pentachlorophenol, ammonia, sulfuric acid, nitroglycerine, calcium cyanamide, lead, ethylene glycol, methanol, copper, aluminum, bis(2-ethylhexyl) adipate, dibutyl phthalate, sodium hydroxide, mercury, and silver
	Fertilizer	Calcium phosphate, calcium sulfate, nitrates, ammonium sulfate, carbonates, potassium, copper, magnesium, molybdenum, boron, and cadmium
	Flocculants	Aluminum
	Foam production	Urethane, formaldehyde, and styrene
	Fungicides	Carbamates, copper sulfate, copper chloride, sulfur, and chromium
	Herbicides	Ammonium thiocyanate, carbanates, organochlorines, organophosphates, arsenic, and mercury
	Paints	Arsenic, barium, cadmium, chromium, cobalt, lead,
	Heavy metals	manganese, mercury, selenium, and zinc
	General	Titanium dioxide
	Solvent	Toluene, oils natural (e.g., pine oil) or synthetic
	Pesticides	Arsenic, lead, organochlorines, and organophosphates
	Active ingredients	Sodium, tetraborate, carbamates, sulfur, and synthetic
	Solvents	pyrethroids
		Xylene, kerosene, methyl isobutyl ketone, amyl acetate, and chlorinated solvents
	Pharmacy	Dextrose and starch
	General/solvents	Acetone, cyclohexane, methylene chloride, ethyl acetate, butyl acetate, methanol, ethanol, isopropanol, butanol, pyridine methyl ethyl ketone, methyl isobutyl ketone, and tetrahydrofuran
	Photography	Hydroquinone, pheidom, sodium carbonate, sodium sulfite, potassium bromide, monomethyl paraaminophenol sulfates, ferricyanide, chromium, silver, thiocyanate, ammonium compounds, sulfur compounds, phosphate, phenylene diamine, ethyl alcohol, thiosulfates, and formaldehyde
	Plastics	Sulfates, carbonates, cadmium, solvents, acrylates, phthalates, and styrene
	Rubber	Carbon black

(Continued)

TABLE 1 (*Continued*) Industries, Land Uses, and Associated Chemicals Contributing to Points, Non-Point Source Pollution

Industry	Type of Chemical	Associated Chemicals
	Soap/detergent General	Potassium compounds, phosphates, ammonia, alcohols, esters, sodium hydroxide, surfactants (sodium lauryl sulfate), and silicate compounds
	Acids	Sulfuric acid and stearic acid
	Oils	Palm, coconut, pine, and tea tree
	Solvents	
	General	Ammonia
	Hydrocarbons	e.g., BTEX (benzene, toluene, ethylbenzene, xylene)
	Chlorinated organics	e.g., trichloroethane, carbon tetrachloride, and methylene chloride
Defense works		See *Explosives* under *Chemicals Manufacture and* Use, Foundries, Engine Works, and Service Stations
Drum reconditioning		See *Chemicals Manufacture and Use*
Dry cleaning		Trichlorethylene and ethane Carbon tetrachloride Perchlorethylene
Electrical		PCBs (transformers and capacitors), solvents, tin, lead, and copper
Engine works	Hydrocarbons Metals Solvents Acids/alkalis Refrigerants	
	Antifreeze	Ethylene glycol, nitrates, phosphates, and silicates
Foundries	Metals	Particularly aluminum, manganese, iron, copper, nickel, chromium, zinc, cadmium and lead and oxides, chlorides, fluorides and sulfates of these metals
	Acids	Phenolics and amines coke/graphite dust
Gas works	Inorganics	Ammonia, cyanide, nitrate, sulfide, and thiocyanate
	Metals	Aluminum, antimony, arsenic, barium, cadmium, chromium, copper, iron, lead, manganese, mercury, nickel, selenium, silver, vanadium, and zinc
	Semivolatiles	Benzene, ethylbenzene, toluene, total xylenes, coal tar, phenolics, and PAHs
Iron and steel works		Metals and oxides of iron, nickel, copper, chromium, magnesium and manganese, and graphite
Landfill sites Marinas		Methane, hydrogen sulfides, heavy metals, and complex acids Engine works, electroplating under metal treatment
	Antifouling paints	Copper, tributyltin (TBT)
Metal treatments	Electroplating metals	Nickel, chromium, zinc, aluminum, copper, lead, cadmium, and tin
	Acids	Sulfuric, hydrochloric, nitric, and phosphoric
	General	Sodium hydroxide, 1,1,1-trichloroethane, tetrachloroethylene, toluene, ethylene glycol, and cyanide compounds
	Liquid carburizing baths	Sodium, cyanide, barium, chloride, potassium chloride, sodium chloride, sodium carbonate, and sodium cyanate
	Mining and extracting industries	Arsenic, mercury, and cyanides and also refer to *Explosives* under *Chemicals Manufacture and Use*

(*Continued*)

TABLE 1 (*Continued*)   Industries, Land Uses, and Associated Chemicals Contributing to Points, Non-Point Source Pollution

Industry	Type of Chemical	Associated Chemicals
	Power stations	Asbestos, PCBs, fly ash, and metals
	Printing shops	Acids, alkalis, solvents, chromium see *Photography* under *Chemicals Manufacture and Use*
Scrap yards	Service stations and fuel storage facilities	Hydrocarbons, metals, and solvents Aliphatic hydrocarbons
		BTEX (i.e., benzene, toluene, ethylbenzene, xylene) PAHs (e.g., benzo(a) pyrene) Phenols Lead
Sheep and cattle dips		Arsenic, organochlorines and organophosphates, carbamates, and synthetic pyrethroids
Smelting and refining		Metals and the fluorides, chlorides and oxides of copper, tin, silver, gold, selenium, lead, and aluminum
Tanning and associated trades	Metals	Chromium, manganese, and aluminum
	General	Ammonium sulfate, ammonia, ammonium nitrate, phenolics (creosote), formaldehyde, and tannic acid
Wood preservation	Metals	Chromium, copper, and arsenic
	General	Naphthalene, ammonia, pentachlorophenol, dibenzofuran, anthracene, biphenyl, ammonium sulfate, quinoline, boron, creosote, and organochlorine pesticides

Source: Barzi et al.[11]

also important sites of PS contamination. The contaminants associated with such activities are listed in Table 1. Contamination at many of these sites appears to have resulted because of lax regulatory measures prior to the establishment of legislation protecting the environment.

Contaminant Interactions in Soil and Water

Inorganic Chemicals

Inorganic contaminant interactions with colloid particulates include: adsorption–desorption at surface sites, precipitation, exchange with clay minerals, binding by organically coated particulate matter or organic colloidal material, or adsorption of contaminant ligand complexes. Depending on the nature of contaminants, these interactions are controlled by solution pH and ionic strength of soil solution, nature of the species, dominant cation, and inorganic and organic ligands present in the soil solution.[2]

Organic Chemicals

The fate and behavior of organic compounds depend on a variety of processes including sorption–desorption, volatilization, chemical and biological degradation, plant uptake, surface runoff, and leaching. Sorption–desorption and degradation (both biotic and abiotic) are perhaps the two most important processes as the bulk of the chemicals is either sorbed by organic and inorganic soil constituents, and chemically or microbially transformed/degraded. The degradation is not always a detoxification process. This is because in some cases the transformation or degradation process leads to intermediate products that are more mobile, more persistent, or more toxic to non-target organisms. The relative importance of these processes is determined by the chemical nature of the compound.

Implications to Soil and Environmental Quality

Considerable amount of literature is available on the effects of contaminants on soil microorganisms and their functions in soil. The negative impacts of contaminants on microbial processes are important from the ecosystem point of view and any such effects could potentially result in a major ecological perturbance. Hence, it is most relevant to examine the effects of contaminants on microbial processes in combination with communities. The most commonly used indicators of metal effects on microflora in soil are: (1) soil respiration, (2) soil nitrification, (3) soil microbial biomass, and (4) soil enzymes.

Contaminants can reach the food chain by way of water, soil, plants, and animals. In addition to the food chain transfer, pollutants may also enter via direct consumption or dust inhalation of soil by children or animals. Accumulation of these pollutants can take place in certain target tissues of the organism depending on the solubility and nature of the compound. For example, DDT and PCBs accumulate in human adipose tissue. Consequently, several of these pollutants have the potential to cause serious abnormalities including cancer and reproductive impairments in animal and human systems.

Sampling for PS Pollution

The aims of the sampling system must be clearly defined before it can be optimized.[3] The type of decision may be to determine land use, how much of an area is to be remediated, or what type of remediation process is required. Because sampling and the associated chemical and statistical analyses are expensive, careful planning of the sampling scheme is therefore a good investment. One of the best ways to achieve this is to use any ancillary data that are available. These data could be in the form of emission history from a stack, old photographs that give details of previous land uses, or agricultural records. Such data can at least give qualitative information.

As discussed before, PS pollution will typically be airborne from a stack, or waterborne from some effluent such as tannery waste, cattle dips, or mine waste. In many cases, the industry will have modified its emissions (e.g., cleaner production) or point of release (increased stack height), hence the current pattern of emission may not be closely related to the historic pattern of pollution. For example, liquid effluent may have been discharged previously into a bay, but that effluent may now be treated and perhaps discharged at some other point. Typically, the aim of a sampling scheme in these situations is to assess the maximum concentrations, the extent of the pollution, and the rate of decline in concentration from the PS. Often the sampling scheme will be used to produce maps of concentration isopleths of the pollutant.

The location of the sampling points would normally be concentrated towards the source of the pollution. A good scheme is to have sufficient samples to accurately assess the maximum pollution, and then space additional samples at increasing intervals. In most cases, the distribution of the pollutant will be asymmetric, with the maximum spread down the slope or down the prevailing wind. In such cases more samples should be placed in the direction of the expected gradient. This is a clear case of when ancillary data can be used effectively. A graph of concentration of the pollutant against the reciprocal of distance from the source is often informative.[4] Sampling depths will depend on both the nature of the pollution and the reason for the investigation. If the pollution is from dust and it is unlikely to be leached, only surface sampling will be required. An example of this is pollution from silver smelting in Wales.[5] In contrast, contamination from organic or mobile inorganic pollutants such as F compounds may migrate well down to the profile and deep sampling may be required.[6,7]

Assessment

In order to assess the impacts of pollution, reliable and effective monitoring techniques are important. Pollution can be assessed and monitored by chemical analyses, toxicity tests, and field surveys. Comparison of contaminant data with an uncontaminated reference site and available databases for baseline concentrations can be useful in establishing the extent of contamination. However, this may not always be possible in the field. Chemical analyses must be used in conjunction with biological assays to reveal site contamination and associated adverse effects. Toxicological assays can also reveal information about synergistic interactions of two or more contaminants present as mixtures in soil, which cannot be measured by chemical assays alone.

Microorganisms serve as rapid detectors of environmental pollution and are thus of importance as pollution indicators. The presence of pollutants can induce alteration of microbial communities and reduction of species diversity, inhibition of certain microbial processes (organic matter breakdown, mineralization of carbon and nitrogen, enzymatic activities, etc.). A measure of the functional diversity of the bacterial flora can be assessed using ecoplates (see http://www.biolog.com/section_4.html). It has been shown that algae are especially sensitive to various organic and inorganic pollutants and thus may serve as a good indicator of pollution.[8] A variety of toxicity tests involving microorganisms, invertebrates, vertebrates, and plants may be used with soil or water samples.[9]

Management and/or Remediation of PS Pollution

The major objective of any remediation process is to (1) reduce the actual or potential environmental threat; and (2) reduce unacceptable risks to man, animals, and the environment to acceptable levels.[10] Therefore, strategies to either manage and/or remediate contaminated sites have been developed largely from application of stringent regulatory measures set up to safeguard ecosystem function as well as to minimize the potential adverse effects of toxic substances on animal and human health.

The available remediation technologies may be grouped into two categories: (1) ex situ techniques that require removal of the contaminated soil or groundwater for treatment either on-site or off-site; and (2) in situ techniques that attempt to remediate without excavation of contaminated soils. Generally, in situ techniques are favored over ex situ techniques because of (1) reduced costs due to elimination or minimization of excavation, transportation to disposal sites, and sometimes treatment itself; (2) reduced health impacts on the public or the workers; and, (3) the potential for remediation of inaccessible sites, e.g., those located at greater depths or under buildings. Although in situ techniques have been successful with organic contaminated sites, the success of in situ strategies with metal contaminants has been limited. Given that organic and inorganic contaminants often occur as a mixture, a combination of more than one strategy is often required to either successfully remediate or manage metal contaminated soils.

Global Challenges and Responsibility

The last 100 years has seen massive industrialization. Indeed such developments were coupled with the rapid increase in world population and the desire to enhance economy and food productivity. While industrialization has led to increased economic activity and much benefit to human race, the lack of regulatory measures and appropriate waste management strategies until early 1980s (including the use of agrochemicals) has resulted in contamination of our biosphere. Continued pollution of the environment through industrial emissions is of global concern. There is, therefore, a need for politicians, regulatory organizations, and scientists to work together to minimize environmental contamination and to remediate contaminated sites. The responsibility to check this pollution lies with every individual and country although the majority of this pollution is due to the industrialized nations. There is a clear need of better coordination of efforts in dealing with numerous forms of PS pollution problems that are being faced globally.

References

1. Mehlum, H.K.; Arnesen, A.K.M.; Singh, B.R. Extractability and plant uptake of heavy metals in alum shale soils. Commun. Soil Sci. Plant Anal. **1998**, *29*, 183–198.
2. McBride, M.B. Reactions controlling heavy metal solubility in soils. Adv. Soil Sci. **1989**, *10*, 1–56.
3. Patil, G.P.; Gore, S.D.; Johnson, G.D. *EPA Observational Economy Series Volume 3: Manual on Statistical Design and Analysis with Composite Samples; Technical Report* No. 96–0501; Center for Statistical Ecology and Environmental Statistics: Pennsylvania State University, 1996.
4. Ward, T.J.; Correll, R.L. Estimating background concentrations of heavy metals in the marine environment, Proceedings of a Bioaccumulation Workshop: Assessment of the Distribution, Impacts and Bioaccumulation of Contaminants in Aquatic Environments, Sydney, 1990; Miskiewicz, A.G., Ed.; Water Board and Australian Marine Science Association: Sydney, 1992, 133–139.
5. Jones, K.C.; Davies, B.E.; Peterson, P.J. Silver in welsh soils: physical and chemical distribution studies. Geoderma **1986**, *37*, 157–174.
6. Barber, C.; Bates, L.; Barron, R.; Allison, H. Assessment of the relative vulnerability of groundwater to pollution: a review and background paper for the conference workshop on vulnerability assessment. J. Aust. Geol. Geophys. **1993**, *14* (2–3), 147–154.
7. Wenzel, W.W.; Blum, W.E.H. Effects of fluorine deposition on the chemistry of acid luvisols. Int. J. Environ. Anal. Chem. **1992**, *46*, 223–231.
8. Megharaj, M.; Singleton, I.; McClure, N.C. Effect of pentachlorophenol pollution towards microalgae and microbial activities in soil from a former timber processing facility. Bull. Environ. Contam. Toxicol. **1998**, *61*, 108–115.
9. Juhasz, A.L.; Megharaj, M.; Naidu, R. Bioavailability: the major challenge (constraint) to bioremediation of organically contaminated soils. In *Remediation Engineering of Contaminated Soils*; Wise, D., Trantolo, D.J., Cichon, E.J., Inyang, H.I., Stottmeister, U., Eds.; Marcel Dekker: New York, 2000; 217–241.
10. Wood, P.A. Remediation methods for contaminated sites. In *Contaminated Land and Its Reclamation*; Hester, R.E., Harrison, R.M., Eds.; Royal Society of Chemistry, Thomas Graham House: Cambridge, U.K., 1997; 47–73.
11. Barzi, F.; Naidu, R.; McLaughlin, M.J. Contaminants and the Australian soil environment. In *Contaminants and the Soil Environment in the Australasia–Pacific Region*; Naidu, R., Kookana, R.S., Oliver, D., Rogers, S., McLaughlin, M.J., Eds.; Kluwer Academic Publishers: Dordrecht, the Netherlands, 1996; 451–484.

54

Radioactivity

Bogdan Skwarzec

Introduction

Radioactivity was discovered at the end of the 19th century by Henri Becquerel, Marie Curie (Polish native name, Maria Sklodowska-Curie), and Pierre Curie. Henri Becquerel found that uranium salts caused fogging of an unexposed photographic plate,[1] and Marie Curie discovered that only certain elements gave off these rays of energy.[2] She named this behavior "radioactivity" (natural radioactivity). A systematic search for the total radioactivity in uranium ores also guided Marie Curie to isolate a new element, polonium, and to separate a new element, radium.[3–7] The two elements have chemical similarity that would otherwise have made them difficult to distinguish from each other. In 1934, Marie Curie's daughter, Irene Joliot-Curie and her husband Frederic Jean Joliot were the first creators of artificial radioactivity. They bombarded boron with alpha particles to make the neutron-poor nitrogen isotope ^{13}N; this isotope emitted positrons. In addition, they bombarded aluminum and magnesium with neutrons to make new radioisotopes.[8]

Radioactive Decay

Radioactive decay is the process by which an unstable atomic nucleus spontaneously loses energy by emitting ionizing particles and radiation. The three main types of radiation were discovered by Ernest Rutherford, the alpha (α), beta (β), and gamma (γ) rays (alpha, beta, and gamma radiation).[9–11] With Ernest Rutherford, he saw that radioactive substances are transformed from one element to another. About 10 years later, he unraveled the rules for the elemental transformations that accompanied radioactive decay, first for α decay and later for β decay. Emission of an α particle changes the emitting atom to an atom of the element two places to the left in the periodic table; emission of a β^- particle changes the emitting atom to an atom of the element one place to the right. These rules taken together are known as the Displacement Law; Kazimierz Fajans published it slightly earlier than did Soddy in 1913.[12] At about the same time, Soddy came to the conclusion that several substances with different radioactive properties and different atomic weights were chemically the same element. He named such substances isotopes.[13] Now, the radioactive principles are named the Soddy–Fajans periodic method.

$$\alpha \text{ decay}: {}_{Z}^{A}X \rightarrow {}_{Z-2}^{A-4}Y + {}_{2}^{4}He$$

$$\beta^{-} \text{ decay}: {}_{Z}^{A}X \rightarrow {}_{Z+1}^{A}Y + {}_{-1}^{0}\beta$$

$$\beta^{+} \text{ decay}: {}_{Z}^{A}X \rightarrow {}_{Z-1}^{A}Y + {}_{+1}^{0}\beta$$

where X and Y are symbols for nuclides, Z is the mass number, and A is the atomic number.

1. Alpha (α) decay is a method of decay in large nuclei. Alpha particles (helium nuclei, He^{2+}), consisting of two neutrons and two protons, are emitted. Because of the particles' relatively high charge, it is heavily ionizing and will cause severe damage if ingested. However, owing to the high mass of the particle, it has little energy and a low range; typically, alpha particles can be stopped with a sheet of paper (or skin).
2. Beta minus (β^{-}) radiation consists of an energetic electron. It is less ionizing than alpha radiation, but more than gamma. The electron can be stopped with a few centimeters of metal. It occurs when a neutron decays into a proton in a nucleus, releasing the beta particle and an antineutrino. Beta-plus (β^{+}) radiation is the emission of positrons. As these are antimatter particles, they annihilate any matter nearby, releasing gamma photons.
3. 3. Gamma (γ) radiation consists of photons with a frequency greater than 10^{19} Hz. Gamma radiation occurs to rid the decaying nucleus of excess energy after it has emitted either alpha or beta radiation.

The activity (A) of radionuclide is lost at time (t) according to the formula

$$A = A_0 \cdot e^{-\lambda \cdot t}$$

where A is the radionuclide activity at time $t = 0$, A_0 is the radionuclide activity at time t, and λ is the decay constant of the radionuclide.

The SI unit of activity is the becquerel (Bq). One becquerel is defined as one transformation (or decay) per second. Another unit of radioactivity is the curie (Ci), which was originally defined as the amount of radium emanation (by gaseous radon-222), in equilibrium with 1 g of pure radium isotope ^{226}Ra. At present, it is equal, by definition, to the activity of any radionuclide decaying with a disintegration rate of 3.7×10^{10} Bq. The activity of a radioactive substance is characterized by its half-time—the time taken for the activity of a given amount of radioactive substance to decay to half of its value.[14,15]

After the discovery of neutron in 1932, Encico Fermi and colleagues studied the results of bombarding uranium with neutron in 1934.[16] The first person that mentioned the idea of nuclear fission in 1934 was Ida Noddack.[17] After Fermi's publication, Lise Meittner, Otto Hahn, and Fritz Strassmann began to perform a similar experiment and discovered nuclear fission of uranium ^{235}U in 1938.[18,19] Also, Józef Rotblat in 1939 published the results of a study about fission of uranium ^{235}U nuclei.[20]

In nuclear physics and nuclear chemistry, nuclear fission is a nuclear reaction in which the nucleus of an atom splits into smaller parts (lighter nuclei), often producing free neutrons and photons (in the form of gamma rays), as well. Fission of heavy elements is an exothermic reaction that can release large amounts of energy both as electromagnetic radiation and as kinetic energy of the fragments (heating the bulk material where fission takes place).

Three heavy radionuclides, natural ^{235}U and artificial ^{239}Pu and ^{233}U, are capable of reactions (nuclear fission) in which an atom's nucleus splits into smaller parts, releasing a large amount of energy in the process. During the fission of ^{235}U, three neutrons are released in addition to the two daughter atoms (see reaction below) (Figure 1).[21]

$$^{235}_{92}U + {}^{1}_{0}n \rightarrow {}^{236}_{92}U \rightarrow {}^{90}_{36}Kr + {}^{143}_{56}Ba + 3{}^{1}_{0}n + 200 \text{ MeV}$$

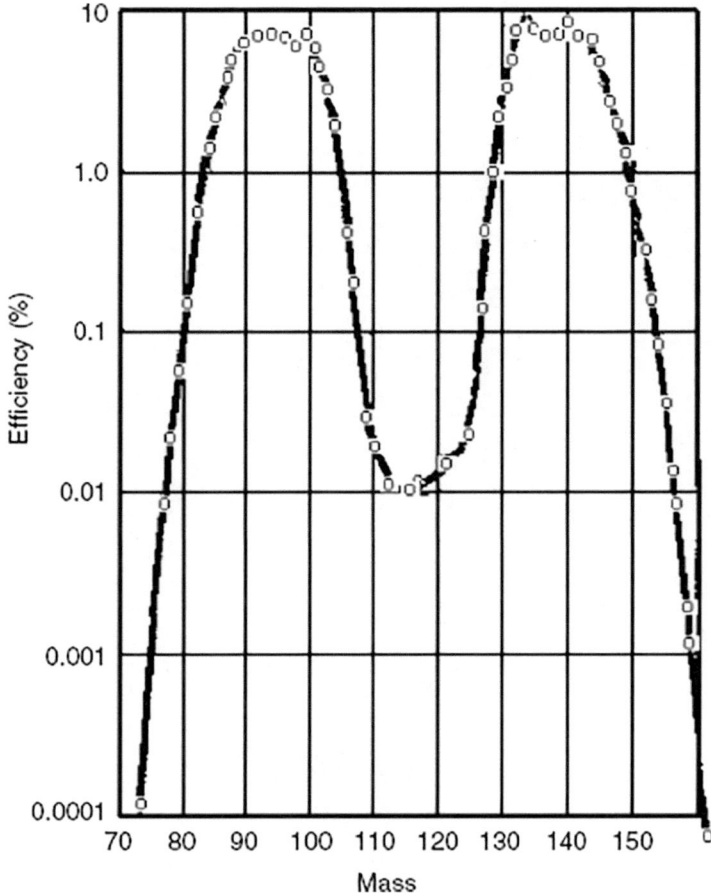

FIGURE 1 Fission yield as a function of mass number for the slow neutron fission of ^{235}U.
Source: Environmental radiochemistry and radiological protection.[21]

Natural and Artificial Radionuclides

Radionuclides present in the natural environment are classified as either of natural or anthropogenic origin. Naturally occurring radionuclides occur in different ecosystems with cosmogenic and primordial providence.[22,23]

1. *Cosmogenic radionuclides:* Cosmic ray–produced radionuclides are generated in the upper-atmosphere gases, e.g., O_2, N_2, and Ar. They are transported to the lower atmosphere and next to the oceans and to the continents. Most of the cosmic radionuclides are produced in very small amounts and only four of them, ^{3}H, ^{7}Be, ^{14}C, and ^{22}Na, constitute significant contributions to the radiation dose to humans. Cosmogenic radionuclides have been measured in humans, topsoil, polar ice, surface rocks, sediments, the biosphere, the ocean floor, and the atmosphere.[24]
2. *Primordial radionuclides:* Among non-series radionuclides of terrestrial origin, only ^{40}K and ^{87}Rb are significant sources of radiation to humans. They are characterized by a long half-time (more than 10^9 years) and small concentrations in crustal rocks (below 1 mBq/kg).[25,26]

The serially occurring radionuclides are contained in four natural decay series—uranium, thorium, actinium, and neptunium—and, except for the actinium series, are named after their parent nuclides (Figures 2–5).

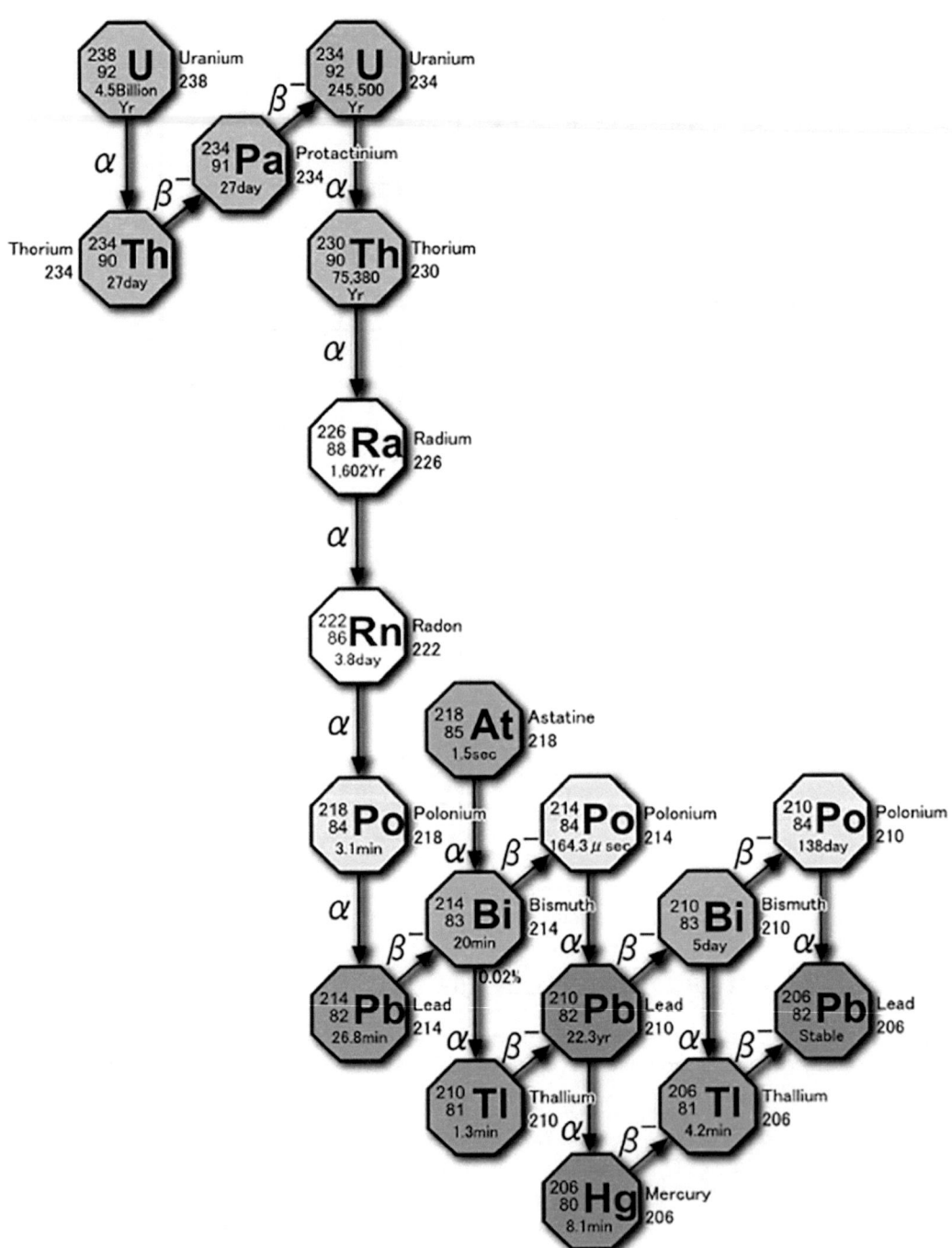

FIGURE 2 Uranium-238 decay series.
Source: Wikipedia, uranium series decay chain, http://upload.wikimedia.org/wikipedia/commons/a/a1/Decay_chain%284n%2B2%2C_Uranium_series%29.PNG.[40]

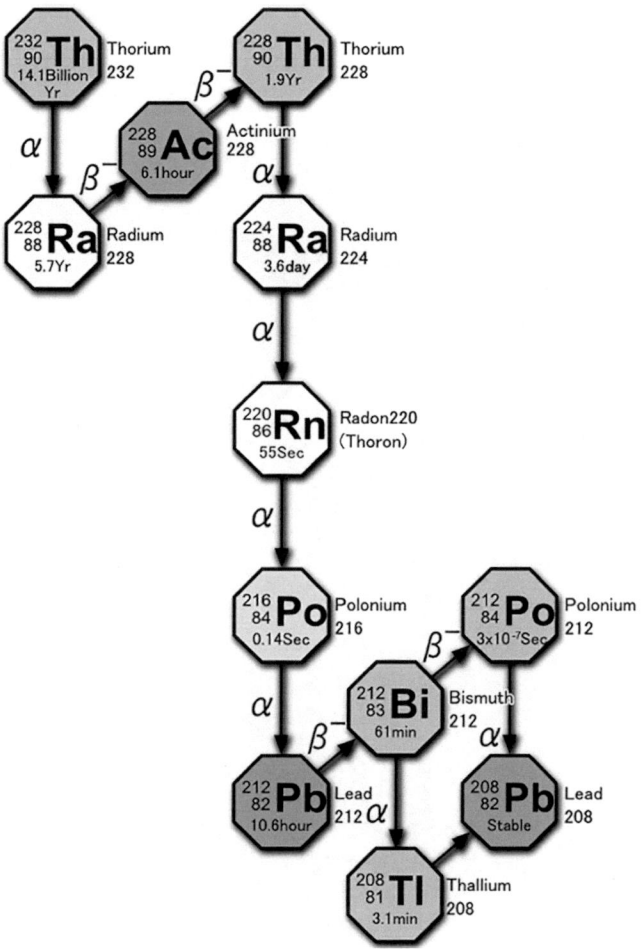

FIGURE 3 Thorium-232 decay series.
Source: Wikipedia, Thorium series decay chain, http://upload.wikimedia.org/wikipedia/commons/1/1c/Decay_chain%284n%2CThorium_series%29.PNG.[41]

Anthropogenic Radionuclides

Anthropogenic-derived radionuclides have been mainly released from several sources since the 1940s. Major sources in the environment are nuclear weapons, nuclear power production, accidents (e.g., the Chernobyl accident in 1986), radioactive waste disposal, solid radioactive waste disposal, and man-made radionuclides as tracers of environmental processes. Fallout from nuclear weapons explosions represents the largest contribution of anthropogenic-derived radionuclides to the natural environment. Anthropogenic radionuclides are divided into three groups:[22,34]

1. *Neutron activation products:* By neutron irradiation of objects, it is possible to induce radioactivity. This activation of stable isotopes enables to create radioisotopes. A lot of artificial radionuclides in the natural environment are produced as a result of the activation process

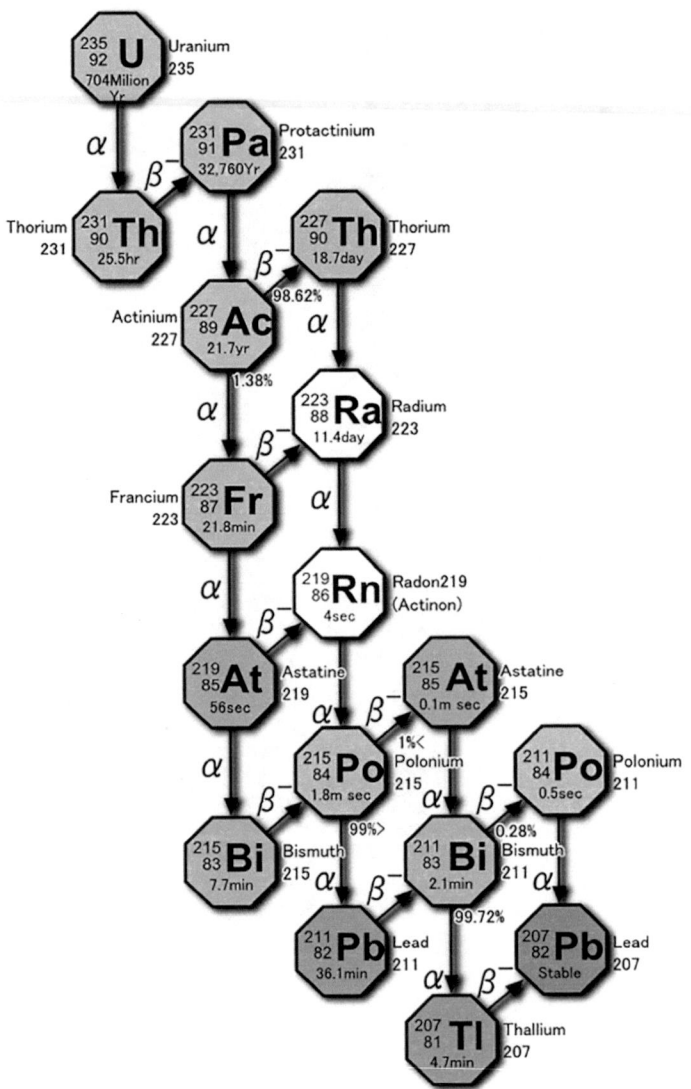

FIGURE 4   Actino-uranium 235U decay series.
Source: Wikipedia, Actinium series decay chain, http://upload.wikimedia.org/wikipedia/commons/1/1e/Decay_
chain%284n%2B3%2C_Actinium_series%29.PNG.[42]

during nuclear weapons tests, the work of reprocessing plants and nuclear reactors used in power plants, as well as in nuclear studies. Owing to the use of new radioanalytical techniques, activation products such as ^{22}Na, ^{51}Cr, ^{54}Mn, ^{65}Zn, ^{110m}Ag, and ^{124}Sb could be detected in the natural environment.[26]

2. *Fission radionuclides:* During the fission of ^{235}U, three neutrons are released in addition to two daughter atoms. In the detonation of a nuclear bomb, radioactive fission products are generated from the primary fission of ^{235}U or ^{239}Pu. The most important radionuclides from two families are ^{90}Sr, ^{95}Zr, ^{131}I, ^{132}I, ^{132}Te, ^{137}Cs,^{140}Ba, and ^{144}Ce. These radionuclides are deposited from the atmosphere to the surface of earth, with the fallout comprising components from the stratosphere (78%), local radioactive pollution (12%), and the troposphere (10%).[27]

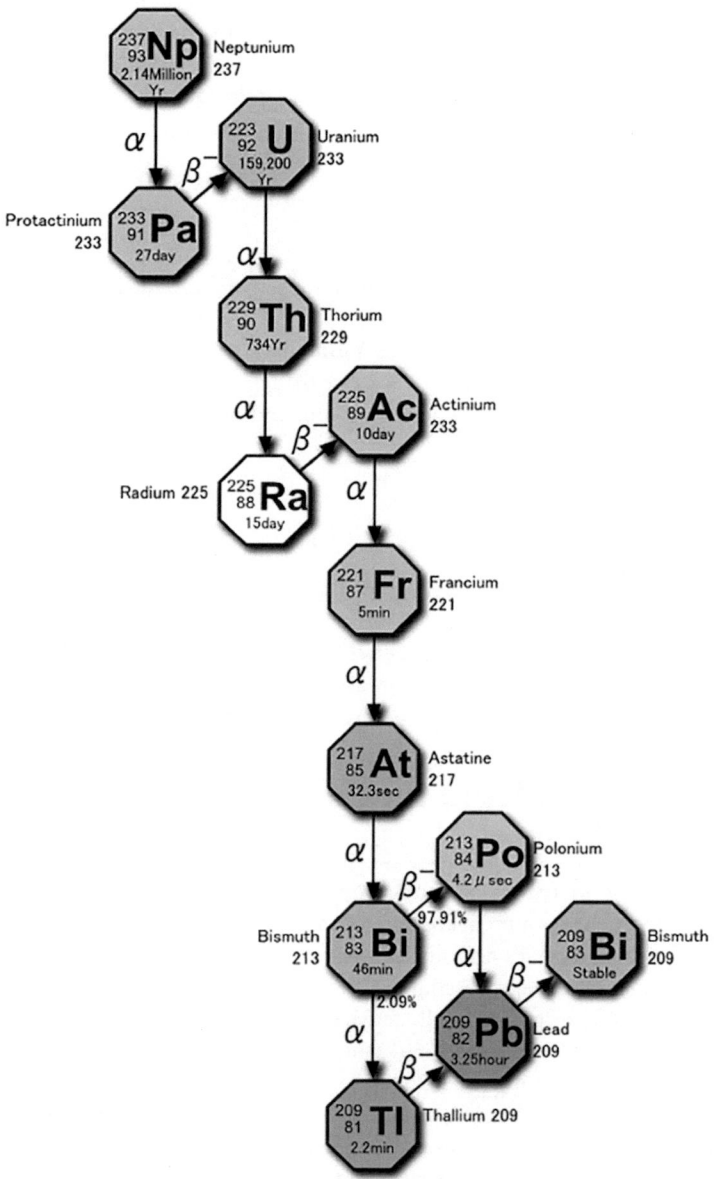

FIGURE 5 Neptunium-237 decay series.
Source: Wikipedia, Neptunium series decay chain, http://upload.wikimedia.org/wikipedia/commons/8/8c/Decay_chain%284n%2B1%2CNeptunium_series%29.PNG.[43]

3. *Transuranic elements:* In chemistry, transuranic elements are chemical elements with atomic numbers greater than 92 (the atomic number of uranium). All transuranic elements are radioactive; 20 transuranic elements have been discovered to date: neptunium (Np), plutonium (Pu), americium (Am), curium (Cm), berkelium (Bk), californium (Cf), einsteinium (Es), fermium (Fm), mendelevium (Md), nobelium (No), lawrencium (Lr), rutherfordium (Rf), dubnium (Db), seaborgium (Sg), bohrium (Bh), hassium (Hs), meitnerium (Mt), darmstadium (Ds), roentgenium (Rg), and copernicium (Cn). Small quantities of neptunium and plutonium are found in nature (in uranium rocks), but most of them are synthesized in nuclear reactors. The most important

sources of transuranic elements (generally, neptunium, plutonium, americium, and curium) in the natural environment are nuclear weapons explosions and nuclear power production as a result of the activation process of uranium, ^{238}U.[21,27,28] The environmental chemistry of some transuranic elements, such as plutonium, is complicated by the fact that solutions of this element can undergo disproportionation, and as a result many different oxidation states can coexist at once.

The most important (long half-time, type of decay, and strong radiotoxicity) natural and anthropogenic radionuclides present in the environment are as follows:[21]

Naturally occurring radionuclides

a. Radionuclides of terrestrial origin—primordial nonseries radionuclide (e.g., ^{40}K and ^{87}Rb)
b. Cosmogenic radionuclides (e.g., ^3H and ^{14}C).
c. Primary radionuclides—primordial series radionuclide: long-lived; have been ubiquitous on Earth since their formation (i.e., ca. 4.5×10^9 years ago). The radionuclides ^{238}U, ^{232}Th, and ^{235}U are the parent members of the uranium, thorium, and actinouranium radioactive decay series, respectively

Anthropogenic Radionuclides

a. Neutron activation products (e.g., ^{54}Mn, ^{55}Fe, ^{60}Co,
b. ^{235}U and ^{239}Pu fission radionuclides (e.g., ^{90}Sr, ^{95}Zr, ^{131}I, ^{132}I, ^{132}Te, ^{137}Cs, and ^{144}Ce)
c. Transuranic elements (e.g., ^{237}Np, ^{238}Pu, ^{239}Pu, ^{240}Pu, ^{241}Pu, ^{241}Am, and ^{243}Am)

Natural and artificial radionuclides in different environmental samples (natural water, sediments, soils, biological organisms) are determined by many radiometric methods, in particular neutron activation analysis (NAA), and alpha, beta, and gamma spectrometry.[29,30]

Sources of Radionuclides in the Environment and Pollution Problem

Radionuclides are present in the whole environment in which we live. The principal sources of radionuclides in the natural environment are the wet and dry atmospheric radioactive fallout of particles from natural rock erosion processes and nuclear weapons tests, as well as release from power plants, nuclear submarines, and nuclear reprocessing facilities. Since April 26, 1986, another source of artificial radionuclides, the Chernobyl-originated radioactive debris, has had to be taken into account.[21] A lot of artificial radionuclides (e.g., 200 kg of ^3H, 1550 kg of ^{14}C, >160 kg of ^{90}Sr, ~350 kg of ^{137}Cs, and >4600 kg of plutonium isotopes—^{238}Pu, ^{240}Pu, and ^{241}Pu) were deposited in different components of the natural environment (seas and oceans, biota, air, and land).[21] The large amount of radioactive waste produced by human utilization of nuclear power is nowadays one of the major problems. The marine environment is especially exposed to radionuclide contamination because many nuclear power plants and nuclear reprocessing facilities are located in coastal areas. In recent decades, the development of new technologies has also resulted in the production of by-products and waste with the so-called technologically enhanced naturally occurring radioactive materials (TENORM).[31] Therefore, human technological activities (e.g., gas, oil, coal, and fertilizers industries) can increase radiation exposure, not only to the persons directly involved in these activities but also to the local or even the whole population.[32] Moreover, coal mining is the source of a huge amount of waste containing large quantities of natural radionuclides, especially polonium, radium, thorium, and uranium. During combustion, some radionuclides are emitted to the atmosphere as gas and radioactive dust; others remain as concentrated ash.[33,34] Phosphate rocks that are used to produce phosphate fertilizers contain natural radioactive elements (polonium, radium, uranium). During the production of phosphate fertilizers, about 10% of the initial ^{226}Ra, 20% of uranium, and about 85% of ^{210}Po is found in the waste phosphogypsum.[35] These radionuclides are leached by rain from phosphogypsum and, as a consequence, in the neighborhood of the phosphate fertilizers plant, their concentration in soil, flora, and water samples is much higher than in non-contaminated areas.[36,37] Radionuclides are strongly accumulated by some species and the bioaccumulation factor values for some radioactive elements (polonium, plutonium, americium) in sea algae, benthic animals, and fish are more than 1.000.[38,39] Some of these organisms are often used as bio-indicators of radioactive pollution of the natural environment.[21] Also, transuranic elements (especially plutonium) belong to the group of radioactivity caused by humans. These radionuclides are important from the radiological point of view due to their high radiotoxicity, long physical lifetime, high chemical reactivity, and long residence in biological systems in the natural environment.[39]

Solution to Radioactive Pollution

A possible solution to radioactive pollution is by the reduction of radioactive emission to the natural environment, change of nuclear technology, and recognition of determination and accumulation processes in living organisms. Plants and animals are capable of accumulating natural and artificial radionuclides from the environment. That is why it is very important to recognize the impact of radionuclides on living organisms and their possible transfer to the human body by way of feeding.[21] Due to the importance of water, air, and food (also cigarette smoking) to human life, their quality must be strictly controlled and monitored. For this reason, studies of food for human consumption must be performed to guarantee that food materials have a low level of radioactivity, both natural and artificial. Especially, long-lived alpha emitters are the most dangerous nuclides in case of ingestion, because the long-term effects of their intake on the human body are the most important from the radiochemical and radiological points of view. A large contamination to the radiation dose received by humans comes from naturally occurring radionuclides accumulated in the body. At the moment, knowledge about accumulation of natural alpha radionuclides by organisms and their ingestion by humans is still very poor.

Conclusion

Radioactivity and radionuclides have been widely applied for more than a hundred years.

Radioactive substances are used to study living organisms, to diagnose and treat diseases (nuclear medicine), to sterilize medical instruments and food, to produce energy for heat and electric power, and to monitor various steps in all types of industrial processes, as well as in research studies (nuclear physics, chemistry, radiology, geochronology and geology, and cosmic research). Many natural and artificial radionuclides are strong radiotoxicants (generally, alpha emitters with a long half-time) to biological organisms. Studies on the bioaccumulation and distribution of radionuclides in different components of the natural environment are very important for radioprotection. Radionuclides used to produce energy in nuclear power plants generate the present and future problems of adequate and responsible utilization of nuclear wastes. For these reasons, the determination and distribution of some important natural and artificial radionuclides in the natural environment should be controlled and monitored.

References

1. Becquerel, H. Sur les radiations invisibles emises par les sels d'uranium. C. R. Acad. Sci. Paris **1896**, *122*, 689–694.
2. Sklodowska-Curie, M. Rayons emis par les composes de l'uranium et du thorium. C. R. Acad. Sci. Paris **1898**, *126*, 1101–1103.
3. Curie, P.; Sklodowska-Curie, M. Sur une substance nouvelle radio-active continue dans la pechblend. C. R. Acad. Sci. Paris **1898**, *127*, 175–178.
4. Sklodowska-Curie, M. Poszukiwania nowego metalu w pechblendzie. Światło **1898**, *1*, 54.
5. Curie, P.; Curie, M.; Bemont, G. Su rune nouvelle substance fortement radioactive, continue dans la pechblende. C. R. Acad. Sci. Paris **1898**, *127*, 1215–1217.

6. Sklodowska-Curie, M. Les rayons de Becquerel et le polonium. Rev. Gen. Sci. Pures Appl. **1899**, *10,* 41–50.

7. Curie, M. Sur le poids atomique du radium. C. R. Acad. Sci. Paris **1902**, *135,* 161–163.

8. Curie, I.; Joliot, M.F. Un nouveau type de radioactivite. C. R. Acad. Sci. Paris **1934**, *198,* 254–256.

9. Rutherford, E.; Soddy F. The cause and nature of radio-activity. Philos. Mag. **1902**, *4,* 370–396, 569–585.

10. Rutherford, E.; Soddy, F. Radioactive change. Philos. Mag. **1903**, *6,* 445457.

11. Rutherford, E. The scattering a, ß and g particles by matter. Philos. Mag. **1911**, *21,* 669–688.

12. Fajans, K. Die stellung der radioelemente im periodischen system. Phys. Z. **1913**, *14,* 136–142.

13. Soddy, F. The radio-element and the periodic law. Chem. News **1913**, *107,* 97–99.

14. Radioactivity. *Encyklopedia Británica Online,* December 18, 2006.

15. Browne, E.; Firestone, F.B. *Table of Radioactive Isotopes,* Shirley, V.S., Ed.; John Willey and Sons: New York, 1986.

16. Fermi, E.; Amaldi, E.; D'Agostino, O.; Rasetti, F.; Segrè, E. Radioacttività provocata da bombardamento di neutroni III. Ric. Sci. **1934**, *5* (1), 452–453.

17. Noddack, I. Über das element 93. Z. Angew. Chem. **1934**, *47* (37), 653–655.

18. Meitner, L; Frisch, O.R. Disintegration of uranium by neutrons: A new type of nuclear reaction. Nature **1939**, *143,* 239–240.

19. Hahn, O.; Strassmann, F. Über den Nachweis und das Verhalten der bei der Bestrahlung des Urans mittels Neutronen entstehenden Erdalkalimetalle. Naturwissenschaften **1939**, *27* (1), 11–15.

20. Rotblat, J. Emission of neutrons accompanying the fission of uranium nuclei. Nature **1939**, *143,* 852.

21. Skwarzec, B. Radiochemia środowiska i ochrona radiologiczna (in Polish), Environmental radiochemistry and radiological protecteon, Wydawnictwo DJ sc., Gdańsk, 2002.

22. Sobkowski, J.; Jelińska-Kaczmarczuk, M. *Chemia Jądrowa;* Wydawnictwo Adamantan: Warszawa, Poland, 2006.

23. Choppin, G.R.; Liljenzin, J.O.; Rydberg, J. *Radiochemistry and Nuclear Chemistry;* Butterworth-Heinemann Ltd.: Oxford, U.K., 1995.

24. Magill, J.; Galy, J. *Radioactivity, Radionuclides, Radiation;* Springer Verlag: Berlin, 2005.

25. Brune, D.; Forkman, B.; Persson, B. *Nuclear Analytical Chemistry;* Verlag Chemie International Inc.: Deerfield Beach, FL. 1984.

26. Holm, E., Ed. *Radioecology, Lectures in Environmental Radioactivity;* World Scientific: Singapore, 1994.

27. MacKenzie, A.B. Environmental radioactivity: Experience from the 20th century—Trends and issues for the 21st century. Sci. Total Environ. **2000**, *249,* 313–329.

28. Hardy, E.P.; Krey, P.W.; Volchok, H.L. Global inventory and distribution of fallout plutonium. Nature **1973**, *241,* 444–445.

29. Skwarzec, B. Determination of radionuclides in the aquatic environment. In *Analytical Measurements in Aquatic Environments;* Namiesnik, J., Szefer, P., Eds.; CRC Press, Taylor and Francis Group: Boca Raton, FL, 2010; 241–258.

30. L'Annunziata, M.F. *Handbook of Radioactivity Analysis,* 2nd Ed.; Academic Press, Oxford, Great Britain, 2003.

31. TENORM Sources, Radiation Protection, U.S. EPA, March 11th, 2009, available at http://www.epa.gov/radiation/ sources.html#summary-table.

32. Bou-Rabee, F.; Al-Zamel, A.; Al-Fares, R.; Bem, H. Technologically enhanced naturally occurring radioactive materials in the oil industry (TENORM). A review. Nukleonika **2009**, *54,* 3–9.

33. Nakaoka, A.; Fukushima, M.; Takagi, S. Environmental effect of natural radionuclides from coal-fired power plants. Health Phys. **1984**, *3,* 407–416.

34. Flues, M.; Morales, M.; Mazilli, B.P. The influence of a coal-plant operation on radionuclides in soil. J. Environ. Radioact. **2002**, *63,* 285–294.

35. Carvalho, F.P.; Oliviera, J.M.; Lopes, I.; Batista, A. Radionuclides from post uranium mining in rivers in Portugal. J. Environ. Radioact. **2007,** *98,* 298–314.

36. Borylo, A.; Nowicki, W.; Skwarzec, B. Isotope of polonium (^{210}Po) and uranium (^{234}U and ^{238}U) in the industrialized area of Wislinka (Northern Poland). Int. J. Environ. Anal. Chem. **2009,** *89,* 677–685.

37. Skwarzec, B.; Borylo, A.; Kosmska, A.; Radziejewska, S. Polonium (^{210}Po) and uranium (^{234}U and ^{238}U) in water, phosphogypsum and their bioaccumulation in plants around phosphogypsum waste heap in Wislinka (northern Poland). Nukleonika **2010,** *55,* 187–193.

38. Skwarzec, B. Polonium, uranium and plutonium in the southern Baltic Sea. Ambio **1997,** *26* (2), 113–117.

39. Coughtrey, P.J.; Jackson, D.; Jones, C.H.; Kene, P.; Thorne, M.C. *Radionuclides Distribution and Transport in Terrestrial and Aquatic Ecosystems. A Critical Review of Data*; A.A. Balkema: Rotterdam, 1984.

40. Wikipedia, uranium series decay chain, available at http://upload.wikimedia.org/wikipedia/commons/a/a1/Decay_chain%284n%2B2%2C_Uranium_series%29.PNG.

41. Wikipedia, thorium series decay chain, available at http://upload.wikimedia.org/wikipedia/commons/1/1c/Decay_chain%284n%2CThorium_series%29.PNG.

42. Wikipedia, actinium series decay chain, available at http://upload.wikimedia.org/wikipedia/commons/1/1e/Decay_chain%284n%2B3%2C_Actinium_series%29.PNG.

43. Wikipedia, neptunium series decay chain, available at http://upload.wikimedia.org/wikipedia/commons/8/8c/Decay_chain%284n%2B1%2CNeptunium_series%29.PNG.

55

Telecouplings

Vilma Sandström

Introduction

Human and natural systems are increasingly influenced by distant actions and globalized processes such as the flows of material, capital, or information through trade or the migration of human or animal populations (Verburg et al., 2015; Liu et al., 2019). These connections affect the understanding and management of global challenges such as climate change, biodiversity conservation, urbanization, and food security by creating a spatial separation between an action and its drivers. Therefore, they have critical implications to wider sustainability agenda such as the efforts to achieve the United Nations' Sustainable Development Goals (United Nations, 2015). Land use is in the core of these efforts. Drivers and impacts of land-use change and the consequent social and ecological impacts operate across geographical scales and over administrative territories. To disentangle the causal relations between an action, such as land-use change, and its underlying drivers and spatial connections, analytical concepts of teleconnection and telecoupling have been introduced. As suggested by the prefix "tele," both of these concepts focus the attention to distal linkages between land systems, broadening the understanding from the proximate drivers and impacts also to processes taking place in geographically or functionally distant areas (Friis et al., 2016).

Teleconnection. The concept of *teleconnection* was first used in physical sciences over a century ago to describe the atmospheric connections and the interrelations of climatic processes over long distances. Ångstrom (1935) was the first to apply the concept when describing the north–south dipole atmospheric anomaly pattern currently known as the North Atlantic Oscillation (Moser and Hart, 2015). The use of the concept was later broadened to include also other socioecological processes and drivers or land system change, such as flows of capital, people or materials, or species dispersal (Seto et al., 2012; Friis et al., 2016).

Telecoupling. The concept of *telecoupling* builds on the concept of teleconnection, expanding its meaning to include not only the links between an action and its distant driver but also the causal relationships, multidirectional flows, and feedback of land-use change in multiple, connected land systems (Friis et al., 2016; Liu et al., 2019). A telecoupling is created when there is a material, energy, or information flow from one system to another (Liu et al., 2013). These coupled systems are characterized by complex interactions of socioeconomic and biophysical elements that impact each other in dynamic, nonlinear, and emergent ways (Friis and Nielsen, 2017; Liu et al., 2007).

Telecoupling Research

Telecoupling research responds to the need to understand and analyze human–environment systems identifying the specific origin, destination, and the actors involved in multiple interactions between systems, in contrast to the traditional place-based studies that often treat factors from outside the system borders as bulk of exogenous variables without addressing the feedback caused by them (Liu et al., 2013). For example, up to 30% of all global biodiversity loss is related to the production of exported commodities (Lenzen et al., 2012). Thus, linking the consumption of imported products in a region to the specific locations of primary production and the consequent ecosystem impacts allows the identification of consumption as an underlying driver of environmental change and the search for leverage points for change (e.g. Sandström et al., 2017).

Interdisciplinarity is inherent in the telecoupling research related to connected human–environment systems covering both social and natural processes. It creates analytical and epistemic differences between the approaches adopted (Friis, 2019). But then, it represents also the flexibility of the concept to adopt to a wide view of perspectives. Therefore, telecoupling can be seen as an umbrella concept covering various approaches to analyze distant interactions among coupled human and natural systems over distances (Liu et al., 2013).

Liu et al. (2013) present a structural framework for analyzing telecouplings between systems. It has been applied and developed by many scholars from wide range of disciplinary backgrounds (see e.g. Friis and Nielsen, 2019; Liu et al., 2019). The framework builds upon the conceptual frameworks of coupled human–environment systems (Turner et al., 2003), socioecological systems (e.g. Young et al., 2006) and coupled human and natural systems (Liu et al., 2007) adding to these the emphasis on the distant interactions between systems (Liu et al., 2013). Figure 1 presents the major components of the analysis: (1) systems, (2) flows, (3) agents, (4) causes, and (5) effects of a telecoupling, with an emphasis on the interactions between these components (Liu et al., 2019). Telecoupled systems are classified as sending (the systems from where the flows move outward), receiving (the systems that receive the flows from the sending systems), or spillover systems (the systems that impact and/or are impacted by the interactions between the sending and receiving systems) (Liu et al., 2013). Areas can also act simultaneously as sending, receiving, or spillover systems. Flows are assisted by agents, such as different actors, people, or animals, and they can cause socioeconomic or environmental effects in any of the systems analyzed (Liu et al., 2013). Defining system boundaries is always context-dependent (Liu et al., 2019). Systems can be separated by spatial distance, often by jurisdictional boundaries or functional distance in terms of governance (Eakin et al., 2014). While the framework of Liu et al. (2013) emphasizes the spatial hierarchy, Eakin et al. (2014) outline telecouplings as results of networked interactions across scales (Friis and Nielsen, 2017) (Figure 1).

In the recent years, the number of studies exploring the telecouplings has increased rapidly (Kapsar et al., 2019). Although the concept of telecoupling was first introduced in land system science, its utility extends also beyond the disciplinary borders. It has been adopted within and across disciplines such as earth science, ecology, and economics as well as environmental and political science (Kapsar et al., 2019). Many of these studies have concentrated on international trade as a driver creating these spatial connections between sometimes seemingly unconnected land systems, for instance, by studying the consumption of imported products and the related displaced land use and its environmental impacts (e.g. Rulli et al., 2019; Sandström et al., 2017). Also, topics such as species invasion and migration (Liu et al., 2014; Hulina et al., 2017), urbanization (Fang and Ren, 2017), tourism (Liu et al., 2015), fisheries management (Carlson et al., 2018), and conservation (Carrasco et al., 2017; Fang and Liu, 2016) have been studied in a telecoupling context.

Conclusions

The analytical concept and framework of telecoupling provide an approach for analyzing the distant interactions of systems in a socioecological context integrating the often isolated disciplinary knowledge into a more comprehensive picture. In doing so, the concept and approach facilitate the broader

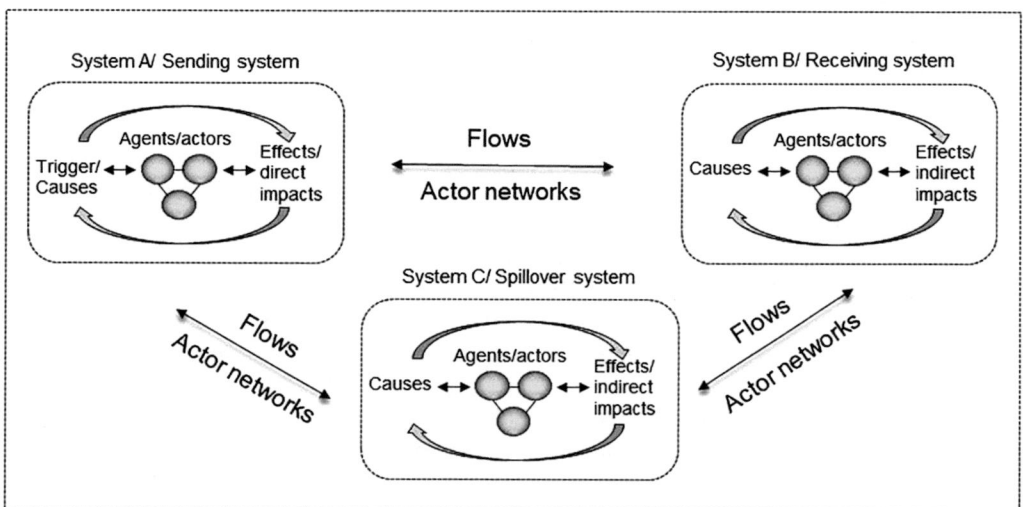

FIGURE 1   Telecoupling framework, adapted from Liu et al. (2013, 2014) and Eakin et al. (2014). The figure presents the main components of analysis as described by Liu et al. (2013): systems, flows, agents, causes, and effects, adapted with the inclusion of actor networks creating a telecoupling (Eakin et al., 2014). A first trigger initiates the telecoupling causing both direct impacts in the same system and indirect impacts in distantly coupled systems together with the feedback processes influencing potentially all systems in question (Eakin et al., 2014). The role of the systems as sending, receiving, or spillover systems can change depending on the perspective (Liu et al., 2014). System boundaries can be determined by spatial hierarchy (Liu et al., 2013) or by scope of governance (Eakin et al., 2014).

understanding of spatial complexity between an action and its drivers and impacts, that is inherent in all human and natural systems in the increasingly globalizing world. The concept has been applied in various temporal and spatial scales. Being relatively novel, the telecoupling framework, approach, and tools are in continuous development to capture and understand all the dimensions involved in and affected by distant interactions.

References

Ångstrom, A. 1935. Teleconnections of climatic changes in present time. *Geografiska Annaler*, 17:243–258.

Carlson, A., Taylor, W., Liu, J., and Orlic, I. 2018. Peruvian anchoveta as a telecoupled fisheries system. *Ecology and Society*, 23(1):35.

Carrasco, L. R., Chan, J., McGrath, F., and Nghiem, L. 2017. Biodiversity conservation in a telecoupled world. *Ecology and Society*, 22(3):24.

Eakin, H., DeFries, R., Kerr, S., Lambin, E. F., Liu, J., Marcotullio, P. J., Messerli, P., Reenberg, A., Rueda, X., Swaffield, S. R., Wicke, B., and Zimmerer, K. 2014. Significance of telecoupling for exploration of land-use change. In Seto, K, &, Reenberg, A. (Eds.) *Rethinking Global Land Use in an Urban Era* (pp. 141–161). Cambridge, MA: MIT Press.

Fang, W., and Liu, J. 2016. Conservation planning beyond giant pandas: the need for an innovative telecoupling framework. *Science China Life Sciences*, 60(5):551–554.

Fang, C., and Ren, Y. 2017. Analysis of emergy-based metabolic efficiency and environmental pressure on the local coupling and telecoupling between urbanization and the eco-environment in the Beijing-Tianjin-Hebei urban agglomeration. *Science China Earth Sciences*, 60(6):1083–1097.

Friis, C., Nielsen, J. Ø., Otero, I., Haberl, H., Niewöhner, J., and Hostert, P. 2016. From teleconnection to telecoupling: taking stock of an emerging framework in land system science. *Journal of Land Use Science*, 11(2):131–153.

Friis, C., and Nielsen, J. 2017. Land-use change in a telecoupled world: the relevance and applicability of the telecoupling framework in the case of banana plantation expansion in Laos. *Ecology and Society*, 22(4):30.

Friis, C. 2019. Telecoupling: a new framework for researching land-use change in a globalised world. In *Telecoupling* (pp. 49–67). Palgrave Macmillan, Cham.

Friis, C., and Nielsen, J. Ø. 2019. Global land-use change through a telecoupling lens: an introduction. In *Telecoupling* (pp. 1–15). Palgrave Macmillan, Cham.

Hulina, J., Bocetti, C., Ii, H., Hull, V., Yang, W., and Liu, J. 2017. Telecoupling framework for research on migratory species in the anthropocene. *Elementa—Science of the Anthropocene*, 5:5. https://doi.org/10.1525/elementa.184.

Kapsar, K. E., Hovis, C. L., Bicudo da Silva, R. F., Buchholtz, E. K., Carlson, A. K., Dou, Y., Du, Y., Furumo, P. R., Li, Y., Torres, A., Yang, D., Wan, H. Y., Zaehringer, J. G., and Liu, J. (2019). Telecoupling research: the first five years. *Sustainability*, 11(4):1033.

Lenzen, M., Moran, D., Kanemoto, K., Foran, B., Lobefaro, L., and Geschke, A. 2012. International trade drives biodiversity threats in developing nations. *Nature*, 486:109–112. http://dx.doi.org/10.1038/nature11145

Liu, J., Dietz, T., Carpenter, S. R., Alberti, M., Folke, C., Moran, E., Pell, A. N., Deadman, P., Kratz, T., Lubchenco, J., Ostrom, E., Uyang, Z., Provencher, W., Redman, C. L., Schneider, S. H., and Taylor, W. 2007. Complexity of coupled human and natural systems. *Science*, 317(5844):1513–1516.

Liu, J., Hull, V., Batistella, M., DeFries, R., Dietz, T., Fu, F., Hertel, T. W., Izaurralde, R. C., Lambin, E. F., Li, S., Martinelli, L. A., McConnell, W. J., Moran, E. F., Naylor, R., Ouyand, Z., Polenske, K. R., Reenberg, A., de Miranda Rocha, G., Simmons, C. S., Verbugr, P. H., Vitousek, P. M., Zhang, F., and Zhu, C. 2013. Framing sustainability in a telecoupled world. *Ecology and Society*, 18(2):26.

Liu, J., Hull, V., Moran, E., Nagendra, H., Swaffield, S., and Turner, B. L. 2014. Applications of the telecoupling framework to land-change science. In K. C. Seto & A. Reenberg (Eds.), *Rethinking Global Land Use in and Urban Age*. MIT Press, Cambridge, MA.

Liu, J., Hull, V., Luo, J., Yang, W., Liu, W, Viña, A., Vogt, C., Xu, Z., Yang, H., Zhang, J., An, L., Chen, X., Li, Z., Ouyang, Z., Xu,, W., and Zhang, H. 2015. Multiple telecouplings and their complex interrelationships. *Ecology and Society* 20(3):44.

Liu, J., Herzberger, A., Kapsar, K., Carlson, A. K., and Connor, T. 2019. What is telecoupling? In: Friis, C., Nielsen, J. (eds) *Telecoupling*. (pp. 19–48). Palgrave Studies in Natural Resource Management. Palgrave Macmillan, Cham.

Moser, S. C., and Hart, J. A. F. 2015. The long arm of climate change: societal teleconnections and the future of climate change impacts studies. *Climatic Change*, 129(1–2):13–26.

Rulli, M. C., Casirati, S., Dell'Angelo, J., Davis, K. F., Passera, C., and D'Odorico, P. 2019. Interdependencies and telecoupling of oil palm expansion at the expense of Indonesian rainforest. *Renewable and Sustainable Energy Reviews*, 105:499–512.

Sandström, V., Kauppi, P. E., Scherer, L., and Kastner, T. 2017. Linking country level food supply to global land and water use and biodiversity impacts: the case of Finland. *Science of the Total Environment*, 575:33–40.

Seto, K. C., Reenberg, A., Boone, C. G., Fragkias, M., Haase, D., Langanke, T., Marcotullio, P., Munroe, D. K., Olah, B., and Simon, D. 2012. Urban land teleconnections and sustainability. *Proceedings of the National Academy of Sciences*, 109(20):7687–7692.

Turner, B. L., Kasperson, R. E., Matson, P. A., McCarthy, J. J., Corell, R. W., Christensen, L., Eckley, N., Kasperson, J. X., Luers, A., Martello, M. L., Polsky, C., Pulsipher, A., and Schiller, A. 2003. A framework for vulnerability analysis in sustainability science. *Proceedings of the National Academy of Sciences*, 100(14):8074–8079.

United Nations. 2015. Transforming our world: the 2030 agenda for sustainable development. Resolution adopted by the General Assembly. https://sustainabledevelopment.un.org/post2015/transformingourworld.

Verburg, P. H., Crossman, N., Ellis, E. C., Heinimann, A., Hostert, P., Mertz, O., Nagendra, H., Sikor, T., Erb, K-H., Golubiewski, N., Grau, R., Grove, M., Konaté, S., Meyfroidt, P., Parker, D. C., Chowdhury, R. R., Shibata, H., Thomson, A., and Zhen, L. 2015. Land system science and sustainable development of the earth system: a global land project perspective. *Anthropocene*, 12:29–41.

Young, O., Lambin, E., Alcock, F., Haberl, H., Karlsson, S., McConnell, W., Myint, T., Pahl-Wostl, C., Polsky, C., Ramakrishnan, P. S., Schroeder, H., Scouvart, M., and Verburg, P H. 2006. A portfolio approach to analyzing complex human-environment interactions: institutions and land change. *Ecology and Society*, 11(2):31.

Index

Page numbers followed by f and t indicate figures and tables, respectively